Student's Solutions Manual to Accompany

Precalculus
A Problems-Oriented Approach
Fourth Edition

David Cohen
Department of Mathematics
University of California
Los Angeles, California

Prepared by
Ross Rueger
Department of Mathematics
College of the Sequoias
Visalia, California

West Publishing Company
Minneapolis/St. Paul   New York   Los Angeles   San Francisco

WEST'S COMMITMENT TO THE ENVIRONMENT
In 1906, West Publishing Company began recycling materials left over from the production of books. This began a tradition of efficient and responsible use of resources. Today, up to 95% of our legal books and 70% of our college and school texts are printed on recycled, acid-free stock. West also recycles nearly 22 million pounds of scrap paper annually—the equivalent of 181,717 trees. Since the 1960s, West has devised ways to capture and recycle waste inks, solvents, oils, and vapors created in the printing process. We also recycle plastics of all kinds, wood, glass, corrugated cardboard, and batteries, and have eliminated the use of Styrofoam book packaging. We at West are proud of the longevity and the scope of commitment to the environment.

Production, Prepress, Printing and Binding by West Publishing Company.

ISBN 0–314–02299–6

Contents

Chapter 12   Additional Topics in Algebra

Appendix

Preface

This <u>Student's Solutions Manual</u> contains complete solutions to all odd-numbered regular and graphing calculator exercises of <u>Precalculus: A Problems-Oriented Approach</u> by David Cohen. It also contains complete solutions to odd and even exercises for each chapter test. I have attempted to format solutions for readability and accuracy, and apologize to you for any errors that you may encounter. If you have any comments, suggestions, error corrections, or alternative solutions please feel free to drop me a note.

Please use this manual with some degree of caution. Be sure that you have attempted a solution, and re-attempted it, before you look it up in this manual. Mathematics can only be learned by **doing**, and not by observing! As you use this manual, do not just read the solution but work it along with the manual, using my solution to check your work. If you use this manual in that fashion then it should be helpful to you in your studying.

I would like to thank a number of people for their assistance in preparing this manual. Thanks go to Peter Marshall, Jane Bass, Stacy Lenzen, and Deanna Quinn at West Educational Publishing for their valuable assistance and support. Special thanks go to Charles Heuer of Concordia College for his meticulous error-checking of my solutions, as well as numerous suggestions for improvement of solutions.

I wish to express my deepest appreciation to David Cohen for continuing his tradition of excellence with this textbook. His addition of TI-81 graphing calculator exercises in this fourth edition will help you in conceptualizing the notion of a function. That notion is the heart of calculus and mathematics in general.

I would especially like to thank my father, who has lived his entire life with pride and dignity. If I can be a fraction of the man that he is then my life will be complete.

Ross Rueger
College of the Sequoias
915 South Mooney Boulevard
Visalia, CA 93277

March, 1993

Chapter One
Algebra Background for Precalculus

1.1 Sets of Real Numbers

1. (a) natural number, integer, rational number
   (b) integer, rational number

3. (a) rational number
   (b) irrational number

5. (a) natural number, integer, rational number
   (b) rational number

7. (a) rational number
   (b) rational number

9. irrational number

11. irrational number

13. Since $\frac{11}{4} \approx 2.75$, we have the following graph:

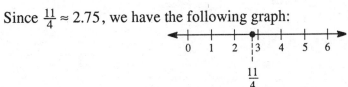

15. Since $1 + \sqrt{2} \approx 2.4$, we have the following graph:

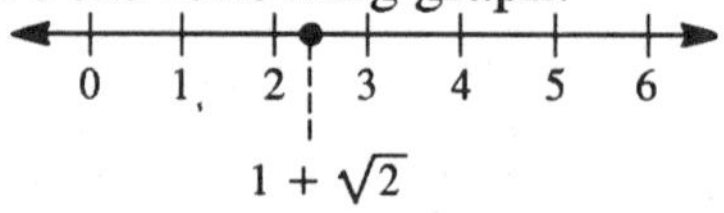

17. Since $\sqrt{2} - 1 \approx 0.4$, we have the following graph:

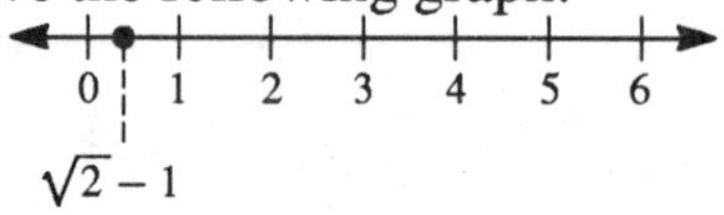

19. Since $\sqrt{2} + \sqrt{3} \approx 3.1$, we have the following graph:

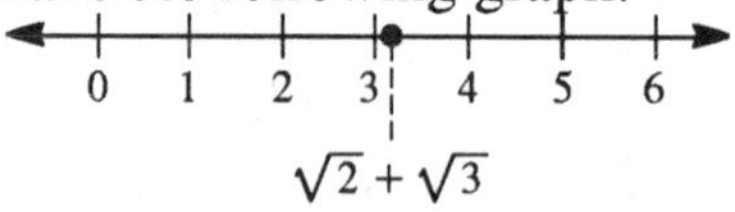

21. Since $\dfrac{1+\sqrt{2}}{2} \approx \dfrac{2.4}{2} \approx 1.2$, we have the following graph:

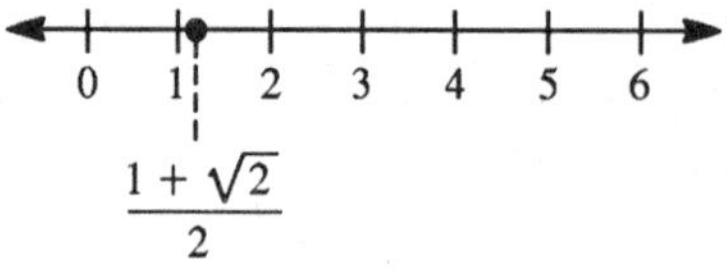

23. We draw the graph:

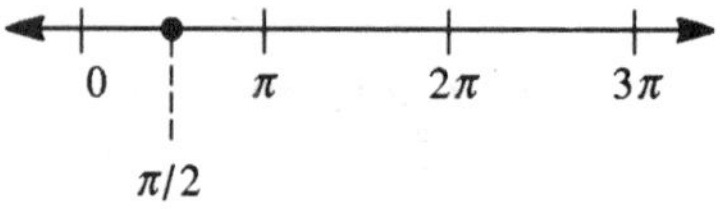

25. We draw the graph:

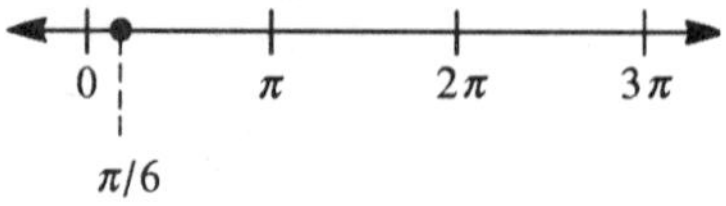

27. We draw the graph:

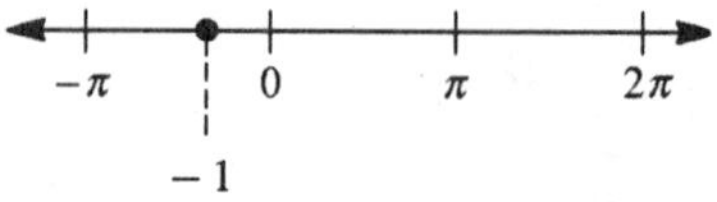

29. We draw the graph:

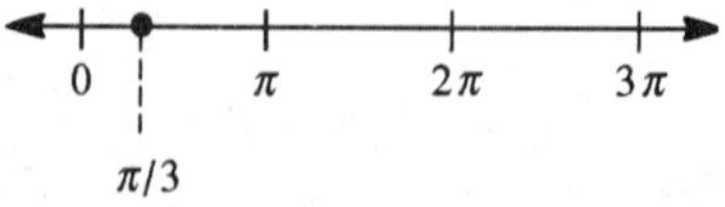

31. Since $1 \approx \frac{\pi}{3}$, we have the following graph:

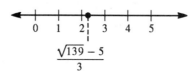

33. Since $\dfrac{\sqrt{139} - 5}{3} \approx 2.26$, we have the following graph:

35. False

37. True (since $-2 = -2$, then it is also true that $-2 \le -2$)

39. False

41. False (since $2\pi \approx 6.2$)

43. True (since $2\sqrt{2} \approx 2.8$)

45. We graph the interval $(2, 5)$:

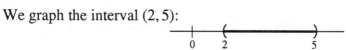

47. We graph the interval $[1, 4]$:

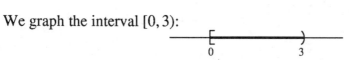

49. We graph the interval $[0, 3)$:

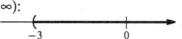

51. We graph the interval $(-3, \infty)$:

53. We graph the interval $[-1, \infty)$:

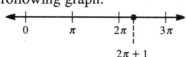

55. We graph the interval $(-\infty, 1)$:

57. We graph the interval $(-\infty, \pi]$:

59. (a) The common value to six decimal places is 3.863703.
(b) The common value (unrounded) to six decimal places is 3.162277.
(c) The common value to six decimal places is 1.847759.

61. (a) Since $(4/3)^4 \approx 3.16$, it agrees with π to one decimal place.
(b) Since $\frac{22}{7} \approx 3.142$, it agrees with π to two decimal places.
(c) Since $\frac{355}{113} \approx 3.1415929$, it agrees with π to six decimal places.
(d) Since $\frac{63}{25}\left(\frac{17+15\sqrt{5}}{7+15\sqrt{5}}\right) \approx 3.1415926538$, it agrees with π to nine decimal places.

63. (a) We need to find two irrational numbers a and b such that their product is rational. If we choose $a = \sqrt{2}$ and $b = \sqrt{8}$, then $ab = \sqrt{2} \cdot \sqrt{8} = \sqrt{16} = 4$, which is rational.

(b) We need to find two irrational numbers a and b such that their product is irrational. If we choose $a = \sqrt{2}$ and $b = \sqrt{3}$, then $ab = \sqrt{2} \cdot \sqrt{3} = \sqrt{6}$, which is irrational.

65. (a) Raising 2 to the 1/2 power results in $2^{1/2} = \sqrt{2}$, which is irrational.
(b) Raising $\sqrt{2}$ to the 2 power results in $\left(\sqrt{2}\right)^2 = 2$, which is rational.

1.2 Absolute Value

1. $|3| = 3$

3. $|-6| = 6$

5. $|-1+3| = |2| = 2$

7. We simplify the expression:
$$\left|-\frac{4}{5}\right| - \frac{4}{5} = \frac{4}{5} - \frac{4}{5} = 0$$

9. We simplify the expression:
$$|-6+2| - |4| = |-4| - |4| = 4 - 4 = 0$$

11. We simplify the expression:
$$\big||-8|+|-9|\big| = |8+9| = |17| = 17$$

13. We simplify the expression:
$$\left|\frac{27-5}{5-27}\right| = \left|\frac{22}{-22}\right| = |-1| = 1$$

15. We simplify the expression:
$$|7(-8)| - |7||-8| = |-56| - 7(8) = 56 - 56 = 0$$

17. Substituting $a = -2$ and $b = 3$:
$$|a-b|^2 = |-2-3|^2 = |-5|^2 = (5)^2 = 25$$

19. Substituting $a = -2$, $b = 3$, and $c = -4$:
$$|c| - |b| - |a| = |-4| - |3| - |-2| = 4 - 3 - 2 = -1$$

21. Substituting $a = -2$, $b = 3$, and $c = -4$:
$$|a+b|^2 - |b+c|^2 = |-2+3|^2 - |3+(-4)|^2 = |1|^2 - |-1|^2 = (1)^2 - (1)^2 = 1 - 1 = 0$$

23. Substituting $a = -2$ and $b = 3$:
$$\frac{a+b+|a-b|}{2} = \frac{-2+3+|-2-3|}{2} = \frac{1+|-5|}{2} = \frac{1+5}{2} = \frac{6}{2} = 3$$

25. Since $\sqrt{2} - 1 > 0$, we have:
$$|\sqrt{2}-1| - 1 = (\sqrt{2}-1) - 1 = \sqrt{2} - 2$$

27. Since $x \geq 3$, then $x - 3 \geq 0$ and thus:
$$|x-3| = x - 3$$

29. Since $t^2 + 1 > 0$, then:
$$|t^2+1| = t^2 + 1$$

31. Since $-\sqrt{3} - 4 < 0$, then:
$$|-\sqrt{3}-4| = -(-\sqrt{3}-4) = \sqrt{3} + 4$$

33. Since $x < 3$, then $x - 3 < 0$ and $x - 4 < 0$, and thus:
$$|x-3| + |x-4| = -(x-3) + [-(x-4)] = -x + 3 - x + 4 = -2x + 7$$

35. Since $3 < x < 4$, then $x - 3 > 0$ and $x - 4 < 0$, and thus:
$$|x-3| + |x-4| = (x-3) + [-(x-4)] = x - 3 - x + 4 = 1$$

37.   Since $-\frac{5}{2} < x < -\frac{3}{2}$, then $x + 1 < 0$ and $x + 3 > 0$, and thus:
$$|x + 1| + 4|x + 3| = [-(x + 1)] + 4(x + 3) = -x - 1 + 4x + 12 = 3x + 11$$

39.   The absolute value equality can be written as $|x - 4| = 8$.

41.   The absolute value equality can be written as $|x - 1| = \frac{1}{2}$.

43.   The absolute value inequality can be written as $|x - 1| \geq \frac{1}{2}$.

45.   The absolute value inequality can be written as $|y - (-4)| < 1$, or $|y + 4| < 1$.

47.   The absolute value inequality can be written as $|y - 0| < 3$, or $|y| < 3$.

49.   The absolute value inequality can be written as $|x^2 - a^2| < M$.

51.   We graph the interval $|x| < 4$:

53.   We graph the interval $|x| > 1$:

55.   We graph the interval $|x - 5| < 3$:

57.   We graph the interval $|x - 3| \leq 4$:

59.   We graph the interval $\left|x + \frac{1}{3}\right| < \frac{3}{2}$:

61.   We graph the interval $|x - 5| \geq 2$:

63. (a) We graph the interval $|x-2|<1$:

(b) We graph the interval $0<|x-2|<1$, noting that $x=2$ is excluded:

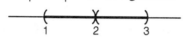

(c) The interval in (b) does not include 2.

65. (a) We know $\min(6,1)=1$, now substituting $a=6$ and $b=1$:
$$\min(6,1)=\frac{6+1-|6-1|}{2}=\frac{7-5}{2}=\frac{2}{2}=1$$
Thus the equation is verified.

(b) We know $\min(1,-6)=-6$, now substituting $a=1$ and $b=-6$:
$$\min(1,-6)=\frac{1+(-6)-|1-(-6)|}{2}=\frac{-5-|1+6|}{2}=\frac{-5-7}{2}=\frac{-12}{2}=-6$$
Thus the equation is verified.

(c) We know $\min(-6,-6)=-6$, now substituting $a=-6$ and $b=-6$:
$$\min(-6,-6)=\frac{-6+(-6)-|-6-(-6)|}{2}=\frac{-12-|-6+6|}{2}=\frac{-12}{2}=-6$$
Thus the equation is verified.

67. Using the triangle inequality twice, we have:
$$|a+b+c|=|a+(b+c)|\le|a|+|b+c|\le|a|+|b|+|c|$$

69. We will consider three cases: $a=b$, $a>b$, and $a<b$.

case 1: If $a=b$, then $\max(a,b)=a$, now verifying:
$$\frac{a+b+|a-b|}{2}=\frac{a+a+|a-a|}{2}=\frac{2a}{2}=a$$
Thus the equation is verified.

case 2: If $a>b$, then $\max(a,b)=a$, and since $a-b>0$, we have:
$$\frac{a+b+|a-b|}{2}=\frac{a+b+a-b}{2}=\frac{2a}{2}=a$$
Thus the equation is verified.

case 3: If $a<b$, then $\max(a,b)=b$, and since $a-b<0$, we have:
$$\frac{a+b+|a-b|}{2}=\frac{a+b+-(a-b)}{2}=\frac{a+b-a+b}{2}=\frac{2b}{2}=b$$
Thus the equation is verified.

71.   (a) Property 1(b).

(b) $a + b \leq |a| + |b|$

(c) Since $(-a) + (-b) \leq |a| + |b|$, then $-(a + b) \leq |a| + |b|$.

(d) Since $a + b \leq |a| + |b|$ and $-(a + b) \leq |a| + |b|$, then $|a + b| \leq |a| + |b|$, since $|a + b|$ is either $a + b$ or $-(a + b)$.

1.3 Polynomials and Factoring

1.   (a) The domain is the set of all real numbers.
(b) The domain is the set of all non-negative real numbers.

3.   (a) The domain is the set of all real numbers.
(b) The domain is the set of all real numbers.

5.   (a) The domain is the set of all non-negative real numbers.
(b) The domain is the set of all positive real numbers except 1.

7.   (a) The degree is 0 and the (nonzero) coefficient is 4.
(b) The degree is 3 and the (nonzero) coefficient is 4.
(c) The degree is 6 and the (nonzero) coefficients are 1, 4, –1, and 2.

9.   We verify special product 1:
$$(A - B)(A + B) = A^2 - AB + AB - B^2 = A^2 - B^2$$
We verify special product 2(a):
$$(A + B)^2 = (A + B)(A + B) = A^2 + AB + AB + B^2 = A^2 + 2AB + B^2$$
We verify special product 2(b):
$$(A - B)^2 = (A - B)(A - B) = A^2 - AB - AB + B^2 = A^2 - 2AB + B^2$$

11.   (a) Using the special product for $(A - B)(A + B)$, we have:
$$(x - y)(x + y) = x^2 - y^2$$

(b) Using the special product for $(A - B)(A + B)$, we have:
$$(x^2 - 5)(x^2 + 5) = (x^2)^2 - (5)^2 = x^4 - 25$$

13.   (a) Using the special product for $(A - B)(A + B)$, we have:
$$(A - 4)(A + 4) = A^2 - 16$$

(b) Using the special product for $(A - B)(A + B)$, we have:
$$[(a + b) - 4][(a + b) + 4] = (a + b)^2 - 16 = a^2 + 2ab + b^2 - 16$$

15.   (a) Using the special product for $(A - B)^2$, we have:
$$(x - 8)^2 = x^2 - 2(x)(8) + 64 = x^2 - 16x + 64$$

(b) Using the special product for $(A - B)^2$, we have:
$$\left(2x^2 - 5\right)^2 = \left(2x^2\right)^2 - 2\left(2x^2\right)(5) + 25 = 4x^4 - 20x^2 + 25$$

17.   (a) Using the special product for $(A + B)^2$, we have:
$$\left(\sqrt{x} + \sqrt{y}\right)^2 = \left(\sqrt{x}\right)^2 + 2\left(\sqrt{x}\right)\left(\sqrt{y}\right) + \left(\sqrt{y}\right)^2 = x + 2\sqrt{xy} + y$$

(b) Using the special product for $(A - B)^2$, we have:
$$\begin{aligned}\left(\sqrt{x+y} - \sqrt{x}\right)^2 &= \left(\sqrt{x+y}\right)^2 - 2\left(\sqrt{x+y}\right)\left(\sqrt{x}\right) + \left(\sqrt{x}\right)^2 \\ &= x + y - 2\sqrt{x^2 + xy} + x \\ &= 2x + y - 2\sqrt{x^2 + xy}\end{aligned}$$

19.   (a) Using the special product for $(A + B)^3$, we have:
$$(a + 1)^3 = a^3 + 3\left(a^2\right)(1) + 3(a)(1)^2 + (1)^3 = a^3 + 3a^2 + 3a + 1$$

(b) Using the special product for $(A - B)^3$, we have:
$$\begin{aligned}\left(3x^2 - 2a^2\right)^3 &= \left(3x^2\right)^3 - 3\left(3x^2\right)^2\left(2a^2\right) + 3\left(3x^2\right)\left(2a^2\right)^2 - \left(2a^2\right)^3 \\ &= 27x^6 - 54x^4a^2 + 36x^2a^4 - 8a^6\end{aligned}$$

21.   (a) Using the special product for $(A + B)\left(A^2 - AB + B^2\right)$, we have:
$$(x + 1)\left(x^2 - x + 1\right) = x^3 + 1$$

(b) Using the special product for $(A + B)\left(A^2 - AB + B^2\right)$, we have:
$$\left(x^2 + 1\right)\left(x^4 - x^2 + 1\right) = \left(x^2\right)^3 + (1)^3 = x^6 + 1$$

23.   (a) Using a difference of squares, we have:
$$x^2 - 64 = (x + 8)(x - 8)$$

(b) Using a common factor of $7x^2$, we have:
$$7x^4 + 14x^2 = 7x^2\left(x^2 + 2\right)$$

(c) Using a common factor of z then a difference of squares, we have:
$$121z - z^3 = z\left(121 - z^2\right) = z(11 + z)(11 - z)$$

(d) Using a difference of squares, we have:
$$a^2b^2 - c^2 = (ab)^2 - (c)^2 = (ab + c)(ab - c)$$

25.   (a) By trial and error, we have:
$$x^2 + 2x - 3 = (x+3)(x-1)$$

(b) By trial and error, we have:
$$x^2 - 2x - 3 = (x-3)(x+1)$$

(c) By trial and error, we find that $x^2 - 2x + 3$ is irreducible.

(d) By trial and error, we have:
$$-x^2 + 2x + 3 = (-x+3)(x+1)$$
Note that we could also use a common factor of -1, then trial and error:
$$-x^2 + 2x + 3 = -1(x^2 - 2x - 3) = -(x-3)(x+1)$$

27.   (a) By the sum of cubes, we have:
$$x^3 + 1 = (x+1)(x^2 - x + 1)$$

(b) Since $6^3 = 216$, we use the sum of cubes to obtain:
$$x^3 + 216 = x^3 + 6^3 = (x+6)(x^2 - 6x + 36)$$

(c) Using a common factor of 8 then a difference of cubes, we have:
$$1000 - 8x^6 = 8(125 - x^6) = 8(5^3 - x^6) = 8(5 - x^2)(25 + 5x^2 + x^4)$$

(d) By the difference of cubes, we have:
$$64a^3x^3 - 125 = (4ax)^3 - 5^3 = (4ax - 5)(16a^2x^2 + 20ax + 25)$$

29.   (a) Using a difference of squares, we have:
$$144 - x^2 = (12 + x)(12 - x)$$

(b) Since this is a sum of squares, the expression $144 + x^2$ is irreducible.

(c) Using a difference of squares, we have:
$$144 - (y-3)^2 = (12 + y - 3)(12 - y + 3) = (9 + y)(15 - y)$$

31.   (a) Using a common factor of h^3 then a difference of squares, we have:
$$h^3 - h^5 = h^3(1 - h^2) = h^3(1 + h)(1 - h)$$

(b) Using a common factor of h^3 then a difference of squares, we have:
$$100h^3 - h^5 = h^3(100 - h^2) = h^3(10 + h)(10 - h)$$

(c) Using a common factor of $(h + 1)^3$ then a difference of squares, we have:

$$100(h+1)^3 - (h+1)^5 = (h+1)^3\left[100 - (h+1)^2\right]$$
$$= (h+1)^3(10 + h + 1)(10 - h - 1)$$
$$= (h+1)^3(11 + h)(9 - h)$$

33.   (a) By trial and error, we have:
$$x^2 - 13x + 40 = (x - 8)(x - 5)$$

(b) By trial and error, we find that $x^2 - 13x - 40$ is irreducible.

35.   (a) By trial and error, we have:
$$x^2 + 5x - 36 = (x + 9)(x - 4)$$

(b) By trial and error, we have:
$$x^2 - 13x + 36 = (x - 9)(x - 4)$$

37.   (a) By trial and error, we have:
$$3x^2 - 22x - 16 = (3x + 2)(x - 8)$$

(b) By trial and error, we find that $3x^2 - x - 16$ is irreducible.

39.   (a) By trial and error, we have:
$$6x^2 + 13x - 5 = (3x - 1)(2x + 5)$$

(b) By trial and error, we have:
$$6x^2 - x - 5 = (6x + 5)(x - 1)$$

41.   (a) Using the special product for $(A + B)^2$, we have:
$$t^4 + 2t^2 + 1 = \left(t^2 + 1\right)^2$$

(b) Using the special product for $(A - B)^2$ and the difference of squares, we have:
$$t^4 - 2t^2 + 1 = \left(t^2 - 1\right)^2 = (t + 1)^2(t - 1)^2$$

(c) By trial and error, we find that $t^4 - 2t^2 - 1$ is irreducible.

43.   (a) Using the common factor x, we have:
$$4x^3 - 20x^2 - 25x = x\left(4x^2 - 20x - 25\right)$$

(b) Using the common factor x and the special product for $(A - B)^2$, we have:
$$4x^3 - 20x^2 + 25x = x\left(4x^2 - 20x + 25\right) = x(2x - 5)^2$$

45. (a) Using grouping, we have:
$$ab - bc + a^2 - ac = b(a-c) + a(a-c) = (a-c)(b+a)$$

  (b) Using grouping, we have:
$$(u+v)x - xy + (u+v)^2 - (u+v)y = x(u+v-y) + (u+v)(u+v-y)$$
$$= (u+v-y)(x+u+v)$$

47. Using grouping, we have:
$$x^2z^2 + xzt + xyz + yt = xz(xz+t) + y(xz+t) = (xz+t)(xz+y)$$

49. Using the special product for $(A-B)^2$, we have:
$$a^4 - 4a^2b^2c^2 + 4b^4c^4 = \left(a^2 - 2b^2c^2\right)^2$$

51. Since this is a sum of squares, $A^2 + B^2$ is irreducible.

53. Since $4^3 = 64$, using the sum of cubes factoring yields:
$$x^3 + 64 = (x+4)\left(x^2 - 4x + 16\right)$$

55. Using the difference of cubes factoring, we have:
$$(x+y)^3 - y^3 = (x+y-y)\left[(x+y)^2 + (x+y)(y) + y^2\right]$$
$$= x\left(x^2 + 2xy + y^2 + xy + y^2 + y^2\right)$$
$$= x\left(x^2 + 3xy + 3y^2\right)$$

57. Using the difference of cubes factoring, then grouping, we have:
$$x^3 - y^3 + x - y = (x-y)\left(x^2 + xy + y^2\right) + (x-y)(1) = (x-y)\left(x^2 + xy + y^2 + 1\right)$$

59. (a) Using the difference of squares factoring (twice), we have:
$$p^4 - 1 = \left(p^2 + 1\right)\left(p^2 - 1\right) = \left(p^2 + 1\right)(p+1)(p-1)$$

  (b) Using the difference of squares factoring (three times), we have:
$$p^8 - 1 = \left(p^4 + 1\right)\left(p^4 - 1\right)$$
$$= \left(p^4 + 1\right)\left(p^2 + 1\right)\left(p^2 - 1\right)$$
$$= \left(p^4 + 1\right)\left(p^2 + 1\right)(p+1)(p-1)$$

61. Using the special product for $(A+B)^3$, we have:
$$x^3 + 3x^2 + 3x + 1 = (x+1)^3$$

Note: If you do not recognize the special product, you can also use grouping:

$$x^3 + 3x^2 + 3x + 1 = (x^3 + 1) + (3x^2 + 3x)$$
$$= (x+1)(x^2 - x + 1) + 3x(x+1)$$
$$= (x+1)(x^2 - x + 1 + 3x)$$
$$= (x+1)(x^2 + 2x + 1)$$
$$= (x+1)(x+1)^2$$
$$= (x+1)^3$$

63. Since this is a sum of squares, $x^2 + 16y^2$ is irreducible.

65. Using the difference of squares factoring, we have:
$$\tfrac{25}{16} - c^2 = \left(\tfrac{5}{4} + c\right)\left(\tfrac{5}{4} - c\right)$$

67. Using the difference of squares factoring (twice), we have:
$$z^4 - \tfrac{81}{16} = \left(z^2 + \tfrac{9}{4}\right)\left(z^2 - \tfrac{9}{4}\right) = \left(z^2 + \tfrac{9}{4}\right)\left(z + \tfrac{3}{2}\right)\left(z - \tfrac{3}{2}\right)$$

69. Using the difference of cubes factoring, we have:
$$\frac{125}{m^3 n^3} - 1 = \left(\frac{5}{mn} - 1\right)\left(\frac{25}{m^2 n^2} + \frac{5}{mn} + 1\right)$$

71. Using the special product for $(A + B)^2$, we have:
$$\tfrac{1}{4}x^2 + xy + y^2 = \left(\tfrac{1}{2}x + y\right)^2$$

73. Using the common factor $x - a$, then a difference of squares, we have:
$$64(x-a)^3 - x + a = (x-a)\left[64(x-a)^2 - 1\right]$$
$$= (x-a)[8(x-a) + 1][8(x-a) - 1]$$
$$= (x-a)(8x - 8a + 1)(8x - 8a - 1)$$

75. Factoring by trial and error, we have:
$$ax^2 + (a+b)x + b = (ax + b)(x+1)$$

77. Using the common factor $(x + 1)^{1/2}$, we have:
$$(x+1)^{1/2} - (x+1)^{3/2} = (x+1)^{1/2}\left[1 - (x+1)^{2/2}\right]$$
$$= (x+1)^{1/2}(1 - x - 1)$$
$$= (x+1)^{1/2}(-x)$$
$$= -x(x+1)^{1/2}$$

79. Using the common factor $(x + 1)^{-3/2}$, we have:
$$(x+1)^{-1/2} - (x+1)^{-3/2} = (x+1)^{-3/2}\left[(x+1)^1 - 1\right]$$
$$= (x+1)^{-3/2}(x)$$
$$= x(x+1)^{-3/2}$$

81. Using the common factor $\frac{1}{3}(2x+3)^{1/2}$, we have:
$$(2x+3)^{1/2} - \tfrac{1}{3}(2x+3)^{3/2} = \tfrac{1}{3}(2x+3)^{1/2}\left[3 - (2x+3)^1\right]$$
$$= \tfrac{1}{3}(2x+3)^{1/2}(3-2x-3)$$
$$= \tfrac{1}{3}(2x+3)^{1/2}(-2x)$$
$$= -\tfrac{2}{3}x(2x+3)^{1/2}$$

83. Factoring the sum of cubes and then grouping, we have:
$$A^3 + B^3 + 3AB(A+B) = (A+B)(A^2 - AB + B^2) + (A+B)(3AB)$$
$$= (A+B)(A^2 - AB + B^2 + 3AB)$$
$$= (A+B)(A^2 + 2AB + B^2)$$
$$= (A+B)(A+B)^2$$
$$= (A+B)^3$$

85. Using the common factor $x(a^2 + x^2)^{-3/2}$, we have:
$$2x(a^2 + x^2)^{-1/2} - x^3(a^2 + x^2)^{-3/2} = x(a^2 + x^2)^{-3/2}\left[2(a^2 + x^2) - x^2\right]$$
$$= x(a^2 + x^2)^{-3/2}(2a^2 + 2x^2 - x^2)$$
$$= x(a^2 + x^2)^{-3/2}(2a^2 + x^2)$$

87. Using grouping (first with fourth term, second with third term), we have:
$$y^4 - p^2q^2 - (p+q)y^3 + pq(p+q)y = (y^2 + pq)(y^2 - pq) - (p+q)(y)(y^2 - pq)$$
$$= (y^2 - pq)(y^2 + pq - py - qy)$$

89. (a) We complete the tables:

x	$\dfrac{x^2-16}{x-4}$
3.9	7.9
3.99	7.99
3.999	7.999
3.9999	7.9999
3.99999	7.99999

x	$\dfrac{x^2-16}{x-4}$
4.1	8.1
4.01	8.01
4.001	8.001
4.0001	8.0001
4.00001	8.00001

(b) As x approaches 4, the value of the expression $\frac{x^2-16}{x-4}$ approaches 8.

(c) Since $x^2 - 16 = (x+4)(x-4)$, then:
$$\frac{x^2 - 16}{x - 4} = \frac{(x+4)(x-4)}{x-4} = x + 4, \text{ if } x \neq 4$$
Thus as x approaches 4, we would expect $x + 4$ to approach 8.

91. Adding and subtracting $16x^2$, we have:
$$x^4 + 64 = x^4 + 16x^2 + 64 - 16x^2$$
$$= \left(x^2 + 8\right)^2 - 16x^2$$
$$= \left(x^2 + 8 + 4x\right)\left(x^2 + 8 - 4x\right)$$
$$= \left(x^2 + 4x + 8\right)\left(x^2 - 4x + 8\right)$$

93. We factor as follows:
$$(x+y)^2 + (x+z)^2 - (z+t)^2 - (y+t)^2$$
$$= \left[(x+y)^2 - (y+t)^2\right] + \left[(x+z)^2 - (z+t)^2\right]$$
$$= \left[x+y-(y+t)\right]\left[x+y+(y+t)\right] + \left[x+z-(z+t)\right]\left[x+z+(z+t)\right]$$
$$= (x-t)(x+2y+t) + (x-t)(x+2z+t)$$
$$= (x-t)(x+2y+t+x+2z+t)$$
$$= (x-t)(2x+2y+2z+2t)$$
$$= 2(x-t)(x+y+z+t)$$

95. We factor as follows:
$$(b-c)^3 + (c-a)^3 + (a-b)^3$$
$$= b^3 - 3b^2c + 3bc^2 - c^3 + c^3 - 3c^2a + 3ca^2 - a^3 + a^3 - 3a^2b + 3ab^2 - b^3$$
$$= -3b^2c + 3bc^2 - 3c^2a + 3ca^2 - 3a^2b + 3ab^2$$
$$= \left(3bc^2 - 3c^2a\right) + \left(3ca^2 - 3b^2c\right) + \left(3ab^2 - 3a^2b\right)$$
$$= 3c^2(b-a) + 3c\left(a^2 - b^2\right) + 3ab(b-a)$$
$$= 3c^2(b-a) - 3c\left(b^2 - a^2\right) + 3ab(b-a)$$
$$= 3(b-a)\left[c^2 - c(b+a) + ab\right]$$
$$= 3(b-a)\left[c^2 - cb - ca + ab\right]$$
$$= 3(b-a)\left[c(c-b) - a(c-b)\right]$$
$$= 3(b-a)\left[(c-b)(c-a)\right]$$
$$= 3(b-a)(c-a)(c-b)$$

1.4 Quadratic Equations

1.   We substitute −3 for x:
$$2(-3)^2 - 6(-3) - 36 = 0$$
$$18 + 18 - 36 = 0, \text{ which is true}$$
So $x = -3$ is a solution.

3.   We substitute $-\frac{1}{4}$ for x:
$$4\left(-\frac{1}{4}\right)^2 - 1 = 0$$
$$4\left(\frac{1}{16}\right) - 1 = 0$$
$$\frac{1}{4} - 1 = 0, \text{ which is false.}$$

5.   We substitute $-1 - \sqrt{7}$ for x:
$$\left(-1 - \sqrt{7}\right)^2 - 2\left(-1 - \sqrt{7}\right) - 6 = 0$$
$$1 + 2\sqrt{7} + 7 + 2 + 2\sqrt{7} - 6 = 0$$
$$4\sqrt{7} + 4 = 0, \text{ which is false}$$

7.   Factoring, we have:
$$x^2 - 5x - 6 = 0$$
$$(x - 6)(x + 1) = 0$$
$$x = 6 \text{ or } x = -1$$

9.   Factoring, we have:
$$x^2 - 100 = 0$$
$$(x - 10)(x + 10) = 0$$
$$x = \pm 10$$

11.   Factoring, we have:
$$25x^2 - 60x + 36 = 0$$
$$(5x - 6)^2 = 0$$
$$5x - 6 = 0$$
$$x = \tfrac{6}{5} \text{ (double root)}$$

13.   Factoring, we have:
$$10z^2 - 13z - 3 = 0$$
$$(5z + 1)(2z - 3) = 0$$
$$z = -\tfrac{1}{5} \text{ or } z = \tfrac{3}{2}$$

15. Factoring, we have:
$$(x+1)^2 - 4 = 0$$
$$(x+1)^2 - 2^2 = 0$$
$$(x+1-2)(x+1+2) = 0$$
$$(x-1)(x+3) = 0$$
$$x = 1 \text{ or } x = -3$$

17. Factoring, we have:
$$x(2x-13) = -6$$
$$2x^2 - 13x + 6 = 0$$
$$(2x-1)(x-6) = 0$$
$$x = \tfrac{1}{2} \text{ or } x = 6$$

19. Factoring, we have:
$$x^3 + x^2 - 156x = 0$$
$$x\left(x^2 + x - 156\right) = 0$$
$$x(x+13)(x-12) = 0$$
$$x = 0, -13, \text{ or } 12$$

21. Factoring by grouping, we have:
$$8y^3 - 4y^2 - 6y + 3 = 0$$
$$4y^2(2y-1) - 3(2y-1) = 0$$
$$(2y-1)\left(4y^2 - 3\right) = 0$$
$$y = \tfrac{1}{2} \text{ or } y = \pm\sqrt{\tfrac{3}{4}} = \pm\tfrac{1}{2}\sqrt{3}$$

23. Multiplying by -1 and factoring by grouping, we have:
$$-y^3 + y^2 + 2y = 2$$
$$y^3 - y^2 - 2y = -2$$
$$y^3 - y^2 - 2y + 2 = 0$$
$$y^2(y-1) - 2(y-1) = 0$$
$$(y-1)\left(y^2 - 2\right) = 0$$
$$y = 1 \text{ or } y = \pm\sqrt{2}$$

25. Using the quadratic formula with $a = 1$, $b = 10$, $c = 9$, we have:
$$x = \frac{-10 \pm \sqrt{100 - 36}}{2} = \frac{-10 \pm \sqrt{64}}{2} = \frac{-10 \pm 8}{2} = -1 \text{ or } -9$$

27. Using the quadratic formula with $a = 1$, $b = -1$, $c = -5$, we have:
$$x = \frac{1 \pm \sqrt{1 - 4(1)(-5)}}{2(1)} = \frac{1 \pm \sqrt{21}}{2}$$

29.   Using the quadratic formula with $a = 2$, $b = 3$, $c = -4$, we have:

$$x = \frac{-3 \pm \sqrt{9 - 4(2)(-4)}}{2(2)} = \frac{-3 \pm \sqrt{41}}{4}$$

31.   Using the quadratic formula with $a = 12$, $b = 32$, $c = 5$, we have:

$$x = \frac{-32 \pm \sqrt{(32)^2 - 4(12)(5)}}{2(12)} = \frac{-32 \pm \sqrt{784}}{24} = \frac{-32 \pm 28}{24}$$

$$x = \frac{-32 + 28}{24} = \frac{-4}{24} = -\frac{1}{6} \text{ or } x = \frac{-32 - 28}{24} = \frac{-60}{24} = -\frac{5}{2}$$

33.   First re-write the equation as $2x^2 - x - 5 = 0$. Using the quadratic formula with $a = 2$, $b = -1$, $c = -5$, we have:

$$x = \frac{1 \pm \sqrt{1 - 4(2)(-5)}}{2(2)} = \frac{1 \pm \sqrt{41}}{4}$$

35.   First re-write the equation as $-6x^2 + 12x + 1 = 0$. Using the quadratic formula with $a = -6$, $b = 12$, $c = 1$, we have:

$$x = \frac{-12 \pm \sqrt{144 - 4(-6)(1)}}{2(-6)} = \frac{-12 \pm \sqrt{168}}{-12} = \frac{-12 \pm 2\sqrt{42}}{-12} = \frac{6 \pm \sqrt{42}}{6}$$

37.   Using the quadratic formula, we have:

$$x = \frac{-3 \pm \sqrt{9 - 4(1)(2.249)}}{2(1)} \approx -1.47, -1.53$$

39.   Using the quadratic formula with $a = 1$, $b = 156$, and $c = 5963$, we have:

$$x = \frac{-156 \pm \sqrt{(156)^2 - 4(1)(5963)}}{2(1)} = \frac{-156 \pm \sqrt{484}}{2} = \frac{-156 \pm 22}{2}$$

$$x = \frac{-156 + 22}{2} = \frac{-134}{2} = -67 \text{ or } x = \frac{-156 - 22}{2} = \frac{-178}{2} = -89$$

41.   Solving by factoring, we have:

$$(3x - 2)^2 = 3x - 2$$
$$9x^2 - 12x + 4 = 3x - 2$$
$$9x^2 - 15x + 6 = 0$$
$$3(3x - 2)(x - 1) = 0$$
$$x = \tfrac{2}{3} \text{ or } x = 1$$

43.   Using the quadratic formula with $a = 3$, $b = 4$, and $c = -3$, we have:

$$x = \frac{-4 \pm \sqrt{(4)^2 - 4(3)(-3)}}{2(3)} = \frac{-4 \pm \sqrt{52}}{6} = \frac{-4 \pm 2\sqrt{13}}{6} = \frac{-2 \pm \sqrt{13}}{3}$$

45. Using the quadratic formula, we have:

$$x^2 = 24$$
$$x^2 - 24 = 0$$
$$x = \frac{0 \pm \sqrt{0 - 4(-24)}}{2} = \frac{\pm\sqrt{96}}{2} = \frac{\pm 4\sqrt{6}}{2} = \pm 2\sqrt{6}$$

47. Using the quadratic formula, we have:

$$x(x-1) = 1$$
$$x^2 - x = 1$$
$$x^2 - x - 1 = 0$$
$$x = \frac{1 \pm \sqrt{1 - 4(-1)}}{2} = \frac{1 \pm \sqrt{5}}{2}$$

49. Multiply by 6 to clear fractions, then use the quadratic formula:

$$6\left(\tfrac{1}{2}x^2 - x - \tfrac{1}{3}\right) = 6(0)$$
$$3x^2 - 6x - 2 = 0$$
$$x = \frac{6 \pm \sqrt{36 - 4(3)(-2)}}{2(3)} = \frac{6 \pm \sqrt{60}}{6} = \frac{6 \pm 2\sqrt{15}}{6} = \frac{3 \pm \sqrt{15}}{3}$$

51. Multiply by $\sqrt{5}$ to simplify the equation:

$$\sqrt{5}\left(2\sqrt{5}x^2 - x - 2\sqrt{5}\right) = \sqrt{5}(0)$$
$$10x^2 - \sqrt{5}x - 10 = 0$$

Using the quadratic formula, we have:

$$x = \frac{\sqrt{5} \pm \sqrt{5 - 4(10)(-10)}}{2(10)} = \frac{\sqrt{5} \pm \sqrt{405}}{20} = \frac{\sqrt{5} \pm 9\sqrt{5}}{20}$$
$$x = \frac{\sqrt{5} + 9\sqrt{5}}{20} = \frac{10\sqrt{5}}{20} = \frac{\sqrt{5}}{2} \quad \text{or } x = \frac{\sqrt{5} - 9\sqrt{5}}{20} = \frac{-8\sqrt{5}}{20} = -\frac{2\sqrt{5}}{5}$$

53. Solving for x by factoring, we have:

$$2y^2x^2 - 3yx + 1 = 0$$
$$(2yx - 1)(yx - 1) = 0$$
$$x = \frac{1}{2y} \quad \text{or } x = \frac{1}{y}$$

55. Solving for x by factoring, we have:

$$(x - p)^2 + (x - q)^2 = p^2 + q^2$$
$$x^2 - 2px + p^2 + x^2 - 2qx + q^2 = p^2 + q^2$$
$$2x^2 - 2px - 2qx = 0$$
$$x^2 - px - qx = 0$$
$$x(x - p - q) = 0$$
$$x = 0 \text{ or } x = p + q$$

57.   Solving for x by factoring, we have:
$$12x^2 = ax + 20a^2$$
$$12x^2 - ax - 20a^2 = 0$$
$$(4x + 5a)(3x - 4a) = 0$$
$$x = -\frac{5a}{4} \text{ or } x = \frac{4a}{3}$$

59.   Solving for r using the quadratic formula, we have:
$$2\pi r^2 + 2\pi rh = 20\pi$$
$$2\pi r^2 + 2\pi rh - 20\pi = 0$$
$$2\pi(r^2 + rh - 10) = 0$$
$$r^2 + rh - 10 = 0$$
$$r = \frac{-h \pm \sqrt{h^2 - 4(-10)}}{2(1)} = \frac{-h \pm \sqrt{h^2 + 40}}{2}$$

61.   Solving for t by factoring, we have:
$$-16t^2 + v_0 t = 0$$
$$t(-16t + v_0) = 0$$
$$t = 0 \text{ or } t = \frac{v_0}{16}$$

63.   Solving for x by factoring:
$$x^3 + bx^2 - 2b^2 x = 0$$
$$x(x^2 + bx - 2b^2) = 0$$
$$x(x + 2b)(x - b) = 0$$
$$x = 0, -2b, b$$

65.   Given $a = 1$, $b = -12$, $c = 16$, we compute the discriminant:
$$b^2 - 4ac = 144 - 4(1)(16) = 80 > 0$$
The equation has two real roots.

67.   Given $a = 4$, $b = -5$, $c = -\frac{1}{2}$, we compute the discriminant:
$$b^2 - 4ac = 25 - 4(4)\left(-\frac{1}{2}\right) = 33 > 0$$
The equation has two real roots.

69.   Given $a = 1$, $b = \sqrt{3}$, $c = \frac{3}{4}$, we compute the discriminant:
$$b^2 - 4ac = 3 - 4(1)\left(\frac{3}{4}\right) = 0$$
The equation has one real root.

71. Given $a = 1$, $b = -\sqrt{5}$, $c = 1$, we compute the discriminant:
$$b^2 - 4ac = 5 - 4(1)(1) = 1 > 0$$
The equation has two real roots.

73. We set the discriminant equal to 0:
$$b^2 - 4ac = 0$$
$$144 - 4(1)(k) = 0$$
$$4k = 144$$
$$k = 36$$

75. We set the discriminant equal to 0:
$$b^2 - 4ac = 0$$
$$k^2 - 4(1)(5) = 0$$
$$k = \pm\sqrt{20} = \pm 2\sqrt{5}$$

77. (a) We solve the equation by squaring each side:
$$\sqrt{x - 8} = 4$$
$$x - 8 = 16$$
$$x = 24$$
Upon checking, we find this is a solution.

(b) Again we solve the equation by squaring each side:
$$\sqrt{x - 8} = 2x + 1$$
$$x - 8 = 4x^2 + 4x + 1$$
$$4x^2 + 3x + 9 = 0$$
Using the quadratic formula with $a = 4$, $b = 3$, and $c = 9$, we have:
$$x = \frac{-3 \pm \sqrt{(3)^2 - 4(4)(9)}}{2(4)} = \frac{-3 \pm \sqrt{-135}}{8}$$
There are no real number solutions to the equation.

79. Squaring both sides yields:
$$\sqrt{1 - 3x} = 2$$
$$1 - 3x = 4$$
$$-3x = 3$$
$$x = -1$$
Upon checking, we find that -1 is the solution.

81. Letting $x^2 = t$ and $x^4 = t^2$, we have:
$$\sqrt{t^2 - 13t + 37} = 1$$
$$t^2 - 13t + 37 = 1$$
$$t^2 - 13t + 36 = 0$$
$$(t - 9)(t - 4) = 0$$
$$t = 4, 9$$
Since $x^2 = t$, then $x^2 = 4$ or $x^2 = 9$, thus $x = \pm 2$ or $x = \pm 3$. Upon checking, we find that ± 2 and ± 3 are solutions.

83. Isolating $\sqrt{1 - 2x}$ and squaring yields:
$$\sqrt{1 - 2x} = 4 - \sqrt{x + 5}$$
$$1 - 2x = 16 - 8\sqrt{x + 5} + x + 5$$
$$8\sqrt{x + 5} = 3x + 20$$
$$64(x + 5) = 9x^2 + 120x + 400$$
$$0 = 9x^2 + 56x + 80$$
$$0 = (x + 4)(9x + 20)$$
$$x = -4, -\tfrac{20}{9}$$
Upon checking, we find that -4 and $-\tfrac{20}{9}$ are solutions.

85. Isolating $\sqrt{3 + 2t}$ and squaring yields:
$$\sqrt{3 + 2t} = 1 - \sqrt{-1 + 4t}$$
$$3 + 2t = 1 - 2\sqrt{-1 + 4t} - 1 + 4t$$
$$2\sqrt{-1 + 4t} = 2t - 3$$
$$4(-1 + 4t) = 4t^2 - 12t + 9$$
$$0 = 4t^2 - 28t + 13$$
$$0 = (2t - 1)(2t - 13)$$
$$t = \tfrac{1}{2}, \tfrac{13}{2}$$
Upon checking, we find that neither of these values are solutions.

87. Isolating $\sqrt{2x + 1}$ and squaring yields:
$$\sqrt{2x + 1} = 1 - \sqrt{x + 4}$$
$$2x + 1 = 1 - 2\sqrt{x + 4} + (x + 4)$$
$$2x + 1 = x + 5 - 2\sqrt{x + 4}$$
$$x - 4 = -2\sqrt{x + 4}$$
Squaring each side again, we have:
$$(x - 4)^2 = 4(x + 4)$$
$$x^2 - 8x + 16 = 4x + 16$$
$$x^2 - 12x = 0$$
$$x(x - 12) = 0$$
$$x = 0, 12$$
Upon checking, we find that neither of these values are solutions.

89. Multiplying through by the least common denominator $x(x + 5)$ gives us:
$$3x + 4(x + 5) = 2x(x + 5)$$
$$7x + 20 = 2x^2 + 10x$$
$$-2x^2 - 3x + 20 = 0$$
$$2x^2 + 3x - 20 = 0$$
$$(2x - 5)(x + 4) = 0$$
$$x = \tfrac{5}{2} \text{ or } x = -4$$
Both of these values check in the original equation.

91. Multiplying through by the least common denominator $6x + 1$ gives us:
$$-6x^2 + 5x - 1 = 0$$
$$6x^2 - 5x + 1 = 0$$
$$(3x - 1)(2x - 1) = 0$$
$$x = \tfrac{1}{3} \text{ or } x = \tfrac{1}{2}$$
Both values check in the original equation.

93. Multiplying through by the least common denominator $(x + 1)(x - 2)(x - 3)$ gives us:
$$3x^2 - 6x - 3 + (5 - 2x)(x + 1) = 0$$
$$3x^2 - 6x - 3 - 2x^2 + 3x + 5 = 0$$
$$x^2 - 3x + 2 = 0$$
$$(x - 2)(x - 1) = 0$$
$$x = 2 \text{ or } x = 1$$
Of these two values, only $x = 1$ checks in the original equation.

95. Multiplying through by the least common denominator $(x - 1)(x + 1)(x - 3)$ gives us:
$$2x(x + 3) - (x - 1)(x + 1) = 0$$
$$2x^2 + 6x - x^2 + 1 = 0$$
$$x^2 + 6x + 1 = 0$$
$$x = \frac{-6 \pm \sqrt{36 - 4}}{2} = \frac{-6 \pm 4\sqrt{2}}{2} = -3 \pm 2\sqrt{2}$$
Both values check in the original equation.

97. Since $x^2 + 4x + 3 = (x + 1)(x + 3)$, multiply through by the least common denominator $(x + 1)(x + 3)$ to yield:
$$(x - 1)(x + 3) - (x + 1)(x + 1) + 4 = 0$$
$$\left(x^2 + 2x - 3\right) - \left(x^2 + 2x + 1\right) + 4 = 0$$
$$-4 + 4 = 0$$
Since this last equation is true for any real number, any value of x (except $x = -1$ or $x = -3$, which yield 0 denominators) is a solution to the equation.

99.   Squaring both sides yields:

$$\sqrt{\sqrt{x}+\sqrt{a}}+\sqrt{\sqrt{x}-\sqrt{a}}=\sqrt{2\sqrt{x}+2\sqrt{b}}$$

$$\left(\sqrt{\sqrt{x}+\sqrt{a}}+\sqrt{\sqrt{x}-\sqrt{a}}\right)^2=\left(\sqrt{2\sqrt{x}+2\sqrt{b}}\right)^2$$

$$\sqrt{x}+\sqrt{a}+2\sqrt{\sqrt{x}+\sqrt{a}}\sqrt{\sqrt{x}-\sqrt{a}}+\sqrt{x}-\sqrt{a}=2\sqrt{x}+2\sqrt{b}$$

$$\sqrt{\sqrt{x}+\sqrt{a}}\sqrt{\sqrt{x}-\sqrt{a}}=\sqrt{b}$$

$$\left(\sqrt{x}+\sqrt{a}\right)\left(\sqrt{x}-\sqrt{a}\right)=b$$

$$x-a=b$$

$$x=a+b$$

Upon checking, we find that this value of x satisfies the original equation. Note: the following fact is useful in carrying out the check. Two non-negative quantities are equal if and only if their squares are equal.

101.   Let $t=\sqrt{x}$. Then the given equation can be written as:

$$\frac{t-a}{t}=\frac{t+a}{t-b}$$

$$t^2+at=t^2-at-bt+ab$$

$$2at+bt=ab$$

$$t(2a+b)=ab$$

$$t=\frac{ab}{2a+b}$$

Thus $x=t^2=\frac{a^2b^2}{(2a+b)^2}$. This value of x checks in the original equation.

103.   Let $t=x^2-x-1$. Then the given equation becomes:

$$\sqrt{t}-\frac{2}{\sqrt{t}}=1$$

$$t-2=\sqrt{t}$$

$$t^2-4t+4=t$$

$$t^2-5t+4=0$$

$$(t-4)(t-1)=0$$

$$t=4 \ \ \text{or} \ \ t=1$$

Upon checking, we find that $t=4$ satisfies the original equation involving t, but $t=1$ does not. With $t=4$ we have:

$$x^2-x-1=4$$

$$x^2-x-5=0$$

Now applying the quadratic formula:

$$x=\frac{1\pm\sqrt{1-4(-5)}}{2}=\frac{1\pm\sqrt{21}}{2}$$

105. Replacing h by 3 and S by 8π in the formula $S = 2\pi r^2 + 2\pi rh$ yields:

$$2\pi r^2 + 2\pi r(3) = 8\pi$$
$$r^2 + 3r = 4$$
$$r^2 + 3r - 4 = 0$$
$$(r + 4)(r - 1) = 0$$
$$r = -4 \text{ or } r = 1$$

We choose the positive root here because $r > 0$. Thus $r = 1$ meter.

107. (a) Replacing t by 10 in the equation yields:

$$P = -(10)^2 + 26(10) + 106 = -100 + 260 + 106 = 266$$

So after 10 days approximately 266 people will have caught the flu.

(b) With $P = 250$, the equation yields:

$$-t^2 + 26t + 106 = 250$$
$$t^2 - 26t + 144 = 0$$
$$(t - 8)(t - 18) = 0$$

Thus $t = 8$ days or $t = 18$ days. We discard the value $t = 18$ because the problem states that $t \le 13$. After 8 days approximately 250 people will have caught the flu.

109. Denoting the lengths of the two pieces by x and $L - x$, we have:

$$\left(\frac{x}{4}\right)^2 + \left(\frac{L-x}{4}\right)^2 = \frac{5L^2}{128}$$
$$\frac{x^2}{16} + \frac{L^2 - 2Lx + x^2}{16} = \frac{5L^2}{128}$$

After multiplying both sides by 128 and then simplifying, we obtain:

$$16x^2 - 16Lx + 3L^2 = 0$$
$$(4x - 3L)(4x - L) = 0$$
$$x = \frac{3L}{4} \text{ or } x = \frac{L}{4}$$

If $x = \frac{3L}{4}$, then $L - x = \frac{L}{4}$. Similarly, if $x = \frac{L}{4}$, then $L - x = \frac{3L}{4}$. So in either case the lengths of the two pieces are $\frac{L}{4}$ inches and $\frac{3L}{4}$ inches.

111. (a) Making the desired substitution $x = y + k$, we have:

$$(y + k)^2 + (y + k) - 1 = 0$$
$$y^2 + 2ky + k^2 + y + k - 1 = 0$$
$$y^2 + (2k + 1)y = 1 - k - k^2$$

(b) When $k = -\frac{1}{2}$, we have:

$$y^2 + (-1 + 1)y = 1 + \frac{1}{2} - \frac{1}{4}$$
$$y^2 = \frac{5}{4}$$

(c) Solving for y, we have $y = \pm\frac{\sqrt{5}}{2}$. Since $x = y + k$, $x = -\frac{1}{2} \pm \frac{\sqrt{5}}{2}$.

113. If the roots are A and B, then:
$$x^2 + Ax + B = (x - A)(x - B)$$
$$x^2 + Ax + B = x^2 - (A + B)x + AB$$
Setting the coefficients equal, we have:

$-(A + B) = A$ $\qquad$ and $\qquad$ $AB = B$

$\quad -A - B = A$ $\qquad\qquad\qquad$ $B(A - 1) = 0$

$\quad 2A + B = 0$ $\qquad\qquad\qquad$ $B = 0$ or $A = 1$

If $B = 0$, then $A = 0$, and if $A = 1$, then $B = -2$. So the values of A and B are:

$A = 0$ and $B = 0$

$A = 1$ and $B = -2$

115. Let $\alpha = \dfrac{-b + \sqrt{b^2 - 4ac}}{2a}$ and $\beta = \dfrac{-b - \sqrt{b^2 - 4ac}}{2a}$.

(a) Working from the left-hand side, we have:
$$\alpha + \beta = \frac{-2b}{2a} = \frac{-b}{a}$$

(b) Working from the left-hand side, we have:
$$\alpha\beta = \frac{b^2 - (b^2 - 4ac)}{4a^2} = \frac{b^2 - b^2 + 4ac}{4a^2} = \frac{4ac}{4a^2} = \frac{c}{a}$$

(c) Using the hint, we have:
$$\alpha^2 + \beta^2 = (\alpha + \beta)^2 - 2\alpha\beta = \left(-\frac{b}{a}\right)^2 - 2\left(\frac{c}{a}\right) = \frac{b^2}{a^2} - \frac{2ac}{a^2} = \frac{b^2 - 2ac}{a^2}$$

(d) Using the hint, we have:
$$\frac{1}{\alpha^2} + \frac{1}{\beta^2} = \frac{\alpha^2 + \beta^2}{\alpha^2\beta^2} = \frac{\alpha^2 + \beta^2}{(\alpha\beta)^2} = \frac{\frac{b^2 - 2ac}{a^2}}{\left(\frac{c}{a}\right)^2} = \frac{b^2 - 2ac}{c^2}$$

117. (a) Let r be one root, thus $r^2 + pr + q = 0$. Since $3r$ will be another root, then $(3r)^2 + p(3r) + q = 0$, thus $9r^2 + 3pr + q = 0$. Therefore:
$$r^2 + pr = 9r^2 + 3pr$$
$$8r^2 + 2pr = 0$$
$$2r(4r + p) = 0$$
$$r = 0 \text{ or } r = -\frac{p}{4}$$

Now $r = 0$ implies $q = 0$, which is impossible. So $r = \frac{-p}{4}$, and:

$$\left(-\frac{p}{4}\right)^2 + p\left(-\frac{p}{4}\right) + q = 0$$

$$\frac{p^2}{16} - \frac{p^2}{4} + q = 0$$

$$\frac{-3p^2}{16} + q = 0$$

$$-3p^2 + 16q = 0$$

$$16q = 3p^2$$

$$\frac{p^2}{q} = \frac{16}{3}$$

(b) Let r be one root, so $r^2 + pr + q = 0$. Since nr will be another root, then $(nr)^2 + p(nr) + q = 0$, so $n^2r^2 + npr + q = 0$. Therefore:

$$r^2 + pr = n^2r^2 + npr$$

$$r + p = n^2r + np \quad \text{(since } r \neq 0)$$

$$p - np = n^2r - r$$

$$p(1 - n) = (n^2 - 1)r$$

$$r = \frac{p(1-n)}{n^2 - 1} = \frac{-p}{n+1}$$

Therefore:

$$\left(\frac{-p}{n+1}\right)^2 + p\left(\frac{-p}{n+1}\right) + q = 0$$

$$\frac{p^2}{(n+1)^2} - \frac{p^2}{n+1} + q = 0$$

$$p^2 - (n+1)p^2 + (n+1)^2q = 0$$

$$(n+1)^2q = np^2$$

$$\frac{p^2}{q} = \frac{(n+1)^2}{n}$$

119. The equation can be solved by means of the quadratic formula. Alternately, we have the following solution by factoring:

$$[abx - (a+b)][(a+b)x + 2] = 0$$

$$abx = a + b \quad \text{or} \quad (a+b)x + 2 = 0$$

$$x = \frac{a+b}{ab} \quad \text{or} \quad x = \frac{-2}{a+b}$$

The two solutions are $\frac{a+b}{ab}$ and $\frac{-2}{a+b}$.

1.5 Inequalities

1.   We solve the inequality:
$$x + 5 < 4$$
$$x < -1$$
The solution set is $(-\infty, -1)$.

3.   We solve the inequality:
$$1 - 3x \le 0$$
$$-3x \le -1$$
$$x \ge \tfrac{1}{3}$$
The solution set is $\left[\tfrac{1}{3}, \infty\right)$.

5.   We solve the inequality:
$$4x + 6 < 3(x - 1) - x$$
$$4x + 6 < 2x - 3$$
$$2x < -9$$
$$x < -\tfrac{9}{2}$$
The solution set is $\left(-\infty, -\tfrac{9}{2}\right)$.

7.   We solve the inequality:
$$1 - 2(t + 3) - t \le 1 - 2t$$
$$1 - 2t - 6 - t \le 1 - 2t$$
$$-6 - t \le 0$$
$$-t \le 6$$
$$t \ge -6$$
The solution set is $[-6, \infty)$.

9.   Multiplying by 15, we solve the inequality:
$$\frac{3x}{5} - \frac{x - 1}{3} < 1$$
$$9x - 5(x - 1) < 15$$
$$4x + 5 < 15$$
$$4x < 10$$
$$x < \tfrac{5}{2}$$
The solution set is $\left(-\infty, \tfrac{5}{2}\right)$.

11. Multiplying by 20, we solve the inequality:
$$\frac{x-1}{4} - \frac{2x+3}{5} \le x$$
$$5(x-1) - 4(2x+3) \le 20x$$
$$5x - 5 - 8x - 12 \le 20x$$
$$-23x \le 17$$
$$x \ge -\tfrac{17}{23}$$
The solution set is $\left[-\tfrac{17}{23}, \infty\right)$.

13. We solve the double inequality:
$$-2 \le x - 6 \le 0$$
$$4 \le x \le 6$$
The solution set is $[4, 6]$.

15. Multiplying by 3, we solve the double inequality:
$$-1 \le \frac{1-4t}{3} \le 1$$
$$-3 \le 1 - 4t \le 3$$
$$-4 \le -4t \le 2$$
$$1 \ge t \ge -\tfrac{1}{2}$$
$$-\tfrac{1}{2} \le t \le 1$$
The solution set is $\left[-\tfrac{1}{2}, 1\right]$.

17. Multiplying by 2, we solve the double inequality:
$$0.99 < \frac{x}{2} - 1 < 0.999$$
$$1.98 < x - 2 < 1.998$$
$$3.98 < x < 3.998$$
The solution set is $(3.98, 3.998)$.

19. (a) The inequality is equivalent to $-\tfrac{1}{2} \le x \le \tfrac{1}{2}$, so the solution set is $\left[-\tfrac{1}{2}, \tfrac{1}{2}\right]$.

(b) The inequality is equivalent to $x > \tfrac{1}{2}$ or $x < -\tfrac{1}{2}$, so the solution set is
$\left(-\infty, -\tfrac{1}{2}\right) \cup \left(\tfrac{1}{2}, \infty\right)$.

21. (a) The inequality is equivalent to $x > 0$ or $x < 0$, so the solution set is $(-\infty, 0) \cup (0, \infty)$.
(b) Since $|x| \ge 0$ for all x, there is no solution to this inequality.

23. (a) We solve the inequality:
$$x - 2 < 1$$
$$x < 3$$
The solution set is $(-\infty, 3)$.

(b) We solve the inequality:
$$|x-2|<1$$
$$-1<x-2<1$$
$$1<x<3$$
The solution set is $(1,3)$.

(c) We solve the inequality:
$$|x-2|>1$$
$$x-2>1 \quad \text{or} \quad x-2<-1$$
$$x>3 \quad \text{or} \quad\quad x<1$$
The solution set is $(-\infty,1)\cup(3,\infty)$.

25. (a) We solve the inequality:
$$1-x\le 5$$
$$-x\le 4$$
$$x\ge -4$$
The solution set is $[-4,\infty)$.

(b) We solve the inequality:
$$|1-x|\le 5$$
$$-5\le 1-x\le 5$$
$$-6\le -x\le 4$$
$$6\ge x\ge -4$$
The solution set is $[-4,6]$.

(c) We solve the inequality:
$$|1-x|>5$$
$$1-x>5 \quad \text{or} \quad 1-x<-5$$
$$-x>4 \quad \text{or} \quad -x<-6$$
$$x<-4 \quad \text{or} \quad\quad x>6$$
The solution set is $(-\infty,-4)\cup(6,\infty)$.

27. (a) We solve the inequality:
$$a-x<c$$
$$-x<c-a$$
$$x>a-c$$
The solution set is $(a-c,\infty)$.

(b) We solve the inequality:
$$|a-x|<c$$
$$-c<a-x<c$$
$$-c-a<-x<c-a$$
$$a+c>x>a-c$$
The solution set is $(a-c,a+c)$.

(c) We solve the inequality:
$$|a - x| \geq c$$

$$
\begin{array}{lll}
a - x \geq c & \text{or} & a - x \leq -c \\
-x \geq c - a & \text{or} & -x \leq -c - a \\
x \leq a - c & \text{or} & x \geq a + c
\end{array}
$$

The solution set is $(-\infty, a - c] \cup [a + c, \infty)$.

29. We solve the inequality:
$$\left| \frac{x - 2}{3} \right| < 4$$

$$-4 < \frac{x - 2}{3} < 4$$
$$-12 < x - 2 < 12$$
$$-10 < x < 14$$

The solution set is $(-10, 14)$.

31. We solve the inequality:
$$\left| \frac{x + 1}{2} - \frac{x - 1}{3} \right| < 1$$

$$-1 < \frac{x + 1}{2} - \frac{x - 1}{3} < 1$$
$$-6 < 3(x + 1) - 2(x - 1) < 6$$
$$-6 < x + 5 < 6$$
$$-11 < x < 1$$

The solution set is $(-11, 1)$.

33. (a) We solve the inequality:
$$\left| (x + h)^2 - x^2 \right| < 3h^2$$
$$\left| x^2 + 2xh + h^2 - x^2 \right| < 3h^2$$
$$\left| 2xh + h^2 \right| < 3h^2$$
$$-3h^2 < 2xh + h^2 < 3h^2$$
$$-4h^2 < 2xh < 2h^2$$

Since $h > 0$, we can divide by $2h$ to obtain $-2h < x < h$. The solution set is $(-2h, h)$.

(b) Since $h < 0$, when we divide the inequality in (a) by $2h$, we reverse the inequalities to obtain $-2h > x > h$. The solution set is $(h, -2h)$.

35.   Since $F = \frac{9}{5}C + 32$ is the conversion from Fahrenheit to Celsius temperature, we substitute and solve the resulting inequality:

$$-280° \leq F \leq 260°$$
$$-280° \leq \frac{9}{5}C + 32 \leq 260°$$
$$-312° \leq \frac{9}{5}C \leq 228°$$
$$-173.3° \leq C \leq 126.7°$$

To the nearest 5° C, the corresponding interval is $-175° \leq C \leq 125°$.

37.   First solve the equation $F = \frac{9}{5}C + 32$ for C. This yields:

$$C = \frac{5F - 160}{9}$$

The following inequalities are then equivalent:

$$-25 \leq C \leq 475$$
$$-25 \leq \frac{5F - 160}{9} \leq 475$$
$$-225 \leq 5F - 160 \leq 4275$$
$$-65 \leq 5F \leq 4435$$
$$-13° \leq F \leq 887°$$

The corresponding range on the Fahrenheit scale is $[-13°, 887°]$.

39.   Using the expression for C found in Exercise 37, we have:

$$5100 \leq C \leq 6500$$
$$5100 \leq \frac{5F - 160}{9} \leq 6500$$
$$45900 \leq 5F - 160 \leq 58500$$
$$46060 \leq 5F \leq 58660$$
$$9212 \leq F \leq 11732$$

Rounding to the nearest 100°, the corresponding range is $[9200°, 11700°]$.

41.   (a) We complete the table:

a	b	$\sqrt{ab}$ (G.M.)	$\frac{a+b}{2}$ (A.M.)	Which is larger, G.M. or A.M.?
1	2	1.4142	1.5	A.M.
1	3	1.7320	2.0	A.M.
1	4	2.0000	2.5	A.M.
2	3	2.4495	2.5	A.M.
3	4	3.4641	3.5	A.M.
5	10	7.0711	7.5	A.M.
9	10	9.4868	9.5	A.M.
99	100	99.4987	99.5	A.M.
999	1000	999.4999	999.5	A.M.

(b) Using the hint and squaring each side, proving the inequality is equivalent to proving the following inequality:

$$ab \le \left(\frac{a+b}{2}\right)^2$$

We note that the inequality $(a - b)^2 \ge 0$ is true for all a and b, thus:

$$(a - b)^2 \ge 0$$
$$a^2 - 2ab + b^2 \ge 0$$
$$a^2 + b^2 \ge 2ab$$

Working from the right-hand side of the original inequality:

$$\left(\frac{a+b}{2}\right)^2 = \frac{a^2 + 2ab + b^2}{4} = \frac{a^2 + b^2}{4} + \frac{ab}{2} \ge \frac{2ab}{4} + \frac{ab}{2} = \frac{ab}{2} + \frac{ab}{2} = ab$$

This proves the required inequality.

43. (a) We complete the table:

x	y	$\dfrac{x}{y}+\dfrac{y}{x}$	True or False $\dfrac{x}{y}+\dfrac{y}{x} \ge 2$
1	1	2.0000	true
2	3	2.1667	true
3	5	2.2667	true
4	7	2.3214	true
5	9	2.3556	true
9	10	2.0111	true
49	50	2.0004	true
99	100	2.0001	true

(b) Using the inequality in Exercise 41(b) and taking $a = \frac{x}{y}$ and $b = \frac{y}{x}$, we have:

$$\sqrt{ab} \le \frac{a+b}{2}$$

$$\sqrt{\frac{x}{y} \cdot \frac{y}{x}} \le \frac{\frac{x}{y}+\frac{y}{x}}{2}$$

$$1 \le \frac{\frac{x}{y}+\frac{y}{x}}{2}$$

$$2 \le \frac{x}{y} + \frac{y}{x}$$

This proves the required inequality.

45.   Since $x(y-3) + 3(x-2) = xy - 3x + 3x - 6 = xy - 6$, then:
$$\begin{aligned}
|xy-6| &= |x(y-3) + 3(x-2)| \\
&\leq |x||y-3| + 3|x-2| \\
&\leq |x| \cdot (0.01) + 3(0.1) \\
&= 0.01|x| + 0.3
\end{aligned}$$

We still need to eliminate the $|x|$ term. Note that:
$$\begin{aligned}
|x-2| &< 0.1 \\
-0.1 &< x - 2 < 0.1 \\
1.9 &< x < 2.1
\end{aligned}$$

Thus $|x| < 2.1$. Now we complete the proof:
$$|xy-6| \leq 0.01|x| + 0.3 \leq (0.01)(2.1) + 0.3 = 0.321$$

47.   (a) Using the result of Exercise 41(b), we have:
$$\sqrt{abcd} = \sqrt{(ab)(cd)} \leq \frac{ab+cd}{2}$$

(b) Working with the squares of each side and using part (a), we have:
$$\begin{aligned}
\left(\sqrt{ab} + \sqrt{cd}\right)^2 &= ab + 2\sqrt{(ad)(bc)} + cd \\
&\leq ab + 2\left(\frac{ad+bc}{2}\right) + cd \\
&= ab + ad + bc + cd \\
&= (a+c)(b+d)
\end{aligned}$$

Therefore $\sqrt{ab} + \sqrt{cd} \leq \sqrt{(a+c)(b+d)}$.

49.   (a) Let l be the length, so $xl = 25$, and $l = \frac{25}{x}$. So the perimeter is given by:
$$P = 2x + 2\left(\frac{25}{x}\right) = 2x + \frac{50}{x}$$

(b) Using the result of Exercise 41(b), we have:
$$\frac{2x + \frac{50}{x}}{2} \geq \sqrt{2x \cdot \frac{50}{x}}$$
$$2x + \frac{50}{x} \geq 2\sqrt{100} = 20$$

(c) We set $P = 20$:
$$\begin{aligned}
2x + \frac{50}{x} &= 20 \\
2x^2 + 50 &= 20x \\
x^2 - 10x + 25 &= 0 \\
(x-5)^2 &= 0 \\
x &= 5
\end{aligned}$$

Since the length is also 5, the rectangle is a square of dimensions 5 ft by 5 ft.

1.6 More on Inequalities

1. We first factor $x^2 + x - 6 = (x + 3)(x - 2)$, so the key numbers are -3 and 2.
Now draw the sign chart:

Interval	Test Number	$x+3$	$x-2$	$(x+3)(x-2)$
$(-\infty, -3)$	-4	neg.	neg.	pos.
$(-3, 2)$	0	pos.	neg.	neg.
$(2, \infty)$	3	pos.	pos.	pos.

So $x^2 + x - 6 < 0$ on the interval $(-3, 2)$.

3. We first factor $x^2 - 11x + 18 = (x - 2)(x - 9)$, so the key numbers are 2 and 9. Now draw the sign chart:

Interval	Test Number	$x-2$	$x-9$	$(x-2)(x-9)$
$(-\infty, 2)$	0	neg.	neg.	pos.
$(2, 9)$	3	pos.	neg.	neg.
$(9, \infty)$	10	pos.	pos.	pos.

So $x^2 - 11x + 18 > 0$ on the intervals $(-\infty, 2) \cup (9, \infty)$.

5. We first simplify and factor the inequality:
$$9x - x^2 \le 20$$
$$-x^2 + 9x - 20 \le 0$$
$$x^2 - 9x + 20 \ge 0$$
$$(x - 4)(x - 5) \ge 0$$
So the key numbers are 4 and 5. Now draw the sign chart:

Interval	Test Number	$x-4$	$x-5$	$(x-4)(x-5)$
$(-\infty, 4)$	0	neg.	neg.	pos.
$(4, 5)$	$\frac{9}{2}$	pos.	neg.	neg.
$(5, \infty)$	6	pos.	pos.	pos.

So $x^2 - 9x + 20 \ge 0$ (and consequently the original inequality is satisfied) on the intervals $(-\infty, 4] \cup [5, \infty)$.

7. We first factor $x^2 - 16 = (x + 4)(x - 4)$, so the key numbers are -4 and 4. Now draw the sign chart:

Interval	Test Number	$x-4$	$x+4$	$(x-4)(x+4)$
$(-\infty, -4)$	-5	neg.	neg.	pos.
$(-4, 4)$	0	neg.	pos.	neg.
$(4, \infty)$	5	pos.	pos.	pos.

So $x^2 - 16 \ge 0$ on the intervals $(-\infty, -4] \cup [4, \infty)$.

9.   We first simplify and factor the inequality:
$$16x^2 + 24x < -9$$
$$16x^2 + 24x + 9 < 0$$
$$(4x + 3)^2 < 0$$
Since $(4x + 3)^2 \geq 0$ for all real values of x, this inequality has no solution.

11.   We first factor $x^3 + 13x^2 + 42x = x(x^2 + 13x + 42) = x(x + 7)(x + 6)$, so the key numbers are 0, –7 and – 6. Now draw the sign chart:

Interval	Test Number	x	$x + 7$	$x + 6$	$x(x + 7)(x + 6)$
$(-\infty, -7)$	-8	neg.	neg.	neg.	neg.
$(-7, -6)$	$-\frac{13}{2}$	neg.	pos.	neg.	pos.
$(-6, 0)$	-1	neg.	pos.	pos.	neg.
$(0, \infty)$	1	pos.	pos.	pos.	pos.

So $x^3 + 13x^2 + 42x > 0$ on the intervals $(-7, -6) \cup (0, \infty)$.

13.   We first simplify and factor the inequality:
$$225x \leq x^3$$
$$225x - x^3 \leq 0$$
$$x(225 - x^2) \leq 0$$
$$x(15 + x)(15 - x) \leq 0$$
So the key numbers are 0, –15, and 15. Now draw the sign chart:

Interval	Test Number	x	$15 - x$	$15 + x$	$x(15 - x)(15 + x)$
$(-\infty, -15)$	-16	neg.	pos.	neg.	pos.
$(-15, 0)$	-1	neg.	pos.	pos.	neg.
$(0, 15)$	1	pos.	pos.	pos.	pos.
$(15, \infty)$	16	pos.	neg.	pos.	neg.

So $225 - x^3 \leq 0$ (and consequently the original inequality is satisfied) on the intervals $[-15, 0] \cup [15, \infty)$.

15.   Since $2x^2 + 1 \geq 1$ for all real numbers x, the inequality is satisfied for the interval $(-\infty, \infty)$.

17. We first factor $12x^3 + 17x^2 + 6x = x(12x^2 + 17x + 6) = x(4x + 3)(3x + 2)$, so the key numbers are $0, -\frac{3}{4}$ and $-\frac{2}{3}$. Now draw the sign chart:

Interval	Test Number x	$4x+3$	$3x+2$	$x(4x+3)(3x+2)$
$\left(-\infty, -\frac{3}{4}\right)$	-1	neg.	neg.	neg.	neg.
$\left(-\frac{3}{4}, -\frac{2}{3}\right)$	$-\frac{17}{24}$	neg.	pos.	neg.	pos.
$\left(-\frac{2}{3}, 0\right)$	$-\frac{1}{3}$	neg.	pos.	pos.	neg.
$(0, \infty)$	1	pos.	pos.	pos.	pos.

So $12x^3 + 17x^2 + 6x < 0$ on the intervals $\left(-\infty, -\frac{3}{4}\right) \cup \left(-\frac{2}{3}, 0\right)$.

19. The key numbers are found by solving the equation $x^2 + x - 1 = 0$. Using the quadratic formula we have:

$$x = \frac{-1 \pm \sqrt{1 - 4(-1)}}{2} = \frac{-1 \pm \sqrt{5}}{2}$$

For purposes of picking appropriate test numbers, note that:

$$\frac{-1 + \sqrt{5}}{2} \approx 0.6 \quad \text{and} \quad \frac{-1 - \sqrt{5}}{2} \approx -1.6$$

We draw the sign chart:

Interval	Test Number	$x+1.6$	$x-0.6$	x^2+x-1
$\left(-\infty, \frac{-1-\sqrt{5}}{2}\right)$	-2	neg.	neg.	pos.
$\left(\frac{-1-\sqrt{5}}{2}, \frac{-1+\sqrt{5}}{2}\right)$	0	pos.	neg.	neg.
$\left(\frac{-1+\sqrt{5}}{2}, \infty\right)$	1	pos.	pos.	pos.

So $x^2 + x - 1 > 0$ on the intervals $\left(-\infty, \frac{-1-\sqrt{5}}{2}\right) \cup \left(\frac{-1+\sqrt{5}}{2}, \infty\right)$.

21. The key numbers are found by using the quadratic formula to solve $x^2 - 8x + 2 = 0$. We have:

$$x = \frac{8 \pm \sqrt{64 - 4(2)}}{2} = \frac{8 \pm \sqrt{56}}{2} = \frac{8 \pm 2\sqrt{14}}{2} = 4 \pm \sqrt{14}$$

For purposes of picking appropriate test numbers, note that $4 + \sqrt{14} \approx 7.7$ and $4 - \sqrt{14} \approx 0.3$. We draw the sign chart:

Interval	Test Number	$x-7.7$	$x-0.3$	x^2-8x+2
$(-\infty, 4-\sqrt{14})$	0	neg.	neg.	pos.
$(4-\sqrt{14}, 4+\sqrt{14})$	4	neg.	pos.	neg.
$(4+\sqrt{14}, \infty)$	8	pos.	pos.	pos.

So $x^2 - 8x + 2 \le 0$ on the interval $\left[4 - \sqrt{14}, 4 + \sqrt{14}\right]$.

23.   The key numbers are -4, -3, and 1. We draw the sign chart:

Interval	Test Number	$x-1$	$x+3$	$x+4$	$(x-1)(x+3)(x+4)$
$(-\infty,-4)$	-5	neg.	neg.	neg.	neg.
$(-4,-3)$	$-\frac{7}{2}$	neg.	neg.	pos.	pos.
$(-3,1)$	0	neg.	pos.	pos.	neg.
$(1,\infty)$	2	pos.	pos.	pos.	pos.

So $(x-1)(x+3)(x+4) \geq 0$ on the intervals $[-4,-3] \cup [1,\infty)$.

25.   The key numbers are -4, -5, and -6. We draw the sign chart:

Interval	Test Number	$x+4$	$x+5$	$x+6$	$(x+4)(x+5)(x+6)$
$(-\infty,-6)$	-7	neg.	neg.	neg.	neg.
$(-6,-5)$	$-\frac{11}{2}$	neg.	neg.	pos.	pos.
$(-5,-4)$	$-\frac{9}{2}$	neg.	pos.	pos.	neg.
$(-4,\infty)$	0	pos.	pos.	pos.	pos.

So $(x+4)(x+5)(x+6) < 0$ on the intervals $(-\infty,-6) \cup (-5,-4)$.

27.   The key numbers are $-\frac{1}{3}$, $\frac{1}{3}$ and 2. We draw the sign chart:

Interval	Test Number	$(x-2)^2$	$(3x+1)^3$	$3x-1$	product
$\left(-\infty,-\frac{1}{3}\right)$	-1	pos.	neg.	neg.	pos.
$\left(-\frac{1}{3},\frac{1}{3}\right)$	0	pos.	pos.	neg.	neg.
$\left(\frac{1}{3},2\right)$	1	pos.	pos.	pos.	pos.
$(2,\infty)$	3	pos.	pos.	pos.	pos.

So $(x-2)^2(3x+1)^3(3x-1) > 0$ on the intervals $\left(-\infty,-\frac{1}{3}\right) \cup \left(\frac{1}{3},2\right) \cup (2,\infty)$.

29.   The key numbers are 3, -1, $-\frac{1}{2}$ and $-\frac{2}{3}$. We draw the sign chart:

Interval	Test Number	$(x-3)^2$	$(x+1)^4$	$(2x+1)^4$	$3x+2$	product
$(-\infty,-1)$	-2	pos.	pos.	pos.	neg.	neg.
$\left(-1,-\frac{2}{3}\right)$	$-\frac{3}{4}$	pos.	pos.	pos.	neg.	neg.
$\left(-\frac{2}{3},-\frac{1}{2}\right)$	$-\frac{7}{12}$	pos.	pos.	pos.	pos.	pos.
$\left(-\frac{1}{2},3\right)$	0	pos.	pos.	pos.	pos.	pos.
$(3,\infty)$	4	pos.	pos.	pos.	pos.	pos.

So $(x-3)^2(x+1)^4(2x+1)^4(3x+2) \le 0$ on the intervals $(-\infty,-1] \cup [-1,-\frac{2}{3}]$, which is equivalent to the interval $(-\infty,-\frac{2}{3}]$. It is also zero at 3 and $-1/2$, so the inequality is satisfied for the values $(-\infty,-\frac{2}{3}] \cup \{3,-\frac{1}{2}\}$.

31. We first simplify and factor the inequality:
$$20 \ge x^2(9-x^2)$$
$$20 \ge 9x^2 - x^4$$
$$x^4 - 9x^2 + 20 \ge 0$$
$$(x^2-4)(x^2-5) \ge 0$$
$$(x-2)(x+2)(x-\sqrt{5})(x+\sqrt{5}) \ge 0$$

The key numbers are $-2, 2, -\sqrt{5}$, and $\sqrt{5}$. We draw the sign chart:

Interval	Test Number	$x-\sqrt{5}$	$x-2$	$x+2$	$x+\sqrt{5}$	x^4-9x^2+20
$(-\infty,-\sqrt{5})$	-3	neg.	neg.	neg.	neg.	pos.
$(-\sqrt{5},-2)$	-2.1	neg.	neg.	neg.	pos.	neg.
$(-2,2)$	0	neg.	neg.	pos.	pos.	pos.
$(2,\sqrt{5})$	2.1	neg.	pos.	pos.	pos.	neg.
$(\sqrt{5},\infty)$	3	pos.	pos.	pos.	pos.	pos.

So $x^4 - 9x^2 + 20 \ge 0$ (and consequently the original inequality is satisfied) on the intervals $(-\infty,-\sqrt{5}] \cup [-2,2] \cup [\sqrt{5},\infty)$.

33. We first simplify and factor the inequality:
$$9(x-4) - x^2(x-4) < 0$$
$$(x-4)(9-x^2) < 0$$
$$(x-4)(3+x)(3-x) < 0$$

The key numbers are 4, -3 and 3. We draw the sign chart:

Interval	Test Number	$x-4$	$3+x$	$3-x$	$(x-4)(3+x)(3-x)$
$(-\infty,-3)$	-4	neg.	neg.	pos.	pos.
$(-3,3)$	0	neg.	pos.	pos.	neg.
$(3,4)$	3.5	neg.	pos.	neg.	pos.
$(4,\infty)$	6	pos.	pos.	neg.	neg.

So $(x+4)(3+x)(3-x) < 0$ (and consequently the original inequality is satisfied) on the intervals $(-3,3) \cup (4,\infty)$.

35. We first simplify and factor the inequality:
$$4(x^2 - 9) - (x^2 - 9)^2 > -5$$
$$4x^2 - 36 - x^4 + 18x^2 - 81 > -5$$
$$-x^4 + 22x^2 - 112 > 0$$
$$x^4 - 22x^2 + 112 < 0$$
$$(x^2 - 8)(x^2 - 14) < 0$$
$$(x - 2\sqrt{2})(x + 2\sqrt{2})(x - \sqrt{14})(x + \sqrt{14}) < 0$$
The key numbers are $2\sqrt{2}, -2\sqrt{2}, \sqrt{14}$, and $-\sqrt{14}$. We draw the sign chart:

Interval	Test Number	$x^2 - 8$	$x^2 - 14$	$x^4 - 22x^2 + 112$
$(-\infty, -\sqrt{14})$	-5	pos.	pos.	pos.
$(-\sqrt{14}, -2\sqrt{2})$	-3	pos.	neg.	neg.
$(-2\sqrt{2}, 2\sqrt{2})$	0	neg.	neg.	pos.
$(2\sqrt{2}, \sqrt{14})$	3	pos.	neg.	neg.
$(\sqrt{14}, \infty)$	5	pos.	pos.	pos.

So $x^4 - 22x^2 + 112 < 0$ (and consequently the original inequality is satisfied) on the intervals $(-\sqrt{14}, -2\sqrt{2}) \cup (2\sqrt{2}, \sqrt{14})$.

37. One of the key numbers is 4. The others are found by solving the equation $2x^2 - 6x - 1 = 0$. With the quadratic formula, we have:
$$x = \frac{6 \pm \sqrt{36 - 4(2)(-1)}}{2(2)} = \frac{6 \pm \sqrt{44}}{4} = \frac{6 \pm 2\sqrt{11}}{4} = \frac{3 \pm \sqrt{11}}{2}$$
Thus, the key numbers are 4 and $\frac{3 \pm \sqrt{11}}{2}$. For purposes of picking appropriate test numbers, note that $\frac{3 + \sqrt{11}}{2} \approx 3.2$ and that $\frac{3 - \sqrt{11}}{2} \approx -0.2$. We draw the sign chart:

Interval	Test Number	$x - 4$	$x - 3.2$	$x + 0.2$	$(x - 4)(2x^2 - 6x - 1)$
$(-\infty, -0.2)$	-1	neg.	neg.	neg.	neg.
$(-0.2, 3.2)$	0	neg.	neg.	pos.	pos.
$(3.2, 4)$	3.5	neg.	pos.	pos.	neg.
$(4, \infty)$	5	pos.	pos.	pos.	pos.

So $(x - 4)(2x^2 - 6x - 1) < 0$ on the intervals $\left(-\infty, \frac{3 - \sqrt{11}}{2}\right) \cup \left(\frac{3 + \sqrt{11}}{2}, 4\right)$.

39. We first factor the inequality:
$$x^3 + 2x^2 - x - 2 > 0$$
$$x^2(x + 2) - (x + 2) > 0$$
$$(x + 2)(x^2 - 1) > 0$$
$$(x + 2)(x + 1)(x - 1) > 0$$

The key numbers are –2, –1 and 1. We draw the sign chart:

Interval	Test Number	$x-1$	$x+1$	$x+2$	$(x-1)(x+1)(x+2)$
$(-\infty,-2)$	-3	neg.	neg.	neg.	neg.
$(-2,-1)$	-1.5	neg.	neg.	pos.	pos.
$(-1,1)$	0	neg.	pos.	pos.	neg.
$(1,\infty)$	2	pos.	pos.	pos.	pos.

So $x^3 + 2x^2 - x - 2 > 0$ on the intervals $(-2,-1)\cup(1,\infty)$.

41. The key numbers are –1 and 1. We draw the sign chart:

Interval	Test Number	$x-1$	$x+1$	$\frac{x-1}{x+1}$
$(-\infty,-1)$	-2	neg.	neg.	pos.
$(-1,1)$	0	neg.	pos.	neg.
$(1,\infty)$	2	pos.	pos.	pos.

Thus the quotient is negative on $(-1,1)$ and it is zero when $x=1$. So $\frac{x-1}{x+1}\le 0$ on the interval $(-1,1]$.

43. The key numbers are 2 and $\frac{3}{2}$. We draw the sign chart:

Interval	Test Number	$2-x$	$3-2x$	$\frac{2-x}{3-2x}$
$(-\infty,\frac{3}{2})$	0	pos.	pos.	pos.
$(\frac{3}{2},2)$	$\frac{7}{4}$	pos.	neg.	neg.
$(2,\infty)$	3	neg.	neg.	pos.

Thus the quotient is positive on $\left(-\infty,\frac{3}{2}\right)\cup(2,\infty)$ and it is zero when $x=2$. So $\frac{2-x}{3-2x}\ge 0$ on the intervals $\left(-\infty,\frac{3}{2}\right)\cup[2,\infty)$.

45. Factoring, we have $\frac{(x+1)(x-9)}{x} < 0$. So the key numbers are –1, 0, and 9. We draw the sign chart:

Interval	Test Number	$x-9$	x	$x+1$	$\frac{(x-9)(x+1)}{x}$
$(-\infty,-1)$	-2	neg.	neg.	neg.	neg.
$(-1,0)$	$-\frac{1}{2}$	neg.	neg.	pos.	pos.
$(0,9)$	1	neg.	pos.	pos.	neg.
$(9,\infty)$	10	pos.	pos.	pos.	pos.

Thus the quotient is negative on the intervals $(-\infty,-1)\cup(0,9)$.

47. Factoring the inequality yields:

$$\frac{2x^3 + 5x^2 - 7x}{3x^2 + 7x + 4} > 0$$

$$\frac{x(2x^2 + 5x - 7)}{(3x + 4)(x + 1)} > 0$$

$$\frac{x(2x + 7)(x - 1)}{(3x + 4)(x + 1)} > 0$$

So the key numbers are $0, -\frac{7}{2}, 1, -\frac{4}{3}$ and -1. We draw the sign chart:

Interval	Test Number	$x-1$	x	$x+1$	$3x+4$	$2x+7$	$\frac{x(2x+7)(x-1)}{(3x+4)(x+1)}$
$\left(-\infty, -\frac{7}{2}\right)$	-4	neg.	neg.	neg.	neg.	neg.	neg.
$\left(-\frac{7}{2}, -\frac{4}{3}\right)$	-2	neg.	neg.	neg.	neg.	pos.	pos.
$\left(-\frac{4}{3}, -1\right)$	$-\frac{7}{6}$	neg.	neg.	neg.	pos.	pos.	neg.
$(-1, 0)$	$-\frac{1}{2}$	neg.	neg.	pos.	pos.	pos.	pos.
$(0, 1)$	$\frac{1}{2}$	neg.	pos.	pos.	pos.	pos.	neg.
$(1, \infty)$	2	pos.	pos.	pos.	pos.	pos.	pos.

Thus the quotient is positive on the intervals $\left(-\frac{7}{2}, -\frac{4}{3}\right) \cup (-1, 0) \cup (1, \infty)$.

49. We first simplify the inequality:

$$\frac{x}{x+1} - 1 > 0$$

$$\frac{x - x - 1}{x + 1} > 0$$

$$\frac{-1}{x+1} > 0$$

This is true whenever $x + 1 < 0$, so $x < -1$. So the solution is $(-\infty, -1)$.

51. We first simplify the inequality:

$$\frac{1}{x} - \frac{1}{x+1} \le 0$$

$$\frac{x + 1 - x}{x(x + 1)} \le 0$$

$$\frac{1}{x(x + 1)} \le 0$$

So the key numbers are 0 and -1. We draw the sign chart:

Interval	Test Number	x	$x+1$	$\frac{1}{x(x+1)}$
$(-\infty, -1)$	-2	neg.	neg.	pos.
$(-1, 0)$	-0.5	neg.	pos.	neg.
$(0, \infty)$	1	pos.	pos.	pos.

So $\frac{1}{x(x+1)} \le 0$ (and consequently the original inequality is satisfied) on the interval $(-1, 0)$.

53. We first simplify the inequality:
$$\frac{x+2}{x+5} - 1 \le 0$$
$$\frac{x+2-x-5}{x+5} \le 0$$
$$\frac{-3}{x+5} \le 0$$
This is true whenever $x + 5 > 0$, so $x > -5$. So the solution is $(-5, \infty)$.

55. We first simplify the inequality:
$$\frac{1}{x-2} - \frac{1}{x-1} - \frac{1}{6} \ge 0$$
$$\frac{6(x-1) - 6(x-2) - (x-2)(x-1)}{6(x-2)(x-1)} \ge 0$$
$$\frac{-x^2 + 3x + 4}{6(x-2)(x-1)} \ge 0$$
$$\frac{x^2 - 3x - 4}{6(x-2)(x-1)} \le 0$$
$$\frac{(x-4)(x+1)}{6(x-2)(x-1)} \le 0$$
So the key numbers are -1, 1, 2, and 4. We draw the sign chart:

Interval	Test Number	$x-4$	$x-2$	$x-1$	$x+1$	$\cdot\frac{(x-4)(x+1)}{6(x-2)(x-1)}$
$(-\infty, -1)$	-2	neg.	neg.	neg.	neg.	pos.
$(-1, 1)$	0	neg.	neg.	neg.	pos.	neg.
$(1, 2)$	$\frac{3}{2}$	neg.	neg.	pos.	pos.	pos.
$(2, 4)$	3	neg.	pos.	pos.	pos.	neg.
$(4, \infty)$	5	pos.	pos.	pos.	pos.	pos.

So the quotient is negative on the intervals $(-1, 1) \cup (2, 4)$ and zero at $x = 4$ and $x = -1$, so our original inequality is satisfied on the intervals $[-1, 1) \cup (2, 4]$.

57. We first simplify the inequality:
$$\frac{1+x}{1-x} - \frac{1-x}{1+x} + 1 < 0$$
$$\frac{(1+x)^2 - (1-x)^2 + (1-x^2)}{(1-x)(1+x)} < 0$$
$$\frac{-x^2 + 4x + 1}{(1-x)(1+x)} < 0$$
$$\frac{x^2 - 4x - 1}{(x-1)(x+1)} < 0$$

The denominator is zero when $x = \pm 1$. Using the quadratic formula, we find the numerator is zero when $x = 2 \pm \sqrt{5}$. Thus, the key numbers are ± 1 and $2 \pm \sqrt{5}$. For purposes of picking appropriate test numbers, note that $2 + \sqrt{5} \approx 4.2$ and that $2 - \sqrt{5} \approx -0.2$. We draw the sign chart:

Interval	Test Number	$x-4.2$	$x-1$	$x+0.2$	$x+1$	$\frac{x^2-4x-1}{(x-1)(x+1)}$
$(-\infty,-1)$	-2	neg.	neg.	neg.	neg.	pos.
$(-1,-0.2)$	-0.5	neg.	neg.	neg.	pos.	neg.
$(-0.2,1)$	0	neg.	neg.	pos.	pos.	pos.
$(1,4.2)$	2	neg.	pos.	pos.	pos.	neg.
$(4.2,\infty)$	5	pos.	pos.	pos.	pos.	pos.

So $\frac{x^2-4x-1}{(x-1)(x+1)} < 0$ (and consequently the original inequality is satisfied) on the intervals $\left(-1, 2-\sqrt{5}\right) \cup \left(1, 2+\sqrt{5}\right)$.

59. We first simplify the inequality:

$$\frac{3-2x}{3+2x} - \frac{1}{x} > 0$$

$$\frac{x(3-2x)-(3+2x)}{x(3+2x)} > 0$$

$$\frac{x-2x^2-3}{x(3+2x)} > 0$$

$$\frac{2x^2-x+3}{x(3+2x)} < 0$$

The denominator is zero when $x = 0$ and $x = -\frac{3}{2}$. By using the quadratic formula, we find that there are no real numbers for which the numerator is zero. Also note that $2x^2 - x + 3$ is positive for all x. Thus 0 and $-\frac{3}{2}$ are the only key numbers. We draw the sign chart:

Interval	Test Number	$2x+3$	x	$\frac{2x^2-x+3}{x(2x+3)}$
$\left(-\infty,-\frac{3}{2}\right)$	-2	neg.	neg.	pos.
$\left(-\frac{3}{2},0\right)$	-1	pos.	neg.	neg.
$(0,\infty)$	1	pos.	pos.	pos.

So $\frac{2x^2-x+3}{x(2x+3)} < 0$ (and consquently the original inequality is satisfied) on the interval $\left(-\frac{3}{2},0\right)$.

61. We first simplify the inequality:

$$1 + \frac{1}{x} - \frac{1}{1+x} \geq 0$$

$$\frac{x(1+x) + (1+x) - x}{x(1+x)} \geq 0$$

$$\frac{x^2 + x + 1}{x(1+x)} \geq 0$$

The only key numbers are 0 and –1. By using the quadratic formula, we find that there are no real numbers for which the numerator $x^2 + x + 1$ is zero. Note that $x^2 + x + 1$ is positive for all values of x. We draw the sign chart:

Interval	Test Number	$x+1$	x	$\frac{x^2+x+1}{x(x+1)}$
$(-\infty, -1)$	-2	neg.	neg.	pos.
$(-1, 0)$	$-\frac{1}{2}$	pos.	neg.	neg.
$(0, \infty)$	1	pos.	pos.	pos.

So $\frac{x^2+x+1}{x(1+x)} \geq 0$ (and consequently the original inequality is satisfied) on the intervals $(-\infty, -1) \cup (0, \infty)$.

63. (a) We solve the inequality:

$$x^2 - 4x - 5 \geq 0$$

$$(x-5)(x+1) \geq 0$$

The key numbers are 5 and –1. We draw the sign chart:

Interval	Test Number	$x-5$	$x+1$	$(x-5)(x+1)$
$(-\infty, -1)$	-2	neg.	neg.	pos.
$(-1, 5)$	0	neg.	pos.	neg.
$(5, \infty)$	6	pos.	pos.	pos.

The solution set for $(x-5)(x+1) \geq 0$ is $(-\infty, -1] \cup [5, \infty)$.

(b) We solve the inequality $\frac{1}{x^2 - 4x - 5} \geq 0$. This will have the same solutions as in (a), except the endpoints $x = -1$ and $x = 5$ are not included. The solution set for this is $(-\infty, -1) \cup (5, \infty)$.

65. The solutions will be real provided the discriminant $b^2 - 4ac$ is non-negative. Thus, we have:

$$b^2 - 4 \geq 0$$

$$(b-2)(b+2) \geq 0$$

The key numbers are ± 2. We draw the sign chart:

Interval	Test Number	$b-2$	$b+2$	$(b-2)(b+2)$
$(-\infty,-2)$	-3	neg.	neg.	pos.
$(-2,2)$	0	neg.	pos.	neg.
$(2,\infty)$	3	pos.	pos.	pos.

The solution set consists of the two intervals $(-\infty,-2]\cup[2,\infty)$. For values of b in either of these two intervals, the equation $x^2 + bx + 1 = 0$ will have real solutions.

67. If $x = 1$ is a solution of $\frac{2a+x}{x-2a} < 1$, then we have:

$$\frac{2a+1}{1-2a} < 1$$

$$\frac{2a+1}{1-2a} - 1 < 0$$

$$\frac{2a+1-(1-2a)}{1-2a} < 0$$

$$\frac{4a}{1-2a} < 0$$

The key numbers are 0 and $\frac{1}{2}$. We draw the sign chart:

Interval	Test Number	$4a$	$1-2a$	$\frac{4a}{1-2a}$
$(-\infty,0)$	-1	neg.	pos.	neg.
$\left(0,\frac{1}{2}\right)$	$\frac{1}{4}$	pos.	pos.	pos.
$\left(\frac{1}{2},\infty\right)$	1	pos.	neg.	neg.

The allowable values of a are those numbers in either of the intervals $(-\infty,0)\cup\left(\frac{1}{2},\infty\right)$.

69. Using the Pythagorean theorem, we find the hypotenuse is $\sqrt{x^2 + (1-x)^2}$, or $\sqrt{2x^2 - 2x + 1}$. If this is less than $\frac{\sqrt{17}}{5}$, then we have:

$$\sqrt{2x^2 - 2x + 1} < \frac{\sqrt{17}}{5}$$

$$\left(5\sqrt{2x^2 - 2x + 1}\right)^2 < \left(\sqrt{17}\right)^2$$

$$50x^2 - 50x + 25 < 17$$

$$50x^2 - 50x + 8 < 0$$

$$25x^2 - 25x + 4 < 0$$

$$(5x-1)(5x-4) < 0$$

The key numbers here are $\frac{1}{5}$ and $\frac{4}{5}$. We draw the sign chart:

Interval	Test Number	$5x-1$	$5x-4$	$(5x-1)(5x-4)$
$\left(-\infty,\frac{1}{5}\right)$	0	neg.	neg.	pos.
$\left(\frac{1}{5},\frac{4}{5}\right)$	$\frac{2}{5}$	pos.	neg.	neg.
$\left(\frac{4}{5},\infty\right)$	1	pos.	pos.	pos.

The solution set is the interval $\left(\frac{1}{5},\frac{4}{5}\right)$.

71.   For the cylinder, we have:
$$\frac{V}{S}=\frac{\pi r^2 h}{2\pi r^2+2\pi rh}=\frac{\pi r^2}{2\pi r^2+2\pi r}=\frac{r}{2r+2}$$
The condition $\frac{V}{S}<\frac{1}{3}$ then becomes:
$$\frac{r}{2r+2}<\frac{1}{3}$$
Since r is positive here, we can multiply through by the positive quantity $3(2r+2)$. This yields:
$$3r<2r+2$$
$$r<2$$
Since $r>0$, the possible values of r are in the interval $(0,2)$.

73.   (a) We complete the table:

x	1	2	5	10	100	200	10,000
$\dfrac{x^2+1000}{2x^2+x}$	333.67	100.40	18.64	5.24	0.55	0.51	0.50

(b) We simplify the inequality:
$$\frac{x^2+1000}{2x^2+x}<\frac{1}{2}$$
$$\frac{x^2+1000}{x(2x+1)}-\frac{1}{2}<0$$
$$\frac{2(x^2+1000)-x(2x+1)}{2x(2x+1)}<0$$
$$\frac{-x+2000}{2x(2x+1)}<0$$

Thus the key numbers are 2000, 0, and $-\frac{1}{2}$. We draw the sign chart:

Interval	Test Number	$2x+1$	x	$-x+2000$	$\frac{-x+2000}{2x(2x+1)}$
$\left(-\infty,-\frac{1}{2}\right)$	-1	neg.	neg.	pos.	pos.
$\left(-\frac{1}{2},0\right)$	$-\frac{1}{4}$	pos.	neg.	pos.	neg.
$(0,2000)$	1	pos.	pos.	pos.	pos.
$(2000,\infty)$	2001	pos.	pos.	neg.	neg.

The solution set for $\frac{x^2+1000}{2x^2+x} < \frac{1}{2}$ is $\left(-\frac{1}{2},0\right) \cup (2000,\infty)$. So the positive real numbers which satisfy the inequality are those in the interval $(2000, \infty)$.

(c) The inequality holds when $n > 2000$, and the smallest natural number fulfilling this condition is $n = 2001$.

75. We multiply this out to obtain:

$$x^2 - 2ax + a^2 - x^2 + 2bx - b^2 > \frac{(a-b)^2}{4}$$

$$(2b - 2a)x + (a^2 - b^2) > \frac{(a-b)^2}{4}$$

$$8(b-a)x + 4(a+b)(a-b) > (a-b)^2$$

Since $a > b$, then $a - b > 0$, so we can divide by $a - b$ and preserve the inequality. Thus:

$$-8x + 4(a+b) > a - b$$
$$-8x + 4a + 4b > a - b$$
$$-8x > -3a - 5b$$
$$x < \frac{3a + 5b}{8}$$

So the solution is $\left(-\infty, \frac{3a+5b}{8}\right)$.

Chapter One Review Exercises

1. Factoring as a difference of squares:
$$a^2 - 16b^2 = a^2 - (4b)^2 = (a - 4b)(a + 4b)$$

3. Factoring as a difference of cubes:
$$8 - (a+1)^3 = \left[2 - (a+1)\right]\left[4 + 2(a+1) + (a+1)^2\right]$$
$$= (1-a)\left(4 + 2a + 2 + a^2 + 2a + 1\right)$$
$$= (1-a)\left(7 + 4a + a^2\right)$$

5. Factoring the common factor x then using the special product for $(A + B)^2$:
$$a^2x^3 + 2ax^2b + b^2x = x\left(a^2x^2 + 2abx + b^2\right) = x(ax + b)^2$$

7.   Factoring by trial and error:
$$8x^2 + 6x + 1 = (4x + 1)(2x + 1)$$

9.   Factoring the common factor a^4x^4 then as a difference of squares (twice):
$$a^4x^4 - x^8a^8 = a^4x^4(1 - x^4a^4)$$
$$= a^4x^4(1 - x^2a^2)(1 + x^2a^2)$$
$$= a^4x^4(1 - xa)(1 + xa)(1 + x^2a^2)$$

11.   Factoring by grouping:
$$8 + 12a + 6a^2 + a^3 = (8 + a^3) + (12a + 6a^2)$$
$$= (2 + a)(4 - 2a + a^2) + 6a(2 + a)$$
$$= (2 + a)(4 - 2a + a^2 + 6a)$$
$$= (2 + a)(4 + 4a + a^2)$$
$$= (2 + a)(2 + a)^2$$
$$= (2 + a)^3$$

13.   Factoring the common factor z^3 then by trial and error:
$$4x^2y^2z^3 - 3xyz^3 - z^3 = z^3(4x^2y^2 - 3xy - 1) = z^3(4xy + 1)(xy - 1)$$

15.   Factoring as a difference of squares then as a sum and difference of cubes:
$$1 - x^6 = (1^3)^2 - (x^3)^2$$
$$= (1 - x^3)(1 + x^3)$$
$$= (1 - x)(1 + x + x^2)(1 + x)(1 - x + x^2)$$

17.   Factoring by grouping the first three terms:
$$a^2x^2 + 2abx + b^2 - 4a^2b^2x^2 = (ax + b)^2 - (2abx)^2 = (ax + b - 2abx)(ax + b + 2abx)$$

19.   Factoring by grouping pairs of terms:
$$a^2 - b^2 + ac - bc + a^2b - b^2a = (a - b)(a + b) + c(a - b) + ab(a - b)$$
$$= (a - b)(a + b + c + ab)$$

21.   Using the special product for $(A - B)(A + B)$, we have:
$$(9x - y)(9x + y) = (9x)^2 - y^2 = 81x^2 - y^2$$

23.   Using the special product for $(A - B)^2$, we have:
$$(2ab - 3)^2 = (2ab)^2 - 2(2ab)(3) + 3^2 = 4a^2b^2 - 12ab + 9$$

25.   Using the special product for $(A + B)^3$, we have:
$$(2a + 3)^3 = (2a)^3 + 3(2a)^2(3) + 3(2a)(3)^2 + 3^3 = 8a^3 + 36a^2 + 54a + 27$$

27.   Using the special product for $(A - B)(A^2 + AB + B^2)$, we have:
$$(1 - 3a)(1 + 3a + 9a^2) = 1^3 - (3a)^3 = 1 - 27a^3$$

29.   Using the special product for $(A - B)(A + B)$, we have:
$$\left(x^{1/2} - y^{1/2}\right)\left(x^{1/2} + y^{1/2}\right) = \left(x^{1/2}\right)^2 - \left(y^{1/2}\right)^2 = x - y$$

31.   Using the special product for $(A + B)^3$, we have:
$$\left(x^{1/3} + y^{1/3}\right)^3 = \left(x^{1/3}\right)^3 + 3\left(x^{1/3}\right)^2\left(y^{1/3}\right) + 3\left(x^{1/3}\right)\left(y^{1/3}\right)^2 + \left(y^{1/3}\right)^3$$
$$= x + 3x^{2/3}y^{1/3} + 3x^{1/3}y^{2/3} + y$$

33.   Using the special product for $(A - B)^2$, we have:
$$\left(x^2 - 3x + 1\right)^2 = \left[x^2 - (3x - 1)\right]^2$$
$$= x^4 - 2x^2(3x - 1) + (3x - 1)^2$$
$$= x^4 - 6x^3 + 2x^2 + 9x^2 - 6x + 1$$
$$= x^4 - 6x^3 + 11x^2 - 6x + 1$$

35.   Using absolute value, the equality is $|x - 6| = 2$.

37.   Using absolute value, the equality is $|a - b| = 3$.

39.   Using absolute value, the inequality is $|x - 0| > 10$, or $|x| > 10$.

41.   Since $\sqrt{6} - 2 > 0$, then:
$$|\sqrt{6} - 2| = \sqrt{6} - 2$$

43.   Since $x^4 + x^2 + 1 > 0$, then:
$$|x^4 + x^2 + 1| = x^4 + x^2 + 1$$

45.   (a) If $x < 2$, then $x - 2 < 0$ and $x - 3 < 0$, so:
$$|x - 2| + |x - 3| = -(x - 2) - (x - 3) = -2x + 5$$

   (b) If $2 < x < 3$, then $x - 2 > 0$ and $x - 3 < 0$, so:
$$|x - 2| + |x - 3| = x - 2 - (x - 3) = x - 2 - x + 3 = 1$$

   (c) If $x > 3$, then $x - 2 > 0$ and $x - 3 > 0$, so:
$$|x - 2| + |x - 3| = x - 2 + x - 3 = 2x - 5$$

47.   We sketch the interval $(3, 5)$:

49. Note that the real numbers must be negative. We sketch the interval $[-5, 0)$:

51. We sketch the interval $[-1, \infty)$:

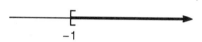

53. We sketch the interval $(3, 9)$:

55. We sketch the intervals $(-\infty, -2]$ and $[0, \infty)$:

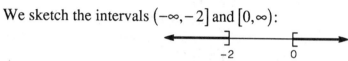

57. (a) We sketch the interval $(-1, 9)$, excluding the point $x = 4$:

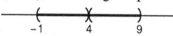

(b) We sketch the interval $(-1, 9)$:

59. We complete the table:

x	$1 + \frac{x^2}{2} + \frac{3x^4}{8}$	$\frac{1}{\sqrt{1-x^2}}$
0.1	1.00504	1.00504
0.01	1.00005	1.00005

61. We solve the equation:
$$12x^2 + 2x - 2 = 0$$
$$6x^2 + x - 1 = 0$$
$$(3x - 1)(2x + 1) = 0$$
$$x = \tfrac{1}{3}, -\tfrac{1}{2}$$

63. Multiplying by 2, we solve the equation:
$$\tfrac{1}{2}x^2 + x - 12 = 0$$
$$x^2 + 2x - 24 = 0$$
$$(x + 6)(x - 4) = 0$$
$$x = -6, 4$$

65.   We solve the equation:
$$\frac{x}{5-x} = \frac{-2}{11-x}$$
$$11x - x^2 = -10 + 2x$$
$$-x^2 + 9x + 10 = 0$$
$$x^2 - 9x - 10 = 0$$
$$(x-10)(x+1) = 0$$
$$x = 10, -1$$

67.   We solve the equation:
$$\frac{1}{3x-7} - \frac{2}{5x-5} - \frac{3}{3x+1} = 0$$
$$(5x-5)(3x+1) - 2(3x-7)(3x+1) - 3(3x-7)(5x-5) = 0$$
$$15x^2 - 10x - 5 - 18x^2 + 36x + 14 - 45x^2 + 150x - 105 = 0$$
$$-48x^2 + 176x - 96 = 0$$
$$3x^2 - 11x + 6 = 0$$
$$(3x-2)(x-3) = 0$$
$$x = \tfrac{2}{3}, 3$$

69.   Factoring yields $x(x^4 - 2x^2 - 2) = 0$, so $x = 0$ is one solution. Using the quadratic formula, we have:
$$x^2 = \frac{2 \pm \sqrt{4 - 4(-2)}}{2} = \frac{2 \pm \sqrt{12}}{2} = \frac{2 \pm 2\sqrt{3}}{2} = 1 \pm \sqrt{3}$$
Note that $1 - \sqrt{3}$ is negative, so no real solutions are obtained from $x^2 = 1 - \sqrt{3}$.
Consequently, $x^2 = 1 + \sqrt{3}$ and $x = \pm\sqrt{1+\sqrt{3}}$. So the solutions are $\pm\sqrt{1+\sqrt{3}}$ and 0.

71.   Squaring, we have:
$$\sqrt{4 - 3x} = 5$$
$$4 - 3x = 25$$
$$-3x = 21$$
$$x = -7$$
This value checks in the original equation.

73.   Squaring, we have:
$$\sqrt{4x+3} = \sqrt{11-8x} - 1$$
$$4x + 3 = 11 - 8x - 2\sqrt{11-8x} + 1$$
$$12x - 9 = -2\sqrt{11-8x}$$
$$144x^2 - 216x + 81 = 44 - 32x$$
$$144x^2 - 184x + 37 = 0$$
$$(4x-1)(36x-37) = 0$$
$$x = \tfrac{1}{4} \text{ or } x = \tfrac{37}{36}$$
Upon checking, we find only $x = \tfrac{1}{4}$ checks in the original equation.

75.   Multiplying by $\sqrt{x+6}\sqrt{x-6}$, we obtain:

$$\frac{2}{\sqrt{x^2-36}}+\frac{1}{\sqrt{x+6}}-\frac{1}{\sqrt{x-6}}=0$$
$$2+\sqrt{x-6}-\sqrt{x+6}=0$$
$$2+\sqrt{x-6}=\sqrt{x+6}$$
$$4+4\sqrt{x-6}+x-6=x+6$$
$$4\sqrt{x-6}=8$$
$$\sqrt{x-6}=2$$
$$x-6=4$$
$$x=10$$

This value checks in the original equation.

77.   Squaring, we have:

$$\sqrt{x+7}-\sqrt{x+2}=\sqrt{x-1}-\sqrt{x-2}$$
$$x+7-2\sqrt{x+7}\sqrt{x+2}+x+2=x-1-2\sqrt{x-1}\sqrt{x-2}+x-2$$
$$-2\sqrt{x+7}\sqrt{x+2}+9=-2\sqrt{x-1}\sqrt{x-2}-3$$
$$-2\sqrt{x+7}\sqrt{x+2}=-2\sqrt{x-1}\sqrt{x-2}-12$$
$$\sqrt{x+7}\sqrt{x+2}=\sqrt{x-1}\sqrt{x-2}+6$$
$$(x+7)(x+2)=(x-1)(x-2)+12\sqrt{x-1}\sqrt{x-2}+36$$
$$x^2+9x+14=x^2-3x+2+12\sqrt{x-1}\sqrt{x-2}+36$$
$$12x-24=12\sqrt{x-1}\sqrt{x-2}$$
$$x-2=\sqrt{x-1}\sqrt{x-2}$$
$$x^2-4x+4=x^2-3x+2$$
$$-x=-2$$
$$x=2$$

This value checks in the original equation.

79.   We solve the inequality:

$$-1<\frac{1-2(1+x)}{3}<1$$
$$-3<1-2-2x<3$$
$$-2<-2x<4$$
$$1>x>-2$$

The solution set is $(-2, 1)$.

81.   The inequality is equivalent to $-\frac{1}{2}\le x\le\frac{1}{2}$, so the solution set is $\left[-\frac{1}{2},\frac{1}{2}\right]$.

83.   We solve the inequality:

$$|x+4|<\tfrac{1}{10}$$
$$-\tfrac{1}{10}<x+4<\tfrac{1}{10}$$
$$-\tfrac{41}{10}<x<-\tfrac{39}{10}$$

The solution set is $\left(-\frac{41}{10},-\frac{39}{10}\right)$.

85. We solve the inequality $|2x-1| \geq 5$:

$$2x - 1 \geq 5 \qquad \text{or} \qquad -(2x-1) \geq 5$$
$$2x \geq 6 \qquad\qquad\qquad 2x - 1 \leq -5$$
$$x \geq 3 \qquad\qquad\qquad\quad 2x \leq -4$$
$$x \leq -2$$

The solution set is $(-\infty, -2] \cup [3, \infty)$.

87. We factor $x^2 + 3x - 40 = (x-5)(x+8)$, so the key numbers are 5 and -8. Now draw the sign chart:

Interval	Test Number	$x+8$	$x-5$	$(x+8)(x-5)$
$(-\infty, -8)$	-9	neg.	neg.	pos.
$(-8, 5)$	0	pos.	neg.	neg.
$(5, \infty)$	6	pos.	pos.	pos.

So the product is negative on the interval $(-8, 5)$.

89. The key numbers are found by using the quadratic formula to solve $x^2 - 6x - 1 = 0$. The result is $x = 3 \pm \sqrt{10}$. For purposes of picking appropriate test numbers, note that $3 + \sqrt{10} \approx 6.2$ and that $3 - \sqrt{10} \approx -0.2$. Now draw the sign chart:

Interval	Test Number	$x - 6.2$	$x + 0.2$	$x^2 - 6x - 1$
$\left(-\infty, 3 - \sqrt{10}\right)$	-1	neg.	neg.	pos.
$\left(3 - \sqrt{10}, 3 + \sqrt{10}\right)$	0	neg.	pos.	neg.
$\left(3 + \sqrt{10}, \infty\right)$	7	pos.	pos.	pos.

So the product is negative on the interval $\left(3 - \sqrt{10}, 3 + \sqrt{10}\right)$.

91. We first factor the inequality:

$$x^4 - 34x^2 + 225 < 0$$
$$\left(x^2 - 9\right)\left(x^2 - 25\right) < 0$$
$$(x-3)(x+3)(x-5)(x+5) < 0$$

The key numbers are -3, 3, -5, and 5. Now draw the sign chart:

Interval	Test Number	$x-5$	$x-3$	$x+3$	$x+5$	$\left(x^2 - 9\right)\left(x^2 - 25\right)$
$(-\infty, -5)$	-6	neg.	neg.	neg.	neg.	pos.
$(-5, -3)$	-4	neg.	neg.	neg.	pos.	neg.
$(-3, 3)$	0	neg.	neg.	pos.	pos.	pos.
$(3, 5)$	4	neg.	pos.	pos.	pos.	neg.
$(5, \infty)$	6	pos.	pos.	pos.	pos.	pos.

So the product is negative on the intervals $(-5, -3) \cup (3, 5)$.

93. The key numbers are 7 and –2. Now draw the sign chart:

Interval	Test Number	$(x+2)^3$	$(x-7)^2$	$\frac{(x-7)^2}{(x+2)^3}$
$(-\infty,-2)$	-3	neg.	pos.	neg.
$(-2,7)$	0	pos.	pos.	pos.
$(7,\infty)$	8	pos.	pos.	pos.

The quotient is positive on the intervals $(-2,7)\cup(7,\infty)$, and zero at $x=7$. The single interval $(-2,\infty)$ satisfies the inequality.

95. We factor the inequality as $\frac{(x-1)(x-9)}{(x+1)(x^2-x+1)}\le 0$. The key numbers are 1, 9, and –1. We draw the sign chart:

Interval	Test Number	$x-9$	$x-1$	$x+1$	$\frac{(x-1)(x-9)}{x^3+1}$
$(-\infty,-1)$	-2	neg.	neg.	neg.	neg.
$(-1,1)$	0	neg.	neg.	pos.	pos.
$(1,9)$	2	neg.	pos.	pos.	neg.
$(9,\infty)$	10	pos.	pos.	pos.	pos.

The quotient is negative on the intervals $(-\infty,-1)\cup(1,9)$, and is zero at $x=1$ and $x=9$. The solution set is $(-\infty,-1)\cup[1,9]$.

97. We first simplify and factor the inequality:
$$\frac{1-2x}{1+2x}-\frac{1}{2}\le 0$$
$$\frac{2(1-2x)-(1+2x)}{2(1+2x)}\le 0$$
$$\frac{-6x+1}{2(1+2x)}\le 0$$

The key numbers are $\frac{1}{6}$ and $-\frac{1}{2}$. Now draw the sign chart:

Interval	Test Number	$2x+1$	$-6x+1$	$\frac{-6x+1}{2(2x+1)}$
$\left(-\infty,-\frac{1}{2}\right)$	-1	neg.	pos.	neg.
$\left(-\frac{1}{2},\frac{1}{6}\right)$	0	pos.	pos.	pos.
$\left(\frac{1}{6},\infty\right)$	1	pos.	neg.	neg.

The quotient is negative on the intervals $\left(-\infty,-\frac{1}{2}\right)\cup\left(\frac{1}{6},\infty\right)$, and zero at $\frac{1}{6}$. The solution set is $\left(-\infty,-\frac{1}{2}\right)\cup\left[\frac{1}{6},\infty\right)$.

99. We first simplify the inequality:

$$\sqrt{x} - \frac{5}{\sqrt{x}} - 4 \le 0$$

$$\frac{x - 5 - 4\sqrt{x}}{\sqrt{x}} \le 0$$

Looking at the denominator, we see that one key number is $x = 0$. The others, if any, are found by solving the equation:

$$x - 5 - 4\sqrt{x} = 0$$

Writing this as $x - 5 = 4\sqrt{x}$, and then squaring both sides, we have:

$$x^2 - 10x + 25 = 16x$$
$$x^2 - 26x + 25 = 0$$
$$(x - 25)(x - 1) = 0$$
$$x = 25 \text{ or } x = 1$$

The value $x = 25$ satisfies the equation $x - 5 - 4\sqrt{x} = 0$, but $x = 1$ is an extraneous root. Thus the key numbers are 0 and 25. Now draw the sign chart:

Interval	Test Number	$\sqrt{x}$	$x - 5 - 4\sqrt{x}$	$\frac{x-5-4\sqrt{x}}{\sqrt{x}}$
$(-\infty, 0)$	-1		undefined	
$(0, 25)$	1	pos.	neg.	neg.
$(25, \infty)$	36	pos.	pos.	pos.

The quotient is negative on the interval $(0, 25)$ and zero at $x = 25$. So the solution set is $(0, 25]$.

101. The discriminant is $b^2 - 4ac = 1 - 4k^2$. We solve for when this discriminant is non-negative:

$$1 - 4k^2 \ge 0$$
$$(1 - 2k)(1 + 2k) \ge 0$$

The key numbers here are $k = \pm\frac{1}{2}$. We draw the sign chart:

Interval	Test Number	$1 + 2k$	$1 - 2k$	$1 - 4k^2$
$\left(-\infty, -\frac{1}{2}\right)$	-1	neg.	pos.	neg.
$\left(-\frac{1}{2}, \frac{1}{2}\right)$	0	pos.	pos.	pos.
$\left(\frac{1}{2}, \infty\right)$	1	pos.	neg.	neg.

The product is positive on $\left(-\frac{1}{2}, \frac{1}{2}\right)$, and zero at $x = -\frac{1}{2}$ and $x = \frac{1}{2}$. So the values of k are chosen from the interval $\left[-\frac{1}{2}, \frac{1}{2}\right]$.

103. The discriminant is:

$$b^2 - 4ac = (k + 1)^2 - 4(2k) = k^2 + 2k + 1 - 8k = k^2 - 6k + 1$$

To find the key numbers for $k^2 - 6k + 1 \ge 0$, we use the quadratic formula:

$$k = \frac{6 \pm \sqrt{36 - 4}}{2} = \frac{6 \pm \sqrt{32}}{2} = \frac{6 \pm 4\sqrt{2}}{2} = 3 \pm 2\sqrt{2}$$

For purposes of picking appropriate test numbers, note that $3 + 2\sqrt{2} \approx 5.8$ and that $3 - 2\sqrt{2} \approx 0.2$. Now draw the sign chart:

Interval	Test Number	$k - 5.8$	$k - 0.2$	$k^2 - 6k + 1$
$(-\infty, 3 - 2\sqrt{2})$	0	neg.	neg.	pos.
$(3 - 2\sqrt{2}, 3 + 2\sqrt{2})$	3	neg.	pos.	neg.
$(3 + 2\sqrt{2}, \infty)$	7	pos.	pos.	pos.

The expression is positive on the interval $(-\infty, 3 - 2\sqrt{2}) \cup (3 + 2\sqrt{2}, \infty)$, and zero at $k = 3 - 2\sqrt{2}$ and $k = 3 + 2\sqrt{2}$. So the values of k are chosen from the interval $(-\infty, 3 - 2\sqrt{2}] \cup [3 + 2\sqrt{2}, \infty)$.

105. Given that the length of a side is $\frac{x}{4}$, we compute the area and the perimeter:

$$\text{area} = \left(\frac{x}{4}\right)^2 = \frac{x^2}{16}$$

$$\text{perimeter} = 4\left(\frac{x}{4}\right) = x$$

If the area is numerically greater than the perimeter, then we have:

$$\frac{x^2}{16} > x$$
$$x^2 > 16x$$
$$x^2 - 16x > 0$$
$$x(x - 16) > 0$$

Since x must be positive, this is satisfied only if $x - 16 > 0$, and hence $x > 16$. We conclude that the area will be numerically greater than the perimeter when $x > 16$ cm, which is the interval $(16 \text{ cm}, \infty)$.

107. (a) With $v_0 = 64$ ft/sec and $h_0 = 0$, the formula is $h = -16t^2 + 64t$. Setting $h = 15$ ft yields:

$$-16t^2 + 64t = 15$$
$$-16t^2 + 64t - 15 = 0$$
$$16t^2 - 64t + 15 = 0$$
$$(4t - 1)(4t - 15) = 0$$
$$t = \tfrac{1}{4} \text{ or } t = \tfrac{15}{4}$$

The height of the ball is 15 ft when $t = \frac{1}{4}$ sec (on the way up) and again when $t = \frac{15}{4}$ sec (on the way down).

(b) If $h > 63$ feet, then we have:

$$-16t^2 + 64t > 63$$
$$16t^2 - 64t + 63 < 0$$
$$(4t - 9)(4t - 7) < 0$$

The key numbers are $\frac{9}{4}$ and $\frac{7}{4}$. Since $t > 0$, we draw the sign chart:

Interval	Test Number	$4t-9$	$4t-7$	$(4t-9)(4t-7)$
$\left(0,\frac{7}{4}\right)$	1	neg.	neg.	pos.
$\left(\frac{7}{4},\frac{9}{4}\right)$	2	neg.	pos.	neg.
$\left(\frac{9}{4},\infty\right)$	3	pos.	pos.	pos.

The product is negative on the interval $\left(\frac{7}{4},\frac{9}{4}\right)$. Thus the height will exceed 63 ft for times between 1.75 sec and 2.25 sec.

109. Since the height is H ft when $t = t_1$ sec and when $t = t_2$ sec, we have the two equations:

$$H = -16t_1^2 + v_0 t_1 + h_0 \qquad \text{(I)}$$
$$H = -16t_2^2 + v_0 t_2 + h_0 \qquad \text{(II)}$$

Subtracting Equation (II) from Equation (I) yields:

$$0 = \left(-16t_1^2 + v_0 t_1 + h_0\right) - \left(-16t_2^2 + v_0 t_2 + h_0\right)$$
$$0 = -16t_1^2 + 16t_2^2 + v_0 t_1 - v_0 t_2$$
$$0 = -16(t_1 - t_2)(t_1 + t_2) + v_0(t_1 - t_2)$$

Now assuming $t_1 \neq t_2$, we can divide both sides by $t_1 - t_2$ to obtain:

$$0 = -16(t_1 + t_2) + v_0$$
$$16(t_1 + t_2) = v_0$$

Chapter One Test

1. (a) Factoring by difference of cubes, we have:

$$64 - (x-2)^3 = \left[4 - (x-2)\right]\left[16 + 4(x-2) + (x-2)^2\right]$$
$$= (4 - x + 2)(16 + 4x - 8 + x^2 - 4x + 4)$$
$$= (6 - x)(12 + x^2)$$

(b) Since this is a sum of squares, $w^2 y^2 + 16$ is irreducible.

2. (a) Using trial and error, $2x^2 + 3x - 8$ is irreducible.

(b) Factoring the common factor z and using trial and error, we have:

$$6x^2 y^2 z + xyz - z = z\left(6x^2 y^2 + xy - 1\right) = z(3xy - 1)(2xy + 1)$$

3. Factoring the common factor $3x(1-x^2)^{-3/2}$, we have:

$$6x(1-x^2)^{-1/2} - 3x^2(1-x^2)^{-3/2} = 3x(1-x^2)^{-3/2}\left[2(1-x^2)-x\right]$$
$$= 3x(1-x^2)^{-3/2}(2-2x^2-x)$$
$$= -3x(1-x^2)^{-3/2}(2x^2+x-2)$$

4. Factoring using grouping, we have:
$$9ax + 6bx - 6ay - 4by = 3x(3a+2b) - 2y(3a+2b) = (3a+2b)(3x-2y)$$

5. Using the special product for $(A+B)^3$, we have:
$$(2a+3b^2)^3 = (2a)^3 + 3(2a)^2(3b^2) + 3(2a)(3b^2)^2 + (3b^2)^3$$
$$= 8a^3 + 36a^2b^2 + 54ab^4 + 27b^6$$

6. If $6 < x < 7$, then $x - 6 > 0$ and $x - 7 < 0$, so:
$$|x-6| + |x-7| = (x-6) - (x-7) = x - 6 - x + 7 = 1$$

7. (a) Since the distance from x to 4 is less than $\frac{1}{10}$, then:
$$4 - \tfrac{1}{10} < x < 4 + \tfrac{1}{10}$$
$$\tfrac{39}{10} < x < \tfrac{41}{10}$$
The interval is $\left(\frac{39}{10}, \frac{41}{10}\right)$.

(b) Since $x \geq -2$, the interval is $[-2, \infty)$.

8. (a) We solve by factoring:
$$x^2 + 4x = 5$$
$$x^2 + 4x - 5 = 0$$
$$(x+5)(x-1) = 0$$
$$x = -5, 1$$

(b) We solve by writing the equation as $x^2 + 4x - 1 = 0$, then using the quadratic formula:
$$x = \frac{-4 \pm \sqrt{(4)^2 - 4(1)(-1)}}{2(1)} = \frac{-4 \pm \sqrt{20}}{2} = \frac{-4 \pm 2\sqrt{5}}{2} = -2 \pm \sqrt{5}$$

9. Isolating $\sqrt{5-2x}$ and squaring, we have:
$$\sqrt{5-2x} - \sqrt{2-x} - \sqrt{3-x} = 0$$
$$\sqrt{5-2x} = \sqrt{2-x} + \sqrt{3-x}$$
$$5 - 2x = 2 - x + 2\sqrt{2-x}\sqrt{3-x} + 3 - x$$
$$0 = 2\sqrt{2-x}\sqrt{3-x}$$

Thus $\sqrt{2-x} = 0$ and consequently $x = 2$ or $\sqrt{3-x} = 0$ and consequently $x = 3$. The value $x = 2$ checks in the original equation, but (considering only the real number system) the value $x = 3$ does not check.

10. We solve by factoring:
$$2x^3 + x^2 - 3x = 0$$
$$x(2x^2 + x - 3) = 0$$
$$x(2x+3)(x-1) = 0$$
$$x = 0, -\tfrac{3}{2}, 1$$

11. We factor by grouping:
$$y^3 + 3y^2 - 8y - 24 = 0$$
$$y^2(y+3) - 8(y+3) = 0$$
$$(y+3)(y^2 - 8) = 0$$
$$y = -3 \quad \text{or} \quad y = \pm\sqrt{8} = \pm 2\sqrt{2}$$

12. (a) Given $|3x - 1| = 2$, either $3x - 1 = 2$ or $3x - 1 = -2$. Therefore:
$$
\begin{array}{ccc}
3x - 1 = 2 & \text{or} & 3x - 1 = -2 \\
3x = 3 & & 3x = -1 \\
x = 1 & & x = -\tfrac{1}{3}
\end{array}
$$
The solutions are 1 and $-\tfrac{1}{3}$.

(b) We solve for x:
$$|x - 8| \le 1$$
$$-1 \le x - 8 \le 1$$
$$7 \le x \le 9$$
Using interval notation, the solution is $[7, 9]$.

13. We solve the inequality:
$$4(1 + x) - 3(2x - 1) \ge 1$$
$$4 + 4x - 6x + 3 \ge 1$$
$$-2x \ge -6$$
$$x \le 3$$
The solution set is $(-\infty, 3]$.

14. Solving the double inequality, we have:
$$\frac{3}{5} < \frac{3 - 2x}{-4} < \frac{4}{5}$$
$$-12 > 15 - 10x > -16$$
$$-27 > -10x > -31$$
$$\tfrac{27}{10} < x < \tfrac{31}{10}$$
The solution set is $\left(\tfrac{27}{10}, \tfrac{31}{10}\right)$.

15. The key numbers are 4 and -8. We draw the sign chart:

Interval	Test Number	$(x+8)^3$	$(x-4)^2$	$(x-4)^2(x+8)^3$
$(-\infty,-8)$	-9	neg.	pos.	neg.
$(-8,4)$	0	pos.	pos.	pos.
$(4,\infty)$	5	pos.	pos.	pos.

The product is positive on $(-8,4)\cup(4,\infty)$ and zero at $x=-8$ and $x=4$. So the inequality is satisfied on $[-8,4]\cup[4,\infty)$, which combine as $[-8,\infty)$.

16. Simplifying the inequality and factoring, we obtain:
$$\frac{(x+1)(x+2)+x(x+2)+x(x+1)}{x(x+1)(x+2)}\geq 0$$
$$\frac{3x^2+6x+2}{x(x+1)(x+2)}\geq 0$$

Three of the key numbers are 0, -1 and -2. The other two are found by using the quadratic formula to solve $3x^2+6x+2=0$. The roots are found to be $\frac{-3\pm\sqrt{3}}{3}$. For purposes of picking appropriate test numbers, note that $\frac{-3+\sqrt{3}}{3}\approx -0.4$ and that $\frac{-3-\sqrt{3}}{3}\approx -1.6$. We draw the sign chart:

Interval	Test Number	x	$x+0.4$	$x+1$	$x+1.6$	$x+2$	$\frac{3x^2+6x+2}{x(x+1)(x+2)}$
$(-\infty,-2)$	-3	neg.	neg.	neg.	neg.	neg.	neg.
$(-2,-1.6)$	-1.8	neg.	neg.	neg.	neg.	pos.	pos.
$(-1.6,-1)$	-1.5	neg.	neg.	neg.	pos.	pos.	neg.
$(-1,-0.4)$	-0.5	neg.	neg.	pos.	pos.	pos.	pos.
$(-0.4,0)$	-0.1	neg.	pos.	pos.	pos.	pos.	neg.
$(0,\infty)$	2	pos.	pos.	pos.	pos.	pos.	pos.

So the solution set is $\left(-2,\frac{-3-\sqrt{3}}{3}\right]\cup\left(-1,\frac{-3+\sqrt{3}}{3}\right]\cup(0,\infty)$.

Chapter Two
Coordinates and Graphs

2.1 Rectangular Coordinates

1. We plot the points:

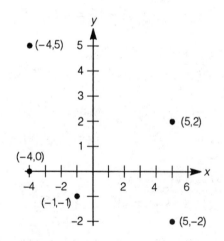

3. (a) We draw the right triangle *PQR*:

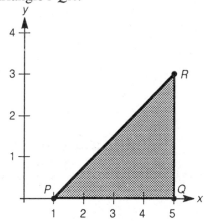

(b) Since the base is $b = 5 - 1 = 4$ and the height is $h = 3 - 0 = 3$, then the area is given by:
$$A = \tfrac{1}{2}bh = \tfrac{1}{2}(4)(3) = 6 \text{ square units}$$

5. (a) Here $(x_1, y_1) = (0,0)$ and $(x_2, y_2) = (-3, 4)$, so by the distance formula:
$$d = \sqrt{(-3-0)^2 + (4-0)^2} = \sqrt{9+16} = \sqrt{25} = 5$$

(b) Here $(x_1, y_1) = (2, 1)$ and $(x_2, y_2) = (7, 13)$, so:
$$d = \sqrt{(7-2)^2 + (13-1)^2} = \sqrt{25+144} = \sqrt{169} = 13$$

7. (a) Here $(x_1, y_1) = (-5, 0)$ and $(x_2, y_2) = (5, 0)$, so:
$$d = \sqrt{[5-(-5)]^2 + (0-0)^2} = \sqrt{100+0} = \sqrt{100} = 10$$

(b) Here $(x_1, y_1) = (0, -8)$ and $(x_2, y_2) = (0, 1)$, so:
$$d = \sqrt{(0-0)^2 + [1-(-8)]^2} = \sqrt{0+81} = \sqrt{81} = 9$$

Note that we really don't need to use the distance formula for either (a) or (b), since in each case one of the coordinates (either *x* or *y*) is the same. Draw quick graphs and you can find the distance by inspection:

(a) This graph indicates the distance is 10:

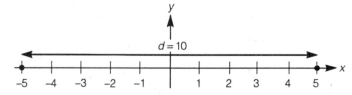

(b) This graph indicates the distance is 9:

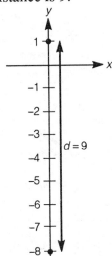

9.   Here $(x_1, y_1) = (1, \sqrt{3})$ and $(x_2, y_2) = (-1, -\sqrt{3})$, so:

$$d = \sqrt{(-1-1)^2 + \left(-\sqrt{3} - \sqrt{3}\right)^2} = \sqrt{(-2)^2 + \left(-2\sqrt{3}\right)^2} = \sqrt{4+12} = 4$$

11. (a) We calculate the distance of each point from the origin:

$$(3,-2): \quad d = \sqrt{(3-0)^2 + (-2-0)^2} = \sqrt{9+4} = \sqrt{13}$$

$$\left(4, \tfrac{1}{2}\right): \quad d = \sqrt{(4-0)^2 + \left(\tfrac{1}{2} - 0\right)^2} = \sqrt{16 + \tfrac{1}{4}} = \sqrt{16.25}$$

So $\left(4, \tfrac{1}{2}\right)$ is farther from the origin.

(b) We calculate the distance of each point from the origin:

$$(-6,7): \quad d = \sqrt{(-6-0)^2 + (7-0)^2} = \sqrt{36+49} = \sqrt{85}$$

$$(9,0): \quad d = \sqrt{(9-0)^2 + (0-0)^2} = \sqrt{81+0} = \sqrt{81}$$

So $(-6,7)$ is farther from the origin.

13. We will graph each triangle and then determine (using the converse of the Pythagorean theorem) whether $a^2 + b^2 = c^2$.

(a)   We graph the points:

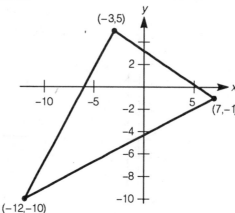

Now calculate the distances:
$$a = \sqrt{(-3-7)^2 + [5-(-1)]^2} = \sqrt{100+36} = \sqrt{136}$$
$$b = \sqrt{[-12-(-3)]^2 + (-10-5)^2} = \sqrt{81+225} = \sqrt{306}$$
$$c = \sqrt{(-12-7)^2 + [-10-(-1)]^2} = \sqrt{361+81} = \sqrt{442}$$
Now check the converse of the Pythagorean theorem:
$$a^2 + b^2 = 136 + 306 = 442 = c^2$$
So the triangle is a right triangle.

(b)   We graph the points:

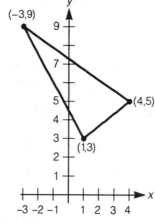

Now calculate the distances:
$$a = \sqrt{(-3-1)^2 + (9-3)^2} = \sqrt{16+36} = \sqrt{52}$$
$$b = \sqrt{(4-1)^2 + (5-3)^2} = \sqrt{9+4} = \sqrt{13}$$
$$c = \sqrt{(-3-4)^2 + (9-5)^2} = \sqrt{49+16} = \sqrt{65}$$
Now check the converse of the Pythagorean theorem:
$$a^2 + b^2 = 52 + 13 = 65 = c^2$$
So the triangle is a right triangle.

(c)   We graph the points:

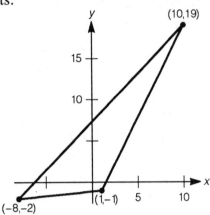

Now calculate the distances:

$$a = \sqrt{(-8-1)^2 + [-2-(-1)]^2} = \sqrt{81+1} = \sqrt{82}$$
$$b = \sqrt{(10-1)^2 + [19-(-1)]^2} = \sqrt{81+400} = \sqrt{481}$$
$$c = \sqrt{[10-(-8)]^2 + [19-(-2)]^2} = \sqrt{324+441} = \sqrt{765}$$

Now check the converse of the Pythagorean theorem:

$$a^2 + b^2 = 82 + 481 = 563 \neq 765 = c^2$$

So the triangle is not a right triangle.

15.   Let $(x_1, y_1) = (1, -4)$, $(x_2, y_2) = (5, 3)$, and $(x_3, y_3) = (13, 17)$, so using the formula from Exercise 14(b) we have:

$$\text{Area} = \left(\tfrac{1}{2}\right)|1(3) - 5(-4) + 5(17) - 13(3) + 13(-4) - 1(17)|$$
$$= \left(\tfrac{1}{2}\right)|3 + 20 + 85 - 39 - 52 - 17|$$
$$= \left(\tfrac{1}{2}\right)|0|$$
$$= 0$$

For the area of the triangle to be 0 it must be that these three points do not form a triangle. The only way that could occur is if the three points are collinear; that is, they all lie on the same line.

17.   The circle is in standard form, so its center is $(3, 1)$ and its radius is $\sqrt{25} = 5$. The y-intercepts are found by setting $x = 0$:

$$9 + (y-1)^2 = 25$$
$$(y-1)^2 = 16$$
$$y - 1 = \pm 4$$
$$y = 1 \pm 4$$
$$y = 5 \text{ or } y = -3$$

19. The circle is in standard form, so its center is $(0,0)$ and its radius is $\sqrt{\sqrt{2}} = \sqrt[4]{2}$. The y-intercepts are found by setting $x = 0$:
$$y^2 = \sqrt{2}$$
$$y = \pm\sqrt{\sqrt{2}} = \pm\sqrt[4]{2}$$

21. We must complete the square:
$$x^2 + 8x + y^2 - 6y = -24$$
$$\left(x^2 + 8x + 16\right) + \left(y^2 - 6y + 9\right) = -24 + 16 + 9$$
$$(x+4)^2 + (y-3)^2 = 1$$
The center is $(-4, 3)$ and the radius is $\sqrt{1} = 1$. The y-intercepts are found by setting $x = 0$:
$$16 + (y-3)^2 = 1$$
$$(y-3)^2 = -15$$
Since this equation has no real solution, there are no y-intercepts.

23. We divide by 9 and complete the square:
$$x^2 + 6x + y^2 - \tfrac{2}{3}y = -\tfrac{64}{9}$$
$$\left(x^2 + 6x + 9\right) + \left(y^2 - \tfrac{2}{3}y + \tfrac{1}{9}\right) = -\tfrac{64}{9} + 9 + \tfrac{1}{9}$$
$$(x+3)^2 + \left(y - \tfrac{1}{3}\right)^2 = 2$$
The center is $\left(-3, \tfrac{1}{3}\right)$ and the radius is $\sqrt{2}$. The y-intercepts are found by setting $x = 0$:
$$9 + \left(y - \tfrac{1}{3}\right)^2 = 2$$
$$\left(y - \tfrac{1}{3}\right)^2 = -7$$
Since this equation has no real solution, there are no y-intercepts.

25. (a) Since P lies on the unit circle, the coordinates of P satisfy the equation $x^2 + y^2 = 1$. We are given that the x-coordinate of P is $\tfrac{3}{5}$. Substituting the value $x = \tfrac{3}{5}$ in the equation $x^2 + y^2 = 1$ gives us:
$$\left(\tfrac{3}{5}\right)^2 + y^2 = 1$$
$$\tfrac{9}{25} + y^2 = 1$$
$$y^2 = \tfrac{16}{25}$$
$$y = \pm\tfrac{4}{5}$$
We want to choose the positive root here because P lies in quadrant I (in which all y-coordinates are positive). Thus, the y-coordinate of P is $\tfrac{4}{5}$.

(b) The solution for the y-coordinates are as in (a), where $y = \pm\tfrac{4}{5}$. We want to choose the negative root here because P lies in quadrant IV (in which all y-coordinates are negative). Thus, the y-coordinate of P is $-\tfrac{4}{5}$.

27. (a) Since P lies on the unit circle, the coordinates of P satisfy the equation $x^2 + y^2 = 1$. We are given that the y-coordinate of P is $-\frac{\sqrt{2}}{2}$. Substituting the value $y = -\frac{\sqrt{2}}{2}$ in the equation $x^2 + y^2 = 1$ gives us:

$$x^2 + \left(-\frac{\sqrt{2}}{2}\right)^2 = 1$$
$$x^2 + \frac{1}{2} = 1$$
$$x^2 = \frac{1}{2}$$
$$x = \pm\frac{\sqrt{2}}{2}$$

We want to choose the negative root here because P lies in quadrant III (in which all x-coordinates are negative). Thus, the x-coordinate of P is $-\frac{\sqrt{2}}{2}$.

(b) The solution for the x-coordinates are as in (a), where $x = \pm\frac{\sqrt{2}}{2}$. We want to choose the positive root here because P lies in quadrant IV (in which all x-coordinates are positive). Thus, the x-coordinate of P is $\frac{\sqrt{2}}{2}$.

29. (a) Since P lies on the unit circle, the coordinates of P satisfy the equation $x^2 + y^2 = 1$. We are given that the x-coordinate of P is $-\frac{2}{3}$. Substituting the value $x = -\frac{2}{3}$ in the equation $x^2 + y^2 = 1$ gives us:

$$\left(-\frac{2}{3}\right)^2 + y^2 = 1$$
$$\frac{4}{9} + y^2 = 1$$
$$y^2 = \frac{5}{9}$$
$$y = \pm\frac{\sqrt{5}}{3}$$

We want to choose the positive root here because P lies in quadrant II (in which all y-coordinates are positive). Thus, the y-coordinate of P is $\frac{\sqrt{5}}{3}$.

(b) The solution for the y-coordinates are as in (a), where $y = \pm\frac{\sqrt{5}}{3}$. We want to choose the negative root here because P lies in quadrant III (in which all y-coordinates are negative). Thus, the y-coordinate of P is $-\frac{\sqrt{5}}{3}$.

31. (a) Since P lies on the unit circle, the coordinates of P satisfy the equation $x^2 + y^2 = 1$. We are given that the x-coordinate of P is 0.652. Substituting the value $x = 0.652$ in the equation $x^2 + y^2 = 1$ gives us:

$$(0.652)^2 + y^2 = 1$$
$$0.425104 + y^2 = 1$$
$$y^2 = 0.574896$$
$$y \approx \pm 0.758$$

We want to choose the positive root here because P lies in quadrant I (in which all y-coordinates are positive). Thus, the y-coordinate of P is (approximately) 0.758.

(b) Substituting the value $y = 0.652$ in the equation $x^2 + y^2 = 1$ gives us:
$$x^2 + (0.652)^2 = 1$$
$$x^2 + 0.425104 = 1$$
$$x^2 = 0.574896$$
$$x \approx \pm 0.758$$
We want to choose the positive root here because P lies in quadrant I (in which all x-coordinates are positive). Thus, the x-coordinate of P is (approximately) 0.758.

33. Since $\overline{PQ}$ is the diameter of the circle, then its midpoint must be the center of the circle. Using the midpoint formula:
$$\text{center} = \left(\tfrac{-4+6}{2}, \tfrac{-2+4}{2}\right) = \left(\tfrac{2}{2}, \tfrac{2}{2}\right) = (1,1)$$
The radius of the circle is the distance from this center $(1,1)$ to point Q (or point P), so using the distance formula:
$$r = \sqrt{(6-1)^2 + (4-1)^2} = \sqrt{25+9} = \sqrt{34}$$
Using the standard form $(x-h)^2 + (y-k)^2 = r^2$ for a circle, the equation is $(x-1)^2 + (y-1)^2 = 34$.

35. (a) Here $(x_1, y_1) = (3, 2)$ and $(x_2, y_2) = (9, 8)$, so by the midpoint formula:
$$M = \left(\frac{3+9}{2}, \frac{2+8}{2}\right) = \left(\tfrac{12}{2}, \tfrac{10}{2}\right) = (6,5)$$

(b) Here $(x_1, y_1) = (-4, 0)$ and $(x_2, y_2) = (5, -3)$, so by the midpoint formula:
$$M = \left(\frac{-4+5}{2}, \frac{0-3}{2}\right) = \left(\tfrac{1}{2}, -\tfrac{3}{2}\right)$$

(c) Here $(x_1, y_1) = (3, -6)$ and $(x_2, y_2) = (-1, -2)$, so by the midpoint formula:
$$M = \left(\frac{3-1}{2}, \frac{-6-2}{2}\right) = \left(\tfrac{2}{2}, -\tfrac{8}{2}\right) = (1, -4)$$

37. Label $C(x, y)$. Since B is the midpoint of $\overline{AC}$, then:
$$\frac{-1+x}{2} = 5 \quad \text{and} \quad \frac{2+y}{2} = -3$$
We solve each of these for x and y:
$$\frac{-1+x}{2} = 5 \quad \text{and} \quad \frac{2+y}{2} = -3$$
$$-1+x = 10 \qquad\qquad 2+y = -6$$
$$x = 11 \qquad\qquad y = -8$$
So the coordinates of C must be $(11, -8)$.

39. (a)   We sketch the parallelogram $ABCD$:

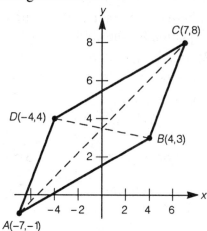

(b)   For AC, let $(x_1, y_1) = (-7, -1)$ and $(x_2, y_2) = (7, 8)$, so the midpoint is:
$$M_1 = \left(\tfrac{-7+7}{2}, \tfrac{-1+8}{2}\right) = \left(0, \tfrac{7}{2}\right)$$
For BD, let $(x_1, y_1) = (4, 3)$ and $(x_2, y_2) = (-4, 4)$, so the midpoint is:
$$M_2 = \left(\tfrac{4-4}{2}, \tfrac{3+4}{2}\right) = \left(0, \tfrac{7}{2}\right)$$

(c)   It appears that the midpoints of the two diagonals of a parallelogram are the same. Stated more concisely, the diagonals of a parallelogram bisect each other.

41. (a)   Since $(AB)^2 = 68$, $(BC)^2 = 40$, and $(AC)^2 = 20$, then:
$$(AB)^2 + (BC)^2 + (AC)^2 = 128$$

(b)   Let the midpoints be X, Y, and Z, so:
$$X = M_{AB} = (5, 2)$$
$$Y = M_{BC} = (6, 4)$$
$$Z = M_{AC} = (2, 3)$$
Then the medians are:
$$AY = \sqrt{(6-1)^2 + (4-1)^2} = \sqrt{25+9} = \sqrt{34}$$
$$BZ = \sqrt{(2-9)^2 + (3-3)^2} = \sqrt{49+0} = \sqrt{49} = 7$$
$$CX = \sqrt{(5-3)^2 + (2-5)^2} = \sqrt{4+9} = \sqrt{13}$$
So the required value is:
$$(AY)^2 + (BZ)^2 + (CX)^2 = 34 + 49 + 13 = 96$$

(c)   This ratio is $\tfrac{128}{96} = \tfrac{4}{3}$.

43. Since the center of the circle is $(3,2)$, we know the equation will take the form $(x-3)^2 + (y-2)^2 = r^2$, where r is the radius of the circle. We can find r since the point $(-2,-10)$ must satisfy the equation for the circle:

$$(-2-3)^2 + (-10-2)^2 = r^2$$
$$25 + 144 = r^2$$
$$169 = r^2$$

The equation of the circle is $(x-3)^2 + (y-2)^2 = 169$.

45. Since the radius is 3, the equation of the circle is $(x-3)^2 + (y-5)^2 = 9$.

47. Using the Pythagorean theorem, we have:

$$a^2 = 1^2 + 1^2 = 2, \text{ so } a = \sqrt{2}$$
$$b^2 = 1^2 + \left(\sqrt{2}\right)^2 = 1 + 2 = 3, \text{ so } b = \sqrt{3}$$
$$c^2 = 1^2 + \left(\sqrt{3}\right)^2 = 1 + 3 = 4, \text{ so } c = 2$$
$$d^2 = 1^2 + 2^2 = 1 + 4 = 5, \text{ so } d = \sqrt{5}$$
$$e^2 = 1^2 + \left(\sqrt{5}\right)^2 = 1 + 5 = 6, \text{ so } e = \sqrt{6}$$
$$f^2 = 1^2 + \left(\sqrt{6}\right)^2 = 1 + 6 = 7, \text{ so } f = \sqrt{7}$$
$$g^2 = 1^2 + \left(\sqrt{7}\right)^2 = 1 + 7 = 8, \text{ so } g = \sqrt{8} = 2\sqrt{2}$$

49. If the circle is concentric with $x^2 - 6x + y^2 - 4y + 4 = 0$, then its center will be the same. We complete the square:

$$x^2 - 6x + y^2 - 4y = -4$$
$$x^2 - 6x + 9 + y^2 - 4y + 4 = -4 + 9 + 4$$
$$(x-3)^2 + (y-2)^2 = 9$$

So the center is $(3,2)$. Since its equation is $(x-3)^2 + (y-2)^2 = r^2$, and $(0,0)$ lies on the circle, we have:

$$(-3)^2 + (-2)^2 = r^2$$
$$13 = r^2$$

So the equation is $(x-3)^2 + (y-2)^2 = 13$.

51. Using the hint and squaring, we have:

$$13 = \sqrt{(12-0)^2 + (t-2)^2}$$
$$169 = 144 + (t-2)^2$$
$$25 = (t-2)^2$$

So either $t - 2 = 5$ and thus $t = 7$, or $t - 2 = -5$ and thus $t = -3$. The values of t are 7 and -3.

53. (a) Call (x, y) the coordinates of B. Clearly $y = c$, since the top and bottom of the parallelogram must be parallel, and thus since the bottom is horizontal then the top must be also. Now compute the lengths OC and OB using the distance formula:

$$OC = \sqrt{(b-0)^2 + (c-0)^2} = \sqrt{b^2 + c^2}$$
$$AB = \sqrt{(x-a)^2 + (c-0)^2} = \sqrt{(x-a)^2 + c^2}$$

Since $OC = AB$, we have:

$$\sqrt{b^2 + c^2} = \sqrt{(x-a)^2 + c^2}$$
$$b^2 + c^2 = (x-a)^2 + c^2$$
$$b^2 = (x-a)^2$$

Taking roots, we have $x - a = \pm b$, or $x = a \pm b$. But clearly $x = a - b$ doesn't make sense, as $x \geq a$ from the figure, so $x = a + b$.

(b) We find the midpoints of the two diagonals $\overline{OB}$ and $\overline{AC}$:

For $O(0,0)$ and $B(a+b, c)$, $M_{\overline{OB}} = \left(\frac{a+b}{2}, \frac{c}{2}\right)$

For $A(a, c)$ and $C(b, c)$, $M_{\overline{AC}} = \left(\frac{a+b}{2}, \frac{c}{2}\right)$

(c) Clearly our two midpoints from (b) are equal. Since the two midpoints on the diagonals are equal they must bisect each other.

55. Using the hint and the lengths of our diagonals from Exercise 54, and squaring each side, we have:

$$\sqrt{(a+b)^2 + c^2} = \sqrt{(b-a)^2 + c^2}$$
$$(a+b)^2 + c^2 = b^2 - 2ab + a^2 + c^2$$
$$4ab = 0$$
$$a = 0 \text{ or } b = 0$$

Now $a = 0$ is impossible, since our figure would no longer be a parallelogram. So $b = 0$. But if $b = 0$, then we must have a rectangle (look at the figure). This proves the desired result.

57. We first compute the left-hand side of the equality:

$$AB = \sqrt{(2-0)^2 + (0-0)^2} = \sqrt{4} = 2$$
$$AC = \sqrt{(2a-0)^2 + (2b-0)^2} = \sqrt{4a^2 + 4b^2}$$

So $AB^2 + AC^2 = 4 + 4a^2 + 4b^2$. For the right-hand side, we first find:

$$M = \left(\frac{2a+a}{2}, \frac{2b+0}{2}\right) = (a+1, b)$$
$$BM = \sqrt{(a+1-2)^2 + (b-0)^2} = \sqrt{(a-1)^2 + b^2}$$
$$AM = \sqrt{(a+1-0)^2 + (b-0)^2} = \sqrt{(a+1)^2 + b^2}$$

Therefore:

$$2\left(BM^2 + AM^2\right) = 2\left[(a-1)^2 + b^2 + (a+1)^2 + b^2\right]$$
$$= 2\left(a^2 - 2a + 1 + b^2 + a^2 + 2a + 1 + b^2\right)$$
$$= 2\left(2a^2 + 2b^2 + 2\right)$$
$$= 4a^2 + 4b^2 + 4$$

So $AB^2 + AC^2 = 2\left(BM^2 + AM^2\right)$.

59. (a) Using the distance formula, we have:

$$PM = \sqrt{\left(\frac{x_1 + x_2}{2} - x_1\right)^2 + \left(\frac{y_1 + y_2}{2} - y_1\right)^2}$$
$$= \sqrt{\left(\frac{x_2 - x_1}{2}\right)^2 + \left(\frac{y_2 - y_1}{2}\right)^2}$$
$$= \frac{\sqrt{\left(x_2 - x_1\right)^2 + \left(y_2 - y_1\right)^2}}{2}$$

$$MQ = \sqrt{\left(x_2 - \frac{x_1 + x_2}{2}\right)^2 + \left(y_2 - \frac{y_1 + y_2}{2}\right)^2}$$
$$= \sqrt{\left(\frac{x_2 - x_1}{2}\right)^2 + \left(\frac{y_2 - y_1}{2}\right)^2}$$
$$= \frac{\sqrt{\left(x_2 - x_1\right)^2 + \left(y_2 - y_1\right)^2}}{2}$$

Thus $PM = MQ$.

(b) Adding the two distances together:

$$PM + MQ = 2(PM) = \sqrt{\left(x_2 - x_1\right)^2 + \left(y_2 - y_1\right)^2} = PQ$$

61. (a) Each side of the outermost quadrilateral is the hypotenuse of the triangle in Figure A, so each side has a length of c. Also, since each angle of the outermost quadrilateral is the sum of the two acute angles from the triangle, then the measure of each angle is $90°$. Since the quadrilateral has angles of $90°$ and sides of equal length, it must be a square.

(b) Each side of the innermost quadrilateral is the difference of the two legs of the triangle in Figure A, so the length is $b - a$.

(c) The area of each triangle is $\frac{1}{2}ab$, and the area of the innermost square is $(b-a)^2$. Since there are four triangles, the figure area is:

$$4\left(\tfrac{1}{2}ab\right) + (b-a)^2 = 2ab + b^2 - 2ab + a^2 = a^2 + b^2$$

Since the figure has an area of c^2, we have the result $a^2 + b^2 = c^2$.

63. (a) We apply the Pythagorean theorem:

$$u^2 + v^2 = \frac{4(m+n)^2}{n^2} + \frac{16m^2}{(m-n)^2}$$

$$= \frac{4(m+n)^2(m-n)^2}{n^2(m-n)^2} + \frac{16m^2n^2}{n^2(m-n)^2}$$

$$= \frac{4(m^2-n^2)^2 + 16m^2n^2}{n^2(m-n)^2}$$

$$= \frac{4m^4 - 8m^2n^2 + 4n^4 + 16m^2n^2}{n^2(m-n)^2}$$

$$= \frac{4m^4 + 8m^2n^2 + 4n^4}{n^2(m-n)^2}$$

$$= \frac{4(m^2+n^2)^2}{n^2(m-n)^2}$$

Taking roots, we have:

$$w = \sqrt{u^2 + v^2} = \sqrt{\frac{4(m^2+n^2)^2}{n^2(m-n)^2}} = \frac{2(m^2+n^2)}{(m-n)n}$$

(b) We compute the right-hand side:

$$u + v + w = \frac{2m+2n}{n} + \frac{4m}{m-n} + \frac{2m^2+2n^2}{(m-n)n}$$

$$= \frac{(2m+2n)(m-n) + 4mn + 2m^2 + 2n^2}{(m-n)n}$$

$$= \frac{2m^2 - 2n^2 + 4mn + 2m^2 + 2n^2}{(m-n)n}$$

$$= \frac{4m^2 + 4mn}{(m-n)n}$$

$$= \frac{4m(m+n)}{(m-n)n}$$

But this is exactly $\frac{1}{2}uv$, thus proving the desired result.

(c) We simply choose m and n such that u, v, and w will be integers, say $m = 3$ and $n = 2$. Then $u = 5$, $v = 12$, and $w = 13$. As a check, we compute:

$$\text{area} = \tfrac{1}{2}(5)(12) = 30$$
$$\text{perimeter} = 5 + 12 + 13 = 30$$

Optional TI-81 Graphing Calculator Exercises for Section 2.1

1.   (a)   Solving for y, we have:
$$x^2 + y^2 = 4$$
$$y^2 = 4 - x^2$$
$$y = \pm\sqrt{4 - x^2}$$
Using ZOOM-6 settings, we draw the graph:

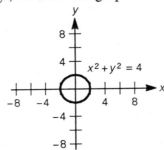

(b)   Using the indicated settings, we draw the graph:

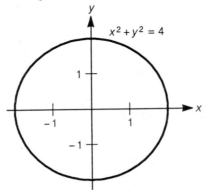

(c)   Using ZOOM-5 settings, we draw the graph:

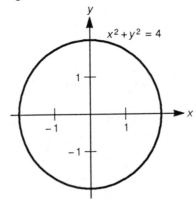

(d)   The graph is identical to that in part (c).

3.   Using $X_{min} = -9$, $X_{max} = 9$, $Y_{min} = -9$, and $Y_{max} = 9$, we draw the graph:

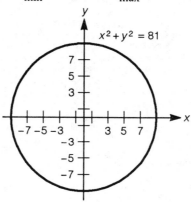

5.   Solving for y, we obtain:
$$(x+3)^2 + (y+2)^2 = 1$$
$$(y+2)^2 = 1 - (x+3)^2$$
$$y+2 = \pm\sqrt{1-(x+3)^2}$$
$$y = -2 \pm \sqrt{1-(x+3)^2}$$
Using $X_{min} = -6$, $X_{max} = 0$, $Y_{min} = -4$, and $Y_{max} = 0$, we draw the graph:

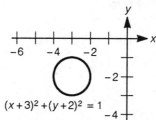

7.   (a)   Using the settings $X_{min} = -6$, $X_{max} = 6$, $Y_{min} = -6$, and $Y_{max} = 6$, we draw the graph:

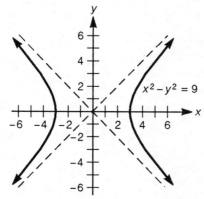

(b)   The curves appear to intersect at $(-3, 0)$ and $(3, 0)$:

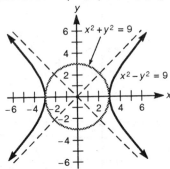

(c)   Solving $x^2 + y^2 = 9$ for y^2 yields $y^2 = 9 - x^2$, now substitute:

$$x^2 - (9 - x^2) = 9$$
$$2x^2 - 9 = 9$$
$$2x^2 = 18$$
$$x^2 = 9$$
$$x = \pm 3$$
$$y = 0$$

Thus the intersection points are indeed $(-3, 0)$ and $(3, 0)$

9.   (a)   Solving for y, we obtain $y = (9 - x^2)^{1/3}$. We graph the curve:

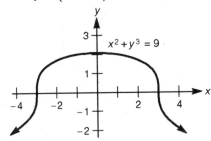

(b)   Solving for y, we obtain $y = \pm\sqrt{9 - x^3}$. We graph each curve:

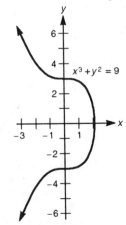

2.2 Graphs and Equations, A Second Look

1. Substituting $x = 8$, we have:
    $$y = \tfrac{1}{2}(8) + 3 = 4 + 3 = 7 \neq 6$$
    The point $(8, 6)$ does not lie on the graph.

3. Substituting $(x, y) = (4, 3)$, we have:
    $$3(4)^2 + (3)^2 = 3(16) + 9 = 48 + 9 = 57 \neq 52$$
    The point $(4, 3)$ does not lie on the graph.

5. Substituting $x = a$, we have:
    $$y = 4a$$
    The point $(a, 4a)$ lies on the graph.

7. Since the denominator cannot equal 0, we must have $x \neq 4$. To assist us in graphing the curve, we simplify when $x \neq 4$:
    $$y = \frac{x^2 + x - 20}{x - 4} = \frac{(x + 5)(x - 4)}{x - 4} = x + 5$$
    Now graph the line $y = x + 5$, leaving an open circle at the point $(4, 9)$:

    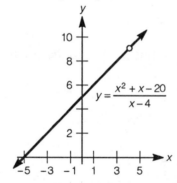

9. Since the denominator cannot equal 0, we must have $x \neq 3$. To assist us in graphing the curve, we simplify when $x \neq 3$:
    $$y = \frac{x^2 - 9}{x - 3} = \frac{(x + 3)(x - 3)}{x - 3} = x + 3$$
    Now graph the line $y = x + 3$, leaving an open circle at the point $(3, 6)$:

    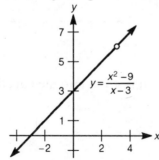

11. Since the denominator cannot equal 0, we must have $x^2 - 1 \neq 0$, so $x \neq \pm 1$. To simplify the fraction we factor the numerator by grouping:
$$-x^3 + 4x^2 + x - 4 = -x^2(x-4) + 1(x-4) = (x-4)\left(1 - x^2\right)$$
Now simplify the fraction when $x \neq \pm 1$:
$$y = \frac{-x^3 + 4x^2 + x - 4}{x^2 - 1} = \frac{(x-4)\left(1 - x^2\right)}{x^2 - 1} = -1(x-4) = -x + 4$$
Now graph the line $y = -x + 4$, leaving open circles at the points $(1, 3)$ and $(-1, 5)$:

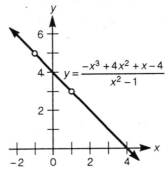

13. To find the x-intercept, we substitute $y = 0$ into the equation $3x + 4y = 12$:
$$3x + 4(0) = 12$$
$$3x = 12$$
$$x = 4$$
To find the y-intercept, we substitute $x = 0$ into the equation $3x + 4y = 12$:
$$3(0) + 4y = 12$$
$$4y = 12$$
$$y = 3$$
Now graph the line:

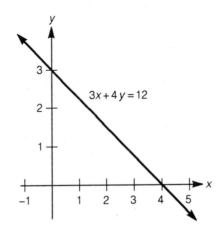

15. To find the x-intercept, we substitute $y = 0$ into the equation $y = 2x - 4$:
$$0 = 2x - 4$$
$$-2x = -4$$
$$x = 2$$

To find the y-intercept, we substitute $x = 0$ into the equation $y = 2x - 4$:
$$y = 2(0) - 4$$
$$y = -4$$
Now graph the line:

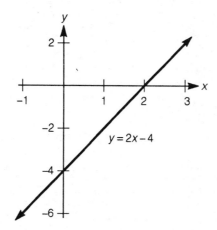

17. To find the x-intercept, we substitute $y = 0$ into the equation $x + y = 1$:
$$x + 0 = 1$$
$$x = 1$$
To find the y-intercept, we substitute $x = 0$ into the equation $x + y = 1$:
$$0 + y = 1$$
$$y = 1$$
Now graph the line:

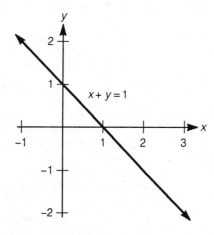

19. (a) To find the y-intercepts, let $x = 0$:
$$y = (0)^2 + 3(0) + 2 = 0 + 0 + 2 = 2$$
To find the x-intercepts, let $y = 0$:
$$x^2 + 3x + 2 = 0$$
$$(x + 2)(x + 1) = 0$$
$$x = -1, -2$$

(b) To find the y-intercepts, let $x = 0$:
$$y = (0)^2 + 2(0) + 3 = 0 + 0 + 3 = 3$$
To find the x-intercepts, we must solve the equation $x^2 + 2x + 3 = 0$. Using the quadratic formula, we obtain:
$$x = \frac{-2 \pm \sqrt{(2)^2 - 4(1)(3)}}{2(1)} = \frac{-2 \pm \sqrt{4 - 12}}{2} = \frac{-2 \pm \sqrt{-8}}{2}$$
Since this equation has no real solutions, there are no x-intercepts.

21. (a) To find the y-intercepts, let $x = 0$:
$$y = (0)^2 + 0 - 1 = 0 + 0 - 1 = -1$$
To find the x-intercepts, we must solve the equation $x^2 + x - 1 = 0$. Using the quadratric formula, we obtain:
$$x = \frac{-1 \pm \sqrt{(1)^2 - 4(1)(-1)}}{2(1)} = \frac{-1 \pm \sqrt{1 + 4}}{2} = \frac{-1 \pm \sqrt{5}}{2}$$

(b) To find the y-intercepts, let $x = 0$:
$$y = (0)^2 + 0 + 1 = 0 + 0 + 1 = 1$$
To find the x-intercepts, we must solve the equation $x^2 + x + 1 = 0$. Using the quadratic formula, we obtain:
$$x = \frac{-1 \pm \sqrt{(1)^2 - 4(1)(1)}}{2(1)} = \frac{-1 \pm \sqrt{1 - 4}}{2} = \frac{-1 \pm \sqrt{-3}}{2}$$
Since this equation has no real solutions, there are no x-intercepts.

23. To find the y-intercepts, let $x = 0$:
$$\frac{0}{2} + \frac{y}{3} = 1$$
$$\frac{y}{3} = 1$$
$$y = 3$$
To find the x-intercepts, let $y = 0$:
$$\frac{x}{2} + \frac{0}{3} = 1$$
$$\frac{x}{2} = 1$$
$$x = 2$$

25. To find the y-intercepts, let $x = 0$:
$$3(0) - 5y = 10$$
$$-5y = 10$$
$$y = -2$$
To find the x-intercepts, let $y = 0$:
$$3x - 5(0) = 10$$
$$3x = 10$$
$$x = \frac{10}{3}$$

27. To find the y-intercepts, let $x = 0$:
$$y = (0)^3 - 8 = 0 - 8 = -8$$
To find the x-intercepts, let $y = 0$:
$$x^3 - 8 = 0$$
$$x^3 = 8$$
$$x = 2$$

29. To find the y-intercepts, let $x = 0$:
$$y = \sqrt{0-1} - 0 + 3 = \sqrt{-1} + 3$$
Since this is not a real number, there are no y-intercepts.
To find the x-intercepts, let $y = 0$:
$$0 = \sqrt{x-1} - x + 3$$
$$x - 3 = \sqrt{x-1}$$
$$(x-3)^2 = x - 1$$
$$x^2 - 6x + 9 = x - 1$$
$$x^2 - 7x + 10 = 0$$
$$(x-2)(x-5) = 0$$
$$x = 2, 5$$
Note that $x = 5$ satisfies the required equation, but $x = 2$ does not. So $x = 5$ is the only x-intercept.

31. To find the y-intercepts, let $x = 0$:
$$y = 11(0) - 2(0)^2 - (0)^3 = 0 - 0 - 0 = 0$$
To find the x-intercepts, let $y = 0$:
$$0 = 11x - 2x^2 - x^3$$
$$0 = x\left(11 - 2x - x^2\right)$$
So $x = 0$ is clearly one x-intercept. We find the other two using the quadratic formula:
$$x = \frac{2 \pm \sqrt{(-2)^2 - 4(11)(-1)}}{2(-1)} = \frac{2 \pm \sqrt{48}}{-2} = -1 \pm 2\sqrt{3} \approx 2.46, -4.46$$

33. To find the y-intercepts, let $x = 0$:
$$y = (0)^4 - 2(0)^2 - 3 = 0 - 0 - 3 = -3$$
To find the x-intercepts, let $y = 0$:
$$0 = x^4 - 2x^2 - 3$$
$$0 = \left(x^2 - 3\right)\left(x^2 + 1\right)$$
Setting the first factor equal to zero yields $x^2 - 3 = 0$, so $x^2 = 3$ and thus $x = \pm\sqrt{3} \approx \pm 1.73$. Setting the second factor equal to zero yields $x^2 + 1 = 0$, so $x^2 = -1$, which has no real solutions.

35. To find the y-intercepts, let $x = 0$:
$$y = |2(0) + 3| - 4 = 3 - 4 = -1$$

To find the x-intercepts, let $y = 0$:

$$|2x+3|-4=0$$
$$|2x+3|=4$$
$$2x+3=4 \quad \text{or} \quad 2x+3=-4$$
$$2x=1 \quad \text{or} \quad 2x=-7$$
$$x=\tfrac{1}{2} \quad \text{or} \quad x=-\tfrac{7}{2}$$

37. (a) We set up a table of values:

x	-3	-2	-1	0	1	2	3
$y=|x|$	3	2	1	0	1	2	3

Now graph the equation:

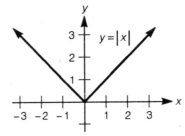

(b) We set up a table of values:

x	-3	-2	-1	0	1	2	3
$y=x^2$	9	4	1	0	1	4	9

Now graph the equation:

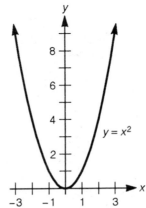

(c) We set up a table of values:

x	-2	-1	0	1	2
$y=x^3$	-8	-1	0	1	8

Now graph the equation:

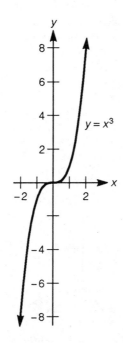

39. We "piece together" the curves $y = |x|$ when $x \le 0$ and $y = x^2$ when $x > 0$:

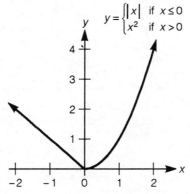

41. (a) We "piece together" the curves $y = \sqrt{x}$ when $0 \le x \le 1$ and $y = \dfrac{1}{x}$ when $1 < x < 2$:

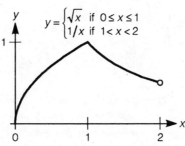

(b)   This graph is identical to that from part (a), except the point $(1, 1)$ is excluded:

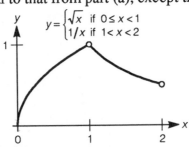

43.   We "piece together" the curves $y = x^3$ when $-1 \leq x < 0$, $y = \sqrt{x}$ when $0 \leq x < 1$, and
$y = \dfrac{1}{x}$ when $1 \leq x \leq 3$:

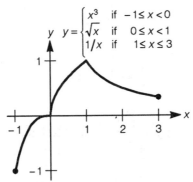

45.   (a)   From the graph it appears that $0° \text{ F}$ corresponds to $-18° \text{ C}$, which is the
(approximate) C-intercept of the graph.

(b)   Substituting $F = 0°$ into the equation and solving for C:
$$\tfrac{9}{5}C + 32 = 0$$
$$\tfrac{9}{5}C = -32$$
$$C = -\tfrac{160}{9} \approx -17.8°$$
This value is consistent with our approximation from part (a).

47.   (a)   Tracing up to the curve from $x = 2$, we get $y = \sqrt{2} \approx 1.4$.
(b)   Tracing up to the curve from $x = 3$, we get $y = \sqrt{3} \approx 1.7$.
(c)   Since $\sqrt{ab} = \sqrt{a} \cdot \sqrt{b}$, then $\sqrt{6} = \sqrt{2 \cdot 3} = \sqrt{2} \cdot \sqrt{3} \approx (1.4)(1.7) \approx 2.4$.

49.   (a)   Tracing up to the curve from $t = 0$, we get $N = 500$ bacteria.

(b)   Since $(0, 500)$ lies on the curve, we find where $(t, 1000)$ would be on the curve,
as the population would now be double. Tracing down from $N = 1000$, we get
$t = 1.5$ hours. So the population will double in 1.5 hours.

(c)   As in (b), we find where $(t, 2500)$ would be on the curve. Tracing down from $N = 2500$, we get $t = 3.5$ hours.

(d)   Between $t = 0$ and $t = 1$, the population has grown from $N = 500$ to $N = 800$, so it has increased at an average rate of 300 bacteria per hour. Between $t = 3$ and $t = 4$, the population has grown from $N = 2000$ to $N = 3000$, so it has increased at an average rate of 1000 bacteria per hour. So the population has increased more rapidly between $t = 3$ and $t = 4$.

51. Using the hint, we factor by grouping:
$$y^2 - x^2 y - x^3 y + x^5 = 0$$
$$y\left(y - x^2\right) - x^3\left(y - x^2\right) = 0$$
$$\left(y - x^2\right)\left(y - x^3\right) = 0$$
$$y = x^2 \text{ or } y = x^3$$
We graph each curve for $-1 \le x \le 1$:

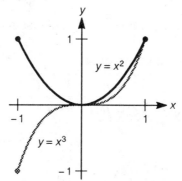

53. Since the y-coordinate for point A is 1.5, we substitute to find x:
$$\tfrac{1}{2}x + 1 = 1.5$$
$$\tfrac{1}{2}x = 0.5$$
$$x = 1$$
Since the y-coordinate for point B is 2.5, we substitute to find x:
$$\tfrac{1}{2}x + 1 = 2.5$$
$$\tfrac{1}{2}x = 1.5$$
$$x = 3$$

55. Since $(x_1, 0)$ and $(x_2, 0)$ satisfy the equation of the circle, we must have:
$$x_1^2 + 2Ax_1 + 0^2 + 2B(0) = C, \text{ so } x_1^2 + 2Ax_1 = C$$
$$x_2^2 + 2Ax_2 + 0^2 + 2B(0) = C, \text{ so } x_2^2 + 2Ax_2 = C$$

Similarly, since $(0, y_1)$ and $(0, y_2)$ satisfy the equation of the circle, we must have:

$$0^2 + 2A(0) + y_1^2 + 2By_1 = C, \text{ so } y_1^2 + 2By_1 = C$$
$$0^2 + 2A(0) + y_2^2 + 2By_2 = C, \text{ so } y_2^2 + 2By_2 = C$$

Since each of these four equations is quadratic, we can solve for x_1, x_2, y_1, and y_2 using the quadratic formula. Assuming x_1 and y_1 are the larger of the intercepts:

$$x_1 = \frac{-(2A) + \sqrt{(2A)^2 - 4(1)(-C)}}{2(1)} = \frac{-2A + 2\sqrt{A^2 + C}}{2} = -A + \sqrt{A^2 + C}$$

$$x_2 = \frac{-(2A) - \sqrt{(2A)^2 - 4(1)(-C)}}{2(1)} = \frac{-2A - 2\sqrt{A^2 + C}}{2} = -A - \sqrt{A^2 + C}$$

$$y_1 = \frac{-(2B) + \sqrt{(2B)^2 - 4(1)(-C)}}{2(1)} = \frac{-2B + 2\sqrt{B^2 + C}}{2} = -B + \sqrt{B^2 + C}$$

$$y_2 = \frac{-(2B) - \sqrt{(2B)^2 - 4(1)(-C)}}{2(1)} = \frac{-2B - 2\sqrt{B^2 + C}}{2} = -B - \sqrt{B^2 + C}$$

Using each of these expressions, we now prove the statements.

(a) Substitute for x_1, x_2, y_1, and y_2 to obtain:

$$\frac{x_1 + x_2}{y_1 + y_2} = \frac{-A + \sqrt{A^2 + C} - A - \sqrt{A^2 + C}}{-B + \sqrt{B^2 + C} - B - \sqrt{B^2 + C}} = \frac{-2A}{-2B} = \frac{A}{B}$$

(b) Substitute for x_1, x_2, y_1, and y_2 to obtain:

$$x_1 x_2 - y_1 y_2$$
$$= \left(-A + \sqrt{A^2 + C}\right)\left(-A - \sqrt{A^2 + C}\right) - \left(-B + \sqrt{B^2 + C}\right)\left(-B - \sqrt{B^2 + C}\right)$$
$$= \left[A^2 - \left(A^2 + C\right)\right] - \left[B^2 - \left(B^2 + C\right)\right]$$
$$= -C + C$$
$$= 0$$

(c) Substitute for x_1, x_2, y_1, and y_2 to obtain:

$$x_1 x_2 + y_1 y_2$$
$$= \left(-A + \sqrt{A^2 + C}\right)\left(-A - \sqrt{A^2 + C}\right) + \left(-B + \sqrt{B^2 + C}\right)\left(-B - \sqrt{B^2 + C}\right)$$
$$= \left[A^2 - \left(A^2 + C\right)\right] + \left[B^2 - \left(B^2 + C\right)\right]$$
$$= -C - C$$
$$= -2C$$

Optional TI-81 Graphing Calculator Exercises for Section 2.2

1.   (a)   Using Zoom-6 settings, we obtain the graph:

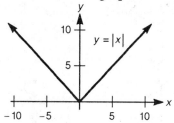

(b)   Using the indicated settings, we obtain the graph:

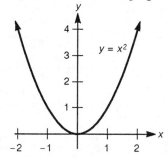

(c)   Using the indicated settings, we obtain the graph:

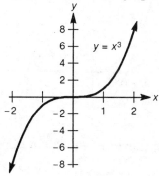

(d)   Using the indicated settings, we obtain the graphs:

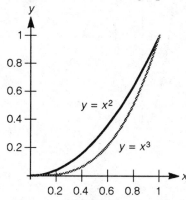

The graphs intersect at $(0,0)$ and $(1,1)$. In the interval $(0,1)$, it appears that x^2 is larger than x^3.

(e) Using the settings $X_{min} = 1$, $X_{max} = 3$, $Y_{min} = 0$, and $Y_{max} = 30$, we obtain the graphs:

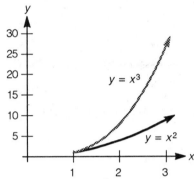

In the interval $(1, \infty)$, it appears that x^3 is larger than x^2.

3.   (a) Following the indicated keystrokes, we obtain the graph:

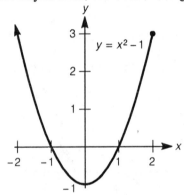

(b) After entering the second function, we obtain the graph:

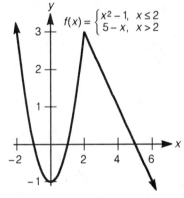

5.   Using the illustrated technique, we obtain the graph:

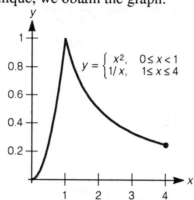

$$y = \begin{cases} x^2, & 0 \le x < 1 \\ 1/x, & 1 \le x \le 4 \end{cases}$$

7.   Using the illustrated technique, we obtain the graph:

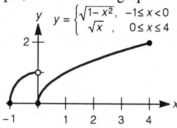

$$y = \begin{cases} \sqrt{1-x^2}, & -1 \le x < 0 \\ \sqrt{x}, & 0 \le x \le 4 \end{cases}$$

9.   Using the illustrated technique, we obtain the graph:

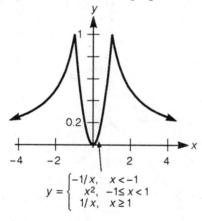

$$y = \begin{cases} -1/x, & x < -1 \\ x^2, & -1 \le x < 1 \\ 1/x, & x \ge 1 \end{cases}$$

11.   We graph each "piece" of the function over the required intervals:

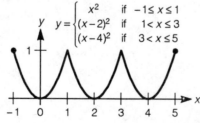

$$y = \begin{cases} x^2 & \text{if } -1 \le x \le 1 \\ (x-2)^2 & \text{if } 1 < x \le 3 \\ (x-4)^2 & \text{if } 3 < x \le 5 \end{cases}$$

2.3 Equations of Lines

1. (a)   Here $(x_1, y_1) = (-3, 2)$ and $(x_2, y_2) = (1, -6)$, so:
$$\text{slope} = \frac{-6 - 2}{1 - (-3)} = -\frac{8}{4} = -2$$

(b)   Here $(x_1, y_1) = (2, -5)$ and $(x_2, y_2) = (4, 1)$, so:
$$\text{slope} = \frac{1 - (-5)}{4 - 2} = \frac{6}{2} = 3$$

(c)   Here $(x_1, y_1) = (-2, 7)$ and $(x_2, y_2) = (1, 0)$, so:
$$\text{slope} = \frac{0 - 7}{1 - (-2)} = -\frac{7}{3}$$

(d)   Here $(x_1, y_1) = (4, 5)$ and $(x_2, y_2) = (5, 8)$, so:
$$\text{slope} = \frac{8 - 5}{5 - 4} = \frac{3}{1} = 3$$

3. (a)   Here $(x_1, y_1) = (1, 1)$ and $(x_2, y_2) = (-1, -1)$, so:
$$\text{slope} = \frac{-1 - 1}{-1 - 1} = \frac{-2}{-2} = 1$$

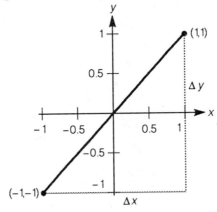

(b)   Here $(x_1, y_1) = (0, 5)$ and $(x_2, y_2) = (-8, 5)$, so:

$$\text{slope} = \frac{5-5}{-8-0} = \frac{0}{-8} = 0$$

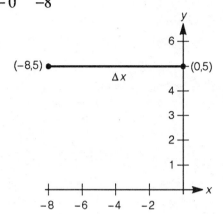

(c)   Here $(x_1, y_1) = (-1, 1)$ and $(x_2, y_2) = (1, -1)$, so:

$$\text{slope} = \frac{-1-1}{1-(-1)} = \frac{-2}{2} = -1$$

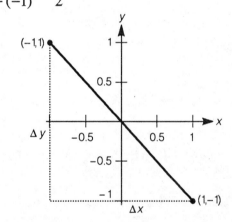

(d)   Here $(x_1, y_1) = (a, b)$ and $(x_2, y_2) = (b, a)$, so:

$$\text{slope} = \frac{a-b}{b-a} = -1$$

5. m_3 is smallest (since it is negative)
m_2 is next (it appears to be near zero)
m_4 is next (it is positive but not as steep as m_1)
m_1 is largest (it is steeper than m_4)

7. We compute the slopes:

$$\text{slope}_{AB} = \frac{\frac{1}{2} - (-2)}{2 - (-8)} = \frac{\frac{5}{2}}{10} = \frac{1}{4}$$

$$\text{slope}_{BC} = \frac{-1 - \frac{1}{2}}{11 - 2} = \frac{-\frac{3}{2}}{9} = -\frac{1}{6}$$

Since these slopes are different the three points cannot be collinear.

9. We compute the slopes:

$$\text{slope}_{AB} = \frac{4 - (-5)}{3 - 0} = \frac{9}{3} = 3$$

$$\text{slope}_{BC} = \frac{-8 - 4}{-1 - 3} = \frac{-12}{-4} = 3$$

Since these slopes are equal the three points are collinear.

11. (a) Here $(x_1, y_1) = (-2, 1)$ and $m = -5$, so by the point-slope formula:

$$y - 1 = -5[x - (-2)]$$
$$y - 1 = -5(x + 2)$$
$$y - 1 = -5x - 10$$
$$y = -5x - 9$$

(b) Here $(x_1, y_1) = (4, -4)$ and $m = 4$, so by the point-slope formula:

$$y - (-4) = 4(x - 4)$$
$$y + 4 = 4x - 16$$
$$y = 4x - 20$$

(c) Here $(x_1, y_1) = \left(-6, -\frac{2}{3}\right)$ and $m = \frac{1}{3}$, so by the point-slope formula:

$$y - \left(-\frac{2}{3}\right) = \left(\frac{1}{3}\right)[x - (-6)]$$
$$y + \frac{2}{3} = \left(\frac{1}{3}\right)(x + 6)$$
$$y + \frac{2}{3} = \frac{1}{3}x + 2$$
$$y = \frac{1}{3}x + \frac{4}{3}$$

(d) Here $(x_1, y_1) = (0, 1)$ and $m = -1$, so by the point-slope formula:

$$y - 1 = -1(x - 0)$$
$$y - 1 = -x$$
$$y = -x + 1$$

Note in (d) that, since the given point is the y-intercept, we could have immediately written $y = -x + 1$ using the slope-intercept formula.

13. (a) First we find the slope:
$$m = \frac{-6-8}{-3-4} = \frac{-14}{-7} = 2$$
Using $(x_1, y_1) = (4, 8)$ in the point-slope formula, we have:
$$y - 8 = 2(x - 4)$$
$$y - 8 = 2x - 8$$
$$y = 2x$$

(b) First we find the slope:
$$m = \frac{-10-0}{3-(-2)} = \frac{-10}{5} = -2$$
Using $(x_1, y_1) = (-2, 0)$ in the point-slope formula, we have:
$$y - 0 = -2[x - (-2)]$$
$$y = -2(x + 2)$$
$$y = -2x - 4$$

(c) First we find the slope:
$$m = \frac{-1-(-2)}{4-(-3)} = \tfrac{1}{7}$$
Using $(x_1, y_1) = (4, -1)$ in the point-slope formula, we have:
$$y - (-1) = \tfrac{1}{7}(x - 4)$$
$$y + 1 = \tfrac{1}{7}x - \tfrac{4}{7}$$
$$y = \tfrac{1}{7}x - \tfrac{11}{7}$$

15. Since vertical lines have the form $x =$ constant, and $(-3, 4)$ is on the line, then the equation is $x = -3$.

17. Since horizontal lines have the form $y =$ constant, and $(-3, 4)$ is on the line, then the equation is $y = 4$.

19. The y-axis is vertical, so its equation must have the form $x =$ constant. Since $(0, 0)$ is on the y-axis, the equation is $x = 0$.

21. (a) Using the slope-intercept formula with $m = -4$ and $b = 7$, we have $y = -4x + 7$.
    (b) Using the slope-intercept formula with $m = 2$ and $b = \tfrac{3}{2}$, we have $y = 2x + \tfrac{3}{2}$.
    (c) Using the slope-intercept formula with $m = -\tfrac{4}{3}$ and $b = 14$, we have $y = -\tfrac{4}{3}x + 14$.

23. (a) We use the point-slope formula:
$$y - (-1) = 4[x - (-3)]$$
$$y + 1 = 4(x + 3)$$
$$y + 1 = 4x + 12$$
$$y = 4x + 11$$
Now draw the graph:

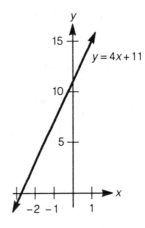

(b) We use the point-slope formula:
$$y - 0 = \tfrac{1}{2}\left(x - \tfrac{5}{2}\right)$$
$$y = \tfrac{1}{2}x - \tfrac{5}{4}$$
Now draw the graph:

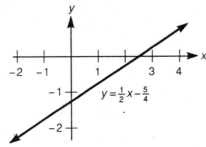

(c) We find the slope between the points $(6, 0)$ and $(0, 5)$:
$$m = \frac{5 - 0}{0 - 6} = -\tfrac{5}{6}$$
Since 5 is the y-intercept, then by the slope-intercept formula:
$$y = -\tfrac{5}{6}x + 5$$

Now draw the graph:

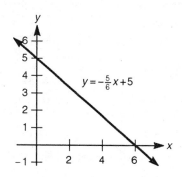

(d)   We use the point-slope formula with the point $(-2, 0)$:

$$y - 0 = \tfrac{3}{4}[x - (-2)]$$
$$y = \tfrac{3}{4}(x + 2)$$
$$y = \tfrac{3}{4}x + \tfrac{3}{2}$$

Now draw the graph:

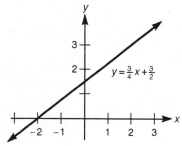

(e)   We first find the slope:

$$m = \frac{6 - 2}{2 - 1} = \tfrac{4}{1} = 4$$

Using the point $(1, 2)$ in the point-slope formula, we have:

$$y - 2 = 4(x - 1)$$
$$y - 2 = 4x - 4$$
$$y = 4x - 2$$

Now draw the graph:

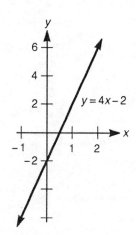

25. We first find the x- and y-intercepts of the circle by letting $x = 0$ and $y = 0$:

$$
\begin{array}{cc}
y = 0 & x = 0 \\
x^2 + 4x + 4 = 0 & y^2 - 4y + 4 = 0 \\
(x+2)^2 = 0 & (y-2)^2 = 0 \\
x = -2 & y = 2
\end{array}
$$

So the line passes through the points $(-2, 0)$ and $(0, 2)$. We find its slope:

$$m = \frac{2-0}{0-(-2)} = \frac{2}{2} = 1$$

So $y = x + 2$ is the equation (in slope-intercept form) of the line. We draw the graph:

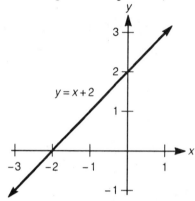

27. If the line is parallel to the x-axis, then its slope is 0. Since this line is of the form $y = $ constant, then for $(-3, 4)$ to lie on the line the equation is $y = 4$, or $y - 4 = 0$ in the desired form. We draw the graph:

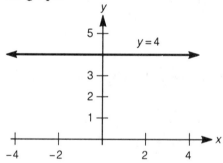

29. (a) If $x = 0$, then $y = 3$ and if $y = 0$, then $x = 5$. So the x-intercept is 5 and the y-intercept is 3. Thus we have:

area $= \frac{1}{2}(5)(3) = \frac{15}{2}$

perimeter $= 5 + 3 + \sqrt{5^2 + 3^2} = 8 + \sqrt{34}$

(b) If $x = 0$, then $y = -3$ and if $y = 0$, then $x = 5$. So the x-intercept is 5 and the y-intercept is -3. Thus we have:

area $= \frac{1}{2}(5)(3) = \frac{15}{2}$

perimeter $= 5 + 3 + \sqrt{5^2 + 3^2} = 8 + \sqrt{34}$

31. (a)   We find the slopes of each:

$$3x - 4y = 12 \qquad\qquad 4x - 3y = 12$$
$$-4y = -3x + 12 \qquad\qquad -3y = -4x + 12$$
$$y = \tfrac{3}{4}x - 3 \qquad\qquad y = \tfrac{4}{3}x - 4$$
$$m = \tfrac{3}{4} \qquad\qquad m = \tfrac{4}{3}$$

The lines are not parallel (slopes aren't the same), the lines are not perpendicular $\left(\tfrac{3}{4} \cdot \tfrac{4}{3} = 1, \text{not} -1\right)$, so they are neither.

(b)   We find the slopes of each:

$$y = 5x - 16 \qquad\qquad y = 5x + 2$$
$$m = 5 \qquad\qquad m = 5$$

Since these slopes are the same, the lines are parallel.

(c)   We find the slopes of each:

$$5x - 6y = 25 \qquad\qquad 6x + 5y = 0$$
$$-6y = -5x + 25 \qquad\qquad 5y = -6x$$
$$y = \tfrac{5}{6}x - \tfrac{25}{6} \qquad\qquad y = -\tfrac{6}{5}x$$
$$m = \tfrac{5}{6} \qquad\qquad m = -\tfrac{6}{5}$$

Since $\left(\tfrac{5}{6}\right)\left(-\tfrac{6}{5}\right) = -1$, the lines are perpendicular.

(d)   We find the slopes of each:

$$y = -\tfrac{2}{3}x - 1 \qquad\qquad y = \tfrac{3}{2}x - 1$$
$$m = -\tfrac{2}{3} \qquad\qquad m = \tfrac{3}{2}$$

Since $\left(-\tfrac{2}{3}\right)\left(\tfrac{3}{2}\right) = -1$, the lines are perpendicular.

(e)   We find the slopes of each:

$$-2x - 5y = 1 \qquad\qquad y - \tfrac{2}{5}x - 4 = 0$$
$$-5y = 2x + 1 \qquad\qquad y = \tfrac{2}{5}x + 4$$
$$y = -\tfrac{2}{5}x - \tfrac{1}{5} \qquad\qquad m = \tfrac{2}{5}$$
$$m = -\tfrac{2}{5}$$

The lines are not parallel (slopes aren't the same), the lines are not perpendicular $\left(-\tfrac{2}{5} \cdot \tfrac{2}{5} = -\tfrac{4}{25}, \text{not} -1\right)$, so they are neither.

(f)   We find the slopes of each:

$$x = 8y + 3 \qquad\qquad 4y - \tfrac{1}{2}x = 32$$
$$8y = x - 3 \qquad\qquad 4y = \tfrac{1}{2}x + 32$$
$$y = \tfrac{1}{8}x - \tfrac{3}{8} \qquad\qquad y = \tfrac{1}{8}x + 8$$
$$m = \tfrac{1}{8} \qquad\qquad m = \tfrac{1}{8}$$

Since these slopes are the same, the lines are parallel.

33. We first find the slope:
$$2x - 5y = 10$$
$$-5y = -2x + 10$$
$$y = \tfrac{2}{5}x - 2$$

So the slope is $\tfrac{2}{5}$. We use the point $(-1, 2)$ in the point-slope formula:
$$y - 2 = \tfrac{2}{5}[x - (-1)]$$
$$y - 2 = \tfrac{2}{5}(x + 1)$$
$$y - 2 = \tfrac{2}{5}x + \tfrac{2}{5}$$
$$y = \tfrac{2}{5}x + \tfrac{12}{5} \qquad \text{[form } y = mx + b\text{]}$$
$$5y = 2x + 12$$
$$2x - 5y + 12 = 0 \qquad \text{[form } Ax + By + C = 0\text{]}$$

35. We first find the slope of $4y - 3x = 1$:
$$4y = 3x + 1$$
$$y = \tfrac{3}{4}x + \tfrac{1}{4}$$

Since this slope is $\tfrac{3}{4}$, the perpendicular line slope is $-\tfrac{4}{3}$. We now use the point $(4, 0)$ in the point-slope formula:
$$y - 0 = -\tfrac{4}{3}(x - 4)$$
$$y = -\tfrac{4}{3}x + \tfrac{16}{3} \qquad \text{[form } y = mx + b\text{]}$$
$$3y = -4x + 16$$
$$4x + 3y - 16 = 0 \qquad \text{[form } Ax + By + C = 0\text{]}$$

37. We first find the slope of $3x - 5y = 25$:
$$-5y = -3x + 25$$
$$y = \tfrac{3}{5}x - 5$$

So the slope is $\tfrac{3}{5}$. We now find the y-intercept of $6x - y + 11 = 0$:
$$y = 6x + 11$$

So the y-intercept is 11. We write the equation:
$$y = \tfrac{3}{5}x + 11 \qquad \text{[form } y = mx + b\text{]}$$
$$5y = 3x + 55$$
$$3x - 5y + 55 = 0 \qquad \text{[form } Ax + By + C = 0\text{]}$$

39. (a) The center is $(0, 0)$ and the radius is 5. We draw the graph:

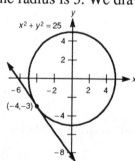

(b) The tangent line will be perpendicular to the radius drawn from the center $(0,0)$ and the point $(-4,-3)$. We find the slope:

$$m = \frac{-3-0}{-4-0} = \frac{-3}{-4} = \frac{3}{4}$$

So the perpendicular line will have slope $= -\frac{4}{3}$. We use the point $(-4,-3)$ in the point-slope formula:

$$y-(-3) = -\tfrac{4}{3}[x-(-4)]$$
$$y+3 = -\tfrac{4}{3}(x+4)$$
$$y+3 = -\tfrac{4}{3}x - \tfrac{16}{3}$$
$$y = -\tfrac{4}{3}x - \tfrac{25}{3}$$

41. Let $(x_1,y_1) = (3,9)$ and $(x_2,y_2) = \left[3+h,(3+h)^2\right]$, so:

$$m = \frac{(3+h)^2-9}{3+h-3} = \frac{9+6h+h^2-9}{h} = \frac{6h+h^2}{h} = 6+h$$

43. Let $(x_1,y_1) = (x,x^3)$ and $(x_2,y_2) = \left[x+h,(x+h)^3\right]$. Therefore the slope is given by:

$$\frac{(x+h)^3-x^3}{x+h-x} = \frac{x^3+3x^3h+3xh^2+h^3-x^3}{h}$$
$$= \frac{3x^2h+3xh^2+h^3}{h}$$
$$= \frac{h(3x^2+3xh+h^2)}{h}$$
$$= 3x^2+3xh+h^2$$

45. Let $(x_1,y_1) = \left(x,\tfrac{1}{x}\right)$ and $(x_2,y_2) = \left(x+h,\tfrac{1}{x+h}\right)$, so:

$$\text{slope} = \frac{\frac{1}{x+h}-\frac{1}{x}}{x+h-x} = \frac{\frac{1}{x+h}-\frac{1}{x}}{h}$$

We multiply the numerator and denominator by $x(x+h)$ to obtain:

$$\text{slope} = \frac{x-x-h}{hx(x+h)} = \frac{-h}{hx(x+h)} = \frac{-1}{x(x+h)}$$

47. We draw the graph:

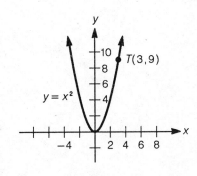

The completed table is:

x	2.5	2.9	2.99	2.999	2.9999
y	6.25	8.41	8.9401	8.994001	8.99940001
Δx	0.5	0.1	0.01	0.001	0.0001
Δy	2.75	0.59	0.0599	0.005999	0.00059999
m	5.5	5.9	5.99	5.999	5.9999

It would appear that as x approaches 3, the slope of these lines (called secant lines) approaches 6. So we would estimate that the slope of the tangent line to the curve $y = x^2$ at $T(3,9)$ is 6.

49. Let the coordinates of P be (x, x^3). Since the slope of the line passing through P and $(1,1)$ is $\frac{3}{4}$, we have:

$$\frac{x^3 - 1}{x - 1} = \frac{3}{4}$$
$$\frac{(x-1)(x^2 + x + 1)}{x - 1} = \frac{3}{4}$$
$$x^2 + x + 1 = \frac{3}{4}$$
$$4x^2 + 4x + 4 = 3$$
$$4x^2 + 4x + 1 = 0$$
$$(2x + 1)^2 = 0$$
$$2x + 1 = 0$$
$$x = -\frac{1}{2}$$
$$y = \left(-\frac{1}{2}\right)^3 = -\frac{1}{8}$$

Thus the point P is $\left(-\frac{1}{2}, -\frac{1}{8}\right)$.

51. Let the coordinates of P be $\left(x, \frac{1}{x}\right)$. Since the slope of the line through P and $\left(2, \frac{1}{2}\right)$ is $-\frac{1}{16}$, we have:

$$\frac{\frac{1}{x} - \frac{1}{2}}{x - 2} = -\frac{1}{16}$$
$$\frac{2 - x}{2x(x - 2)} = -\frac{1}{16}$$
$$\frac{-1}{2x} = -\frac{1}{16}$$
$$2x = 16$$
$$x = 8$$

So the point P is $\left(8, \frac{1}{8}\right)$.

53. We use the point-slope formula to find the equation of the line:

$$y - 6 = -5(x - 3)$$
$$y - 6 = -5x + 15$$
$$y = -5x + 21$$

To find the x-intercept, let $y = 0$:

$$-5x + 21 = 0$$
$$5x = 21$$
$$x = \tfrac{21}{5}$$

To find the y-intercept, let $x = 0$, so $y = 21$. The area of the triangle is given by:

$$\text{Area} = \tfrac{1}{2}(\text{base})(\text{height}) = \tfrac{1}{2} \cdot \tfrac{21}{5} \cdot 21 = 44.1 \text{ square units}$$

55. (a) We draw the graph:

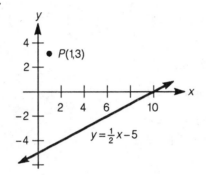

(b) Since the slope must be -2, we use $(1, 3)$ in the point-slope formula:

$$y - 3 = -2(x - 1)$$
$$y - 3 = -2x + 2$$
$$y = -2x + 5$$

(c) Set the two equations equal:

$$-2x + 5 = \tfrac{1}{2}x - 5$$
$$-4x + 10 = x - 10$$
$$-5x = -20$$
$$x = 4$$
$$y = -2(4) + 5 = -3$$

The intersection point is $(4, -3)$.

(d) We find the distance from $(1, 3)$ to $(4, -3)$:

$$d = \sqrt{(4 - 1)^2 + (-3 - 3)^2} = \sqrt{9 + 36} = \sqrt{45} = 3\sqrt{5}$$

57. (a) Since the slope is m and the point $(2, 1)$ lies on the line, we use the point-slope formula to obtain $y - 1 = m(x - 2)$.

(b) For the x-intercept, let $y = 0$:
$$-1 = m(x - 2)$$
$$-1 = mx - 2m$$
$$mx = 2m - 1$$
$$x = \frac{2m - 1}{m}$$
For the y-intercept, let $x = 0$:
$$y - 1 = m(-2)$$
$$y - 1 = -2m$$
$$y = -2m + 1$$

(c) The area is given by:
$$\text{Area} = \tfrac{1}{2}(\text{base})(\text{height}) = \frac{1}{2}\left(\frac{2m-1}{m}\right)(-2m+1)$$

(d) Since the area is 4, we have:
$$\frac{1}{2}\left(\frac{2m-1}{m}\right)(-2m+1) = 4$$
$$(2m - 1)(-2m + 1) = 8m$$
$$-4m^2 + 4m - 1 = 8m$$
$$4m^2 + 4m + 1 = 0$$

(e) We solve for m:
$$(2m + 1)^2 = 0$$
$$2m + 1 = 0$$
$$2m = -1$$
$$m = -\tfrac{1}{2}$$

59. (a) Since $\frac{\Delta x}{\Delta P} = -\frac{2}{3}$, we have the equation:
$$x = 280 - \tfrac{2}{3}(P - 195) = 410 - \tfrac{2}{3}P$$

(b) If $P = 270$, we have:
$$x = 410 - \tfrac{2}{3}(270) = 410 - 180 = 230$$
So 230 units can be sold in a month.

(c) If $x = 205$, we have:
$$205 = 410 - \tfrac{2}{3}P$$
$$\tfrac{2}{3}P = 205$$
$$P = \tfrac{3}{2}(205) = 307.5$$
The price would be \$307.50 per unit.

61. (a) Since A and C lie on the line $y = m_1 x + b_1$, and the x-coordinates of A and C are 0 and 1, respectively, then the y-coordinates will be b_1 and $m_1 + b_1$, so $A = (0, b_1)$ and $C = (1, m_1 + b_1)$. Similarly, the points B and D are $B = (0, b_2)$ and $D = (1, m_2 + b_2)$.

(b) We verify the results:
$$AB = \sqrt{(0-0)^2 + (b_1 - b_2)^2} = \sqrt{(b_1 - b_2)^2} = |b_1 - b_2|,$$
which is $b_1 - b_2$ since $b_1 > b_2$
$$CD = \sqrt{(1-1)^2 + [(m_1 + b_1) - (m_2 + b_2)]^2}$$
$$= \sqrt{[(m_1 + b_1) - (m_2 + b_2)]^2}$$
$$= |(m_1 + b_1) - (m_2 + b_2)|$$
$$= m_1 + b_1 - (m_2 + b_2) \text{ since } m_1 + b_1 > m_2 + b_2$$

(c) Since $AB = CD$, we have:
$$b_1 - b_2 = m_1 + b_1 - m_2 - b_2$$
$$0 = m_1 - m_2$$
$$m_1 = m_2$$

63. Multiplying both numerator and denominator by -1, we obtain:
$$\frac{y_2 - y_1}{x_2 - x_1} \cdot \frac{-1}{-1} = \frac{y_1 - y_2}{x_1 - x_2}$$
This identity tells us that, in calculating slope, either point can be chosen as the starting point.

65. (a) Since the dotted lines are parallel to the coordinate axes, we can label points as shown in the figure:

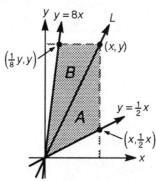

The areas of triangles A and B can be found by subtracting areas of the right triangles formed by the lines and the coordinate axes:
$$\text{Area}_A = \tfrac{1}{2}(x)(y) - \tfrac{1}{2}(x)\left(\tfrac{1}{2}x\right) = \tfrac{1}{2}xy - \tfrac{1}{4}x^2$$
$$\text{Area}_B = \tfrac{1}{2}(y)(x) - \tfrac{1}{2}(y)\left(\tfrac{1}{8}y\right) = \tfrac{1}{2}xy - \tfrac{1}{16}y^2$$

Since these two areas must be equal, we have:

$$\tfrac{1}{2}xy - \tfrac{1}{4}x^2 = \tfrac{1}{2}xy - \tfrac{1}{16}y^2$$
$$-\tfrac{1}{4}x^2 = -\tfrac{1}{16}y^2$$
$$y^2 = 4x^2$$
$$y = 2x$$

Note that we can choose the positive root since L has a positive slope.

(b)   Again we label points as indicated in the figure:

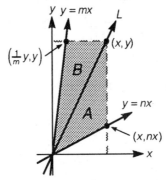

As in part (a), we obtain expressions for the areas of each triangle:

$$\text{Area}_A = \tfrac{1}{2}(x)(y) - \tfrac{1}{2}(x)(nx) = \tfrac{1}{2}xy - \tfrac{1}{2}nx^2$$
$$\text{Area}_B = \tfrac{1}{2}(y)(x) - \tfrac{1}{2}(y)\left(\tfrac{1}{m}y\right) = \tfrac{1}{2}xy - \tfrac{1}{2m}y^2$$

Since these two areas must be equal, we have:

$$\tfrac{1}{2}xy - \tfrac{1}{2}nx^2 = \tfrac{1}{2}xy - \tfrac{1}{2m}y^2$$
$$-\tfrac{1}{2}nx^2 = -\tfrac{1}{2m}y^2$$
$$y^2 = mnx^2$$
$$y = \sqrt{mn}\, x$$

Again note that we can choose the positive root since L has a positive slope.

Optional TI-81 Graphing Calculator Exercises for Section 2.3

1. (a) Since the lines pass through the origin, the y-intercepts are 0. So the equations of the four lines are $y = x$, $y = 2x$, $y = 3x$, and $y = 10x$. We sketch the graphs:

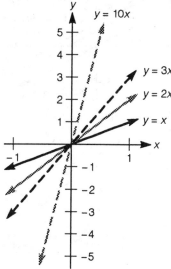

As the positive slopes increase, the lines become steeper.

(b) Since the lines pass through the origin, the y-intercepts are 0. So the equations of the four lines are $y = -x$, $y = -2x$, $y = -3x$, and $y = -10x$. We sketch the graphs:

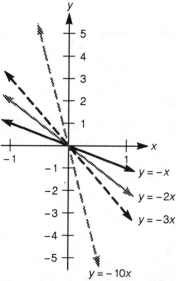

As the negative slopes decrease, the lines become steeper.

3.   We graph the three parallel lines:

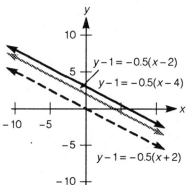

5.   (a)   Solving for y, we have:

$$\frac{x}{2}+\frac{y}{3}=1$$
$$3x+2y=6$$
$$2y=-3x+6$$
$$y=-\tfrac{3}{2}x+3$$

Now graph the line:

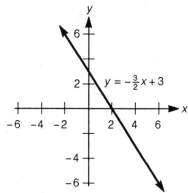

Note that the x-intercept is 2 and the y-intercept is 3.

(b)   Solving for y, we have:

$$\frac{x}{-2}+\frac{y}{-3}=1$$
$$3x+2y=-6$$
$$2y=-3x-6$$
$$y=-\tfrac{3}{2}x-3$$

Now graph the line:

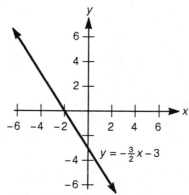

Note that the x-intercept is -2 and the y-intercept is -3.

(c)   Solving for y, we have:

$$\frac{x}{6}+\frac{y}{5}=1$$
$$5x+6y=30$$
$$6y=-5x+30$$
$$y=-\frac{5}{6}x+5$$

Now graph the line:

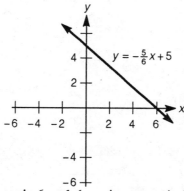

Note that the x-intercept is 6 and the y-intercept is 5.

(d)   Solving for y, we have:

$$\frac{x}{-6}+\frac{y}{-5}=1$$
$$5x+6y=-30$$
$$6y=-5x-30$$
$$y=-\frac{5}{6}x-5$$

Now graph the line:

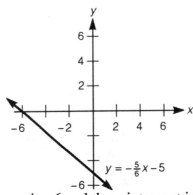

$$y = -\tfrac{5}{6}x - 5$$

Note that the x-intercept is -6 and the y-intercept is -5.

(e)   It appears that $\tfrac{x}{a} + \tfrac{y}{b} = 1$ is a line with an x-intercept of a and y-intercept of b.

7.   (a)   We draw the triangle with vertices $A(-4,0), B(2,0)$, and $C(0,6)$:

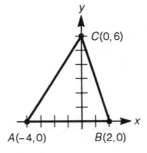

(b)   Using the midpoint formula:
$$M_{AB} = \left(\tfrac{-4+2}{2}, \tfrac{0+0}{2}\right) = \left(-\tfrac{2}{2}, \tfrac{0}{2}\right) = (-1,0)$$
$$M_{BC} = \left(\tfrac{2+0}{2}, \tfrac{0+6}{2}\right) = \left(\tfrac{2}{2}, \tfrac{6}{2}\right) = (1,3)$$
$$M_{AC} = \left(\tfrac{-4+0}{2}, \tfrac{0+6}{2}\right) = \left(-\tfrac{4}{2}, \tfrac{6}{2}\right) = (-2,3)$$
Now add the medians to the graph of the triangle:

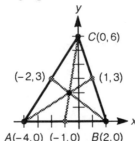

The medians appear to intersect at the point $(-1,2)$.

(c)   For the median from A to side BC, the line passes through the points $(-4,0)$ and $(1,3)$. Finding the slope:

$$m = \frac{3-0}{1+4} = \frac{3}{5}$$

Using the point-slope formula with the point $(1,3)$:

$$y - 3 = \tfrac{3}{5}(x-1)$$
$$y - 3 = \tfrac{3}{5}x - \tfrac{3}{5}$$
$$y = \tfrac{3}{5}x + \tfrac{12}{5}$$

For the median from B to side AC, the line passes through the points $(2,0)$ and $(-2,3)$. Finding the slope:

$$m = \frac{3-0}{-2-2} = -\frac{3}{4}$$

Using the pont-slope formula with the point $(2,0)$:

$$y - 0 = -\tfrac{3}{4}(x-2)$$
$$y = -\tfrac{3}{4}x + \tfrac{3}{2}$$

For the median from C to side AB, the line passes through the points $(0,6)$ and $(-1,0)$. Finding the slope:

$$m = \frac{0-6}{-1-0} = 6$$

Since the y-intercept is 6, its equation is $y = 6x + 6$.

Solving any pair of simultaneous equations will yield the coordinates of the centroid. Using $y = -\tfrac{3}{4}x + \tfrac{3}{2}$ and $y = 6x + 6$, we solve the system:

$$-\tfrac{3}{4}x + \tfrac{3}{2} = 6x + 6$$
$$-3x + 6 = 24x + 24$$
$$-27x = 18$$
$$x = -\tfrac{2}{3}$$
$$y = 6\left(-\tfrac{2}{3}\right) + 6 = 2$$

The exact coordinates of the centroid are $\left(-\tfrac{2}{3}, 2\right)$, which agrees with our estimate from part (b), to the nearest integer.

2.4 Symmetry and Graphs

1.   (a)   The reflection of $\overline{AB}$ about the x-axis is given by the graph:

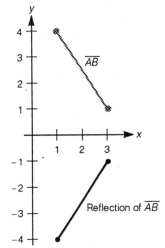

(b)   The reflection of $\overline{AB}$ about the y-axis is given by the graph:

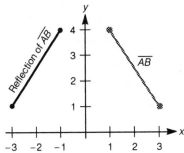

(c)   The reflection of $\overline{AB}$ about the origin is given by the graph:

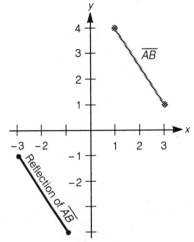

(d)   The reflection of $\overline{AB}$ about the line $y = x$ is given by the graph:

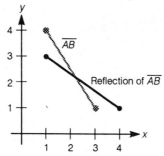

3.   (a)   The reflection of $\overline{AB}$ about the x-axis is given by the graph:

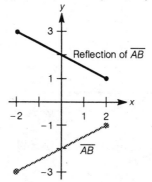

(b)   The reflection of $\overline{AB}$ about the y-axis is given by the graph:

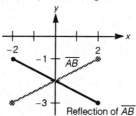

(c)   The reflection of $\overline{AB}$ about the origin is given by the graph:

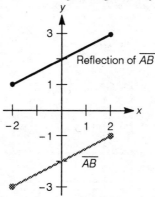

(d) The reflection of $\overline{AB}$ about the line $y = x$ is given by the graph:

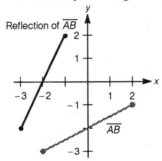

5. (a) The reflection of $\overline{AB}$ about the x-axis is given by the graph:

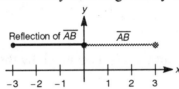

(b) The reflection of $\overline{AB}$ about the y-axis is given by the graph:

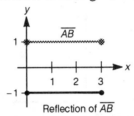

(c) The reflection of $\overline{AB}$ about the origin is given by the graph:

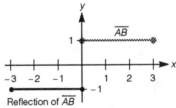

(d) The reflection of $\overline{AB}$ about the line $y = x$ is given by the graph:

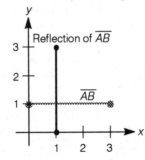

7. We complete the table:

	Symmetric about the x - axis	Symmetric about the y - axis	Symmetric about the origin
$y = x^2$	no	yes	no
$y = x^3$	no	no	yes
$y = \sqrt{x}$	no	no	no

9. The x-intercepts are 2 and -2, and the y-intercept is 4. The graph is symmetric about the y-axis.

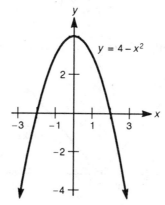

11. There are no x- or y-intercepts. The graph is symmetric about the origin.

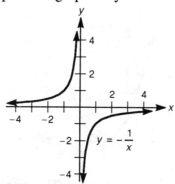

13. The x- and y-intercepts are both 0. The graph is symmetric about the y-axis.

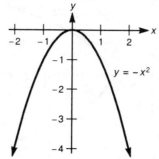

15. There are no x- or y-intercepts. The graph is symmetric about the origin.

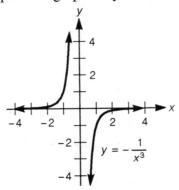

17. The x- and y-intercepts are both 0. The graph is symmetric about the y-axis.

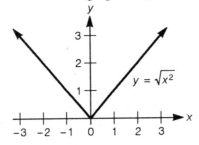

19. The x- and y-intercepts are both 1. The graph does not possess any of the three types of symmetry.

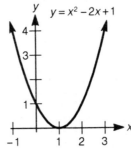

21. The x-intercept is 2 and there is no y-intercept. The graph is symmetric about the x-axis.

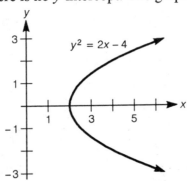

23. To find the *x*-intercepts, we use the quadratic formula:

$$x = \frac{-1 \pm \sqrt{(1)^2 - 4(2)(-4)}}{2(2)} = \frac{-1 \pm \sqrt{33}}{4}$$

The *y*-intercept is −4. The graph does not possess any of the three types of symmetry.

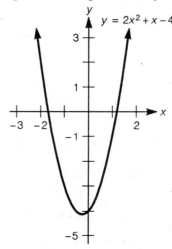

25. (a) The *x*- and *y*-intercepts are both 2. The graph does not possess any of the three types of symmetry.

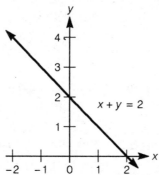

(b) The *x*-intercepts are ±2 and the *y*-intercept is 2. The graph is symmetric about the *y*-axis.

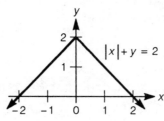

27. (a) The x-intercept for each graph is the same. Setting $y = 0$, we obtain:

$$\tfrac{3}{4}x - 2 = 0$$
$$\tfrac{3}{4}x = 2$$
$$x = \tfrac{8}{3}$$

The y-intercept for $y = \tfrac{3}{4}x - 2$ is -2, so the y-intercept for $y = \left|\tfrac{3}{4}x - 2\right|$ is $|-2| = 2$.

(b) The two graphs are identical on the interval $\left[\tfrac{8}{3}, \infty\right)$.

(c) The graph of $y = \left|\tfrac{3}{4}x - 2\right|$ can be obtained by reflecting $y = \tfrac{3}{4}x - 2$ about the x-axis for the interval $\left(-\infty, \tfrac{8}{3}\right)$. For the interval $\left[\tfrac{8}{3}, \infty\right)$, no reflection is necessary.

29. Let the points $P = (7, -1)$ and $Q = (-1, 7)$. We must first show that the line segment $\overline{PQ}$ is perpendicular to the line $y = x$. Line segment $\overline{PQ}$ has a slope of:

$$m = \frac{7 - (-1)}{-1 - 7} = \frac{8}{-8} = -1$$

Since $y = x$ has a slope of 1, and $1(-1) = -1$, then the two lines are perpendicular. Next, we must show that P and Q are equidistant from $y = x$. We find the midpoint of line segment $\overline{PQ}$:

$$M = \left(\tfrac{7-1}{2}, \tfrac{-1+7}{2}\right) = (3, 3)$$

Since $(3, 3)$ lies on the line $y = x$, and since $\overline{PQ}$ is perpendicular to $y = x$, and $PM = QM$, then by the definition of symmetry P and Q are symmetric about the line $y = x$.

31. Let the points $P = (a, b)$ and $Q = (b, a)$. We must first show that the line segment $\overline{PQ}$ is perpendicular to the line $y = x$. Line segment $\overline{PQ}$ has a slope of:

$$m = \frac{a - b}{b - a} = -1$$

Since $y = x$ has a slope of 1, and $1(-1) = -1$, then the two lines are perpendicular. Next, we must show that P and Q are equidistant from $y = x$. We find the midpoint of line segment $\overline{PQ}$:

$$M = \left(\tfrac{a+b}{2}, \tfrac{b+a}{2}\right) = \left(\tfrac{a+b}{2}, \tfrac{a+b}{2}\right)$$

Since $\left(\tfrac{a+b}{2}, \tfrac{a+b}{2}\right)$ lies on the line $y = x$, and since $\overline{PQ}$ is perpendicular to $y = x$, and $PM = QM$, then by the definition of symmetry P and Q are symmetric about the line $y = x$.

33. Let's pick the points $P(0,3)$, $Q(2,4)$, $R(4,5)$. Then the reflected points will be $P'(3,0)$, $Q'(4,2)$, and $R'(5,4)$. We will show that P', Q', and R' are collinear by computing the slopes:

$$m_{P'Q'} = \frac{2-0}{4-3} = \frac{2}{1} = 2$$

$$m_{Q'R'} = \frac{4-2}{5-4} = \frac{2}{1} = 2$$

So P', Q', and R' all are collinear (lie on the same line). There are two ways to find the equation of the line.

1st way: We use the point $(3,0)$ and $m=2$ in the point-slope formula:
$$y-0 = 2(x-3)$$
$$y = 2x-6$$

2nd way: We realize that this line must be the inverse function of $y = \frac{1}{2}x + 3$.

Exchange x and y and solve for y:
$$x = \frac{1}{2}y + 3$$
$$x - 3 = \frac{1}{2}y$$
$$y = 2x - 6$$

In either case, we obtain the same equation.

35. The x-intercepts are found by setting $y = 0$:
$$x^3 - 27x = 0$$
$$x(x - 3\sqrt{3})(x + 3\sqrt{3}) = 0$$
$$x = 0, \pm 3\sqrt{3} \approx 0, \pm 5.2$$

When $x = 0$, $y = (0)^3 - 27(0) = 0$, so the y-intercept is 0. Using the point $(-x, -y)$, we find the graph is symmetric with respect to the origin:
$$-y = (-x)^3 - 27(-x)$$
$$-y = -x^3 + 27x$$
$$y = x^3 - 27x$$

We set up a table of values:

x	0	0.5	1	1.5	2	2.5	3	3.5	4	4.5	5	5.5	6
y	0	−13.4	−26	−37.1	−46	−51.9	−54	−51.6	−44	−30.4	−10	17.9	54

Now graph the curve:

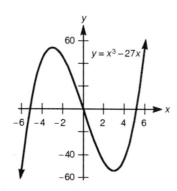

37. Let $P = (8, 2)$ and $Q = (4, 8)$. We first find the slope m_1 of the line segment $\overline{PQ}$ perpendicular to the line $y = mx + b$:

$$m_1 = \frac{8 - 2}{4 - 8} = -\frac{6}{4} = -\frac{3}{2}$$

So $-\frac{3}{2}m = -1$, and thus $m = \frac{2}{3}$. So $y = \frac{2}{3}x + b$.

Next, we must find the value of b so that P and Q are equidistant from $y = \frac{2}{3}x + b$. Call the point $A\left(a, \frac{2}{3}a + b\right)$ on $y = \frac{2}{3}x + b$. We use the distance formula:

$$PA = \sqrt{(a - 8)^2 + \left(\frac{2}{3}a + b - 2\right)^2}$$

$$QA = \sqrt{(a - 4)^2 + \left(\frac{2}{3}a + b - 2\right)^2}$$

Now, since $PA = QA$, we have:

$$\sqrt{(a - 8)^2 + \left(\frac{2}{3}a + b - 2\right)^2} = \sqrt{(a - 4)^2 + \left(\frac{2}{3}a + b - 8\right)^2}$$

$$(a - 8)^2 + \left(\frac{2}{3}a + b - 2\right)^2 = (a - 4)^2 + \left(\frac{2}{3}a + b - 8\right)^2$$

$$a^2 - 16a + 64 + \left(\frac{2}{3}a + b\right)^2 - 4\left(\frac{2}{3}a + b\right) + 4 = a^2 - 8a + 16 + \left(\frac{2}{3}a + b\right)^2 - 16\left(\frac{2}{3}a + b\right) + 64$$

$$-16a - 4\left(\frac{2}{3}a + b\right) + 4 = -8a + 16 - 16\left(\frac{2}{3}a + b\right)$$

$$12\left(\frac{2}{3}a + b\right) = 8a + 12$$

$$8a + 12b = 8a + 12$$

$$12b = 12$$

$$b = 1$$

So $y = \frac{2}{3}x + 1$ is the line of symmetry, and so $m = \frac{2}{3}$ and $b = 1$.

39. (a) Let $P = (a, b)$ and $Q = (x, y)$. Q is the point we are trying to find. Since $\overline{PQ}$ is perpendicular to $y = 3x$, which has a slope of 3, then we have:

$$\frac{y - b}{x - a} = -\frac{1}{3}$$

$$3y - 3b = -x + a$$

$$x + 3y = a + 3b$$

Now call $C(c, 3c)$ a point on the line $y = 3x$:

$$\frac{3c - b}{c - a} = -\frac{1}{3}$$

$$9c - 3b = -c + a$$

$$10c = a + 3b$$

$$c = \frac{a + 3b}{10}$$

Since $PC = QC$, then by the distance formula we have:

$$\sqrt{(a-c)^2 + (b-3c)^2} = \sqrt{(x-c)^2 + (y-3c)^2}$$

$$\left(a - \tfrac{a+3b}{10}\right)^2 + \left(b - \tfrac{3a+9b}{10}\right)^2 = \left(x - \tfrac{a+3b}{10}\right)^2 + \left(y - \tfrac{3a+9b}{10}\right)^2$$

Recall that $x + 3y = a + 3b$, and thus $3x + 9y = 3a + 9b$. Substituting these values into the right-hand side, we obtain:

$$\left(a - \tfrac{a+3b}{10}\right)^2 + \left(b - \tfrac{3a+9b}{10}\right)^2 = \left(x - \tfrac{x+3y}{10}\right)^2 + \left(y - \tfrac{3x+9y}{10}\right)^2$$

$$\left(\tfrac{9a-3b}{10}\right)^2 + \left(\tfrac{b-3a}{10}\right)^2 = \left(\tfrac{9x-3y}{10}\right)^2 + \left(\tfrac{y-3x}{10}\right)^2$$

$$81a^2 - 54ab + 9b^2 + b^2 - 6ab + 9a^2 = 81x^2 - 54xy + 9y^2 + y^2 - 6xy + 9x^2$$

$$90a^2 - 60ab + 10b^2 = 90x^2 - 60xy + 10y^2$$

$$9a^2 - 6ab + b^2 = 9x^2 - 6xy + y^2$$

$$(3a - b)^2 = (3x - y)^2$$

$$|3a - b| = |3x - y|$$

So, either $3x - y = 3a - b$ or $3x - y = -(3a - b) = -3a + b$.

Case 1:

    $x + 3y = a + 3b$, so $x = a + 3b - 3y$

    $3x - y = 3a - b$

    Substituting:

        $3(a + 3b - 3y) - y = 3a - b$

        $3a + 9b - 9y - y = 3a - b$

                $-10y = -10b$

                   $y = b$

                   $x = a$

But if $(x, y) = (a, b)$, then the point lies **on** the line $y = 3x$.

Note: An alternate approach is that if $3x - y = 3a - b$, then $\dfrac{y-b}{x-a} = 3$. But this is impossible, since $\dfrac{y-b}{x-a} = -\tfrac{1}{3}$.

Case 2:

    $x + 3y = a + 3b$, so $x = a + 3b - 3y$

    $3x - y = -3a + b$

    Substituting:

        $3(a + 3b - 3y) - y = -3a + b$

        $3a + 9b - 9y - y = -3a + b$

                $-10y = -6a - 8b$

                   $y = \dfrac{3a + 4b}{5}$

Since $x = a + 3b - 3y$, we have:

$$x = a + 3b - 3\left(\frac{3a+4b}{5}\right)$$

$$= a + 3b - \frac{9a+12b}{5}$$

$$= \frac{5a + 15b - 9a - 12b}{5}$$

$$= \frac{-4a + 3b}{5}$$

So the reflected point (x, y) is $\left(\frac{3b-4a}{5}, \frac{3a+4b}{5}\right)$.

(b) Calling P and Q as in (a), we have:

$$\frac{y-b}{x-a} = -\frac{1}{m}$$

$$my - mb = -x + a$$

$$x + my = a + mb$$

Let $C(c, mc)$ be a point on the line $y = mx$:

$$\frac{mc-b}{c-a} = -\frac{1}{m}$$

$$m^2c - mb = a - c$$

$$\left(m^2 + 1\right)c = a + mb$$

$$c = \frac{a+mb}{m^2+1}$$

Now since $PC = QC$, we have:

$$\sqrt{(a-c)^2 + (b-mc)^2} = \sqrt{(x-c)^2 + (y-mc)^2}$$

$$\left(a - \frac{a+mb}{m^2+1}\right)^2 + \left(b - \frac{am+m^2b}{m^2+1}\right)^2 = \left(x - \frac{a+mb}{m^2+1}\right)^2 + \left(y - \frac{am+m^2b}{m^2+1}\right)^2$$

Recall that $x + my = a + mb$, and thus $mx + m^2y = am + m^2b$. Substituting these values into the right-hand side, we obtain:

$$\left(a - \frac{a+mb}{m^2+1}\right)^2 + \left(b - \frac{am+m^2b}{m^2+1}\right)^2 = \left(x - \frac{x+my}{m^2+1}\right)^2 + \left(y - \frac{mx+m^2y}{m^2+1}\right)^2$$

$$\left(\frac{m^2a-mb}{m^2+1}\right)^2 + \left(\frac{b-am}{m^2+1}\right)^2 = \left(\frac{m^2x-my}{m^2+1}\right)^2 + \left(\frac{y-mx}{m^2+1}\right)^2$$

$$m^2a^2 - 2abm + b^2 = m^2x^2 - 2mxy + y^2$$

$$(ma - b)^2 = (mx - y)^2$$

$$|ma - b| = |mx - y|$$

So, either $mx - y = ma - b$ or $mx - y = -(ma + b) = -ma + b$.

Case 1:

$x + my = a + mb$, so $x = a + mb - my$

$mx - y = ma - b$

Substituting:

$$m(a + mb - my) - y = ma - b$$

$$ma + m^2b - m^2y - y = ma - b$$

$$(m^2 + 1)b = (m^2 + 1)y$$

$$b = y$$

$$a = x$$

But, if $(x, y) = (a, b)$, then the point lies **on** the line $y = mx$.

Note: An alternate approach is that if $mx - y = ma - b$, then $\dfrac{y - b}{x - a} = m$.

But this is impossible, since $\dfrac{y - b}{x - a} = -\dfrac{1}{m}$.

Case 2:

$x + my = a + mb$, so $x = a + mb - my$

$mx - y = -ma + b$

Substituting:

$$m(a + mb - my) - y = -ma + b$$

$$ma + m^2b - m^2y - y = -ma + b$$

$$-(m^2 + 1)y = -2ma + b(1 - m^2)$$

$$y = \frac{2ma + (m^2 + 1)b}{m^2 + 1}$$

Since $x = a + mb - my$, we have:

$$x = a + mb - \frac{2m^2a + m(m^2 - 1)b}{m^2 + 1}$$

$$= \frac{am^2 + a + m^3b + mb - 2m^2a - m^3b + mb}{m^2 + 1}$$

$$= \frac{2mb - (m^2 - 1)a}{m^2 + 1}$$

So the reflected point (x, y) is:

$$\left(\frac{2mb - (m^2 - 1)a}{m^2 + 1}, \frac{2ma + (m^2 - 1)b}{m^2 + 1} \right)$$

Note: Using $m = 3$ and our answer from (a) verifies this formula.

Optional TI-81 Graphing Calculator Exercises for Section 2.4

1.   We draw the graph:

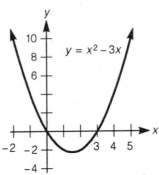

The graph does not possess any of the three types of symmetry.

3.   We draw the graph:

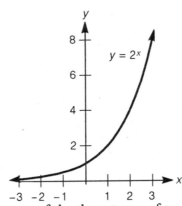

The graph does not possess any of the three types of symmetry.

5.   We draw the graph:

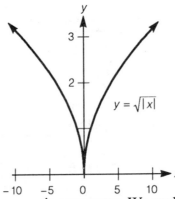

The graph appears to possess y-axis symmetry. We replace (x,y) with $(-x,y)$:

$$y = \sqrt{|-x|} = \sqrt{|x|}$$

So the graph does possess y-axis symmetry.

7.   We draw the graph:

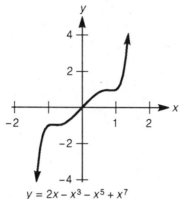

$$y = 2x - x^3 - x^5 + x^7$$

The graph appears to have origin symmetry. We replace (x, y) with $(-x, -y)$:
$$-y = 2(-x) - (-x)^3 - (-x)^5 + (-x)^7$$
$$-y = -2x + x^3 + x^5 - x^7$$
$$y = 2x - x^3 - x^5 + x^7$$
So the graph does possess origin symmetry.

9.   Using the indicated settings, we draw the graph:

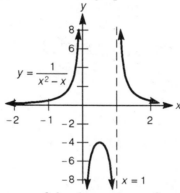

$$y = \frac{1}{x^2 - x}$$

The graph does not possess any of the three types of symmetry.

11.   (a)   We draw the graph:

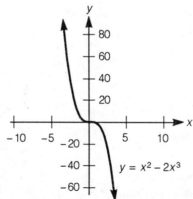

$$y = x^2 - 2x^3$$

The graph "appears" to possess origin symmetry (see part (b)).

(b)   Using the indicated settings, we draw the graph:

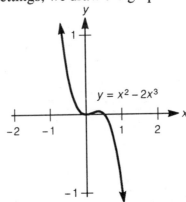

$y = x^2 - 2x^3$

Note that the graph no longer appears to possess origin symmetry.

(c)   We replace (x, y) with $(-x, -y)$:
$$-y = (-x)^2 - 2(-x)^3$$
$$-y = x^2 + 2x^3$$
$$y = -x^2 - 2x^3$$
The graph does not possess origin symmetry.

Chapter Two Review Exercises

1.   We find the slope:
$$m = \frac{6-2}{-6-(-4)} = \frac{4}{-2} = -2$$
Using the point $(-4, 2)$ in the point-slope formula:
$$y - 2 = -2[x - (-4)]$$
$$y - 2 = -2(x + 4)$$
$$y - 2 = -2x - 8$$
$$y = -2x - 6$$

3.   Using the point-slope formula:
$$y - (-3) = \tfrac{1}{4}[x - (-2)]$$
$$y + 3 = \tfrac{1}{4}(x + 2)$$
$$y + 3 = \tfrac{1}{4}x + \tfrac{1}{2}$$
$$y = \tfrac{1}{4}x - \tfrac{5}{2}$$

5.   We find the slope between the points $(-4, 0)$ and $(0, 8)$:
$$m = \frac{8-0}{0-(-4)} = \tfrac{8}{4} = 2$$
Using the slope-intercept formula, we have $y = 2x + 8$.

7. Since the line is parallel to the x-axis, its equation will be of the form $y = $ constant (horizontal line). Since $(0, -2)$ is on the line, its equation is $y = -2$.

9. We find the slope of $x + y + 1 = 0$ by writing it in slope-intercept form, and obtain $y = -x - 1$. So its slope is -1, and thus the perpendicular slope would be 1. We use the point $(1, 2)$ in the point-slope formula:
$$y - 2 = 1(x - 1)$$
$$y - 2 = x - 1$$
$$y = x + 1$$

11. We find the center of each circle by completing the square:
$$x^2 + 4x + y^2 + 2y = 0$$
$$\left(x^2 + 4x + 4\right) + \left(y^2 + 2y + 1\right) = 0 + 4 + 1$$
$$(x + 2)^2 + (y + 1)^2 = 5$$
$$\text{center: } (-2, -1)$$

$$x^2 - 4x + y^2 - 16y = 0$$
$$\left(x^2 - 4x + 4\right) + \left(y^2 - 16y + 64\right) = 0 + 4 + 64$$
$$(x - 2)^2 + (y - 8)^2 = 68$$
$$\text{center: } (2, 8)$$
We find the slope between $(-2, -1)$ and $(2, 8)$:
$$m = \frac{8 - (-1)}{2 - (-2)} = \frac{9}{4}$$
We use the point $(2, 8)$ in the point-slope formula:
$$y - 8 = \tfrac{9}{4}(x - 2)$$
$$4y - 32 = 9x - 18$$
$$-9x + 4y - 14 = 0$$
$$9x - 4y + 14 = 0$$

13. We find the midpoint of the line segment joining $(-2, -3)$ and $(6, -5)$:
$$M = \left(\frac{-2 + 6}{2}, \frac{-3 - 5}{2}\right) = \left(\tfrac{4}{2}, -\tfrac{8}{2}\right) = (2, -4)$$
Now find the slope between $(0, 0)$ and $(2, -4)$:
$$m = \frac{-4 - 0}{2 - 0} = \frac{-4}{2} = -2$$
Using the slope-intercept formula, we obtain $y = -2x$, or $2x + y = 0$.

15. First find the center of the circle by completing the square:
$$x^2 - 6x + y^2 + 8y = 0$$
$$\left(x^2 - 6x + 9\right) + \left(y^2 + 8y + 16\right) = 0 + 9 + 16$$
$$(x - 3)^2 + (y + 4)^2 = 25$$

So the center is $(3, -4)$. Now find the slope of the radius drawn from $(3, -4)$ to $(0, 0)$:

$$m = \frac{0 - (-4)}{0 - 3} = \frac{4}{-3} = -\frac{4}{3}$$

So the slope of the perpendicular tangent line is $\frac{3}{4}$. Using the slope-intercept formula, we obtain $y = \frac{3}{4}x$, or $3x - 4y = 0$.

17. Call the x-intercept a. Then the y-intercept is $2 - a$, and we find the slope from each to $(2, -1)$:

$$m = \frac{-1 - 0}{2 - a} = \frac{-1}{2 - a}$$

$$m = \frac{-1 - (2 - a)}{2 - 0} = \frac{-1 - 2 + a}{2} = \frac{a - 3}{2}$$

Since these two slopes must be equal, we have:

$$\frac{-1}{2 - a} = \frac{a - 3}{2}$$
$$-2 = (2 - a)(a - 3)$$
$$-2 = -a^2 + 5a - 6$$
$$0 = a^2 - 5a + 4$$
$$0 = (a - 4)(a - 1)$$
$$a = 1 \text{ or } a = 4$$

When $a = 1$, we have $m = \frac{1-3}{2} = \frac{-2}{2} = -1$. When $a = 4$, we have $m = \frac{4-3}{2} = \frac{1}{2}$. Using the point $(2, -1)$ and each of these slopes in the point-slope formula:

$$y - (-1) = -1(x - 2) \qquad\qquad y - (-1) = \tfrac{1}{2}(x - 2)$$
$$y + 1 = -x + 2 \qquad\qquad\quad 2y + 2 = x - 2$$
$$x + y - 1 = 0 \qquad\qquad\qquad x - 2y - 4 = 0$$

Both of these lines satisfy the given conditions.

19. Using the distance formula with $(x_1, y_1) = (-1, 2)$ and $(x_2, y_2) = (4, -10)$:

$$d = \sqrt{[4 - (-1)]^2 + (-10 - 2)^2} = \sqrt{25 + 144} = \sqrt{169} = 13$$

21. We find the distance of each from the origin:

$$d_1 = \sqrt{(15 - 0)^2 + (6 - 0)^2} = \sqrt{225 + 36} = \sqrt{261}$$
$$d_2 = \sqrt{(16 - 0)^2 + (2 - 0)^2} = \sqrt{256 + 4} = \sqrt{260}$$

The point $(15, 6)$ is farther from the origin.

23. For x-axis symmetry, we replace (x, y) with $(x, -y)$:
$$-y = x^4 - 2x^2$$
$$y = -x^4 + 2x^2$$
The graph is not symmetric about the x-axis.
For y-axis symmetry, we replace (x, y) with $(-x, y)$:
$$y = (-x)^4 - 2(-x)^2 = x^4 - 2x^2$$
The graph is symmetric about the y-axis.

For origin symmetry, we replace (x, y) with $(-x, -y)$:
$$-y = (-x)^4 - 2(-x)^2$$
$$-y = x^4 - 2x^2$$
$$y = -x^4 + 2x^2$$
The graph is not symmetric about the origin.

25. For x-axis symmetry, we replace (x, y) with $(x, -y)$:
$$-y = x^3 + 5x$$
$$y = -x^3 - 5x$$
The graph is not symmetric about the x-axis.
For y-axis symmetry, we replace (x, y) with $(-x, y)$:
$$y = (-x)^3 + 5(-x) = -x^3 - 5x$$
The graph is not symmetric about the y-axis.
For origin symmetry, we replace (x, y) with $(-x, -y)$:
$$-y = (-x)^3 + 5(-x)$$
$$-y = -x^3 - 5x$$
$$y = x^3 + 5x$$
The graph is symmetric about the origin.

27. For x-axis symmetry, we replace (x, y) with $(x, -y)$:
$$(-y)^2 = [x + (-y)]^4$$
$$y^2 = (x - y)^4$$
The graph is not symmetric about the x-axis.
For y-axis symmetry, we replace (x, y) with $(-x, y)$:
$$y^2 = (-x + y)^4$$
The graph is not symmetric about the y-axis.
For origin symmetry, we replace (x, y) with $(-x, -y)$:
$$(-y)^2 = [-x + (-y)]^4$$
$$y^2 = (-x - y)^4$$
$$y^2 = (x + y)^4$$
The graph is symmetric about the origin.

29. For x-axis symmetry, we replace (x, y) with $(x, -y)$:
$$-y = 3x - \tfrac{1}{x}$$
$$y = -3x + \tfrac{1}{x}$$
The graph is not symmetric about the x-axis.
For y-axis symmetry, we replace (x, y) with $(-x, y)$:
$$y = 3(-x) - \tfrac{1}{-x} = -3x + \tfrac{1}{x}$$
The graph is not symmetric about the y-axis.

For origin symmetry, we replace (x, y) with $(-x, -y)$:
$$-y = 3(-x) - \frac{1}{-x}$$
$$-y = -3x + \frac{1}{x}$$
$$y = 3x - \frac{1}{x}$$
The graph is symmetric about the origin.

31. For x-axis symmetry, we replace (x, y) with $(x, -y)$:
$$-y = 2^x - 2^{-x}$$
$$y = -2^x + 2^{-x}$$
The graph is not symmetric about the x-axis.
For y-axis symmetry, we replace (x, y) with $(-x, y)$:
$$y = 2^{-x} - 2^{-(-x)} = 2^{-x} - 2^x$$
The graph is not symmetric about the y-axis.
For origin symmetry, we replace (x, y) with $(-x, -y)$:
$$-y = 2^{-x} - 2^{-(-x)}$$
$$-y = 2^{-x} - 2^x$$
$$y = 2^x - 2^{-x}$$
The graph is symmetric about the origin.

33. The graph is symmetric about the x-axis only.

35. The graph is symmetric about the x-axis, the y-axis, the origin, and the line $y = x$.

37. The graph is symmetric about the x-axis only.

39. The graph is symmetric about the x-axis only.

41. Using the distance formula and substituting $y = -\frac{2}{3}x + 1$, we have:
$$\sqrt{x^2 + \left(-\frac{2}{3}x + 1\right)^2} = \sqrt{10}$$
$$x^2 + \frac{4}{9}x^2 - \frac{4}{3}x + 1 = 10$$
$$\frac{13}{9}x^2 - \frac{4}{3}x - 9 = 0$$
$$13x^2 - 12x - 81 = 0$$
$$(13x + 27)(x - 3) = 0$$
$$x = -\frac{27}{13}, 3$$
$$y = \frac{31}{13}, -1$$
The two points are $\left(-\frac{27}{13}, \frac{31}{13}\right)$ and $(3, -1)$.

43.   Since the sum of the *x*- and *y*-coordinates is 2 and $y = x^2$, we have:

$$x + x^2 = 2$$
$$x^2 + x - 2 = 0$$
$$(x + 2)(x - 1) = 0$$
$$x = -2, 1$$
$$y = 4, 1$$

The two points are $(-2, 4)$ and $(1, 1)$.

45.   We draw the graph:

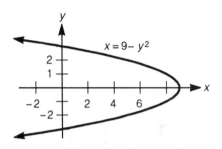

47.   This is a circle with center = $(0, 0)$ and radius = 1. We draw the graph:

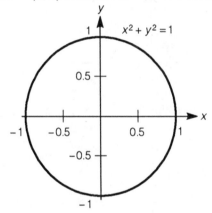

49.   We draw the graph:

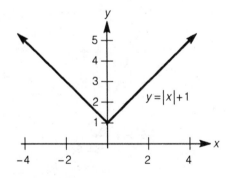

51. We draw the graph:

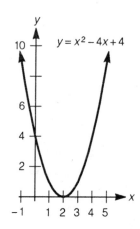

$y = x^2 - 4x + 4$

53. We draw the graph:

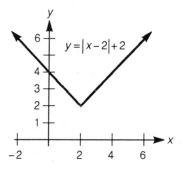

$y = |x - 2| + 2$

55. We write the line in slope-intercept form:
$$3x + y = 0$$
$$y = -3x$$
Now draw the graph:

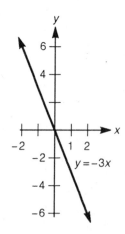

$y = -3x$

57. Complete the square:
$$x^2 - 4x + y^2 + 6y = 0$$
$$\left(x^2 - 4x + 4\right) + \left(y^2 + 6y + 9\right) = 4 + 9$$
$$(x-2)^2 + (y+3)^2 = 13$$

This is a circle with center $= (2,-3)$ and radius $= \sqrt{13}$.

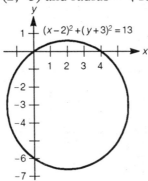

59. As long as $x \neq 4$, we can simplify:
$$y = \frac{x^2 - 16}{x - 4} = \frac{(x+4)(x-4)}{(x-4)} = x + 4$$

This is a line in slope-intercept form. Now draw the graph:

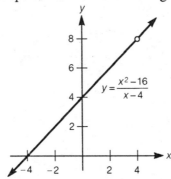

61. Setting each factor equal to 0, we have $4x - y + 4 = 0$ and thus $y = 4x + 4$, or $4x + y - 4 = 0$ and thus $y = -4x + 4$. So the graph consists of two lines:

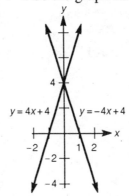

63. We set up the equation involving slope:
$$\frac{t-1}{5-2} = 6$$
$$\frac{t-1}{3} = 6$$
$$t-1 = 18$$
$$t = 19$$

65. The slope between the points $(-2, -8)$ and (x, x^3) is given by:
$$m = \frac{x^3 - (-8)}{x - (-2)} = \frac{x^3 + 8}{x + 2} = \frac{(x+2)(x^2 - 2x + 4)}{x + 2} = x^2 - 2x + 4$$

67. The midpoint of the hypotenuse BC would be:
$$M = \left(\frac{0 + 2c}{2}, \frac{2b + 0}{2}\right) = \left(\frac{2c}{2}, \frac{2b}{2}\right) = (c, b)$$
We compute the distances:
$$MA = \sqrt{(c-0)^2 + (b-0)^2} = \sqrt{c^2 + b^2}$$
$$MB = \sqrt{(c-0)^2 + (b-2b)^2} = \sqrt{c^2 + b^2}$$
$$MC = \sqrt{(c-2c)^2 + (b-0)^2} = \sqrt{c^2 + b^2}$$
These distances are all the same.

69. (a)   We compute the distance:
$$d = \sqrt{(5-2)^2 + (-6-5)^2} = \sqrt{9 + 121} = \sqrt{130}$$

(b)   We compute the slope:
$$m = \frac{-6-5}{5-2} = -\tfrac{11}{3}$$

(c)   We compute the midpoint:
$$M = \left(\frac{2+5}{2}, \frac{5-6}{2}\right) = \left(\tfrac{7}{2}, -\tfrac{1}{2}\right)$$

71. (a)   We compute the distance:
$$d = \sqrt{\left(-\tfrac{\sqrt{3}}{2} - \tfrac{\sqrt{3}}{2}\right)^2 + \left(-\tfrac{1}{2} - \tfrac{1}{2}\right)^2} = \sqrt{\left(-\sqrt{3}\right)^2 + (-1)^2} = \sqrt{3 + 1} = \sqrt{4} = 2$$

(b)   We compute the slope:
$$m = \frac{-\tfrac{1}{2} - \tfrac{1}{2}}{-\tfrac{\sqrt{3}}{2} - \tfrac{\sqrt{3}}{2}} = \frac{-1}{-\sqrt{3}} = \frac{1}{\sqrt{3}} = \frac{\sqrt{3}}{3}$$

(c)   We compute the midpoint:
$$M = \left(\frac{\frac{\sqrt{3}}{2} - \frac{\sqrt{3}}{2}}{2}, \frac{\frac{1}{2} - \frac{1}{2}}{2} \right) = (0,0)$$

73. We find the slope:
$$m = \frac{1-2}{4-1} = -\tfrac{1}{3}$$
Using $(1,2)$ in the point-slope formula:
$$y - 2 = -\tfrac{1}{3}(x-1)$$
$$y - 2 = -\tfrac{1}{3}x + \tfrac{1}{3}$$
$$y = -\tfrac{1}{3}x + \tfrac{7}{3}$$
So the x- and y-intercepts are $\tfrac{7}{3}$ (if $x = 0$) and 7 (if $y = 0$). Then the area is given by:
$$\text{Area} = \tfrac{1}{2}(\text{base})(\text{height}) = \tfrac{1}{2} \cdot 7 \cdot \tfrac{7}{3} = \tfrac{49}{6}$$

75. (a)   We find each midpoint:
$$M_1 = \left(\frac{-5+7}{2}, \frac{3+7}{2} \right) = (1,5)$$
$$M_2 = \left(\frac{3+7}{2}, \frac{1+7}{2} \right) = (5,4)$$
$$M_3 = \left(\frac{-5+3}{2}, \frac{3+1}{2} \right) = (-1,2)$$
We now use the result from Exercise 74 to find the desired points. Let P_1, P_2, and P_3 be the desired points on medians CM_1, AM_2, and BM_3. Then:
$$P_1 = \left(\tfrac{1}{3}(3) + \tfrac{2}{3}(1), \tfrac{1}{3}(1) + \tfrac{2}{3}(5) \right) = \left(1 + \tfrac{2}{3}, \tfrac{1}{3} + \tfrac{10}{3} \right) = \left(\tfrac{5}{3}, \tfrac{11}{3} \right)$$
$$P_2 = \left(\tfrac{1}{3}(-5) + \tfrac{2}{3}(5), \tfrac{1}{3}(3) + \tfrac{2}{3}(4) \right) = \left(-\tfrac{5}{3} + \tfrac{10}{3}, 1 + \tfrac{8}{3} \right) = \left(\tfrac{5}{3}, \tfrac{11}{3} \right)$$
$$P_3 = \left(\tfrac{1}{3}(7) + \tfrac{2}{3}(-1), \tfrac{1}{3}(7) + \tfrac{2}{3}(2) \right) = \left(\tfrac{7}{3} - \tfrac{2}{3}, \tfrac{7}{3} + \tfrac{4}{3} \right) = \left(\tfrac{5}{3}, \tfrac{11}{3} \right)$$
They all intersect at the same point.

(b)   We find each midpoint:
$$M_1 = \left(\frac{0+2b}{2}, \frac{0+2c}{2} \right) = (b,c)$$
$$M_1 = \left(\frac{2b+2a}{2}, \frac{2c+0}{2} \right) = (a+b,c)$$
$$M_3 = \left(\frac{0+2a}{2}, \frac{0+0}{2} \right) = (a,0)$$

We now use the result from Exercise 74 to find the desired points. Let P_1, P_2, and P_3 be the desired points on medians CM_3, BM_1, and AM_2. Then:

$$P_1 = \left(\tfrac{1}{3}(2b) + \tfrac{2}{3}(a), \tfrac{1}{3}(2c) + \tfrac{2}{3}(0)\right) = \left(\frac{2b}{3} + \frac{2a}{3}, \frac{2c}{3}\right)$$

$$P_2 = \left(\tfrac{1}{3}(2a) + \tfrac{2}{3}(b), \tfrac{1}{3}(0) + \tfrac{2}{3}(c)\right) = \left(\frac{2a}{3} + \frac{2b}{3}, \frac{2c}{3}\right)$$

$$P_3 = \left(\tfrac{1}{3}(0) + \tfrac{2}{3}(a+b), \tfrac{1}{3}(0) + \tfrac{2}{3}(c)\right) = \left(\frac{2a}{3} + \frac{2b}{3}, \frac{2c}{3}\right)$$

Again all these intersect at the same point. All medians of a triangle intersect at a point which is $\frac{2}{3}$ of the distance from each vertex to the midpoint on the opposite side.

77. We first complete the square to find the center and radius:
$$x^2 - 4x + y^2 + 2y = 15$$
$$\left(x^2 - 4x + 4\right) + \left(y^2 + 2y + 1\right) = 15 + 4 + 1$$
$$(x-2)^2 + (y+1)^2 = 20$$

So the center is $(2,-1)$ and the radius is $\sqrt{20} = 2\sqrt{5}$. Since $(1,-2)$ is the midpoint of the chord, the chord must be perpendicular to the radius line drawn to that point. We find the distance a of the line segment joining $(2,-1)$ and $(1,-2)$:
$$a = \sqrt{(1-2)^2 + [-2-(-1)]^2} = \sqrt{(-1)^2 + (-1)^2} = \sqrt{1+1} = \sqrt{2}$$
We can find d (half of the length of the chord) by the Pythagorean theorem:
$$d^2 + \left(\sqrt{2}\right)^2 = \left(\sqrt{20}\right)^2$$
$$d^2 + 2 = 20$$
$$d^2 = 18$$
$$d = \sqrt{18} = 3\sqrt{2}$$
The length of the chord is $2d = 2(3\sqrt{2}) = 6\sqrt{2}$.

79. Using the hint, we have:
$$m^2 = (a + c - 0)^2 + (b - 0)^2$$
$$m^2 = a^2 + 2ac + c^2 + b^2$$
$$s^2 = (2a - 0)^2 + (2b - 0)^2 = 4a^2 + 4b^2$$
$$t^2 = (2c)^2 = 4c^2$$
$$u^2 = (2a - 2c)^2 + (2b - 0)^2 = 4a^2 - 8ac + 4c^2 + 4b^2$$
Working from the right-hand side, we have:
$$\tfrac{1}{2}\left(s^2 + t^2\right) - \tfrac{1}{4}u^2 = \tfrac{1}{2}\left(4a^2 + 4b^2 + 4c^2\right) - \tfrac{1}{4}\left(4a^2 - 8ac + 4c^2 + 4b^2\right)$$
$$= 2a^2 + 2b^2 + 2c^2 - a^2 + 2ac - c^2 - b^2$$
$$= a^2 + b^2 + c^2 + 2ac$$
$$= m^2$$

81. (a) The reflection of $\overline{AB}$ about the line $y = x$ is given by the graph:

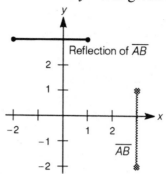

(b) The reflection of $\overline{AB}$ about the x-axis is given by the graph:

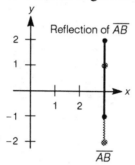

(c) The reflection of $\overline{AB}$ about the y-axis is given by the graph:

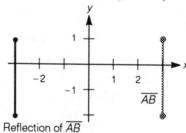

(d) The reflection of $\overline{AB}$ about the origin is given by the graph:

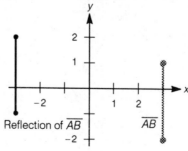

83. Here $(x_0, y_0) = (1, 2)$, $m = \frac{1}{2}$, and $b = -5$:
$$d = \frac{2 - \frac{1}{2}(1) - (-5)}{\sqrt{1 + \left(\frac{1}{2}\right)^2}} = \frac{2 - \frac{1}{2} + 5}{\sqrt{1 + \frac{1}{4}}} = \frac{\frac{13}{2}}{\sqrt{\frac{5}{4}}} = \frac{13\sqrt{5}}{5}$$

85. Here $(x_0, y_0) = (-1, -3)$, $A = 2$, $B = 3$ and $C = -6$:
$$d = \frac{|2(-1) + 3(-3) + (-6)|}{\sqrt{2^2 + 3^2}} = \frac{|-2 - 9 - 6|}{\sqrt{13}} = \frac{17}{\sqrt{13}} = \frac{17\sqrt{13}}{13}$$

87. **(a)** If the center is (h, k) and the circle is tangent to the coordinate axes, then the points of tangency are $(h, 0)$ and $(0, k)$. Since the distances from the center to the coordinate axes are the same, $h = k = r$ (the radius) and the center must be (r, r). We write the line in slope-intercept form:
$$3x + 4y = 12$$
$$4y = -3x + 12$$
$$y = -\tfrac{3}{4}x + 3$$

We find the distance from (r, r) to this line, using the fact that this distance is also r:
$$r = \frac{\left|r + \frac{3}{4}r - 3\right|}{\sqrt{1 + (3/4)^2}} = \frac{\left|\frac{7}{4}r - 3\right|}{5/4}$$

So we have the equation:
$$\tfrac{5}{4}r = \left|\tfrac{7}{4}r - 3\right|$$

Thus we have:

$$\tfrac{5}{4}r = \tfrac{7}{4}r - 3 \qquad \text{or} \qquad -\tfrac{5}{4}r = \tfrac{7}{4}r - 3$$
$$5r = 7r - 12 \qquad\qquad\qquad -5r = 7r - 12$$
$$-2r = -12 \qquad\qquad\qquad -12r = -12$$
$$r = 6 \qquad\qquad\qquad\qquad r = 1$$

Clearly $r = 6$ corresponds to a circle above the line, so $r = 1$. Thus the equation of the circle is $(x - 1)^2 + (y - 1)^2 = 1$.

(b) We have the coordinates $S(1, 0)$ and $U(0, 1)$. We must find the coordinates of T. We know the slope of the center to $3x + 4y = 12$ is $4/3$, so we use the point $(1, 1)$ in the point-slope formula:
$$y - 1 = \tfrac{4}{3}(x - 1)$$
$$y - 1 = \tfrac{4}{3}x - \tfrac{4}{3}$$
$$y = \tfrac{4}{3}x - \tfrac{1}{3}$$

We now find the intersection point of this line with our given line:

$$\frac{4}{3}x - \frac{1}{3} = -\frac{3}{4}x + 3$$
$$16x - 4 = -9x + 36$$
$$25x = 40$$
$$x = \frac{8}{5}$$

Substituting, we find $y = 9/5$, so we have the point $T(8/5, 9/5)$. We now can find the required equations. For $\overline{AT}$ the slope is $m = \dfrac{9/5 - 0}{8/5 - 0} = \dfrac{9}{8}$, so using the point $(0,0)$ in the point-slope equation yields:

$$y - 0 = \frac{9}{8}(x - 0)$$
$$y = \frac{9}{8}x$$

For $\overline{BU}$ the slope is $m = \dfrac{0 - 1}{4 - 0} = -\dfrac{1}{4}$, so using the point $(0, 1)$ in the point-slope equation yields:

$$y - 1 = -\frac{1}{4}(x - 0)$$
$$y = -\frac{1}{4}x + 1$$

For $\overline{CS}$ the slope is $m = \dfrac{3 - 0}{0 - 1} = -3$, so using the point $(1,0)$ in the point-slope equation yields:

$$y - 0 = -3(x - 1)$$
$$y = -3x + 3$$

(c) To find the required intersection points, we set the equations equal. For $\overline{AT}$ and $\overline{CS}$ we have:

$$\frac{9}{8}x = -3x + 3$$
$$\frac{33}{8}x = 3$$
$$x = \frac{24}{33} = \frac{8}{11}$$
$$y = \frac{9}{11}$$

The intersection point is $(8/11, 9/11)$. For $\overline{AT}$ and $\overline{BU}$ we have:

$$\frac{9}{8}x = -\frac{1}{4}x + 1$$
$$9x = -2x + 8$$
$$11x = 8$$
$$x = \frac{8}{11}$$
$$y = \frac{9}{11}$$

The intersection point is $(8/11, 9/11)$. We observe that the point of intersection is the same in both cases.

Chapter Two Test

1. We first find the slope:
$$m = \frac{8-(-2)}{3-1} = \frac{10}{2} = 5$$
Now using the point $(3,8)$ in the point-slope formula, we have:
$$y - 8 = 5(x-3)$$
$$y - 8 = 5x - 15$$
$$y = 5x - 7$$

2. We first write $5x + 6y = 30$ in slope-intercept form:
$$5x + 6y = 30$$
$$6y = -5x + 30$$
$$y = -\tfrac{5}{6}x + 5$$
So the perpendicular slope is $\tfrac{6}{5}$. Now using the point $(2,-1)$ in the point-slope formula, we have:
$$y - (-1) = \tfrac{6}{5}(x-2)$$
$$y + 1 = \tfrac{6}{5}x - \tfrac{12}{5}$$
$$y = \tfrac{6}{5}x - \tfrac{17}{5}$$

3. We first find the midpoint of the line segment joining the points $(-5,-2)$ and $(3,8)$:
$$M = \left(\frac{-5+3}{2}, \frac{-2+8}{2}\right) = \left(-\tfrac{2}{2}, \tfrac{6}{2}\right) = (-1,3)$$
Since the line passes through the origin, its slope is given by:
$$m = \frac{3-0}{-1-0} = \frac{3}{-1} = -3$$
So the equation is $y = -3x$.

4. We first complete the square to find the center of the circle:
$$x^2 - 8x + y^2 + 10y = 0$$
$$x^2 - 8x + 16 + y^2 + 10y + 25 = 16 + 25$$
$$(x-4)^2 + (y+5)^2 = 41$$
So the center of the circle is $(4,-5)$. The slope from this center to the origin is therefore:
$$m = \frac{-5-0}{4-0} = -\tfrac{5}{4}$$
So the tangent line slope, which is perpendicular to this radius slope, is $\tfrac{4}{5}$. So the equation of the tangent line is $y = \tfrac{4}{5}x$.

5. We compute the distance from each point to the origin:
$$d_1 = \sqrt{(3-0)^2 + (9-0)^2} = \sqrt{9+81} = \sqrt{90}$$
$$d_2 = \sqrt{(5-0)^2 + (8-0)^2} = \sqrt{25+64} = \sqrt{89}$$
So $(3,9)$ is farther from the origin.

6. **(a)** To test for symmetry about the x-axis, we replace (x,y) with $(x,-y)$:
$$-y = x^3 + 5x$$
$$y = -x^3 - 5x$$
Since the equation is changed, there is no x-axis symmetry. To test for symmetry about the y-axis, we replace (x,y) with $(-x,y)$:
$$y = (-x)^3 + 5(-x) = -x^3 - 5x$$
Since the equation is changed, there is no y-axis symmetry. To test for symmetry about the origin, we replace (x,y) with $(-x,-y)$:
$$-y = (-x)^3 + 5(-x)$$
$$-y = -x^3 - 5x$$
$$y = x^3 + 5x$$
Since the equation is unchanged, there is origin symmetry.

(b) To test for symmetry about the x-axis, we replace (x,y) with $(x,-y)$:
$$-y = 3^x + 3^{-x}$$
$$y = -3^x - 3^{-x}$$
Since the equation is changed, there is no x-axis symmetry. To test for symmetry about the y-axis, we replace (x,y) with $(-x,y)$:
$$y = 3^{-x} + 3^{-(-x)} = 3^{-x} + 3^x$$
Since the equation is unchanged, there is y-axis symmetry. To test for symmetry about the origin, we replace (x,y) with $(-x,-y)$:
$$-y = 3^{-x} + 3^{-(-x)}$$
$$-y = 3^{-x} + 3^x$$
$$y = -3^{-x} - 3^x$$
Since the equation is changed, there is no origin symmetry.

(c) To test for symmetry about the x-axis, we replace (x,y) with $(x,-y)$:
$$(-y)^2 = 5x^2 + x$$
$$y^2 = 5x^2 + x$$
Since the equation is unchanged, there is x-axis symmetry. To test for symmetry about the y-axis, we replace (x,y) with $(-x,y)$:
$$y^2 = 5(-x)^2 + (-x) = 5x^2 - x$$
Since the equation is changed, there is no y-axis symmetry. To test for symmetry about the origin, we replace (x,y) with $(-x,-y)$:
$$(-y)^2 = 5(-x)^2 + (-x)$$
$$y^2 = 5x^2 - x$$
Since the equation is changed, there is no origin symmetry.

7. To find the x-intercept, let $y = 0$:
$$3x = 15$$
$$x = 5$$

To find the y-intercept, let $x = 0$:
$$-5y = 15$$
$$y = -3$$
Now draw the graph:

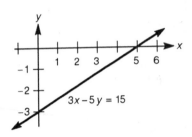

8.   Dividing by 2, we complete the square:
$$x^2 - 5x + y^2 + 3y = \tfrac{17}{2}$$
$$x^2 - 5x + \tfrac{25}{4} + y^2 + 3y + \tfrac{9}{4} = \tfrac{17}{2} + \tfrac{25}{4} + \tfrac{9}{4}$$
$$\left(x - \tfrac{5}{2}\right)^2 + \left(y + \tfrac{3}{2}\right)^2 = 17$$

This is the graph of a circle centered at $\left(\tfrac{5}{2}, -\tfrac{3}{2}\right)$ with radius $\sqrt{17}$.
To find the x-intercepts, let $y = 0$:
$$\left(x - \tfrac{5}{2}\right)^2 + \tfrac{9}{4} = 17$$
$$\left(x - \tfrac{5}{2}\right)^2 = \tfrac{59}{4}$$
$$x - \tfrac{5}{2} = \tfrac{\pm\sqrt{59}}{2}$$
$$x = \tfrac{5 \pm \sqrt{59}}{2}$$
To find the y-intercepts, let $x = 0$:
$$\tfrac{25}{4} + \left(y + \tfrac{3}{2}\right)^2 = 17$$
$$\left(y + \tfrac{3}{2}\right)^2 = \tfrac{43}{4}$$
$$y + \tfrac{3}{2} = \tfrac{\pm\sqrt{43}}{2}$$
$$y = \tfrac{-3 \pm \sqrt{43}}{2}$$
Now draw the graph:

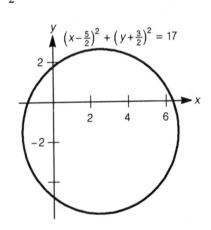

9. To find the *x*-intercept, let $y = 0$:
$$0 = |x+1| + 2$$
$$-2 = |x+1|$$
Since the equation has no solution, there is no *x*-intercept.
To find the *y*-intercept, let $x = 0$:
$$y = |0+1| + 2 = 1 + 2 = 3$$
Now draw the graph:

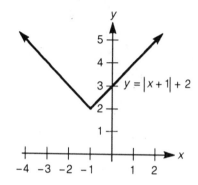

10. The slope of the line passing through $(-4, 3)$ and $(t, 13)$ is given by:
$$m = \frac{13-3}{t-(-4)} = \frac{10}{t+4}$$
Setting the slope $m = 2$, we have:
$$\frac{10}{t+4} = 2$$
$$10 = 2t + 8$$
$$2 = 2t$$
$$1 = t$$

11. Substituting $y = x^2$ into $x + y = 12$, we have:
$$x + x^2 = 12$$
$$x^2 + x - 12 = 0$$
$$(x+4)(x-3) = 0$$
$$x = -4, 3$$
$$y = 16, 9$$
The points are $(-4, 16)$ and $(3, 9)$.

12. Substituting $x = -\frac{1}{2}$ into $y = 4x^2 - 8x$:
$$y = 4\left(-\tfrac{1}{2}\right)^2 - 8\left(-\tfrac{1}{2}\right) = 4\left(\tfrac{1}{4}\right) + 4 = 1 + 4 = 5$$
So $\left(-\tfrac{1}{2}, 5\right)$ lies on the graph of $y = 4x^2 - 8x$.

13. **(a)**   We sketch the graph, noting that $x \neq -1/2$:

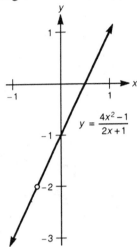

$$y = \frac{4x^2 - 1}{2x + 1}$$

(b)   We sketch the graph:

$$y = \begin{cases} \sqrt{1-x^2} & \text{if} \quad -1 \leq x \leq 0 \\ |x| & \text{if} \quad 0 < x \leq 2 \end{cases}$$

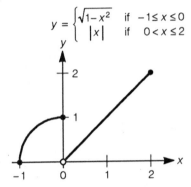

(c)   We sketch the graph:

$$y = \begin{cases} \sqrt{x} & \text{if} \quad 0 \leq x \leq 1 \\ 1/x & \text{if} \quad 1 < x \leq 4 \end{cases}$$

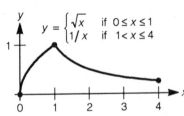

14. **(a)**   The reflection of $\overline{AB}$ about the x-axis is given by the graph:

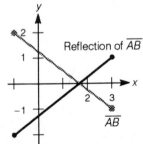

Reflection of $\overline{AB}$

$\overline{AB}$

(b) The reflection of $\overline{AB}$ about the y-axis is given by the graph:

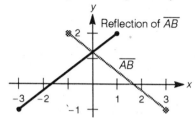

(c) The reflection of $\overline{AB}$ about the origin is given by the graph:

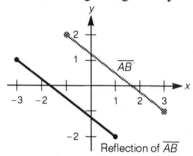

(d) The reflection of $\overline{AB}$ about the line $y = x$ is given by the graph:

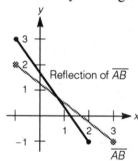

15. We first find the slope of the line:
$$m = \frac{4-2}{1-6} = -\frac{2}{5}$$
Using the point-slope formula with the point $(1, 4)$, we have:
$$y - 4 = -\frac{2}{5}(x - 1)$$
$$5y - 20 = -2x + 2$$
$$2x + 5y = 22$$
Now we find the x- and y-intercepts:

let $x = 0$	let $y = 0$
$5y = 22$	$2x = 22$
$y = \frac{22}{5}$	$x = 11$

So the area is given by:
$$\text{Area} = \tfrac{1}{2}(\text{base})(\text{height}) = \tfrac{1}{2} \cdot 11 \cdot \tfrac{22}{5} = \tfrac{121}{5}$$

Chapter Three
Functions

3.1 The Definition of a Function

1. (a) Since $-5x + 1$ is defined for all real numbers, the domain is all real numbers, or $(-\infty, \infty)$.

   (b) For $\sqrt{-5x+1}$ to be defined, we must solve $-5x + 1 \geq 0$:
   $$-5x \geq -1$$
   $$x \leq \tfrac{1}{5}$$
   So the domain is $\left(-\infty, \tfrac{1}{5}\right]$.

   (c) For the expression to be defined, we must be sure $\left|-5x+1\right| > 0$. Since $\left|-5x+1\right| \geq 0$ because it is an absolute value, we must solve $-5x + 1 = 0$:
   $$-5x = -1$$
   $$x = \tfrac{1}{5}$$
   So the domain is all real numbers except $\tfrac{1}{5}$, or $\left(-\infty, \tfrac{1}{5}\right) \cup \left(\tfrac{1}{5}, \infty\right)$.

3. (a) Since $x^2 - 10x + 24$ is defined for all real numbers, the domain is all real numbers, or $(-\infty, \infty)$.

(b)   We must be sure the quantity inside the radical is non-negative:
$$x^2 - 10x + 24 \geq 0$$
$$(x-4)(x-6) \geq 0$$
The key numbers are 4 and 6. We draw the sign chart:

Interval	Test Number	$x-4$	$x-6$	$(x-4)(x-6)$
$(-\infty, 4)$	2	neg.	neg.	pos.
$(4,6)$	5	pos.	neg.	neg.
$(6,\infty)$	8	pos.	pos.	pos.

So the expression is positive on the intervals $(-\infty,4) \cup (6,\infty)$, and it is 0 at $x = 4$ and $x = 6$. Thus the domain is $(-\infty,4] \cup [6,\infty)$.

(c)   We must make sure the quantity inside the radical is non-negative:
$$\frac{x}{x^2 - 10x + 24} \geq 0$$
$$\frac{x}{(x-4)(x-6)} \geq 0$$
The key numbers are 0, 4, and 6. We draw the sign chart:

Interval	Test Number	x	$x-4$	$x-6$	$\dfrac{x}{(x-4)(x-6)}$
$(-\infty, 0)$	-2	neg.	neg.	neg.	neg.
$(0,4)$	2	pos.	neg.	neg.	pos.
$(4,6)$	5	pos.	pos.	neg.	neg.
$(6,\infty)$	6	pos.	pos.	pos.	pos.

So the expression is positive on the intervals $(0,4) \cup (6,\infty)$, and it is 0 at $x = 0$. Thus the domain is $[0,4) \cup (6,\infty)$.

5.   (a)   The expression is undefined when its denominator is equal to 0, which occurs when $x = 4$. So the domain is all real numbers except 4, or $(-\infty,4) \cup (4,\infty)$.

(b)   The expression is undefined when its denominator is equal to 0, so we must solve $x^2 - 4 = 0$:
$$x^2 - 4 = 0$$
$$(x+2)(x-2) = 0$$
$$x = -2, 2$$
So the domain is all real numbers except -2 and 2, or $(-\infty,-2) \cup (-2,2) \cup (2,\infty)$.

(c) We must be sure the quantity inside the radical is non-negative:

$$\frac{x^2+4}{x^2-4} \geq 0$$

$$\frac{x^2+4}{(x+2)(x-2)} \geq 0$$

The key numbers are –2 and 2. We draw the sign chart:

Interval	Test Number	$x+2$	$x-2$	x^2+4	$\dfrac{x^2+4}{(x+2)(x-2)}$
$(-\infty,-2)$	-3	neg.	neg.	pos.	pos.
$(-2,2)$	0	pos.	neg.	pos.	neg.
$(2,\infty)$	3	pos.	pos.	pos.	pos.

So the expression is positive on the intervals $(-\infty,-2) \cup (2,\infty)$, and it is never equal to 0. Thus the domain is $(-\infty,-2) \cup (2,\infty)$.

7. (a) We must be sure the quantity inside the radical is non-negative:

$$x^2 - 4x - 5 \geq 0$$
$$(x-5)(x+1) \geq 0$$

The key numbers are –1 and 5. We draw the sign chart:

Interval	Test Number	$x-5$	$x+1$	$(x-5)(x+1)$
$(-\infty,-1)$	-2	neg.	neg.	pos.
$(-1,5)$	0	neg.	pos.	neg.
$(5,\infty)$	6	pos.	pos.	pos.

So the expression is positive on the intervals $(-\infty,-1) \cup (5,\infty)$, and it is 0 at $x=-1$ and $x=5$. Thus the domain is $(-\infty,-1] \cup [5,\infty)$.

(b) We must solve $x^2 - 4x - 5 > 0$, which has the same solutions as in part (a) but excluding the endpoints. So the domain is $(-\infty,-1) \cup (5,\infty)$.

(c) There are no restrictions on $\sqrt[3]{x^2-4x-5}$, so the domain is all real numbers, or $(-\infty,\infty)$.

9. (a) To find the range, we first solve for x:

$$y = \frac{x+3}{x-5}$$
$$y(x-5) = x+3$$
$$yx - 5y = x+3$$
$$yx - x = 5y+3$$
$$x(y-1) = 5y+3$$
$$x = \frac{5y+3}{y-1}$$

So the range is all real numbers except 1, or $(-\infty,1) \cup (1,\infty)$.

(b) To find the range, we first solve for x:
$$y = \frac{x-5}{x+3}$$
$$y(x+3) = x-5$$
$$yx + 3y = x - 5$$
$$yx - x = -3y - 5$$
$$x(y-1) = -3y - 5$$
$$x = \frac{-3y-5}{y-1} = \frac{3y+5}{1-y}$$
So the range is all real numbers except 1, or $(-\infty,1) \cup (1,\infty)$.

11. (a) To find the range, we first solve for x:
$$y = x^2 + 4$$
$$y - 4 = x^2$$
$$x = \pm\sqrt{y-4}$$
In order for the radical to be defined, we must have $y - 4 \geq 0$, so $y \geq 4$. So the range is $[4,\infty)$. Another approach is to note that since x^2 will assume values from 0 to ∞, then $x^2 + 4$ will assume values from 4 to ∞. So the range is $[4,\infty)$.

(b) To find the range, we first solve for x:
$$y = x^3 + 4$$
$$y - 4 = x^3$$
$$x = \sqrt[3]{y-4}$$
Since the cube root is defined for all real numbers, the range is all real numbers. Another approach is to note that since x^3 will assume values from $-\infty$ to ∞, then so will $x^3 + 4$. So the range is all real numbers, or $(-\infty,\infty)$.

13. Rules f, g, F and H are functions. Rule h is not a function since $h(x) = 1$ and $h(x) = 2$, which violates the definition of a function. Rule G is not a function, since $G(y)$ has not been assigned a value.

15. (a) The range of f is $\{1,2,3\}$, the range of g is $\{2,3\}$, the range of F is $\{1\}$, and the range of H is $\{1,2\}$.

(b) The range of g is $\{i,j\}$, the range of F is $\{i,j\}$, and the range of G is $\{k\}$.

17. (a) $y = (x-3)^2$
(b) $y = x^2 - 3$
(c) $y = (3x)^2$
(d) $y = 3x^2$

19. (a) We compute $f(1)$:
$$f(1) = (1)^2 - 3(1) + 1 = 1 - 3 + 1 = -1$$

(b) We compute $f(0)$:
$$f(0) = (0)^2 - 3(0) + 1 = 0 - 0 + 1 = 1$$

(c) We compute $f(-1)$:
$$f(-1) = (-1)^2 - 3(-1) + 1 = 1 + 3 + 1 = 5$$

(d) We compute $f\left(\frac{3}{2}\right)$:
$$f\left(\frac{3}{2}\right) = \left(\frac{3}{2}\right)^2 - 3\left(\frac{3}{2}\right) + 1 = \frac{9}{4} - \frac{9}{2} + 1 = -\frac{5}{4}$$

(e) We compute $f(z)$:
$$f(z) = (z)^2 - 3(z) + 1 = z^2 - 3z + 1$$

(f) We compute $f(x + 1)$:
$$\begin{aligned} f(x+1) &= (x+1)^2 - 3(x+1) + 1 \\ &= x^2 + 2x + 1 - 3x - 3 + 1 \\ &= x^2 - x - 1 \end{aligned}$$

(g) We compute $f(a + 1)$:
$$\begin{aligned} f(a+1) &= (a+1)^2 - 3(a+1) + 1 \\ &= a^2 + 2a + 1 - 3a - 3 + 1 \\ &= a^2 - a - 1 \end{aligned}$$

(h) We compute $f(-x)$:
$$f(-x) = (-x)^2 - 3(-x) + 1 = x^2 + 3x + 1$$

(i) Using our result from (a):
$$|f(1)| = |-1| = 1$$

(j) We compute $f(\sqrt{3})$:
$$f(\sqrt{3}) = (\sqrt{3})^2 - 3(\sqrt{3}) + 1 = 3 - 3\sqrt{3} + 1 = 4 - 3\sqrt{3}$$

(k) We compute $f(1 + \sqrt{2})$:
$$f(1+\sqrt{2}) = (1+\sqrt{2})^2 - 3(1+\sqrt{2}) + 1 = 1 + 2\sqrt{2} + 2 - 3 - 3\sqrt{2} + 1 = 1 - \sqrt{2}$$

(l) We compute $|1 - f(2)|$:
$$|1 - f(2)| = |1 - [(2)^2 - 3(2) + 1]| = |1 - [4 - 6 + 1]| = |1 - (-1)| = |1 + 1| = 2$$

21. (a) We compute $f(2x)$:
$$f(2x) = 3(2x)^2 = 3(4x^2) = 12x^2$$

(b) We compute $2f(x)$:
$$2f(x) = 2(3x^2) = 6x^2$$

(c) We compute $f(x^2)$:
$$f(x^2) = 3(x^2)^2 = 3x^4$$

(d) We compute $[f(x)]^2$:
$$[f(x)]^2 = (3x^2)^2 = 9x^4$$

(e) We compute $f\left(\dfrac{x}{2}\right)$:
$$f\left(\tfrac{x}{2}\right) = 3\left(\tfrac{x}{2}\right)^2 = 3 \cdot \frac{x^2}{4} = \tfrac{3}{4}x^2$$

(f) We compute $\dfrac{f(x)}{2}$:
$$\frac{f(x)}{2} = \frac{3x^2}{2} = \tfrac{3}{2}x^2$$

23. (a) We compute $H(0)$:
$$H(0) = 1 - 2(0)^2 = 1 - 0 = 1$$

(b) We compute $H(2)$:
$$H(2) = 1 - 2(2)^2 = 1 - 2(4) = 1 - 8 = -7$$

(c) We compute $H(\sqrt{2})$:
$$H(\sqrt{2}) = 1 - 2(\sqrt{2})^2 = 1 - 2(2) = 1 - 4 = -3$$

(d) We compute $H\left(\tfrac{5}{6}\right)$:
$$H\left(\tfrac{5}{6}\right) = 1 - 2\left(\tfrac{5}{6}\right)^2 = 1 - 2\left(\tfrac{25}{36}\right) = 1 - \tfrac{25}{18} = -\tfrac{7}{18}$$

(e) We compute $H(x+1)$:
$$\begin{aligned}
H(x+1) &= 1 - 2(x+1)^2 \\
&= 1 - 2(x^2 + 2x + 1) \\
&= 1 - 2x^2 - 4x - 2 \\
&= -2x^2 - 4x - 1
\end{aligned}$$

(f)   We compute $H(x+h)$:
$$H(x+h)=1-2(x+h)^2=1-2\left(x^2+2xh+h^2\right)=1-2x^2-4xh-2h^2$$

(g)   Using our result from part (f), we have:
$$\begin{aligned}H(x+h)-H(x)&=\left(1-2x^2-4xh-2h^2\right)-\left(1-2x^2\right)\\&=1-2x^2-4xh-2h^2-1+2x^2\\&=-4xh-2h^2\end{aligned}$$

(h)   Using our result from part (g), we have:
$$\frac{H(x+h)-H(x)}{h}=\frac{-4xh-2h^2}{h}=\frac{h(-4x-2h)}{h}=-4x-2h$$

25.   (a)   For the domain, we must exclude those values which make $x-2=0$, or $x=2$. So the domain is all real numbers except 2, or $(-\infty,2)\cup(2,\infty)$. For the range, we solve for x:
$$\begin{aligned}y&=\frac{2x-1}{x-2}\\y(x-2)&=2x-1\\yx-2y&=2x-1\\yx-2x&=2y-1\\x(y-2)&=2y-1\\x&=\frac{2y-1}{y-2}\end{aligned}$$
Since the denominator cannot be zero, $y=2$ is excluded. So the range is all real numbers except 2, or $(-\infty,2)\cup(2,\infty)$.

(b)   We compute $R(0)$:
$$R(0)=\frac{2(0)-1}{0-2}=\frac{-1}{-2}=\tfrac{1}{2}$$

(c)   We compute $R\left(\tfrac{1}{2}\right)$:
$$R\left(\tfrac{1}{2}\right)=\frac{2\left(\tfrac{1}{2}\right)-1}{\tfrac{1}{2}-2}=\frac{1-1}{-\tfrac{3}{2}}=0$$

(d)   We compute $R(-1)$:
$$R(-1)=\frac{2(-1)-1}{-1-2}=\frac{-2-1}{-3}=\frac{-3}{-3}=1$$

(e) We compute $R(x^2)$:

$$R(x^2) = \frac{2(x^2)-1}{x^2-2} = \frac{2x^2-1}{x^2-2}$$

(f) We compute $R(\frac{1}{x})$:

$$R(\tfrac{1}{x}) = \frac{2(\frac{1}{x})-1}{\frac{1}{x}-2} = \frac{\frac{2}{x}-1}{\frac{1}{x}-2} = \frac{2-x}{1-2x}$$

(g) We compute $R(a)$:

$$R(a) = \frac{2(a)-1}{a-2} = \frac{2a-1}{a-2}$$

(h) We compute $R(x-1)$:

$$R(x-1) = \frac{2(x-1)-1}{(x-1)-2} = \frac{2x-2-1}{x-3} = \frac{2x-3}{x-3}$$

27. (a) We compute the indicated values:

$$d(1) = -16(1)^2 + 96(1) = -16 + 96 = 80$$
$$d(\tfrac{3}{2}) = -16(\tfrac{3}{2})^2 + 96(\tfrac{3}{2}) = -16(\tfrac{9}{4}) + 144 = 108$$
$$d(2) = -16(2)^2 + 96(2) = -64 + 192 = 128$$
$$d(t_0) = -16t_0^2 + 96t_0$$

(b) We set $d(t) = 0$:

$$-16t^2 + 96t = 0$$
$$-16t(t-6) = 0$$
$$t = 0, 6$$

(c) We set $d(t) = 1$:

$$-16t^2 + 96t = 1$$
$$16t^2 - 96t + 1 = 0$$

Using the quadratic formula, we have:

$$t = \frac{96 \pm \sqrt{(-96)^2 - 4(16)(1)}}{2(16)}$$
$$= \frac{96 \pm \sqrt{9216 - 64}}{32}$$
$$= \frac{96 \pm \sqrt{9152}}{32}$$
$$= \frac{96 \pm 8\sqrt{143}}{32}$$
$$= \frac{12 \pm \sqrt{143}}{4}$$

29. We compute the given values:
$$g(3) = |3 - 4| = |-1| = 1$$
$$g(x + 4) = |x + 4 - 4| = |x|$$

31. (a) We first compute:
$$f(x + h) = (x + h)^2 = x^2 + 2xh + h^2$$
Therefore:
$$\frac{f(x + h) - f(x)}{h} = \frac{(x^2 + 2xh + h^2) - (x^2)}{h} = \frac{2xh + h^2}{h} = \frac{h(2x + h)}{h} = 2x + h$$

(b) We first compute:
$$f(x + h) = 2(x + h)^2 - 3(x + h) + 1$$
$$= 2(x^2 + 2xh + h^2) - 3x - 3h + 1$$
$$= 2x^2 + 4xh + 2h^2 - 3x - 3h + 1$$
Therefore:
$$\frac{f(x + h) - f(x)}{h} = \frac{(2x^2 + 4xh + 2h^2 - 3x - 3h + 1) - (2x^2 - 3x + 1)}{h}$$
$$= \frac{2x^2 + 4xh + 2h^2 - 3x - 3h + 1 - 2x^2 + 3x - 1}{h}$$
$$= \frac{4xh + 2h^2 - 3h}{h}$$
$$= \frac{h(4x + 2h - 3)}{h}$$
$$= 4x + 2h - 3$$

(c) We first compute:
$$f(x + h) = (x + h)^3 = x^3 + 3x^2h + 3xh^2 + h^3$$
Therefore:
$$\frac{f(x + h) - f(x)}{h} = \frac{(x^3 + 3x^2h + 3xh^2 + h^3) - (x^3)}{h}$$
$$= \frac{3x^2h + 3xh^2 + h^3}{h}$$
$$= \frac{h(3x^2 + 3xh + h^2)}{h}$$
$$= 3x^2 + 3xh + h^2$$

33. (a)   We compute the difference quotient:

$$\frac{f(x)-f(a)}{x-a}=\frac{\frac{x}{x-1}-\frac{a}{a-1}}{x-a}\cdot\frac{(x-1)(a-1)}{(x-1)(a-1)}$$

$$=\frac{x(a-1)-a(x-1)}{(x-a)(x-1)(a-1)}$$

$$=\frac{ax-x-ax+a}{(x-a)(x-1)(a-1)}$$

$$=\frac{-(x-a)}{(x-a)(x-1)(a-1)}$$

$$=\frac{-1}{(x-1)(a-1)}$$

(b)   Using our result from (a) with $a = 3$, we have:

$$\frac{f(x)-f(3)}{x-3}=\frac{-1}{(x-1)(3-1)}=\frac{-1}{2(x-1)}$$

(c)   We compute the difference quotient:

$$\frac{f(x+h)-f(x)}{h}=\frac{\frac{x+h}{x+h-1}-\frac{x}{x-1}}{h}\cdot\frac{(x-1)(x+h-1)}{(x-1)(x+h-1)}$$

$$=\frac{(x+h)(x-1)-x(x+h-1)}{h(x-1)(x+h-1)}$$

$$=\frac{x^2+hx-x-h-x^2-xh+x}{h(x-1)(x+h-1)}$$

$$=\frac{-h}{h(x-1)(x+h-1)}$$

$$=\frac{-1}{(x-1)(x+h-1)}$$

(d)   Using our result from (c) with $x = 3$, we have:

$$\frac{f(3+h)-f(3)}{h}=\frac{-1}{(3-1)(3+h-1)}=\frac{-1}{2(2+h)}$$

35. (a)   We set $f(x_0) = g(x_0)$:

$$4x_0-3=8-x_0$$
$$5x_0=11$$
$$x_0=\tfrac{11}{5}$$

(b)   We set $f(x_0) = g(x_0)$:
$$x_0^2 - 4 = 4 - x_0^2$$
$$2x_0^2 = 8$$
$$x_0^2 = 4$$
$$x_0 = \pm 2$$

(c)   We set $f(x_0) = g(x_0)$:
$$x_0^2 = x_0^3$$
$$x_0^3 - x_0^2 = 0$$
$$x_0^2(x_0 - 1) = 0$$
$$x_0 = 0, 1$$

(d)   We set $f(x_0) = g(x_0)$:
$$2x_0^2 - x_0 = 3$$
$$2x_0^2 - x_0 - 3 = 0$$
$$(2x_0 - 3)(x_0 + 1) = 0$$
$$x_0 = \tfrac{3}{2}, -1$$

37. (a)   We compute $A(1)$ and $A(0)$:
$$A(1) = 1000\left(1 + \tfrac{0.12}{4}\right)^{4(1)} \approx \$1125.51$$
$$A(0) = 1000\left(1 + \tfrac{0.12}{4}\right)^{4(0)} = \$1000.00$$
So $A(1) - A(0) \approx 1125.51 - 1000 = \$125.51$.

(b)   We compute $A(10)$ and $A(9)$:
$$A(10) = 1000\left(1 + \tfrac{0.12}{4}\right)^{4(10)} \approx \$3262.04$$
$$A(9) = 1000\left(1 + \tfrac{0.12}{4}\right)^{4(9)} \approx \$2898.28$$
So $A(10) - A(9) \approx 3262.04 - 2898.28 = \$363.76$.

39. (a)   We complete the table:

n	2	3	4	5	6	7	8
$g(n)$	1.4142	1.4422	1.4142	1.3797	1.3480	1.3205	1.2968

(b)   Computing $g(15) = 1.19786$ and $g(14) = 1.20744$, so 15 is the smallest natural number n such that $g(n) < 1.2$.

41. Let $a = 1$ and $b = 2$. Then:
$$f(a+b) = f(3) = 3^2 - 1 = 8$$
$$f(a) = f(1) = 1^2 - 1 = 0$$
$$f(b) = f(2) = 2^2 - 1 = 3$$
So $f(a+b) \neq f(a) + f(b)$.

43. Let $a = 2$. Then:
$$f\left(\tfrac{1}{a}\right) = f\left(\tfrac{1}{2}\right) = \left(\tfrac{1}{2}\right)^2 - 1 = \tfrac{1}{4} - 1 = -\tfrac{3}{4}$$
$$\frac{1}{f(a)} = \frac{1}{f(2)} = \frac{1}{2^2 - 1} = \tfrac{1}{3}$$
So $f\left(\tfrac{1}{a}\right) \neq \dfrac{1}{f(a)}$.

45. (a) We compute $f(a), f(2a)$ and $f(3a)$:
$$f(a) = \frac{a-a}{a+a} = \frac{0}{2a} = 0$$
$$f(2a) = \frac{2a-a}{2a+a} = \frac{a}{3a} = \tfrac{1}{3}$$
$$f(3a) = \frac{3a-a}{3a+a} = \frac{2a}{4a} = \tfrac{1}{2}$$
Since $\tfrac{1}{2} \neq 0 + \tfrac{1}{3}$, then $f(3a) \neq f(a) + f(2a)$.

   (b) We compute $f(5a)$:
$$f(5a) = \frac{5a-a}{5a+a} = \frac{4a}{6a} = \tfrac{2}{3}$$
Since $f(2a) = \tfrac{1}{3}$, then $f(5a) = 2f(2a)$.

47. We simplify $\phi\left(y^2\right)$ and $\left[\phi(y)\right]^2$:
$$\phi\left(y^2\right) = 2\left(y^2\right) - 3 = 2y^2 - 3$$
$$\left[\phi(y)\right]^2 = (2y-3)^2 = 4y^2 - 12y + 9$$
So $\phi\left(y^2\right) \neq \left[\phi(y)\right]^2$.

49. We simplify $f(ax + b) = 2(ax + b) + 3 = 2ax + 2b + 3$. Since $f(ax + b) = x$, we have $2ax + 2b + 3 = x$. Since a and b are constants, we must have:

$$2ax = x \qquad\qquad 2b + 3 = 0$$
$$2a = 1 \qquad\qquad 2b = -3$$
$$a = \tfrac{1}{2} \qquad\qquad b = -\tfrac{3}{2}$$

So $a = \tfrac{1}{2}$ and $b = -\tfrac{3}{2}$.

51. We first compute:

$$f(x+y) = \frac{(x+y)-x}{(x+y)+y} = \frac{y}{x+2y} \text{ and } f(x-y) = \frac{(x-y)-x}{(x-y)+y} = \frac{-y}{x}$$

Therefore:

$$f(x+y) + f(x-y) = \frac{y}{x+2y} - \frac{y}{x}$$
$$= \frac{yx - y(x+2y)}{x(x+2y)}$$
$$= \frac{yx - yx - 2y^2}{x^2 + 2xy}$$
$$= \frac{-2y^2}{x^2 + 2xy}$$

53. We compute $F\left(\frac{ax+b}{cx-a}\right)$:

$$F\left(\frac{ax+b}{cx-a}\right) = \frac{a\left(\frac{ax+b}{cx-a}\right) + b}{c\left(\frac{ax+b}{cx-a}\right) - a}$$
$$= \frac{a(ax+b) + b(cx-a)}{c(ax+b) - a(cx-a)}$$
$$= \frac{a^2x + ab + bcx - ab}{acx + bc - acx + a^2}$$
$$= \frac{a^2x + bcx}{a^2 + bc}$$
$$= \frac{x(a^2 + bc)}{a^2 + bc}$$
$$= x$$

55. Since $g(1) = (1)^2 - 3(1)k - 4 = 1 - 3k - 4 = -3k - 3$, then if $g(1) = -2$, we have:

$$-3k - 3 = -2$$
$$-3k = 1$$
$$k = -\tfrac{1}{3}$$

57. (a) We have $L(1) = 0$, since $2^0 = 1$.

(b) We have $L(2) = 1$, since $2^1 = 2$.

(c) We have $L(4) = 2$, since $2^2 = 4$.

(d) We have $L(64) = 6$, since $2^6 = 64$.

(e) We have $L\left(\tfrac{1}{2}\right) = -1$, since $2^{-1} = \tfrac{1}{2}$.

(f) We have $L\left(\tfrac{1}{4}\right) = -2$, since $2^{-2} = \tfrac{1}{4}$.

(g) We have $L\left(\tfrac{1}{64}\right) = -6$, since $2^{-6} = \tfrac{1}{64}$.

(h) We have $L(\sqrt{2}) = \tfrac{1}{2}$, since $2^{1/2} = \sqrt{2}$.

59. Actually, we already know the answer to this question. Since $\frac{-b+\sqrt{b^2-4ac}}{2a}$ is one of the roots to the quadratic equation $q(x) = 0$, we know $q\left(\frac{-b+\sqrt{b^2-4ac}}{2a}\right) = 0$. Let's check our answer manually:

$$q\left(\frac{-b+\sqrt{b^2-4ac}}{2a}\right) = a\left(\frac{-b+\sqrt{b^2-4ac}}{2a}\right)^2 + b\left(\frac{-b+\sqrt{b^2-4ac}}{2a}\right) + c$$

$$= a\left(\frac{b^2 - 2b\sqrt{b^2-4ac} + b^2 - 4ac}{4a^2}\right) + \left(\frac{-b^2 + b\sqrt{b^2-4ac}}{2a}\right) + c$$

$$= \frac{2b^2 - 4ac - 2b\sqrt{b^2-4ac}}{4a} + \frac{-b^2 + b\sqrt{b^2-4ac}}{2a} + c$$

$$= \frac{b^2 - 2ac - b\sqrt{b^2-4ac} - b^2 + b\sqrt{b^2-4ac}}{2a} + c$$

$$= \frac{-2ac}{2a} + c$$

$$= -c + c$$

$$= 0$$

61. (a) We compute $f[f(x)]$:

$$f[f(x)] = f\left(\frac{3x-4}{x-3}\right)$$

$$= \frac{3\left(\frac{3x-4}{x-3}\right) - 4}{\left(\frac{3x-4}{x-3}\right) - 3}$$

$$= \frac{3(3x-4) - 4(x-3)}{(3x-4) - 3(x-3)}$$

$$= \frac{9x - 12 - 4x + 12}{3x - 4 - 3x + 9}$$

$$= \frac{5x}{5}$$

$$= x$$

(b) Since $f[f(x)] = x$, then $f\left[f\left(\frac{22}{7}\right)\right] = \frac{22}{7}$.

63. The problem here is the word "nearest." $G(4) = 3$, but also $G(4) = 5$, since both 3 and 5 are equally "near" 4. So G is not a function, since it assigns more than one value to $x = 4$. To alter the definition of G, one could define G to assign the closest prime number less than or equal to x (or, for that matter, greater than or equal to x). This would provide G with a way of "deciding" between 3 and 5, in the previous example. Why do we need the "or equal to" portion of the definition? [Hint: How would you define $G(5)$ otherwise?]

65. We compute the values:
$$P(1) = (1)^2 - 1 + 17 = 17$$
$$P(2) = (2)^2 - 2 + 17 = 19$$
$$P(3) = (3)^2 - 3 + 17 = 23$$
$$P(4) = (4)^2 - 4 + 17 = 29$$
The first natural number x such that $P(x)$ is not prime is $x = 17$. We compute:
$$P(17) = (17)^2 - 17 + 17 = 17^2$$

67. (a)   We compute the function values:
$$S(0) = \frac{3^0 - 3^{-0}}{2} = \frac{1-1}{2} = \frac{0}{2} = 0$$
$$C(0) = \frac{3^0 + 3^{-0}}{2} = \frac{1+1}{2} = \frac{2}{2} = 1$$
$$S(1) = \frac{3^1 - 3^{-1}}{2} = \frac{3 - 1/3}{2} = \frac{8/3}{2} = \frac{4}{3}$$
$$C(1) = \frac{3^1 + 3^{-1}}{2} = \frac{3 + 1/3}{2} = \frac{10/3}{2} = \frac{5}{3}$$

(b)   Working from the left-hand side:
$$[C(x)]^2 - [S(x)]^2 = \left(\frac{3^x + 3^{-x}}{2}\right)^2 - \left(\frac{3^x - 3^{-x}}{2}\right)^2$$
$$= \frac{3^{2x} + 2 \cdot 3^0 + 3^{-2x}}{4} - \frac{3^{2x} - 2 \cdot 3^0 + 3^{-2x}}{4}$$
$$= \frac{3^{2x} + 2 + 3^{-2x} - 3^{2x} + 2 - 3^{-2x}}{4}$$
$$= \frac{4}{4}$$
$$= 1$$

(c)   Working from the left-hand sides:
$$S(-x) = \frac{3^{-x} - 3^{-(-x)}}{2} = \frac{3^{-x} - 3^x}{2} = -\frac{3^x - 3^{-x}}{2} = -S(x)$$
$$C(-x) = \frac{3^{-x} + 3^{-(-x)}}{2} = \frac{3^{-x} + 3^x}{2} = \frac{3^x + 3^{-x}}{2} = C(x)$$

(d)   Working from the right-hand side:

$S(x)C(y)+C(x)S(y)$

$$= \left(\frac{3^x - 3^{-x}}{2}\right)\left(\frac{3^y + 3^{-y}}{2}\right) + \left(\frac{3^x + 3^{-x}}{2}\right)\left(\frac{3^y - 3^{-y}}{2}\right)$$

$$= \frac{3^x 3^y - 3^{-x} 3^y + 3^x 3^{-y} - 3^{-x} 3^{-y}}{4} + \frac{3^x 3^y + 3^{-x} 3^y - 3^x 3^{-y} - 3^{-x} 3^{-y}}{4}$$

$$= \frac{3^{x+y} - 3^{y-x} + 3^{x-y} - 3^{-(x+y)} + 3^{x+y} + 3^{y-x} - 3^{x-y} - 3^{-(x+y)}}{4}$$

$$= \frac{2[3^{x+y} - 3^{-(x+y)}]}{4}$$

$$= \frac{3^{x+y} - 3^{-(x+y)}}{2}$$

$$= S(x+y)$$

(e)   Working from the right-hand side:

$C(x)C(y)+S(x)S(y)$

$$= \left(\frac{3^x + 3^{-x}}{2}\right)\left(\frac{3^y + 3^{-y}}{2}\right) + \left(\frac{3^x - 3^{-x}}{2}\right)\left(\frac{3^y - 3^{-y}}{2}\right)$$

$$= \frac{3^x 3^y + 3^{-x} 3^y + 3^x 3^{-y} + 3^{-x} 3^{-y}}{4} + \frac{3^x 3^y - 3^{-x} 3^y - 3^x 3^{-y} + 3^{-x} 3^{-y}}{4}$$

$$= \frac{3^{x+y} + 3^{y-x} + 3^{x-y} + 3^{-(x+y)} + 3^{x+y} - 3^{y-x} - 3^{x-y} + 3^{-(x+y)}}{4}$$

$$= \frac{2(3^{x+y} + 3^{-(x+y)})}{4}$$

$$= \frac{3^{x+y} + 3^{-(x+y)}}{2}$$

$$= C(x+y)$$

(f)   Working from the right-hand sides:

$$2S(x)C(x) = 2 \cdot \left(\frac{3^x - 3^{-x}}{2}\right)\left(\frac{3^x + 3^{-x}}{2}\right)$$

$$= \frac{3^x 3^x - 3^{-x} 3^x + 3^x 3^{-x} - 3^{-x} 3^{-x}}{2}$$

$$= \frac{3^{2x} - 3^{-2x}}{2}$$

$$= S(2x)$$

$[C(x)]^2 + [S(x)]^2$

$$= \left(\frac{3^x + 3^{-x}}{2}\right)^2 + \left(\frac{3^x - 3^{-x}}{2}\right)^2$$

$$= \frac{3^x 3^x + 3^x 3^{-x} + 3^x 3^{-x} + 3^{-x} 3^{-x}}{4} + \frac{3^x 3^x - 3^x 3^{-x} - 3^{-x} 3^x + 3^{-x} 3^{-x}}{4}$$

$$= \frac{3^{2x} + 2 + 3^{-2x} + 3^{2x} - 2 + 3^{-2x}}{4}$$

$$= \frac{2(3^{2x} + 3^{-2x})}{4}$$

$$= \frac{3^{2x} + 3^{-2x}}{2}$$

$$= C(2x)$$

(g) Working from the right-hand side:

$$3S(x) + 4[S(x)]^3 = 3\left(\frac{3^x - 3^{-x}}{2}\right) + 4\left(\frac{3^x - 3^{-x}}{2}\right)^3$$

$$= \tfrac{3}{2}(3^x - 3^{-x}) + \tfrac{4}{8}(3^x - 3^{-x})(3^{2x} - 2 + 3^{-2x})$$

$$= \tfrac{3}{2}(3^x - 3^{-x}) + \tfrac{1}{2}(3^{3x} - 2 \cdot 3^x + 3^{-x} - 3^x + 2 \cdot 3^{-x} - 3^{-3x})$$

$$= \tfrac{3}{2}3^x - \tfrac{3}{2}3^{-x} + \tfrac{1}{2}3^{3x} - \tfrac{3}{2}3^x + \tfrac{3}{2}3^{-x} - \tfrac{1}{2}3^{-3x}$$

$$= \tfrac{1}{2}(3^{3x} - 3^{-3x})$$

$$= S(3x)$$

(h) Since $S(x) + C(x) = \dfrac{3^x - 3^{-x}}{2} + \dfrac{3^x + 3^{-x}}{2} = \dfrac{2 \cdot 3^x}{2} = 3^x$, then:

$$[S(x) + C(x)]^2 = (3^x)^2 = 3^{2x}$$

We simplify the right-hand side:

$$S(2x) + C(2x) = \frac{3^{2x} - 3^{-2x}}{2} + \frac{3^{2x} + 3^{-2x}}{2} = \frac{2 \cdot 3^{2x}}{2} = 3^{2x}$$

So $[S(x) + C(x)]^2 = S(2x) + C(2x)$.

69. (a) Since $\phi(1) = 1$, $\phi(2) = 1$, $\phi(3) = 2$, and $\phi(6) = 2$, then:
$\phi(1) + \phi(2) + \phi(3) + \phi(6) = 1 + 1 + 2 + 2 = 6$

(b) Since $\phi(1) = 1$, $\phi(3) = 2$, $\phi(5) = 4$, and $\phi(15) = 8$, then:
$\phi(1) + \phi(3) + \phi(5) + \phi(15) = 1 + 2 + 4 + 8 = 15$

3.2 The Graph of a Function

1.   The domain is $[-4, 2]$ and the range is $[-3, 3]$.

3.   The domain is $[-4, -1) \cup (-1, 4]$ and the range is $[-2, 3)$.

5.   The domain is $[-4, 4]$ and the range is $[-2, 2)$.

7.   The domain is $[-4, 3]$ and the range is $\{2\}$.

9.   (a) Since $(-5, 1)$ lies on the graph of F, then $F(-5) = 1$.
      (b) Since $(2, -3)$ lies on the graph of F, then $F(2) = -3$.
      (c) Since $F(1) \approx -2$, then $F(1)$ is negative, not positive.
      (d) Since $(2, -3)$ lies on the graph of F, then $F(x) = -3$ when $x = 2$.
      (e) Since $F(2) = -3$ and $F(-2) = -2$, then $F(2) - F(-2) = -3 - (-2) = -1$.

11.   (a) positive
      (b) $f(-2) = 4; f(1) = 1; f(2) = 2; f(3) = 0$
      (c) $f(2)$, since $f(2) > 0$ and $f(4) < 0$
      (d) $f(4) - f(1) = -2 - 1 = -3$
      (e) $|f(4) - f(1)| = |-3| = 3$
      (f) domain $= [-2, 4]$; range $= [-2, 4]$

13.   (a) $f(-2) = 0$ and $g(-2) = 1$, so $g(-2)$ is larger.

      (b) $f(0) - g(0) = 2 - (-3) = 2 + 3 = 5$

      (c) We compute the three values:
$$f(1) - g(1) = 1 - (-1) = 2$$
$$f(2) - g(2) = 1 - 0 = 1$$
$$f(3) - g(3) = 4 - 1 = 3$$
      So $f(2) - g(2)$ is the smallest.

      (d) Since $f(1) = 1$, we look for where $g(x) = 1$. This occurs at $x = -2$ or $x = 3$.

      (e) Since $(3, 4)$ is a point on the graph of f, then 4 is in the range of f.

15. We complete the table:

Function	$\lvert x \rvert$	x^2	x^3
Domain	$(-\infty,\infty)$	$(-\infty,\infty)$	$(-\infty,\infty)$
Range	$[0,\infty)$	$[0,\infty)$	$(-\infty,\infty)$
Turning Point	$(0,0)$	$(0,0)$	none
Maximum Value	none	none	none
Minimum Value	0	0	none
Interval(s) Where Increasing	$(0,\infty)$	$(0,\infty)$	$(-\infty,\infty)$
Intervals(s) Where Decreasing	$(-\infty,0)$	$(-\infty,0)$	none

17. (a) The range is $[-1,1]$.
(b) The maximum value is 1 (occurring at $x = 1$).
(c) The minimum value is -1 (occurring at $x = 3$).
(d) The function is increasing on the intervals $[0,1]$ and $[3,4]$.
(e) The function is decreasing on the interval $[1,3]$.

19. (a) The range is $[-3,0]$.
(b) The maximum value is 0 (occurring at $x = 0$ and $x = 4$).
(c) The minimum value is -3 (occurring at $x = 2$).
(d) The function is increasing on the interval $[2,4]$.
(e) The function is decreasing on the interval $[0,2]$.

21. (a) Since any vertical line drawn intersects the graph in at most one point, this is the graph of a function.

(b) Since a vertical line can be drawn which intersects the graph in two points, this cannot be the graph of a function.

(c) Since a vertical line can be drawn which intersects the graph in two points, this cannot be the graph of a function.

(d) Since any vertical line drawn intersects the graph in only one point, this is the graph of a function.

23. We graph the function:

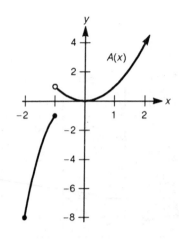

25. We graph the function:

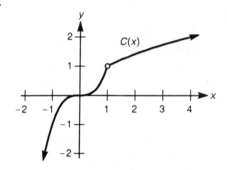

27. (a) We graph the function:

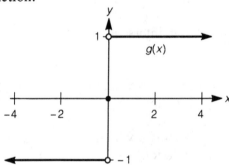

(b) We graph the function:

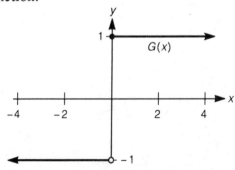

29. We graph the function:

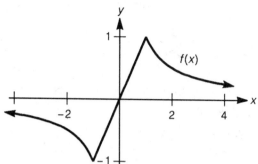

31. We graph the function:

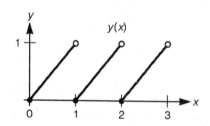

33. The domain is the set of all real numbers, or $(-\infty, \infty)$, and the range is the number 3, or $\{3\}$.

35. Since $f(3) = 9$ and $f(4) = 16$, we find the slope:
$$m = \frac{f(4) - f(3)}{4 - 3} = \frac{16 - 9}{4 - 3} = \frac{7}{1} = 7$$
This represents the average rate of change of f on the interval $[3, 4]$.

37. Since $f(3) = 3^2 + 2(3) = 15$ and $f(5) = 5^2 + 2(5) = 35$, the average rate of change is given by:
$$\frac{f(5) - f(3)}{5 - 3} = \frac{35 - 15}{5 - 3} = \frac{20}{2} = 10$$

39. Since $g(-1) = 2(-1)^2 - 4(-1) = 6$ and $g(3) = 2(3)^2 - 4(3) = 6$, the average rate of change is given by:
$$\frac{g(3) - g(-1)}{3 - (-1)} = \frac{6 - 6}{3 + 1} = 0$$

41. Since $h(5) = 2(5) - 6 = 4$ and $h(12) = 2(12) - 6 = 18$, the average rate of change is given by:
$$\frac{h(12) - h(5)}{12 - 5} = \frac{18 - 4}{12 - 5} = \frac{14}{7} = 2$$

43. Since $f(a) = \frac{3}{a}$ and $f(b) = \frac{3}{b}$, the average rate of change is given by:

$$\frac{f(b)-f(a)}{b-a} = \frac{\frac{3}{b}-\frac{3}{a}}{b-a} \cdot \frac{ab}{ab} = \frac{3a-3b}{ab(b-a)} = \frac{3(a-b)}{ab(b-a)} = -\frac{3}{ab}$$

45. Since $f(a) = -2a^3$ and $f(b) = -2b^3$, the average rate of change is given by:

$$\frac{f(b)-f(a)}{b-a} = \frac{-2b^3 + 2a^3}{b-a}$$

$$= \frac{2\left(a^3 - b^3\right)}{b-a}$$

$$= \frac{2(a-b)\left(a^2 + ab + b^2\right)}{b-a}$$

$$= -2\left(a^2 + ab + b^2\right)$$

47. (a)   Since $G(0) = 22$ and $G(3) = 23$, the average rate of change is given by:

$$\frac{G(3)-G(0)}{3-0} = \frac{23-22}{3-0} = \frac{1}{3}\,°\text{C/min}$$

(b)   Since $G(3) = 23$ and $G(6) = 27$, the average rate of change is given by:

$$\frac{G(6)-G(3)}{6-3} = \frac{27-23}{6-3} = \frac{4}{3}\,°\text{C/min}$$

(c)   Since $G(6) = 27$ and $G(8) = 28$, the average rate of change is given by:

$$\frac{G(8)-G(6)}{8-6} = \frac{28-27}{8-6} = \frac{1}{2}\,°\text{C/min}$$

49. Since $f(1) = 1$ and $f(b) = \frac{1}{b}$, the average rate of change is given by:

$$\frac{f(b)-f(1)}{b-1} = \frac{\frac{1}{b}-1}{b-1} \cdot \frac{b}{b} = \frac{1-b}{b(b-1)} = -\frac{1}{b}$$

Since this average rate of change is equal to $-\frac{1}{5}$, we have:

$$-\frac{1}{b} = -\frac{1}{5}$$

$$b = 5$$

51. (a)   To simplify calculations we compute the average rate of change on the interval $[1,b]$:

$$\frac{s(b)-s(1)}{b-1} = \frac{16b^2 - 16}{b-1} = \frac{16(b+1)(b-1)}{b-1} = 16(b+1)$$

We can complete the table by evaluating this expression at $b = 1.1$, $b = 1.01$, $b = 1.001$, $b = 1.0001$, and $b = 1.00001$. The resulting values are displayed in the table:

Interval	[1,1.1]	[1,1.01]	[1,1.001]	[1,1.0001]	[1,1.00001]
$\Delta s/\Delta t$	33.6	32.16	32.016	32.0016	32.00016

(b) The average velocity $\Delta s/\Delta t$ seems to be approaching 32 ft/sec.

53. (a) Using the points $(0,0)$ and $(1,4)$:
$$m = \frac{4-0}{1-0} = \frac{4}{1} = 4\,°\text{F/hour}$$

(b) Using the points $(1,4)$ and $(2,2)$:
$$m = \frac{2-4}{2-1} = \frac{-2}{1} = -2\,°\text{F/hour}$$

(c) Using the points $(0,0)$ and $(3,0)$:
$$m = \frac{0-0}{3-0} = \frac{0}{3} = 0\,°\text{F/hour}$$

(d) Using the points $(0,0)$ and $(4,4)$:
$$m = \frac{4-0}{4-0} = \frac{4}{4} = 1\,°\text{F/hour}$$

55. (a) We complete the table:

t	0	0.25	0.5	0.75	1	1.25	1.5	1.75	2	2.25	2.5	2.75	3
$S(t)$	0	0.25	0.5	0.75	1	1	1	1	1	1	1	1	1

Now graph the function $S(t)$ on the interval $0 \le t \le 3$:

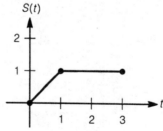

(b) We complete the table:

t	3	3.25	3.5	3.75	4	4.25	4.5	4.75	5
$S(t)$	1	0.75	0.5	0.25	0	−0.25	−0.5	−0.75	−1

Now graph the function $S(t)$ on the interval $3 \le t \le 5$:

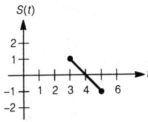

(c) We draw the graph:

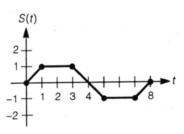

(d) We draw the graph:

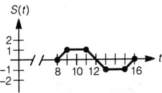

The graph is identical to that of part (c) except for the t-values. This is an example of a periodic function, that is, one which repeats its values over a specified period. In this example, the period is $t = 8$.

57. (a) We complete the table:

x	0	0.1	0.5	0.9	1.0
$[x]$	0	0	0	0	1

(b) We graph $y = [x]$ on the interval $-2 \le x < 3$:

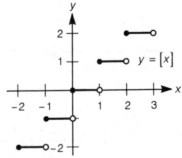

(c) The range is all integers.

Optional TI-81 Graphing Calculator Exercises for Section 3.2

1. (a) We graph the function:

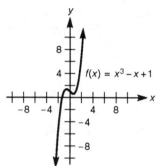

There appear to be two turning points.

(b) Using the settings $X_{min} = -2$, $X_{max} = 2$, $Y_{min} = -2$, and $Y_{max} = 2$:

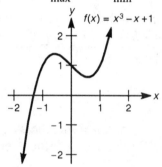

(c) The turning points are approximately $(0.568, 0.615)$ and $(-0.568, 1.385)$.

(d) Our result agrees to one decimal place with $x = \pm 1/\sqrt{3} \approx \pm 0.577$.

(e) f is increasing on the intervals $\left(-\infty, -1/\sqrt{3}\right)$ and $\left(1/\sqrt{3}, \infty\right)$, while f is decreasing on the interval $\left(-1/\sqrt{3}, 1/\sqrt{3}\right)$.

3. (a) We graph the function:

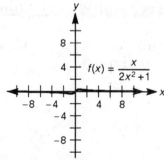

There appear to be two turning points.

(b)   Using the settings $X_{min} = -2$, $X_{max} = 2$, $Y_{min} = -0.5$, and $Y_{max} = 0.5$:

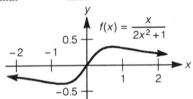

(c)   The turning points are approximately $(0.695, 0.353)$ and $(-0.695, -0.353)$.

(d)   Our result agrees to one decimal place with $x = \pm\sqrt{2}/2$.

(e) f is increasing on the interval $\left(-\sqrt{2}/2, \sqrt{2}/2\right)$, while f is decreasing on the intervals $\left(-\infty, -\sqrt{2}/2\right)$ and $\left(\sqrt{2}/2, \infty\right)$.

5.   (a)   Since $b = m + h$ and $a = m - h$, the average rate of change is given by:
$$\frac{f(b) - f(a)}{b - a} = \frac{f(m+h) - f(m-h)}{(m+h) - (m-h)} = \frac{f(m+h) - f(m-h)}{2h}$$

(b)   The result 3 is confirmed.

7.   (a)   Using the TI-81 with $m = 2.5$ and $h = 0.5$, the average rate of change is 13.
(b)   Using the TI-81 with $m = 1/3$ and $h = 1/3$, the average rate of change is 0.

9.   (a)   Using the TI-81 with $m = 0$ and $h = 1$, the average rate of change is approximately 1.1750.

(b)   Using the TI-81 with $m = 0$ and $h = 0.1$, the average rate of change is approximately 1.0016.

(c)   Using the TI-81 with $m = 0$ and $h = 0.01$, the average rate of change is approximately 0.9999.

3.3 Techniques in Graphing

1.   (a) C   (b) F   (c) I   (d) A   (e) J   (f) K
     (g) D   (h) B   (i) E   (j) H   (k) G

3. This graph will be that of $y = x^3$ translated down 3 units:

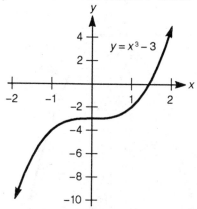

5. This graph will be that of $y = x^2$ translated to the left 4 units:

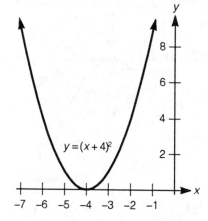

7. This graph will be that of $y = x^2$ translated to the right 4 units:

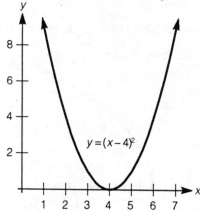

9.   This graph will be that of $y = x^2$ reflected across the x-axis:

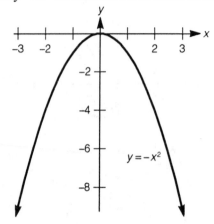

11.   This graph will be that of $y = x^2$ translated to the right 3 units, then reflected across the x-axis:

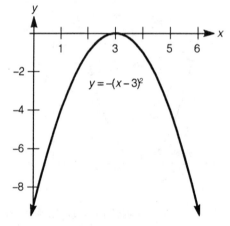

13.   This graph will be that of $y = \sqrt{x}$ translated to the right 3 units:

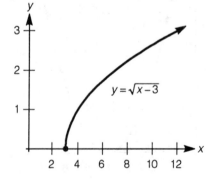

15. This graph will be that of $y = \sqrt{x}$ translated to the left 1 unit, then reflected across the x-axis:

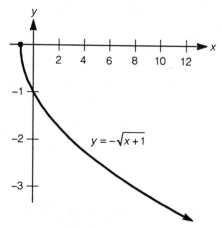

$$y = -\sqrt{x+1}$$

17. This graph will be that of $y = \dfrac{1}{x}$ translated to the left 2 units, then translated up 2 units:

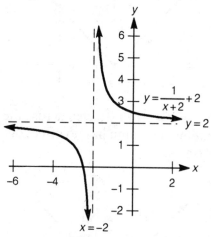

$$y = \frac{1}{x+2} + 2$$

$$y = 2$$

$$x = -2$$

19. This graph will be that of $y = x^3$ translated to the right 2 units:

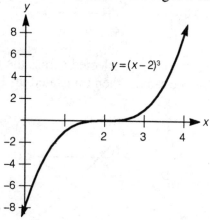

$$y = (x-2)^3$$

21. This graph will be that of $y = x^3$ reflected across the x-axis, then translated up 4 units:

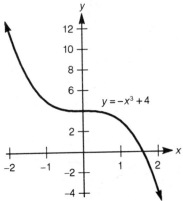

23. (a) This graph will be that of $y = |x|$ translated to the left 4 units:

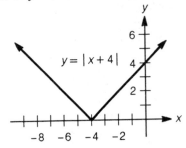

(b) This graph will be that of $y = |x|$ reflected across the y-axis, then translated to the right 4 units. Note that the reflection has no effect on the graph, since $y = |x|$ is symmetric about the y-axis:

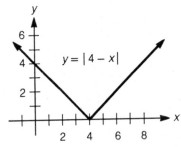

(c) This graph will be that of $y = |x|$ reflected across the y-axis, translated to the right 4 units, reflected across the x-axis, then translated up 1 unit:

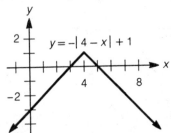

25. This is $f(x) = |x|$ translated to the right 5 units:

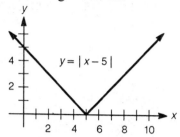

27. This is $f(x) = |x|$ reflected across the y-axis, then translated to the right 5 units. Note that the reflection has no effect on the graph, since $y = |x|$ is symmetric about the y-axis:

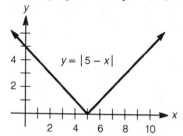

29. This is $f(x) = |x|$ translated to the right 5 units, reflected across the x-axis, then translated up 1 unit:

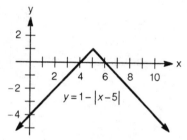

31. This is $F(x) = \dfrac{1}{x}$ translated to the left 3 units:

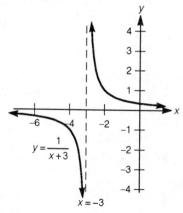

33. This is $F(x) = \dfrac{1}{x}$ translated to the left 3 units, then reflected across the x-axis:

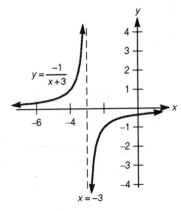

$$y = \frac{-1}{x+3}$$

$$x = -3$$

35. This is $g(x) = \sqrt{1-x^2}$ translated to the right 2 units:

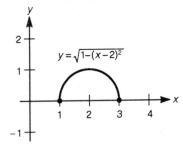

$$y = \sqrt{1-(x-2)^2}$$

37. This is $g(x) = \sqrt{1-x^2}$ translated to the right 2 units, reflected across the x-axis, then translated up 1 unit:

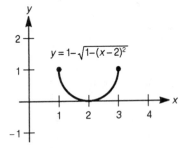

$$y = 1 - \sqrt{1-(x-2)^2}$$

39. This is $g(x) = \sqrt{1-x^2}$ reflected across the y-axis, then translated to the right 2 units:

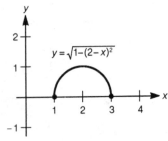

$$y = \sqrt{1-(2-x)^2}$$

41. (a) This is $y = f(x)$ reflected across the x-axis:

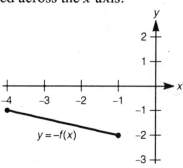

(b) This is $y = f(x)$ reflected across the y-axis:

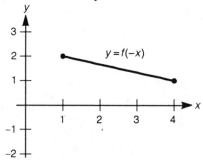

(c) This is $y = f(x)$ reflected across the y-axis, then reflected across the x-axis:

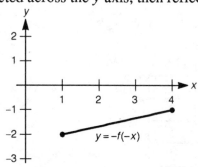

43. (a) This is $y = g(x)$ reflected across the y-axis:

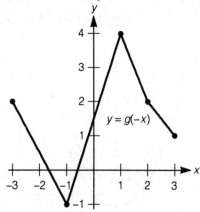

(b)   This is $y = g(x)$ reflected across the x-axis:

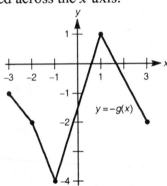

(c)   This is $y = g(x)$ reflected across the y-axis, then reflected across the x-axis:

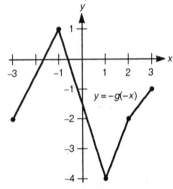

45. (a)   We complete the table:

x	x^2	$x^2 - 1$	$x^2 + 1$
0	0	−1	1
±1	1	0	2
±2	4	3	5
±3	9	8	10

(b)   Notice that the graph of $y = x^2 - 1$ is a vertical displacement down one unit (from $y = x^2$), while the graph of $y = x^2 + 1$ is a vertical displacement up one unit (from $y = x^2$):

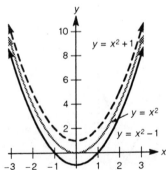

47. (a)   We complete the table:

x	$\sqrt{x}$	$-\sqrt{x}$
0	0.0	0.0
1	1.0	−1.0
2	1.4	−1.4
3	1.7	−1.7
4	2.0	−2.0
5	2.2	−2.2

(b)   Notice that the graph of $y = -\sqrt{x}$ is a reflection of $y = \sqrt{x}$ across the x-axis:

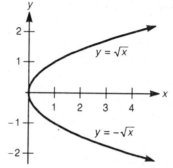

49. (a)   The point is $(a + 3, b)$, since $f(a + 3 - 3) = f(a) = b$.
   (b)   The point is $(a, b - 3)$, since $f(a) - 3 = b - 3$.
   (c)   The point is $(a + 3, b - 3)$, since $f(a + 3 - 3) - 3 = f(a) - 3 = b - 3$.
   (d)   The point is $(a, -b)$, since $-f(a) = -b$.
   (e)   The point is $(-a, b)$, since $f[-(-a)] = f(a) = b$.
   (f)   The point is $(-a, -b)$, since $-f[-(-a)] = -f(a) = -b$.
   (g)   The point is $(-a + 3, b)$, since $f[3 - (-a + 3)] = f(3 + a - 3) = f(a) = b$.
   (h)   The point is $(-a + 3, -b + 1)$, since $-f[3 - (-a + 3)] + 1 = -f(a) + 1 = -b + 1$.

51. We work from the right-hand side of the equality:
$$\frac{1}{x-1} + 1 = \frac{1 + x - 1}{x - 1} = \frac{x}{x - 1}$$
Since $g(x) = \frac{1}{x-1} + 1$, then we graph $y = \frac{1}{x}$ displaced 1 unit to the right and 1 unit up.
See graph:

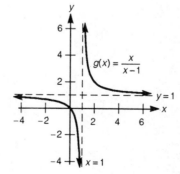

53. We rationalize the denominator (which is valid for $0 < |x| \le 1$):

$$\frac{x^2}{1-\sqrt{1-x^2}} \cdot \frac{1+\sqrt{1-x^2}}{1+\sqrt{1-x^2}} = \frac{x^2\left(1+\sqrt{1-x^2}\right)}{1-1+x^2} = \frac{x^2\left(1+\sqrt{1-x^2}\right)}{x^2} = 1+\sqrt{1-x^2}$$

Now, since $0 \le \sqrt{1-x^2} < 1$, then $1 \le 1+\sqrt{1-x^2} < 2$. So the range is $[1,2)$.

55. (a) Call $y = f(x)$. We replace x by $-x$ and y by $-y$:

$$-y = f(-x)$$
$$-y = -f(x) \qquad \text{[since } f(-x) = -f(x)\text{]}$$
$$y = f(x)$$

So the resulting equation is identical to the original equation, and thus the graph of $y = f(x)$ is symmetric about the origin.

(b) (i) We compute $f(-x)$:

$$f(-x) = (-x)^3 = -x^3 = -f(x)$$

(ii) We compute $f(-x)$:

$$f(-x) = -2(-x)^5 + 4(-x)^3 - (-x)$$
$$= 2x^5 - 4x^3 + x$$
$$= -\left(-2x^5 + 4x^3 - x\right)$$
$$= -f(x)$$

(iii) We compute $f(-x)$:

$$f(-x) = \frac{|-x|}{(-x)+(-x)^7} = \frac{|x|}{-x-x^7} = -\frac{|x|}{x+x^7} = -f(x)$$

57. (a) Since $y = f(-x)$ is a reflection across the y-axis, then the new function will have these intervals reversed with a change in sign:

         increasing: $(-\infty, -4) \cup (-2, \infty)$
         decreasing: $(-4, -2)$

(b) Since $y = -f(x)$ is a reflection across the x-axis, then the new function will have these intervals reversed:

         increasing: $(-\infty, 2) \cup (4, \infty)$
         decreasing: $(2, 4)$

Optional TI-81 Graphing Calculator Exercises for Section 3.3

1. (a)   We graph the two functions using the settings $X_{min} = -10$, $X_{max} = 10$, $Y_{min} = 0$, and $Y_{max} = 10$:

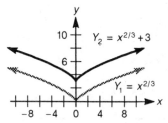

   (b)   We graph the two functions using the settings $X_{min} = -10$, $X_{max} = 10$, $Y_{min} = -5$, and $Y_{max} = 5$:

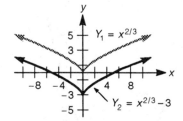

3. (a)   We graph the two functions using the settings $X_{min} = -4$, $X_{max} = 4$, $Y_{min} = -4$, and $Y_{max} = 4$:

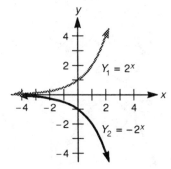

   (b)   We graph the two functions using the same settings:

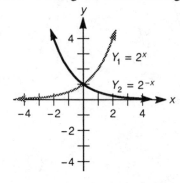

5.   (a)   We draw the graph:

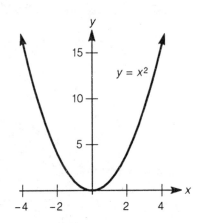

(b)   This will be a translation of $y = x^2$ to the left 3 units:

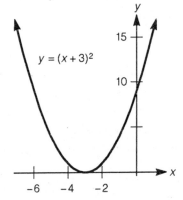

(c)   This will be a translation of $y = x^2$ to the left 3 units, then a reflection across the x-axis:

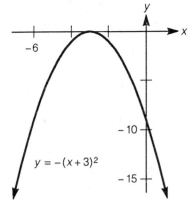

(d) This will be a translation of $y = x^2$ to the left 3 units, then a reflection across the y-axis:

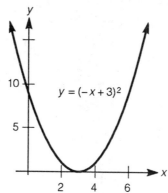

$y = (-x + 3)^2$

7. (a) We draw the graph:

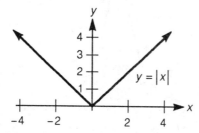

$y = |x|$

(b) This will be a translation of $y = |x|$ to the left 4 units:

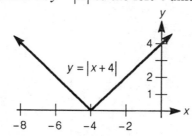

$y = |x + 4|$

(c) This will be a translation of $y = |x|$ to the left 4 units, then a reflection across the y-axis:

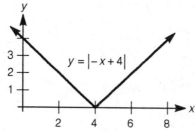

$y = |-x + 4|$

(d)   This will be a translation of $y = |x|$ to the left 4 units, then a reflection across the y-axis, then a translation down 2 units:

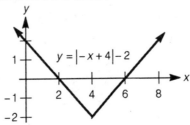

9.   (a)   We draw the graph:

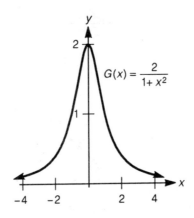

(b)   This will be a translation of $G(x)$ to the left 3 units. We compute $G(x + 3)$:
$$G(x+3) = \frac{2}{1+(x+3)^2}$$
Now draw the graph:

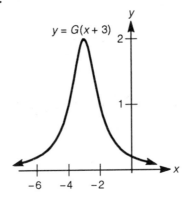

(c)   This will be a translation of $G(x)$ to the right 3 units. We compute $G(x - 3)$:
$$G(x-3) = \frac{2}{1+(x-3)^2}$$

Now draw the graph:

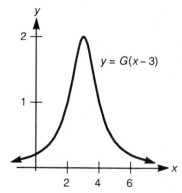

(d) This will be a reflection of $G(x)$ across the x-axis, then a translation up 2 units. We compute $-G(x) + 2$:

$$-G(x) + 2 = \frac{-2}{1 + x^2} + 2$$

Now draw the graph:

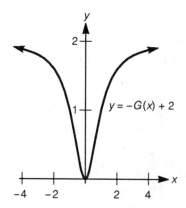

3.4 Methods of Combining Functions

1. (a) We compute $(f + g)(x)$:
$$(f + g)(x) = f(x) + g(x) = (2x - 1) + (x^2 - 3x - 6) = x^2 - x - 7$$

(b) We compute $(f - g)(x)$:
$$(f - g)(x) = f(x) - g(x) = (2x - 1) - (x^2 - 3x - 6) = -x^2 + 5x + 5$$

(c) Using our answer from part (b), we have:
$$(f - g)(0) = -(0)^2 + 5(0) + 5 = 0 + 0 + 5 = 5$$

3. (a) We compute $(m-f)(x)$:
$$(m-f)(x) = m(x) - f(x)$$
$$= (x^2 - 9) - (2x - 1)$$
$$= x^2 - 9 - 2x + 1$$
$$= x^2 - 2x - 8$$

   (b) We compute $(f-m)(x)$:
$$(f-m)(x) = f(x) - m(x)$$
$$= (2x - 1) - (x^2 - 9)$$
$$= 2x - 1 - x^2 + 9$$
$$= -x^2 + 2x + 8$$

5. (a) We compute $(fk)(x)$:
$$(fk)(x) = f(x)k(x) = (2x - 1)(2) = 4x - 2$$

   (b) We compute $(kf)(x)$:
$$(kf)(x) = k(x)f(x) = 2(2x - 1) = 4x - 2$$

   (c) Using our results from parts (a) and (b):
$$(fk)(1) - (kf)(2) = [4(1) - 2] - [4(2) - 2] = 2 - 6 = -4$$

7. (a) We compute $(f/m)(x) - (m/f)(x)$:
$$\frac{f}{m}(x) - \frac{m}{f}(x) = \frac{f(x)}{m(x)} - \frac{m(x)}{f(x)}$$
$$= \frac{[f(x)]^2 - [m(x)]^2}{f(x)m(x)}$$
$$= \frac{(2x - 1)^2 - (x^2 - 9)^2}{(x^2 - 9)(2x - 1)}$$
$$= \frac{(4x^2 - 4x + 1) - (x^4 - 18x^2 + 81)}{2x^3 - x^2 - 18x + 9}$$
$$= \frac{-x^4 + 22x^2 - 4x - 80}{2x^3 - x^2 - 18x + 9}$$

   (b) Using our result from part (a):
$$\frac{f}{m}(0) - \frac{m}{f}(0) = \frac{-0^4 + 22(0)^2 - 4(0) - 80}{2(0)^3 - 0^2 - 18(0) + 9} = -\frac{80}{9}$$

9. (a) We compute $[m \cdot (k - h)](x)$:

$$\begin{aligned}
[m \cdot (k - h)](x) &= m(x)[k(x) - h(x)] \\
&= (x^2 - 9)(2 - x^3) \\
&= 2x^2 - 18 - x^5 + 9x^3 \\
&= -x^5 + 9x^3 + 2x^2 - 18
\end{aligned}$$

(b) We compute $(mk)(x) - (mh)(x)$:

$$\begin{aligned}
(mk)(x) - (mh)(x) &= m(x)k(x) - m(x)h(x) \\
&= (x^2 - 9)(2) - (x^2 - 9)(x^3) \\
&= 2x^2 - 18 - x^5 + 9x^3 \\
&= -x^5 + 9x^3 + 2x^2 - 18
\end{aligned}$$

(c) Using our result from part (b):

$$\begin{aligned}
(mk)(-1) - (mh)(-1) &= -(-1)^5 + 9(-1)^3 + 2(-1)^2 - 18 \\
&= -(-1) + 9(-1) + 2(1) - 18 \\
&= 1 - 9 + 2 - 18 \\
&= -24
\end{aligned}$$

11. (a) We compute $(f \circ g)(x)$:

$$(f \circ g)(x) = f[g(x)] = f(-2x - 5) = 3(-2x - 5) + 1 = -6x - 15 + 1 = -6x - 14$$

(b) Using our result from part (a):

$$(f \circ g)(10) = -6(10) - 14 = -60 - 14 = -74$$

(c) We compute $(g \circ f)(x)$:

$$(g \circ f)(x) = g[f(x)] = g(3x + 1) = -2(3x + 1) - 5 = -6x - 2 - 5 = -6x - 7$$

(d) Using our result from part (a):

$$(g \circ f)(10) = -6(10) - 7 = -60 - 7 = -67$$

13. (a) We find the indicated compositions:

$$\begin{aligned}
(f \circ g)(x) &= f[g(x)] = f(1 + x) = 1 - (1 + x) = 1 - 1 - x = -x \\
(f \circ g)(-2) &= -(-2) = 2 \\
(g \circ f)(x) &= g[f(x)] = g(1 - x) = 1 + (1 - x) = 2 - x \\
(g \circ f)(-2) &= 2 - (-2) = 2 + 2 = 4
\end{aligned}$$

(b) We find the indicated compositions:

$$\begin{aligned}
(f \circ g)(x) &= f[g(x)] \\
&= f(2 - 3x) \\
&= (2 - 3x)^2 - 3(2 - 3x) - 4 \\
&= 4 - 12x + 9x^2 - 6 + 9x - 4 \\
&= 9x^2 - 3x - 6 \\
(f \circ g)(-2) &= 9(-2)^2 - 3(-2) - 6 = 36 + 6 - 6 = 36
\end{aligned}$$

$$(g \circ f)(x) = g[f(x)]$$
$$= g(x^2 - 3x - 4)$$
$$= 2 - 3(x^2 - 3x - 4)$$
$$= 2 - 3x^2 + 9x + 12$$
$$= -3x^2 + 9x + 14$$
$$(g \circ f)(-2) = -2(-2)^2 + 9(-2) + 14 = -12 - 18 + 14 = -16$$

(c) We find the indicated compositions:

$$(f \circ g)(x) = f[g(x)] = f(1 - x^4) = \frac{1 - x^4}{3}$$
$$(f \circ g)(-2) = \frac{1 - (-2)^4}{3} = \frac{1 - 16}{3} = \frac{-15}{3} = -5$$
$$(g \circ f)(x) = g[f(x)] = g\left(\frac{x}{3}\right) = 1 - \left(\frac{x}{3}\right)^4 = 1 - \frac{x^4}{81}$$
$$(g \circ f)(-2) = 1 - \frac{(-2)^4}{81} = 1 - \frac{16}{81} = \frac{65}{81}$$

(d) We find the indicated compositions:

$$(f \circ g)(x) = f[g(x)] = f(x^2 + 1) = 2^{x^2 + 1}$$
$$(f \circ g)(-2) = 2^{(-2)^2 + 1} = 2^{4+1} = 2^5 = 32$$
$$(g \circ f)(x) = g[f(x)] = g(2^x) = (2^x)^2 + 1 = 2^{2x} + 1$$
$$(g \circ f)(-2) = 2^{2(-2)} + 1 = 2^{-4} + 1 = \frac{1}{16} + 1 = \frac{17}{16}$$

(e) We find the indicated compositions:

$$(f \circ g)(x) = f[g(x)] = f(3x^5 - 4x^2) = 3x^5 - 4x^2$$
$$(f \circ g)(-2) = 3(-2)^5 - 4(-2)^2 = -96 - 16 = -112$$
$$(g \circ f)(x) = g[f(x)] = g(x) = 3x^5 - 4x^2$$
$$(g \circ f)(-2) = 3(-2)^5 - 4(-2)^2 = -96 - 16 = -112$$

(f) We find the indicated compositions:

$$(f \circ g)(x) = f[g(x)] = f\left(\frac{x+4}{3}\right) = 3\left(\frac{x+4}{3}\right) - 4 = x + 4 - 4 = x$$
$$(f \circ g)(-2) = -2$$
$$(g \circ f)(x) = g[f(x)] = g(3x - 4) = \frac{3x - 4 + 4}{3} = \frac{3x}{3} = x$$
$$(g \circ f)(-2) = -2$$

15. (a)   We compute $(F \circ G)(x)$:

$$
\begin{aligned}
(F \circ G)(x) &= F[G(x)] \\
&= F\left(\tfrac{x+1}{x-1}\right) \\
&= \frac{3\left(\frac{x+1}{x-1}\right) - 4}{3\left(\frac{x+1}{x-1}\right) + 3} \\
&= \frac{3(x+1) - 4(x-1)}{3(x+1) + 3(x-1)} \\
&= \frac{3x + 3 - 4x + 4}{3x + 3 + 3x - 3} \\
&= \frac{-x + 7}{6x}
\end{aligned}
$$

(b)   Using our result from part (a):

$$F[G(t)] = \frac{-t + 7}{6t}$$

(c)   Using our result from part (a):

$$(F \circ G)(2) = F[G(2)] = \frac{-2 + 7}{6(2)} = \tfrac{5}{12}$$

(d)   We compute $(G \circ F)(x)$:

$$
\begin{aligned}
(G \circ F)(x) &= G[F(x)] \\
&= G\left(\tfrac{3x-4}{3x+3}\right) \\
&= \frac{\frac{3x-4}{3x+3} + 1}{\frac{3x-4}{3x+3} - 1} \\
&= \frac{(3x - 4) + 1(3x + 3)}{(3x - 4) - 1(3x + 3)} \\
&= \frac{3x - 4 + 3x + 3}{3x - 4 - 3x - 3} \\
&= \frac{6x - 1}{-7} \\
&= \frac{1 - 6x}{7}
\end{aligned}
$$

(e)   Using our result from part (d):

$$G[F(y)] = \frac{1 - 6y}{7}$$

(f)   Using our result from part (d):

$$(G \circ F)(2) = G[F(2)] = \frac{1 - 6(2)}{7} = -\tfrac{11}{7}$$

17. (a) We compute $M(7)$ and $M[M(7)]$:

$$M(7) = \frac{2(7)-1}{7-2} = \frac{14-1}{5} = \frac{13}{5}$$

$$M[M(7)] = M\left(\tfrac{13}{5}\right) = \frac{2\left(\tfrac{13}{5}\right)-1}{\tfrac{13}{5}-2} = \frac{\tfrac{26}{5}-1}{\tfrac{3}{5}} = \frac{\tfrac{21}{5}}{\tfrac{3}{5}} = 7$$

(b) We compute $(M \circ M)(x)$:

$$
\begin{aligned}
(M \circ M)(x) &= M[M(x)] \\
&= M\left(\tfrac{2x-1}{x-2}\right) \\
&= \frac{2\left(\tfrac{2x-1}{x-2}\right)-1}{\left(\tfrac{2x-1}{x-2}\right)-2} \\
&= \frac{2(2x-1)-1(x-2)}{(2x-1)-2(x-2)} \\
&= \frac{4x-2-x+2}{2x-1-2x+4} \\
&= \frac{3x}{3} \\
&= x
\end{aligned}
$$

(c) Using our result from part (b), we have $(M \circ M)(7) = M[M(7)] = 7$. This agrees with our answer from part (a).

19. (a) $f[g(3)] = f(0) = 1$
(b) $g[f(3)] = g(4) = -3$
(c) $f[h(3)] = f(2) = -1$
(d) $(h \circ g)(2) = h[g(2)] = h(1) = 2$
(e) $h\{f[g(3)]\} = h[f(0)] = h(1) = 2$
(f) $(g \circ f \circ h \circ f)(2) = (g \circ f \circ h)(-1) = (g \circ f)(3) = g(4) = -3$

21. (a) We compute each composition:

$$(T \circ I)(x) = T[I(x)] = T(x) = 4x^3 - 3x^2 + 6x - 1$$
$$(I \circ T)(x) = I[T(x)] = I\left(4x^3 - 3x^2 + 6x - 1\right) = 4x^3 - 3x^2 + 6x - 1$$

(b) We compute each composition:

$$(G \circ I)(x) = G[I(x)] = G(x) = ax^2 + bx + c$$
$$(I \circ G)(x) = I[G(x)] = I\left(ax^2 + bx + c\right) = ax^2 + bx + c$$

(c) In general, given any function $f(x)$ and $I(x) = x$, then $(f \circ I)(x) = f(x)$ and $(I \circ f)(x) = f(x)$. The function $I(x) = x$ is called the identity function.

23. We compute the compositions:

$$(f \circ g)(0) = f[g(0)] = f(3) = 1$$
$$(f \circ g)(1) = f[g(1)] = f(2) = 3$$
$$(f \circ g)(2) = f[g(2)] = f(0) = 2$$
$$(f \circ g)(3) = f[g(3)] = f(4) = \text{undefined}$$
$$(f \circ g)(4) = f[g(4)] = f(-1) = 2$$

Thus we have the table:

x	0	1	2	3	4
$(f \circ g)(x)$	1	3	2	undef.	2

We compute the compositions:

$$(g \circ f)(-1) = g[f(-1)] = g(2) = 0$$
$$(g \circ f)(0) = g[f(0)] = g(2) = 0$$
$$(g \circ f)(1) = g[f(1)] = g(0) = 3$$
$$(g \circ f)(2) = g[f(2)] = g(3) = 4$$
$$(g \circ f)(3) = g[f(3)] = g(1) = 2$$
$$(g \circ f)(4) = g[f(4)] = \text{undefined}$$

Thus we have the table:

x	−1	0	1	2	3	4
$(g \circ f)(x)$	0	0	3	4	2	undef.

25. (a) We compute $(f \circ g)(x)$:

$$(f \circ g)(x) = f[g(x)] = f(3x - 4) = 2(3x - 4) + 1 = 6x - 8 + 1 = 6x - 7$$

We draw the graph:

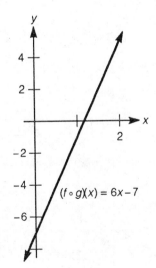

$(f \circ g)(x) = 6x - 7$

(b) We compute $(g \circ f)(x)$:

$$(g \circ f)(x) = g[f(x)] = g(2x + 1) = 3(2x + 1) - 4 = 6x + 3 - 4 = 6x - 1$$

We draw the graph:

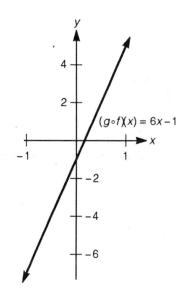

27. (a)   The domain is $[0, \infty)$ and the range is $[-3, \infty)$:

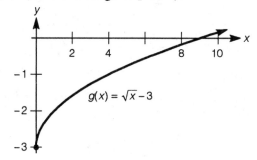

(b)   The domain is $(-\infty, \infty)$ and the range is $(-\infty, \infty)$:

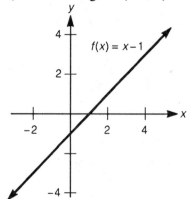

(c)   We first compute $(f \circ g)(x)$:
$$(f \circ g)(x) = f[g(x)] = f(\sqrt{x} - 3) = (\sqrt{x} - 3) - 1 = \sqrt{x} - 4$$

The domain is $[0, \infty)$ and the range is $[-4, \infty)$:

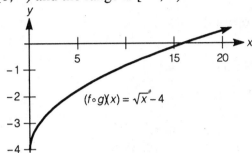

(d) We first compute $g[f(x)]$:
$$g[f(x)] = g(x-1) = \sqrt{x-1} - 3$$
The domain is $[1, \infty)$.

(e) We draw the graph:

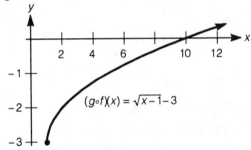

29. Let $f(x) = x^4$ and $g(x) = 3x - 1$. Then $C(x) = (f \circ g)(x)$, since:
$$(f \circ g)(x) = f[g(x)] = f(3x-1) = (3x-1)^4$$

31. (a) Let $f(x) = \sqrt[3]{x}$ and $g(x) = 3x + 4$. Then $F(x) = (f \circ g)(x)$, since:
$$(f \circ g)(x) = f[g(x)] = f(3x+4) = \sqrt[3]{3x+4}$$

(b) Let $f(x) = |x|$ and $g(x) = 2x - 3$. Then $G(x) = (f \circ g)(x)$, since:
$$(f \circ g)(x) = f[g(x)] = f(2x-3) = |2x-3|$$

(c) Let $f(x) = x^5$ and $g(x) = ax + b$. Then $H(x) = (f \circ g)(x)$, since:
$$(f \circ g)(x) = f[g(x)] = f(ax+b) = (ax+b)^5$$

(d) Let $f(x) = \frac{1}{x}$ and $g(x) = \sqrt{x}$. Then $T(x) = (f \circ g)(x)$, since:
$$(f \circ g)(x) = f[g(x)] = f(\sqrt{x}) = \frac{1}{\sqrt{x}}$$

33. (a) We have $f(x) = (b \circ c)(x)$, since:
$$(b \circ c)(x) = b[c(x)] = b(2x+1) = \sqrt[3]{2x+1}$$

(b)   We have $g(x) = (a \circ d)(x)$, since:
$$(a \circ d)(x) = a[d(x)] = a(x^2) = \frac{1}{x^2}$$

(c)   We have $h(x) = (c \circ d)(x)$, since:
$$(c \circ d)(x) = c[d(x)] = c(x^2) = 2x^2 + 1$$

(d)   We have $K(x) = (c \circ b)(x)$, since:
$$(c \circ b)(x) = c[b(x)] = c(\sqrt[3]{x}) = 2\sqrt[3]{x} + 1$$

(e)   We have $l(x) = (c \circ a)(x)$, since:
$$(c \circ a)(x) = c[a(x)] = c\left(\tfrac{1}{x}\right) = 2\left(\tfrac{1}{x}\right) + 1 = \frac{2}{x} + 1$$

(f)   We have $m(x) = (a \circ c)(x)$, since:
$$(a \circ c)(x) = a[c(x)] = a(2x + 1) = \frac{1}{2x + 1}$$

(g)   We have $n(x) = (b \circ d)(x)$, since:
$$(b \circ d)(x) = b[d(x)] = b(x^2) = \sqrt[3]{x^2} = x^{2/3}$$
Note that we could also use $n(x) = (d \circ b)(x)$, since:
$$(d \circ b)(x) = d[b(x)] = d(\sqrt[3]{x}) = \left(\sqrt[3]{x}\right)^2 = x^{2/3}$$

35.   We first compute $(C \circ f)(t)$:
$$(C \circ f)(t) = C[f(t)] = C\left(\frac{1}{t^2 + 1}\right) = 2\pi\left(\frac{1}{t^2 + 1}\right) = \frac{2\pi}{t^2 + 1}$$
When $t = 3$, we have:
$$(C \circ f)(3) = \frac{2\pi}{3^2 + 1} = \frac{2\pi}{10} = \frac{\pi}{5} \text{ ft}$$

37.   (a)   We compute $(C \circ f)(t)$:
$$(C \circ f)(t) = C[f(t)] = C(5t) = 100 + 90(5t) - (5t)^2 = 100 + 450t - 25t^2$$

(b)   When $t = 3$ hr, we have:
$$C[f(3)] = 100 + 450(3) - 25(3)^2 = \$1225$$

(c)   When $t = 6$ hr, we have:
$$C[f(6)] = 100 + 450(6) - 25(6)^2 = \$1900$$
No, the cost is not twice as much for 6 hours.

39. Call $y = f(x)$. Since $(g \circ f)(x) = g[f(x)] = g(y)$, then:

$$g(y) = x + 5$$
$$4y - 1 = x + 5$$
$$4y = x + 6$$
$$y = \frac{x+6}{4}$$

So $f(x) = \frac{x+6}{4}$.

41. We set $f[g(x)] = x$:

$$f(ax + b) = x$$
$$-2(ax + b) + 1 = x$$
$$-2ax - 2b + 1 = x$$

Since a and b are constants, we can equate components:

$$-2a = 1 \qquad \text{and} \qquad -2b + 1 = 0$$
$$a = -\tfrac{1}{2} \qquad\qquad\qquad -2b = -1$$
$$\qquad\qquad\qquad\qquad b = \tfrac{1}{2}$$

So $a = -\tfrac{1}{2}$ and $b = \tfrac{1}{2}$.

43. (a) We compute the difference quotient:

$$\frac{f[g(x)] - f[g(a)]}{g(x) - g(a)} = \frac{f(2x-1) - f(2a-1)}{(2x-1) - (2a-1)}$$
$$= \frac{(2x-1)^2 - (2a-1)^2}{2x - 1 - 2a + 1}$$
$$= \frac{[(2x-1) + (2a-1)][(2x-1) - (2a-1)]}{2x - 2a}$$
$$= \frac{(2x + 2a - 2)(2x - 2a)}{2x - 2a}$$
$$= 2x + 2a - 2$$

(b) We compute the difference quotient:

$$\frac{f[g(x)] - f[g(a)]}{x - a} = \frac{(2x + 2a - 2)(2x - 2a)}{x - a} = \frac{4(x + a - 1)(x - a)}{x - a} = 4x + 4a - 4$$

45. (a) We compute $(g \circ h \circ f)(x)$:

$$(g \circ h \circ f)(x) = g\{h[f(x)]\} = g[h(x^2)] = g\left(\frac{x^2}{2}\right) = \frac{x^2}{2} + 1$$

(b) We compute $(h \circ f \circ g)(x)$:

$$(h \circ f \circ g)(x) = h\{f[g(x)]\} = h[f(x+1)] = h[(x+1)^2] = \frac{(x+1)^2}{2}$$

(c)   We compute $(g \circ f \circ h)(x)$:

$$(g \circ f \circ h)(x) = g\{f[h(x)]\} = g\left[f\left(\tfrac{x}{2}\right)\right] = g\left[\left(\tfrac{x}{2}\right)^2\right] = g\left(\frac{x^2}{4}\right) = \frac{x^2}{4} + 1$$

(d)   We compute $(f \circ h \circ g)(x)$:

$$(f \circ h \circ g)(x) = f\{h[g(x)]\} = f[h(x+1)] = f\left(\frac{x+1}{2}\right) = \left(\frac{x+1}{2}\right)^2 = \frac{(x+1)^2}{4}$$

(e)   We compute $(h \circ g \circ f)(x)$:

$$(h \circ g \circ f)(x) = h\{g[f(x)]\} = h\left[g(x^2)\right] = h(x^2+1) = \frac{x^2+1}{2}$$

47. (a)   We have $p(x) = (g \circ f \circ h)(x)$, since:
$$(g \circ f \circ h)(x) = g\{f[h(x)]\} = g[f(3x)] = g\left[(3x)^2\right] = g(9x^2) = 1 - 9x^2$$

(b)   We have $q(x) = (h \circ g \circ f)(x)$, since:
$$(h \circ g \circ f)(x) = h\{g[f(x)]\} = h\left[g(x^2)\right] = h(1-x^2) = 3(1-x^2) = 3 - 3x^2$$

(c)   We have $r(x) = (f \circ g \circ h)(x)$, since:
$$(f \circ g \circ h)(x) = f\{g[h(x)]\} = f[g(3x)] = f(1-3x) = (1-3x)^2 = 1 - 6x + 9x^2$$

(d)   We have $s(x) = (h \circ f \circ g)(x)$, since:
$$
\begin{aligned}
(h \circ f \circ g)(x) &= h\{f[g(x)]\} \\
&= h[f(1-x)] \\
&= h\left[(1-x)^2\right] \\
&= h(1 - 2x + x^2) \\
&= 3(1 - 2x + x^2) \\
&= 3 - 6x + 3x^2
\end{aligned}
$$

49. The y-coordinate of P is $f(a)$. Since this is the x-coordinate of Q, the y-coordinate of Q is $f(f(a))$. Since this is the x-coordinate of R, the y-coordinate of R is $f(f(f(a)))$.

51. (a)   The first six iterates are as follows:
$$
\begin{aligned}
f(16) &= \sqrt{16} = 4 \\
f(4) &= \sqrt{4} = 2 \\
f(2) &= \sqrt{2} \approx 1.41 \\
f(1.41) &= \sqrt{1.41} \approx 1.19 \\
f(1.19) &= \sqrt{1.19} \approx 1.09 \\
f(1.09) &= \sqrt{1.09} \approx 1.04
\end{aligned}
$$

(b)   The next six iterates are as follows:
$$f(1.04) = \sqrt{1.04} \approx 1.022$$
$$f(1.022) = \sqrt{1.022} \approx 1.011$$
$$f(1.011) = \sqrt{1.011} \approx 1.005$$
$$f(1.005) = \sqrt{1.005} \approx 1.003$$
$$f(1.003) = \sqrt{1.003} \approx 1.001$$
$$f(1.001) = \sqrt{1.001} \approx 1.001$$
The iterates seem to be approaching 1.

53. (a)   We have the following compositions:

$(i \circ i)(x) = i(x)$ $(a \circ i)(x) = a(x)$

$(i \circ a)(x) = i(-x) = -x = a(x)$ $(a \circ a)(x) = a(-x) = x = i(x)$

$(i \circ b)(x) = i\left(\dfrac{1}{x}\right) = \dfrac{1}{x} = b(x)$ $(a \circ b)(x) = a\left(\dfrac{1}{x}\right) = -\dfrac{1}{x} = c(x)$

$(i \circ c)(x) = i\left(-\dfrac{1}{x}\right) = -\dfrac{1}{x} = c(x)$ $(a \circ c)(x) = a\left(-\dfrac{1}{x}\right) = \dfrac{1}{x} = b(x)$

$(b \circ i)(x) = b(x)$ $(c \circ i)(x) = c(x)$

$(b \circ a)(x) = b(-x) = -\dfrac{1}{x} = c(x)$ $(c \circ a)(x) = c(-x) = \dfrac{1}{x} = b(x)$

$(b \circ b)(x) = b\left(\dfrac{1}{x}\right) = x = i(x)$ $(c \circ b)(x) = c\left(\dfrac{1}{x}\right) = -x = a(x)$

$(b \circ c)(x) = b\left(-\dfrac{1}{x}\right) = -x = a(x)$ $(c \circ c)(x) = c\left(-\dfrac{1}{x}\right) = x = i(x)$

So we have the composition table:

$\circ$	i	a	b	c
i	i	a	b	c
a	a	i	c	b
b	b	c	i	a
c	c	b	a	i

(b)   yes

(c) $a^2 = i;\ b^2 = i;\ c^2 = i;\ c^3 = c \circ c^2 = c \circ i = c$

(d)   Since $(a \circ b) \circ c = (c) \circ c = i$ and $a \circ (b \circ c) = a \circ (a) = i$, then $(a \circ b) \circ c = a \circ (b \circ c)$.

(e)   Since $i^2 = a^2 = b^2 = c^2 = i$, then the range of f is $\{i\}$.

(f)   We compute:
$$f(a \circ b) = f(c) = c^2 = i$$
$$[f(a)] \circ [f(b)] = a^2 \circ b^2 = i \circ i = i,$$
So $f(a \circ b) = [f(a)] \circ [f(b)]$.

(g)   We compute:

$$f(a \circ b \circ c) = f(a \circ a) = f(i) = i^2 = i$$
$$[f(a)] \circ [f(b)] \circ [f(c)] = a^2 \circ b^2 \circ c^2 = i \circ (i \circ i) = i,$$

So $f(a \circ b \circ c) = [f(a)] \circ [f(b)] \circ [f(c)]$.

Optional TI-81 Graphing Calculator Exercises for Section 3.4

1.   Using the illustrated procedure, the completed table is:

x	0.1	0.2	0.3
$f(g(x))$	194.672	148.176	109.744

3.   (a)   Entering $Y_1 = -x^2 + 5x + 1$ and $Y_2 = 3x^2 - x + 4$, we complete the table:

x	7	8	9
$f(g(x))$	524	1614	3714

(b)   Entering $Y_1 = 3x^2 - x + 4$ and $Y_2 = -x^2 + 5x + 1$, we complete the table:

x	7	8	9
$g(f(x))$	-20015	-34403	-55453

5.   (a)   The successive iterates are 2.25, 2.236111111, and 2.236067978. This last iterate is the target value, which is reached in three iterations.

(b)   The successive iterates are 2.333333333, 2.238095238, 2.236068896, and 2.236067978. This last iterate is the target value, which is reached in four iterations.

(c)   The successive iterates are 12.6, 6.498412698, 3.63391568, 2.504920969, 2.250495964, 2.236114227, and 2.236067978. This last iterate is the target value, which is reached in seven iterations.

(d)   The results are equal to $\sqrt{5}$.

7.   (a)   The successive iterates are 3.733333333, 3.206363379, 3.110265742, 3.107235463, and 3.107232506. This last iterate is the target value, which is reached in five iterations.

(b)   The successive iterates are 16.6826667, 11.15770876, 7.518797361, 5.189421557, 3.830946132, 3.235341778, 3.112238272, 3.107240553, and 3.107232506. This last iterate is the target value, which is reached in nine iterations.

(c)   The results are equal to $\sqrt[3]{30}$.

3.5 Inverse Functions

1. (a) We must show that $f[g(x)] = x$ and $g[f(x)] = x$:

$$f[g(x)] = f\left(\tfrac{x}{3}\right) = 3\left(\tfrac{x}{3}\right) = x$$
$$g[f(x)] = g(3x) = \tfrac{3x}{3} = x$$

So $f(x)$ and $g(x)$ are inverse functions.

(b) We must show that $f[g(x)] = x$ and $g[f(x)] = x$:

$$f[g(x)] = f\left(\tfrac{x+1}{4}\right) = 4\left(\tfrac{x+1}{4}\right) - 1 = x + 1 - 1 = x$$
$$g[f(x)] = g(4x - 1) = \tfrac{(4x-1)+1}{4} = \tfrac{4x}{4} = x$$

So $f(x)$ and $g(x)$ are inverse functions.

(c) We must show that $g[h(x)] = x$ and $h[g(x)] = x$:

$$g[h(x)] = g\left(x^2\right) = \sqrt{x^2} = x, \quad \text{since } x \geq 0$$
$$h[g(x)] = h(\sqrt{x}) = \left(\sqrt{x}\right)^2 = x$$

So $g(x)$ and $h(x)$ are inverse functions.

3. (a) Since $f[f^{-1}(x)] = x$ for all x in the domain of f^{-1}, then $f[f^{-1}(4)] = 4$ since the domain of f^{-1} is $(-\infty, \infty)$.

(b) Since $f^{-1}[f(x)] = x$ as long as $f(x)$ is in the domain of f^{-1}, then $f^{-1}[f(-1)] = -1$ since the domain of f^{-1} is $(-\infty, \infty)$.

(c) Since $f[f^{-1}(x)] = x$ for all x in the domain of f^{-1}, then $f[f^{-1}(\sqrt{2})] = \sqrt{2}$ since the domain of f^{-1} is $(-\infty, \infty)$.

(d) Since $f[f^{-1}(x)] = x$ for all x in the domain of f^{-1}, then $f[f^{-1}(t + 1)] = t + 1$ since the domain of f^{-1} is $(-\infty, \infty)$.

(e) We first compute $f(0)$:

$$f(0) = (0)^3 + 2(0) + 1 = 0 + 0 + 1 = 1$$

Since the inverse function interchanges inputs and outputs, and $f(0) = 1$, then $f^{-1}(1) = 0$.

(f) We first compute $f(-1)$:

$$f(-1) = (-1)^3 + 2(-1) + 1 = -1 - 2 + 1 = -2$$

Since the inverse function interchanges inputs and outputs, and $f(-1) = -2$, then $f^{-1}(-2) = -1$.

5. (a) Let $y = 3x - 1$. We switch the roles of x and y and solve the resulting equation for y:

$$x = 3y - 1$$
$$3y = x + 1$$
$$y = \frac{x+1}{3}$$

So the inverse is $f^{-1}(x) = \frac{x+1}{3}$.

(b) We verify that $f[f^{-1}(x)] = x$ and $f^{-1}[f(x)] = x$:

$$f\left[f^{-1}(x)\right] = f\left(\tfrac{x+1}{3}\right) = 3\left(\tfrac{x+1}{3}\right) - 1 = x + 1 - 1 = x$$
$$f^{-1}\left[f(x)\right] = f^{-1}(3x - 1) = \frac{(3x - 1) + 1}{3} = \frac{3x}{3} = x$$

(c) The graphs of each line are given below. Note the symmetry of the two lines about the line $y = x$:

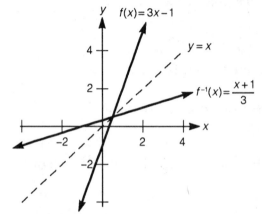

7. (a) Let $y = \sqrt{x-1}$. We switch the roles of x and y and solve the resulting equation for y:

$$x = \sqrt{y - 1}$$
$$x^2 = y - 1$$
$$y = x^2 + 1$$

So the inverse is $f^{-1}(x) = x^2 + 1$ for $x \geq 0$.

(b) We verify that $f[f^{-1}(x)] = x$ and $f^{-1}[f(x)] = x$:

$$f\left[f^{-1}(x)\right] = f\left(x^2 + 1\right) = \sqrt{\left(x^2 + 1\right) - 1} = \sqrt{x^2} = x \quad \text{(since } x \geq 0)$$
$$f^{-1}\left[f(x)\right] = f^{-1}\left(\sqrt{x - 1}\right) = \left(\sqrt{x - 1}\right)^2 + 1 = x - 1 + 1 = x$$

(c)   The graphs of each curve are given below. Note the symmetry of the curves about the line $y = x$:

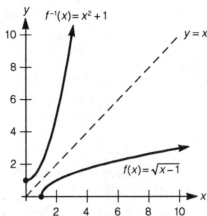

9.   (a)   Since the denominator is 0 when $x = 3$, the domain of f is the set of all real numbers except 3, or $(-\infty, 3) \cup (3, \infty)$. To find the range of f, we first solve for x:

$$y = \frac{x+2}{x-3}$$
$$y(x-3) = x+2$$
$$yx - 3y = x+2$$
$$yx - x = 3y+2$$
$$x(y-1) = 3y+2$$
$$x = \frac{3y+2}{y-1}$$

Since the denominator is 0 when $y = 1$, the range of f is the set of all real numbers except 1, or $(-\infty, 1) \cup (1, \infty)$.

(b)   Let $y = \dfrac{x+2}{x-3}$. We switch the roles of x and y and solve the resulting equation for y:

$$x = \frac{y+2}{y-3}$$
$$x(y-3) = y+2$$
$$xy - 3x = y+2$$
$$xy - y = 3x+2$$
$$y(x-1) = 3x+2$$
$$y = \frac{3x+2}{x-1}$$

So the inverse is $f^{-1}(x) = \dfrac{3x+2}{x-1}$.

(c)   Since the denominator is 0 when $x = 1$, the domain of f^{-1} is the set of all real numbers except 1, or $(-\infty, 1) \cup (1, \infty)$. To find the range of f^{-1}, we first solve for x:

$$y = \frac{3x+2}{x-1}$$
$$y(x-1) = 3x+2$$
$$yx - y = 3x + 2$$
$$yx - 3x = y + 2$$
$$x(y-3) = y + 2$$
$$x = \frac{y+2}{y-3}$$

Since the denominator is 0 when $y = 3$, the range of f^{-1} is the set of all real numbers except 3, or $(-\infty, 3) \cup (3, \infty)$. We observe that the domain of f is equal to the range of f^{-1}, and that the range of f is equal to the domain of f^{-1}.

11.   Let $y = 2x^3 + 1$. We switch the roles of x and y and solve the resulting equation for y:

$$x = 2y^3 + 1$$
$$2y^3 = x - 1$$
$$y^3 = \frac{x-1}{2}$$
$$y = \sqrt[3]{\frac{x-1}{2}}$$

So the inverse is $f^{-1}(x) = \sqrt[3]{\frac{x-1}{2}}$.

13.   We compute $f\left[f^{-1}(x)\right]$ and $f^{-1}\left[f(x)\right]$:

$$f\left[f^{-1}(x)\right] = f\left(\frac{x+4}{3}\right) = 3\left(\frac{x+4}{3}\right) - 4 = x + 4 - 4 = x$$
$$f^{-1}\left[f(x)\right] = f^{-1}(3x - 4) = \frac{(3x-4)+4}{3} = \frac{3x}{3} = x$$

This verfies that f and f^{-1} are inverse functions. We sketch the graph:

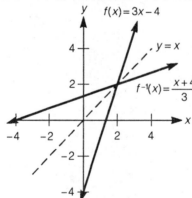

15. (a) Let $y = (x - 3)^3 - 1$. We switch the roles of x and y and solve the resulting equation for y:

$$x = (y - 3)^3 - 1$$
$$(y - 3)^3 = x + 1$$
$$y - 3 = \sqrt[3]{x + 1}$$
$$y = \sqrt[3]{x + 1} + 3$$

So the inverse is $f^{-1}(x) = \sqrt[3]{x + 1} + 3$.

(b) Note that $f(x)$ and $f^{-1}(x)$ are symmetric about the line $y = x$:

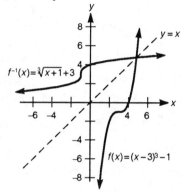

17. (a) The graph of $y = g^{-1}(x)$ is a reflection of $g(x)$ across the line $y = x$:

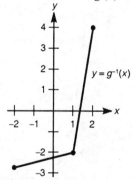

(b) The graph of $y = g^{-1}(x) - 1$ is a displacement of $g^{-1}(x)$ down 1 unit:

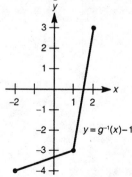

(c)   The graph of $y = g^{-1}(x - 1)$ is a displacement of $g^{-1}(x)$ to the right 1 unit:

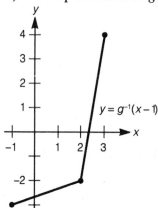

(d)   The graph of $y = g^{-1}(-x)$ is a reflection of $g^{-1}(x)$ across the y-axis:

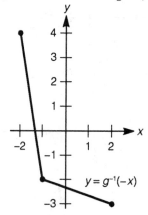

(e)   The graph of $y = -g^{-1}(x)$ is a reflection of $g^{-1}(x)$ across the x-axis:

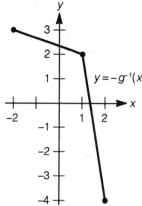

(f) The graph of $y = -g^{-1}(-x)$ is a reflection of $g^{-1}(x)$ across the y-axis and across the x-axis:

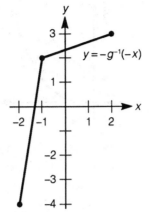

19. The graph of $y = x^2 + 1$ fails the horizontal line test, so it is not one-to-one:

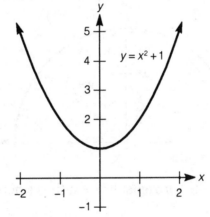

21. The graph of $y = \dfrac{1}{x}$ passes the horizontal line test, so it is one-to-one:

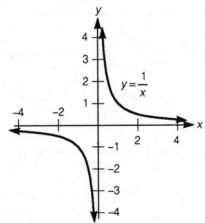

23. The graph of $y = x^3$ passes the horizontal line test, so it is one-to-one:

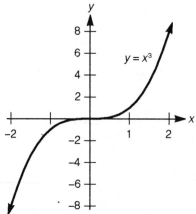

25. The graph of $y = \sqrt{1 - x^2}$ fails the horizontal line test, so it is not one-to-one:

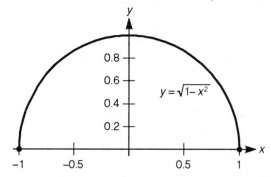

27. The graph of $g(x) = 5$ fails the horizontal line test (it <u>is</u> a horizontal line), so it is not one-to-one:

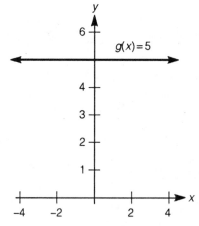

29. The graph of $f(x)$ passes the horizontal line test, so it is one-to-one:

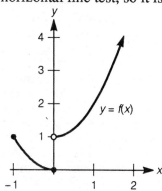

31. We find $f[f(x)]$:
$$f[f(x)] = f\left(\tfrac{3x-2}{5x-3}\right)$$
$$= \frac{3\left(\tfrac{3x-2}{5x-3}\right) - 2}{5\left(\tfrac{3x-2}{5x-3}\right) - 3}$$
$$= \frac{3(3x-2) - 2(5x-3)}{5(3x-2) - 3(5x-3)}$$
$$= \frac{9x - 6 - 10x + 6}{15x - 10 - 15x + 9}$$
$$= \frac{-x}{-1}$$
$$= x$$
Thus $f^{-1}(x) = f(x)$.

33. (a) We solve the equation:
$$7 + f^{-1}(x-1) = 9$$
$$f^{-1}(x-1) = 2$$
$$x - 1 = f(2)$$
$$x - 1 = 6$$
$$x = 7$$

(b) We solve the equation:
$$4 + f(x+3) = -3$$
$$f(x+3) = -7$$
$$x + 3 = f^{-1}(-7)$$
$$x + 3 = 0$$
$$x = -3$$

35. We solve the equation:

$$f^{-1}\left(\frac{t+1}{t-2}\right) = 12$$

$$\frac{t+1}{t-2} = f(12)$$

$$\frac{t+1}{t-2} = 13$$

$$t+1 = 13t - 26$$

$$27 = 12t$$

$$t = \tfrac{27}{12} = \tfrac{9}{4}$$

37. (a) Let $y = \sqrt{x}$. We switch the roles of x and y and solve the resulting equation for y:

$$x = \sqrt{y}$$

$$x^2 = y$$

So the inverse function is $f^{-1}(x) = x^2$. Since $x = \sqrt{y}$, then $x \geq 0$. So the domain of $f^{-1}(x)$ is $[0, \infty)$.

(b) (i)   Since $2 = \sqrt{4}$, then $(4, 2)$ lies on the graph of f.

(ii)   Since $(4, 2)$ lies on the graph of f, then $(2, 4)$ lies on the graph of f^{-1}.

(iii)   Since $\sqrt{5} = \sqrt{5}$, then $(5, \sqrt{5})$ lies on the graph of f.

(iv)   Since $(5, \sqrt{5})$ lies on the graph of f, then $(\sqrt{5}, 5)$ lies on the graph of f^{-1}.

(v)   Since $f(a) = f(a)$, then $(a, f(a))$ lies on the graph of f.

(vi)   Since $(a, f(a))$ lies on the graph of f, then $(f(a), a)$ lies on the graph of f^{-1}.

(vii)   Since the graph of f^{-1} is all points $(x, f^{-1}(x))$, then $(b, f^{-1}(b))$ lies on the graph of f^{-1}.

(viii)   Since $b = f[f^{-1}(b)]$, then $(f^{-1}(b), b)$ lies on the graph of f.

39. The coordinates of A are $(a, f(a))$. Since B lies on the line $y = x$, and its y-coordinate is $f(a)$, the coordinates of B are $(f(a), f(a))$. The y-coordinate of C is also $f(a)$, and so its x-coordinate is $f(f(a))$, since $f^{-1}(f(f(a))) = f(a)$, so the coordinates of C are $(f(f(a)), f(a))$. Since D is the reflection of C across $y = x$, the coordinates of D are $(f(a), f(f(a)))$.

Optional TI-81 Graphing Calculator Exercises for Section 3.5

1. We draw the graphs, noting the symmetry about $y = x$:

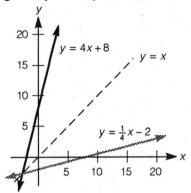

3. We draw the graphs, noting the symmetry about $y = x$:

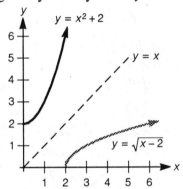

5. We draw the graphs, noting the symmetry about $y = x$:

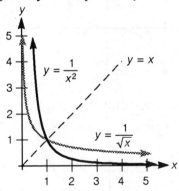

7.   The function is not one-to-one. We draw the graph:

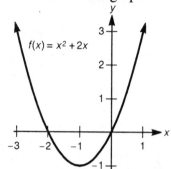

9.   The function is not one-to-one. We draw the graph:

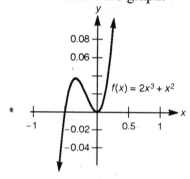

11. (a)   The function is one-to-one. We draw the graph:

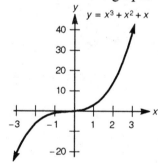

(b)   The function is one-to-one. We draw the graph:

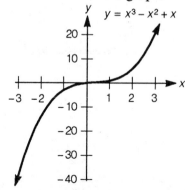

(c)   The function is not one-to-one. We draw the graph:

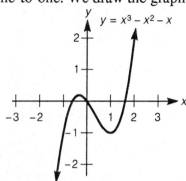

13. We first find $F^{-1}(x)$. Let $y = F(x)$, then switch x and y and solve the resulting equation for y:

$$x = \sqrt{1 - y^2}$$
$$x^2 = 1 - y^2$$
$$y^2 = 1 - x^2$$
$$y = -\sqrt{1 - x^2}$$

Note that $y \le 0$ since $x \le 0$ in the original function $F(x)$. So $F^{-1}(x) = -\sqrt{1 - x^2}$ with domain $0 \le x \le 1$. We draw the graph using Zoom-5 to observe the symmetry:

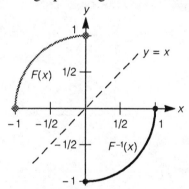

15. (a)   We graph the function, noting that it passes the horizontal line test:

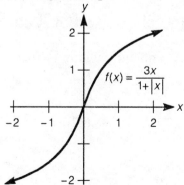

(b)   If $x \geq 0$, then $y = \frac{3x}{1+x}$ since $|x| = x$. Switching the roles of x and y, and solving the resulting equation for y:

$$\frac{3y}{1+y} = x$$
$$3y = x + xy$$
$$y(3 - x) = x$$
$$y = \frac{x}{3-x} \quad \text{(for } x < 3\text{)}$$

If $x < 0$, then $y = \frac{3x}{1-x}$ since $|x| = -x$. Switching the roles of x and y, and solving the resulting equation for y:

$$\frac{3y}{1-y} = x$$
$$3y = x - xy$$
$$y(3 + x) = x$$
$$y = \frac{x}{3+x} \quad \text{(for } x > -3\text{)}$$

Thus the inverse function is:

$$f^{-1}(x) = \begin{cases} x/(3-x) & \text{if } 0 \leq x < 3 \\ x/(3+x) & \text{if } -3 < x < 0 \end{cases}$$

Now graph both functions, noting the symmetry about $y = x$:

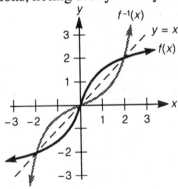

Chapter Three Review Exercises

1.   (a)   We must guarantee that the quantity within the radical is non-negative:

$$15 - 5x \geq 0$$
$$-5x \geq -15$$
$$x \leq 3$$

So the domain is $(-\infty, 3]$.

(b)   We solve $y = \frac{3+x}{2x-5}$ for x:
$$y(2x-5) = 3+x$$
$$2xy - 5y = 3+x$$
$$2xy - x = 5y+3$$
$$x(2y-1) = 5y+3$$
$$x = \frac{5y+3}{2y-1}$$

So the range is all real numbers except $\frac{1}{2}$, or $\left(-\infty,\frac{1}{2}\right)\cup\left(\frac{1}{2},\infty\right)$.

3.   Yes. Let $f(x) = ax + b$ and $g(x) = cx + d$, then:
$$(f \circ g)(x) = f(cx+d) = a(cx+d)+b = acx+(ad+b)$$
So $(f \circ g)(x)$ is a linear function.

5.   (a)   We compute the difference quotient:
$$\frac{F(x)-F(a)}{x-a} = \frac{\frac{1}{x}-\frac{1}{a}}{x-a}\cdot\frac{ax}{ax} = \frac{a-x}{ax(x-a)} = -\frac{1}{ax}$$

(b)   We first compute:
$$g(x+h) = (x+h)-2(x+h)^2 = x+h-2x^2-4xh-2h^2$$
So we have:
$$\frac{g(x+h)-g(x)}{h} = \frac{\left(x+h-2x^2-4xh-2h^2\right)-\left(x-2x^2\right)}{h}$$
$$= \frac{x+h-2x^2-4xh-2h^2-x+2x^2}{h}$$
$$= \frac{h-4xh-2h^2}{h}$$
$$= 1-4x-2h$$

7.   (a)   We switch the roles of x and y and solve the resulting equation for y:
$$x = \frac{1-5y}{3y}$$
$$3xy = 1-5y$$
$$3xy+5y = 1$$
$$y(3x+5) = 1$$
$$y = \frac{1}{3x+5}$$
So $g^{-1}(x) = \frac{1}{3x+5}$.

(b)   We graph f^{-1}:

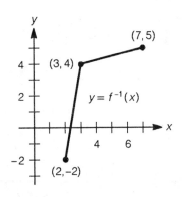

9.   (a)   We displace $y = |x|$ two units to the left and 3 units down. The x-intercepts are -5 and 1, and the y-intercept is -1:

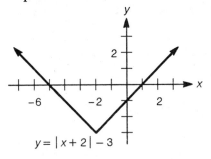

(b)   We displace $y = \frac{1}{x}$ two units to the left and 1 unit down. The x-intercept is -1 and the y-intercept is $-\frac{1}{2}$:

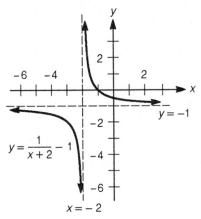

11.   (a)   We compute $f(-1)$:
$$f(-1) = 3(-1)^2 - 2(-1) = 3 + 2 = 5$$

(b)   We compute $f(1-\sqrt{2})$:

$$\begin{aligned}
f(1-\sqrt{2}) &= 3(1-\sqrt{2})^2 - 2(1-\sqrt{2}) \\
&= 3(1-2\sqrt{2}+2) - 2 + 2\sqrt{2} \\
&= 9 - 6\sqrt{2} - 2 + 2\sqrt{2} \\
&= 7 - 4\sqrt{2}
\end{aligned}$$

13.   The slope is given by:

$$m = \frac{(5+h)^2 - 25}{5+h-5} = \frac{25+10h+h^2-25}{h} = \frac{10h+h^2}{h} = 10+h$$

Using functional notation, we have $m(h) = 10 + h$.

15.   The graph of $y = f(-x)$ will result in a reflection across the y-axis:

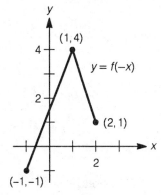

17.   (a)   Since the graph of $y = f(x) + 1$ is obtained by translating the graph of $y = f(x)$ up one unit, then the point (a,b) is translated to the point $(a, b+1)$. Thus E is the correct answer.

(b)   Since the graph of $y = f(x+1)$ is obtained by translating the graph of $y = f(x)$ to the left one unit, then the point (a,b) is translated to the point $(a-1,b)$. Thus C is the correct answer.

(c)   Since the graph of $y = f(x-1) + 1$ is obtained by translating the graph of $y = f(x)$ to the right one unit and up one unit, then the point (a,b) is translated to the point $(a+1, b+1)$. Thus L is the correct answer.

(d)   Since the graph of $y = f(-x)$ is obtained by reflecting the graph of $y = f(x)$ across the y-axis, then the point (a,b) is reflected to the point $(-a,b)$. Thus A is the correct answer.

(e)   Since the graph of $y = -f(x)$ is obtained by reflecting the graph of $y = f(x)$ across the x-axis, then the point (a,b) is reflected to the point $(a,-b)$. Thus J is the correct answer.

(f)   Since the graph of $y = -f(-x)$ is obtained by reflecting the graph of $y = f(x)$ across the x-axis and across the y-axis, then the point (a, b) is reflected to the point $(-a, -b)$. Thus G is the correct answer.

(g)   Since the graph of $y = f^{-1}(x)$ is obtained by reflecting the graph of $y = f(x)$ across the line $y = x$, then the point (a, b) is reflected to the point (b, a). Thus B is the correct answer.

(h)   Since the graph of $y = f^{-1}(x) + 1$ is obtained by reflecting the graph of $y = f(x)$ across the line $y = x$, then translating up one unit, then the point (a, b) is reflected to the point (b, a) then translated to the point $(b, a + 1)$. Thus M is the correct answer.

(i)   Since the graph of $y = f^{-1}(x - 1)$ is obtained by reflecting the graph of $y = f(x)$ across the line $y = x$, then translating to the right one unit, then the point (a, b) is reflected to the point (b, a) then translated to the point $(b + 1, a)$. Thus K is the correct answer.

(j)   Since the graph of $y = f^{-1}(-x) + 1$ is obtained by reflecting the graph of $y = f(x)$ across the line $y = x$, then across the y-axis, then translating up one unit, then the point (a, b) is reflected to the point (b, a), then reflected to the point $(-b, a)$, then translated to the point $(-b, a + 1)$. Thus D is the correct answer.

(k)   Since the graph of $y = -f^{-1}(x)$ is obtained by reflecting the graph of $y = f(x)$ across the line $y = x$ then across the x-axis, then the point (a, b) is reflected to the point (b, a) then reflected to the point $(b, -a)$. Thus I is the correct answer.

(l)   Since the graph of $y = -f^{-1}(-x) + 1$ is obtained by reflecting the graph of $y = f(x)$ across the line $y = x$, the x-axis, and the y-axis, then translating up one unit, then the point (a, b) is reflected to the point (b, a), then reflected to the point $(-b, a)$ and then $(-b, -a)$, and finally translated to the point $(-b, -a + 1)$. Thus H is the correct answer.

(m)   Since the graph of $y = 1 - f^{-1}(x)$ is obtained by reflecting the graph of $y = f(x)$ across the line $y = x$ and across the x-axis, then translated up one unit, then the point (a, b) is reflected to the point (b, a), then reflected to the point $(b, -a)$, then translated to the point $(b, -a + 1)$. Thus N is the correct answer.

(n)   Since the graph of $y = f(1 - x)$ is obtained by reflecting the graph of $y = f(x)$ across the y-axis, then translating to the right one unit, then the point (a, b) is reflected to the point $(-a, b)$ then translated to the point $(-a + 1, b)$. Thus F is the correct answer.

19. To find the x-intercepts, we let $y = 0$:

$$-(x-1)^2 + 2 = 0$$
$$-(x-1)^2 = -2$$
$$x - 1 = \pm\sqrt{2}$$
$$x = 1 \pm \sqrt{2}$$

To find the y-intercept, we let $x = 0$:

$$y = -(0-1)^2 + 2 = -1 + 2 = 1$$

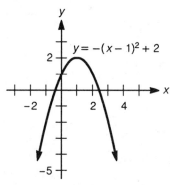

21. To find the x-intercepts, we let $y = 0$:

$$\frac{1}{x+1} = 0$$
$$1 = 0$$

Clearly this is impossible, so there are no x-intercepts. To find the y-intercept, we let $x = 0$:

$$y = \frac{1}{0+1} = 1$$

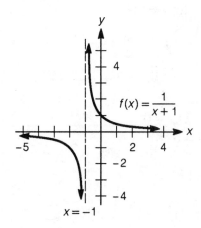

23. To find the x-intercepts, we let $y = 0$:

$$|x+3| = 0$$
$$x = -3$$

To find the y-intercept, we let $x = 0$:
$$y = |0 + 3| = 3$$

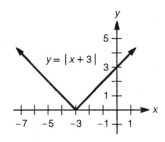

25.   To find the x-intercepts, we let $y = 0$:
$$\sqrt{1 - x^2} = 0$$
$$1 - x^2 = 0$$
$$-x^2 = -1$$
$$x = \pm 1$$
To find the y-intercept, we let $x = 0$:
$$y = \sqrt{1 - 0} = 1$$

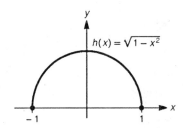

27.   To find the x-intercepts, we let $y = 0$:
$$1 - (x + 1)^3 = 0$$
$$-(x + 1)^3 = -1$$
$$(x + 1)^3 = 1$$
$$x + 1 = 1$$
$$x = 0$$
To find the y-intercept, we let $x = 0$:
$$y = 1 - (0 + 1)^3 = 1 - 1 = 0$$

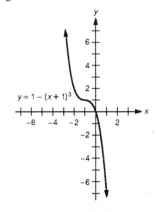

29. The x-intercept is 0 and the y-intercept is 0:

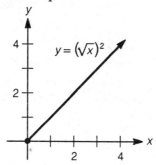

31. We first compute $(f \circ g)(x)$:

$$(f \circ g)(x) = f(\sqrt{x-1}) = -(\sqrt{x-1})^2 = 1 - x \text{ for } x \geq 1$$

The x-intercept is 1 and there is no y-intercept:

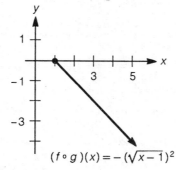

33. The x-intercepts are -1 and 0, and the y-intercept is 0:

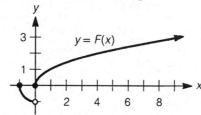

35. There are no x- or y-intercepts:

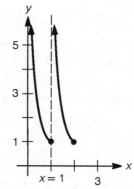

37. To find the inverse function, we switch the roles of x and y and solve the resulting equation for y:

$$x = \frac{y+1}{2}$$
$$2x = y+1$$
$$2x-1 = y$$

So $f^{-1}(x) = 2x - 1$. To find the x-intercepts, we let $y = 0$:

$$2x-1 = 0$$
$$2x = 1$$
$$x = \tfrac{1}{2}$$

To find the y-intercept, we let $x = 0$:

$$y = 2(0) - 1 = -1$$

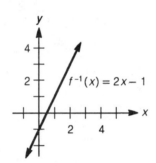

39. We compute $\left(f \circ f^{-1}\right)(x) = x$ for $x \geq 0$. The x-intercept is 0 and the y-intercept is 0:

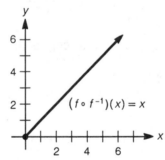

41. We must make sure that $x^2 - 9 \neq 0$. We find the points to exclude:

$$x^2 - 9 = 0$$
$$x^2 = 9$$
$$x = \pm 3$$

So the domain is all real numbers except 3 and –3, or $(-\infty, -3) \cup (-3, 3) \cup (3, \infty)$.

43. We must be sure that the quantity inside the radical is non-negative:

$$8 - 2x \geq 0$$
$$-2x \geq -8$$
$$x \leq 4$$

So the domain is $(-\infty, 4]$.

45. We must be sure that the quantity inside the radical is non-negative, so $|2-5x| \geq 0$. But this is true for all real numbers, so the domain is all real numbers, or $(-\infty, \infty)$.

47. We must be sure that the quantity inside the radical is non-negative:
$$x^2 - 2x - 3 \geq 0$$
$$(x-3)(x+1) \geq 0$$
The key numbers are -1 and 3. We draw the sign chart:

Interval	Test Number	$x-3$	$x+1$	$(x-3)(x+1)$
$(-\infty, -1)$	-2	neg.	neg.	pos.
$(-1, 3)$	0	neg.	pos.	neg.
$(3, \infty)$	4	pos.	pos.	pos.

So the product is positive on the intervals $(-\infty, -1) \cup (3, \infty)$, and it is zero when $x = -1$ and $x = 3$. So the domain is $(-\infty, -1] \cup [3, \infty)$.

49. We must be sure that $x \neq 0$. So the domain is all real numbers except 0, or $(-\infty, 0) \cup (0, \infty)$.

51. We solve for x:
$$y = \frac{x+4}{3x-1}$$
$$y(3x-1) = x+4$$
$$3xy - y = x+4$$
$$3xy - x = y+4$$
$$x(3y-1) = y+4$$
$$x = \frac{y+4}{3y-1}$$
Now $3y - 1 = 0$ when $y = \frac{1}{3}$, so the range is all real numbers except $\frac{1}{3}$, or $\left(-\infty, \frac{1}{3}\right) \cup \left(\frac{1}{3}, \infty\right)$.

53. We first compute $(f \circ g)(x)$:
$$(f \circ g)(x) = f(3x+4) = \frac{1}{3x+4}$$
We solve $y = \frac{1}{3x+4}$ for x:
$$y = \frac{1}{3x+4}$$
$$y(3x+4) = 1$$
$$3xy + 4y = 1$$
$$3xy = 1 - 4y$$
$$x = \frac{1-4y}{3y}$$
Now $3y = 0$ when $y = 0$, so the range is all real numbers except 0, or $(-\infty, 0) \cup (0, \infty)$.

55. Since the range of f^{-1} is the domain of f, we must exclude the values of x where $3x - 6 = 0$, or $x = 2$. So the range is all real numbers except 2, or $(-\infty, 2) \cup (2, \infty)$.

57. We have $a(x) = (f \circ g)(x)$, since:
$$(f \circ g)(x) = f[g(x)] = f(x - 1) = \frac{1}{x - 1}$$

59. We have $c(x) = (G \circ g)(x)$, since:
$$(G \circ g)(x) = G[g(x)] = G(x - 1) = \sqrt{x - 1}$$

61. We have $A(x) = (g \circ f \circ G)(x)$, since:
$$(g \circ f \circ G)(x) = (g \circ f)(\sqrt{x}) = g\left(\frac{1}{\sqrt{x}}\right) = \frac{1}{\sqrt{x}} - 1$$

63. We have $C(x) = (g \circ G \circ G)(x)$, since:
$$(g \circ G \circ G)(x) = (g \circ G)(\sqrt{x}) = g\left(\sqrt{\sqrt{x}}\right) = g\left(\sqrt[4]{x}\right) = \sqrt[4]{x} - 1$$

65. Using $f(x)$, we compute $f(-3)$:
$$f(-3) = (-3)^2 - (-3) = 9 + 3 = 12$$

67. Using $F(x)$, we compute $F\left(\frac{3}{4}\right)$:
$$F\left(\tfrac{3}{4}\right) = \frac{\frac{3}{4} - 3}{\frac{3}{4} + 4} = \frac{3 - 3(4)}{3 + 4(4)} = \frac{3 - 12}{3 + 16} = -\frac{9}{19}$$

69. Using $f(x)$, we compute $f(-t)$:
$$f(-t) = (-t)^2 - (-t) = t^2 + t$$

71. Using $f(x)$, we compute $f(x - 2)$:
$$f(x - 2) = (x - 2)^2 - (x - 2) = x^2 - 4x + 4 - x + 2 = x^2 - 5x + 6$$

73. Using $g(x)$, we first compute $g(2)$ and $g(0)$:
$$g(2) = 1 - 2(2) = 1 - 4 = -3$$
$$g(0) = 1 - 2(0) = 1 - 0 = 1$$
So $g(2) - g(0) = -3 - 1 = -4$

75. Using $f(x)$, we first compute $f(1)$ and $f(3)$:
$$f(1) = (1)^2 - 1 = 1 - 1 = 0$$
$$f(3) = (3)^2 - 3 = 9 - 3 = 6$$
So $|f(1) - f(3)| = |0 - 6| = |-6| = 6$.

77. Using $f(x)$, we compute $f\left(x^2\right)$:

$$f\left(x^2\right)=\left(x^2\right)^2-x^2=x^4-x^2$$

79. We compute $[f(x)][g(x)]$:

$$[f(x)][g(x)]=\left(x^2-x\right)(1-2x)=x^2-2x^3-x+2x^2=-2x^3+3x^2-x$$

81. We compute $f[g(x)]$:

$$f[g(x)]=f(1-2x)=(1-2x)^2-(1-2x)=1-4x+4x^2-1+2x=4x^2-2x$$

83. We compute $(g\circ f)(x)$:

$$(g\circ f)(x)=g[f(x)]=g\left(x^2-x\right)=1-2\left(x^2-x\right)=2-2x^2+2x=-2x^2+2x+1$$

85. We compute $(F\circ g)(x)$:

$$(F\circ g)(x)=F[g(x)]=F(1-2x)=\frac{(1-2x)-3}{(1-2x)+4}=\frac{-2x-2}{-2x+5}\text{ or }\frac{2x+2}{2x-5}$$

87. We find the difference quotient:

$$\frac{f(x+h)-f(x)}{h}=\frac{2xh+h^2-h}{h}=\frac{h(2x+h-1)}{h}=2x+h-1$$

89. Let $y=\dfrac{x-3}{x+4}$. To find $F^{-1}(x)$, we switch the roles of x and y and solve the resulting equation for y:

$$x=\frac{y-3}{y+4}$$
$$x(y+4)=y-3$$
$$xy+4x=y-3$$
$$xy-y=-4x-3$$
$$y(x-1)=-4x-3$$
$$y=\frac{-4x-3}{x-1}$$

So $F^{-1}(x)=\dfrac{-4x-3}{x-1}=\dfrac{4x+3}{1-x}$.

91. $F^{-1}[F(x)]=x$ for $x\neq 4$, by definition of $F^{-1}(x)$.

93. The value is x, since $\left(g\circ g^{-1}\right)(x)=x$ for all x.

95. Since $g^{-1}(x) = \dfrac{1-x}{2}$, we have:

$$g^{-1}(-x) = \dfrac{1-(-x)}{2} = \dfrac{1+x}{2}$$

97. Since $\frac{22}{7}$ is in the domain of $F(x)$, then $F^{-1}\left[F\left(\frac{22}{7}\right)\right] = \frac{22}{7}$.

99. Since $f(0) = -2$, it is negative.

101. Since the point $(-3, -1)$ lies on the graph, then $f(-3) = -1$.

103. Since $f(0) = -2$ and $f(8) = -1$, then $f(0) - f(8) = -2 - (-1) = -2 + 1 = -1$.

105. The coordinates of the turning points are $(0, -2)$ and $(5, 1)$.

107. $f(x)$ is decreasing on the intervals $[-6, 0]$ and $[5, 8]$.

109. Since $|x| \le 2$ corresponds to the interval $-2 \le x \le 2$, the largest value of $f(x)$ is 0, occurring at $x = 2$.

111. Since $f(x)$ is not a one-to-one function (it does not pass the horizontal line test) then it does not possess an inverse function.

113. From the graph, we note that $f(x) = g(x)$ when $x = 4$.

115. (a) From the graph, we note that $f(x) = 0$ when $x = 10$.
(b) From the graph, we note that $g(x) = 0$ when $x = 0$.

117. (a) $(f + g)(8) = f(8) + g(8) = 1 + 4 = 5$
(b) $(f - g)(8) = f(8) - g(8) = 1 - 4 = -3$
(c) $fg(8) = f(8) \cdot g(8) = 1 \cdot 4 = 4$
(d) $(f / g)(x) = \dfrac{f(8)}{g(8)} = \frac{1}{4}$

119. We compute each value:
$$(f \circ f)(10) = f[f(10)] = f(0) = 5$$
$$(g \circ g)(10) = g[g(10)] = g(3) = 1$$
So $(f \circ f)(10)$ is larger.

121. We note that $f(x) \ge 3$ for $0 \le x \le 4$, which is the interval $[0, 4]$.

123. The maximum point of $g(x)$ is $(6, 5)$, so the largest number in the range of g is 5.

125. We note that $g(x)$ is decreasing when $1 < x < 3$ and $6 < x < 10$, which are the intervals $(1,3) \cup (6,10)$.

127. We consider two cases:

If $4 < x < 5$, then $f(x) > f(5)$ and $x < 5$, thus $f(x) - f(5) > 0$ while $x - 5 < 0$. Thus $\dfrac{f(x) - f(5)}{x - 5} < 0$.

If $5 < x < 7$, then $f(x) < f(5)$ and $x > 5$, thus $f(x) - f(5) < 0$ while $x - 5 > 0$. Thus $\dfrac{f(x) - f(5)}{x - 5} < 0$.

Therefore, the quantity is negative for all x-values in the interval $(4,7)$.

Chapter Three Test

1. We must be sure the quantity inside the radical is non-negative:
$$x^2 - 5x - 6 \geq 0$$
$$(x - 6)(x + 1) \geq 0$$
The key numbers are -1 and 6. We draw the sign chart:

Interval	Test Number	$x - 6$	$x + 1$	$(x - 6)(x + 1)$
$(-\infty, -1)$	-2	neg.	neg.	pos.
$(-1, 6)$	0	neg.	pos.	neg.
$(6, \infty)$	8	pos.	pos.	pos.

The product is positive on the intervals $(-\infty, -1) \cup (6, \infty)$ and 0 at $x = -1$ and $x = 6$. So the domain is $(-\infty, -1] \cup [6, \infty)$.

2. Let $y = \frac{2x-8}{3x+5}$. We solve for x:
$$\frac{2x - 8}{3x + 5} = y$$
$$2x - 8 = 3yx + 5y$$
$$2x - 3yx = 5y + 8$$
$$x(2 - 3y) = 5y + 8$$
$$x = \frac{5y + 8}{2 - 3y}$$
So $y \neq \frac{2}{3}$. The range is $\left(-\infty, \frac{2}{3}\right) \cup \left(\frac{2}{3}, \infty\right)$.

3. (a) We compute $(f - g)(x)$:
$$(f - g)(x) = f(x) - g(x) = (2x^2 - 3x) - (2 - x) = 2x^2 - 3x - 2 + x = 2x^2 - 2x - 2$$

(b)   We compute $(f \circ g)(x)$:
$$\begin{aligned}(f \circ g)(x) &= f[g(x)] \\ &= f(2-x) \\ &= 2(2-x)^2 - 3(2-x) \\ &= 8 - 8x + 2x^2 - 6 + 3x \\ &= 2x^2 - 5x + 2 \end{aligned}$$

(c)   Using our result from part (b):
$$f[g(-4)] = 2(-4)^2 - 5(-4) + 2 = 32 + 20 + 2 = 54$$

4.   We compute the difference quotient:
$$\frac{f(t) - f(a)}{t-a} = \frac{\frac{2}{t} - \frac{2}{a}}{t-a} \cdot \frac{at}{at} = \frac{2a - 2t}{at(t-a)} = \frac{-2(t-a)}{at(t-a)} = -\frac{2}{at}$$

5.   We first compute $g(x + h)$:
$$g(x+h) = 2(x+h)^2 - 5(x+h) = 2x^2 + 4xh + 2h^2 - 5x - 5h$$
Now compute the difference quotient:
$$\begin{aligned}\frac{g(x+h) - g(x)}{h} &= \frac{2x^2 + 4xh + 2h^2 - 5x - 5h - 2x^2 + 5x}{h} \\ &= \frac{4xh + 2h^2 - 5h}{h} \\ &= 4x + 2h - 5\end{aligned}$$

6.   Let $y = \frac{-4x}{6x+1}$. We switch the roles of x and y, then solve the resulting equation for y:
$$\frac{-4y}{6y+1} = x$$
$$-4y = 6xy + x$$
$$-x = 6xy + 4y$$
$$-x = y(6x + 4)$$
$$\frac{-x}{6x+4} = y$$
So $g^{-1}(x) = -\frac{x}{6x+4}$.

7.   Since $f^{-1}(x)$ will be the line segment joining the points $(1, -3)$ and $(6, 5)$, then $y = -f^{-1}(x)$ will be the line segment joining the points $(1, 3)$ and $(6, -5)$. We compute the slope:
$$m = \frac{-5 - 3}{6 - 1} = -\frac{8}{5}$$

8.   (a)   The x-intercepts are 4 and 2, and the y-intercept is –2:

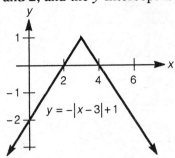

(b)   The x-intercept is $-\frac{5}{2}$ and the y-intercept is $-\frac{5}{3}$:

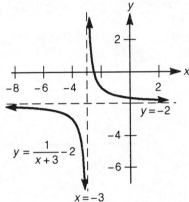

9.   (a)   The range of g is [–3, 1].
(b)   The turning point has coordinates (–2, 1).
(c)   The minimum value of g is –3, which occurs where x = 2.
(d)   The maximum value of g is 1, occurring at x = –2.
(e)   For –2 < x < 2, g is decreasing. So g is decreasing on (–2, 2).
(f)   On the interval [–4, 2], we have:
$$\frac{\Delta g}{\Delta x} = \frac{-3-(-1)}{2-(-4)} = \frac{-2}{6} = -\frac{1}{3}$$

10.   (a)   We compute $f\left(-\frac{3}{2}\right)$:
$$f\left(-\tfrac{3}{2}\right) = \left(-\tfrac{3}{2}\right)^{2} - 3\left(-\tfrac{3}{2}\right) - 1 = \tfrac{9}{4} + \tfrac{9}{2} - 1 = \tfrac{23}{4}$$

(b)   We compute $f\left(\sqrt{3}-2\right)$:
$$\begin{aligned}
f\left(\sqrt{3}-2\right) &= \left(\sqrt{3}-2\right)^{2} - 3\left(\sqrt{3}-2\right) - 1 \\
&= 3 - 4\sqrt{3} + 4 - 3\sqrt{3} + 6 - 1 \\
&= 12 - 7\sqrt{3}
\end{aligned}$$

11. The domain is all real numbers except -1, or $(-\infty, -1) \cup (-1, \infty)$. We draw the graph:

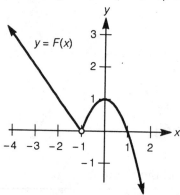

12. Since $y = f(-x)$ is a reflection of $y = f(x)$ across the y-axis, it will be the line segment joining the points $(-1, 3)$ and $(-5, -2)$:

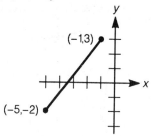

13. We solve the equation:
$$5 + f(4t - 3) = 2$$
$$f(4t - 3) = -3$$
$$4t - 3 = f^{-1}(-3)$$
$$4t - 3 = 1$$
$$4t = 4$$
$$t = 1$$

14. Since $f(1) = 1$ and $f(b) = \dfrac{1}{b}$, we have:
$$\frac{\Delta f}{\Delta x} = \frac{f(b) - f(1)}{b - 1} = \frac{\frac{1}{b} - 1}{b - 1} \cdot \frac{b}{b} = \frac{1 - b}{b(b - 1)} = -\frac{1}{b}$$
Therefore:
$$-\frac{1}{b} = -\frac{1}{10}$$
$$b = 10$$

15. (a)   Since $F(t_1) = 0.16t_1^2 - 1.6t_1 + 35$ and $F(5) = 0.16(5)^2 - 1.6(5) + 35 = 31$, the average value is given by:

$$\frac{\Delta F}{\Delta t} = \frac{F(5) - F(t_1)}{5 - t_1}$$

$$= \frac{31 - \left(0.16t_1^2 - 1.6t_1 + 35\right)}{5 - t_1}$$

$$= \frac{-0.16t_1^2 + 1.6t_1 - 4}{5 - t_1}$$

$$= \frac{0.16t_1^2 - 1.6t_1 + 4}{t_1 - 5}$$

$$= \frac{\left(t_1 - 5\right)\left(0.16t_1 - 0.8\right)}{t_1 - 5}$$

$$= 0.16t_1 - 0.8$$

(b)   Using the result from part (a) where $t_1 = 4$, we have:

$$\frac{\Delta F}{\Delta t} = 0.16(4) - 0.8 = -0.16 \text{ °C/hour}$$

Chapter Four
Polynomial and Rational Functions.
Applications to Optimization

4.1 Linear Functions

1. We first find the slope between the points $(-1, 0)$ and $(5, 4)$:
$$m = \frac{4 - 0}{5 - (-1)} = \frac{4}{6} = \frac{2}{3}$$
Now use the point $(-1, 0)$ in the point-slope formula:
$$y - 0 = \frac{2}{3}[x - (-1)]$$
$$y = \frac{2}{3}(x + 1)$$
$$y = \frac{2}{3}x + \frac{2}{3}$$
Using functional notation we have $f(x) = \frac{2}{3}x + \frac{2}{3}$.

3. We first find the slope between the points $(0, 0)$ and $(1, \sqrt{2})$:
$$m = \frac{\sqrt{2} - 0}{1 - 0} = \sqrt{2}$$
Now, since $(0, 0)$ is the y-intercept, we use the slope-intercept formula to write $y = \sqrt{2}x$.
Using functional notation we have $g(x) = \sqrt{2}x$.

5. We find the slope of $x - y = 1$:
$$-y = -x + 1$$
$$y = x - 1$$

233

So the parallel slope is 1. Use the point $\left(\tfrac{1}{2}, -3\right)$ in the point-slope formula:
$$y - (-3) = 1\left(x - \tfrac{1}{2}\right)$$
$$y + 3 = x - \tfrac{1}{2}$$
$$y = x - \tfrac{7}{2}$$
Using functional notation we have $f(x) = x - \tfrac{7}{2}$.

7.   We set $x = 0$ to find the y-intercepts of the circle:
$$0 - 0 + y^2 - 3 = 0$$
$$y^2 = 3$$
$$y = \pm\sqrt{3}$$
So a horizontal line passing through $\left(0, \sqrt{3}\right)$ is $y = \sqrt{3}$. Using functional notation we have $f(x) = \sqrt{3}$.

9.   Since the points $(-1, 2)$ and $(0, 4)$ lie on the graph of the inverse function, then $(2, -1)$ and $(4, 0)$ must lie on the graph of the function. We find the slope:
$$m = \frac{0 - (-1)}{4 - 2} = \tfrac{1}{2}$$
We use the point $(4, 0)$ in the point-slope formula:
$$y - 0 = \tfrac{1}{2}(x - 4)$$
$$y = \tfrac{1}{2}x - 2$$
Using functional notation we have $f(x) = \tfrac{1}{2}x - 2$.

11.   We compute $(f \circ g)(x)$:
$$(f \circ g)(x) = f[g(x)] = f(1 - 2x) = 3(1 - 2x) - 4 = 3 - 6x - 4 = -6x - 1$$
So $f \circ g$ is a linear function (it is in standard form).

13.   Call $V(t)$ the value of the machine after t years. When $t = 0$ we have $V = 20,000$ and when $t = 8$ we have $V = 1,000$. We find the slope of the line between the points $(0, 20000)$ and $(8, 1000)$:
$$m = \frac{1000 - 20000}{8 - 0} = \frac{-19000}{8} = -2375$$
Since $(0, 20000)$ is the y-intercept, we use the slope-intercept formula to write $V = -2375t + 20000$. Using functional notation we have $V(t) = -2375t + 20000$.

15.   (a)   Call $V(t)$ the value of the machine after t years. Now $V = 60,000$ when $t = 0$ and $V = 0$ when $t = 5$. We find the slope of the line between $(0, 60000)$ and $(5, 0)$:
$$m = \frac{0 - 60000}{5 - 0} = \frac{-60000}{5} = -12000$$
Since $(0, 60000)$ is the y-intercept, we use the slope-intercept formula to write $V = -12000t + 60000$. Using functional notation we have $V(t) = -12000t + 60000$.

(b) The completed schedule is:

End of Year	Yearly Depreciation	Accumulated Depreciation	Value V
0	0	0	60,000
1	12,000	12,000	48,000
2	12,000	24,000	36,000
3	12,000	36,000	24,000
4	12,000	48,000	12,000
5	12,000	60,000	0

17. (a) We let $x = 10$, so $C(10) = 450 + 8x = 450 + 8(10) = \$530$.

(b) We let $x = 11$, so $C(11) = 450 + 8(11) = \$538$.

(c) There are two ways to find the marginal cost. One way is to recognize that the marginal cost will be the slope of the line, which is \$8/fan. Another way would be to use the definition of the marginal cost: it is cost of producing the <u>next</u> unit. Since $C(10)$ is the cost of producing 10 fans and $C(11)$ is the cost of producing 11 fans, then the marginal cost would be $C(11) - C(10) = 538 - 530 = \$8$/fan. We get the same answer using either approach.

19. (a) We first compute $C(n + 1)$:
$$C(n+1) = 400 + 50(n+1) = 400 + 50n + 50 = 450 + 50n$$
Therefore:
$$C(n+1) - C(n) = (450 + 50n) - (400 + 50n) = 450 + 50n - 400 - 50n = 50$$

(b) The marginal cost, which represents the cost to produce the next unit, is the slope of the line. So the marginal cost is \$50/player.

(c) The answers are the same. Notice that (a) directly computed marginal cost from its definition.

21. Since the velocity will be the slope of the distance line, we compute the slope of each line.

(a) Let $(x_1, y_1) = (1, 4)$ and $(x_2, y_2) = (6, 8)$. Then:
$$m = \frac{8-4}{6-1} = \frac{4}{5}$$
So the velocity is $\frac{4}{5}$ ft/sec.

(b)   Let $(x_1, y_1) = (2, 4)$ and $(x_2, y_2) = (5, 4)$. Then:
$$m = \frac{4-4}{5-2} = \frac{0}{3} = 0$$
So the velocity is 0 cm/sec.

(c)   Let $(x_1, y_1) = (0, 0)$ and $(x_2, y_2) = (2, 16)$. Then:
$$m = \frac{16-0}{2-0} = \frac{16}{2} = 8$$
So the velocity is 8 mph.

23.   (a)   Since the velocity of A is 3 units/sec and the velocity of B is 20 units/sec, then B is traveling faster.

(b)   When $t = 0$, A is at $x = 100$ and B is at $x = -36$. So A is farther to the right.

(c)   We set the two x-coordinates equal:
$$3t + 100 = 20t - 36$$
$$136 = 17t$$
$$8 = t$$
When $t = 8$ sec, A and B have the same x-coordinate.

25.   (a)   We plot the points:

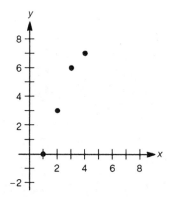

(b)   The slope appears to be approximately 2.5 and the y-intercept is -2:

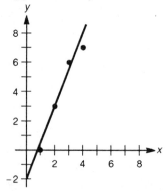

(c) We graph the regression line $y = 2.4x - 2$:

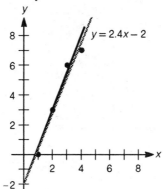

27. (a) We let $x = 2000$:
$$y = 37025(2000) - 70226566 = 3823434$$
It would appear that Los Angeles will not reach a population of 4 million by the year 2000.

(b) We let $x = 2010$:
$$y = 37025(2010) - 70226566 = 4195000$$
The population in Los Angeles will be approximately 4,195,000 in the year 2010.

29. (a) We let $x = 1996$:
$$y = (0.3428)(1996) - 680.6114 = 3.6174$$
The value of the toy imports from China in 1996 will be approximately 3.6174 billion dollars.

(b) We let $x = 2000$:
$$y = (0.3428)(2000) - 680.6114 = 4.9886$$
The value of the toy imports will not exceed 5 billion dollars by the year 2000.

31. (a) The first three iterates are 2.5, 3.25, and 3.625.

(b) The y-coordinate of point A is $f(1) = 2.5$. The x-coordinate of point B is 2.5, so the y-coordinate of point B is $f(2.5) = 3.25$. The x-coordinate of point C is 3.25, so the y-coordinate of point C is $f(3.25) = 3.625$.

(c) At the intersection point the y-coordinates must be equal, so:
$$x = \tfrac{1}{2}x + 2$$
$$2x = x + 4$$
$$x = 4$$
Since $y = x$, the y-coordinate is also 4.

(d) The fourth through tenth iterates are 3.813, 3.906, 3.953, 3.977, 3.988, 3.994, and 3.997. These iterates appear to be approaching 4.

33. (a) The average rate of change is given by:
$$\frac{f(b)-f(a)}{b-a}=\frac{(Ab+B)-(Aa+B)}{b-a}=\frac{Ab-Aa}{b-a}=\frac{A(b-a)}{b-a}=A$$

(b) Since $(f\circ f)(x)=f(Ax+B)=A(Ax+B)+B=A^2x+AB+B$, the average rate of change is given by:
$$\frac{(f\circ f)(b)-(f\circ f)(a)}{b-a}=\frac{\left(A^2b+AB+B\right)-\left(A^2a+AB+B\right)}{b-a}$$
$$=\frac{A^2b-A^2a}{b-a}$$
$$=\frac{A^2(b-a)}{b-a}$$
$$=A^2$$

(c) Since $(g\circ f)(x)=g(Ax+B)=C(Ax+B)+D=ACx+BC+D$, the average rate of change is given by:
$$\frac{(g\circ f)(b)-(g\circ f)(a)}{b-a}=\frac{(ACb+BC+D)-(ACa+BC+D)}{b-a}$$
$$=\frac{ACb-ACa}{b-a}$$
$$=\frac{AC(b-a)}{b-a}$$
$$=AC$$

(d) Since $(f\circ g)(x)=f(Cx+D)=A(Cx+D)+B=ACx+AD+B$, the average rate of change is given by:
$$\frac{(f\circ g)(b)-(f\circ g)(a)}{b-a}=\frac{(ACb+AD+B)-(ACa+AD+B)}{b-a}$$
$$=\frac{ACb-ACa}{b-a}$$
$$=\frac{AC(b-a)}{b-a}$$
$$=AC$$

35. (a) Since $(f\circ g)(x)=ACx+AD+B$, we must find $(f\circ g)^{-1}$ by switching the roles of x and y, then solving for y:
$$ACy+AD+B=x$$
$$ACy=x-AD-B$$
$$y=\frac{x}{AC}-\frac{D}{C}-\frac{B}{AC}$$

So $(f \circ g)^{-1} = \dfrac{x}{AC} - \dfrac{D}{C} - \dfrac{B}{AC}$. The average rate of change is given by:

$$\frac{(f \circ g)^{-1}(b) - (f \circ g)^{-1}(a)}{b - a} = \frac{\left(\frac{b}{AC} - \frac{D}{C} - \frac{B}{AC}\right) - \left(\frac{a}{AC} - \frac{D}{C} - \frac{B}{AC}\right)}{b - a}$$

$$= \frac{\frac{b}{AC} - \frac{a}{AC}}{b - a}$$

$$= \frac{\frac{1}{AC}(b - a)}{b - a}$$

$$= \frac{1}{AC}$$

(b)   Since $(g \circ f)(x) = ACx + BC + D$, we must find $(g \circ f)^{-1}$ by switching the roles of x and y, then solving for y:

$$ACy + BC + D = x$$
$$ACy = x - BC - D$$
$$y = \frac{x}{AC} - \frac{B}{A} - \frac{D}{AC}$$

So $(g \circ f)^{-1} = \dfrac{x}{AC} - \dfrac{B}{A} - \dfrac{D}{AC}$. The average rate of change is given by:

$$\frac{(g \circ f)^{-1}(b) - (g \circ f)^{-1}(a)}{b - a} = \frac{\left(\frac{b}{AC} - \frac{B}{A} - \frac{D}{AC}\right) - \left(\frac{a}{AC} - \frac{B}{A} - \frac{D}{AC}\right)}{b - a}$$

$$= \frac{\frac{b}{AC} - \frac{a}{AC}}{b - a}$$

$$= \frac{\frac{1}{AC}(b - a)}{b - a}$$

$$= \frac{1}{AC}$$

37.   (a)   Using $f(x) = mx$, we have:

$$f(a + b) = m(a + b) = ma + mb = f(a) + f(b)$$

(b)   Using $f(x) = mx$, we have:

$$f(ax) = m(ax) = a(mx) = af(x)$$

39.   We first simplify $(f \circ f)(x)$:

$$(f \circ f)(x) = f(mx + b) = m(mx + b) + b = m^2 x + (mb + b)$$

Since $(f \circ f)(x) = 9x + 4$, and since m and b are constants, we must have the two equations (equating x-coefficients and constants):

$$m^2 = 9 \quad \text{and} \quad mb + b = 4$$

Since $m > 0$, then $m = 3$ and:
$$3b + b = 4$$
$$4b = 4$$
$$b = 1$$
So $f(x) = 3x + 1$.

41. (a) We compute the sums:
$$\sum x = 1 + 2 + 3 + 4 = 10$$
$$\sum y = 0 + 3 + 6 + 7 = 16$$

(b) We compute the sums:
$$\sum x^2 = 1 + 4 + 9 + 16 = 30$$
$$\sum xy = 0 + 6 + 18 + 28 = 52$$

(c) We multiply the first equation by -3:
$$-12b - 30m = -48$$
$$10b + 30m = 52$$
Adding, we obtain:
$$-2b = 4$$
$$b = -2$$
Substituting into the first equation:
$$4b + 10m = 16$$
$$-8 + 10m = 16$$
$$10m = 24$$
$$m = 2.4$$
So the regression line is $y = 2.4x - 2$.

43. We first do the computations:
$$\sum x = 1 + 2 + 3 + 4 + 5 = 15$$
$$\sum y = 2 + 3 + 9 + 9 + 11 = 34$$
$$\sum x^2 = 1 + 4 + 9 + 16 + 25 = 55$$
$$\sum xy = 2 + 6 + 27 + 36 + 55 = 126$$
So the system of equations becomes:
$$5b + 15m = 34$$
$$15b + 55m = 126$$
We multiply the first equation by -3:
$$-15b - 45m = -102$$
$$15b + 55m = 126$$
Adding, we obtain:
$$10m = 24$$
$$m = 2.4$$

Substituting into the first equation:
$$5b + 36 = 34$$
$$5b = -2$$
$$b = -0.4$$
So the regression line is $y = 2.4x - 0.4$.

45. We first do the computations:
$$\sum x = 520 + 740 + 560 + 610 + 650 = 3080$$
$$\sum y = 81 + 98 + 83 + 88 + 95 = 445$$
$$\sum x^2 = 270400 + 547600 + 313600 + 372100 + 422500 = 1926200$$
$$\sum xy = 42120 + 72520 + 46480 + 53680 + 61750 = 276550$$
So the system of equations becomes:
$$5b + 3080m = 445$$
$$3080b + 1926200m = 276550$$
Divide the first equation by 5 and the second equation by 10:
$$b + 616m = 89$$
$$308b + 192620m = 27655$$
We multiply the first equation by -308:
$$-308b - 189728m = -27412$$
$$308b + 192620m = 27655$$
Adding, we obtain:
$$2892m = 243$$
$$m \approx 0.084$$
Substituting into the first equation:
$$5b + 258.8 = 445$$
$$5b = 186.2$$
$$b \approx 37.241$$
So the regression line is $y = 0.084x + 37.241$.

Optional TI-81 Graphing Calculator Exercises for Section 4.1

1.   Using the indicated settings, we obtain the graph:

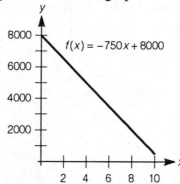

The y-coordinate is 4250, which matches the result from the text.

3.   (a)   We graph the two functions:

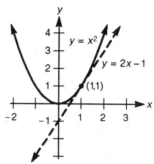

(b)   We complete the tables:

x	0.9	0.99	0.999
$2x-1$	0.8	0.98	0.998
x^2	0.81	0.9801	0.998001

x	1.1	1.01	1.001
$2x-1$	1.2	1.02	1.002
x^2	1.21	1.0201	1.002001

4.2 Quadratic Functions

1.   The vertex is $(-2,0)$, the axis of symmetry is $x = -2$, the minimum value is 0, the x-intercept is -2, and the y-intercept is 4. We draw the graph:

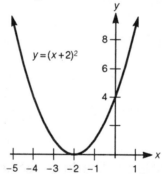

3.   The vertex is $(-2,0)$, the axis of symmetry is $x = -2$, the minimum value is 0, the x-intercept is -2, and the y-intercept is 8. We draw the graph:

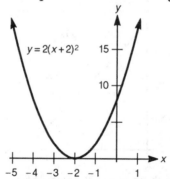

5. The vertex is $(-2, 4)$, the axis of symmetry is $x = -2$, the maximum value is 4, the x-intercepts are $-2 \pm \sqrt{2}$, and the y-intercept is -4. We draw the graph:

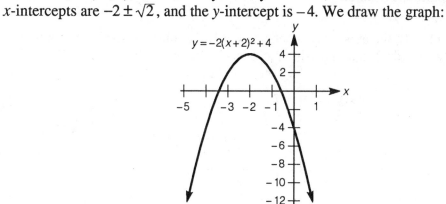

7. We first complete the square:
    $$f(x) = x^2 - 4x = \left(x^2 - 4x + 4\right) - 4 = (x-2)^2 - 4$$
    The vertex is $(2, -4)$, the axis of symmetry is $x = 2$, the minimum value is -4, the x-intercepts are 0 and 4, and the y-intercept is 0. We draw the graph:

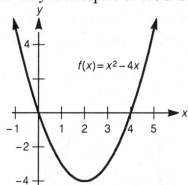

9. The vertex is $(0, 1)$, the axis of symmetry is $x = 0$, the maximum value is 1, the x-intercepts are ± 1, and the y-intercept is 1. We draw the graph:

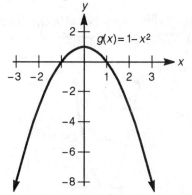

11. We first complete the square:
$$y = x^2 - 2x - 3 = \left(x^2 - 2x + 1\right) - 4 = (x - 1)^2 - 4$$
The vertex is $(1, -4)$, the axis of symmetry is $x = 1$, the minimum value is -4, the x-intercepts are 3 and -1, and the y-intercept is -3. We draw the graph:

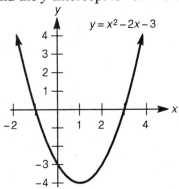

13. We first complete the square:
$$y = -x^2 + 6x + 2 = -\left(x^2 - 6x\right) + 2 = -\left(x^2 - 6x + 9\right) + 11 = -(x - 3)^2 + 11$$
The vertex is $(3, 11)$, the axis of symmetry is $x = 3$, the maximum value is 11, the x-intercepts are $3 \pm \sqrt{11}$, and the y-intercept is 2. We draw the graph:

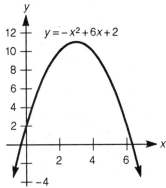

15. The vertex is $(0, 0)$, the axis of symmetry is $t = 0$, the minimum value is 0, the t-intercept is 0, and the s-intercept is 0. We draw the graph:

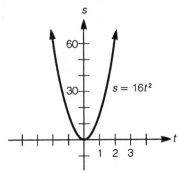

17. We first complete the square:
$$s = -9t^2 + 3t + 2 = -9\left(t^2 - \tfrac{1}{3}t\right) + 2 = -9\left(t^2 - \tfrac{1}{3}t + \tfrac{1}{36}\right) + \tfrac{9}{4} = -9\left(t - \tfrac{1}{6}\right)^2 + \tfrac{9}{4}$$

The vertex is $\left(\tfrac{1}{6}, \tfrac{9}{4}\right)$, the axis of symmetry is $t = \tfrac{1}{6}$, the maximum value is $\tfrac{9}{4}$, the t-intercepts are $-\tfrac{1}{3}$ and $\tfrac{2}{3}$, and the s-intercept is 2. We draw the graph:

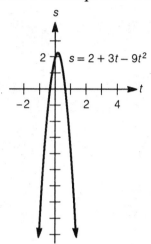

19. We complete the square:
$$y = 2x^2 - 4x + 11 = 2\left(x^2 - 2x\right) + 11 = 2\left(x^2 - 2x + 1\right) + 9 = 2(x - 1)^2 + 9$$
Since the vertex is $(1, 9)$ and the parabola will be pointed up, $x = 1$ will yield a minimum output value.

21. We complete the square:
$$g(x) = -6x^2 + 18x = -6\left(x^2 - 3x\right) = -6\left(x^2 - 3x + \tfrac{9}{4}\right) + \tfrac{27}{2} = -6\left(x - \tfrac{3}{2}\right)^2 + \tfrac{27}{2}$$
Since the vertex is $\left(\tfrac{3}{2}, \tfrac{27}{2}\right)$ and the parabola will be pointed down, $x = \tfrac{3}{2}$ will yield a maximum output value.

23. Since the vertex is $(0, -10)$ and the parabola will be pointed up, $x = 0$ will yield a minimum output value.

25. We complete the square:
$$y = x^2 - 8x + 3 = \left(x^2 - 8x + 16\right) - 13 = (x - 4)^2 - 13$$
Since the vertex is $(4, -13)$ and the parabola will be pointed up, the function has a minimum value of -13.

27. We complete the square:
$$y = -2x^2 - 3x + 2 = -2\left(x^2 + \tfrac{3}{2}x\right) + 2 = -2\left(x^2 + \tfrac{3}{2}x + \tfrac{9}{16}\right) + 2 + \tfrac{9}{8} = -2\left(x + \tfrac{3}{4}\right)^2 + \tfrac{25}{8}$$
Since the vertex is $\left(-\tfrac{3}{4}, \tfrac{25}{8}\right)$ and the parabola will be pointed down, the function has a maximum value of $\tfrac{25}{8}$.

29. We do not need to complete the square, since the vertex is $(0, 1000)$ and the parabola will be pointed down. So the function has a maximum value of 1000.

31. We find the vertex of the parabola by completing the square:
$$y = x^2 - 6x + 13 = \left(x^2 - 6x + 9\right) + 4 = (x - 3)^2 + 4$$
So the vertex is $(3, 4)$. We now use the distance formula with the points $(0, 0)$ and $(3, 4)$:
$$d = \sqrt{(3 - 0)^2 + (4 - 0)^2} = \sqrt{9 + 16} = \sqrt{25} = 5$$
So the vertex is 5 units from the origin.

33. We compute $(f \circ g)(x)$:
$$(f \circ g)(x) = 2\left(x^2 + 4x + 1\right) - 3 = 2x^2 + 8x + 2 - 3 = 2x^2 + 8x - 1$$
So $f \circ g$ is a quadratic function.

35. We compute $(g \circ h)(x)$:
$$\begin{aligned} (g \circ h)(x) &= \left(1 - 2x^2\right)^2 + 4\left(1 - 2x^2\right) + 1 \\ &= 1 - 4x^2 + 4x^4 + 4 - 8x^2 + 1 \\ &= 4x^4 - 12x^2 + 6 \end{aligned}$$
So $g \circ h$ is neither linear nor quadratic.

37. We compute $(f \circ f)(x)$:
$$(f \circ f)(x) = 2(2x - 3) - 3 = 4x - 6 - 3 = 4x - 9$$
So $f \circ f$ is a linear function.

39. (a) We first complete the square on $x^2 - 6x + 73$:
$$x^2 - 6x + 73 = \left(x^2 - 6x + 9\right) + 73 - 9 = (x - 3)^2 + 64$$
So $f(x) = \sqrt{(x - 3)^2 + 64}$. This would achieve a minimum value at $\left(3, \sqrt{64}\right) = (3, 8)$.

   (b) Here $g(x) = \sqrt[3]{(x - 3)^2 + 64}$, which would achieve a minimum value at $\left(3, \sqrt[3]{64}\right) = (3, 4)$.

   (c) We complete the square on $x^4 - 6x^2 + 73$:
$$x^4 - 6x^2 + 73 = \left(x^4 - 6x^2 + 9\right) + 73 - 9 = \left(x^2 - 3\right)^2 + 64$$
So $h(x) = (x^2 - 3)^2 + 64$, which would achieve a minimum value at $\left(\pm\sqrt{3}, 64\right)$.

41. (a) We complete the square on $-x^2 + 4x + 12$:
$$-x^2 + 4x + 12 = -\left(x^2 - 4x\right) + 12 = -\left(x^2 - 4x + 4\right) + 16 = -(x - 2)^2 + 16$$
So $f(x) = \sqrt{-(x - 2)^2 + 16}$, which has a maximum value at $\left(2, \sqrt{16}\right) = (2, 4)$.

   (b) Now $g(x) = \sqrt[3]{-(x - 2)^2 + 16}$, which has a maximum value at $\left(2, \sqrt[3]{16}\right) = \left(2, 2\sqrt[3]{2}\right)$.

(c) Here $h(x) = -\left(x^2 - 2\right)^2 + 16$, which has a maximum value at $\left(\pm\sqrt{2}, 16\right)$.

43. The average rate of change is given by:

$$\frac{f(x_2) - f(x_1)}{x_2 - x_1} = \frac{\left(ax_2^2 + bx_2 + c\right) - \left(ax_1^2 + bx_1 + c\right)}{x_2 - x_1}$$

$$= \frac{ax_2^2 + bx_2 - ax_1^2 - bx_1}{x_2 - x_1}$$

$$= \frac{a\left(x_2^2 - x_1^2\right) + b\left(x_2 - x_1\right)}{x_2 - x_1}$$

$$= \frac{a\left(x_2 + x_1\right)\left(x_2 - x_1\right) + b\left(x_2 - x_1\right)}{x_2 - x_1}$$

$$= \frac{\left(x_2 - x_1\right)\left(ax_1 + ax_2 + b\right)}{x_2 - x_1}$$

$$= ax_1 + ax_2 + b$$

45. We complete the square:

$$y = ax^2 + bx + c$$

$$= a\left(x^2 + \frac{b}{a}x\right) + c$$

$$= a\left(x^2 + \frac{b}{a}x + \frac{b^2}{4a^2}\right) + c - \frac{ab^2}{4a^2}$$

$$= a\left(x + \frac{b}{2a}\right)^2 + \frac{4ac - b^2}{4a}$$

$$= a\left(x + \frac{b}{2a}\right)^2 - \frac{D}{4a}, \text{ where } D = b^2 - 4ac$$

So the vertex is $\left(-\dfrac{b}{2a}, -\dfrac{D}{4a}\right)$, where $D = b^2 - 4ac$.

47. (a) We compute $f\left(\dfrac{-b}{2a} + h\right)$:

$$f\left(\frac{-b}{2a} + h\right) = a\left(\frac{-b}{2a} + h\right)^2 + b\left(\frac{-b}{2a} + h\right) + c$$

$$= a\left(\frac{b^2}{4a^2} - \frac{bh}{a} + h^2\right) - \frac{b^2}{2a} + bh + c$$

$$= \frac{b^2}{4a} - bh + ah^2 - \frac{b^2}{2a} + bh + c$$

$$= ah^2 - \frac{b^2}{4a} + c$$

$$= ah^2 - \frac{b^2 - 4ac}{4a}$$

$$= ah^2 - \frac{D}{4a}, \text{ where } D = b^2 - 4ac$$

(b) We compute $f\left(\dfrac{-b}{2a}-h\right)$:

$$f\left(\frac{-b}{2a}-h\right)=a\left(\frac{-b}{2a}-h\right)^2+b\left(\frac{-b}{2a}-h\right)+c$$

$$=a\left(\frac{b^2}{4a^2}+\frac{bh}{a}+h^2\right)-\frac{b^2}{2a}-bh+c$$

$$=\frac{b^2}{4a}+bh+ah^2-\frac{b^2}{2a}-bh+c$$

$$=ah^2-\frac{b^2}{4a}+c$$

$$=ah^2-\frac{b^2-4ac}{4a}$$

$$=ah^2-\frac{D}{4a},\ \text{where } D=b^2-4ac$$

49. (a) The y-coordinate of point A is $f(x_0)$, the first iterate of x_0. The x-coordinate of point B is $f(x_0)$, so the y-coordinate of point B is $f(f(x_0))$, the second iterate of x_0. The x-coordinate of point C is $f(f(x_0))$, so the y-coordinate of point C is $f(f(f(x_0)))$, the third iterate of x_0. The x-coordinate of point D is $f(f(f(x_0)))$, so the y-coordinate of point D is $f(f(f(f(x_0))))$, the fourth iterate of x_0.

(b) The first four iterates of $x_0=0.2$ are -0.56, -0.2864, -0.518, and -0.332.

51. Since the vertex is $(2,2)$, we know its equation will be $y=A(x-2)^2+2$. Since $(0,0)$ must lie on this parabola, we substitute the points to find A:

$$y=A(x-2)^2+2$$
$$0=A(0-2)^2+2$$
$$0=4A+2$$
$$-2=4A$$
$$A=-\tfrac{1}{2}$$

So the parabola is $y=-\tfrac{1}{2}(x-2)^2+2$.

53. Since the vertex is $(3,-1)$, we know its equation will be $y=A(x-3)^2-1$. Since $(1,0)$ must lie on this parabola, we substitute the points to find A:

$$y=A(x-3)^2-1$$
$$0=A(1-3)^2-1$$
$$0=4A-1$$
$$1=4A$$
$$A=\tfrac{1}{4}$$

So the parabola is $y=\tfrac{1}{4}(x-3)^2-1$.

55. We find the vertex of the parabola by completing the square:
$$y = 2x^2 + 12x + 14 = 2(x^2 + 6x) + 14 = 2(x^2 + 6x + 9) - 4 = 2(x+3)^2 - 4$$
Since the vertex is $(-3, -4)$, the circle would have an equation of $(x + 3)^2 + (y + 4)^2 = r^2$.

Since $(0, 0)$ lies on the circle, we have:
$$(0+3)^2 + (0+4)^2 = r^2$$
$$r^2 = 9 + 16$$
$$r^2 = 25$$
$$r = 5, \text{ since } r > 0$$

So the circle is $(x + 3)^2 + (y + 4)^2 = 25$.

57. (a) We first complete the square to find the vertex:
$$y = p(x^2 + x) + r = p\left(x^2 + x + \tfrac{1}{4}\right) + r - \frac{p}{4} = p\left(x + \tfrac{1}{2}\right)^2 + \frac{4r - p}{4}$$
If the vertex lies on the x-axis, then $\frac{4r-p}{4} = 0$, so $4r - p = 0$, thus $p = 4r$.

(b) If $p = 4r$, then $y = 4rx^2 + 4rx + r$. We complete the square:
$$y = 4r(x^2 + x) + r = 4r\left(x^2 + x + \tfrac{1}{4}\right) + r - r = 4r\left(x + \tfrac{1}{2}\right)^2$$
But then the vertex is $\left(-\tfrac{1}{2}, 0\right)$, and thus it lies on the x-axis.

59. If $a + bi$ and $a - bi$ are roots of the equation, then we can write the equation in factored form as:
$$[x - (a + bi)][x - (a - bi)] = 0$$
$$x^2 - (a + bi)x - (a - bi)x + (a + bi)(a - bi) = 0$$
$$x^2 - 2ax + (a^2 + b^2) = 0$$

We complete the square:
$$f(x) = (x^2 - 2ax) + (a^2 + b^2) = (x^2 - 2ax + a^2) + (a^2 + b^2 - a^2) = (x - a)^2 + b^2$$

So the vertex is (a, b^2).

Optional TI-81 Graphing Calculator Exercises for Section 4.2

1.   (a)   We graph all four functions:

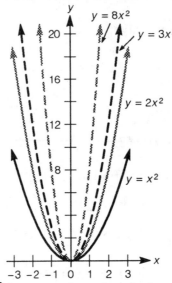

As the coefficient of x^2 increases from 1 to 8, the graph "narrows" and produces larger y-coordinates.

(b)   The graph should be "flatter" than the others:

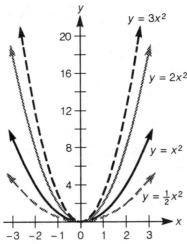

3. (a) Using the indicated settings, we graph both functions:

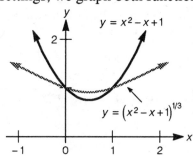

(b) When $x \approx 0.5$, the value of y is approximately 0.91.

(c) Since $\sqrt[3]{\frac{3}{4}} \approx 0.90856$, the approximation from part (b) is close.

4.3 Applied Functions: Setting Up Equations

1. (a) Call P the perimeter and A the area. We are asked to come up with a formula for A in terms of x. Since $P = 16$ and $P = 2x + 2l$, we have:
$$2x + 2l = 16$$
$$2l = 16 - 2x$$
$$l = 8 - x$$
Now $A = xl = x(8 - x) = 8x - x^2$. Using functional notation we have $A(x) = 8x - x^2$.

(b) Since $A = 85$ and $A = xl$, we have:
$$xl = 85$$
$$l = \frac{85}{x}$$
Now $P = 2x + 2l = 2x + 2\left(\frac{85}{x}\right) = 2x + \frac{170}{x}$. Using functional notation we have $P(x) = 2x + \frac{170}{x}$.

3. (a) We first draw the figure:

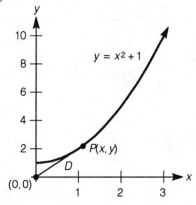

Call D the distance from $P(x,y)$ to the origin. We are asked to come up with a formula for D in terms of x. By the distance formula, we have:
$$D = \sqrt{(x-0)^2 + (y-0)^2} = \sqrt{x^2 + y^2}$$
Since $P(x,y)$ lies on the curve, then $y = x^2 + 1$. Substituting this for y in our equation for D, we have:
$$D = \sqrt{x^2 + y^2} = \sqrt{x^2 + (x^2 + 1)^2} = \sqrt{x^2 + x^4 + 2x^2 + 1} = \sqrt{x^4 + 3x^2 + 1}$$
Using functional notation we have $D(x) = \sqrt{x^4 + 3x^2 + 1}$.

(b)   Let m denote the slope of the line segment from the origin to $P(x,y)$. Then:
$$m = \frac{y-0}{x-0} = \frac{y}{x}$$
Substituting $y = x^2 + 1$ in for y in this equation, we have:
$$m = \frac{y}{x} = \frac{x^2 + 1}{x}$$
Using functional notation we have $m(x) = \frac{x^2+1}{x}$.

5.   (a)   We first draw the figure:

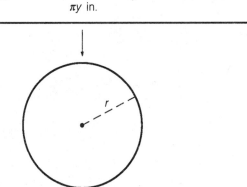

πy in.

Call A the area of the circle and r its radius. We are asked to come up with a formula for A in terms of y. Since πy is the circumference C of the circle, and $C = 2\pi r$, we have:
$$2\pi r = \pi y$$
$$r = \frac{y}{2}$$
Now $A = \pi r^2 = \pi\left(\frac{y}{2}\right)^2 = \frac{\pi y^2}{4}$. Using functional notation we have $A(y) = \frac{\pi y^2}{4}$.

(b) We draw the figure:

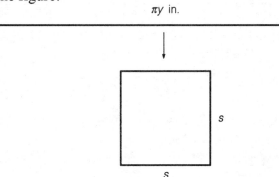

πy in.

s

s

Let A denote the area of the square, and s the length of the side. Since the perimeter P is πy and $P = 4s$, we have:

$$4s = \pi y$$

$$s = \frac{\pi y}{4}$$

Now $A = s^2 = \left(\frac{\pi y}{4}\right)^2 = \frac{\pi^2 y^2}{16}$. Using functional notation we have $A(y) = \frac{\pi^2 y^2}{16}$.

7. (a) Let the two numbers be x and $16 - x$. Then the product P would be:

$$P = x(16 - x) = 16x - x^2$$

Using functional notation we have $P(x) = 16x - x^2$.

(b) Since the two numbers are x and $16 - x$, then the sum of squares S would be:

$$S = (x)^2 + (16 - x)^2 = x^2 + 256 - 32x + x^2 = 2x^2 - 32x + 256$$

Using functional notation we have $S(x) = 2x^2 - 32x + 256$.

(c) There are two ways to set this up. Since the two numbers are x and $16 - x$, then the difference of the cubes D could be:

$$D = (x)^3 - (16 - x)^3 \text{ or } D = (16 - x)^3 - x^3$$

Using functional notation we have:

$$D(x) = (x)^3 - (16 - x)^3 \text{ or } D(x) = (16 - x)^3 - x^3$$

(d) Let A denote the average of the two numbers. Since the two numbers are x and $16 - x$, we have:

$$A = \frac{x + 16 - x}{2} = \frac{16}{2} = 8$$

So $A(x) = 8$. Notice that the average does not depend on what the two numbers are!

9. Let R be the revenue, x be the number of units sold, and p be the demand (price). Then:

$$R = xp = x\left(-\tfrac{1}{4}x + 8\right) = -\tfrac{1}{4}x^2 + 8x$$

Using functional notation we have $R(x) = -\tfrac{1}{4}x^2 + 8x$.

11. (a)   We complete the table:

x	1	2	3	4	5	6	7
$P(x)$	17.88	19.49	20.83	21.86	22.49	22.58	21.75

(b)   The largest value for $P(x)$ is 22.58, corresponding to $x = 6$.

(c)   We compute $P(4\sqrt{2})$:
$$P(4\sqrt{2}) = 2(4\sqrt{2}) + 2\sqrt{64 - (4\sqrt{2})^2}$$
$$= 8\sqrt{2} + 2\sqrt{64 - 32}$$
$$= 8\sqrt{2} + 2\sqrt{32}$$
$$= 8\sqrt{2} + 8\sqrt{2}$$
$$= 16\sqrt{2}$$
$$\approx 22.63$$
This is indeed larger than any of our table values.

13. (a)   We first draw the figure:

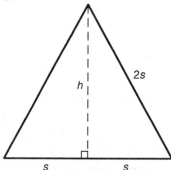

Let h denote the height and $2s$ denote the sides. Note that the height (called the altitude) bisects the base into the lengths of s and s. We are asked to find h in terms of s, so we use the Pythagorean theorem on the right triangle:
$$h^2 + s^2 = (2s)^2$$
$$h^2 + s^2 = 4s^2$$
$$h^2 = 3s^2$$
$$h = \sqrt{3s^2} = \sqrt{3}s$$
Using functional notation we have $h(s) = \sqrt{3}s$.

(b)   Let A denote the area of the triangle. Then:
$$A = \tfrac{1}{2}(\text{base})(\text{height}) = \tfrac{1}{2}(2s)(\sqrt{3}s) = \sqrt{3}s^2$$
Using functional notation we have $A(s) = \sqrt{3}s^2$.

(c)   If each side is 8 cm, then:
$$2s = 8$$
$$s = 4$$

Using the function from (a), we have:
$$h(4) = \sqrt{3} \cdot 4 = 4\sqrt{3} \text{ cm}$$

(d) If each side is 5 in., then:
$$2s = 5$$
$$s = \tfrac{5}{2}$$
Using the function from (b), we have:
$$A\left(\tfrac{5}{2}\right) = \sqrt{3}\left(\tfrac{5}{2}\right)^2 = \frac{25\sqrt{3}}{4} \text{ in.}^2$$

15. Let h be the height, r be the radius, and V be the volume. We know that:
$$V = \pi r^2 h$$
We are also given that $h = 2r$, so we substitute into the formula for V:
$$V = \pi r^2 (2r) = 2\pi r^3$$
Using functional notation we have $V(r) = 2\pi r^3$.

17. (a) Let h be the height, r be the radius and V be the volume. We know that $V = 12\pi$ and $V = \pi r^2 h$, so:
$$\pi r^2 h = 12\pi$$
$$h = \frac{12\pi}{\pi r^2} = \frac{12}{r^2}$$
Using functional notation we have $h(r) = \frac{12}{r^2}$.

(b) Let S be the total surface area. Then:
$$S = 2\pi r^2 + 2\pi r h = 2\pi r^2 + 2\pi r\left(\frac{12}{r^2}\right) = 2\pi r^2 + \frac{24\pi}{r}$$
Using functional notation we have $S(r) = 2\pi r^2 + \frac{24\pi}{r}$.

19. We solve $S = 4\pi r^2$ for r:
$$4\pi r^2 = S$$
$$r^2 = \frac{S}{4\pi}$$
$$r = \sqrt{\frac{S}{4\pi}}$$
Now since $V = \frac{4}{3}\pi r^3$, we have:
$$V = \frac{4}{3}\pi\left(\sqrt{\frac{S}{4\pi}}\right)^3 = \frac{4\pi S\sqrt{S}}{3(4\pi)\sqrt{4\pi}} = \frac{S\sqrt{S}}{3\sqrt{4\pi}} = \frac{S\sqrt{S\pi}}{6\pi}$$
Using functional notation we have $V(S) = \frac{S\sqrt{S\pi}}{6\pi}$.

21. We draw a figure:

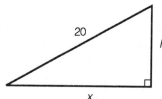

Let A be the area of the triangle and let x and h be its two legs. By the Pythagorean theorem, we have:

$$x^2 + h^2 = 20^2$$
$$h^2 = 400 - x^2$$
$$h = \sqrt{400 - x^2}$$

Therefore:

$$A = \tfrac{1}{2}(\text{base})(\text{height}) = \tfrac{1}{2}(x)(h) = \tfrac{1}{2}x\sqrt{400 - x^2}$$

Using functional notation we have $A(x) = \tfrac{1}{2}x\sqrt{400 - x^2}$.

23. We draw the figure:

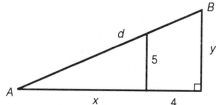

Using similar triangles, we have:

$$\frac{x}{5} = \frac{x+4}{y}$$
$$xy = 5(x+4)$$
$$y = \frac{5(x+4)}{x}$$

Therefore:

$$d^2 = (x+4)^2 + y^2$$
$$= (x+4)^2 + \left(\frac{5(x+4)}{x}\right)^2$$
$$= (x+4)^2 + \frac{25(x+4)^2}{x^2}$$
$$= \frac{x^2(x+4)^2 + 25(x+4)^2}{x^2}$$
$$= \frac{(x+4)^2(x^2+25)}{x^2}$$

So $d = \dfrac{(x+4)\sqrt{x^2+25}}{x}$. Using functional notation we have $d(x) = \dfrac{(x+4)\sqrt{x^2+25}}{x}$.

25. We substitute into $A(x) = 50x - x^2$:

x	5	10	20	24	24.8	24.9	25	25.1	25.2	45
$A(x)$	225	400	600	624	624.96	624.99	625	624.99	624.96	225

So $x = 25$ yields the largest area. Since $L = 50 - x = 50 - 25 = 25$, then $L = 25$ is the corresponding value.

27. (a) We substitute into $A(x) = 8x - \frac{1}{2}x^3$:

Table 1:

x	1	2	3	4
A	7.5	12	10.5	0

Here $x = 2$ yields the largest area.

Table 2:

x	1.75	2.00	2.25	2.50	2.75
A	11.3203	12.0000	12.3047	12.1875	11.6016

Here $x = 2.25$ yields the largest area.

Table 3:

x	2.15	2.20	2.25	2.30	2.35
A	12.2308	12.2760	12.3047	12.3165	12.3111

Here $x = 2.30$ yields the largest area.

(b) Since $x = \frac{4\sqrt{3}}{3} = 2.309$ to four significant places, then $x = 2.30$ is the closest x-value. This yields an area of 12.3168. Notice that $x = 2.30$ agrees with this to five significant digits.

29. For the square, the side is $\frac{x}{4}$. For the rectangle which has a perimeter of $3 - x$, we have:

$$2w + 2l = 3 - x$$
$$2 \cdot \tfrac{1}{2}l + 2l = 3 - x$$
$$3l = 3 - x$$
$$l = 1 - \tfrac{1}{3}x$$
$$w = \tfrac{1}{2} - \tfrac{1}{6}x$$

So the combined area of the square and rectangle is:

$$A = \left(\tfrac{x}{4}\right)^2 + \left(1 - \tfrac{1}{3}x\right)\left(\tfrac{1}{2} - \tfrac{1}{6}x\right) = \tfrac{1}{16}x^2 + \tfrac{1}{2} - \tfrac{1}{3}x + \tfrac{1}{18}x^2 = \tfrac{17}{144}x^2 - \tfrac{1}{3}x + \tfrac{1}{2}$$

Using functional notation we have $A(x) = \frac{17}{144}x^2 - \frac{1}{3}x + \frac{1}{2}$.

31. (a) Since $V = \frac{1}{3}\pi r^2 h$ and $h = \sqrt{3}r$, we have:

$$V = \tfrac{1}{3}\pi r^2 (\sqrt{3}r) = \tfrac{\sqrt{3}}{3}\pi r^3$$

Using functional notation we have $V(r) = \frac{\sqrt{3}}{3}\pi r^3$.

(b) Since $S = \pi r\sqrt{r^2 + h^2}$ and $h = \sqrt{3}r$, we have:

$$S = \pi r\sqrt{r^2 + h^2} = \pi r\sqrt{r^2 + \left(\sqrt{3}r\right)^2} = \pi r\sqrt{r^2 + 3r^2} = \pi r\sqrt{4r^2} = \pi r(2r) = 2\pi r^2$$

Using functional notation we have $S(r) = 2\pi r^2$.

33. (a) Since $V = \frac{1}{3}\pi r^2 h$ and $S = \pi r\sqrt{r^2 + h^2}$, and $V = S$, we have:

$$\frac{1}{3}\pi r^2 h = \pi r\sqrt{r^2 + h^2}$$

$$rh = 3\sqrt{r^2 + h^2}$$

$$r^2 h^2 = 9\left(r^2 + h^2\right)$$

$$r^2 h^2 = 9r^2 + 9h^2$$

$$r^2 h^2 - 9r^2 = 9h^2$$

$$r^2\left(h^2 - 9\right) = 9h^2$$

$$r^2 = \frac{9h^2}{h^2 - 9}$$

$$r = \sqrt{\frac{9h^2}{h^2 - 9}} = \frac{3h}{\sqrt{h^2 - 9}}$$

Using functional notation we have $r(h) = \dfrac{3h}{\sqrt{h^2 - 9}}$.

(b) After squaring in (a), we had:

$$r^2 h^2 = 9r^2 + 9h^2$$

$$r^2 h^2 - 9h^2 = 9r^2$$

$$h^2\left(r^2 - 9\right) = 9r^2$$

$$h^2 = \frac{9r^2}{r^2 - 9}$$

Taking roots:

$$h = \sqrt{\frac{9r^2}{r^2 - 9}} = \frac{3r}{\sqrt{r^2 - 9}}$$

Using functional notation we have $h(r) = \dfrac{3r}{\sqrt{r^2 - 9}}$.

35. Let x be the length of wire used for the circle, so $14 - x$ is the length of wire used for the square. For the circle, we have $2\pi r = x$, so $r = \frac{x}{2\pi}$ and thus the area is given by:

$$\pi r^2 = \pi\left(\frac{x}{2\pi}\right)^2 = \frac{x^2}{4\pi}$$

For the square, we have $4s = 14 - x$, so $s = \frac{14-x}{4}$ and thus the area is given by:

$$\left(\frac{14-x}{4}\right)^2 = \frac{(14-x)^2}{16}$$

So the total combined area is:

$$A = \frac{x^2}{4\pi} + \frac{(14-x)^2}{16} = \frac{4x^2 + \pi(14-x)^2}{16\pi}$$

Using functional notation we have $A(x) = \dfrac{4x^2 + \pi(14-x)^2}{16\pi}$.

37. The perimeter of each semi-circle is $\frac{1}{2}(2\pi r) = \pi r$, so the total perimeter P is given by:

$$P = \pi r + \pi r + l + l = 2\pi r + 2l, \text{ where } l \text{ is the length of the rectangle}$$

Since $P = \frac{1}{4}$, we have:

$$2\pi r + 2l = \tfrac{1}{4}$$
$$2l = \tfrac{1}{4} - 2\pi r$$
$$2l = \frac{1 - 8\pi r}{4}$$
$$l = \frac{1 - 8\pi r}{8}$$

We now find the area A. The area of each semicircle is $\frac{1}{2}\pi r^2$, and the area of the rectangle is length • width, where $w = 2r$, so:

$$A = \tfrac{1}{2}\pi r^2 + \tfrac{1}{2}\pi r^2 + lw$$
$$= \pi r^2 + \left(\frac{1 - 8\pi r}{8}\right)(2r)$$
$$= \pi r^2 + \frac{r - 8\pi r^2}{4}$$
$$= \frac{4\pi r^2 + r - 8\pi r^2}{4}$$
$$= \frac{r - 4\pi r^2}{4}$$

Using functional notation we have $A(r) = \dfrac{r(1 - 4\pi r)}{4}$.

39. We draw the figure:

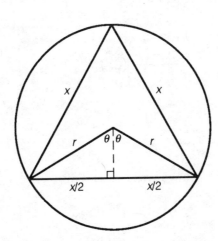

Using geometry, we see that:

$2\theta = \frac{1}{3}(360°)$

$2\theta = 120°$

$\theta = 60°$

Since $\theta = 60°$, then:

$\dfrac{x}{2} = \dfrac{\sqrt{3}}{2}r$

$x = \sqrt{3}r$

$r = \dfrac{x}{\sqrt{3}}$

So the area of the circle A is:

$$A = \pi r^2 = \pi\left(\frac{x}{\sqrt{3}}\right)^2 = \frac{\pi x^2}{3}$$

Using functional notation we have $A(x) = \frac{\pi}{3}x^2$.

41. Since $V = l \bullet w \bullet h$ and $l = 8 - 2w$, $w = 6 - 2x$, $h = x$, we have:

$$V = (8 - 2x)(6 - 2x)(x) = \left(48 - 28x + 4x^2\right)(x) = 4x^3 - 28x^2 + 48x$$

Using functional notation we have $V(x) = 4x^3 - 28x^2 + 48x$.

43. (a) The area of the window would be:

$A = \frac{1}{2}\left(\pi r^2\right) + lw$

It remains to find l and w in terms of r. We see that $w = 2r$, and the perimeter $P = 32$, so:

$P = \frac{1}{2}(2\pi r) + 2l + w$

Therefore:

$\frac{1}{2}(2\pi r) + 2l + w = 32$

$\pi r + 2l + 2r = 32$

$2l = 32 - \pi r - 2r$

$l = \dfrac{32 - \pi r - 2r}{2}$

We now find the area:

$A = \frac{1}{2}\left(\pi r^2\right) + lw$

$= \frac{1}{2}\left(\pi r^2\right) + \left(\dfrac{32 - \pi r - 2r}{2}\right)(2r)$

$= \dfrac{\pi r^2}{2} + 32r - \pi r^2 - 2r^2$

$= 32r - 2r^2 - \dfrac{\pi r^2}{2}$

Using functional notation we write $A(r) = 32r - 2r^2 - \frac{\pi r^2}{2}$.

(b)   We have $A(r) = -\left(\frac{4+\pi}{2}\right)r^2 + 32r$, which will open downward. Since $A(0) = 0$, it does pass through the origin. We complete the square:

$$A(r) = -\left(\frac{4+\pi}{2}\right)\left(r^2 - \frac{64}{4+\pi}r\right)$$

$$= -\left(\frac{4+\pi}{2}\right)\left[r^2 - \frac{64}{4+\pi}r + \left(\frac{32}{4+\pi}\right)^2\right] + \left(\frac{4+\pi}{2}\right)\left(\frac{32}{4+\pi}\right)^2$$

$$= -\left(\frac{4+\pi}{2}\right)\left(r - \frac{32}{4+\pi}\right)^2 + \frac{512}{4+\pi}$$

So the vertex is $\left(\frac{32}{4+\pi}, \frac{512}{4+\pi}\right)$.

45. (a)   We use the Pythagorean theorem:

$$3^2 + y^2 = z^2$$

Taking roots:

$$z = \sqrt{y^2 + 9}$$

Now $s = \frac{y}{z} = \frac{y}{\sqrt{y^2+9}}$, so $s\sqrt{y^2+9} = y$. Squaring, we obtain:

$$s^2(y^2 + 9) = y^2$$
$$s^2y^2 + 9s^2 = y^2$$
$$y^2 - s^2y^2 = 9s^2$$
$$y^2(1 - s^2) = 9s^2$$
$$y^2 = \frac{9s^2}{1 - s^2}$$
$$y = \frac{3s}{\sqrt{1 - s^2}}$$

Using functional notation we have $y(s) = \frac{3s}{\sqrt{1-s^2}}$.

(b)   From part (a) we had $s(y) = \frac{y}{\sqrt{y^2+9}}$.

(c)   Since $s = \frac{y}{z}$, then $z = \frac{y}{s}$. Using our result from (a), we have:

$$z = \frac{y}{s} = \frac{\frac{3s}{\sqrt{1-s^2}}}{s} = \frac{3}{\sqrt{1 - s^2}}$$

Using functional notation we have $z(s) = \frac{3}{\sqrt{1-s^2}}$.

(d)   Using our answer from (c), we have:

$$z = \frac{3}{\sqrt{1 - s^2}}$$
$$z\sqrt{1 - s^2} = 3$$

Squaring each side, we obtain:
$$z^2(1-s^2) = 9$$
$$z^2 - z^2 s^2 = 9$$
$$-z^2 s^2 = 9 - z^2$$
$$s^2 = \frac{z^2 - 9}{z^2}$$
$$s = \frac{\sqrt{z^2 - 9}}{z}$$

Using functional notation we have $s(z) = \frac{\sqrt{z^2-9}}{z}$.

47. (a) Using the points $(0, -1)$ and (a, a^2), we have:
$$m(a) = \frac{a^2 - (-1)}{a - 0} = \frac{a^2 + 1}{a}$$

(b) If x_0 is the x-intercept, then the area of the triangle A is:
$$A = \tfrac{1}{2}(\text{base})(\text{height}) = \tfrac{1}{2}(a - x_0)(a^2)$$

To find the x-intercept, we must find the equation of the line. We use $m = \frac{a^2+1}{a}$ (from (a) above) and $(0, -1)$ in the slope-intercept formula to obtain:
$$y = \frac{a^2 + 1}{a} x - 1$$
We find x_0 by letting $y = 0$:
$$0 = \frac{a^2 + 1}{a} x_0 - 1$$
$$\frac{a^2 + 1}{a} x_0 = 1$$
$$x_0 = \frac{a}{a^2 + 1}$$
Therefore:
$$A = \tfrac{1}{2}(a - x_0)a^2$$
$$= \tfrac{1}{2}\left(a - \frac{a}{a^2 + 1}\right)a^2$$
$$= \frac{a^2}{2}\left[\frac{a(a^2 + 1) - a}{a^2 + 1}\right]$$
$$= \frac{a^2(a^3 + a - a)}{2(a^2 + 1)}$$
$$= \frac{a^2(a^3)}{2(a^2 + 1)}$$
$$= \frac{a^5}{2(a^2 + 1)}$$

49. We re-draw the figure (differently):

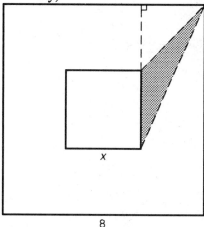

We extend the triangle to form a right triangle as pictured. Now find the areas of the large and small right triangles:

$$A_{\text{large}} = \tfrac{1}{2}(\text{base})(\text{height})$$

$$= \tfrac{1}{2}\left(x + \frac{8-x}{2}\right)\left(\frac{8-x}{2}\right)$$

$$= \tfrac{1}{2}\left(\frac{2x+8-x}{2}\right)\left(\frac{8-x}{2}\right)$$

$$= \frac{(x+8)(8-x)}{8}$$

$$= \frac{64-x^2}{8}$$

$$A_{\text{small}} = \tfrac{1}{2}(\text{base})(\text{height}) = \tfrac{1}{2}\left(\frac{8-x}{2}\right)\left(\frac{8-x}{2}\right) = \frac{64-16x+x^2}{8}$$

Therefore:

$$A = A_{\text{large}} - A_{\text{small}}$$

$$= \frac{64-x^2}{8} - \frac{64-16x+x^2}{8}$$

$$= \frac{64-x^2-64+16x-x^2}{8}$$

$$= \frac{16x-2x^2}{8}$$

$$= \frac{8x-x^2}{4}$$

Using functional notation we write $A(x) = \dfrac{8x-x^2}{4}$.

Note: A much easier approach is to realize that the altitude of the triangle need not lie on the triangle. That is:

$$\text{Area} = \tfrac{1}{2}(\text{base})(\text{altitude}) = \tfrac{1}{2}(x)\left(\frac{8-x}{2}\right) = \frac{8x - x^2}{4}$$

Both approaches are correct.

51. We draw the diagram:

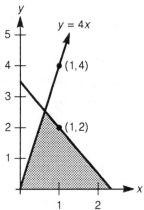

We are asked to find the area A of the shaded triangle. Since the line has slope m and passes through $(1,2)$, then by the point-slope formula, we have:

$$y - 2 = m(x - 1)$$
$$y - 2 = mx - m$$
$$y = mx + (2 - m)$$

Its x-intercept is where $y = 0$:

$$0 = mx + (2 - m)$$
$$mx = m - 2$$
$$x = \frac{m - 2}{m}$$

This is the base of a triangle. To find its height, we must find the value of y where this line and $y = 4x$ intersect. We set the two y-values equal:

$$mx + 2 - m = 4x$$
$$mx - 4x = m - 2$$
$$x(m - 4) = m - 2$$
$$x = \frac{m - 2}{m - 4}$$

Since this point lies on $y = 4x$, its y-coordinate is:

$$y = 4x = 4\left(\frac{m - 2}{m - 4}\right)$$

Finally, we find the area:

$$A = \tfrac{1}{2}(\text{base})(\text{height})$$

$$= \tfrac{1}{2}\left(\frac{m-2}{m}\right)(4)\left(\frac{m-2}{m-4}\right)$$

$$= 2\frac{(m-2)^2}{m(m-4)}$$

$$= \frac{2(m^2-4m+4)}{m^2-4m}$$

$$= \frac{2m^2-8m+8}{m^2-4m}$$

Using functional notation we have $A(m) = \dfrac{2m^2-8m+8}{m^2-4m}$.

53. We draw a figure:

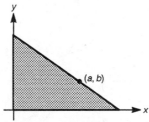

The line has the equation $y - b = m(x - a)$. Since the base and height are the x- and y-intercepts, respectively, we find each intercept:

base: $y = 0$
$$-b = m(x-a)$$
$$-b = mx - ma$$
$$mx = ma - b$$
$$x = \frac{ma-b}{m}$$

height: $x = 0$
$$y - b = m(-a)$$
$$y = b - ma$$

So the area of the triangle is $A = \tfrac{1}{2}\left(\frac{ma-b}{m}\right)(b-ma) = \frac{(ma-b)^2}{-2m}$.

4.4 Maximum and Minimum Problems

1. Call the two numbers x and y. Then $x + y = 5$, so $y = 5 - x$. So the product can be written as:

$$P = xy = x(5-x) = 5x - x^2$$

We now complete the square:

$$P = -\left(x^2 - 5x\right) = -\left(x^2 - 5x + \tfrac{25}{4}\right) + \tfrac{25}{4} = -\left(x - \tfrac{5}{2}\right)^2 + \tfrac{25}{4}$$

Since this is a parabola opening downward, it will have a maximum value of $\tfrac{25}{4}$.

3. Call the two numbers x and y. Then $y - x = 1$, so $y = x + 1$. The sum of their squares can be written as:
$$S = x^2 + y^2 = x^2 + (x+1)^2 = x^2 + x^2 + 2x + 1 = 2x^2 + 2x + 1$$
We now complete the square:
$$S = 2(x^2 + x) + 1 = 2(x^2 + x + \tfrac{1}{4}) + 1 - \tfrac{1}{2} = 2(x + \tfrac{1}{2})^2 + \tfrac{1}{2}$$
Since this is a parabola opening upward, it will have a minimum value of $\tfrac{1}{2}$.

5. Let w and l be the width and length, respectively. Since $P = 2w + 2l$, then:
$$2w + 2l = 25$$
$$2l = 25 - 2w$$
$$l = \frac{25 - 2w}{2}$$
So the area is given by:
$$A = wl = w\left(\frac{25 - 2w}{2}\right) = \tfrac{1}{2}(-2w^2 + 25)$$
We now complete the square:
$$A = -\left(w^2 - \tfrac{25}{2}w\right) = -\left(w^2 - \tfrac{25}{2}w + \tfrac{625}{16}\right) = -\left(w - \tfrac{25}{4}\right)^2 + \tfrac{625}{16}$$
This is a parabola opening downward, so it will achieve a maximum value when $w = \tfrac{25}{4}$.
We find l:
$$l = \frac{25 - 2\left(\tfrac{25}{4}\right)}{2} = \frac{25 - \tfrac{25}{2}}{2} = \tfrac{25}{4}$$
So the largest such rectangle is a square of dimensions $\tfrac{25}{4}$ m by $\tfrac{25}{4}$ m.

7. Let x and y be the lengths of the two shorter sides, so $x + y = 100$, and $y = 100 - x$. Then the area is given by:
$$A = \tfrac{1}{2}xy = \tfrac{1}{2}x(100 - x) = \tfrac{1}{2}(-x^2 + 100x)$$
We now complete the square:
$$A = -\tfrac{1}{2}(x^2 - 100x) = -\tfrac{1}{2}(x^2 - 100x + 2500) + 1250 = -\tfrac{1}{2}(x - 50)^2 + 1250$$
This is a parabola opening downward, so it will achieve a maximum value of 1250 in.^{2}.

9. Let x and y be the two numbers, so $x + y = 6$ and thus $y = 6 - x$.

(a) We have:
$$T = x^2 + y^2 = x^2 + (6 - x)^2 = x^2 + 36 - 12x + x^2 = 2x^2 - 12x + 36$$
We now complete the square:
$$T = 2(x^2 - 6x) + 36 = 2(x^2 - 6x + 9) + 36 - 18 = 2(x - 3)^2 + 18$$
This is a parabola opening upward, so it will have a minimum value of 18.

(b) We have:
$$S = x + y^2 = x + (6 - x)^2 = x + 36 - 12x + x^2 = x^2 - 11x + 36$$

We now complete the square:

$$S = \left(x^2 - 11x\right) + 36 = \left(x^2 - 11x + \tfrac{121}{4}\right) + 36 - \tfrac{121}{4} = \left(x - \tfrac{11}{2}\right)^2 + \tfrac{23}{4}.$$

This is a parabola opening upward, so it will have a minimum value of $\tfrac{23}{4}$.

(c) We have:

$$U = x + 2y^2 = x + 2(6 - x)^2 = x + 72 - 24x + 2x^2 = 2x^2 - 23x + 72$$

We now complete the square:

$$U = 2\left(x^2 - \tfrac{23}{2}x\right) + 72 = 2\left(x^2 - \tfrac{23}{2}x + \tfrac{529}{16}\right) + 72 - \tfrac{529}{8} = 2\left(x - \tfrac{23}{4}\right)^2 + \tfrac{47}{8}$$

This is a parabola opening upward, so it will have a minimum value of $\tfrac{47}{8}$.

(d) We have:

$$\begin{aligned}
V &= x + (2y)^2 \\
&= x + 4y^2 \\
&= x + 4(6 - x)^2 \\
&= x + 144 - 48x + 4x^2 \\
&= 4x^2 - 47x + 144
\end{aligned}$$

We now complete the square:

$$V = 4\left(x^2 - \tfrac{47}{4}x\right) + 144 = 4\left(x^2 - \tfrac{47}{4}x + \tfrac{2209}{64}\right) + 144 - \tfrac{2209}{16} = 4\left(x - \tfrac{47}{8}\right)^2 + \tfrac{95}{16}$$

This is a parabola opening upward, so it will have a minimum value of $\tfrac{95}{16}$.

11. (a) We compute the heights:

$$h(1) = -16(1)^2 + 32(1) = -16 + 32 = 16 \text{ ft}$$

$$h\left(\tfrac{3}{2}\right) = -16\left(\tfrac{3}{2}\right)^2 + 32\left(\tfrac{3}{2}\right) = -16\left(\tfrac{9}{4}\right) + 48 = -36 + 48 = 12 \text{ ft}$$

(b) We complete the square:

$$h = -16t^2 + 32t = -16\left(t^2 - 2t\right) = -16\left(t^2 - 2t + 1\right) + 16 = -16(t - 1)^2 + 16$$

This is a parabola opening downward, so it will have a maximum height of 16 ft, attained after 1 second.

(c) We set $h = 7$:

$$7 = -16t^2 + 32t$$
$$16t^2 - 32t + 7 = 0$$
$$(4t - 7)(4t - 1) = 0$$
$$t = \tfrac{7}{4}, \tfrac{1}{4}$$

So $h = 7$ when $t = \tfrac{7}{4}$ sec or $t = \tfrac{1}{4}$ sec.

13. Every point on the given curve has coordinates of the form $\left(x, \sqrt{x - 2} + 1\right)$, and using the distance formula gives:

$$d = \sqrt{(4 - x)^2 + \left(1 - \sqrt{x - 2} - 1\right)^2} = \sqrt{16 - 8x + x^2 + x - 2} = \sqrt{x^2 - 7x + 14}$$

We complete the square:
$$\left(x^2 - 7x\right) + 14 = \left(x^2 - 7x + \tfrac{49}{4}\right) + 14 - \tfrac{49}{4} = \left(x - \tfrac{7}{2}\right)^2 + \tfrac{7}{4}$$

This is a parabola opening upward which will achieve a minimum value of $\sqrt{\tfrac{7}{4}} = \tfrac{\sqrt{7}}{2}$ at $x = \tfrac{7}{2}$. Then:
$$y = \sqrt{\tfrac{7}{2} - 2} + 1 = \sqrt{\tfrac{3}{2}} + 1 = \tfrac{2 + \sqrt{6}}{2}$$

So the point is $\left(\tfrac{7}{2}, \tfrac{2+\sqrt{6}}{2}\right)$ and the distance is $\tfrac{\sqrt{7}}{2}$.

15. (a) We must find the value of x such that $x - x^2$ is as large as possible. Call $f(x) = -x^2 + x$. We complete the square:
$$f(x) = -\left(x^2 - x\right) = -\left(x^2 - x + \tfrac{1}{4}\right) + \tfrac{1}{4} = -\left(x - \tfrac{1}{2}\right)^2 + \tfrac{1}{4}$$

This is a parabola opening downward, so it will achieve a maximum value when $x = \tfrac{1}{2}$. So the number is $\tfrac{1}{2}$.

(b) We must find the value of x such that $x - 2x^2$ is as large as possible. Call $f(x) = -2x^2 + x$. We complete the square:
$$f(x) = -2\left(x^2 - \tfrac{1}{2}x\right) = -2\left(x^2 - \tfrac{1}{2}x + \tfrac{1}{16}\right) + \tfrac{1}{8} = -2\left(x - \tfrac{1}{4}\right)^2 + \tfrac{1}{8}$$

This is a parabola opening downward, so it will achieve a maximum value when $x = \tfrac{1}{4}$. So the number is $\tfrac{1}{4}$.

17. If we choose x for the depth of the pasture, then $500 - 2x$ is the length parallel to the river. The area of the pasture will then be given by:
$$A = x(500 - 2x) = -2x^2 + 500x$$

We complete the square:
$$A = -2\left(x^2 - 250x\right) = -2\left(x^2 - 250x + 125^2\right) + 2(125)^2 = -2(x - 125)^2 + 31250$$

This is a parabola opening downward, so it will achieve a maximum value at $x = 125$. Then the length is $500 - 2(125) = 500 - 250 = 250$. So the dimensions are 125 ft by 250 ft.

19. We have:
$$\begin{aligned}
R - C &= \left(0.4x^2 + 10x + 5\right) - \left(0.5x^2 + 2x + 101\right) \\
&= 0.4x^2 + 10x + 5 - 0.5x^2 - 2x - 101 \\
&= -0.1x^2 + 8x - 96
\end{aligned}$$

We now complete the square:
$$\begin{aligned}
R - C &= -0.1\left(x^2 - 80x\right) - 96 \\
&= -0.1\left(x^2 - 80x + 1600\right) - 96 + 160 \\
&= -0.1(x - 40)^2 + 64
\end{aligned}$$

This is a parabola opening downward, so it will achieve a maximum value when $x = 40$.

21. Recall that revenue R is $x \cdot p$. So:
$$R = x\left(-\tfrac{1}{4}x + 30\right) = -\tfrac{1}{4}x^2 + 30x$$
We complete the square:
$$R = -\tfrac{1}{4}x^2 + 30x$$
$$= -\tfrac{1}{4}\left(x^2 - 120x\right)$$
$$= -\tfrac{1}{4}\left(x^2 - 120x + 3600\right) + 900$$
$$= -\tfrac{1}{4}(x - 60)^2 + 900$$
This is a parabola opening downward, so it will achieve a maximum value at $x = 60$. The maximum revenue is \$900. The corresponding unit price p is:
$$p = -\tfrac{1}{4}(60) + 30 = -15 + 30 = \$15$$

23. (a) To use max/min methods, we need to substitute in the quantity $x^2 + y^2$ and write it strictly in terms of x or y. So take $2x + 3y = 6$ and solve for y:
$$3y = 6 - 2x$$
$$y = \frac{6 - 2x}{3}$$
Then substitute, and the quantity $x^2 + y^2$ becomes:
$$Q = x^2 + \left(\frac{6 - 2x}{3}\right)^2 = x^2 + \frac{36 - 24x + 4x^2}{9} = \tfrac{13}{9}x^2 - \tfrac{8}{3}x + 4$$
We now complete the square:
$$Q = \tfrac{13}{9}\left(x^2 - \tfrac{24}{13}x\right) + 4 = \tfrac{13}{9}\left(x^2 - \tfrac{24}{13}x + \tfrac{144}{169}\right) + 4 - \tfrac{13}{9}\left(\tfrac{144}{169}\right) = \tfrac{13}{9}\left(x - \tfrac{12}{13}\right)^2 + \tfrac{36}{13}$$
This is a parabola opening up, so it will achieve a minimum value of $\tfrac{36}{13}$.

(b) The equation of a circle with its center at the origin is $x^2 + y^2 = r^2$ where r is the radius. The line $2x + 3y = 6$ will intersect the circle in two points wherever r is sufficiently large. As we reduce r, we gradually reach a position where the circle and line are tangent and this is the minimum value of r or $\sqrt{x^2 + y^2}$. In this case, it is $\sqrt{\tfrac{36}{13}} = \tfrac{6\sqrt{13}}{13}$. This is the square root of the answer from (a).

25. (a) We substitute $y = 15 - x$:
$$Q = x^2 + y^2 = x^2 + (15 - x)^2 = x^2 + 225 - 30x + x^2 = 2x^2 - 30x + 225$$
We now complete the square:
$$Q = 2x^2 - 30x + 225$$
$$= 2\left(x^2 - 15x\right) + 225$$
$$= 2\left(x^2 - 15x + \tfrac{225}{4}\right) + 225 - \tfrac{225}{2}$$
$$= 2\left(x - \tfrac{15}{2}\right)^2 + \tfrac{225}{2}$$
This is a parabola opening upward, so it will achieve a minimum value of $\tfrac{225}{2}$.

(b)   We substitute $y = C - x$:
$$Q = x^2 + y^2 = x^2 + (C - x)^2 = x^2 + C^2 - 2Cx + x^2 = 2x^2 - 2Cx + C^2$$
We now complete the square:
$$\begin{aligned} Q &= 2x^2 - 2Cx + C^2 \\ &= 2(x^2 - Cx) + C^2 \\ &= 2\left(x^2 - Cx + \tfrac{C^2}{4}\right) + C^2 - \tfrac{C^2}{2} \\ &= 2\left(x - \tfrac{C}{2}\right)^2 + \tfrac{C^2}{2} \end{aligned}$$
This is a parabola opening upward, so it will achieve a minimum value of $\tfrac{C^2}{2}$.
When $C = 15$, the result from (a) is verified.

27.   Let the other two sides of each of the four triangles be t and $1 - t$, respectively. Then the area of the square will be a minimum when the area of these triangles is a maximum. Now write an expression for the total area of the four triangles:
$$A = 4\left(\tfrac{1}{2}\right)(t)(1 - t) = 2t - 2t^2 = -2t^2 + 2t$$
We complete the square:
$$A = -2t^2 + 2t = -2(t^2 - t) = -2\left(t^2 - t + \tfrac{1}{4}\right) + \tfrac{1}{2} = -2\left(t - \tfrac{1}{2}\right)^2 + \tfrac{1}{2}$$
This is a parabola opening downward, so it will achieve a maximum area of $\tfrac{1}{2}$ when $t = \tfrac{1}{2}$. Since the large square has area $= 1$, then the smaller square has area $1 - A = 1 - \tfrac{1}{2} = \tfrac{1}{2}$. We use the Pythagorean theorem to find x:
$$x^2 = t^2 + (1 - t)^2 = \tfrac{1}{4} + \tfrac{1}{4} = \tfrac{1}{2}$$
So $x = \sqrt{\tfrac{1}{2}} = \tfrac{\sqrt{2}}{2}$.

29.   We draw the figure:

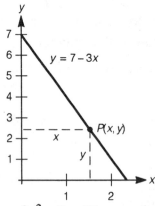

We have $A = xy = x(7 - 3x) = -3x^2 + 7x$. We complete the square:
$$A = -3\left(x^2 - \tfrac{7}{3}x\right) = -3\left(x^2 - \tfrac{7}{3}x + \tfrac{49}{36}\right) + \tfrac{49}{12} = -3\left(x - \tfrac{7}{6}\right)^2 + \tfrac{49}{12}$$
Since this parabola opens downward, the largest possible area is $\tfrac{49}{12}$.

31. We draw the figure:

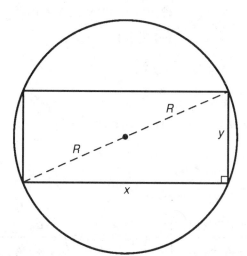

Using the Pythagorean theorem, we obtain $y = \sqrt{4R^2 - x^2}$. The area A of the rectangle is:

$$A = xy = x\sqrt{4R^2 - x^2}$$

Therefore:

$$A^2 = x^2(4R^2 - x^2) = 4R^2x^2 - x^4$$

In order to maximize the expression $4R^2x^2 - x^4$, first let $t = x^2$, so that the expression becomes $4R^2t - t^2$. We complete the square:

$$
\begin{aligned}
A^2 &= -t^2 + 4R^2t \\
&= -(t^2 - 4R^2t) \\
&= -(t^2 - 4R^2t + 4R^4) + 4R^4 \\
&= -(t - 2R^2)^2 + 4R^4
\end{aligned}
$$

This parabola opens downward, so the maximum value of A^2 is $4R^4$. Thus the maximum area is $\sqrt{4R^4} = 2R^2$.

33. Let x = east-west dimension, and y = north-south dimension. So the cost is given by $C = 12(2x) + 8(2y) = 24x + 16y$. Since this cost is \$4800, we have $24x + 16y = 4800$, so:

$$y = \frac{4800 - 24x}{16} = \frac{600 - 3x}{2}$$

Now the area is $A = xy = x\left(\dfrac{600 - 3x}{2}\right) = -\frac{3}{2}x^2 + 300x$. So $A(x) = -\frac{3}{2}x^2 + 300x$.

This will be a parabola opening downward, so it will have a maximum value. We complete the square:

$$
\begin{aligned}
A(x) &= -\tfrac{3}{2}x^2 + 300x \\
&= -\tfrac{3}{2}(x^2 - 200x) \\
&= -\tfrac{3}{2}(x^2 - 200x + 100^2) + 15000 \\
&= -\tfrac{3}{2}(x - 100)^2 + 15000
\end{aligned}
$$

So $x = 100$ will maximize area, which is 15,000 yd^2. We find y:
$$y = \frac{600 - 3(100)}{2} = \frac{600 - 300}{2} = 150 \text{ yd}$$
So the dimensions are 100 yd by 150 yd.

35. The given function can be rewritten:
$$y = (a_1 + a_2)x^2 - 2(a_1x_1 + a_2x_2)x + (a_1x_1^2 + a_2x_2^2)$$
We complete the square:
$$y = (a_1 + a_2)\left[x^2 - \frac{2(a_1x_1 + a_2x_2)}{a_1 + a_2} \right] + (a_1x_1^2 + a_2x_2^2)$$
$$= (a_1 + a_2)\left(x - \frac{a_1x_1 + a_2x_2}{a_1 + a_2} \right)^2 + (a_1x_1^2 + a_2x_2^2) - \frac{(a_1x_1 + a_2x_2)^2}{a_1 + a_2}$$
Since a_1 and a_2 are both positive, then $a_1 + a_2 > 0$ and thus this parabola opens upward.
So the minimum must occur where $x = \frac{a_1x_1 + a_2x_2}{a_1 + a_2}$.

37. (a) We have $\frac{\Delta p}{\Delta x} = \frac{10}{-5} = -2$. Also $p = 200$ when $x = 150$. We use the point-slope formula
with the point $(150, 200)$:
$$p - 200 = -2(x - 150)$$
$$p - 200 = -2x + 300$$
$$p = -2x + 500$$
Using functional notation we have $p(x) = -2x + 500$.

   (b) Since $R = xp$, we have $R = x(-2x + 500) = -2x^2 + 500x$. We complete the square:
$$R = -2x^2 + 500x$$
$$= -2(x^2 - 250x)$$
$$= -2(x^2 - 250x + 15625) + 31250$$
$$= -2(x - 125)^2 + 31250$$
Using functional notation we have $R(x) = -2(x - 125)^2 + 31250$. Since this
parabola opens downward, we have a maximum revenue of \$31,250 when $x = 125$.
We find $p = -2(125) + 500 = \$250$.

39. Let $x = t^2$, so $f(x) = x - x^2 = -x^2 + x$. We complete the square:
$$f(x) = -(x^2 - x) = -\left(x^2 - x + \tfrac{1}{4}\right) + \tfrac{1}{4} = -\left(x - \tfrac{1}{2}\right)^2 + \tfrac{1}{4}$$
So $f(x)$ has a maximum value when $x = \tfrac{1}{2}$. Then $t^2 = \tfrac{1}{2}$, so $t = \pm\frac{\sqrt{2}}{2}$.

41. Let $x = t^2$. Then $y = -t^4 + 6t^2 - 6 = -x^2 + 6x - 6$. We complete the square:
$$y = -x^2 + 6x - 6 = -(x^2 - 6x) - 6 = -(x^2 - 6x + 9) - 6 + 9 = -(x - 3)^2 + 3$$

So $x = 3$ will yield the largest output. Since $x = t^2$, we have:

$$t^2 = 3$$
$$t = \pm\sqrt{3}$$

So $t = \sqrt{3}$ or $t = -\sqrt{3}$ will yield the largest output.

43. (a) For the circle, we have $2\pi r = x$, so $r = \frac{x}{2\pi}$, and thus the area is:

$$A = \pi r^2 = \pi \left(\frac{x}{2\pi}\right)^2 = \frac{x^2}{4\pi}$$

For the square, we have $4s = 16 - x$, so $s = \frac{16-x}{4}$, and thus the area is:

$$A = s^2 = \left(\frac{16-x}{4}\right)^2 = \frac{256 - 32x + x^2}{16} = 16 - 2x + \tfrac{1}{16}x^2$$

So the total combined area is given by:

$$A(x) = \frac{x^2}{4\pi} + 16 - 2x + \tfrac{1}{16}x^2 = \tfrac{4+\pi}{16\pi}x^2 - 2x + 16$$

(b) This is a parabola opening upward, so it will have a minimum value at:

$$x = \frac{2}{2 \cdot \frac{4+\pi}{16\pi}} = \frac{16\pi}{4+\pi}$$

(c) This ratio is given by:

$$\frac{\frac{16\pi}{4+\pi}}{16 - \frac{16\pi}{4+\pi}} \cdot \frac{4+\pi}{4+\pi} = \frac{16\pi}{64 + 16\pi - 16\pi} = \frac{16\pi}{64} = \frac{\pi}{4}$$

45. (a) For the circle, we have $2\pi r = x$, so $r = \frac{x}{2\pi}$, and thus the area is:

$$A = \pi r^2 = \pi \left(\frac{x}{2\pi}\right)^2 = \frac{x^2}{4\pi}$$

For the square, we have $4s = L - x$, so $s = \frac{L-x}{4}$, and thus the area is:

$$A = \left(\frac{L-x}{4}\right)^2 = \frac{L^2 - 2Lx + x^2}{16} = \tfrac{1}{16}x^2 - \frac{L}{8}x + \frac{L^2}{16}$$

So the total combined area is given by:

$$A(x) = \left(\tfrac{1}{4\pi} + \tfrac{1}{16}\right)x^2 - \frac{L}{8}x + \frac{L^2}{16} = \frac{4+\pi}{16\pi}x^2 - \frac{L}{8}x + \frac{L^2}{16}$$

(b) This will have a minimum value when $x = -\frac{b}{2a}$, so:

$$x = \frac{\frac{L}{8}}{\frac{4+\pi}{8\pi}} \cdot \frac{8\pi}{8\pi} = \frac{\pi L}{4+\pi}$$

(c)   This ratio is:

$$\frac{\frac{\pi L}{4+\pi}}{L-\frac{\pi L}{4+\pi}} \cdot \frac{4+\pi}{4+\pi} = \frac{\pi L}{4L+\pi L - \pi L} = \frac{\pi L}{4L} = \frac{\pi}{4}$$

It is interesting to note that this ratio does not depend on L, the length of the wire.

47.   We re-draw the figure and label additional sides:

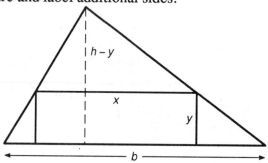

By similar triangles:

$$\frac{h-y}{h} = \frac{x}{b}$$
$$bh - by = xh$$
$$by = bh - xh$$
$$y = h - \frac{h}{b}x$$

So the area of the rectangle is given by:

$$A = xy = x\left(h - \frac{h}{b}x\right) = -\frac{h}{b}x^2 + hx$$

This is a parabola opening downward, so it will have a maximum value when:

$$x = \frac{-h}{2\left(-\frac{h}{b}\right)} = \frac{b}{2}$$

So the maximum area is given by:

$$A\left(\frac{b}{2}\right) = -\frac{h}{b} \cdot \frac{b^2}{4} + h \cdot \frac{b}{2} = -\frac{hb}{4} + \frac{hb}{2} = \frac{hb}{4}$$

Since the triangle has an area of $\frac{hb}{2}$, then the desired ratio is:

$$\frac{\frac{hb}{2}}{\frac{hb}{4}} = 2$$

49. Call h the height of the triangle (see figure), and call y the indicated value:

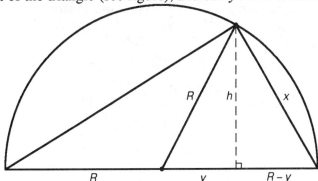

By the Pythagorean theorem, we have the following relationships:

$$y^2 + h^2 = R^2 \quad \text{and} \quad (R-y)^2 + h^2 = x^2$$
$$y^2 = R^2 - h^2$$
$$y = \sqrt{R^2 - h^2}$$

Substituting, we have:

$$\left(R - \sqrt{R^2 - h^2}\right)^2 + h^2 = x^2$$
$$R^2 - 2R\sqrt{R^2 - h^2} + R^2 - h^2 + h^2 = x^2$$
$$2R^2 - 2R\sqrt{R^2 - h^2} = x^2$$

We solve for h:

$$-2R\sqrt{R^2 - h^2} = x^2 - 2R^2$$
$$\sqrt{R^2 - h^2} = R - \frac{x^2}{2R}$$
$$R^2 - h^2 = R^2 - x^2 + \frac{x^4}{4R^2}$$
$$-h^2 = \frac{x^4}{4R^2} - x^2$$
$$h^2 = -\frac{x^4}{4R^2} + x^2$$

We now use the hint. Since the area of the triangle is given by $A = \frac{1}{2}(2R)(h) = Rh$, we maximize:

$$A^2 = R^2 h^2 = R^2\left(\frac{-x^4}{4R^2} + x^2\right) = -\tfrac{1}{4}x^4 + R^2 x^2$$

This will have a maximum value when:

$$x^2 = \frac{-b}{2a} = \frac{-R^2}{-\frac{1}{2}} = 2R^2$$

Therefore we have:

$$A^2 = -\tfrac{1}{4}\left(2R^2\right)^2 + R^2\left(2R^2\right) = -R^4 + 2R^4 = R^4$$

So the maximum value is $A^2 = R^4$, or $A = R^2$. Thus the shaded area is:

$$\tfrac{1}{2}\left(\pi R^2\right) - R^2 = \frac{\pi R^2}{2} - R^2 = R^2\left(\frac{\pi - 2}{2}\right)$$

51. We wish to minimize the sum $x + \dfrac{1}{x}$. Completing the square:

$$x + \frac{1}{x} = \left(\sqrt{x} - \sqrt{\frac{1}{x}} \right)^2 + 2$$

The smallest possible value of 2 occurs if:

$$\sqrt{x} = \sqrt{\frac{1}{x}}$$

$$x = \frac{1}{x}$$

$$x^2 = 1$$

$$x = 1$$

If both numbers are 1, the minimum sum of 2 occurs.

53. Using the hint, we have:

$$G(x) = \frac{(a+x)(b+x)}{x}$$

$$= \frac{ab + (a+b)x + x^2}{x}$$

$$= \frac{ab}{x} + (a+b) + x$$

$$= (a+b) + \left(\frac{ab}{x} + x \right)$$

We can rewrite $G(x)$ as:

$$G(x) = \left(\sqrt{x} - \sqrt{\frac{ab}{x}} \right)^2 + \left(a + b - 2\sqrt{ab} \right) = \left(\sqrt{x} - \sqrt{\frac{ab}{x}} \right)^2 + \left(\sqrt{a} - \sqrt{b} \right)^2$$

Since $G(x)$ is the sum of two squares, we have $G(x) \geq 0$. Thus the minimum value occurs when:

$$\sqrt{x} - \sqrt{\frac{ab}{x}} = 0$$

$$\sqrt{x} = \frac{\sqrt{ab}}{\sqrt{x}}$$

$$x = \sqrt{ab}$$

So the minimum value is:

$$G\left(\sqrt{ab} \right) = (a+b) + \left(\frac{ab}{\sqrt{ab}} + \sqrt{ab} \right) = a + b + 2\sqrt{ab} = \left(\sqrt{a} + \sqrt{b} \right)^2$$

This proves the desired result.

55. (a) The largest value of $f(x)$ is approximately 1.143.

(b) Since $f(x) = \frac{1}{2x^2 - x + 1}$, its maximum value will be the same as the minimum value of $2x^2 - x + 1$. This occurs when:

$$x = \frac{-b}{2a} = \frac{1}{4}$$

Thus the maximum value of $f(x)$ is:

$$f\left(\tfrac{1}{4}\right) = \frac{1}{2\left(\tfrac{1}{4}\right)^2 - \tfrac{1}{4} + 1} = \frac{1}{\tfrac{1}{8} - \tfrac{1}{4} + 1} = \frac{1}{\tfrac{7}{8}} = \tfrac{8}{7}$$

So $f(x) \le \tfrac{8}{7}$ for every value of x.

Optional TI-81 Graphing Calculator Exercises for Section 4.4

1. (a) The value was obtained by completing the square, and noting that the vertex was $(4.5, 20.25)$.

(b) We graph P using the settings $X_{min} = 0$, $X_{max} = 10$, $Y_{min} = 0$, and $Y_{max} = 24$:

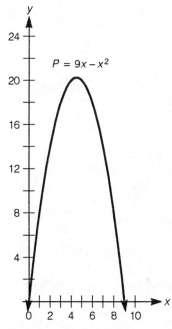

(c) Using the TRACE key, the maximum value is 20.249. This value is within 0.001 of the actual value 20.25.

3.   (a)   Using the settings $X_{min} = 0$, $X_{max} = 1$, $Y_{min} = 0.5$, and $Y_{max} = 1$:

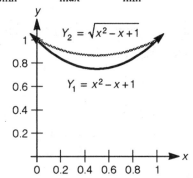

(b)   Using the TRACE key, the minimum values are attained when $x \approx 0.505$. This value is within 0.005 of the actual value of 1/2.

5.   (a)   Using the settings $X_{min} = 0$, $X_{max} = 3$, $Y_{min} = 0$, and $Y_{max} = 3$, we sketch the graphs:

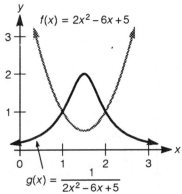

The statement appears to be valid.

(b)   The maximum value of g is approximately 1.998.

(c)   We first find the minimum value of f:

$$f(x) = 2\left(x^2 - 3x\right) + 5 = 2\left(x^2 - 3x + \tfrac{9}{4}\right) + 5 - \tfrac{9}{2} = 2\left(x - \tfrac{3}{2}\right)^2 + \tfrac{1}{2}$$

Since the minimum value of f is $\tfrac{1}{2}$, the maximum value of g is $\dfrac{1}{1/2} = 2$. Our estimate is within 0.002 of this actual value.

4.5 Polynomial Functions

1. This graph has 4 turning points, but a polynomial function of degree 3 can have at most 2 turning points.

3. As $|x|$ gets very large, our function should be similar to $f(x) = a_3x^3$. But $f(x)$ does not have a parabolic shape like the given graph.

5. As $|x|$ gets very large with x negative, then the graph should resemble $2x^5$. But the y-values of $2x^5$ are always negative when x is negative, contrary to the given graph.

7. This graph has a corner, which cannot occur in the graph of a polynomial function.

9. This is $y = x^2$ translated 2 units to the right and 1 unit up. There are no x-intercepts and the y-intercept is 5. We draw the graph:

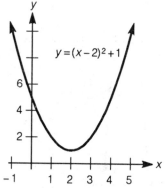

11. This is $y = x^4$ reflected across the x-axis and translated 1 unit to the right. The x-intercept is 1 and the y-intercept is -1. We draw the graph:

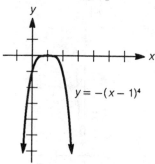

13.   This is $y = x^3$ translated 4 units to the right and 2 units down. The x-intercept is $4 + \sqrt[3]{2}$ and the y-intercept is -66. We draw the graph:

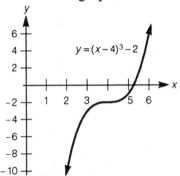

15.   This is $y = 2x^4$ translated 5 units to the right and reflected across the x-axis. The x-intercept is -5 and the y-intercept is -1250. We draw the graph:

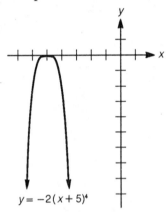

17.   This is $y = \frac{1}{2}x^5$ translated 1 unit to the left. The x-intercept is -1 and the y-intercept is $\frac{1}{2}$. We draw the graph:

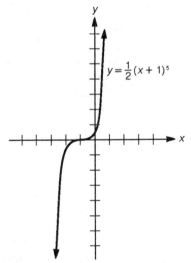

19. This is $y = x^3$ translated 1 unit to the right and 1 unit down, then reflected across the x-axis. The x- and y-intercepts are both 0. We draw the graph:

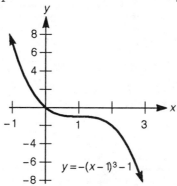

21. (a) The x-intercepts are 2, 1, and –1, and the y-intercept is 2. The signs of y are given by:

Interval	$y = (x-2)(x-1)(x+1)$
$(-\infty,-1)$	neg.
$(-1,1)$	pos.
$(1,2)$	neg.
$(2,\infty)$	pos.

We sketch the excluded regions:

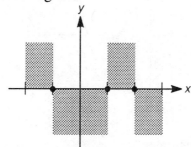

(b) We now graph the function:

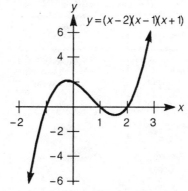

23. (a)   The x-intercepts are 0, 2, and 1, and the y-intercept is 0. The signs of y are given by:

Interval	$y = 2x(x-2)(x-1)$
$(-\infty,0)$	neg.
$(0,1)$	pos.
$(1,2)$	neg.
$(2,\infty)$	pos.

We sketch the excluded regions:

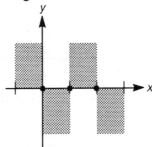

(b)   We now graph the function:

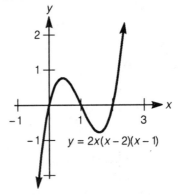

$$y = 2x(x-2)(x-1)$$

25. (a)   We factor to find the x-intercepts:

$$y = x\left(x^2 - 4x - 5\right) = x(x-5)(x+1)$$

The x-intercepts are 0, 5, and –1, and the y-intercept is 0. The signs of y are given by:

Interval	$y = x(x-5)(x+1)$
$(-\infty,-1)$	neg.
$(-1,0)$	pos.
$(0,5)$	neg.
$(5,\infty)$	pos.

We sketch the excluded regions:

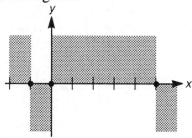

(b) We now graph the function:

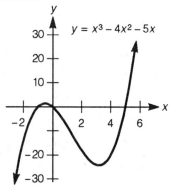

27. (a) We factor (by grouping) to find the x-intercepts:

$$y = x^2(x+3) - 4(x+3) = (x+3)(x^2-4) = (x+3)(x+2)(x-2)$$

The x-intercepts are –3, –2, and 2, and the y-intercept is –12. The signs of y are given by:

Interval	$y = (x+3)(x+2)(x-2)$
$(-\infty, -3)$	neg.
$(-3, -2)$	pos.
$(-2, 2)$	neg.
$(2, \infty)$	pos.

We sketch the excluded regions:

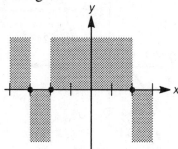

(b)   We now graph the function:

$$y = x^3 + 3x^2 - 4x - 12$$

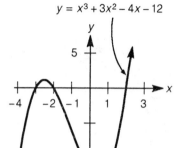

29. (a)   The x-intercepts are 0 and -2, and the y-intercept is 0. The signs of y are given by:

Interval	$y = x^3(x+2)$
$(-\infty, -2)$	pos.
$(-2, 0)$	neg.
$(0, \infty)$	pos.

We sketch the excluded regions:

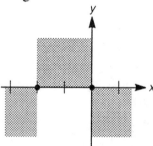

(b)   Near $x = 0$, we have the approximation $y \approx x^3(0 + 2) = 2x^3$. So in the immediate vicinity of $x = 0$, the graph of y resembles $y = 2x^3$:

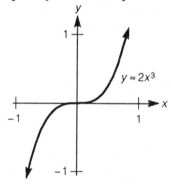

(c)   We now graph the function:

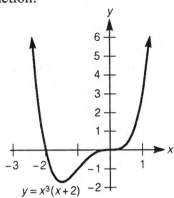

$$y = x^3(x+2)$$

31. (a)   The x-intercepts are 1 and 4, and the y-intercept is 128. The signs of y are given by:

Interval	$y = 2(x-1)(x-4)^3$
$(-\infty, 1)$	pos.
$(1, 4)$	neg.
$(4, \infty)$	pos.

We sketch the excluded regions:

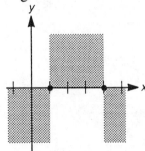

(b)   Near $x = 4$, we have the approximation $y \approx 2(4-1)(x-4)^3 = 6(x-4)^3$. So in the immediate vicinity of $x = 4$, the graph of y resembles $y = 6(x-4)^3$:

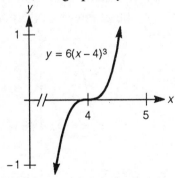

$$y = 6(x-4)^3$$

(c)   We now graph the function:

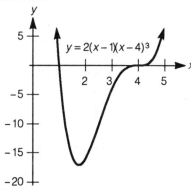

33. (a)   The x-intercepts are -1, 1, and 3, and the y-intercept is 3. The signs of y are given by:

Interval	$y = (x+1)^2(x-1)(x-3)$
$(-\infty,-1)$	pos.
$(-1,1)$	pos.
$(1,3)$	neg.
$(3,\infty)$	pos.

We sketch the excluded regions:

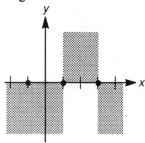

(b)   Near $x = -1$, we have the approximation $y \approx (x+1)^2(-2)(-4) = 8(x+1)^2$. So in the immediate vicinity of $x = -1$, the graph of y resembles $y = 8(x+1)^2$:

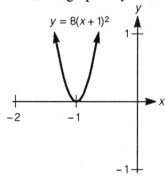

(c)   We now graph the function:

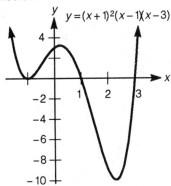

$$y = (x+1)^2(x-1)(x-3)$$

35. (a) The x-intercepts are 0, 4, and –2, and the y-intercept is 0. The signs of y are given by:

Interval	$y = -x^3(x-4)(x+2)$
$(-\infty, -2)$	pos.
$(-2, 0)$	neg.
$(0, 4)$	pos.
$(4, \infty)$	neg.

We sketch the excluded regions:

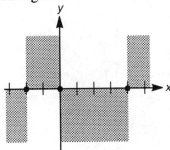

(b)   Near $x = 0$, we have the approximation $y \approx -x^3(0-4)(0+2) = 8x^3$. So in the immediate vicinity of $x = 0$, the graph resembles $y = 8x^3$:

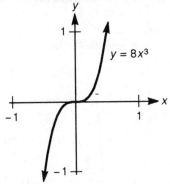

$$y = 8x^3$$

(c)   We now graph the function:

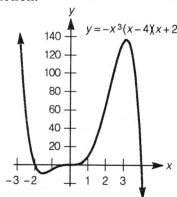

$$y = -x^3(x-4)(x+2)$$

37.   (a)   The x-intercepts are 0, 2, and –2, and the y-intercept is 0. The signs of y are given by:

Interval	$y = -4x(x-2)^2(x+2)^3$
$(-\infty,-2)$	neg.
$(-2,0)$	pos.
$(0,2)$	neg.
$(2,\infty)$	neg.

We sketch the excluded regions:

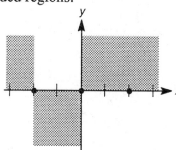

(b)   Near $x = 2$, we have the approximation $y \approx -4(2)(x-2)^2(2+2)^3 = -512(x-2)^2$. So in the immediate vicinity of $x = 2$, the graph resembles $y = -512(x-2)^2$:

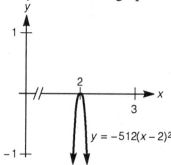

$$y = -512(x-2)^2$$

Near $x = -2$, we have the approximation $y \approx -4(-2)(-2-2)^2(x+2)^3 = 128(x+2)^3$.

So in the immediate vicinity of $x = -2$, the graph resembles $y = 128(x + 2)^3$:

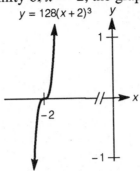

$y = 128(x+2)^3$

(c)   We now graph the function:

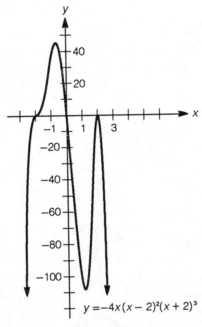

$y = -4x(x - 2)^2(x + 2)^3$

39. From left to right, they are $f(x) = x$, $g(x) = x^2$, $h(x) = x^3$, $F(x) = x^4$, $G(x) = x^5$, and $H(x) = x^6$.

41. We must find where $0 \le H(x) < 0.1$:
$$0 \le x^6 < 0.1$$
$$0 \le x < 0.68$$
When x lies in the interval $[0, 0.68)$, then $H(x)$ will lie in the interval $[0, 0.1)$.

43. We must find where:
$$g(t) - F(t) = 0.26$$
$$t^2 - t^4 = 0.26$$
$$t^4 - t^2 + 0.26 = 0$$

Using the quadratic formula:
$$t^2 = \frac{1 \pm \sqrt{1 - 4(0.26)}}{2} = \frac{1 \pm \sqrt{1 - 1.04}}{2} = \frac{1 \pm \sqrt{-0.04}}{2}$$
Since this equation has no real solutions, there is no such value of t.

45. We set the two y-coordinates equal:
$$x = \tfrac{1}{100} x^2$$
$$0 = \tfrac{1}{100} x^2 - x$$
$$0 = \tfrac{1}{100}\left(x^2 - 100x\right)$$
$$0 = \tfrac{1}{100} x(x - 100)$$
The graph intersects at the origin but also at the point $(100, 100)$.

47. (a) We first factor to obtain:
$$D(x) = x^2\left(1 - x^2\right) = x^2(1 + x)(1 - x)$$
So the x-intercepts are 0, -1, and 1. We draw the graph:

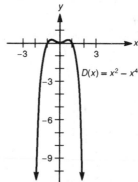

(b) We complete the square:
$$D(x) = x^2 - x^4 = -\left(x^4 - x^2\right) = -\left(x^4 - x^2 + \tfrac{1}{4}\right) + \tfrac{1}{4} = -\left(x^2 - \tfrac{1}{2}\right)^2 + \tfrac{1}{4}$$
The turning points are at $\left(\pm\tfrac{1}{\sqrt{2}}, \tfrac{1}{4}\right) = \left(\pm\tfrac{\sqrt{2}}{2}, \tfrac{1}{4}\right)$, which yield maximum values.
Note that the graph also has a turning point at $(0,0)$, which is a minimum value.

(c) We graph the two functions:

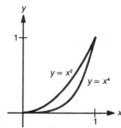

From part (b), we know that the maximum vertical distance between the two curves is $\tfrac{1}{4}$.

49. (a)   Drawing a diagonal between two corners where the cylinder touches the circle yields the right triangle:

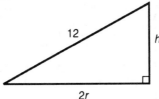

Using the Pythagorean theorem:
$$(2r)^2 + h^2 = 12^2$$
$$4r^2 + h^2 = 144$$
$$h^2 = 144 - 4r^2$$
$$h = \sqrt{144 - 4r^2} = 2\sqrt{36 - r^2}$$

(b)   We compute the volume:
$$V = \pi r^2 h = \pi r^2 \left(2\sqrt{36 - r^2}\right) = 2\pi r^2 \sqrt{36 - r^2}$$

(c)   We must have $36 - r^2 > 0$, so $0 < r < 6$. So the domain is $(0, 6)$.

(d)   We complete the table:

r	$f(r)$
0.0	0
0.5	88
1.0	1382
1.5	6745
2.0	20213
2.5	45878
3.0	86339
3.5	140700
4.0	202129
4.5	254971
5.0	271414
5.5	207720
6.0	0

We graph $f(r) = 4\pi^2 r^4 (36 - r^2)$:

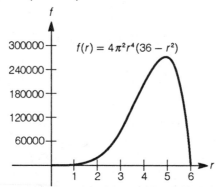

We can approximate the maximum value to be 273000. Thus the maximum possible volume is approximately $\sqrt{273000} \approx 522$ cm^3.

Optional TI-81 Graphing Calculator Exercises for Section 4.5

1. (a) We graph Y_1 using the standard viewing rectangle:

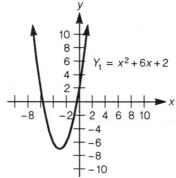

Now graph Y_1 using the settings $X_{min} = -1000$, $X_{max} = 1000$, $Y_{min} = -100,000$, and $Y_{max} = 100,000$:

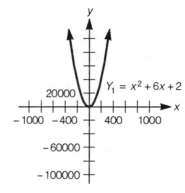

Adding the curve Y_2, we note the two curves appear identical:

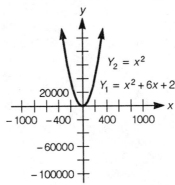

(b)   We graph Y_1 using the standard viewing rectangle:

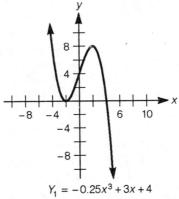

We graph Y_1 and $Y_2 = -0.25x^3$ using the settings $X_{min} = -100$, $X_{max} = 100$, $Y_{min} = -1000$, and $Y_{max} = 1000$. Note that the curves appear identical:

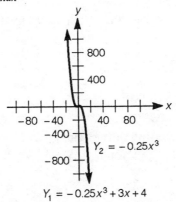

(c)   We graph Y_1 using the settings $X_{min} = -6$, $X_{max} = 8$, $Y_{min} = -2000$, and $Y_{max} = 2000$:

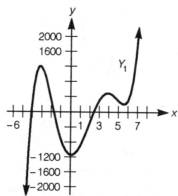

Now sketch Y_1 and $Y_2 = 2x^5$ using the settings $X_{min} = -250$, $X_{max} = 250$, $Y_{min} = -10^{11}$, and $Y_{max} = 10^{11}$:

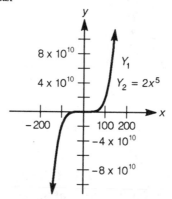

3.   Let $P(x, x^3)$ represent a point on the curve. The distance from this point to the point $(4, 5)$ is given by:

$$D(x) = \sqrt{(x-4)^2 + \left(x^3 - 5\right)^2}$$
$$= \sqrt{x^2 - 8x + 16 + x^6 - 10x^3 + 25}$$
$$= \sqrt{x^6 - 10x^3 + x^2 - 8x + 41}$$

The minimum point on the resulting curve occurs when $x \approx 1.738$, and $y \approx 5.250$. The point $(1.738, 5.250)$ is closest to the point $(4, 5)$.

5.   Let x be the length of the squares cut out. The box will have dimensions of $9 - 2x$ by $12 - 2x$ by x, so the volume is given by:

$$V(x) = x(9 - 2x)(12 - 2x) = x\left(108 - 42x + 4x^2\right) = 4x^3 - 42x^2 + 108x$$

The maximum point on the resulting curve occurs when $x \approx 1.697$, with a maximum possible volume of 81.872 cubic inches.

4.6 Rational Functions

1. For the domain, we must exclude those values of x where:

$$4x - 12 = 0$$
$$4x = 12$$
$$x = 3$$

So the domain is all real numbers except 3, or $(-\infty, 3) \cup (3, \infty)$.
For the x-intercepts, we must find where:

$$3x + 15 = 0$$
$$3x = -15$$
$$x = -5$$

So the x-intercept is –5. For the y-intercept, we let $x = 0$ to obtain:

$$y = \frac{0 + 15}{0 - 12} = -\frac{5}{4}$$

So the y-intercept is $-\frac{5}{4}$.

3. For the domain, we must exclude those values of x where:

$$x^2 - x - 6 = 0$$
$$(x - 3)(x + 2) = 0$$
$$x = 3, -2$$

So the domain is all real numbers except 3 and –2, or $(-\infty, -2) \cup (-2, 3) \cup (3, \infty)$.
For the x-intercepts, we must find where:

$$x^2 - 8x - 9 = 0$$
$$(x - 9)(x + 1) = 0$$
$$x = 9, -1$$

So the x-intercepts are 9 and –1. For the y-intercept, we let $x = 0$ to obtain:

$$y = \frac{-9}{-6} = \frac{3}{2}$$

So the y-intercept is $\frac{3}{2}$.

5. For the domain, we must exclude those values of x where:

$$x^6 = 0$$
$$x = 0$$

So the domain is all real numbers except 0, or $(-\infty, 0) \cup (0, \infty)$.
For the x-intercepts, we must find where:

$$\left(x^2 - 4\right)\left(x^3 - 1\right) = 0$$

$$x^2 - 4 = 0 \qquad \text{or} \qquad x^3 - 1 = 0$$
$$x^2 = 4 \qquad\qquad\qquad x^3 = 1$$
$$x = \pm 2 \qquad\qquad\qquad x = 1$$

So the x-intercepts are –2, 2, and 1. For the y-intercept, we let $x = 0$. But $x = 0$ is not in the domain of the function, so there are no y-intercepts.

7.   There is no x-intercept, the y-intercept is $\frac{1}{4}$, the horizontal asymptote is $y = 0$, and the vertical asymptote is $x = -4$. We draw the graph:

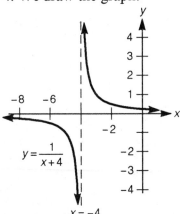

9.   There is no x-intercept, the y-intercept is $\frac{3}{2}$, the horizontal asymptote is $y = 0$, and the vertical asymptote is $x = -2$. We draw the graph:

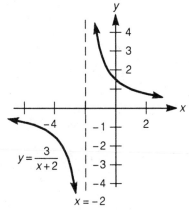

11.   There is no x-intercept, the y-intercept is $\frac{2}{3}$, the horizontal asymptote is $y = 0$, and the vertical asymptote is $x = 3$. We draw the graph:

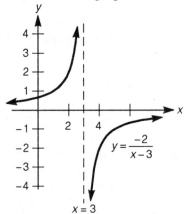

13. Using long division, we have:

$$\frac{x-3}{x-1} = 1 - \frac{2}{x-1}$$

The x-intercept is 3, the y-intercept is 3, the horizontal asymptote is $y = 1$, and the vertical asymptote is $x = 1$. We draw the graph:

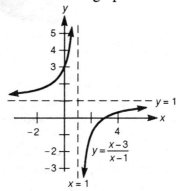

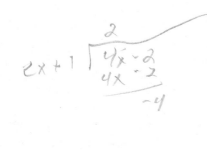

15. Using long division, we have:

$$\frac{4x-2}{2x+1} = 2 - \frac{4}{2x+1}$$

The x-intercept is $\frac{1}{2}$, the y-intercept is -2, the horizontal asymptote is $y = 2$, and the vertical asymptote is $x = -\frac{1}{2}$. We draw the graph:

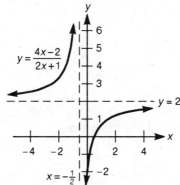

17. There is no x-intercept, the y-intercept is $\frac{1}{4}$, the horizontal asymptote is $y = 0$, and the vertical asymptote is $x = 2$. We draw the graph:

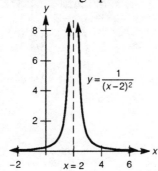

19. There is no x-intercept, the y-intercept is 3, the horizontal asymptote is $y = 0$, and the vertical asymptote is $x = -1$. We draw the graph:

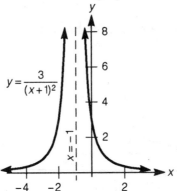

21. There is no x-intercept, the y-intercept is $\frac{1}{8}$, the horizontal asymptote is $y = 0$, and the vertical asymptote is $x = -2$. We draw the graph:

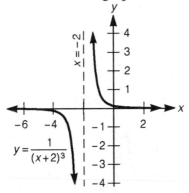

23. There is no x-intercept, the y-intercept is $-\frac{4}{125}$, the horizontal asymptote is $y = 0$, and the vertical asymptote is $x = -5$. We draw the graph:

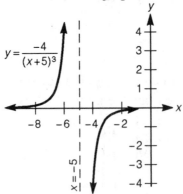

25. The x- and y-intercepts are 0, the horizontal asymptote is $y = 0$, and the vertical asymptotes are $x = -2$ and $x = 2$. We draw the graph:

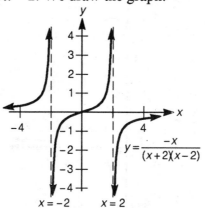

27. The x- and y-intercepts are 0, the horizontal asymptote is $y = 0$, and the vertical asymptotes are $x = -3$ and $x = 1$. We draw the graph:

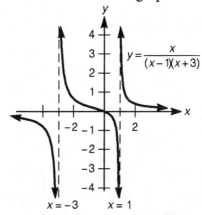

29. (a) The x-intercepts are 2 and 4, there is no y-intercept, the horizontal asymptote is $y = 1$, and the vertical asymptotes are $x = 0$ and $x = 1$. We draw the graph:

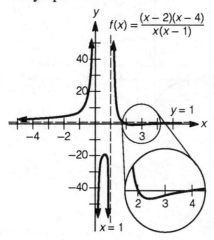

(b)   The x-intercepts are 2 and 4, there is no y-intercept, the horizontal asymptote is $y = 1$, and the vertical asymptotes are $x = 0$ and $x = 3$. We draw the graph:

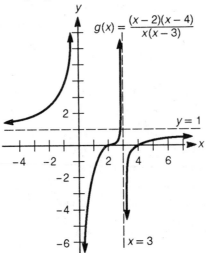

31.   We find where the curve crosses the horizontal asymptote:

$$\frac{(x-4)(x+2)}{(x-1)(x-3)} = 1$$
$$(x-4)(x+2) = (x-1)(x-3)$$
$$x^2 - 2x - 8 = x^2 - 4x + 3$$
$$-2x - 8 = -4x + 3$$
$$2x = 11$$
$$x = \frac{11}{2}$$

The x-intercepts are -2 and 4, the y-intercept is $-\frac{8}{3}$, the horizontal asymptote is $y = 1$, and the vertical asymptotes are $x = 1$ and $x = 3$. We draw the graph:

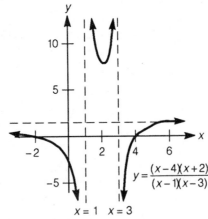

33. We find where the curve crosses the horizontal asymptote:

$$\frac{(x+1)^2}{(x-1)(x-3)} = 1$$
$$(x+1)^2 = (x-1)(x-3)$$
$$x^2 + 2x + 1 = x^2 - 4x + 3$$
$$2x + 1 = -4x + 3$$
$$6x = 2$$
$$x = \tfrac{1}{3}$$

The x-intercept is -1, the y-intercept is $\tfrac{1}{3}$, the horizontal asymptote is $y = 1$, and the vertical asymptotes are $x = 1$ and $x = 3$. We draw the graph:

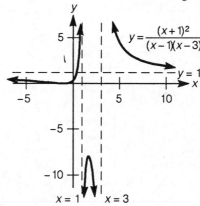

35. (a) For x near -2, we have:

$$y \approx \frac{(-5)(x+2)}{(-1)(-4)} = -\tfrac{5}{4}(x+2)$$

(b) For x near -1, we have.

$$y \approx \frac{(-4)(1)}{(x+1)(-3)} = \frac{\tfrac{4}{3}}{x+1}$$

(c) For x near 2, we have:

$$y \approx \frac{(-1)(4)}{(3)(x-2)} = \frac{-\tfrac{4}{3}}{x-2}$$

37. (a) For $x \neq -3$, we have:

$$y = \frac{x^2 - 9}{x+3} = \frac{(x+3)(x-3)}{x+3} = x - 3$$

So this is the graph of $y = x - 3$, without the point at $(-3, -6)$:

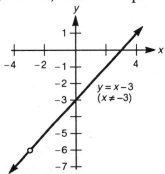

(b)   For $x \neq 3$, we have:
$$y = \frac{x^2 - 5x + 6}{x^2 - 2x - 3} = \frac{(x-2)(x-3)}{(x+1)(x-3)} = \frac{x-2}{x+1}$$

So this is the graph of $y = \frac{x-2}{x+1}$, without the point at $\left(3, \frac{1}{4}\right)$:

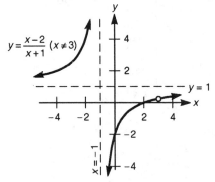

(c)   For $x \neq 1, 2, 3$, we have:
$$y = \frac{(x-1)(x-2)(x-3)}{(x-1)(x-2)(x-3)(x-4)} = \frac{1}{x-4}$$

So this is the graph of $y = \frac{1}{x-4}$, without the points at $\left(1, -\frac{1}{3}\right), \left(2, -\frac{1}{2}\right), (3, -1)$:

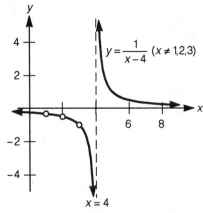

39. The horizontal asymptote is $y = 0$ and the vertical asymptote is $x = 3$. We find the value of k where $k = \frac{x}{(x-3)^2}$:

$$k(x - 3)^2 = x$$
$$kx^2 - 6kx + 9k = x$$
$$kx^2 - (6k + 1)x + 9k = 0$$

Since this equation must have only one solution, we set the discriminant equal to zero:

$$[-(6k + 1)]^2 - 4k(9k) = 0$$
$$(6k + 1)^2 - 36k^2 = 0$$
$$(6k + 1 + 6k)(6k + 1 - 6k) = 0$$
$$(12k + 1)(1) = 0$$
$$k = -\tfrac{1}{12}$$

Thus $y = -\tfrac{1}{12}$. We find x:

$$-\tfrac{1}{12} = \frac{x}{(x - 3)^2}$$
$$-(x - 3)^2 = 12x$$
$$(x - 3)^2 = -12x$$
$$x^2 - 6x + 9 = -12x$$
$$x^2 + 6x + 9 = 0$$
$$(x + 3)^2 = 0$$
$$x = -3$$

So the low point is $\left(-3, -\tfrac{1}{12}\right)$. We draw the graph:

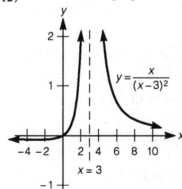

41. (a) We use long division:

$$
\begin{array}{r}
x + 4 \\
x - 3 \overline{\smash{)}\ x^2 + x - 6} \\
\underline{x^2 - 3x} \\
4x - 6 \\
\underline{4x - 12} \\
6
\end{array}
$$

So $\dfrac{x^2 + x - 6}{x - 3} = (x + 4) + \dfrac{6}{x - 3}.$

(b)   We complete the tables:

x	$x+4$	$\dfrac{x^2+x-6}{x-3}$
10	14	14.8571
100	104	104.0619
1000	1004	1004.0060

x	$x+4$	$\dfrac{x^2+x-6}{x-3}$
−10	−6	−6.4615
−100	−96	−96.0583
−1000	−996	−996.0600

(c)   The vertical asymptote is $x = 3$, the x-intercepts are −3 and 2, and the y-intercept is 2.

(d)   We draw the graph:

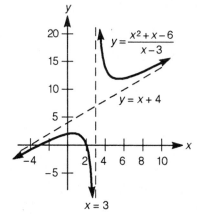

(e)   We find where:

$$\frac{x^2+x-6}{x-3} = k$$
$$x^2+x-6 = kx - 3k$$
$$x^2 + (1-k)x + (3k-6) = 0$$

Setting the discriminant equal to 0, we have:

$$(1-k)^2 - 4(1)(3k-6) = 0$$
$$1 - 2k + k^2 - 12k + 24 = 0$$
$$k^2 - 14k + 25 = 0$$
$$(k-7)^2 = -25 + 49$$
$$(k-7)^2 = 24$$
$$k - 7 = \pm\sqrt{24}$$
$$k = 7 \pm 2\sqrt{6}$$

So either $y = 7 + 2\sqrt{6}$ or $y = 7 - 2\sqrt{6}$. For each of these values, we find x:

$$\frac{x^2 + x - 6}{x - 3} = 7 + 2\sqrt{6}$$

$$x^2 + x - 6 = (7 + 2\sqrt{6})x - 21 - 6\sqrt{6}$$

$$x^2 + (-6 - 2\sqrt{6})x + (15 + 6\sqrt{6}) = 0$$

$$(x - (3 + \sqrt{6}))^2 = 0$$

$$x = 3 + \sqrt{6}$$

So one point is $(3 + \sqrt{6}, 7 + 2\sqrt{6})$. For $y = 7 - 2\sqrt{6}$, we find x:

$$\frac{x^2 + x - 6}{x - 3} = 7 - 2\sqrt{6}$$

$$x^2 + x - 6 = (7 - 2\sqrt{6})x - 21 + 6\sqrt{6}$$

$$x^2 + (-6 + 2\sqrt{6})x + (15 - 6\sqrt{6}) = 0$$

$$(x - (3 - \sqrt{6}))^2 = 0$$

$$x = 3 - \sqrt{6}$$

So the other point is $(3 - \sqrt{6}, 7 - 2\sqrt{6})$.

43. Note that:

$$\frac{-x^2 + 1}{x} = -x + \frac{1}{x}$$

Thus $y = -x$ is a slant asymptote. We sketch the graph:

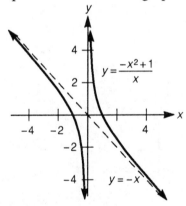

Optional TI-81 Graphing Calculator Exercises for Section 4.6

1. The equation of the line can be found using the point-slope formula:

$$y - 1 = m(x - 2)$$

$$y = mx - 2m + 1$$

We now find the x- and y-intercepts. If $x = 0$, then $y = -2m + 1$, and if $y = 0$, we have:

$$0 = mx - 2m + 1$$
$$2m - 1 = mx$$
$$\frac{2m - 1}{m} = x$$

So the area of the triangle is given by:

$$A(m) = \tfrac{1}{2} \cdot (-2m + 1)\left(\frac{2m - 1}{m}\right) = \frac{-4m^2 + 4m - 1}{2m}$$

The graph of $A(m)$ appears as:

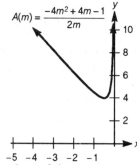

This function has a minimum value of 4 when $m \approx -0.5$.

3.   Let the coordinates of P be $P(x, x^2)$. We find the slope between P and $(0, -1)$:

$$m(x) = \frac{x^2 - (-1)}{x - 0} = \frac{x^2 + 1}{x}$$

The graph of $m(x)$ appears as:

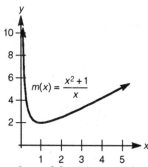

This function has a minimum value of 2 when $x \approx 1$. Thus the point P is $P(1, 1)$. We draw the parabola and the line through P and $(0, -1)$:

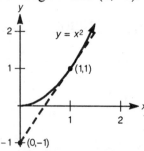

The line appears to be tangent to the parabola at $(1, 1)$.

5.   (a)   Since the area is 6 m², then $xy = 6$, so $y = \frac{6}{x}$ where x is the width and y is the length. Thus the perimeter is given by:

$$P(x) = 2x + 2\left(\frac{6}{x}\right) = 2x + \frac{12}{x} = \frac{2x^2 + 12}{x}$$

(b)   The graph of $P(x)$ appears as:

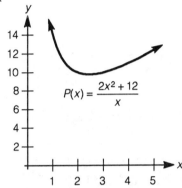

This function has a minimum value of 9.80 when $x \approx 2.45$. Substituting into $y = \frac{6}{x}$, we obtain $y \approx 2.45$. So the length and width are approximately 2.45 m.

7.   (a)   Since $V = \pi r^2 h$ is the formula for volume, we have:

$$\pi r^2 h = 54\pi$$
$$h = \frac{54}{r^2}$$

(b)   Since the total surface area is given by $S = 2\pi r^2 + 2\pi r h$, we have:

$$S(r) = 2\pi r^2 + 2\pi r \cdot \frac{54}{r^2} = 2\pi r^2 + \frac{108\pi}{r} = \frac{2\pi r^3 + 108\pi}{r}$$

(c)   The graph of $S(r)$ appears as:

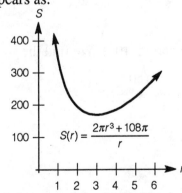

This function has a minimum value of 169.65 when $r = 3$ in.

(d)   Using $h = \frac{54}{r^2}$, we obtain $h = \frac{54}{3^2} = 6$ in. The value of h is double that of r.

Chapter Four Review Exercises

1. We find the slope between the points $(1, -2)$ and $(-2, -11)$:
$$m = \frac{-11-(-2)}{-2-1} = \frac{-9}{-3} = 3$$
Now use the point-slope formula:
$$y - (-2) = 3(x-1)$$
$$y + 2 = 3x - 3$$
$$y = 3x - 5$$
So $G(x) = 3x - 5$, and thus $G(0) = -5$.

3. We first compute the revenue:
$$R(x) = xp = x\left(-\tfrac{1}{8}x + 100\right) = -\tfrac{1}{8}x^2 + 100x$$
Now complete the square:
$$R(x) = -\tfrac{1}{8}\left(x^2 - 800x\right) = -\tfrac{1}{8}\left(x^2 - 800x + 160000\right) + 20000 = -\tfrac{1}{8}(x - 400)^2 + 20000$$
So the maximum possible revenue is \$20,000.

5. We graph the function:

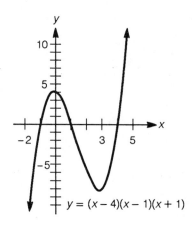

$$y = (x - 4)(x - 1)(x + 1)$$

7. We have the points $(0, 1000)$ and $(5, 100)$. We find the slope:
$$m = \frac{100 - 1000}{5 - 0} = \frac{-900}{5} = -180$$
So $V(t) = -180t + 1000$.

9. The x-intercept is $-\frac{5}{3}$, the y-intercept is $\frac{5}{2}$, the horizontal asymptote is $y = 3$, and the vertical asymptote is $x = -2$. We graph the function:

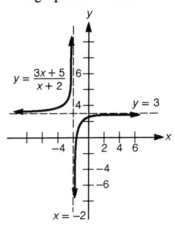

11. We graph the function:

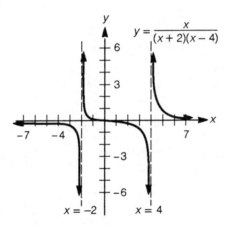

13. We draw the figure:

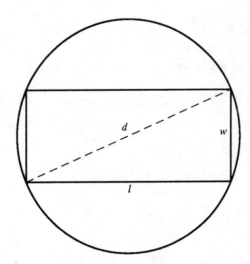

We can find r, since the circumference of the circle is 12 cm:

$$2\pi r = 12, \text{ so } r = \frac{6}{\pi} \text{ and thus } d = \frac{12}{\pi}$$

Using the Pythagorean theorem:

$$w^2 + l^2 = \left(\frac{12}{\pi}\right)^2$$

$$w^2 + l^2 = \frac{144}{\pi^2}$$

$$l^2 = \frac{144}{\pi^2} - w^2$$

$$l = \sqrt{\frac{144 - \pi^2 w^2}{\pi^2}} = \frac{\sqrt{144 - \pi^2 w^2}}{\pi}$$

Since the perimeter is given by $P = 2w + 2l$, we have:

$$P(w) = 2w + \frac{2\sqrt{144 - \pi^2 w^2}}{\pi}$$

15. We find the slope between the points $(3, 5)$ and $(-2, 0)$:

$$m = \frac{0 - 5}{-2 - 3} = \frac{-5}{-5} = 1$$

We use the point $(3, 5)$ in the point-slope formula:

$$y - 5 = 1(x - 3)$$
$$y - 5 = x - 3$$
$$y = x + 2$$

Using functional notation we have $f(x) = x + 2$.

17. We find the slope of the line $3x - 8y = 16$:

$$-8y = -3x + 16$$
$$y = \tfrac{3}{8}x - 2$$

We use $m = \tfrac{3}{8}$ and the point $(4, -1)$ in the point-slope formula:

$$y - (-1) = \tfrac{3}{8}(x - 4)$$
$$y + 1 = \tfrac{3}{8}x - \tfrac{3}{2}$$
$$y = \tfrac{3}{8}x - \tfrac{5}{2}$$

Using functional notation we have $f(x) = \tfrac{3}{8}x - \tfrac{5}{2}$.

19. If the graph of the inverse function passes through $(2, 1)$, then $(1, 2)$ must lie on the graph of the function. We find the slope between the points $(1, 2)$ and $(-3, 5)$:

$$m = \frac{5 - 2}{-3 - 1} = \frac{3}{-4} = -\frac{3}{4}$$

We use the point $(1, 2)$ in the point-slope formula:

$$y - 2 = -\tfrac{3}{4}(x - 1)$$
$$y - 2 = -\tfrac{3}{4}x + \tfrac{3}{4}$$
$$y = -\tfrac{3}{4}x + \tfrac{11}{4}$$

Using functional notation we have $f(x) = -\tfrac{3}{4}x + \tfrac{11}{4}$.

21. We complete the square:
$$y = x^2 + 2x - 3 = (x^2 + 2x + 1) - 4 = (x+1)^2 - 4$$
The vertex is $(-1, -4)$, the x-intercepts are 1 and -3, and the y-intercept is -3. We graph the function:

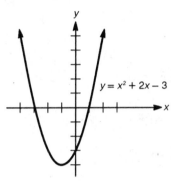

23. We complete the square:
$$y = -x^2 + 2\sqrt{3}x + 3 = -(x^2 - 2\sqrt{3}x) + 3 = -(x^2 - 2\sqrt{3}x + 3) + 6 = -(x - \sqrt{3})^2 + 6$$
The vertex is $(\sqrt{3}, 6)$, the x-intercepts are $\sqrt{3} + \sqrt{6}$ and $\sqrt{3} - \sqrt{6}$, and the y-intercept is 3. We graph the function:

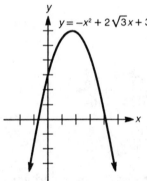

25. We complete the square:
$$y = -3x^2 + 12x = -3(x^2 - 4x) = -3(x^2 - 4x + 4) + 12 = -3(x - 2)^2 + 12$$
The vertex is $(2, 12)$, the x-intercepts are 0 and 4, and the y-intercept is 0. We draw the graph:

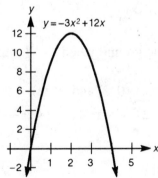

27. We find the two vertices:

$$y = x^2 - 4x + 6$$
$$y - 6 = x^2 - 4x$$
$$y - 6 + 4 = x^2 - 4x + 4$$
$$y - 2 = (x - 2)^2$$
$$\text{vertex: } (2, 2)$$

$$y = -x^2 - 4x - 5$$
$$y + 5 = -(x^2 + 4x)$$
$$y + 5 - 4 = -(x^2 + 4x + 4)$$
$$y + 1 = -(x + 2)^2$$
$$\text{vertex: } (-2, -1)$$

We now find the distance between $(2, 2)$ and $(-2, -1)$:
$$d = \sqrt{(-2 - 2)^2 + (-1 - 2)^2} = \sqrt{16 + 9} = \sqrt{25} = 5$$

29. Call x and y the two numbers, so $x + y = \sqrt{3}$, and thus $y = \sqrt{3} - x$.
We find their product:
$$P = xy = x(\sqrt{3} - x) = -x^2 + \sqrt{3}x$$
Now complete the square:
$$P = -\left(x^2 - \sqrt{3}x\right) = -\left(x^2 - \sqrt{3}x + \tfrac{3}{4}\right) + \tfrac{3}{4} = -\left(x - \tfrac{\sqrt{3}}{2}\right)^2 + \tfrac{3}{4}$$
The maximum product is $\tfrac{3}{4}$.

31. (a) We complete the square:
$$h = -16t^2 + v_0 t$$
$$= -16\left(t^2 - \frac{v_0}{16}t\right)$$
$$= -16\left(t^2 - \frac{v_0}{16}t + \frac{v_0^2}{1024}\right) + \frac{v_0^2}{64}$$
$$= -16\left(t - \frac{v_0}{32}\right)^2 + \frac{v_0^2}{64}$$

So the maximum height of $\frac{v_0^2}{64}$ ft. is obtained when $t = \frac{v_0}{32}$ sec.

(b) The object will strike the ground when $h(t) = 0$:
$$-16t^2 + v_0 t = 0$$
$$t\left(-16t + v_0\right) = 0$$
$$t = 0 \text{ or } t = \frac{v_0}{16}$$

So the object will strike the ground when $t = \frac{v_0}{16}$ sec.

33. (a) We find the distance between $(0, 2)$ and (x, x^2):
$$d = \sqrt{(x - 0)^2 + (x^2 - 2)^2} = \sqrt{x^2 + x^4 - 4x^2 + 4} = \sqrt{x^4 - 3x^2 + 4}$$

(b)   We want to minimize d. This will occur at the same x-coordinate as the minimum of d^2. We complete the square:

$$d^2 = x^4 - 3x^2 + 4 = \left(x^4 - 3x^2 + \tfrac{9}{4}\right) + 4 - \tfrac{9}{4} = \left(x^2 - \tfrac{3}{2}\right)^2 + \tfrac{7}{4}$$

So the minimum occurs where $x^2 = \tfrac{3}{2}$, so $x = \pm\sqrt{\tfrac{3}{2}} = \tfrac{\pm\sqrt{6}}{2}$. Since the point is in the second quadrant, we know $x < 0$, and thus the point on the parabola is $\left(\tfrac{-\sqrt{6}}{2}, \tfrac{3}{2}\right)$.

35. We find the distance between $(2, 0)$ and $\left(x, \tfrac{4}{3}x + b\right)$:

$$d = \sqrt{(x-2)^2 + \left(\tfrac{4}{3}x + b\right)^2}$$
$$= \sqrt{x^2 - 4x + 4 + \tfrac{16}{9}x^2 + \tfrac{8b}{3}x + b^2}$$
$$= \sqrt{\tfrac{25}{9}x^2 + \left(\tfrac{8b}{3} - 4\right)x + \left(4 + b^2\right)}$$

We complete the square on d^2:

$$d^2 = \tfrac{25}{9}\left[x^2 + \tfrac{9}{25}\left(\tfrac{8b}{3} - 4\right)x\right] + \left(4 + b^2\right)$$
$$= \tfrac{25}{9}\left[x^2 + \tfrac{12}{25}(2b - 3)x\right] + \left(4 + b^2\right)$$
$$= \tfrac{25}{9}\left[x^2 + \tfrac{12}{25}(2b - 3)x + \tfrac{36}{625}(2b - 3)^2\right] + \left(4 + b^2\right) - \tfrac{4}{25}(2b - 3)^2$$
$$= \tfrac{25}{9}\left[x + \tfrac{6}{25}(2b - 3)\right]^2 + \frac{25\left(4 + b^2\right) - 4\left(4b^2 - 12b + 9\right)}{25}$$
$$= \tfrac{25}{9}\left[x + \tfrac{6}{25}(2b - 3)\right]^2 + \frac{(3b + 8)^2}{25}$$

Since the minimum distance is 5, then $5 = \left|\tfrac{3b+8}{5}\right|$. We now solve for b:

$$|3b + 8| = 25$$

$$
\begin{array}{lll}
3b + 8 = 25 & \text{or} & 3b + 8 = -25 \\
3b = 17 & & 3b = -33 \\
b = \tfrac{17}{3} & & b = -11
\end{array}
$$

37. Since $x + y = \sqrt{2}$, then $y = \sqrt{2} - x$. Then:

$$s = x^2 + y^2 = x^2 + \left(\sqrt{2} - x\right)^2 = x^2 + 2 - 2\sqrt{2}x + x^2 = 2x^2 - 2\sqrt{2}x + 2$$

We complete the square:

$$s = 2\left(x^2 - \sqrt{2}x\right) + 2 = 2\left(x^2 - \sqrt{2}x + \tfrac{1}{2}\right) + 2 - 1 = 2\left(x - \tfrac{\sqrt{2}}{2}\right)^2 + 1$$

So the minimum value of s is 1.

39. Let x and h be the two legs. We have:

$$x^2 + h^2 = 15^2$$
$$h^2 = 225 - x^2$$
$$h = \sqrt{225 - x^2}$$

So $A = \tfrac{1}{2}(\text{base})(\text{height}) = \tfrac{1}{2}x\sqrt{225 - x^2}$. We find $A^2 = \tfrac{1}{4}x^2\left(225 - x^2\right) = -\tfrac{1}{4}x^4 + \tfrac{225}{4}x^2$.

We now complete the square on A^2:
$$A^2 = -\tfrac{1}{4}\left(x^4 - 225x^2\right) = -\tfrac{1}{4}\left[x^4 - 225x^2 + \left(\tfrac{225}{2}\right)^2\right] + \left(\tfrac{225}{4}\right)^2 = -\tfrac{1}{4}\left(x^2 - \tfrac{225}{2}\right)^2 + \left(\tfrac{225}{4}\right)^2$$
So the maximum of $A^2 = \left(\tfrac{225}{4}\right)^2$, thus $A = \tfrac{225}{4}$ cm^2.

41. We factor $f(x)$ as:
$$f(x) = x^2 - \left(a^2 + 2a\right)x + 2a^3 = \left(x - a^2\right)\left(x - 2a\right)$$
So the x-intercepts are a^2 and $2a$. Since $2a > a^2$ when $0 < a < 2$, then the distance between $(a^2, 0)$ and $(2a, 0)$ will be $D = 2a - a^2 = -a^2 + 2a$.
We now complete the square:
$$D = -\left(a^2 - 2a\right) = -\left(a^2 - 2a + 1\right) + 1 = -(a-1)^2 + 1$$
So D is maximum when $a = 1$.

43. We compute R:
$$R = xp = x\left(160 - \tfrac{1}{5}x\right) = -\tfrac{1}{5}x^2 + 160x$$
We now complete the square:
$$R = -\tfrac{1}{5}\left(x^2 - 800x\right) = -\tfrac{1}{5}\left(x^2 - 800x + 160000\right) + 32000 = -\tfrac{1}{5}(x - 400)^2 + 32000$$
So $x = 400$ units will maximize the revenue. Then $p = 160 - \tfrac{1}{5}(400) = \$80$.

45. The x-intercepts are -4 and 2 and the y-intercept is -8. We draw the graph:

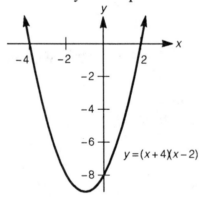

47. The x-intercept is -5 and the y-intercept is -125. We draw the graph:

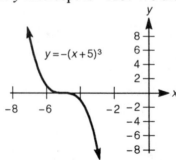

49. The x-intercepts are -1 and 0, and the y-intercept is 0. We draw the graph:

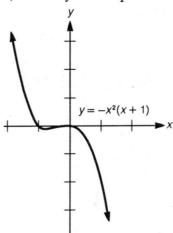

51. The x-intercepts are 0, 2, and -2, and the y-intercept is 0. We draw the graph:

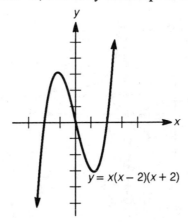

53. The x-intercept is $-\frac{1}{3}$, there is no y-intercept, the horizontal asymptote is $y = 3$, and the vertical asymptote is $x = 0$. We draw the graph:

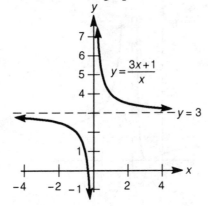

55. There is no x-intercept, the y-intercept is -1, the horizontal asymptote is $y = 0$, and the vertical asymptote is $x = 1$. We draw the graph:

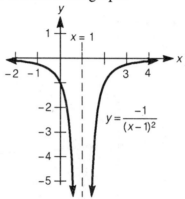

57. The x-intercept is 2, the y-intercept is $\frac{2}{3}$, the horizontal asymptote is $y = 1$, and the vertical asymptote is $x = 3$. We draw the graph:

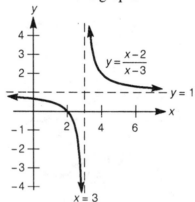

59. The x-intercept is 1, the y-intercept is $\frac{1}{4}$, the horizontal asymptote is $y = 1$, and the vertical asymptote is $x = 2$. We draw the graph:

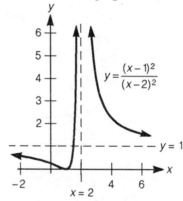

61. (a) If $b = 1$, then $f(x) = x^2 + 2x + 1 = (x + 1)^2$, so the vertex is $(-1, 0)$. This point is 1 unit from the origin.

(b) If $b = 2$, then $f(x) = x^2 + 4x + 1$. We complete the square:
$$f(x) = (x^2 + 4x + 4) + 1 - 4 = (x+2)^2 - 3$$
So the vertex is $(-2, -3)$. By the distance formula:
$$d = \sqrt{(-2-0)^2 + (-3-0)^2} = \sqrt{4+9} = \sqrt{13}$$

(c) We complete the square:
$$f(x) = x^2 + 2bx + 1 = (x^2 + 2bx + b^2) + 1 - b^2 = (x+b)^2 + 1 - b^2$$
So the vertex is $(-b, 1 - b^2)$. By the distance formula:
$$d = \sqrt{(-b)^2 + (1-b^2)^2} = \sqrt{b^2 + 1 - 2b^2 + b^4} = \sqrt{b^4 - b^2 + 1}$$
We now complete the square on d^2:
$$d^2 = b^4 - b^2 + 1 = (b^4 - b^2 + \tfrac{1}{4}) + 1 - \tfrac{1}{4} = (b^2 - \tfrac{1}{2})^2 + \tfrac{3}{4}$$
So d^2 (and thus d) is minimized when $b^2 = \tfrac{1}{2}$, thus $b = \pm\frac{1}{\sqrt{2}} = \frac{\pm\sqrt{2}}{2}$.

63. We need to find the vertex of the parabola. Complete the square:
$$y = x^2 - 2x + k$$
$$y - k = x^2 - 2x$$
$$y - k + 1 = x^2 - 2x + 1$$
$$y = (x-1)^2 + (k-1)$$
Since the vertex is $(1, k-1)$ and the parabola is opening upward, then $k - 1 = 5$, so $k = 6$.

65. We solve for x:
$$y = \frac{(x-1)(x-3)}{x-4}$$
$$y(x-4) = (x-1)(x-3)$$
$$yx - 4y = x^2 - 4x + 3$$
$$0 = x^2 - (4+y)x + (4y+3)$$
Using the quadratic formula:
$$x = \frac{4+y \pm \sqrt{(4+y)^2 - 4(4y+3)}}{2}$$
$$= \frac{4+y \pm \sqrt{16+8y+y^2 - 16y - 12}}{2}$$
$$= \frac{4+y \pm \sqrt{y^2 - 8y + 4}}{2}$$
So we must make sure that $y^2 - 8y + 4 \geq 0$.
We find the key numbers by using the quadratic formula:
$$y = \frac{8 \pm \sqrt{64-16}}{2} = \frac{8 \pm 4\sqrt{3}}{2} = 4 \pm 2\sqrt{3}$$
From a sign chart, we see that the range is $y \leq 4 - 2\sqrt{3}$ or $y \geq 4 + 2\sqrt{3}$. We write this as $(-\infty, 4 - 2\sqrt{3}] \cup [4 + 2\sqrt{3}, \infty)$.

67. We first draw a figure:

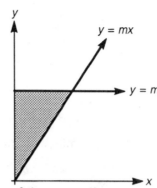

We find the intersection point of these two lines:

$$mx = m$$
$$x = 1$$
$$y = m$$

So the point is $(1, m)$. Since these are the base and height, respectively, of the triangle, we have:

$$A = \tfrac{1}{2}(1)(m) = \frac{m}{2}$$

Using functional notation we have $A(m) = \frac{m}{2}$.

69. We re-draw the figure and label essential parts:

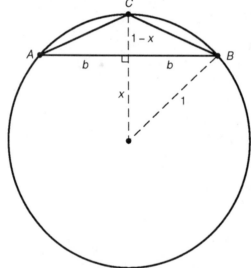

By the labeled parts of the figure, we have:

$$x^2 + b^2 = 1^2, \text{ so } b = \sqrt{1 - x^2}$$

Thus the base of the triangle is $2b = 2\sqrt{1 - x^2}$ and the height is $1 - x$, thus the area is given by:

$$A = \tfrac{1}{2} \cdot 2\sqrt{1 - x^2} \cdot (1 - x) = (1 - x)\sqrt{1 - x^2}$$

Using functional notation we have $A(x) = (1 - x)\sqrt{1 - x^2}$.

71. (a) Substituting $x = 1990$, we have:
$$y = 9.243(1990) - 18044.316 = 349.254$$
The production is approximately 349.254 quadrillion Btu.

(b) Our estimate is too high, by approximately 4.19 quadrillion Btu.

73. (a) We draw the scatter diagram (not to scale):

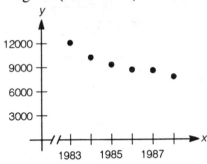

(b) We draw the best-fit line (not to scale):

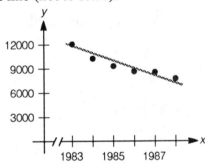

The slope appears to be approximately –770 and the y-intercept appears to be approximately 1.5 million.

(c) We let $x = 1995$:
$$f(1995) = (-771.4)(1995) + 1541090 = 2147$$
We would estimate that approximately 2100 refugees would arrive in 1995.

(d) We switch the roles of x and y, and solve the resulting equation for y:
$$x = -771.4y + 1541090$$
$$771.4y = 1541090 - x$$
$$y = \frac{1541090 - x}{771.4}$$
So $f^{-1}(x) = \frac{1541090 - x}{771.4}$.

(e) We find $f^{-1}(1000) = \frac{1541090 - 1000}{771.4} \approx 1996$. The number of refugees should dip below 1000 in the year 1996.

75.   (a)   The graphs are reflections of each other about the y-axis:

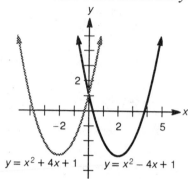

$y = x^2 + 4x + 1$ $y = x^2 - 4x + 1$

(b)   We compute $f(-x)$:
$$f(-x) = a(-x)^2 + b(-x) + c = ax^2 - bx + c$$

(c)   Reversing the sign of b will reflect the graph of $f(x) = ax^2 + bx + c$ across the y-axis.

Chapter Four Test

1.   We first find the slope between $(-2, -4)$ and $(5, 1)$:
$$m = \frac{-4 - 1}{-2 - 5} = \frac{-5}{-7} = \frac{5}{7}$$
Using the point-slope formula, we have:
$$y - 1 = \tfrac{5}{7}(x - 5)$$
$$y - 1 = \tfrac{5}{7}x - \tfrac{25}{7}$$
$$y = \tfrac{5}{7}x - \tfrac{18}{7}$$
So $L(x) = \tfrac{5}{7}x - \tfrac{18}{7}$, and thus $L(0) = -\tfrac{18}{7}$.

2.   (a)   We complete the square:
$$F(x) = -2x^2 + 4x = -2(x^2 - 2x) = -2(x^2 - 2x + 1) + 2 = -2(x - 1)^2 + 2$$
So the graph of $F(x)$ is a parabola pointing downward with vertex $(1, 2)$, and thus it is increasing on the interval $(-\infty, 1)$. The maximum value of $F(x)$ is 2.

(b)   Since $9t^4 \geq 0$ and $6t^2 \geq 0$, then $G(t)$ will achieve a minimum value of 2 when $t = 0$.

3.   (a)   The x-intercepts are 3 and -4, and the y-intercept is -48. The signs of $f(x)$ are given by:

Interval	$f(x) = (x - 3)(x + 4)^2$
$(-\infty, -4)$	neg.
$(-4, 3)$	neg.
$(3, \infty)$	pos.

We sketch the excluded regions for $f(x)$:

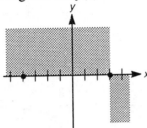

(b)   For $x \approx -4$, we have $f(x) \approx (-4-3)(x+4)^2 = -7(x+4)^2$. So $f(x)$ is approximately $-7(x+4)^2$ when x is close to -4. We sketch $y = -7(x+4)^2$:

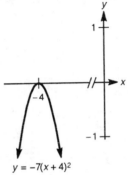

$$y = -7(x+4)^2$$

(c)   We graph the function:

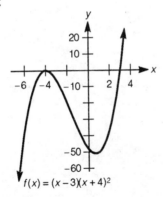

$$f(x) = (x-3)(x+4)^2$$

4.   (a)   We sketch the graph:

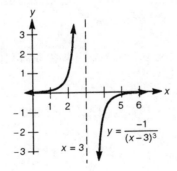

$$y = \frac{-1}{(x-3)^3}$$

$$x = 3$$

(b) We sketch the graph:

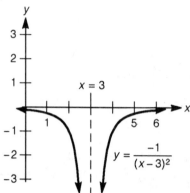

5.   We first complete the square:
$$y = -x^2 + 7x + 6 = -\left(x^2 - 7x\right) + 6 = -\left(x^2 - 7x + \tfrac{49}{4}\right) + \tfrac{73}{4} = -\left(x - \tfrac{7}{2}\right)^2 + \tfrac{73}{4}$$
The turning point (vertex) is $\left(\tfrac{7}{2}, \tfrac{73}{4}\right)$, the x-intercepts are $\tfrac{7 \pm \sqrt{73}}{2}$, the y-intercept is 6, and the axis of symmetry is $x = \tfrac{7}{2}$. We draw the graph:

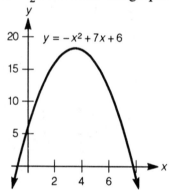

6.   We set up the revenue function:
$$R = xp = x\left(-\tfrac{1}{6}x + 80\right) = -\tfrac{1}{6}x^2 + 80x$$
Now complete the square:
$$R = -\tfrac{1}{6}\left(x^2 - 480x\right) = -\tfrac{1}{6}\left(x^2 - 480x + 57600\right) + 9600 = -\tfrac{1}{6}\left(x - 240\right)^2 + 9600$$
The maximum possible revenue is \$9600, which occurs when $x = 240$.
Thus $p = -\tfrac{1}{6}(240) + 80 = \$40 \,/\, \text{unit}$.

7.   We find the slope between the points $(0, 14000)$ and $(10, 750)$:
$$m = \frac{750 - 14000}{10 - 0} = \frac{-13250}{10} = -1325$$
So the value of the machine is $V(t) = -1325t + 14000$.

8. Since $y = -\frac{1}{2}(3-x)^2$ will be a reflection of $y = -\frac{1}{2}(x+3)^2$ across the y-axis, we have the graph:

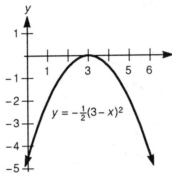

The x-intercept is 3 and the y-intercept is $-\frac{9}{2}$.

9. The x-intercept is $\frac{3}{2}$, the y-intercept is -3, the vertical asymptote is $x = -1$, and the horizontal asymptote is $y = 2$. We draw the graph:

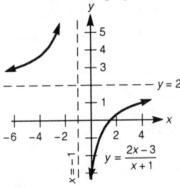

10. (a) Using the distance formula, we have:
$$L(x) = \sqrt{(x+1)^2 + (y-3)^2}$$
$$= \sqrt{(x+1)^2 + (3x-1-3)^2}$$
$$= \sqrt{x^2 + 2x + 1 + 9x^2 - 24x + 16}$$
$$= \sqrt{10x^2 - 22x + 17}$$

(b) We complete the square inside the radical:
$$10x^2 - 22x + 17 = 10(x^2 - 2.2x) + 17$$
$$= 10(x^2 - 2.2x + 1.21) + 4.9$$
$$= 10(x - 1.1)^2 + 4.9$$

So the length will be a minimum when L^2 is a minimum, which occurs when $x = 1.1$.

11. (a)   The vertical asymptotes are $x = 3$ and $x = -3$, and the horizontal asymptote is $y = 1$.

(b)   Near $x = 0$, $f(x)$ will look like:
$$y = \frac{x(0-2)}{0-9} = \tfrac{2}{9}x$$
Near $x = 2$, $f(x)$ will look like:
$$y = \frac{2(x-2)}{4-9} = -\tfrac{2}{5}(x-2)$$
We sketch the graph:

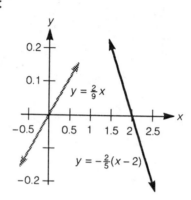

(c)   Near $x = 3$, $f(x)$ will look like:
$$y = \frac{3(3-2)}{(3+3)(x-3)} = \frac{1}{2(x-3)}$$
Near $x = -3$, $f(x)$ will look like:
$$y = \frac{-3(-3-2)}{(x+3)(-3-3)} = -\frac{5}{2(x+3)}$$
We sketch the graph:

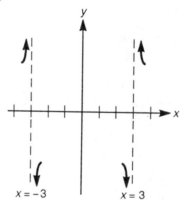

(d)   We graph the function:

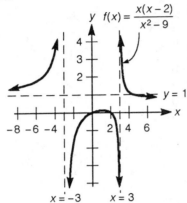

$$y \quad f(x) = \frac{x(x-2)}{x^2-9}$$

12.   Let w and l represent the width and length, respectively. We draw the figure:

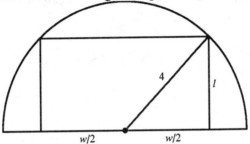

From the Pythagorean theorem, we have:

$$\left(\frac{w}{2}\right)^2 + l^2 = 4^2$$

$$\frac{w^2}{4} + l^2 = 16$$

$$l^2 = \frac{64 - w^2}{4}$$

$$l = \frac{\sqrt{64 - w^2}}{2}$$

So the area of the rectangle is given by:

$$A(w) = w \cdot l = w \cdot \frac{\sqrt{64 - w^2}}{2} = \tfrac{1}{2} w \sqrt{64 - w^2}$$

13.   (a)   As $|x|$ increases in size for positive values of x, $-x^3$ should be large negative values, which the function is not.

(b)   This graph has four turning points, and a polynomial function with highest degree term $-x^3$ can have at most two turning points.

14.   (a)   We sketch the scatter diagram (not to scale):

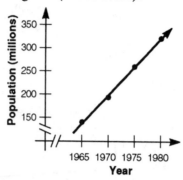

     (b)   Using $x = 1985$, we have:
$$y = -23757 + 12.16(1985) = 380.6$$
Our estimate is 381 million. Our estimate is slightly higher than the actual amount.

Chapter Five
Exponential and Logarithmic Functions

5.1 Exponential Functions

1. (a) Since $2^{10} \approx 10^3$, then $2^{30} = \left(2^{10}\right)^3 \approx \left(10^3\right)^3 = 10^9$.

   (b) Since $2^{10} \approx 10^3$, then $2^{50} = \left(2^{10}\right)^5 \approx \left(10^3\right)^5 = 10^{15}$.

3. Using the property $\left(b^m\right)^n = b^{mn}$, we have:
$$\left(5^{\sqrt{3}}\right)^{\sqrt{3}} = 5^{\sqrt{3} \cdot \sqrt{3}} = 5^3 = 125$$

5. Using the property $b^n b^m = b^{n+m}$, we have:
$$\left(4^{1+\sqrt{2}}\right)\left(4^{1-\sqrt{2}}\right) = 4^{1+\sqrt{2}+1-\sqrt{2}} = 4^2 = 16$$

7. Using the property $\dfrac{b^m}{b^n} = b^{m-n}$, we have:
$$\frac{2^{4+\pi}}{2^{1+\pi}} = 2^{4+\pi-1-\pi} = 2^3 = 8$$

9. Using the property $\left(b^m\right)^n = b^{mn}$, we have:

$$\left(\sqrt{5}^{\sqrt{2}}\right)^2 = \sqrt{5}^{2\sqrt{2}} = \left(5^{1/2}\right)^{2\sqrt{2}} = 5^{\sqrt{2}}$$

11. (a)   We solve for x:

$$3^x = 27$$
$$3^x = 3^3$$
$$x = 3$$

(b)   We solve for t:

$$9^t = 27$$
$$\left(3^2\right)^t = 3^3$$
$$3^{2t} = 3^3$$
$$2t = 3$$
$$t = \tfrac{3}{2}$$

(c)   We solve for y:

$$3^{1-2y} = \sqrt{3}$$
$$3^{1-2y} = 3^{1/2}$$
$$1 - 2y = \tfrac{1}{2}$$
$$-2y = -\tfrac{1}{2}$$
$$y = \tfrac{1}{4}$$

(d)   We solve for z:

$$3^z = 9\sqrt{3}$$
$$3^z = 3^2 \cdot 3^{1/2}$$
$$3^z = 3^{5/2}$$
$$z = \tfrac{5}{2}$$

13. The domain is all real numbers, or $(-\infty, \infty)$.

15. Since $2^{x-1} \neq 0$ (even if $x = 1$), then the domain is all real numbers, or $(-\infty, \infty)$.

17. We graph $y = 2^x$ and $y = 2^{-x}$:

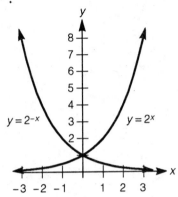

19. We graph $y = 3^x$ and $y = -3^x$ (note that $y = -3^x$ is a reflection of $y = 3^x$ about the x-axis):

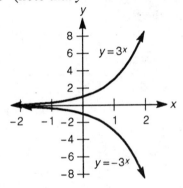

21. We graph $y = 2^x$ and $y = 3^x$:

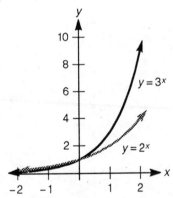

23. We graph $y = \left(\tfrac{1}{2}\right)^x = 2^{-x}$ and $y = \left(\tfrac{1}{3}\right)^x = 3^{-x}$:

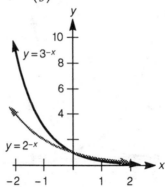

25. The domain is $(-\infty, \infty)$, the range is $(-\infty, 1)$, the x- and y-intercepts are 0, and the asymptote is $y = 1$. We graph the function:

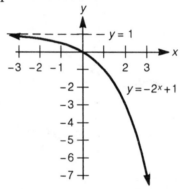

27. The domain is $(-\infty, \infty)$, the range is $(1, \infty)$, there is no x-intercept, the y-intercept is 2, and the asymptote is $y = 1$. We graph the function:

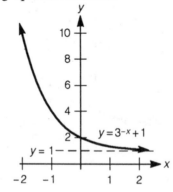

29. The domain is $(-\infty, \infty)$, the range is $(0, \infty)$, there is no x-intercept, the y-intercept is $\frac{1}{2}$, and the asymptote is $y = 0$. We graph the function:

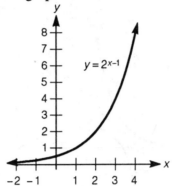

31. The domain is $(-\infty, \infty)$, the range is $(1, \infty)$, there is no x-intercept, the y-intercept is 4, and the asymptote is $y = 1$. We graph the function:

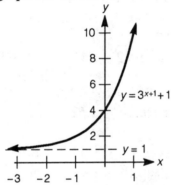

33. We solve for x:

$$3x(10^x) + 10^x = 0$$
$$10^x(3x + 1) = 0$$
$$x = -\tfrac{1}{3} \text{ (since } 10^x \neq 0)$$

35. We solve for x:

$$3(3^x) - 5x(3^x) + 2x^2(3^x) = 0$$
$$3^x(3 - 5x + 2x^2) = 0$$
$$3^x(3 - 2x)(1 - x) = 0$$
$$x = \tfrac{3}{2}, 1 \text{ (since } 3^x \neq 0)$$

37. We simplify the difference quotient:

$$\frac{f(x+h) - f(x)}{h} = \frac{2^{x+h} - 2^x}{h} = \frac{2^x 2^h - 2^x}{h} = \frac{2^x(2^h - 1)}{h} = 2^x\left(\frac{2^h - 1}{h}\right)$$

39. (a)   We know the graph of $y = 2^x$ and the graph of g, the inverse of f, should contain these points with the x and y coordinates interchanged. So $g(x)$ will be the reflection of $f(x)$ across the line $y = x$:

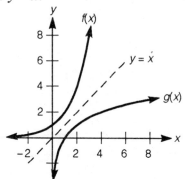

(b)   The domain is $(0, \infty)$, the range is $(-\infty, \infty)$, the x-intercept is 1, there is no y-intercept, and the asymptote is $x = 0$.

41. (a)   From the graph, we have $2^{1/2} \approx 1.4$.

(b)   Using a calculator, we have $\sqrt{2} \approx 1.41$.

43. (a)   From the graph, we have $2^{3/5} = 2^{0.6} \approx 1.5$.

(b)   Using a calculator, we have $2^{3/5} \approx 1.52$.

45. (a)   From the graph, we have $\sqrt{3} = 3^{1/2} \approx 1.7$.

(b)   Using a calculator, we have $\sqrt{3} \approx 1.73$.

47. (a)   From the graph, we have $5^{3/10} = 5^{0.3} \approx 1.6$.

(b)   Using a calculator, we have $5^{3/10} \approx 1.62$.

49.   From the graph, we have $x \approx 0.3$.

51.   From the graph, we have $x \approx 0.7$.

53. (a)   Since $10^0 = 1$, the entry in the table corresponding to $x = 1$ is 0. Since $10^1 = 10$, the entry corresponding to $x = 10$ is 1.

(b)   Since $10^{0.3} \approx 2$, we have $\left(10^{0.3}\right)^2 \approx 2^2$. That is, $10^{0.6} \approx 4$. Therefore the entry in the table corresponding to $x = 4$ is 0.6. Similarly, by cubing both sides of the approximation $10^{0.3} \approx 2$, we obtain $10^{0.9} \approx 8$. Thus the entry in the table corresponding to $x = 8$ is 0.9.

(c)   Using the hint that is given, we have $5 \approx \dfrac{10}{10^{0.3}} = 10^{0.7}$. Thus, the entry in the table corresponding to $x = 5$ is 0.7.

(d)   We have $7^2 \approx (5)(10) \approx \left(10^{0.7}\right)\left(10^1\right) = 10^{1.7}$. Therefore, $\left(7^2\right)^{1/2} \approx \left(10^{1.7}\right)^{1/2}$, or $7 \approx 10^{0.85}$. Thus the power to which 10 must be raised to yield 7 is approximately 0.85.

(e)   We have $3^4 \approx (8)(10) \approx \left(10^{0.9}\right)\left(10^1\right) = 10^{1.9}$. Therefore $\left(3^4\right)^{1/4} \approx \left(10^{1.9}\right)^{1/4} \approx 10^{0.475}$. Thus the power to which 10 must be raised to yield 3 is approximately 0.48.

(f)   We have $6 = (2)(3) \approx \left(10^{0.3}\right)\left(10^{0.48}\right) = 10^{0.78}$. Thus the power to which 10 must be raised to yield 6 is approximately 0.78. We also have:
$$9 = (3)(3) \approx \left(10^{0.475}\right)\left(10^{0.475}\right) = 10^{0.95}$$
Thus the power to which 10 must be raised to yield 9 is approximately 0.95. We complete the table:

x	$\log_{10} x$
1	0.00
2	0.30
3	0.48
4	0.60
5	0.70
6	0.78
7	0.85
8	0.90
9	0.95
10	1.00

5.2 The Exponential Function $y = e^x$

1.   The domain is $(-\infty, \infty)$, the range is $(0, \infty)$, there is no x-intercept, the y-intercept is 1, and the asymptote is $y = 0$. We draw the graph:

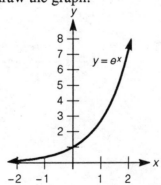

3.   The domain is $(-\infty, \infty)$, the range is $(-\infty, 0)$, there is no x-intercept, the y-intercept is -1, and the asymptote is $y = 0$. We draw the graph:

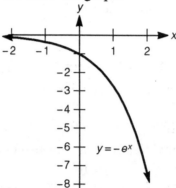

5.   The domain is $(-\infty, \infty)$, the range is $(1, \infty)$, there is no x-intercept, the y-intercept is 2, and the asymptote is $y = 1$. We draw the graph:

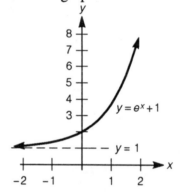

7.   The domain is $(-\infty, \infty)$, the range is $(1, \infty)$, there is no x-intercept, the y-intercept is $e + 1$, and the asymptote is $y = 1$. We draw the graph:

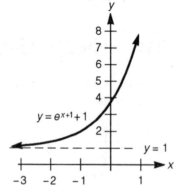

9. The domain is $(-\infty, \infty)$, the range is $(-\infty, 0)$, there is no x-intercept, the y-intercept is $-\frac{1}{e^2}$, and the asymptote is $y = 0$. We draw the graph:

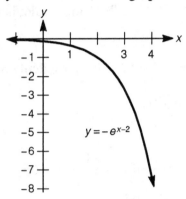

11. The domain is $(-\infty, \infty)$, the range is $(-\infty, e)$, the x-intercept is 1, the y-intercept is $e - 1$, and the asymptote is $y = e$. We draw the graph:

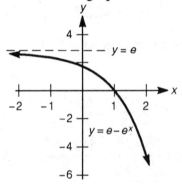

13. We graph the three functions:

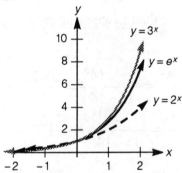

15. False (since $e \approx 2.7$)

17. False (since $\sqrt{2.7} > 1$)

19. True

21. False (since $2.7^{-1} > 0$)

23. From the graph, we have $e^{0.2} \approx 1.2$. Using a calculator, we have $e^{0.2} \approx 1.22$.

25. From the graph, we have $e^{4/5} = e^{0.8} \approx 2.2$. Using a calculator, we have $e^{4/5} = e^{0.8} \approx 2.23$.

27. From the graph, we have $\sqrt{e} = e^{0.5} \approx 1.6$. Using a calculator, we have $\sqrt{e} = e^{0.5} \approx 1.65$.

29. The value of x is approximately -0.2.

31. (a) On the interval $[2, 3]$:

$$\frac{\Delta f}{\Delta x} = \frac{f(3) - f(2)}{3 - 2} = \frac{3^4 - 2^4}{3 - 2} = \frac{81 - 16}{1} = 65$$

$$\frac{\Delta g}{\Delta x} = \frac{g(3) - g(2)}{3 - 2} = \frac{2^3 - 2^2}{3 - 2} = \frac{8 - 4}{1} = 4$$

$$\frac{\Delta h}{\Delta x} = \frac{h(3) - h(2)}{3 - 2} = \frac{e^3 - e^2}{1} \approx 12.70$$

Note that $\Delta f / \Delta x$ is 5 times as large as $\Delta h / \Delta x$.

(b) On the interval $[8, 9]$:

$$\frac{\Delta f}{\Delta x} = \frac{f(9) - f(8)}{9 - 8} = \frac{9^4 - 8^4}{9 - 8} = \frac{6561 - 4096}{1} = 2465$$

$$\frac{\Delta g}{\Delta x} = \frac{g(9) - g(8)}{9 - 8} = \frac{2^9 - 2^8}{9 - 8} = \frac{512 - 256}{1} = 256$$

$$\frac{\Delta h}{\Delta x} = \frac{h(9) - h(8)}{9 - 8} = \frac{e^9 - e^8}{1} \approx 5122.13$$

Note that $\Delta h / \Delta x$ is more than twice as large as $\Delta f / \Delta x$.

(c) On the interval $[14, 15]$:

$$\frac{\Delta f}{\Delta x} = \frac{f(15) - f(14)}{15 - 14} = \frac{15^4 - 14^4}{15 - 14} = \frac{50625 - 38416}{1} = 12,209$$

$$\frac{\Delta g}{\Delta x} = \frac{g(15) - g(14)}{15 - 14} = \frac{2^{15} - 2^{14}}{15 - 14} = \frac{32768 - 16384}{1} = 16,384$$

$$\frac{\Delta h}{\Delta x} = \frac{h(15) - h(14)}{15 - 14} = \frac{e^{15} - e^{14}}{1} \approx 2,066,413$$

Note that $\Delta h / \Delta x$ is 126 times as large as $\Delta g / \Delta x$.

33. (a) Based on the graph, $e^\pi \approx 23.4$ while $\pi^e \approx 22.2$, so e^π is slightly larger.
(b) Since $e^\pi \approx 23.14$ while $\pi^e \approx 22.46$, e^π is larger.

35. We simplify the difference quotient:

$$\frac{E(x+h) - E(x)}{h} = \frac{e^{x+h} - e^x}{h} = \frac{e^x e^h - e^x}{h} = \frac{e^x (e^h - 1)}{h} = e^x \left(\frac{e^h - 1}{h} \right)$$

37. (a) The domain is all real numbers, or $(-\infty, \infty)$.

(b) We evaluate when $x = 0$:

$$C(0) = \frac{e^0 + e^0}{2} = \frac{1+1}{2} = 1$$

(c) We compute $C(-x)$:

$$C(-x) = \frac{e^{-x} + e^x}{2} = C(x)$$

The graph must be symmetric about the y-axis.

(d) We draw the graph:

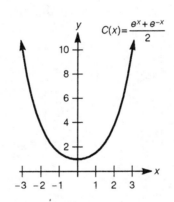

39. (a) We complete the table:

x	1000	10^4	10^5	10^6	10^7
$\left(1 + \frac{2}{x}\right)^x$	7.374312	7.387579	7.388908	7.389041	7.389055

(b) The value of e^2 is approximately 7.389056. Note that the values in the table are approaching e^2.

41. (a) We draw the graphs, noting that $L(x)$ will be the reflection of $f(x)$ across $y = x$:

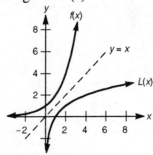

(b) The domain is $(0, \infty)$, the range is $(-\infty, \infty)$, the x-intercept is 1, and the asymptote is $x = 0$.

(c) (i) This will be a reflection of $L(x)$ across the x-axis. The x-intercept is 1 and the asymptote is $x = 0$.

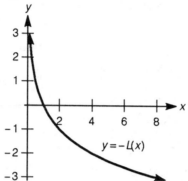

(ii) This will be a reflection of $L(x)$ across the y-axis. The x-intercept is -1 and the asymptote is $x = 0$:

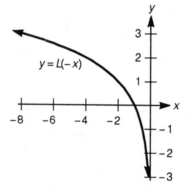

(iii) This will be a displacement of $L(x)$ to the right one unit. The x-intercept is 2 and the asymptote is $x = 1$:

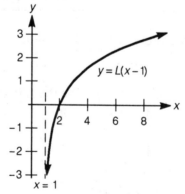

43. **(a)** We first compute $[C(x)]^2$ and $[S(x)]^2$:

$$[C(x)]^2 = \left(\frac{e^x + e^{-x}}{2}\right)^2 = \frac{e^{2x} + 2 + e^{-2x}}{4}$$

$$[S(x)]^2 = \left(\frac{e^x - e^{-x}}{2}\right)^2 = \frac{e^{2x} - 2 + e^{-2x}}{4}$$

Therefore:

$$[C(x)]^2 - [S(x)]^2 = \frac{e^{2x} + 2 + e^{-2x} - e^{2x} + 2 - e^{-2x}}{4} = \frac{4}{4} = 1$$

(b) Working from the left-hand side, we have:

$$S(-x) = \frac{e^{-x} - e^{-(-x)}}{2} = \frac{e^{-x} - e^x}{2} = -\frac{e^x - e^{-x}}{2} = -S(x)$$

(c). Working from the left-hand side, we have:

$$C(-x) = \frac{e^{-x} + e^{-(-x)}}{2} = \frac{e^{-x} + e^x}{2} = \frac{e^x + e^{-x}}{2} = C(x)$$

(d) Working from the right-hand side, we have:

$$S(x)C(y) + C(x)S(y)$$

$$= \left(\frac{e^x - e^{-x}}{2}\right)\left(\frac{e^y + e^{-y}}{2}\right) + \left(\frac{e^x + e^{-x}}{2}\right)\left(\frac{e^y - e^{-y}}{2}\right)$$

$$= \frac{e^{x+y} - e^{-x+y} + e^{x-y} - e^{-x-y} + e^{x+y} + e^{-x+y} - e^{x-y} - e^{-x-y}}{4}$$

$$= \frac{2e^{x+y} - 2e^{-x-y}}{4}$$

$$= \frac{e^{x+y} - e^{-(x+y)}}{2}$$

$$= S(x + y)$$

(e) Working from the right-hand side, we have:

$$C(x)C(y) + S(x)S(y)$$

$$= \left(\frac{e^x + e^{-x}}{2}\right)\left(\frac{e^y + e^{-y}}{2}\right) + \left(\frac{e^x - e^{-x}}{2}\right)\left(\frac{e^y - e^{-y}}{2}\right)$$

$$= \frac{e^{x+y} + e^{-x+y} + e^{x-y} + e^{-x-y} + e^{x+y} - e^{-x+y} - e^{x-y} + e^{-x-y}}{4}$$

$$= \frac{2e^{x+y} + 2e^{-x-y}}{4}$$

$$= \frac{e^{x+y} + e^{-(x+y)}}{2}$$

$$= C(x + y)$$

(f)   Working from the right-hand side, we have:

$$2S(x)C(x) = 2\left(\frac{e^x - e^{-x}}{2}\right)\left(\frac{e^x + e^{-x}}{2}\right) = \frac{e^{2x} - e^{-2x}}{2} = S(2x)$$

(g)   Working from the right-hand side, we have:

$$[C(x)]^2 + [S(x)]^2 = \left(\frac{e^x + e^{-x}}{2}\right)^2 + \left(\frac{e^x - e^{-x}}{2}\right)^2$$

$$= \frac{e^{2x} + 2 + e^{-2x} + e^{2x} - 2 + e^{-2x}}{4}$$

$$= \frac{2e^{2x} + 2e^{-2x}}{4}$$

$$= \frac{e^{2x} + e^{-2x}}{2}$$

$$= C(2x)$$

Optional TI-81 Graphing Calculator Exercises for Sections 5.1 and 5.2

1.   (a)   Using the indicated settings, we draw the graphs:

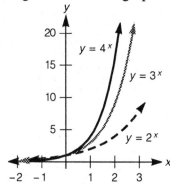

(b)   It will be much "narrower". We draw the graphs:

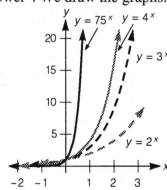

3.   (a)   The x-intercept is 2, the y-intercept is $\frac{8}{9}$, and the asymptote is $y = 1$.

  (b)   We graph the function:

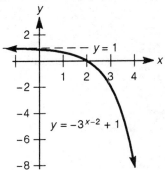

5.   (a)   There is no x-intercept, the y-intercept is 5, and the asymptote is $y = 4$.

  (b)   We graph the function:

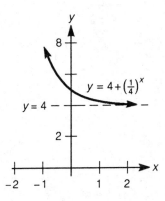

7.   (a)   The x-intercept is $1 - \log_{10} 2$, the y-intercept is 0.4, and the asymptote is $y = 0.5$.

  (b)   We graph the function:

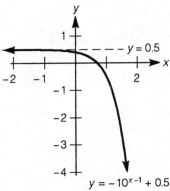

9.   (a)   We graph the function:

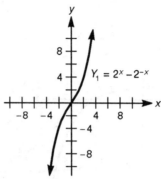

(b)   Adding $y = x^3$, we draw the graph:

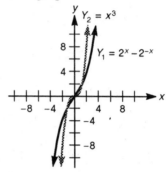

11.   We sketch the graphs when $x < 0$:

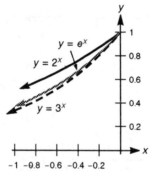

Now sketch the graphs when $x > 0$:

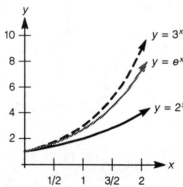

13. The graph is a reflection of $y = e^x$ across the y-axis:

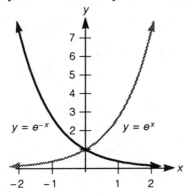

15. The graph is a reflection of $y = e^x$ across the x- and y-axes.

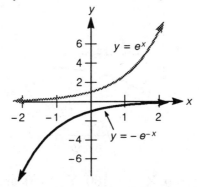

17. (a) The graph is a translation of $y = e^x$ to the right 1 unit:

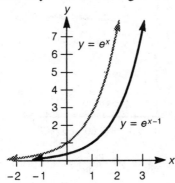

(b) The graph is a translation of $y = e^x$ to the right 1 unit, then a reflection across the
y-axis:

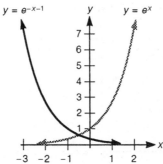

19. (a)   We graph both functions:

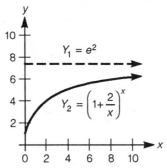

(b)   Using the settings $X_{min} = 0$, $X_{max} = 100$, $Y_{min} = 0$, and $Y_{max} = 8$:

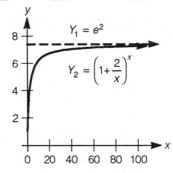

21. Using the settings $X_{min} = 0$, $X_{max} = 50$, $Y_{min} = 0$, and $Y_{max} = 0.5$:

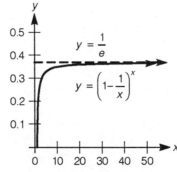

5.3 Logarithmic Functions

1. (a) No, because a horizontal line can intersect the parabola twice.
    (b) Yes, because any horizontal line intersects the line only once.
    (c) Yes, because any horizontal line intersects the curve only once.

3. (a) Let $y = \frac{2x-1}{3x+4}$. We switch the roles of x and y and solve the resulting equation for y:

$$x = \frac{2y-1}{3y+4}$$
$$x(3y+4) = 2y-1$$
$$3xy + 4x = 2y-1$$
$$3xy - 2y = -1 - 4x$$
$$y(3x-2) = -1 - 4x$$
$$y = \frac{-1-4x}{3x-2} = \frac{4x+1}{2-3x}$$

So $f^{-1}(x) = \frac{4x+1}{2-3x}$.

   (b) We compute $\frac{1}{f(x)}$:

$$\frac{1}{f(x)} = \frac{3x+4}{2x-1}$$

   (c) We compute $f^{-1}(0)$:

$$f^{-1}(0) = \frac{4(0)+1}{2-3(0)} = \tfrac{1}{2}$$

   (d) We compute $\frac{1}{f(0)}$:

$$\frac{1}{f(0)} = \frac{3(0)+4}{2(0)-1} = \frac{4}{-1} = -4$$

5. Since $y = f^{-1}(x)$ would join the points $(-2, 3)$ and $(5, -1)$, then $y = f^{-1}(x-1)$ will join the points $(-1, 3)$ and $(6, -1)$.

7. (a) $\log_3 9 = 2$
    (b) $\log_{10} 1000 = 3$
    (c) $\log_7 343 = 3$
    (d) $\log_2 \sqrt{2} = 1/2$

9. (a) $2^5 = 32$
    (b) $10^0 = 1$
    (c) $e^{1/2} = \sqrt{e}$
    (d) $3^{-4} = 1/81$
    (e) $t^v = u$

11. Since $\log_5 30$ represents the power to which 5 must be raised to get 30, it is clearly greater than 2, since $5^2 = 25$. But $\log_8 60$ is less than 2, since $8^2 = 64$. Hence $\log_5 30$ is larger.

13. (a) Since $\log_9 27$ is the power to which 9 must be raised to get 27, we can see it is between 1 and 2, since $9^1 = 9$ while $9^2 = 81$. To find it let $\log_9 27 = n$ then $9^n = 27$ in exponential form and $3^{2n} = 3^3$. So $2n = 3$ and $n = \frac{3}{2}$.

(b) If $\log_4 \frac{1}{32} = n$ then $4^n = \frac{1}{32}$, thus $2^{2n} = 2^{-5}$. So $2n = -5$, and $n = -\frac{5}{2}$. So $\log_4 \frac{1}{32} = -\frac{5}{2}$.

(c) Follow the same steps. If $\log_5 5\sqrt{5} = n$, then:
$$5^n = 5\sqrt{5}$$
$$5^n = 5^{3/2}$$
$$n = \frac{3}{2}$$

15. (a) Writing $\log_4 x = -2$ in expontential form, we have $x = 4^{-2} = \frac{1}{16}$.

(b) Writing $\ln x = -2$ in exponential form, we have $x = e^{-2} \approx 0.14$.

17. (a) We must have $5x > 0$, so $x > 0$. So the domain is $(0, \infty)$.

(b) We must have $3 - 4x > 0$, so $3 > 4x$ and $x < \frac{3}{4}$. So the domain is $\left(-\infty, \frac{3}{4}\right)$.

(c) We must have $x^2 > 0$, so $x \neq 0$. So the domain is all real numbers except 0, or $(-\infty, 0) \cup (0, \infty)$.

(d) We must have $x > 0$. So the domain is $(0, \infty)$.

(e) We solve the inequality:
$$x^2 - 25 > 0$$
$$(x + 5)(x - 5) > 0$$
The key numbers are -5 and 5. We construct the sign chart:

Inteval	Test Number	$x + 5$	$x - 5$	$(x + 5)(x - 5)$
$(-\infty, -5)$	-6	neg.	neg.	pos.
$(-5, 5)$	0	pos.	neg.	neg.
$(5, \infty)$	6	pos.	pos.	pos.

So the product is positive on the intervals $(-\infty, -5) \cup (5, \infty)$, which is the domain.

Note: This inequality can also be solved without using a sign chart:

$$x^2 - 25 > 0$$
$$x^2 > 25$$
$$|x| > 5$$
$$x > 5 \text{ or } x < -5$$

19. The coordinates for each point are:

A: $(0, 1)$
B: $(1, 0)$
C: $(4, \log_2 4) = (4, 2)$
D: $(2, 4)$

21. (a) The domain is $(0, \infty)$, the range is $(-\infty, \infty)$, the x-intercept is 1, there is no y-intercept, and the asymptote is $x = 0$:

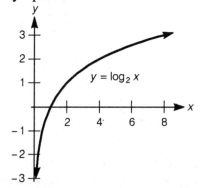

(b) The domain is $(0, \infty)$, the range is $(-\infty, \infty)$, the x-intercept is 1, there is no y-intercept, and the asymptote is $x = 0$:

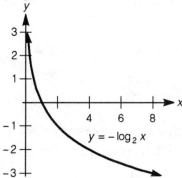

(c)   The domain is $(-\infty, 0)$, the range is $(-\infty, \infty)$, the x-intercept is -1, there is no y-intercept, and the asymptote is $x = 0$:

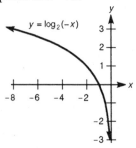

(d)   The domain is $(-\infty, 0)$, the range is $(-\infty, \infty)$, the x-intercept is -1, there is no y-intercept, and the asymptote is $x = 0$:

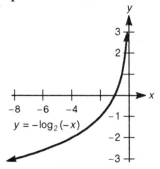

23.   We translate $y = \log_3 x$ two units to the right, reflect across the x-axis, and translate 1 unit up. The domain is $(2, \infty)$, the range is $(-\infty, \infty)$, the x-intercept is 5, there is no y-intercept, and the asymptote is $x = 2$:

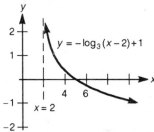

25.   We translate $y = \ln x$ to the left e units. The domain is $(-e, \infty)$, the range is $(-\infty, \infty)$, the x-intercept is $-e + 1$, the y-intercept is 1, and the asymptote is $x = -e$:

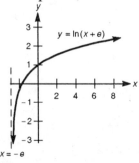

27. (a) Writing $x = \ln e^4$ in exponential form:
$$e^x = e^4$$
$$x = 4$$

(b) Writing $x = \ln \dfrac{1}{e}$ in exponential form:

$$e^x = \frac{1}{e}$$
$$e^x = e^{-1}$$
$$x = -1$$

(c) Writing $x = \ln \sqrt{e}$ in exponential form:
$$e^x = \sqrt{e}$$
$$e^x = e^{1/2}$$
$$x = \tfrac{1}{2}$$

29. We solve for x:
$$10^x = 25$$
$$x = \log_{10} 25 \approx 1.40$$

31. We solve for x:
$$10^{(x^2)} = 40$$
$$x^2 = \log_{10} 40$$
$$x = \pm\sqrt{\log_{10} 40} = \pm\sqrt{1 + \log_{10} 4} \approx \pm 1.27$$

33. We solve for t:
$$e^{2t+3} = 10$$
$$2t + 3 = \ln 10$$
$$2t = -3 + \ln 10$$
$$t = \frac{-3 + \ln 10}{2} \approx -0.35$$

35. We solve for t:
$$e^{1-4t} = 12.405$$
$$1 - 4t = \ln 12.405$$
$$-4t = -1 + \ln 12.405$$
$$t = \frac{1 - \ln 12.405}{4} \approx -0.38$$

37. Since $\sqrt{5000} \approx 70$ while $\ln 5000 \approx 9$, then A is the curve $g(x) = \ln x$ and B is the curve $f(x) = \sqrt{x}$.

39. Since $\sqrt[10]{7 \times 10^{15}} \approx 38$ while $\ln\left(7 \times 10^{15}\right) \approx 36$, then A is the curve $g(x) = \ln x$ and B is the curve $f(x) = \sqrt[10]{x}$.

41. Since $\log_3 3 = 1$ while $\log_2 3 > 1$, the upper curve must be $y = \log_2 x$ and the lower curve must be $y = \log_3 x$.

43. Let $y = e^{x+1}$. We switch the roles of x and y and solve the resulting equation for y:
$$x = e^{y+1}$$
$$\ln x = \ln\left(e^{y+1}\right)$$
$$\ln x = y + 1$$
$$y = -1 + \ln x$$
$$f^{-1}(x) = -1 + \ln x$$
The x-intercept is e and the asymptote is $x = 0$:

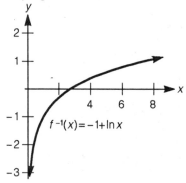

45. We graph the region:

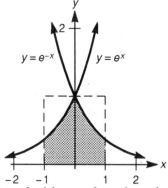

The area is less than the rectangle (shown dotted on the graph), which has an area of two square units.

47. We solve for x:
$$\left(e^x\right)^2 - 5e^x - 6 = 0$$
$$\left(e^x - 6\right)\left(e^x + 1\right) = 0$$
$$e^x = 6, -1$$
Since $e^x > 0$, then $e^x \neq -1$. Then $e^x = 6$, so $x = \ln 6$.

49. Using the approximation $2^{10} \approx 10^3$, then writing $\log_2 x = 100$ in exponential form:
$$x = 2^{100} = \left(2^{10}\right)^{10} \approx \left(10^3\right)^{10} = 10^{30}$$

51. (a) We compute $\text{pH} = -\log_{10}\left(3 \times 10^{-4}\right) \approx 3.5$. This would be an acid.
(b) We compute $\text{pH} = -\log_{10}(1) = 0$. This would be an acid.

53. (a) We have $P(10) = 4$, since 2, 3, 5, 7 do not exceed 10.
We have $P(18) = 7$, since 2, 3, 5, 7, 11, 13, 17 do not exceed 18.
We have $P(19) = 8$, since 2, 3, 5, 7, 11, 13, 17, 19 do not exceed 19.

(b) We complete the table:

x	$P(x)$	$\dfrac{x}{\ln x}$	$\dfrac{P(x)}{x/\ln x}$
10^2	25	22	1.151
10^4	1229	1086	1.132
10^6	78498	72382	1.084
10^8	5761455	5428681	1.061
10^9	50847534	48254942	1.054
10^{10}	455052512	434294482	1.048

(c) We complete the table:

x	$P(x)$	$\dfrac{x}{\ln x - 1.08366}$	$\dfrac{P(x)}{x/(\ln x - 1.08366)}$
10^2	25	28	0.8804
10^4	1229	1231	0.9988
10^6	78498	78543	0.9994
10^8	5761455	5768004	0.9989
10^9	50847534	50917519	0.9986
10^{10}	455052512	455743004	0.9985

55. Taking logarithms (base e), we must find a natural number n such that:
$$\ln(e^n) < \ln(10^{12}) < \ln(e^{n+1})$$
$$n < 12\ln 10 < n+1$$
$$n < 27.63 < n+1$$
So $n = 27$.

57. The easiest approach to use is to convert the equation to exponential form:
$$\ln y = x, \text{ so } y = e^x$$
Thus the graph is identical to $y = e^x$:

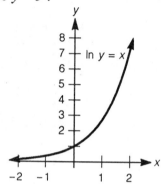

Optional TI-81 Graphing Calculator Exercises for Section 5.3

1. (a) Since $y = e^x$ and $y = \ln x$ are inverse functions, we would expect their graphs to be reflections across the line $y = x$.

   (b) Using ZOOM-5 settings to preserve a "square" graph, we obtain:

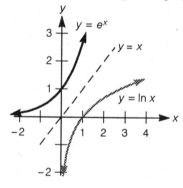

3.   (a)   For $F(x) = e^x$, the domain is $(-\infty, \infty)$:

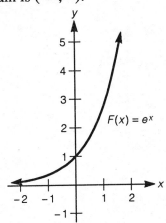

(b)   For $G(x) = \ln x$, the domain is $(0, \infty)$:

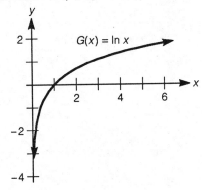

(c)   For $(G \circ F)(x) = \ln e^x = x$, the domain is $(-\infty, \infty)$:

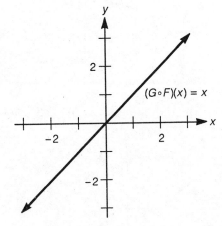

(d)   For $(F \circ G)(x) = e^{\ln x} = x$ for $x > 0$, the domain is $(0, \infty)$:

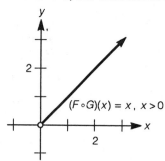

(e)   For $(G \circ H)(x) = \ln(x - 2)$, the domain is $(2, \infty)$:

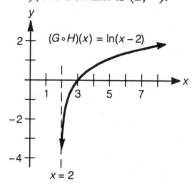

(f)   For $(H \circ G)(x) = \ln x - 2$, the domain is $(0, \infty)$:

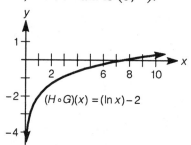

(g)   For $(G \circ K)(x) = \ln x^2$, the domain is $(-\infty, 0) \cup (0, \infty)$:

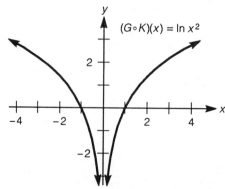

(h)   For $(K \circ G)(x) = (\ln x)^2$, the domain is $(0, \infty)$:

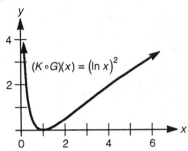

5.   (a)   We graph the function:

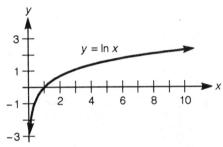

(b)   We graph both functions:

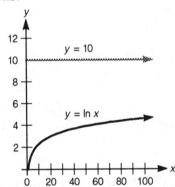

(c)   We graph both functions:

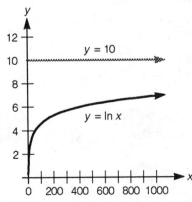

(d)   We graph both functions:

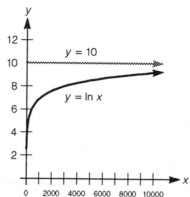

(e)   We graph both functions:

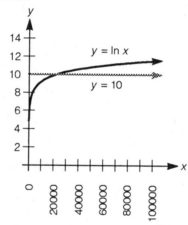

It appears $\ln x = 10$ when $x \approx 22{,}000$. To find the exact value, we write $\ln x = 10$ in exponential form to obtain $x = e^{10} \approx 22026$.

5.4 Properties of Logarithms

1.   Using properties for logarithms, we have:
$$\log_{10} 70 - \log_{10} 7 = \log_{10} \tfrac{70}{7} = \log_{10} 10 = 1$$

3.   Using properties for logarithms, we have:
$$\log_7 \sqrt{7} = \log_7\left(7^{1/2}\right) = \tfrac{1}{2}$$

5.   Using properties for logarithms, we have:
$$\log_3 108 + \log_3 \tfrac{3}{4} = \log_3\left(\tfrac{108 \bullet 3}{4}\right) = \log_3 81 = \log_3 3^4 = 4$$

7.   Using properties for logarithms, we have:
$$-\tfrac{1}{2} + \ln \sqrt{e} = -\tfrac{1}{2} + \ln e^{1/2} = -\tfrac{1}{2} + \tfrac{1}{2} = 0$$

9. Using properties for logarithms, we have:
$$2^{\log_2 5} - 3\log_5 \sqrt[3]{5} = 5 - 3\log_5 5^{1/3} = 5 - 3\left(\tfrac{1}{3}\right) = 5 - 1 = 4$$

11. Using properties for logarithms, we have:
$$\log_{10} 30 + \log_{10} 2 = \log_{10}(30 \cdot 2) = \log_{10} 60$$

13. Using properties for logarithms, we have:
$$\log_5 6 + \log_5 \tfrac{1}{3} + \log_5 10 = \log_5\left(6 \cdot \tfrac{1}{3} \cdot 10\right) = \log_5 20$$

15. (a) Using properties for logarithms, we have:
$$\ln 3 - 2\ln 4 + \ln 32 = \ln 3 - \ln\left(4^2\right) + \ln 32 = \ln\left(\frac{3 \cdot 32}{4^2}\right) = \ln 6$$

(b) Using properties for logarithms, we have:
$$\begin{aligned}
\ln 3 - 2(\ln 4 + \ln 32) &= \ln 3 - 2[\ln(4 \cdot 32)] \\
&= \ln 3 - 2\ln 128 \\
&= \ln 3 - \ln\left(128^2\right) \\
&= \ln\left(\frac{3}{128^2}\right) \\
&= \ln\left(\tfrac{3}{16384}\right)
\end{aligned}$$

17. Using properties for logarithms, we have:
$$\begin{aligned}
\log_b 4 + 3\left[\log_b(1+x) - \tfrac{1}{2}\log_b(1-x)\right] &= \log_b 4 + 3\left(\log_b(1+x) - \log_b\sqrt{1-x}\right) \\
&= \log_b 4 + \log_b(1+x)^3 - \log_b(1-x)^{3/2} \\
&= \log_b\left[\frac{4(1+x)^3}{(1-x)^{3/2}}\right]
\end{aligned}$$

19. Using properties for logarithms, we have:
$$\begin{aligned}
4\log_{10} 3 - 6\log_{10}\left(x^2+1\right) + \tfrac{1}{2}\left[\log_{10}(x+1) - 2\log_{10} 3\right] \\
= \log_{10} 3^4 - \log_{10}\left(x^2+1\right)^6 + \tfrac{1}{2}\left[\log_{10}(x+1) - \log_{10} 3^2\right] \\
= \log_{10} 81 - \log_{10}\left(x^2+1\right)^6 + \tfrac{1}{2}\log_{10}\left(\frac{x+1}{9}\right) \\
= \log_{10} 81 - \log_{10}\left(x^2+1\right)^6 + \log_{10}\left(\frac{x+1}{9}\right)^{1/2} \\
= \log_{10}\left[\frac{81\frac{\sqrt{x+1}}{3}}{\left(x^2+1\right)^6}\right] \\
= \log_{10}\left[\frac{27\sqrt{x+1}}{\left(x^2+1\right)^6}\right]
\end{aligned}$$

21. (a) Using properties for logarithms, we have:

$$\log_{10}\left(\frac{x^2}{1+x^2}\right) = \log_{10}x^2 - \log_{10}(1+x^2) = 2\log_{10}x - \log_{10}(1+x^2)$$

(b) Using properties for logarithms, we have:

$$\ln\left(\frac{x^2}{\sqrt{1+x^2}}\right) = \ln x^2 - \ln\sqrt{1+x^2} = 2\ln x - \tfrac{1}{2}\ln(1+x^2)$$

23. (a) Using properties for logarithms, we have:

$$\log_{10}\sqrt{9-x^2} = \tfrac{1}{2}\log_{10}(9-x^2)$$
$$= \tfrac{1}{2}\log_{10}[(3+x)(3-x)]$$
$$= \tfrac{1}{2}\log_{10}(3+x) + \tfrac{1}{2}\log_{10}(3-x)$$

(b) Using properties for logarithms, we have:

$$\ln\left[\frac{\sqrt{4-x^2}}{(x-1)(x+1)^{3/2}}\right] = \tfrac{1}{2}\ln(4-x^2) - \ln(x-1) - \ln(x+1)^{3/2}$$

$$= \tfrac{1}{2}\ln[(2+x)(2-x)] - \ln(x-1) - \tfrac{3}{2}\ln(x+1)$$
$$= \tfrac{1}{2}\ln(2+x) + \tfrac{1}{2}\ln(2-x) - \ln(x-1) - \tfrac{3}{2}\ln(x+1)$$

25. (a) Using properties for logarithms, we have:

$$\log_b\sqrt{\frac{x}{b}} = \tfrac{1}{2}\log_b\frac{x}{b} = \tfrac{1}{2}\log_b x - \tfrac{1}{2}\log_b b = \tfrac{1}{2}\log_b x - \tfrac{1}{2}$$

(b) Using properties for logarithms, we have:

$$2\ln\sqrt{(1+x^2)(1+x^4)(1+x^6)} = \ln[(1+x^2)(1+x^4)(1+x^6)]$$
$$= \ln(1+x^2) + \ln(1+x^4) + \ln(1+x^6)$$

27. (a) Using properties for logarithms, we have:

$$\log_{10}(AB^2C^3) = \log_{10}A + \log_{10}B^2 + \log_{10}C^3$$
$$= \log_{10}A + 2\log_{10}B + 3\log_{10}C$$
$$= a + 2b + 3c$$

(b) Using properties for logarithms, we have:

$$\log_{10}10\sqrt{A} = \log_{10}10 + \log_{10}\sqrt{A} = 1 + \tfrac{1}{2}\log_{10}A = 1 + \tfrac{1}{2}a$$

(c) Using properties for logarithms, we have:

$$\log_{10}\sqrt{10ABC} = \tfrac{1}{2}\log_{10}(10ABC)$$
$$= \tfrac{1}{2}\log_{10}10 + \tfrac{1}{2}\log_{10}A + \tfrac{1}{2}\log_{10}B + \tfrac{1}{2}\log_{10}C$$
$$= \tfrac{1}{2}(1) + \tfrac{1}{2}a + \tfrac{1}{2}b + \tfrac{1}{2}c$$
$$= \tfrac{1}{2}(1 + a + b + c)$$

(d) Using properties for logarithms, we have:

$$\log_{10}\!\left(\tfrac{10A}{\sqrt{BC}}\right) = \log_{10}(10A) - \tfrac{1}{2}\log_{10}(BC)$$
$$= \log_{10}10 + \log_{10}A - \tfrac{1}{2}\log_{10}B - \tfrac{1}{2}\log_{10}C$$
$$= 1 + a - \tfrac{1}{2}b - \tfrac{1}{2}c$$

29. (a) Using properties for logarithms, we have:

$$\ln(ex) = \ln e + \ln x = 1 + t$$

(b) Using properties for logarithms, we have:

$$\ln(xy) - \ln(x^2) = \ln\frac{xy}{x^2} = \ln\frac{y}{x} = \ln y - \ln x = u - t$$

(c) Using properties for logarithms, we have:

$$\ln\sqrt{xy} + \ln\frac{x}{e} = \tfrac{1}{2}(\ln x + \ln y) + \ln x - \ln e = \tfrac{1}{2}(t + u) + t - 1 = \tfrac{3}{2}t + \tfrac{1}{2}u - 1$$

(d) Using properties for logarithms, we have:

$$\ln(e^2 x\sqrt{y}) = \ln e^2 + \ln x + \ln\sqrt{y} = 2\ln e + \ln x + \tfrac{1}{2}\ln y = 2 + t + \tfrac{1}{2}u$$

31. (a) To find the x-intercept, we set $y = 0$:

$$0 = 2^x - 5$$
$$2^x = 5$$
$$\ln 2^x = \ln 5$$
$$x\ln 2 = \ln 5$$
$$x = \frac{\ln 5}{\ln 2} \approx 2.32$$

Now sketch the graph:

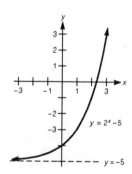

(b)   To find the x-intercept, we set $y = 0$:

$$0 = 2^{x/2} - 5$$
$$2^{x/2} = 5$$
$$\ln 2^{x/2} = \ln 5$$
$$\frac{x}{2}\ln 2 = \ln 5$$
$$x = \frac{2\ln 5}{\ln 2} \approx 4.64$$

Now sketch the graph:

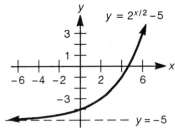

33.   Taking the natural log of each side, we have:

$$\ln 5 = \ln 2 + \ln e^{2x-1}$$
$$\ln 5 = \ln 2 + (2x - 1)$$
$$\ln 5 = \ln 2 + 2x - 1$$
$$2x = \ln 5 - \ln 2 + 1$$
$$x = \frac{\ln 5 - \ln 2 + 1}{2}$$

35.   Taking the natural logarithm of each side, we have:

$$2^x = 13$$
$$\ln 2^x = \ln 13$$
$$x\ln 2 = \ln 13$$
$$x = \frac{\ln 13}{\ln 2}$$

37.   Taking the natural logarithm of each side, we have:

$$10^x = e$$
$$\ln 10^x = \ln e$$
$$x\ln 10 = 1$$
$$x = \frac{1}{\ln 10}$$

39. Solving for x, we have:

$$\log_9(x+1) = \tfrac{1}{2} + \log_9 x$$
$$\log_9(x+1) - \log_9 x = \tfrac{1}{2}$$
$$\log_9\left(\frac{x+1}{x}\right) = \tfrac{1}{2}$$
$$9^{1/2} = \frac{x+1}{x}$$
$$3 = \frac{x+1}{x}$$
$$3x = x+1$$
$$2x = 1$$
$$x = \tfrac{1}{2}$$

41. Solving for x, we have:

$$\log_{10}(2x+4) + \log_{10}(x-2) = 1$$
$$\log_{10}[(2x+4)(x-2)] = 1$$
$$(2x+4)(x-2) = 10^1$$
$$2x^2 - 8 = 10$$
$$2x^2 = 18$$
$$x^2 = 9$$
$$x = \pm 3$$

Since $\log_{10}(-2)$ is undefined, $x = -3$ is an extraneous root. So the solution is $x = 3$.

43. Solving for x, we have:

$$\log_{10}(x+3) - \log_{10}(x-2) = 2$$
$$\log_{10}\left(\frac{x+3}{x-2}\right) = 2$$
$$10^2 = \frac{x+3}{x-2}$$
$$100(x-2) = x+3$$
$$100x - 200 = x+3$$
$$99x = 203$$
$$x = \frac{203}{99}$$

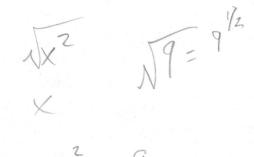

45. Solving for x, we have:

$$\log_b(x+1) = 2\log_b(x-1)$$
$$\log_b(x+1) = \log_b(x-1)^2$$
$$x+1 = (x-1)^2$$
$$x+1 = x^2 - 2x + 1$$
$$x^2 - 3x = 0$$
$$x(x-3) = 0$$
$$x = 0, 3$$

Since $\log_b(-1)$ is undefined, $x = 0$ is an extraneous root. So the solution is $x = 3$.

47. Solving for x, we have:
$$\log_{10}(x-6) + \log_{10}(x+3) = 1$$
$$\log_{10}[(x-6)(x+3)] = 1$$
$$10^1 = (x-6)(x+3)$$
$$x^2 - 3x - 18 = 10$$
$$x^2 - 3x - 28 = 0$$
$$(x-7)(x+4) = 0$$
$$x = 7, -4$$
Since $\log_{10}(-10)$ is undefined, $x = -4$ is an extraneous root. So the solution is $x = 7$.

49. (a) Solving for x, we have:
$$\log_{10} x - y = \log_{10}(3x - 1)$$
$$\log_{10} x - \log_{10}(3x - 1) = y$$
$$\log_{10}\left(\frac{x}{3x-1}\right) = y$$
$$10^y = \frac{x}{3x-1}$$
$$10^y(3x - 1) = x$$
$$3(10^y)x - 10^y = x$$
$$3(10^y)x - x = 10^y$$
$$x[3(10^y) - 1] = 10^y$$
$$x = \frac{10^y}{3(10^y) - 1}$$

(b) Solving for x, we have:
$$\log_{10}(x - y) = \log_{10}(3x - 1)$$
$$x - y = 3x - 1$$
$$-2x = y - 1$$
$$x = \frac{y-1}{-2} \text{ or } \frac{1-y}{2}$$

51. Using the change of base formula, we have:
$$\log_2 5 = \frac{\log_{10} 5}{\log_{10} 2}$$

53. Using the change of base formula, we have:
$$\ln 3 = \log_e 3 = \frac{\log_{10} 3}{\log_{10} e}$$

55. Using the change of base formula, we have:
$$\log_b 2 = \frac{\log_{10} 2}{\log_{10} b}$$

57. Using the change of base formula, we have:
$$\log_{10} 6 = \frac{\ln 6}{\ln 10}$$

59. Using the change of base formula, we have:
$$\log_{10} e = \frac{\ln e}{\ln 10} = \frac{1}{\ln 10}$$

61. Using the change of base formula, we have:
$$\log_{10}(\log_{10} x) = \log_{10}\left(\frac{\ln x}{\ln 10}\right) = \frac{\ln\left(\frac{\ln x}{\ln 10}\right)}{\ln 10} = \frac{\ln(\ln x) - \ln(\ln 10)}{\ln 10}$$

63. (a) true
(b) true
(c) true
(d) false, since $\ln x^3 = 3 \ln x$, not $\ln 3x$
(e) true
(f) false, since $\ln 2x^3 = \ln 2 + 3 \ln x \neq 3 \ln 2x = \ln(2x)^3$
(g) true
(h) false, since $\log_5 24$ is between 1 and 2, not 5^1 and 5^2
(i) true
(j) false, since $\log_5 24$ is close to 2 ($2 = \log_5 25$)
(k) false, since the domain is $(0, \infty)$
(l) true
(m) true

65. (a) We have:
$$\log_{10} \pi^7 \approx 3.48005$$
$$7 \log_{10} \pi \approx 7(0.49715) = 3.48005$$
So $\log_{10} \pi^7 = 7 \log_{10} \pi$.

(b) We have:
$$\log_b P^n = \log_{10} \pi^7 \approx 3.48005$$
$$\left(\log_b P\right)^n = \left(\log_{10} \pi\right)^7 \approx 0.00751$$
So $\log_b \pi^7 \neq \left(\log_b \pi\right)^7$.

(c) We have:
$$b^{\log_b P} = 10^{\log_{10} 1776} = 10^{3.24944} = 1776$$
So $10^{\log_{10} 1776} = 1776$.

(d)   We have:

$$\ln 2 + \ln 3 + \ln 4 \approx 0.69315 + 1.09861 + 1.38629 \approx 3.17805$$

$$\ln 24 \approx 3.17805$$

So $\ln 2 + \ln 3 + \ln 4 = \ln 24$.

(e)   We have:

$$\log_{10} A + \log_{10} B + \log_{10} C = \log_{10} 11 + \log_{10} 12 + \log_{10} 13$$
$$\approx 1.04139 + 1.07918 + 1.11394$$
$$= 3.23451$$
$$\log_{10}(ABC) = \log_{10}(11 \cdot 12 \cdot 13) = \log_{10} 1716 \approx 3.23451$$

So $\log_{10} A + \log_{10} B + \log_{10} C = \log_{10}(ABC)$.

(f)   We compute $f[g(2345.6)]$:

$$f[g(2345.6)] = f(\ln 2345.6) \approx f(7.76030) \approx e^{7.76030} \approx 2345.6$$

(g)   We compute $g[f(0.123456)]$:

$$g[f(0.123456)] = g\left(10^{0.123456}\right) = \log_{10}(1.32879) = 0.123456$$

67.   We complete the table:

x	0.1	0.05	0.005	0.0005
$\ln(1+x)$	0.095310	0.048790	0.004987	0.000499

Note that the values of $\ln(1+x)$ are close to x, when x is close to 0.

69.   Since $(-2, 324)$ and $\left(\frac{1}{2}, \frac{4}{3}\right)$ lie on the curve $y = ae^{bx}$, we have:

$$324 = ae^{-2b} \quad \text{and} \quad \tfrac{4}{3} = ae^{\frac{1}{2}b}$$

Dividing, we have:

$$\frac{324}{\frac{4}{3}} = \frac{ae^{-2b}}{ae^{\frac{1}{2}b}}$$

$$243 = e^{-2.5b}$$

$$\ln 243 = -2.5b$$

$$b = -\frac{\ln 243}{2.5}$$

Substituting into $324 = ae^{-2b}$, we have:

$$324 = ae^{(2\ln 243)/2.5}$$

$$324 = ae^{0.8\ln 243}$$

$$324 = a\left(e^{\ln 243}\right)^{0.8}$$

$$324 = a \cdot 243^{0.8}$$

$$a = \frac{324}{243^{0.8}} = \tfrac{324}{81} = 4$$

71. (a) We solve the equation:

$$2^x - 3 = 0$$
$$2^x = 3$$
$$\log_{10} 2^x = \log_{10} 3$$
$$x \log_{10} 2 = \log_{10} 3$$
$$x = \frac{\log_{10} 3}{\log_{10} 2}$$

(b) We have $\dfrac{\log_{10} 3}{\log_{10} 2} \approx 1.585$ and $\dfrac{\ln 3}{\ln 2} \approx 1.585$.

(c) Using the change of base formula, we have:

$$\frac{\log_{10} 3}{\log_{10} 2} = \frac{\dfrac{\ln 3}{\ln 10}}{\dfrac{\ln 2}{\ln 10}} = \frac{\ln 3}{\ln 2}$$

73. Working from the left-hand side, we have:

$$\log_b\left(\frac{\sqrt{3}+\sqrt{2}}{\sqrt{3}-\sqrt{2}}\right) = \log_b\left(\frac{\sqrt{3}+\sqrt{2}}{\sqrt{3}-\sqrt{2}} \cdot \frac{\sqrt{3}+\sqrt{2}}{\sqrt{3}+\sqrt{2}}\right)$$

$$= \log_b\left[\frac{\left(\sqrt{3}+\sqrt{2}\right)^2}{3-2}\right]$$

$$= \log_b\left(\sqrt{3}+\sqrt{2}\right)^2$$

$$= 2\log_b\left(\sqrt{3}+\sqrt{2}\right)$$

75. Using properties of logarithms, we have:

$$b^{3\log_b x} = b^{\log_b x^3} = x^3$$

77. Using properties of logarithms, we have:

$$\log_2 \sqrt[5]{4\sqrt{2}} = \tfrac{1}{5}\log_2 4\sqrt{2} = \tfrac{1}{5}\log_2\left(2^2 \cdot 2^{1/2}\right) = \tfrac{1}{5}\log_2\left(2^{5/2}\right) = \tfrac{1}{5} \cdot \tfrac{5}{2} = \tfrac{1}{2}$$

79. Solving for x, we have:
$$3\ln x = \alpha + 3\ln\beta$$
$$3\ln x - 3\ln\beta = \alpha$$
$$3(\ln x - \ln\beta) = \alpha$$
$$3\ln\frac{x}{\beta} = \alpha$$
$$\ln\frac{x}{\beta} = \frac{\alpha}{3}$$
$$\frac{x}{\beta} = e^{\alpha/3}$$
$$x = \beta e^{\alpha/3}$$

81. Using the change of base formula, we have:
$$\log_b a = \frac{\log_a a}{\log_a b} = \frac{1}{\log_a b}$$

83. Using the change of base formula, we have:
$$\log_{ab} x = \frac{\log_a x}{\log_a ab} = \frac{\log_a x}{\log_a a + \log_a b} = \frac{\log_a x}{1 + \log_a b}$$
Therefore:
$$\frac{\log_a x}{\log_{ab} x} = \frac{\log_a x}{\frac{\log_a x}{1 + \log_a b}} = 1 + \log_a b$$

85. We work from the right-hand side:
$$\tfrac{1}{2}(\log a + \log b) = \tfrac{1}{2}\log(ab) = \log\sqrt{ab}$$
Proving the desired equality is equivalent to proving:
$$\tfrac{1}{3}(a + b) = \sqrt{ab}, \text{ since log is a one-to-one function.}$$
If a and b are both positive, we can square each side:
$$\tfrac{1}{9}(a^2 + 2ab + b^2) = ab$$
We now work with the left-hand side:
$$\tfrac{1}{9}(a^2 + 2ab + b^2) = \tfrac{1}{9}(7ab + 2ab) \text{ by our assumption } a^2 + b^2 = 7ab$$
$$= \tfrac{1}{9}(9ab)$$
$$= ab$$

87. Let $y = \ln\left(x + \sqrt{x^2 + 1}\right)$. We interchange the roles of x and y, then solve for y:
$$x = \ln\left(y + \sqrt{y^2 + 1}\right)$$
$$e^x = y + \sqrt{y^2 + 1}$$
$$e^x - y = \sqrt{y^2 + 1}$$

Squaring each side:
$$e^{2x} - 2ye^x + y^2 = y^2 + 1$$
$$e^{2x} - 2ye^x = 1$$
$$e^{2x} - 1 = 2ye^x$$
$$\frac{e^{2x} - 1}{2e^x} = y$$

So $f^{-1}(x) = \dfrac{e^{2x} - 1}{2e^x}$.

89. We complete the table:

x	10^2	10^3	10^6	10^{20}	10^{50}	10^{99}
y	0.4234	0.6589	0.9654	1.3428	1.5573	1.6918

As x increases in size, so does $\ln x$, and thus $\ln(\ln x)$, and finally $\ln[\ln(\ln x)]$. The table is misleading--it only indicates that y increases in size much slower! An actual proof of the result is more complicated: To show every positive number y is in the range, we need to show $\ln[\ln(\ln x)] = y$ has a solution for each such y. Let $\ln[\ln(\ln x)] = y$, then:
$$\ln[\ln(\ln x)] = y$$
$$\ln(\ln x) = e^y$$
$$\ln x = e^{e^y}$$
$$x = e^{e^{e^y}}$$

5.5 Compound Interest

1. For annual compounding of money we use the formula $A = P(1 + r)^t$. We find A when $P = \$800$, $r = 0.06$ and $t = 4$ yrs:
$$A = 800(1 + 0.06)^4 = 800(1.06)^4 \approx \$1009.98$$

3. For annual compounding of money we use the formula $A = P(1 + r)^t$. We find r when $A = \$6000$, $P = \$4000$ and $t = 5$:
$$6000 = 4000(1 + r)^5$$
$$1.5 = (1 + r)^5$$
$$(1.5)^{1/5} = 1 + r$$
$$r = (1.5)^{1/5} - 1 \approx 0.0845$$
So the interest rate is 8.45%.

5. In the first bank, we have $P = \$500$, $r = 0.05$, and $t = 4$:
$$A = 500(1 + 0.05)^4 = 500(1.05)^4 \approx 607.75$$
We now deposit $P = \$607.75$, $r = 0.06$, and $t = 4$:
$$A = 607.75(1 + 0.06)^4 = 607.75(1.06)^4 \approx 767.27$$
So the new balance will be $767.27.

7.   (a)   We use $A = P(1 + r)^t$ with $P = 1000$, $r = 0.07$, and $t = 20$:
$$A = 1000(1 + 0.07)^{20} = 1000(1.07)^{20} \approx 3869.68$$
The new balance is \$3869.68.

(b)   We use $A = P\left(1 + \frac{r}{N}\right)^t$ with $P = 1000$, $r = 0.07$, $N = 4$, and $t = 20$:
$$A = 1000\left(1 + \frac{0.07}{4}\right)^{4(20)} = 1000(1.0175)^{80} \approx 4006.39$$
The new balance is \$4006.39.

9.   For compounding quarterly we use the formula:
$$A = P\left(1 + \frac{r}{N}\right)^{Nt}$$
So here $P = \$100$, $r = 6\%$ and we have $N = 4$ compoundings per year. We find the value of t for which $A \geq \$120$:
$$120 \leq 100\left(1 + \frac{0.06}{4}\right)^{4t}$$
$$1.2 \leq (1.015)^{4t}$$
$$\ln 1.2 \leq 4t \ln 1.015$$
$$4t \geq \frac{\ln 1.2}{\ln 1.015}$$
$$t \geq \frac{\ln 1.2}{4 \ln 1.015} \approx 3.06$$
This is slightly over 3 years, and so 13 quarters will be required.

11.   We use $A = P\left(1 + \frac{r}{N}\right)^{Nt}$ where $r = 0.055$, $N = 2$, $A = 6000$, and $t = 10$:
$$6000 = P\left(1 + \frac{0.55}{2}\right)^{2(10)}$$
$$6000 = P(1.0275)^{20}$$
$$P = \frac{6000}{(1.0275)^{20}} \approx 3487.50$$
You must deposit a principal of \$3487.50.

13.   We use $A = Pe^{rt}$ where $A = 5000$, $t = 10$ and $r = 0.065$:
$$5000 = Pe^{(0.065)(10)}$$
$$5000 = Pe^{0.65}$$
$$P = \frac{5000}{e^{0.65}} \approx 2610.23$$
A principal of \$2610.23 will grow to \$5000 in 10 yrs.

15.   Since the effective rate is $r = 0.06$, we have:
$$A = P(1 + 0.06)^1 = 1.06P$$
The nominal rate r would yield a balance of:
$$A = Pe^{r(1)} = Pe^r$$

Setting these equal:
$$Pe^r = P(1.06)$$
$$e^r = 1.06$$
$$r = \ln 1.06 \approx 0.0583$$
So the nominal rate is 5.83%.

17. Let's take the 6% investment first:
$$A = 10000(1 + 0.06)^5 = 10000(1.3382) \approx \$13382.26$$
The second choice will be:
$$A = 10000e^{0.05(5)} = 10000e^{0.25} \approx \$12840.25, \text{ considerably less.}$$

19. (a) We have $T_2 \approx \frac{0.7}{r} = \frac{0.7}{0.05} = 14$ yrs.

   (b) We have $T_2 = \frac{\ln 2}{r} = \frac{\ln 2}{0.05} = 13.86$ yrs.

   (c) Here $d_1 = 13.86$, $d_2 = 14$, so $d = |13.86 - 14| = 0.14$.
   This represents $\frac{0.14}{13.86}(100) \approx 1.01\%$ of the actual doubling time.

21. We compute:
$$A = 1000e^{(0.08)(300)} = 1000e^{24} = 1000\left(2.65 \times 10^{10}\right) = \$2.65 \times 10^{13}$$
That's \$26.5 trillion, a nice inheritance.

23. (a) We have $T_2 \approx \frac{0.7}{0.05} = 14$ yrs.

   (b) We sketch the graph:

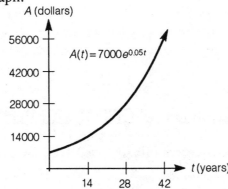

25. (a) For account #1, we have $P = 2000$, $r = 0.04$, and $N = 4$, so:
$$A_1(t) = 2000\left(1 + \frac{0.04}{4}\right)^{4t} = 2000(1.01)^{4t}$$
For account #2, we have $P = 2000$ and $r = 0.04$, so:
$$A_2(t) = 2000e^{0.04t}$$

Now complete the table:

t (years)	1	2	3	4	10
A (account #1)	2081	2166	2254	2345	2978
A (account #2)	2082	2167	2255	2347	2984

(b) For account #1, we find when $A_1(t) = 4000$:

$$4000 = 2000(1.01)^{4t}$$
$$2 = (1.01)^{4t}$$
$$\ln 2 = 4t \ln 1.01$$
$$t = \frac{\ln 2}{4 \ln 1.01} \approx 17.4 \text{ years}$$

For account #2, we find when $A_2(t) = 4000$:

$$4000 = 2000 e^{0.04t}$$
$$2 = e^{0.04t}$$
$$\ln 2 = 0.04t$$
$$t = \frac{\ln 2}{0.04} \approx 17.3 \text{ years}$$

The doubling time for account #1 is 17.4 years while that of account #2 is 17.3 years.

(c) Using the table values:

$$\frac{\Delta A}{\Delta t} = \frac{A_1(3) - A_1(2)}{3 - 2} = \frac{2254 - 2166}{3 - 2} = \$88 \text{ per year}$$

(d) Using the table values:

$$\frac{\Delta A}{\Delta t} = \frac{A_2(3) - A_2(2)}{3 - 2} = \frac{2255 - 2167}{3 - 2} = \$88 \text{ per year}$$

5.6 Exponential Growth and Decay

1. (a) Since 2000 bacteria are present initially, we have $N_0 = 2000$. Now $N = 3800$ when $t = 2$, so:

$$3800 = 2000 e^{k \cdot 2}$$
$$1.9 = e^{2k}$$
$$\ln 1.9 = 2k$$
$$k = \frac{\ln 1.9}{2} \approx 0.3209$$

(b)   Using the formula $N(t) = 2000e^{0.3209t}$, we find $N(5)$:

$$N(5) = 2000e^{(0.3209)(5)} \approx 9951 \text{ bacteria}$$

(c)   We find when $N = 10000$:

$$10000 = 2000e^{0.3209t}$$
$$5 = e^{0.3209t}$$
$$\ln 5 = 0.3209t$$
$$t = \frac{\ln 5}{0.3209} \approx 5.0 \text{ hours}$$

3.   Since $N(0) = 2000$, we have $N_0 = 2000$. Using $N(t) = 2000e^{kt}$, we find when $N(3) = 3400$:

$$3400 = 2000e^{k \cdot 3}$$
$$1.7 = e^{3k}$$
$$\ln 1.7 = 3k$$
$$k = \frac{\ln 1.7}{3} \approx 0.1769$$

5.   The growth problems are solved by assuming an exponential growth rate:
$$N = N_0 e^{kt}$$
First we find the percent distribution by dividing the total population into each of the two regions:
$$\frac{1.224}{5.420} \approx 22.6\% \text{ for more developed regions}$$
Applying the growth formula to the total world population, we have:
$$N = N_0 e^{kt} = 5.420e^{0.017(8)} = 5.420e^{0.136} \approx 6.210 \text{ billion}$$
The other calculations are similar to these two. The completed table is:

	1992 Population (billions)	% Population in 1992	Growth Rate (% per year)	Year 2000 Population (billions)	% of World Population in 2000
World	5.420	100	1.7	6.210	100
More Dev.	1.224	22.6	0.5	1.274	20.50
Less Dev.	4.196	77.4	2.0	4.924	79.29

Note that some discrepency in the last two columns is due to round-off error.

7. (a) We complete the table:

Country	1992 Population (millions)	Relative Growth Rate (percent per year)	Year 2000 Population (millions)	Percent Increase in Population
Iraq	18.2	3.7	24.5	34.62
United Kingdom	57.8	0.3	59.2	2.42

(b) We solve the equation:

$$18.2e^{0.037t} = 57.8e^{0.003t}$$
$$\frac{e^{0.037t}}{e^{0.003t}} = \frac{57.8}{18.2}$$
$$e^{0.034t} \approx 3.18$$
$$0.034t \approx \ln 3.18$$
$$t \approx \frac{\ln 3.18}{0.034} \approx 34$$

Iraq's population will be equal to United Kingdom's in 34 years, which is the year 2026.

9. (a) For Cyprus, we use $N_0 = 0.7$ million, $k = 0.011$, and $t = 8$:

$$N(8) = 0.7e^{(0.011)(8)} \approx 0.8 \text{ million}$$

For Gaza, we use $N_0 = 0.7$ million, $k = 0.046$, and $t = 8$:

$$N(8) = 0.7e^{(0.046)(8)} \approx 1.0 \text{ million}$$

(b) For Cyprus, we find when $N = 1$ million:

$$1 = 0.7e^{0.011t}$$
$$1.43 = e^{0.011t}$$
$$\ln 1.43 = 0.011t$$
$$t = \frac{\ln 1.43}{0.011} \approx 33 \text{ years}$$

Cyprus will reach 1 million people in 33 years, which is the year 2025. At that time, the population of Gaza will be:

$$N(33) = 0.7e^{(0.046)(33)} \approx 3.2 \text{ million}$$

11. (a) We use $N_0 = 1.4$, $t = 25$, and $N = 2$:

$$2 = 1.4e^{k \cdot 25}$$
$$1.43 = e^{25k}$$
$$\ln 1.43 = 25k$$
$$k = \frac{\ln 1.43}{25} \approx 0.0143$$

(b)   We use $N_0 = 1.4$, $t = 108$, and $N = 2$:

$$2 = 1.4e^{k \cdot 108}$$
$$1.43 = e^{108k}$$
$$\ln 1.43 = 108k$$
$$k = \frac{\ln 1.43}{108} \approx 0.0033$$

13. (a)   Use $N = N_0 e^{kt}$:

$$62947714 = 23191876 e^{k(50)}$$
$$2.7142 = e^{50k}$$
$$\ln 2.7142 = 50k$$
$$k = \frac{\ln 2.7142}{50} \approx 0.0200$$

(b)   In 1950, we have:

$$N = (62947714)e^{(0.02)(50)} \approx 170853155$$

Your answer may differ slightly due to round-off error.

(c)   The actual growth over that period was slower than that predicted from (b) with a constant growth rate.

15. (a)   Use $N = N_0 e^{kt}$:

$$13479142 = 12588066 e^{k(10)}$$
$$1.0708 = e^{10k}$$
$$\ln 1.0708 = 10k$$
$$k = \frac{\ln 1.0708}{10} \approx 0.00684$$

The growth rate is 0.684%.

(b)   Use $N = N_0 e^{kt}$:

$$14830192 = 13479142 e^{k(10)}$$
$$1.1002 = e^{10k}$$
$$\ln 1.1002 = 10k$$
$$k = \frac{\ln 1.1002}{10} \approx 0.00955$$

The growth rate is 0.955%.

(c)   Use $N = N_0 e^{kt}$, $N_0 = 14830192$, $k = 0.00955$, and $t = 40$:

$$N = 14830192 e^{0.00955(40)} \approx 21731258$$

(d)   Our prediction is higher than the actual population.

17. (a) We complete the table:

Region	1990 Population (millions)	Growth Rate (%)	2025 Population
North America	275.2	0.7	351.6
Soviet Union	291.3	0.7	372.2
Europe	499.5	0.2	535.7
Nigeria	113.3	3.1	335.3

(b) It will be 335.3 mil – 113.3 mil = 222.0 million.

(c) The net increases are given by:

North America:	351.6 mil – 275.2 mil = 76.4 mil
Soviet Union:	372.2 mil – 291.3 mil = 80.9 mil
Europe:	535.7 mil – 499.5 mil = 36.2 mil
combined:	193.5 million

(d) For Nigeria, our results support this projection.

19. (a) We complete the table:

t(seconds)	0	550	1100	1650	2200
N(grams)	8	4	2	1	0.5

(b) We complete the table:

t(years)	0	4.9×10^9	9.8×10^9	14.7×10^9	19.6×10^9
N(grams)	10	5	2.5	1.25	0.625

21. Given the decay law $N = N_0 e^{kt}$, and $N = \frac{1}{2} N_0$ when $t = 8$, we have:

$$\frac{1}{2} N_0 = N_0 e^{8k}$$

$$\frac{1}{2} = e^{8k}$$

$$\ln \frac{1}{2} = 8k$$

$$k = \frac{\ln \frac{1}{2}}{8} \approx -0.0866$$

So when $t = 7$ and $N_0 = 1$, we have:

$$N(7) = 1e^{(-0.0866)(7)} \approx 0.55 \text{ g}$$

23. (a) Using $N = \frac{1}{2}N_0$ when $t = 14.9$ hours, we have:

$$\frac{1}{2}N_0 = N_0 e^{14.9k}$$
$$\frac{1}{2} = e^{14.9k}$$
$$\ln\frac{1}{2} = 14.9k$$
$$k = \frac{\ln\frac{1}{2}}{14.9} \approx -0.0465$$

Now using $t = 48$ hours and $N_0 = 40$ g, we have:

$$N(48) = 40e^{(-0.0465)(48)} \approx 4.29 \text{ g}$$

Approximately 4.29 g of the sample will remain after 48 hours.

(b) We find when $N(t) = 1$:

$$1 = 40e^{-0.0465t}$$
$$\frac{1}{40} = e^{-0.0465t}$$
$$\ln\frac{1}{40} = -0.0465t$$
$$t = \frac{\ln\frac{1}{40}}{-0.0465} \approx 79 \text{ hours}$$

The isotope will decay to 1 gram in approximately 79 hours.

25. (a) We draw the graph:

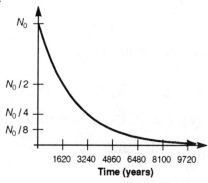

(b) We draw the graph:

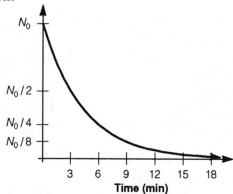

27. (a) We first find the decay constant k:
$$\tfrac{1}{2}N_0 = N_0 e^{13k}$$
$$\tfrac{1}{2} = e^{13k}$$
$$\ln\tfrac{1}{2} = 13k$$
$$k = \frac{\ln\tfrac{1}{2}}{13} \approx -0.0533$$
Now using $N_0 = 2$ g and $t = 5$ years, we have:
$$N(5) = 2e^{(-0.0533)(5)} \approx 1.53 \text{ grams}$$

   (b) We find when $N(t) = 0.2$ g:
$$0.2 = 2e^{-0.0533t}$$
$$0.1 = e^{-0.0533t}$$
$$\ln 0.1 = -0.0533t$$
$$t = \frac{\ln 0.1}{-0.0533} \approx 43 \text{ years}$$

29. (a) The decay law is $N = N_0 e^{kt}$, where $k = \frac{\ln 0.5}{7340}$. We want to determine t when $\frac{N_0}{1000} = N_0 e^{kt}$. After dividing by N_0, and then taking the logarithm of both sides, we obtain $\ln 0.001 = kt$. Thus $t = \frac{\ln 0.001}{k}$, where $k = \frac{\ln 0.5}{7340}$. A calculator then shows that it is approximately 73,000 yrs.

   (b) The answer 73,000 yrs is approximately 10 half-lives.

   (c) After one half-life, we have $N = \frac{N_0}{2}$. After two half-lives, we have $N = \frac{N_0}{2^2}$. After three half-lives, we have $N = \frac{N_0}{2^3}$. Continuing the pattern, we see that after 10 half-lives, $N = \frac{N_0}{2^{10}} \approx \frac{N_0}{1000}$. This agrees with the result in part (b).

31. (a) We have $k = \frac{\ln 0.5}{28} \approx -0.0248$.

   (b) We solve for t:
$$\frac{N_0}{1000} = N_0 e^{kt}$$
$$0.001 = e^{-0.0248t}$$
$$\ln 0.001 = -0.0248t$$
$$t = \frac{\ln 0.001}{-0.0248} \approx 279 \text{ years}$$

(c) Since $2^{10} \approx 1000$, then after 10 half-lives, it should be reduced to $\frac{N_0}{1000}$. Since each half-life is 28 years, this is approximately 280 years. Note that this is close to our answer from (b).

33. (a) Let T represent the half-life. Since $T = \dfrac{\ln \frac{1}{2}}{k} = \dfrac{-\ln 2}{k}$, then $k = \dfrac{-\ln 2}{T}$. Therefore:

$$\mu = \frac{-1}{k} = \frac{-1}{\frac{-\ln 2}{T}} = \frac{T}{\ln 2}$$

(b) The mean life is $\mu = \dfrac{10}{\ln 2} \approx 14.4$ seconds.

(c) After 3 mean lives we have $3(14.4) = 43.2$ seconds passed. Since $k = \dfrac{-\ln 2}{10}$, we have:

$$N(43.2) = 1e^{\left(\frac{-\ln 2}{10}\right)(43.2)} \approx 0.05$$

Approximately 0.05 g of the sample remains.

35. (a) We use $N = N_0 e^{kt}$ with $N = 10 \times 10^9$, $N_0 = 3.6 \times 10^9$, and $k = 0.02$:

$$10 \times 10^9 = (3.6 \times 10^9)e^{0.02t}$$
$$\frac{10}{3.6} = e^{0.02t}$$
$$\ln \frac{10}{3.6} = 0.02t$$
$$t = \frac{\ln \frac{10}{3.6}}{0.02} \approx 51 \text{ years}$$

So 51 years after 1969, or 2020, the carrying capacity will be reached.

(b) We use $N = N_0 e^{kt}$ with $N = 10 \times 10^9$, $N_0 = 4.043 \times 10^9$, and $k = 0.015$:

$$10 \times 10^9 = (4.043 \times 10^9)e^{0.015t}$$
$$\frac{10}{4.043} = e^{0.015t}$$
$$\ln \frac{10}{4.043} = 0.015t$$
$$t = \frac{\ln \frac{10}{4.043}}{0.015} \approx 60 \text{ years}$$

So 60 years after 1975, or 2035, the carrying capacity will be reached.

(c)   We use $N = N_0 e^{kt}$ with $N = 10 \times 10^9$, $N_0 = 4.134 \times 10^9$, and $k = 0.02$:

$$10 \times 10^9 = (4.134 \times 10^9) e^{0.02t}$$

$$\frac{10}{4.134} = e^{0.02t}$$

$$\ln \frac{10}{4.134} = 0.02t$$

$$t = \frac{\ln \frac{10}{4.134}}{0.02} \approx 44 \text{ years}$$

So 44 years after 1975, or 2019, the carrying capacity will be reached.

37. (a)   We compute T:

$$T = \frac{\ln\left(\frac{Ak}{A_0} + 1\right)}{k} = \frac{\ln\left(\frac{161241(0.01)}{250} + 1\right)}{0.01} \approx 201 \text{ years}$$

(b)   We compute T:

$$T = \frac{\ln\left(\frac{Ak}{A_0} + 1\right)}{k} = \frac{\ln\left(\frac{161241(0.02)}{250} + 1\right)}{0.02} \approx 132 \text{ years}$$

39. (a)   Solving for E, we have:

$$\log_{10} E = 11.4 + 1.5M$$

$$E = 10^{11.4 + 1.5M} = 10^{11.4} \cdot 10^{1.5M}$$

(b)   We compute the ratio:

$$\frac{E_1}{E_2} = \frac{10^{11.4} \cdot 10^{1.5M_1}}{10^{11.4} \cdot 10^{1.5M_2}} = 10^{1.5(M_1 - M_2)} = 10^{1.5(8.6-5.9)} = 10^{4.05} \approx 11220$$

41. We compute k:

$$k = \frac{\ln 0.5}{4.7 \times 10^{10}} \approx -1.4748 \times 10^{-11}$$

43. We compute T:

$$T = \frac{\ln\left(\frac{N_s}{N_r} + 1\right)}{-k} = \frac{\ln(0.0636 + 1)}{-(-1.4748 \times 10^{-11})} \approx 4.181 \times 10^9 \text{ yrs}$$

The rock is approximately 4.181 billion years old.

45. We compute T:

$$T = \frac{5730 \ln \frac{N}{920}}{\ln \frac{1}{2}} = \frac{5730 \ln \frac{141}{920}}{\ln \frac{1}{2}} \approx 15505 \text{ yrs}$$

The two paintings are 15,505 years old.

47. We compute T:

$$T = \frac{5730 \ln \frac{N}{920}}{\ln \frac{1}{2}} = \frac{5730 \ln \frac{226}{920}}{\ln \frac{1}{2}} \approx 11605 \text{ yrs}$$

This is 11,500 years (to the nearest 500 years).

49. (a) We compute $\dfrac{\Delta N}{\Delta t}$ over the interval $[t, t + \Delta t]$:

$$\frac{\Delta N}{\Delta t} = \frac{N(t + \Delta t) - N(t)}{t + \Delta t - t}$$

$$= \frac{N_0 e^{k(t + \Delta t)} - N_0 e^{kt}}{\Delta t}$$

$$= \frac{N_0 e^{kt} e^{k\Delta t} - N_0 e^{kt}}{\Delta t}$$

$$= \frac{N_0 e^{kt} \left(e^{k\Delta t} - 1 \right)}{\Delta t}$$

$$= \frac{N \left(e^{k\Delta t} - 1 \right)}{\Delta t}$$

(b) Substituting, we have:

$$\frac{\Delta N}{\Delta t} = \frac{N \left(e^{k\Delta t} - 1 \right)}{\Delta t} \approx \frac{N(1 + k\Delta t - 1)}{\Delta t} = \frac{kN\Delta t}{\Delta t} = kN$$

Thus $\dfrac{\Delta N}{\Delta t} \approx kN$ when Δt is close to 0.

Chapter Five Review Exercises

1. If $x = \log_5 126$, then $5^x = 126$. Since $5^3 = 125$, then $x > 3$.
If $y = \log_{10} 999$, then $10^y = 999$. Since $10^3 = 1000$, then $y < 3$. So $\log_5 126$ is larger.

3. Since $N(t) = N_0 e^{kt}$ and $N_0 = 8000$, then $N(t) = 8000 e^{kt}$. Since $N(4) = 10000$, we can find k:

$$N(4) = 10000$$
$$8000 e^{4k} = 10000$$
$$e^{4k} = 1.25$$
$$k = \frac{\ln 1.25}{4}$$

We now find t such that $N(t) = 12000$:
$$N(t) = 12000$$
$$8000e^{kt} = 12000$$
$$e^{kt} = 1.5$$
$$kt = \ln 1.5$$
$$t = \frac{\ln 1.5}{k} = \frac{\ln 1.5}{\frac{\ln 1.25}{4}} = \frac{4\ln 1.5}{\ln 1.25} \text{ hours}$$

5.   We graph $f(x)$:

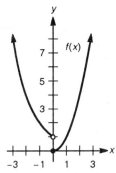

f is not one-to-one since it does not pass the horizontal line test.

7.   We solve for x:
$$\ln(x+1) - 1 = \ln(x-1)$$
$$\ln(x+1) - \ln(x-1) = 1$$
$$\ln\frac{x+1}{x-1} = 1$$
$$\frac{x+1}{x-1} = e$$
$$x + 1 = ex - e$$
$$e + 1 = ex - x$$
$$e + 1 = x(e-1)$$
$$x = \frac{e+1}{e-1}$$

9.   For $y = e^x$, the domain is $(-\infty, \infty)$ and the range is $(0, \infty)$. For $y = \ln x$, the domain is $(0, \infty)$ and the range is $(-\infty, \infty)$. We graph both functions:

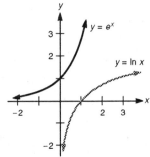

11. We simplify the log:

$$\log_9 \tfrac{1}{27} = \log_9(3^{-3}) = \log_9(9^{1/2})^{-3} = \log_9 9^{-3/2} = -\tfrac{3}{2}$$

13. Using the decay equation $N = N_0 e^{kt}$, we find k when $t = 13$ and $N = \tfrac{1}{2}N_0$:

$$\tfrac{1}{2}N_0 = N_0 e^{13k}$$
$$\tfrac{1}{2} = e^{13k}$$
$$\ln\tfrac{1}{2} = 13k$$
$$k = \frac{\ln\tfrac{1}{2}}{13} \approx -0.05$$

15. We solve for x:

$$5e^{2-x} = 12$$
$$e^{2-x} = \tfrac{12}{5}$$
$$2 - x = \ln\tfrac{12}{5}$$
$$-x = -2 + \ln\tfrac{12}{5}$$
$$x = 2 - \ln\tfrac{12}{5}$$

17. We use $N = N_0 e^{kt}$ where $N_0 = 2$ and $k = 0.02$, so $N(t) = 2e^{0.02t}$. We find t when $N = 3$:

$$3 = 2e^{0.02t}$$
$$1.5 = e^{0.02t}$$
$$\ln 1.5 = 0.02t$$
$$t = \frac{\ln 1.5}{0.02} \approx 20 \text{ yrs}$$

We would expect the population to reach 3 million in the year 2010.

19. We use $A = Pe^{rt}$ where $P = \$1000$ and $r = 0.10$, so $A(t) = 1000e^{0.1t}$. The doubling time is approximately $\tfrac{70}{10} = 7$ years. We sketch the graph:

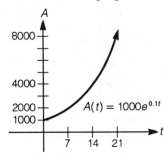

21. The horizontal asymptote is $y = 0$, there is no vertical asymptote, there is no x-intercept, and the y-intercept is 1. We draw the graph:

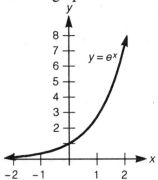

23. There is no horizontal asymptote, the vertical asymptote is $x = 0$, the x-intercept is 1, and there is no y-intercept. We draw the graph:

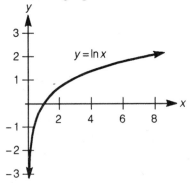

25. The horizontal asymptote is $y = 1$, there is no vertical asymptote, there is no x-intercept, and the y-intercept is 3. We draw the graph:

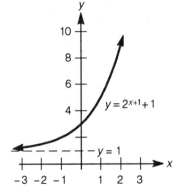

27. The horizontal asymptote is $y = 0$, there is no vertical asymptote, there is no x-intercept, and the y-intercept is 1. We draw the graph:

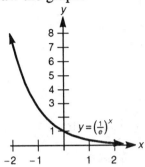

29. The horizontal asymptote is $y = 1$, there is no vertical asymptote, there is no x-intercept, and the y-intercept is $e + 1$. We draw the graph:

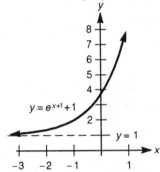

31. There are no asymptotes, and the x- and y-intercepts are both 0. We draw the graph:

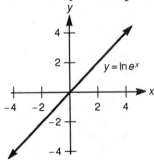

33. Using properties of logarithms, we solve:

$$\log_4 x + \log_4(x - 3) = 1$$
$$\log_4[x(x - 3)] = 1$$
$$x(x - 3) = 4^1$$
$$x^2 - 3x = 4$$
$$x^2 - 3x - 4 = 0$$
$$(x - 4)(x + 1) = 0$$
$$x = 4, -1$$

Since $\log_4(-1)$ is undefined, $x = -1$ is an extraneous root. So the solution is $x = 4$.

35. Using properties of logarithms, we solve:
$$\ln x + \ln(x+2) = \ln 15$$
$$\ln[x(x+2)] = \ln 15$$
$$x(x+2) = 15$$
$$x^2 + 2x = 15$$
$$x^2 + 2x - 15 = 0$$
$$(x+5)(x-3) = 0$$
$$x = -5, 3$$

Since $\ln(-5)$ is undefined, $x = -5$ is an extraneous root. So the solution is $x = 3$.

37. Using properties of logarithms, we solve:
$$\log_2 x + \log_2(3x+10) - 3 = 0$$
$$\log_2[x(3x+10)] = 3$$
$$x(3x+10) = 2^3$$
$$3x^2 + 10x = 8$$
$$3x^2 + 10x - 8 = 0$$
$$(3x-2)(x+4) = 0$$
$$x = \tfrac{2}{3}, -4$$

Since $\log_2(-4)$ is undefined, $x = -4$ is an extraneous root. So the solution is $x = \tfrac{2}{3}$.

39. We solve for x:
$$3\log_9 x = \tfrac{1}{2}$$
$$\log_9 x = \tfrac{1}{6}$$
$$x = 9^{1/6} = \sqrt[6]{9} = \sqrt[3]{3}$$

41. We solve for x:
$$e^{1-5x} = 3\sqrt{e}$$
$$1 - 5x = \ln 3\sqrt{e}$$
$$1 - 5x = \ln 3 + \tfrac{1}{2}\ln e$$
$$1 - 5x = \ln 3 + \tfrac{1}{2}$$
$$-5x = \ln 3 - \tfrac{1}{2}$$
$$x = \frac{1 - 2\ln 3}{10}$$

43. Using properties of logarithms, we solve:
$$\log_{10} x - 2 = \log_{10}(x-2)$$
$$\log_{10} x - \log_{10}(x-2) = 2$$
$$\log_{10} \frac{x}{x-2} = 2$$
$$\frac{x}{x-2} = 100$$
$$x = 100(x-2)$$
$$x = 100x - 200$$
$$200 = 99x$$
$$x = \tfrac{200}{99}$$

45. Using properties of logarithms, we solve:
$$\ln(x+2) = \ln x + \ln 2$$
$$\ln(x+2) = \ln 2x$$
$$x+2 = 2x$$
$$x = 2$$

47. Since $\ln(x^4) = 4\ln x$ is an identity, the solution is all real numbers $x > 0$, or $(0, \infty)$.

49. Using properties of logs, we have:
$$\log_{10} x = \ln x$$
$$\frac{\ln x}{\ln 10} = \ln x$$
$$\ln x = (\ln 10)(\ln x)$$
$$\ln x - (\ln 10)(\ln x) = 0$$
$$\ln x (1 - \ln 10) = 0$$
$$\ln x = 0, \text{ so } x = 1$$

51. Using properties of logs, we have:
$$\log_{10} \sqrt{10} = \log_{10}\left(10^{1/2}\right) = \tfrac{1}{2}$$

53. Using properties of logs, we have:
$$\ln\left(\sqrt[5]{e}\right) = \ln\left(e^{1/5}\right) = \tfrac{1}{5}$$

55. Using properties of logs, we have:
$$\log_{10} \pi - \log_{10} 10\pi = \log_{10} \pi - \left(\log_{10} 10 + \log_{10} \pi\right)$$
$$= \log_{10} \pi - \log_{10} 10 - \log_{10} \pi$$
$$= -\log_{10} 10$$
$$= -1$$

57. Using properties of logs, we have:
$$10^{\log_{10} 16} = 16$$

59.   We evaluate $\ln\left(e^4\right) = 4$.

61.   Using properties of logs, we have:
$$\log_{12} 2 + \log_{12} 18 + \log_{12} 4 = \log_{12}(2 \cdot 18 \cdot 4) = \log_{12} 144 = \log_{12}\left(12^2\right) = 2$$

63.   Using properties of logs, we have:
$$\frac{\ln 100}{\ln 10} = \frac{\ln\left(10^2\right)}{\ln 10} = \frac{2\ln 10}{\ln 10} = 2$$

65.   Using properties of logs, we have:
$$\begin{aligned}
\log_2 \sqrt[7]{16\sqrt[3]{2\sqrt{2}}} &= \tfrac{1}{7}\log_2 16\sqrt[3]{2\sqrt{2}} \\
&= \tfrac{1}{7}\left(\log_2 16 + \log_2 \sqrt[3]{2\sqrt{2}}\right) \\
&= \tfrac{1}{7}\left[\log_2\left(2^4\right) + \tfrac{1}{3}\log_2 2\sqrt{2}\right] \\
&= \tfrac{1}{7}\left[4 + \tfrac{1}{3}\left(\log_2 2 + \log_2 \sqrt{2}\right)\right] \\
&= \tfrac{1}{7}\left[4 + \tfrac{1}{3}\left(1 + \log_2\left(2^{1/2}\right)\right)\right] \\
&= \tfrac{1}{7}\left(4 + \tfrac{1}{3} \cdot \tfrac{3}{2}\right) \\
&= \tfrac{1}{7}\left(4 + \tfrac{1}{2}\right) \\
&= \tfrac{1}{7} \cdot \tfrac{9}{2} \\
&= \tfrac{9}{14}
\end{aligned}$$

67.   Using properties of logs, we have:
$$\begin{aligned}
\log_{10}\left(A^2 B^3 \sqrt{C}\right) &= \log_{10} A^2 + \log_{10} B^3 + \log_{10} \sqrt{C} \\
&= 2\log_{10} A + 3\log_{10} B + \tfrac{1}{2}\log_{10} C \\
&= 2a + 3b + \tfrac{c}{2}
\end{aligned}$$

69.   Using properties of logs, we have:
$$\begin{aligned}
16\log_{10} \sqrt{A}\,\sqrt[4]{B} &= 16\left(\log_{10} \sqrt{A} + \log_{10} \sqrt[4]{B}\right) \\
&= 16\left(\tfrac{1}{2}\log_{10} A + \tfrac{1}{4}\log_{10} B\right) \\
&= 8\log_{10} A + 4\log_{10} B \\
&= 8a + 4b
\end{aligned}$$

71.   Since $\log_{10} 100 = 2$ and $\log_{10} 1000 = 3$, then $\log_{10} 209$ lies between 2 and 3.

73.   Since $\log_6 36 = 2$ and $\log_6 216 = 3$, then $\log_6 100$ lies between 2 and 3.

75.   Since $\log_{10} 0.010 = -2$ and $\log_{10} 0.001 = -3$, then $\log_{10} 0.003$ lies between -2 and -3.

77. (a)   We graph $y = \ln(x + 2)$ and $y = \ln(-x) - 1$:

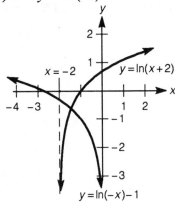

The graph shows these two curves intersect in the third quadrant.

(b)   Setting the y-coordinates equal, we have:
$$\ln(x + 2) = \ln(-x) - 1$$
$$\ln(x + 2) - \ln(-x) = -1$$
$$\ln\frac{x + 2}{-x} = -1$$
$$\frac{x + 2}{-x} = e^{-1}$$
$$e(x + 2) = -x$$
$$ex + 2e = -x$$
$$ex + x = -2e$$
$$x(e + 1) = -2e$$
$$x = \frac{-2e}{e + 1} \approx -1.46$$

79.   We know $N = N_0 e^{kt}$, and $N = \frac{1}{2} N_0$ when $t = T$:
$$\tfrac{1}{2} N_0 = N_0 e^{kT}$$
$$\tfrac{1}{2} = e^{kT}$$
$$\ln \tfrac{1}{2} = kT$$
$$k = \frac{\ln \tfrac{1}{2}}{T}$$

81.   Since 4 half-lives have passed, we will have $\left(\frac{1}{2}\right)^4 N_0 = \frac{1}{16} N_0$ of the substance left, or 6.25% remaining.

83. We have $k = \dfrac{\ln \frac{1}{2}}{d}$ and $N_0 = b$. We want to find the value of t when $N = c$:

$$N = N_0 e^{kt}$$
$$c = b e^{kt}$$
$$\frac{c}{b} = e^{kt}$$
$$\ln \frac{c}{b} = kt$$
$$\ln \frac{c}{b} = \frac{\ln \frac{1}{2}}{d} \cdot t$$
$$t = \frac{d \ln \frac{c}{b}}{\ln \frac{1}{2}} \text{ days}$$

85. Using properties of logs, we have:

$$\log_{10} 8 + \log_{10} 3 - \log_{10} 12 = \log_{10}\left(\frac{8 \cdot 3}{12}\right) = \log_{10} 2$$

87. Using properties of logs, we have:

$$\ln 5 - 3\ln 2 + \ln 16 = \ln 5 - \ln 2^3 + \ln 16 = \ln 5 - \ln 8 + \ln 16 = \ln\left(\frac{5 \cdot 16}{8}\right) = \ln 10$$

89. Using properties of logs, we have:
$$a \ln x + b \ln y = \ln x^a + \ln y^b = \ln x^a y^b$$

91. Using properties of logs, we have:
$$\ln \sqrt{(x-3)(x+4)} = \tfrac{1}{2}\ln[(x-3)(x+4)] = \tfrac{1}{2}\ln(x-3) + \tfrac{1}{2}(x+4)$$

93. Using properties of logs, we have:

$$\log_{10} \frac{x^3}{\sqrt{1+x}} = \log_{10} x^3 - \log_{10} \sqrt{1+x}$$
$$= \log_{10} x^3 - \log_{10}(1+x)^{1/2}$$
$$= 3\log_{10} x - \tfrac{1}{2}\log_{10}(1+x)$$

95. Using properties of logs, we have:

$$\log_{10} \sqrt[3]{\frac{x}{100}} = \log_{10}\left(\frac{x}{100}\right)^{1/3}$$
$$= \tfrac{1}{3}\log_{10} \frac{x}{100}$$
$$= \tfrac{1}{3}\log_{10} x - \tfrac{1}{3}\log_{10} 100$$
$$= \tfrac{1}{3}\log_{10} x - \tfrac{2}{3}$$

97. Using properties of logs, we have:
$$\ln\left(\frac{1+2e}{1-2e}\right)^3 = 3\ln\frac{1+2e}{1-2e} = 3\ln(1+2e) - 3\ln(1-2e)$$

99. We use $B = P(1+r)^t$, where $P = A$, $B = 2A$, and $r = \frac{R}{100}$:
$$2A = A\left(1+\frac{R}{100}\right)^t$$
$$2 = \left(1+\frac{R}{100}\right)^t$$
$$\ln 2 = t\ln\left(1+\frac{R}{100}\right)$$
$$t = \frac{\ln 2}{\ln\left(1+\frac{R}{100}\right)} \text{ years}$$

101. The balance after 1 year will be $A = P\left(1+\frac{0.095}{12}\right)^{12}$. We must find the effective interest rate r where $A = P(1+r)^1$. We set these equal:
$$P(1+r) = P\left(1+\frac{0.095}{12}\right)^{12}$$
$$1+r = \left(1+\frac{0.095}{12}\right)^{12}$$
$$r = \left(1+\frac{0.095}{12}\right)^{12} - 1 \approx 0.0992$$
So the effective interest rate is 9.92%.

103. We use $A = Pe^{rt}$ where $P = D$, $A = 2D$, and $r = \frac{R}{100}$:
$$2D = De^{(R/100)t}$$
$$2 = e^{(R/100)t}$$
$$\ln 2 = \frac{R}{100}(t)$$
$$t = \frac{100\ln 2}{R} \text{ years}$$

105. (a) We use $A = P\left(1+\frac{r}{N}\right)^{Nt}$ where $P = \$660$, $r = 0.055$, $N = 4$, and $A = \$1000$:
$$1000 = 660\left(1+\frac{0.055}{4}\right)^{4t}$$
$$\frac{50}{33} = (1.01375)^{4t}$$
$$\ln\frac{50}{33} = 4t\ln 1.01375$$
$$t = \frac{\ln\frac{50}{33}}{4\ln 1.01375} \approx 7.61 \text{ yrs}$$
So the balance will reach $1000 after $7\frac{3}{4}$ years.

(b)   We use $A = P\left(1+\frac{r}{N}\right)^{Nt}$ where $P = D, r = \frac{R}{100}, N = 4,$ and $A = nD$:

$$nD = D\left(1+\frac{R}{400}\right)^{t}$$

$$n = \left(1+\frac{R}{400}\right)^{t}$$

$$\ln n = t\ln\left(1+\frac{R}{400}\right)$$

$$t = \frac{\ln n}{\ln\left(1+\frac{R}{400}\right)} \text{ years}$$

107. (a) The domain is $x > 0$, or $(0, \infty)$.

  (b) We must make sure $\log_{10} x \geq 0$, so $x \geq 1$. So the domain is $x \geq 1$, or $[1, \infty)$.

109. (a) We must make sure that the expression in the log is non-zero:

$$x^2 - 2x - 15 \neq 0$$
$$(x-5)(x+3) \neq 0$$
$$x \neq 5, -3$$

So the domain is all real numbers except 5 and –3, or $(-\infty, -3) \cup (-3, 5) \cup (5, \infty)$.

  (b) We must make sure that the expression in the log is positive:

$$x^2 - 2x - 15 > 0$$
$$(x-5)(x+3) > 0$$

The key numbers are –3 and 5. We draw a sign chart:

Interval	Test Number	$x-5$	$x+3$	$(x-5)(x+3)$
$(-\infty, -3)$	-5	neg.	neg.	pos.
$(-3, 5)$	0	neg.	pos.	neg.
$(5, \infty)$	6	pos.	pos.	pos.

So the product is positive on the intervals $(-\infty, -3) \cup (5, \infty)$. Thus the domain is $(-\infty, -3) \cup (5, \infty)$.

111. We solve for x:

$$y = \frac{e^x + 1}{e^x - 1}$$
$$y(e^x - 1) = e^x + 1$$
$$ye^x - y = e^x + 1$$
$$ye^x - e^x = y + 1$$
$$e^x(y-1) = y + 1$$
$$e^x = \frac{y+1}{y-1}$$
$$x = \ln\frac{y+1}{y-1}$$

We must solve the inequality $\frac{y+1}{y-1} > 0$. The key numbers are -1 and 1. We draw a sign chart:

Interval	Test Number	$y+1$	$y-1$	$\frac{y+1}{y-1}$
$(-\infty, -1)$	-4	neg.	neg.	pos.
$(-1, 1)$	0	pos.	neg.	neg.
$(1, \infty)$	4	pos.	pos.	pos.

The quotient is positive on the intervals $(-\infty, -1) \cup (1, \infty)$. So the range is $(-\infty, -1) \cup (1, \infty)$.

113. Using properties of logs, we have:
$$\ln 0.5 = \ln(2^{-1}) = -\ln 2 \approx -0.7$$

115. Using properties of logs, we have:
$$\ln \tfrac{1}{9} = \ln(3^{-2}) = -2\ln 3 \approx -2(1.1) = -2.2$$

117. Using properties of logs, we have:
$$\begin{aligned}
\ln 72 &= \ln(3^2 \cdot 2^3) \\
&= \ln(3^2) + \ln(2^3) \\
&= 2\ln 3 + 3\ln 2 \\
&\approx 2(1.1) + 3(0.7) \\
&= 2.2 + 2.1 \\
&= 4.3
\end{aligned}$$

119. Let $x = e$, then $\ln x = 1$. From the graph, we see $x \approx 2.7$.

121. Using properties of logs, we have:
$$\log_2 3 = \frac{\ln 3}{\ln 2} \approx \frac{1.1}{0.7} \approx 1.6$$

123. (a) The corresponding x-coordinate when $y = 3.000$ is $x \approx 1.0986$.

(b) The percent error is given by:
$$\% \text{ error} = \frac{|1.0986 - \ln 3|}{\ln 3} \cdot 100 = 0.00112\%$$

(c) We estimate the quantities:
$$\begin{aligned}
\ln \sqrt{3} &= \tfrac{1}{2}\ln 3 \approx \tfrac{1}{2}(1.0986) \approx 0.5493 \\
\ln 9 &= \ln(3^2) = 2\ln 3 \approx 2(1.0986) \approx 2.1972 \\
\ln \tfrac{1}{3} &= \ln(3^{-1}) = -\ln 3 \approx -1.0986
\end{aligned}$$

Chapter Five Test

1.   The domain is $(-\infty, \infty)$, the range is $(-3, \infty)$, the x-intercept is $-\frac{\ln 3}{\ln 2}$, the y-intercept is -2, and the asymptote is $y = -3$. We draw the graph:

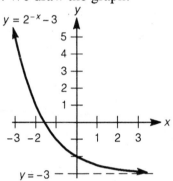

2.   Using $N_0 = 6000$, $N = 6200$, and $t = 1$, we find the growth constant k:
$$6200 = 6000e^{k \cdot 1}$$
$$1.033 = e^k$$
$$k = \ln 1.033 \approx 0.03279$$
We now find the value of t when $N = 10000$:
$$10000 = 6000e^{kt}$$
$$\tfrac{5}{3} = e^{kt}$$
$$kt = \ln \tfrac{5}{3}$$
$$t = \frac{\ln \tfrac{5}{3}}{\ln 1.033} \approx 16 \text{ hrs}$$

3.   Since $2^4 = 16$, then $\log_2 17 > 4$. Since $3^4 = 81$, then $\log_3 80 < 4$. So $\log_2 17$ is larger.

4.   Using the change-of-base formula, we have:
$$\log_2 15 = \frac{\ln 15}{\ln 2}$$

5.   Since $2^{10} \approx 10^3$, we have:
$$2^{40} = \left(2^{10}\right)^4 \approx \left(10^3\right)^4 = 10^{12}$$

6.   Using $A = P(1 + r)^t$ for interest compounded annually, with $P = \$9500$, $A = \$12,000$ and $r = 0.06$, we have:
$$12000 = 9500(1.06)^t$$
$$1.2632 = (1.06)^t$$
$$\ln 1.2632 = t \ln 1.06$$
$$t = \frac{\ln 1.2632}{\ln 1.06} \approx 4 \text{ yrs}$$

7. (a) The identity is valid only if $\ln x$ is defined, which is the interval $(0, \infty)$.

    (b) We graph the two functions:

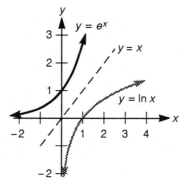

8. Using properties of logs, we have:

$$\log_{10} \frac{A^3}{\sqrt{B}} = \log_{10} A^3 - \log_{10} \sqrt{B} = 3\log_{10} A - \tfrac{1}{2}\log_{10} B = 3a - \tfrac{1}{2}b$$

9. (a) Using properties of logs, we have:
    $$\log_5 \tfrac{1}{\sqrt{5}} = \log_5 5^{-1/2} = -\tfrac{1}{2}$$

    (b) Using properties of logs, we have:
    $$\ln e^2 + \ln 1 - e^{\ln 3} = 2 + 0 - 3 = -1$$

10. Factoring, we have:
$$e^x\left(x^3 - 4x\right) = 0$$
$$xe^x\left(x^2 - 4\right) = 0$$
$$xe^x(x+2)(x-2) = 0$$
$$x = 0, -2, 2 \quad (\text{since } e^x \neq 0)$$

11. (a) Using the formula $T = \dfrac{\ln \frac{1}{2}}{k}$, we have:

$$4 = \frac{\ln \frac{1}{2}}{k}$$
$$k = \frac{\ln \frac{1}{2}}{4} \approx -0.1733$$

    (b) Using $N = N_0 e^{kt}$ with $N_0 = 2$, $k = -0.1733$, and $t = 10$:
    $$N = 2e^{-0.1733(10)} \approx 0.35 \text{ grams}$$

12. Using properties of logs, we have:
$$2\ln x - \ln\sqrt[3]{x^2+1} = \ln x^2 - \ln(x^2+1)^{1/3} = \ln\frac{x^2}{(x^2+1)^{1/3}}$$

13. We solve for x:
$$-2e^{3x-1} = 9$$
$$e^{3x-1} = -4.5$$
But this is impossible $(e^{3x-1} > 0$ for all $x)$, so there is no solution.

14. Let $y = \log_{10}(x-1)$. Switching the roles of x and y and solving the resulting equation for y yields:
$$x = \log_{10}(y-1)$$
$$10^x = y-1$$
$$y = 10^x + 1$$
So $g^{-1}(x) = 10^x + 1$. The range of $g^{-1}(x)$ is $(1, \infty)$.

15. (a)   Using the formula $T = \frac{\ln 2}{r}$, we have:
$$T = \frac{\ln 2}{0.06} \approx 12 \text{ years}$$

   (b)   We graph $A = 12000e^{0.06t}$:

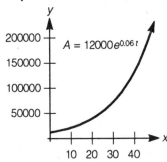

16. (a)   Since $\ln x$ must be defined, this is true on the interval $(0, \infty)$. Note that $\ln(x^2) = 2\ln x$ is an identity on the interval.

   (b)   We solve for x:
$$(\ln x)^2 = 2\ln x$$
$$(\ln x)^2 - 2\ln x = 0$$
$$(\ln x)(\ln x - 2) = 0$$
$$\ln x = 0 \quad \text{or} \quad \ln x = 2$$
$$x = 1 \qquad\qquad x = e^2$$

17. (a)   We graph $f(x)$:

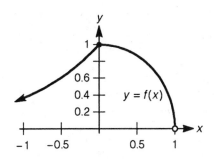

(b)   By the horizontal line test, $f(x)$ is not one-to-one.

18.   Using properties of logs, we solve for x:

$$\tfrac{1}{2} - \log_{16}(x-3) = \log_{16} x$$
$$\tfrac{1}{2} = \log_{16} x + \log_{16}(x-3)$$
$$\tfrac{1}{2} = \log_{16}[x(x-3)]$$
$$16^{1/2} = x(x-3)$$
$$4 = x^2 - 3x$$
$$0 = x^2 - 3x - 4$$
$$0 = (x-4)(x+1)$$
$$x = -1, 4$$

Since $\log_{10}(-1)$ is undefined, $x = -1$ is an extraneous root. So the solution is $x = 4$.

Chapter Six
Trigonometric Functions of Angles

6.1 Trigonometric Functions of Acute Angles

1. (a) We use the definitions:

$$\sin\theta = \frac{\text{opposite}}{\text{hypotenuse}} = \frac{15}{17} \qquad \cos\theta = \frac{\text{adjacent}}{\text{hypotenuse}} = \frac{8}{17}$$

$$\tan\theta = \frac{\text{opposite}}{\text{adjacent}} = \frac{15}{8} \qquad \cot\theta = \frac{\text{adjacent}}{\text{opposite}} = \frac{8}{15}$$

$$\sec\theta = \frac{\text{hypotenuse}}{\text{adjacent}} = \frac{17}{8} \qquad \csc\theta = \frac{\text{hypotenuse}}{\text{opposite}} = \frac{17}{15}$$

(b) We use the definitions:

$$\sin\beta = \frac{\text{opposite}}{\text{hypotenuse}} = \frac{8}{17} \qquad \cos\beta = \frac{\text{adjacent}}{\text{hypotenuse}} = \frac{15}{17}$$

$$\tan\beta = \frac{\text{opposite}}{\text{adjacent}} = \frac{8}{15} \qquad \cot\beta = \frac{\text{adjacent}}{\text{opposite}} = \frac{15}{8}$$

$$\sec\beta = \frac{\text{hypotenuse}}{\text{adjacent}} = \frac{17}{15} \qquad \csc\beta = \frac{\text{hypotenuse}}{\text{opposite}} = \frac{17}{8}$$

3.   (a)   We use the definitions:

$$\sin\theta = \frac{\text{opposite}}{\text{hypotenuse}} = \frac{3}{3\sqrt{5}} = \frac{1}{\sqrt{5}} = \frac{\sqrt{5}}{5}$$

$$\cos\theta = \frac{\text{adjacent}}{\text{hypotenuse}} = \frac{6}{3\sqrt{5}} = \frac{2}{\sqrt{5}} = \frac{2\sqrt{5}}{5}$$

$$\tan\theta = \frac{\text{opposite}}{\text{adjacent}} = \frac{3}{6} = \tfrac{1}{2}$$

$$\cot\theta = \frac{\text{adjacent}}{\text{opposite}} = \frac{6}{3} = 2$$

$$\sec\theta = \frac{\text{hypotenuse}}{\text{adjacent}} = \frac{3\sqrt{5}}{6} = \frac{\sqrt{5}}{2}$$

$$\csc\theta = \frac{\text{hypotenuse}}{\text{opposite}} = \frac{3\sqrt{5}}{3} = \sqrt{5}$$

(b)   We use the definitions:

$$\sin\beta = \frac{\text{opposite}}{\text{hypotenuse}} = \frac{6}{3\sqrt{5}} = \frac{2}{\sqrt{5}} = \frac{2\sqrt{5}}{5}$$

$$\cos\beta = \frac{\text{adjacent}}{\text{hypotenuse}} = \frac{3}{3\sqrt{5}} = \frac{1}{\sqrt{5}} = \frac{\sqrt{5}}{5}$$

$$\tan\beta = \frac{\text{opposite}}{\text{adjacent}} = \frac{6}{3} = 2$$

$$\cot\beta = \frac{\text{adjacent}}{\text{opposite}} = \frac{3}{6} = \tfrac{1}{2}$$

$$\sec\beta = \frac{\text{hypotenuse}}{\text{adjacent}} = \frac{3\sqrt{5}}{3} = \sqrt{5}$$

$$\csc\beta = \frac{\text{hypotenuse}}{\text{opposite}} = \frac{3\sqrt{5}}{6} = \frac{\sqrt{5}}{2}$$

5.   We first draw $\triangle ABC$ and label $AC = 3$ and $BC = 2$:

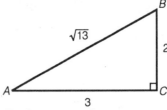

We can find AB by using the Pythagorean theorem:

$$(AC)^2 + (BC)^2 = (AB)^2$$
$$(3)^2 + (2)^2 = (AB)^2$$
$$9 + 4 = (AB)^2$$
$$\sqrt{13} = AB$$

(a) We find $\cos A$, $\sin A$ and $\tan A$:

$$\cos A = \frac{\text{adjacent}}{\text{hypotenuse}} = \frac{3}{\sqrt{13}} = \frac{3\sqrt{13}}{13}$$

$$\sin A = \frac{\text{opposite}}{\text{hypotenuse}} = \frac{2}{\sqrt{13}} = \frac{2\sqrt{13}}{13}$$

$$\tan A = \frac{\text{opposite}}{\text{adjacent}} = \frac{2}{3}$$

(b) We find $\sec B$, $\csc B$ and $\cot B$:

$$\sec B = \frac{\text{hypotenuse}}{\text{adjacent}} = \frac{\sqrt{13}}{2}$$

$$\csc B = \frac{\text{hypotenuse}}{\text{opposite}} = \frac{\sqrt{13}}{3}$$

$$\cot B = \frac{\text{adjacent}}{\text{opposite}} = \frac{2}{3}$$

7. We draw a sketch with $AB = 13$ and $BC = 5$:

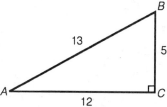

We can find AC by using the Pythagorean theorem:

$$(AC)^2 + (BC)^2 = (AB)^2$$
$$(AC)^2 + (5)^2 = (13)^2$$
$$(AC)^2 + 25 = 169$$
$$(AC)^2 = 144$$
$$AC = 12$$

Now we find the six trigonometric functions of angle B:

$$\sin B = \frac{\text{opposite}}{\text{hypotenuse}} = \frac{12}{13} \qquad \cos B = \frac{\text{adjacent}}{\text{hypotenuse}} = \frac{5}{13}$$

$$\tan B = \frac{\text{opposite}}{\text{adjacent}} = \frac{12}{5} \qquad \cot B = \frac{\text{adjacent}}{\text{opposite}} = \frac{5}{12}$$

$$\sec B = \frac{\text{hypotenuse}}{\text{adjacent}} = \frac{13}{5} \qquad \csc B = \frac{\text{hypotenuse}}{\text{opposite}} = \frac{13}{12}$$

9.   We draw a sketch where $AC = 1$ and $BC = \frac{3}{4}$:

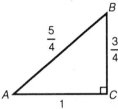

We can find AB by using the Pythagorean theorem:

$$(AC)^2 + (BC)^2 = (AB)^2$$
$$(1)^2 + \left(\tfrac{3}{4}\right)^2 = (AB)^2$$
$$1 + \tfrac{9}{16} = (AB)^2$$
$$\tfrac{25}{16} = (AB)^2$$
$$\tfrac{5}{4} = AB$$

(a)   We find $\sin B$ and $\cos A$:

$$\sin B = \frac{\text{opposite}}{\text{hypotenuse}} = \frac{1}{\frac{5}{4}} = \frac{4}{5}$$

$$\cos A = \frac{\text{adjacent}}{\text{hypotenuse}} = \frac{1}{\frac{5}{4}} = \frac{4}{5}$$

(b)   We find $\sin A$ and $\cos B$:

$$\sin A = \frac{\text{opposite}}{\text{hypotenuse}} = \frac{\frac{3}{4}}{\frac{5}{4}} = \frac{3}{5}$$

$$\cos B = \frac{\text{adjacent}}{\text{hypotenuse}} = \frac{\frac{3}{4}}{\frac{5}{4}} = \frac{3}{5}$$

(c)   We first find $\tan A$ and $\tan B$:

$$\tan A = \frac{\text{opposite}}{\text{adjacent}} = \frac{\frac{3}{4}}{1} = \frac{3}{4}$$

$$\tan B = \frac{\text{opposite}}{\text{adjacent}} = \frac{1}{\frac{3}{4}} = \frac{4}{3}$$

Now find $(\tan A)(\tan B)$:

$$(\tan A)(\tan B) = \tfrac{3}{4} \cdot \tfrac{4}{3} = 1$$

11. We draw a sketch where $AB = 25$ and $AC = 24$:

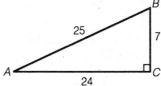

We can find BC by using the Pythagorean theorem:

$$(AC)^2 + (BC)^2 = (AB)^2$$
$$(24)^2 + (BC)^2 = (25)^2$$
$$576 + (BC)^2 = 625$$
$$(BC)^2 = 49$$
$$BC = 7$$

(a)   We find $\cos A$, $\sin A$ and $\tan A$:

$$\cos A = \frac{\text{adjacent}}{\text{hypotenuse}} = \frac{24}{25}$$

$$\sin A = \frac{\text{opposite}}{\text{hypotenuse}} = \frac{7}{25}$$

$$\tan A = \frac{\text{opposite}}{\text{adjacent}} = \frac{7}{24}$$

(b)   We find $\cos B$, $\sin B$ and $\tan B$:

$$\cos B = \frac{\text{adjacent}}{\text{hypotenuse}} = \frac{7}{25}$$

$$\sin B = \frac{\text{opposite}}{\text{hypotenuse}} = \frac{24}{25}$$

$$\tan B = \frac{\text{opposite}}{\text{adjacent}} = \frac{24}{7}$$

(c)   Using the values obtained in parts (a) and (b):
$$(\tan A)(\tan B) = \frac{7}{24} \cdot \frac{24}{7} = 1$$

13. The calculated values for $\theta = 65°$ are approximately:

$\sin \theta \approx 0.906$ $\cos \theta \approx 0.423$ $\tan \theta \approx 2.145$

15. The calculated values for $\theta = 38.5°$ are approximately:

$\sin \theta \approx 0.623$ $\cos \theta \approx 0.783$ $\tan \theta \approx 0.795$

17. The calculated values for $\theta = 80.06°$ are approximately:

$\sin \theta \approx 0.985$ $\cos \theta \approx 0.173$ $\tan \theta \approx 5.706$

19. The calculated values for $\theta = 20°$ are approximately:

$$\sec\theta \approx 1.064 \qquad \csc\theta \approx 2.924 \qquad \cot\theta \approx 2.747$$

21. The calculated values for $\theta = 17.5°$ are approximately:

$$\sec\theta \approx 1.049 \qquad \csc\theta \approx 3.326 \qquad \cot\theta \approx 3.172$$

23. The calculated values for $\theta = 1°$ are approximately:

$$\sec\theta \approx 1.000 \qquad \csc\theta \approx 57.299 \qquad \cot\theta \approx 57.290$$

25. We will compute each side of the equation:

$$\cos 60° = \tfrac{1}{2}$$

$$\cos^2 30° - \sin^2 30° = \left(\tfrac{\sqrt{3}}{2}\right)^2 - \left(\tfrac{1}{2}\right)^2 = \tfrac{3}{4} - \tfrac{1}{4} = \tfrac{2}{4} = \tfrac{1}{2}$$

So $\cos 60° = \cos^2 30° - \sin^2 30°$.

27. We will compute the left-hand side of the equation:

$$\sin^2 30° + \sin^2 45° + \sin^2 60° = \left(\tfrac{1}{2}\right)^2 + \left(\tfrac{\sqrt{2}}{2}\right)^2 + \left(\tfrac{\sqrt{3}}{2}\right)^2 = \tfrac{1}{4} + \tfrac{2}{4} + \tfrac{3}{4} = \tfrac{6}{4} = \tfrac{3}{2}$$

So $\sin^2 30° + \sin^2 45° + \sin^2 60° = \tfrac{3}{2}$.

29. We will compute each side of the equation:

$$2\sin 30° \cos 30° = 2\left(\tfrac{1}{2}\right)\left(\tfrac{\sqrt{3}}{2}\right) = \tfrac{2\sqrt{3}}{4} = \tfrac{\sqrt{3}}{2}$$

$$\sin 60° = \tfrac{\sqrt{3}}{2}$$

So $2\sin 30° \cos 30° = \sin 60°$.

31. We will compute each side of the equation:

$$\sin 30° = \tfrac{1}{2}$$

$$\sqrt{\tfrac{1}{2}(1 - \cos 60°)} = \sqrt{\tfrac{1}{2}\left(1 - \tfrac{1}{2}\right)} = \sqrt{\tfrac{1}{2}\left(\tfrac{1}{2}\right)} = \sqrt{\tfrac{1}{4}} = \tfrac{1}{2}$$

So $\sin 30° = \sqrt{\tfrac{1}{2}(1 - \cos 60°)}$.

33. We will compute each side of the equation:

$$\tan 30° = \frac{1}{\sqrt{3}} = \frac{\sqrt{3}}{3}$$

$$\frac{\sin 60°}{1 + \cos 60°} = \frac{\tfrac{\sqrt{3}}{2}}{1 + \tfrac{1}{2}} = \frac{\tfrac{\sqrt{3}}{2}}{\tfrac{3}{2}} = \frac{\sqrt{3}}{3}$$

So $\tan 30° = \dfrac{\sin 60°}{1 + \cos 60°}$.

35. We will compute each side of the equation:
$$1 + \tan^2 45° = 1 + (1)^2 = 1 + 1 = 2$$
$$\sec^2 45° = \left(\frac{\sqrt{2}}{1}\right)^2 = 2$$
So $1 + \tan^2 45° = \sec^2 45°$.

37. (a) Using the Texas Instruments TI-81, the approximate values are:
$$\cos 30° \approx 0.8660254038$$
$$\cos 45° \approx 0.7071067812$$

(b) Using the values given in Table 1, the approximate values are:
$$\cos 30° = \frac{\sqrt{3}}{2} \approx 0.8660254038$$
$$\cos 45° = \frac{\sqrt{2}}{2} \approx 0.7071067812$$
Note that these results agree with those in part (a).

39. (a) $\cos \theta$ is larger
(b) $\sec \beta$ is larger

41. Extending AB to D an equal distance to AB guarantees $\triangle DBC$ congruent to $\triangle ABC$. Now $\triangle ADC$ is equilateral as each angle is $60°$. By construction $AD = 2AB$ and since the triangle is equilateral, $AC = AD$, hence $AC = 2AB$.

43. The values are equal. Computed to ten decimal places, the values are:

$\sin 3° \approx 0.0523359562$	$\sin 6° \approx 0.1045284633$
$\sin 9° \approx 0.1564344650$	$\sin 12° \approx 0.2079116908$
$\sin 15° \approx 0.2588190451$	$\sin 18° \approx 0.3090169944$

6.2 Algebra and the Trigonometric Functions

1. (a) Combining like terms, we have:
$$-SC + 12SC = 11SC$$

(b) Combining as in (a), we have:
$$-\sin \theta \cos \theta + 12 \sin \theta \cos \theta = 11 \sin \theta \cos \theta$$

3. (a) Combining like terms, we have:
$$4C^3 S - 12 C^3 S = -8C^3 S$$

(b) Combining as in (a), we have:
$$4 \cos^3 \theta \sin \theta - 12 \cos^3 \theta \sin \theta = -8 \cos^3 \theta \sin \theta$$

5. (a) Squaring, we have:
$$(1 + T)^2 = 1 + 2T + T^2$$

(b) Squaring as in (a), we have:
$$(1+\tan\theta)^2 = 1 + 2\tan\theta + \tan^2\theta$$

7. (a) Using the distributive property, we have:
$$(T+3)(T-2) = T^2 + 3T - 2T - 6 = T^2 + T - 6$$

(b) Using the distributive property as in (a), we have:
$$(\tan\theta + 3)(\tan\theta - 2) = \tan^2\theta + 3\tan\theta - 2\tan\theta - 6 = \tan^2\theta + \tan\theta - 6$$

9. (a) Factoring then simplifying, we have:
$$\frac{S-C}{C-S} = \frac{-1(C-S)}{C-S} = -1$$

(b) Factoring then simplifying as in (a), we have:
$$\frac{\sin\theta - \cos\theta}{\cos\theta - \sin\theta} = \frac{-1(\cos\theta - \sin\theta)}{\cos\theta - \sin\theta} = -1$$

11. (a) Obtaining common denominators then adding, we have:
$$C + \frac{2}{S} = \frac{CS}{S} + \frac{2}{S} = \frac{CS+2}{S}$$

(b) Obtaining common denominators then adding as in (a), we have:
$$\cos A + \frac{2}{\sin A} = \frac{\cos A\sin A}{\sin A} + \frac{2}{\sin A} = \frac{\cos A\sin A + 2}{\sin A}$$

13. (a) The expression factors as:
$$T^2 + 8T - 9 = (T-1)(T+9)$$

(b) Factoring as in (a):
$$\tan^2\beta + 8\tan\beta - 9 = (\tan\beta - 1)(\tan\beta + 9)$$

15. (a) Factoring as a difference of squares:
$$4C^2 - 1 = (2C+1)(2C-1)$$

(b) Factoring as in (a):
$$4\cos^2 B - 1 = (2\cos B + 1)(2\cos B - 1)$$

17. (a) Factoring the greatest common factor:
$$9S^2T^3 + 6ST^2 = 3ST^2(3ST+2)$$

(b) Factoring as in (a):
$$9\sec^2 B\tan^3 B + 6\sec B\tan^2 B = 3\sec B\tan^2 B(3\sec B\tan B + 2)$$

19. Since $\sin\theta = \dfrac{\text{opposite}}{\text{hypotenuse}} = \frac{3}{4}$, we can construct the triangle:

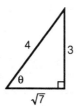

Using the Pythagorean theorem:
$$a^2 + 3^2 = 4^2$$
$$a^2 + 9 = 16$$
$$a^2 = 7$$
$$a = \sqrt{7}$$

So the six trigonometric functions are:

$$\sin\theta = \frac{\text{opposite}}{\text{hypotenuse}} = \frac{3}{4} \qquad\qquad \cos\theta = \frac{\text{adjacent}}{\text{hypotenuse}} = \frac{\sqrt{7}}{4}$$

$$\tan\theta = \frac{\text{opposite}}{\text{adjacent}} = \frac{3}{\sqrt{7}} = \frac{3\sqrt{7}}{7} \qquad\qquad \cot\theta = \frac{\text{adjacent}}{\text{opposite}} = \frac{\sqrt{7}}{3}$$

$$\sec\theta = \frac{\text{hypotenuse}}{\text{adjacent}} = \frac{4}{\sqrt{7}} = \frac{4\sqrt{7}}{7} \qquad\qquad \csc\theta = \frac{\text{hypotenuse}}{\text{opposite}} = \frac{4}{3}$$

21. Since $\cos\beta = \dfrac{\text{adjacent}}{\text{hypotenuse}} = \dfrac{\sqrt{3}}{5}$, we can construct the triangle:

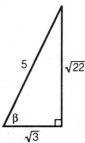

Using the Pythagorean theorem:. we have:
$$\left(\sqrt{3}\right)^2 + b^2 = 5^2$$
$$3 + b^2 = 25$$
$$b^2 = 22$$
$$b = \sqrt{22}$$

So the six trigonometric functions are:

$$\sin\beta = \frac{\text{opposite}}{\text{hypotenuse}} = \frac{\sqrt{22}}{5}$$

$$\cos\beta = \frac{\text{adjacent}}{\text{hypotenuse}} = \frac{\sqrt{3}}{5}$$

$$\tan\beta = \frac{\text{opposite}}{\text{adjacent}} = \frac{\sqrt{22}}{\sqrt{3}} = \frac{\sqrt{66}}{3}$$

$$\cot\beta = \frac{\text{adjacent}}{\text{opposite}} = \frac{\sqrt{3}}{\sqrt{22}} = \frac{\sqrt{66}}{22}$$

$$\sec\beta = \frac{\text{hypotenuse}}{\text{adjacent}} = \frac{5}{\sqrt{3}} = \frac{5\sqrt{3}}{3}$$

$$\csc\beta = \frac{\text{hypotenuse}}{\text{opposite}} = \frac{5}{\sqrt{22}} = \frac{5\sqrt{22}}{22}$$

23. Since $\sin A = \dfrac{\text{opposite}}{\text{hypotenuse}} = \frac{5}{13}$, we can construct the triangle:

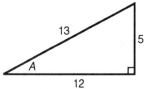

Using the Pythagorean theorem, we have:

$$a^2 + 5^2 = 13^2$$
$$a^2 + 25 = 169$$
$$a^2 = 144$$
$$a = 12$$

So the six trigonometric functions are:

$$\sin A = \frac{\text{opposite}}{\text{hypotenuse}} = \frac{5}{13}$$

$$\cos A = \frac{\text{adjacent}}{\text{hypotenuse}} = \frac{12}{13}$$

$$\tan A = \frac{\text{opposite}}{\text{adjacent}} = \frac{5}{12}$$

$$\cot A = \frac{\text{adjacent}}{\text{opposite}} = \frac{12}{5}$$

$$\sec A = \frac{\text{hypotenuse}}{\text{adjacent}} = \frac{13}{12}$$

$$\csc A = \frac{\text{hypotenuse}}{\text{opposite}} = \frac{13}{5}$$

25. Since $\tan B = \dfrac{\text{opposite}}{\text{adjacent}} = \frac{4}{3}$, we can construct the triangle:

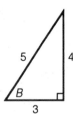

Using the Pythagorean theorem, we have:

$$3^2 + 4^2 = c^2$$
$$9 + 16 = c^2$$
$$25 = c^2$$
$$5 = c$$

So the six trigonometric functions are:

$$\sin B = \frac{\text{opposite}}{\text{hypotenuse}} = \frac{4}{5} \qquad\qquad \cos B = \frac{\text{adjacent}}{\text{hypotenuse}} = \frac{3}{5}$$

$$\tan B = \frac{\text{opposite}}{\text{adjacent}} = \frac{4}{3} \qquad\qquad \cot B = \frac{\text{adjacent}}{\text{opposite}} = \frac{3}{4}$$

$$\sec B = \frac{\text{hypotenuse}}{\text{adjacent}} = \frac{5}{3} \qquad\qquad \csc B = \frac{\text{hypotenuse}}{\text{opposite}} = \frac{5}{4}$$

27. Since $\sec C = \dfrac{\text{hypotenuse}}{\text{adjacent}} = \dfrac{3}{2}$, we can construct the triangle:

Using the Pythagorean theorem, we have:

$$2^2 + b^2 = 3^2$$
$$4 + b^2 = 9$$
$$b^2 = 5$$
$$b = \sqrt{5}$$

So the six trigonometric functions are:

$$\sin C = \frac{\text{opposite}}{\text{hypotenuse}} = \frac{\sqrt{5}}{3} \qquad\qquad \cos C = \frac{\text{adjacent}}{\text{hypotenuse}} = \frac{2}{3}$$

$$\tan C = \frac{\text{opposite}}{\text{adjacent}} = \frac{\sqrt{5}}{2} \qquad\qquad \cot C = \frac{\text{adjacent}}{\text{opposite}} = \frac{2}{\sqrt{5}} = \frac{2\sqrt{5}}{5}$$

$$\sec C = \frac{\text{hypotenuse}}{\text{adjacent}} = \frac{3}{2} \qquad\qquad \csc C = \frac{\text{hypotenuse}}{\text{opposite}} = \frac{3}{\sqrt{5}} = \frac{3\sqrt{5}}{5}$$

29. Since $\cot \alpha = \dfrac{\text{adjacent}}{\text{opposite}} = \dfrac{\sqrt{3}}{3}$, we can construct the triangle:

Using the Pythagorean theorem, we have:

$$\left(\sqrt{3}\right)^2 + (3)^2 = c^2$$
$$3 + 9 = c^2$$
$$12 = c^2$$
$$c = \sqrt{12} = 2\sqrt{3}$$

So the six trigonometric functions are:

$$\sin \alpha = \frac{\text{opposite}}{\text{hypotenuse}} = \frac{3}{2\sqrt{3}} = \frac{3\sqrt{3}}{6} = \frac{\sqrt{3}}{2} \qquad \cos \alpha = \frac{\text{adjacent}}{\text{hypotenuse}} = \frac{\sqrt{3}}{2\sqrt{3}} = \frac{1}{2}$$

$$\tan \alpha = \frac{\text{opposite}}{\text{adjacent}} = \frac{3}{\sqrt{3}} = \frac{3\sqrt{3}}{3} = \sqrt{3} \qquad \cot \alpha = \frac{\text{adjacent}}{\text{opposite}} = \frac{\sqrt{3}}{3}$$

$$\sec \alpha = \frac{\text{hypotenuse}}{\text{adjacent}} = \frac{2\sqrt{3}}{\sqrt{3}} = 2 \qquad \csc \alpha = \frac{\text{hypotenuse}}{\text{opposite}} = \frac{2\sqrt{3}}{3}$$

31. From the identity $\sin^2 \theta + \cos^2 \theta = 1$ we can find $\sin \theta$:

$$\sin^2 \theta + (0.4626)^2 = 1$$
$$\sin^2 \theta + 0.21399876 = 1$$
$$\sin^2 \theta = 0.78600124$$
$$\sin \theta \approx 0.887$$

Using trigonometric identities, we compute:

$$\tan \theta = \frac{\sin \theta}{\cos \theta} \approx \frac{0.887}{0.4626} \approx 1.916$$

$$\cot \theta = \frac{\cos \theta}{\sin \theta} \approx \frac{0.4626}{0.887} \approx 0.522$$

$$\sec \theta = \frac{1}{\cos \theta} = \frac{1}{0.4626} \approx 2.162$$

$$\csc \theta = \frac{1}{\sin \theta} \approx \frac{1}{0.887} \approx 1.128$$

33. Since $\tan \theta = \frac{\text{opposite}}{\text{adjacent}} = \frac{1.1998}{1}$, we can construct the triangle:

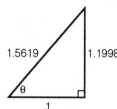

Using the Pythagorean theorem, we have:

$$1^2 + (1.1998)^2 = c^2$$
$$1 + 1.43952004 = c^2$$
$$2.43952004 = c^2$$
$$c \approx 1.5619$$

So the remaining trigonometric functions are:

$$\sin\theta = \frac{\text{opposite}}{\text{hypotenuse}} \approx \frac{1.1998}{1.5619} \approx 0.768$$

$$\cos\theta = \frac{\text{adjacent}}{\text{hypotenuse}} \approx \frac{1}{1.5619} \approx 0.640$$

$$\cot\theta = \frac{1}{\tan\theta} = \frac{1}{1.1998} \approx 0.833$$

$$\sec\theta = \frac{1}{\cos\theta} \approx \frac{1}{0.640} \approx 1.562$$

$$\csc\theta = \frac{1}{\sin\theta} \approx \frac{1}{0.768} \approx 1.302$$

35. By factoring the numerator, we have:

$$\frac{\sin^2 A - \cos^2 A}{\sin A - \cos A} = \frac{(\sin A - \cos A)(\sin A + \cos A)}{\sin A - \cos A} = \sin A + \cos A$$

37. By using the identities for $\csc\theta$ and $\sec\theta$, we have:

$$\sin^2\theta\cos\theta\csc^3\theta\sec\theta = \sin^2\theta \cdot \cos\theta \cdot \frac{1}{\sin^3\theta} \cdot \frac{1}{\cos\theta} = \frac{1}{\sin\theta} = \csc\theta$$

39. By using the identity for $\cot B$, we have:

$$\cot B \sin^2 B \cot B = \frac{\cos B}{\sin B} \cdot \sin^2 B \cdot \frac{\cos B}{\sin B} = \cos^2 B$$

41. By factoring the numerator, we have:

$$\frac{\cos^2 A + \cos A - 12}{\cos A - 3} = \frac{(\cos A - 3)(\cos A + 4)}{\cos A - 3} = \cos A + 4$$

43. By using the identities for $\tan\theta$ and $\sec\theta$, we have:

$$\frac{\tan\theta}{\sec\theta - 1} + \frac{\tan\theta}{\sec\theta + 1} = \frac{\frac{\sin\theta}{\cos\theta}}{\frac{1}{\cos\theta} - 1} + \frac{\frac{\sin\theta}{\cos\theta}}{\frac{1}{\cos\theta} + 1}$$

$$= \frac{\sin\theta}{1 - \cos\theta} + \frac{\sin\theta}{1 + \cos\theta}$$

$$= \frac{\sin\theta(1 + \cos\theta) + \sin\theta(1 - \cos\theta)}{1 - \cos^2\theta}$$

$$= \frac{2\sin\theta}{\sin^2\theta}$$

$$= \frac{2}{\sin\theta}$$

$$= 2\csc\theta$$

45. By using the identities for $\sec A$, $\csc A$, $\tan A$ and $\cot A$, we have:

$$\sec A \csc A - \tan A - \cot A = \frac{1}{\cos A} \cdot \frac{1}{\sin A} - \frac{\sin A}{\cos A} - \frac{\cos A}{\sin A}$$

$$= \frac{1 - \sin^2 A - \cos^2 A}{\sin A \cos A}$$

$$= \frac{1 - \left(\sin^2 A + \cos^2 A\right)}{\sin A \cos A}$$

$$= \frac{1 - 1}{\sin A \cos A}$$

$$= 0$$

47. By using the identities for $\cot \theta$, $\tan \theta$, $\csc \theta$ and $\sec \theta$, we have:

$$\frac{\cot^2 \theta}{\csc^2 \theta} + \frac{\tan^2 \theta}{\sec^2 \theta} = \frac{\frac{\cos^2 \theta}{\sin^2 \theta}}{\frac{1}{\sin^2 \theta}} + \frac{\frac{\sin^2 \theta}{\cos^2 \theta}}{\frac{1}{\cos^2 \theta}} = \cos^2 \theta + \sin^2 \theta = 1$$

49. By using the identity $\cos(90° - \theta) = \sin \theta$, we have:

$$\frac{\cos(90° - \theta)}{\cos \theta} = \frac{\sin \theta}{\cos \theta} = \tan \theta$$

51. By using the identities $\cos(90° - A) = \sin A$ and $\sin(90° - A) = \cos A$, we have:

$$\frac{\cos^2(90° - A)}{\sin^2(90° - A)} - \frac{1}{\cos^2 A} = \frac{\sin^2 A}{\cos^2 A} - \frac{1}{\cos^2 A}$$

$$= \frac{\sin^2 A - 1}{\cos^2 A}$$

$$= -\frac{1 - \sin^2 A}{\cos^2 A}$$

$$= -\frac{\cos^2 A}{\cos^2 A}$$

$$= -1$$

53. By using the identity $\sin(90° - \theta) = \cos \theta$, we have:

$$1 - \sin(90° - \theta)\cos \theta = 1 - \cos \theta \cdot \cos \theta = 1 - \cos^2 \theta = \sin^2 \theta$$

55. By using only algebraic simplification, we have:

$$\frac{\frac{\cos \theta + 1}{\cos \theta} + 1}{\frac{\cos \theta - 1}{\cos \theta} - 1} = \frac{\cos \theta + 1 + \cos \theta}{\cos \theta - 1 - \cos \theta} = \frac{2\cos \theta + 1}{-1} = -2\cos \theta - 1$$

57. We solve the first equation for A:
$$A\sin\theta + \cos\theta = 1$$
$$A\sin\theta = 1 - \cos\theta$$
$$A = \frac{1 - \cos\theta}{\sin\theta}$$

Now solve the second equation for B:
$$B\sin\theta - \cos\theta = 1$$
$$B\sin\theta = 1 + \cos\theta$$
$$B = \frac{1 + \cos\theta}{\sin\theta}$$

Now compute the product AB:
$$AB = \frac{1 - \cos\theta}{\sin\theta} \cdot \frac{1 + \cos\theta}{\sin\theta} = \frac{1 - \cos^2\theta}{\sin^2\theta} = \frac{\sin^2\theta}{\sin^2\theta} = 1$$

59. Recall that $\sin^2\theta + \cos^2\theta = 1$, so we have the system of equations:
$$a\sin^2\theta + b\cos^2\theta = 1$$
$$\sin^2\theta + \cos^2\theta = 1$$

Multiply the second equation by $-b$:
$$a\sin^2\theta + b\cos^2\theta = 1$$
$$-b\sin^2\theta - b\cos^2\theta = -b$$

Adding, we obtain:
$$(a - b)\sin^2\theta = 1 - b$$
$$\sin^2\theta = \frac{1 - b}{a - b}$$

So, we have:
$$\cos^2\theta = 1 - \sin^2\theta = 1 - \frac{1 - b}{a - b} = \frac{(a - b) - (1 - b)}{a - b} = \frac{a - 1}{a - b}$$

Using the identity for $\tan\theta$, we have:
$$\tan^2\theta = \frac{\sin^2\theta}{\cos^2\theta} = \frac{\frac{1-b}{a-b}}{\frac{a-1}{a-b}} = \frac{1 - b}{a - 1} = \frac{b - 1}{1 - a}$$

This proves the desired results.

61. Solving the identity $\sin^2\beta + \cos^2\beta = 1$ for $\cos\beta$ yields $\cos\beta = \sqrt{1-\sin^2\beta}$, so:

$$\cos\beta = \sqrt{1-\sin^2\beta}$$

$$= \sqrt{1 - \frac{m^4 - 2m^2n^2 + n^4}{\left(m^2+n^2\right)^2}}$$

$$= \frac{\sqrt{m^4 + 2m^2n^2 + n^4 - m^4 + 2m^2n^2 - n^4}}{m^2+n^2}$$

$$= \frac{\sqrt{4m^2n^2}}{m^2+n^2}$$

$$= \frac{2mn}{m^2+n^2} \quad \text{(since } m > 0, n > 0\text{)}$$

Now using the identity for $\tan\beta$, we have:

$$\tan\beta = \frac{\sin\beta}{\cos\beta} = \frac{\dfrac{m^2-n^2}{m^2+n^2}}{\dfrac{2mn}{m^2+n^2}} = \frac{m^2-n^2}{2mn}$$

6.3 Right-Triangle Applications

1. We draw the figure:

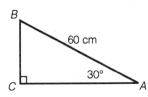

Since $\sin 30° = \dfrac{BC}{60}$, we have:

$$BC = 60\sin 30° = 60 \cdot \tfrac{1}{2} = 30 \text{ cm}$$

Since $\cos 30° = \dfrac{AC}{60}$, we have:

$$AC = 60\cos 30° = 60 \cdot \tfrac{\sqrt{3}}{2} = 30\sqrt{3} \text{ cm}$$

3. We draw the figure:

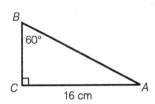

Since $\sin 60° = \dfrac{16}{AB}$, we have:

$$AB = \frac{16}{\sin 60°} = \frac{16}{\frac{\sqrt{3}}{2}} = \frac{32}{\sqrt{3}} = \frac{32\sqrt{3}}{3} \text{ cm}$$

Since $\tan 60° = \dfrac{16}{BC}$, we have:

$$BC = \frac{16}{\tan 60°} = \frac{16}{\sqrt{3}} = \frac{16\sqrt{3}}{3} \text{ cm}$$

5. We draw the figure:

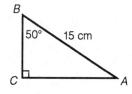

Since $\sin 50° = \dfrac{AC}{15}$, we have:

$$AC = 15\sin 50° \approx 11.5 \text{ cm}$$

Since $\cos 50° = \dfrac{BC}{15}$, we have:

$$BC = 15\cos 50° \approx 9.6 \text{ cm}$$

7. We draw a figure:

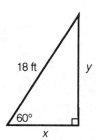

(a) We are asked to find y:

$$\sin 60° = \tfrac{y}{18}$$

$$y = 18\sin 60° = 18\left(\tfrac{\sqrt{3}}{2}\right) = 9\sqrt{3} \text{ ft}$$

Using a calculator, this is approximately 15.59 ft.

(b) We are asked to find x:

$$\cos 60° = \tfrac{x}{18}$$

$$x = 18\cos 60° = 18\left(\tfrac{1}{2}\right) = 9 \text{ ft}$$

9. Using the sine function, we have that $\sin(\angle SEM) = \dfrac{MS}{SE}$, so:

$MS = SE\sin(\angle SEM) = 93\sin 21.16° \approx 34$

Thus the distance MS is approximately 34 million miles.

11. We draw a figure:

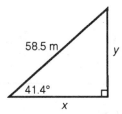

We first find x and y:

$\cos 41.4° = \dfrac{x}{58.5}$, so $x = 58.5\cos 41.4° \approx 43.9$ m

$\sin 41.4° = \dfrac{y}{58.5}$, so $y = 58.5\sin 41.4° \approx 38.7$ m

So the total length of fencing required is:

43.9 m $+ 38.7$ m $+ 58.5$ m ≈ 141.1 m

13. **(a)** We first draw the triangle:

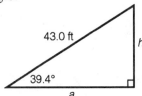

Now find h:

$\sin 39.4° = \dfrac{h}{43.0}$, so $h = 43.0\sin 39.4° \approx 27.3$ ft

(b) We first find a:

$\cos 39.4° = \dfrac{a}{43.0}$, so $a = 43.0\cos 39.4° \approx 33.2$ ft

Now the gable has a base of $2(33.2) = 66.4$ ft and a height of 27.3 ft, so its area is given by:

$\frac{1}{2}(\text{base})(\text{height}) \approx \frac{1}{2}(66.4)(27.3) \approx 906.9$ ft^2

Note: This calculation was done with the full calculator approximation, not the values rounded to one decimal place.

15. Using the formula $A = \frac{1}{2}ab\sin\theta$, we have:

$A = \frac{1}{2}(2)(3)\sin 30° = 1.5$ in.^{2}

17. We start by finding the measure of a central angle of a triangle drawn from the center out to two adjacent vertices:

Since the polygon is 7-sided, we have $7\theta = 360°$, so $\theta = \frac{360°}{7}$. The area of this triangle is thus:
$$\tfrac{1}{2}ab\sin\theta = \tfrac{1}{2}(1)(1)\sin\frac{360°}{7} = \tfrac{1}{2}\sin\frac{360°}{7}$$
The area of the septagon (7-sided figure) is therefore:
$$7\left(\tfrac{1}{2}\sin\frac{360°}{7}\right) = \tfrac{7}{2}\sin\frac{360°}{7} \approx 2.736 \text{ square units}$$

19. We start by finding the measure of a central angle of a triangle drawn from the center out to two adjacent vertices:

Since the polygon is 8-sided, we have $8\theta = 360°$, so $\theta = 45°$. The area of this triangle is thus:
$$\tfrac{1}{2}ab\sin\theta = \tfrac{1}{2}(1)(1)\sin 45° = \frac{\sqrt{2}}{4}$$
The area of the octagon (8-sided figure) is therefore:
$$8\left(\frac{\sqrt{2}}{4}\right) = 2\sqrt{2}$$
The shaded area is obtained by subtracting this area from the circular area, therefore:
$$\text{Area} = \pi(1)^2 - 2\sqrt{2} = \pi - 2\sqrt{2} \approx 0.313 \text{ square units}$$

21. We start by finding the measure of a central angle of a triangle drawn from the center out to two adjacent vertices:

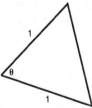

Since the polygon is 6-sided, we have $6\theta = 360°$, so $\theta = 60°$. The area of this triangle is thus:
$$\tfrac{1}{2}ab\sin\theta = \tfrac{1}{2}(1)(1)\sin 60° = \frac{\sqrt{3}}{4}$$
Since the shaded area is comprised of four such triangles, the shaded area is:
$$4\left(\frac{\sqrt{3}}{4}\right) = \sqrt{3} \approx 1.732 \text{ square units}$$

23. Using the hint, we draw the figure:

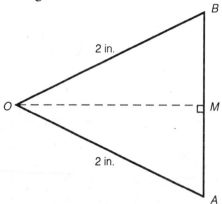

Now $\angle AOB = \frac{360°}{5} = 72°$, so $\angle AOM = 36°$. We find AM:

$$\sin 36° = \frac{AM}{2}$$

$$AM = 2\sin 36°$$

Thus $AB = 4\sin 36°$, and since there are five sides, the perimeter is:

$$5(AB) = 5(4\sin 36°) = 20\sin 36° \text{ inches}$$

25. We draw the figure:

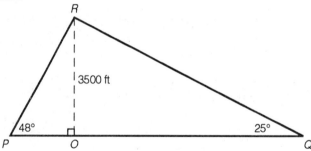

Now compute:

$$\tan 48° = \frac{3500}{PO} \quad \text{and} \quad \tan 25° = \frac{3500}{OQ}$$

So $PO = \dfrac{3500}{\tan 48°}$ and $OQ = \dfrac{3500}{\tan 25°}$. Thus $PQ = \dfrac{3500}{\tan 48°} + \dfrac{3500}{\tan 25°} \approx 10,660$ ft.

27. We draw a figure:

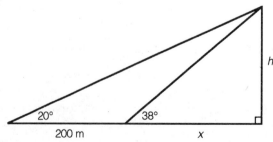

Now $\tan 38° = \dfrac{h}{x}$, so $x = \dfrac{h}{\tan 38°}$. Also $\tan 20° = \dfrac{h}{200+x}$, so $h = (200+x)\tan 20°$.
Substituting, we have:

$$h = \left(200 + \frac{h}{\tan 38°}\right)\tan 20°$$
$$h\tan 38° = 200\tan 38°\tan 20° + h\tan 20°$$
$$h(\tan 38° - \tan 20°) = 200\tan 38°\tan 20°$$
$$h = \frac{200\tan 38°\tan 20°}{\tan 38° - \tan 20°} \approx 136 \text{ m}$$

29. We draw the figure:

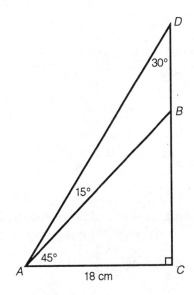

Now compute:

$$\tan 45° = \frac{BC}{18}, \text{ so } BC = 18\tan 45°$$
$$\tan 60° = \frac{CD}{18}, \text{ so } CD = 18\tan 60°$$

Therefore:

$$BD = CD - BC = 18(\tan 60° - \tan 45°) = 18(\sqrt{3} - 1) \text{ cm}$$

31. (a) First, note that $\angle BOA = 90° - \theta$ since it forms a right angle with θ. Also, $\angle OAB = \theta$ since both angles are complemetary to the same angle ($\angle AOB$). Now, $\angle BAP = 90° - \theta$ since it forms a right angle with $\angle OAB$. Finally, $\angle BPA = \theta$ since $\angle BPA$ and $\angle OAB$ are both complementary to the same angle ($\angle BAP$).

(b) Using $\triangle AOP$, we have:

$$\sin\theta = \frac{AO}{OP} = \frac{AO}{1}, \text{ so } AO = \sin\theta$$
$$\cos\theta = \frac{AP}{OP} = \frac{AP}{1}, \text{ so } AP = \cos\theta$$

Using $\triangle AOB$, we have:
$$\sin\theta = \frac{OB}{OA} = \frac{OB}{\sin\theta}, \text{ so } OB = \sin^2\theta$$
Using $\triangle ABP$, we have:
$$\cos\theta = \frac{BP}{AP} = \frac{BP}{\cos\theta}, \text{ so } BP = \cos^2\theta$$

33. (a) Examine the similar triangles having $\overline{AB}$ and $\overline{BC}$ as hypotenuses, and notice that θ is also the angle at B in the smaller triangle. Thus:
$$\sin\theta = \frac{5}{BC}, \text{ so } BC = \frac{5}{\sin\theta}$$

(b) Using the smaller triangle:
$$\cos\theta = \frac{4}{AB}, \text{ so } AB = \frac{4}{\cos\theta}$$

(c) Since $AC = AB + BC$, then:
$$AC = \frac{4}{\cos\theta} + \frac{5}{\sin\theta} = 4\sec\theta + 5\csc\theta$$

35. First observe that the figure $x^2 + y^2 = 1$ is a circle with a radius of one. Since OA, OD and OF all represent the radius, they are each equal to 1. In each case, we will look for a trigonometric relationship involving the required segment:

(a) $\sin\theta = \dfrac{DE}{OD}$, so $DE = \sin\theta$

(b) $\cos\theta = \dfrac{OE}{OD}$, so $OE = \cos\theta$

(c) $\tan\theta = \dfrac{CF}{OF}$, so $CF = \tan\theta$

(d) $\sec\theta = \dfrac{OC}{OF}$, so $OC = \sec\theta$

Going to $\triangle OAB$, $\angle ABO = \theta$, thus:

(e) $\cot\theta = \dfrac{AB}{OA}$, so $AB = \cot\theta$

(f) $\csc\theta = \dfrac{OB}{OA}$, so $OB = \csc\theta$

37. (a) We can use $\sin\theta$ to set up a trigonometric relationship:

$$\sin\theta = \frac{r}{PS+r}$$
$$PS\sin\theta + r\sin\theta = r$$
$$PS\sin\theta = r - r\sin\theta$$
$$PS\sin\theta = r(1-\sin\theta)$$
$$r = \left(\frac{\sin\theta}{1-\sin\theta}\right)PS$$

(b) Using a calculator, we have:

$$r = \left(\frac{\sin 0.257°}{1-\sin 0.257°}\right)(238,857) \approx 1080 \text{ miles}$$

39. (a) We draw the figure:

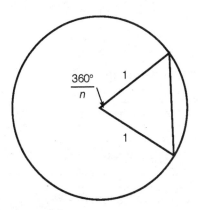

The area of the triangle is given by:
$$\tfrac{1}{2}ab\sin\theta = \tfrac{1}{2}(1)(1)\sin\tfrac{360°}{n} = \tfrac{1}{2}\sin\tfrac{360°}{n}$$

Since there are n congruent triangles in the n-gon, then its area is $\tfrac{n}{2}\sin\tfrac{360°}{n}$.

(b) We complete the table:

n	5	10	50	100	1,000	5,000	10,000
A_n	2.38	2.94	3.1333	3.1395	3.141572	3.1415918	3.1415924

(c) As n gets larger, A_n becomes closer to the area of the circle, which is π.

41.   We draw the figure:

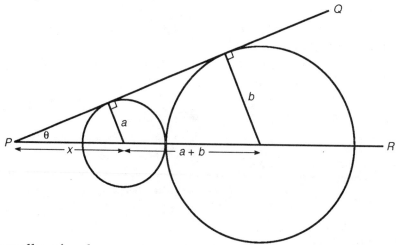

Using the smaller triangle:

$$\sin \theta = \frac{a}{x}, \text{ so } x = \frac{a}{\sin \theta}$$

Using the larger triangle and substituting for x:

$$\sin \theta = \frac{b}{x + a + b}$$

$$\sin \theta = \frac{b}{\frac{a}{\sin \theta} + a + b}$$

$$\sin \theta = \frac{b \sin \theta}{a + (a+b)\sin \theta}$$

$$a \sin \theta + (a+b)\sin^2 \theta = b \sin \theta$$

$$a + (a+b)\sin \theta = b \quad (\text{since } \theta \neq 0 \text{ and } \theta \neq 180°, \text{ then } \sin \theta \neq 0)$$

$$(a+b)\sin \theta = b - a$$

$$\sin \theta = \frac{b-a}{a+b}$$

Since $\cos^2 \theta + \sin^2 \theta = 1$, then:

$$\cos \theta = \sqrt{1 - \sin^2 \theta}$$

$$= \sqrt{1 - \frac{(b-a)^2}{(a+b)^2}}$$

$$= \sqrt{\frac{a^2 + 2ab + b^2 - b^2 + 2ab - a^2}{(a+b)^2}}$$

$$= \sqrt{\frac{4ab}{(a+b)^2}}$$

$$= \frac{2\sqrt{ab}}{a+b}$$

43. We draw the figure:

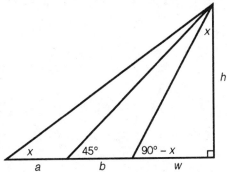

We have the following relationships from the three triangles:

small: $\tan x = \dfrac{w}{h}$

middle: $\tan 45° = \dfrac{h}{b+w}$, so $h = b+w$ and thus $w = h-b$

large: $\tan x = \dfrac{h}{a+b+w}$

Substituting, we have:

$$\frac{w}{h} = \frac{h}{a+b+w}$$

Since $w = h - b$, we have:

$$\frac{h-b}{h} = \frac{h}{a+b+h-b}$$

$$\frac{h-b}{h} = \frac{h}{h+a}$$

$$(h-b)(h+a) = h^2$$

$$h^2 - bh + ah - ab = h^2$$

$$ah - bh = ab$$

$$h(a-b) = ab$$

$$h = \frac{ab}{a-b}$$

So the tower is $\dfrac{ab}{a-b}$ feet high.

6.4 Trigonometric Functions of Angles

1.   (a)   The reference angle for 110° is 70°:

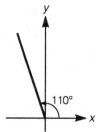

(b)   The reference angle for –110° is 70°:

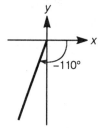

3.   (a)   The reference angle for 200° is 20°:

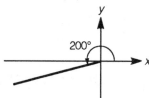

(b)   The reference angle for –200° is 20°:

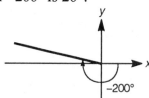

5.   (a)   The reference angle for 300° is 60°:

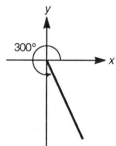

(b)   The reference angle for –300° is 60°:

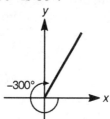

7.   (a)   The reference angle for 60° is 60°:

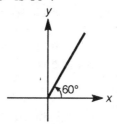

(b)   The reference angle for –60° is 60°:

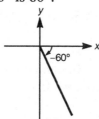

9.   Since 270° corresponds to the point $(0, -1)$ on the unit circle, then using $x = 0$ and $y = -1$ in the definitions yields:

$$\cos 270° = x = 0$$

$$\sin 270° = y = -1$$

$$\tan 270° = \frac{y}{x} = \frac{-1}{0}, \text{ which is undefined}$$

$$\sec 270° = \frac{1}{x} = \frac{1}{0}, \text{ which is undefined}$$

$$\csc 270° = \frac{1}{y} = \frac{1}{-1} = -1$$

$$\cot 270° = \frac{x}{y} = \frac{0}{-1} = 0$$

11.   Since –270° corresponds to the point $(0, 1)$ on the unit circle, then using $x = 0$ and $y = 1$ in the definitions yields:

$$\cos(-270°) = x = 0$$

$$\sin(-270°) = y = 1$$

$$\tan(-270°) = \frac{y}{x} = \frac{1}{0}, \text{ which is undefined}$$

$$\sec(-270°) = \frac{1}{x} = \frac{1}{0}, \text{ which is undefined}$$

$$\csc(-270°) = \frac{1}{y} = \frac{1}{1} = 1$$

$$\cot(-270°) = \frac{x}{y} = \frac{0}{1} = 0$$

13. Since $810°$ results in the same point as $90°$ on the unit circle, which corresponds to the point $(0, 1)$, then using $x = 0$ and $y = 1$ in the definitions yields:

$\cos 810° = x = 0$ $\sec 810° = \dfrac{1}{x} = \dfrac{1}{0}$, which is undefined

$\sin 810° = y = 1$ $\csc 810° = \dfrac{1}{y} = \dfrac{1}{1} = 1$

$\tan 810° = \dfrac{y}{x} = \dfrac{1}{0}$, which is undefined $\cot 810° = \dfrac{x}{y} = \dfrac{0}{1} = 0$

15. Using Figure 8, we have $\sin 10° \approx 0.2$ and $\sin(-10°) \approx -0.2$. Using a calculator, we have $\sin 10° \approx 0.17$ and $\sin(-10°) \approx -0.17$.

17. Using Figure 8, we have $\cos 80° \approx 0.2$ and $\cos(-80°) \approx 0.2$. Using a calculator, we have $\cos 80° \approx 0.17$ and $\cos(-80°) \approx 0.17$.

19. Using Figure 8, we have $\sin 120° \approx 0.9$ and $\sin(-120°) \approx -0.9$. Using a calculator, we have $\sin 120° \approx 0.87$ and $\sin(-120°) \approx -0.87$.

21. Using Figure 8, we have $\sin 150° = 0.5$ and $\sin(-150°) = -0.5$. Using a calculator, we have $\sin 150° = 0.5$ and $\sin(-150°) = -0.5$.

23. Using Figure 8, we have $\cos 220° \approx -0.8$ and $\cos(-220°) \approx -0.8$. Using a calculator, we have $\cos 220° \approx -0.77$ and $\cos(-220°) \approx -0.77$.

25. Using Figure 8, we have $\cos 310° \approx 0.6$ and $\cos(-310°) \approx 0.6$. Using a calculator, we have $\cos 310° \approx 0.64$ and $\cos(-310°) \approx 0.64$.

27. Since $40° + 360°$ is one full revolution on the unit circle past $40°$, then $\sin(40° + 360°) = \sin 40°$. Using Figure 8, we have $\sin(40° + 360°) \approx 0.6$. Using a calculator, we have $\sin(40° + 360°) \approx 0.64$.

29. (a)   The reference angle for $315°$ is $45°$, and $\cos 45° = \frac{\sqrt{2}}{2}$. Since $\cos \theta$ is the x-coordinate, and $\theta = 315°$ lies in the fourth quadrant where the x-coordinates are positive, then $\cos 315° = \frac{\sqrt{2}}{2}$.

    (b)   The reference angle for $-315°$ is $45°$, and $\cos 45° = \frac{\sqrt{2}}{2}$. Since $\cos \theta$ is the x-coordinate, and $\theta = -315°$ lies in the first quadrant where the x-coordinates are positive, then $\cos(-315°) = \frac{\sqrt{2}}{2}$.

    (c)   The reference angle for $315°$ is $45°$, and $\sin 45° = \frac{\sqrt{2}}{2}$. Since $\sin \theta$ is the y-coordinate, and $\theta = 315°$ lies in the fourth quadrant where the y-coordinates are negative, then $\sin 315° = -\frac{\sqrt{2}}{2}$.

(d)   The reference angle for $-315°$ is $45°$, and $\sin 45° = \frac{\sqrt{2}}{2}$. Since $\sin\theta$ is the y-coordinate, and $\theta = -315°$ lies in the first quadrant where the y-coordinates are positive, then $\sin(-315°) = \frac{\sqrt{2}}{2}$.

31. (a)   The reference angle for $300°$ is $60°$, and $\cos 60° = \frac{1}{2}$. Since $\cos\theta$ is the x-coordinate, and $\theta = 300°$ lies in the fourth quadrant where the x-coordinates are positive, then $\cos 300° = \frac{1}{2}$.

(b)   The reference angle for $-300°$ is $60°$, and $\cos 60° = \frac{1}{2}$. Since $\cos\theta$ is the x-coordinate, and $\theta = -300°$ lies in the first quadrant where the x-coordinates are positive, then $\cos(-300°) = \frac{1}{2}$.

(c)   The reference angle for $300°$ is $60°$, and $\sin 60° = \frac{\sqrt{3}}{2}$. Since $\sin\theta$ is the y-coordinate, and $\theta = 300°$ lies in the fourth quadrant where the y-coordinates are negative, then $\sin 300° = -\frac{\sqrt{3}}{2}$.

(d)   The reference angle for $-300°$ is $60°$, and $\sin 60° = \frac{\sqrt{3}}{2}$. Since $\sin\theta$ is the y-coordinate, and $\theta = -300°$ lies in the first quadrant where the y-coordinates are positive, then $\sin(-300°) = \frac{\sqrt{3}}{2}$.

33. (a)   The reference angle for $210°$ is $30°$, and $\cos 30° = \frac{\sqrt{3}}{2}$. Since $\cos\theta$ is the x-coordinate, and $\theta = 210°$ lies in the third quadrant where the x-coordinates are negative, then $\cos 210° = -\frac{\sqrt{3}}{2}$.

(b)   The reference angle for $-210°$ is $30°$, and $\cos 30° = \frac{\sqrt{3}}{2}$. Since $\cos\theta$ is the x-coordinate, and $\theta = -210°$ lies in the second quadrant where the x-coordinates are negative, then $\cos(-210°) = -\frac{\sqrt{3}}{2}$.

(c)   The reference angle for $210°$ is $30°$, and $\sin 30° = \frac{1}{2}$. Since $\sin\theta$ is the y-coordinate, and $\theta = 210°$ lies in the third quadrant where the y-coordinates are negative, then $\sin 210° = -\frac{1}{2}$.

(d)   The reference angle for $-210°$ is $30°$, and $\sin 30° = \frac{1}{2}$. Since $\sin\theta$ is the y-coordinate, and $\theta = -210°$ lies in the second quadrant where the y-coordinates are positive, then $\sin(-210°) = \frac{1}{2}$.

35. (a) The reference angle for 390° is 30° (390° = 30° + 360°), and $\cos 30° = \frac{\sqrt{3}}{2}$. Since $\cos\theta$ is the x-coordinate, and $\theta = 390°$ lies in the first quadrant where the x-coordinates are positive, then $\cos 390° = \frac{\sqrt{3}}{2}$.

(b) The reference angle for −390° is 30° (−390° = −30° − 360°), and $\cos 30° = \frac{\sqrt{3}}{2}$. Since $\cos\theta$ is the x-coordinate, and $\theta = −390°$ lies in the fourth quadrant where the x-coordinates are positive, then $\cos(−390°) = \frac{\sqrt{3}}{2}$.

(c) The reference angle for 390° is 30°, and $\sin 30° = \frac{1}{2}$. Since $\sin\theta$ is the y-coordinate, and $\theta = 390°$ lies in the first quadrant where the y-coordinates are positive, then $\sin 390° = \frac{1}{2}$.

(d) The reference angle for −390° is 30°, and $\sin 30° = \frac{1}{2}$. Since $\sin\theta$ is the y-coordinate, and $\theta = −390°$ lies in the fourth quadrant where the y-coordinates are negative, then $\sin(−390°) = −\frac{1}{2}$.

37. (a) The reference angle for 600° is 60° (600° = 240° + 360°), and $\sec 60° = 2$. Since $\sec\theta$ is the reciprocal of the x-coordinate, and $\theta = 600°$ lies in the third quadrant where the x-coordinates are negative, then $\sec 600° = −2$.

(b) The reference angle for −600° is 60° (−600° = −240° − 360°), and $\csc 60° = \frac{2\sqrt{3}}{3}$. Since $\csc\theta$ is the reciprocal of the y-coordinate, and $\theta = −600°$ lies in the second quadrant where the y-coordinates are positive, then $\csc(−600°) = \frac{2\sqrt{3}}{3}$.

(c) The reference angle for 600° is 60°, and $\tan 60° = \sqrt{3}$. Since $\tan\theta = \frac{y}{x}$, and $\theta = 600°$ lies in the third quadrant where both the x- and y-coordinates are negative, then $\tan 600° = \sqrt{3}$.

(d) The reference angle for −600° is 60°, and $\cot 60° = \frac{\sqrt{3}}{3}$. Since $\cot\theta = \frac{x}{y}$ and $\theta = −600°$ lies in the second quadrant where the y-coordinates are positive and the x-coordinates are negative, then $\cot(−600°) = −\frac{\sqrt{3}}{3}$.

39. We complete the table:

θ	$\sin\theta$	$\cos\theta$	$\tan\theta$
$0°$	0	1	0
$30°$	$\frac{1}{2}$	$\frac{\sqrt{3}}{2}$	$\frac{\sqrt{3}}{3}$
$45°$	$\frac{\sqrt{2}}{2}$	$\frac{\sqrt{2}}{2}$	1
$60°$	$\frac{\sqrt{3}}{2}$	$\frac{1}{2}$	$\sqrt{3}$
$90°$	1	0	undefined
$120°$	$\frac{\sqrt{3}}{2}$	$-\frac{1}{2}$	$-\sqrt{3}$
$135°$	$\frac{\sqrt{2}}{2}$	$-\frac{\sqrt{2}}{2}$	-1
$150°$	$\frac{1}{2}$	$-\frac{\sqrt{3}}{2}$	$-\frac{\sqrt{3}}{3}$
$180°$	0	-1	0

41. We draw the figure:

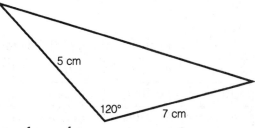

Now using the area formula, we have:
$$\text{Area} = \tfrac{1}{2}ab\sin\theta = \tfrac{1}{2}(7)(5)\sin120° = \left(\tfrac{35}{2}\right)\left(\tfrac{\sqrt{3}}{2}\right) = \tfrac{35\sqrt{3}}{4}\,\text{cm}^2$$

43. We draw the figure:

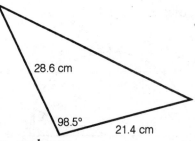

Now using the area formula, we have:
$$\text{Area} = \tfrac{1}{2}ab\sin\theta = \tfrac{1}{2}(21.4)(28.6)\sin98.5° \approx 302.7\ \text{cm}^2$$

45. We start by finding the measure of a central angle of a triangle drawn from the center out to two adjacent vertices:

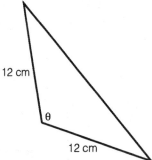

Since the triangle is 3-sided, we have $3\theta = 360°$, so $\theta = 120°$. The area of this triangle is thus:

$$\tfrac{1}{2}ab\sin\theta = \tfrac{1}{2}(12)(12)\sin 120° = 72 \cdot \tfrac{\sqrt{3}}{2} = 36\sqrt{3}$$

The area of the shaded figure (3 triangles) is therefore:

$$3\left(36\sqrt{3}\right) = 108\sqrt{3} \approx 187.06 \text{ cm}^2$$

47. (a) We complete the table:

Terminal side of angle θ lies in

	Quadrant I	Quadrant II	Quadrant III	Quadrant IV
$\sin\theta$	positive	positive	negative	negative
$\cos\theta$	positive	negative	negative	positive
$\tan\theta$	positive	negative	positive	negative

(b) It works!

49. (a) A sketch of $\sin 10°$ indicates that $\sin 10° \approx 0.2$:

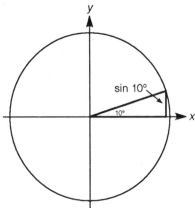

A sketch of sin 70° indicates that sin 70° ≈ 0.9:

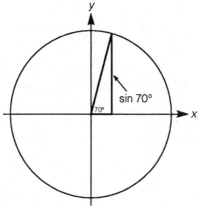

Clearly sin 70° is larger.

(b)   A sketch of cos 10° indicates that cos 10° ≈ 1.0:

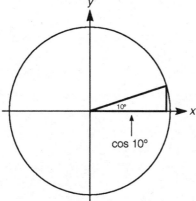

A sketch of cos 70° indicates that cos 70° ≈ 0.3:

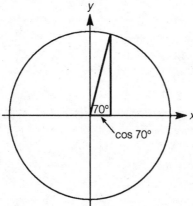

Clearly cos 10° is larger.

51. (a)   A sketch of cos 140° indicates that cos 140° ≈ −0.8:

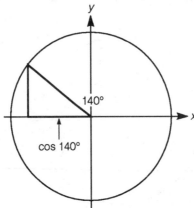

A sketch of cos 100° indicates that cos 100° ≈ −0.2:

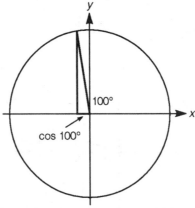

Clearly cos 100° is larger.

(b)   A sketch of sin 140° indicates that sin 140° ≈ 0.6:

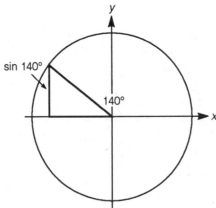

A sketch of sin 100° indicates that sin 100° ≈ 1.0:

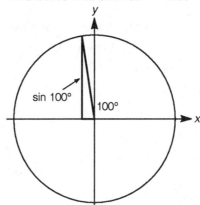

Clearly sin 100° is larger.

53. (a) We sketch the figure showing the obtuse angle 180° − θ in standard position:

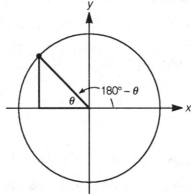

Note that the terminal side of the angle 180° − θ lies in the second quadrant.

(b) The reference angle for 180° − θ is θ.

(c) Since 180° − θ lies in the second quadrant where all y-coordinates are positive, then sin(180°−θ) = sin θ.

55. (a) In △ADC, $\sec\alpha = \dfrac{AC}{AD} = \dfrac{AC}{1} = AC$.

Similarly, in △ADB we have $\sec\beta = \dfrac{AB}{AD} = \dfrac{AB}{1} = AB$.

(b) We compute the areas:

Area △ADC $= \frac{1}{2}(AC)(AD)\sin\alpha = \frac{1}{2}\sec\alpha\sin\alpha$

Area △ADB $= \frac{1}{2}(AD)(AB)\sin\beta = \frac{1}{2}\sec\beta\sin\beta$

Area △ABC $= \frac{1}{2}(AC)(AB)\sin(\alpha+\beta) = \frac{1}{2}\sec\alpha\sec\beta\sin(\alpha+\beta)$

(c)   Since the sum of the areas of the two smaller triangles equals the area of $\triangle ABC$, we have:

$$\tfrac{1}{2}\sec\alpha\sec\beta\sin(\alpha+\beta)=\tfrac{1}{2}\sec\alpha\sin\alpha+\tfrac{1}{2}\sec\beta\sin\beta$$

Now multiply both sides of this equation by the quantity $2\cos\alpha\cos\beta$. This yields:

$$\sin(\alpha+\beta)=\sin\alpha\cos\beta+\cos\alpha\sin\beta$$

(d)   Using the hint, we have:

$$\begin{aligned}
\sin 75°&=\sin(30°+45°)\\
&=\sin 30°\cos 45°+\cos 30°\sin 45°\\
&=\frac{1}{2}\cdot\frac{\sqrt{2}}{2}+\frac{\sqrt{3}}{2}\cdot\frac{\sqrt{2}}{2}\\
&=\frac{\sqrt{2}+\sqrt{6}}{4}
\end{aligned}$$

(e)   We compute $\sin 75°=\dfrac{\sqrt{2}+\sqrt{6}}{4}\approx\dfrac{1.4+2.4}{4}=\dfrac{3.8}{4}<1$.

But $\sin 30°+\sin 45°\approx 0.5+0.7=1.2>1$. Since $\sin 75°$ is less than 1, whereas $\sin 30°+\sin 45°$ is greater than 1, the quantities cannot be equal. We can also argue with exact values (not approximations):

$$\sin 30°+\sin 45°=\frac{1}{2}+\frac{\sqrt{2}}{2}=\frac{2+2\sqrt{2}}{4}$$

$$\sin 75°=\frac{\sqrt{2}+\sqrt{6}}{4}$$

So, if these were to be equal then $2+2\sqrt{2}=\sqrt{2}+\sqrt{6}$, thus $2+\sqrt{2}=\sqrt{6}$.
But $2+\sqrt{2}>3$ while $\sqrt{6}<3$, thus they cannot be equal.

(f)   We compute $\sin 105°\approx 0.9659$, $\sin 45°\approx 0.7071$, and $\sin 60°\approx 0.8660$.
So $\sin 45°+\sin 60°\approx 1.5731\neq\sin 105°$.
Note: An argument similar to (e) using exact values can be made here.

57.   (a) $h(60°)=\tan 60°=\sqrt{3}$

(b) $h(2\cdot 60°)=\tan 120°=-\sqrt{3}$

(c) $h\left(\frac{60°}{2}\right)=\tan 30°=\frac{\sqrt{3}}{3}$

59.   (a)   Using a calculator, we have:

$$\frac{f(27°)}{g(27°)}=\frac{\sin 27°}{\cos 27°}\approx 0.510$$

(b) $h(27°)=\tan 27°\approx 0.510$

(c) $\left[f(27°)\right]^2+\left[g(27°)\right]^2=\sin^2 27°+\cos^2 27°=1$
Note that a calculator is not necessary, since $\sin^2\theta+\cos^2\theta=1$ for all angles θ.

6.5 Trigonometric Identities

1. (a) Substituting $\sin\theta = \frac{1}{5}$ into $\cos^2\theta = 1 - \sin^2\theta$ yields:

$$\cos^2\theta = 1 - \left(\frac{1}{5}\right)^2 = 1 - \frac{1}{25} = \frac{24}{25}$$

Since the terminal side of θ lies in quadrant 2, $\cos\theta < 0$ and thus:

$$\cos\theta = -\sqrt{\frac{24}{25}} = -\frac{2\sqrt{6}}{5}$$

Now compute the remaining four trigonometric values:

$$\tan\theta = \frac{\sin\theta}{\cos\theta} = \frac{\frac{1}{5}}{-\frac{2\sqrt{6}}{5}} = -\frac{1}{2\sqrt{6}} = -\frac{\sqrt{6}}{12}$$

$$\cot\theta = \frac{\cos\theta}{\sin\theta} = \frac{-\frac{2\sqrt{6}}{5}}{\frac{1}{5}} = -2\sqrt{6}$$

$$\sec\theta = \frac{1}{\cos\theta} = \frac{1}{-\frac{2\sqrt{6}}{5}} = -\frac{5}{2\sqrt{6}} = -\frac{5\sqrt{6}}{12}$$

$$\csc\theta = \frac{1}{\sin\theta} = \frac{1}{\frac{1}{5}} = 5$$

(b) Substituting $\sin\theta = -\frac{1}{5}$ into $\cos^2\theta = 1 - \sin^2\theta$ yields:

$$\cos^2\theta = 1 - \left(\frac{1}{5}\right)^2 = 1 - \frac{1}{25} = \frac{24}{25}$$

Since the terminal side of θ lies in quadrant 3, $\cos\theta < 0$ and thus:

$$\cos\theta = -\sqrt{\frac{24}{25}} = -\frac{2\sqrt{6}}{5}$$

Now compute the remaining four trigonometric values:

$$\tan\theta = \frac{\sin\theta}{\cos\theta} = \frac{-\frac{1}{5}}{-\frac{2\sqrt{6}}{5}} = \frac{1}{2\sqrt{6}} = \frac{\sqrt{6}}{12}$$

$$\cot\theta = \frac{\cos\theta}{\sin\theta} = \frac{-\frac{2\sqrt{6}}{5}}{-\frac{1}{5}} = 2\sqrt{6}$$

$$\sec\theta = \frac{1}{\cos\theta} = \frac{1}{-\frac{2\sqrt{6}}{5}} = -\frac{5}{2\sqrt{6}} = -\frac{5\sqrt{6}}{12}$$

$$\csc\theta = \frac{1}{\sin\theta} = \frac{1}{-\frac{1}{5}} = -5$$

3. (a) Substituting $\cos\theta = \frac{5}{13}$ into $\sin^2\theta = 1 - \cos^2\theta$ yields:

$$\sin^2\theta = 1 - \left(\frac{5}{13}\right)^2 = 1 - \frac{25}{169} = \frac{144}{169}$$

Since the terminal side of θ lies in quadrant 1, $\sin\theta > 0$ and thus:
$$\sin\theta = \sqrt{\tfrac{144}{169}} = \tfrac{12}{13}$$
Now compute the remaining four trigonometric values:

$$\tan\theta = \frac{\sin\theta}{\cos\theta} = \frac{\tfrac{12}{13}}{\tfrac{5}{13}} = \tfrac{12}{5} \qquad\qquad \cot\theta = \frac{\cos\theta}{\sin\theta} = \frac{\tfrac{5}{13}}{\tfrac{12}{13}} = \tfrac{5}{12}$$

$$\sec\theta = \frac{1}{\cos\theta} = \frac{1}{\tfrac{5}{13}} = \tfrac{13}{5} \qquad\qquad \csc\theta = \frac{1}{\sin\theta} = \frac{1}{\tfrac{12}{13}} = \tfrac{13}{12}$$

(b) Substituting $\cos\theta = -\tfrac{5}{13}$ into $\sin^2\theta = 1 - \cos^2\theta$ yields:
$$\sin^2\theta = 1 - \left(-\tfrac{5}{13}\right)^2 = 1 - \tfrac{25}{169} = \tfrac{144}{169}$$
Since the terminal side of θ lies in quadrant 3, $\sin\theta < 0$ and thus:
$$\sin\theta = -\sqrt{\tfrac{144}{169}} = -\tfrac{12}{13}$$
Now compute the remaining four trigonometric values:

$$\tan\theta = \frac{\sin\theta}{\cos\theta} = \frac{-\tfrac{12}{13}}{-\tfrac{5}{13}} = \tfrac{12}{5} \qquad\qquad \cot\theta = \frac{\cos\theta}{\sin\theta} = \frac{-\tfrac{5}{13}}{-\tfrac{12}{13}} = \tfrac{5}{12}$$

$$\sec\theta = \frac{1}{\cos\theta} = \frac{1}{-\tfrac{5}{13}} = -\tfrac{13}{5} \qquad\qquad \csc\theta = \frac{1}{\sin\theta} = \frac{1}{-\tfrac{12}{13}} = -\tfrac{13}{12}$$

5. Since $\csc A = -3$, then $\sin A = -\tfrac{1}{3}$. Substituting into $\cos^2 A = 1 - \sin^2 A$ yields:
$$\cos^2 A = 1 - \left(-\tfrac{1}{3}\right)^2 = 1 - \tfrac{1}{9} = \tfrac{8}{9}$$
Since the terminal side of A lies in quadrant 4, $\cos A > 0$ and thus:
$$\cos A = \sqrt{\tfrac{8}{9}} = \tfrac{2\sqrt{2}}{3}$$
Now compute the remaining three trigonometric values:

$$\tan A = \frac{\sin A}{\cos A} = \frac{-\tfrac{1}{3}}{\tfrac{2\sqrt{2}}{3}} = -\frac{1}{2\sqrt{2}} = -\frac{\sqrt{2}}{4}$$

$$\cot A = \frac{\cos A}{\sin A} = \frac{\tfrac{2\sqrt{2}}{3}}{-\tfrac{1}{3}} = -2\sqrt{2}$$

$$\sec A = \frac{1}{\cos A} = \frac{1}{\tfrac{2\sqrt{2}}{3}} = \frac{3}{2\sqrt{2}} = \frac{3\sqrt{2}}{4}$$

7. Since $\sec B = -\tfrac{3}{2}$, then $\cos B = -\tfrac{2}{3}$. Substituting into $\sin^2 B = 1 - \cos^2 B$ yields:
$$\sin^2 B = 1 - \left(-\tfrac{2}{3}\right)^2 = 1 - \tfrac{4}{9} = \tfrac{5}{9}$$
Since the terminal side of B lies in quadrant 3, $\sin B < 0$ and thus:
$$\sin B = -\sqrt{\tfrac{5}{9}} = -\tfrac{\sqrt{5}}{3}$$

Now compute the remaining three trigonometric values:

$$\tan B = \frac{\sin B}{\cos B} = \frac{-\frac{\sqrt{5}}{3}}{-\frac{2}{3}} = \frac{\sqrt{5}}{2}$$

$$\cot B = \frac{\cos B}{\sin B} = \frac{-\frac{2}{3}}{-\frac{\sqrt{5}}{3}} = \frac{2}{\sqrt{5}} = \frac{2\sqrt{5}}{5}$$

$$\csc B = \frac{1}{\sin B} = \frac{1}{-\frac{\sqrt{5}}{3}} = -\frac{3}{\sqrt{5}} = -\frac{3\sqrt{5}}{5}$$

9. Substituting $\cos\theta = \dfrac{t}{3}$ into $\sin^2\theta = 1 - \cos^2\theta$ yields:

$$\sin^2\theta = 1 - \left(\frac{t}{3}\right)^2 = 1 - \frac{t^2}{9} = \frac{9 - t^2}{9}$$

Since the terminal side of θ lies in quadrant 4, $\sin\theta < 0$ and thus:

$$\sin\theta = -\sqrt{\frac{9 - t^2}{9}} = -\frac{\sqrt{9 - t^2}}{3}$$

Now compute the remaining four trigonometric values:

$$\tan\theta = \frac{\sin\theta}{\cos\theta} = \frac{-\frac{\sqrt{9-t^2}}{3}}{\frac{t}{3}} = -\frac{\sqrt{9 - t^2}}{t}$$

$$\cot\theta = \frac{\cos\theta}{\sin\theta} = \frac{\frac{t}{3}}{-\frac{\sqrt{9-t^2}}{3}} = -\frac{t}{\sqrt{9-t^2}} = -\frac{t\sqrt{9-t^2}}{9 - t^2}$$

$$\sec\theta = \frac{1}{\cos\theta} = \frac{1}{\frac{t}{3}} = \frac{3}{t}$$

$$\csc\theta = \frac{1}{\sin\theta} = \frac{1}{-\frac{\sqrt{9-t^2}}{3}} = -\frac{3}{\sqrt{9-t^2}} = -\frac{3\sqrt{9 - t^2}}{9 - t^2}$$

11. Substituting $\sin\theta = -3u$ into $\cos^2\theta = 1 - \sin^2\theta$ yields:

$$\cos^2\theta = 1 - (-3u)^2 = 1 - 9u^2$$

Since the terminal side of θ lies in quadrant 3, $\cos\theta < 0$ and thus:

$$\cos\theta = -\sqrt{1 - 9u^2}$$

Now compute the remaining four trigonometric values:

$$\tan\theta = \frac{\sin\theta}{\cos\theta} = \frac{-3u}{-\sqrt{1-9u^2}} = \frac{3u\sqrt{1-9u^2}}{1 - 9u^2}$$

$$\cot\theta = \frac{\cos\theta}{\sin\theta} = \frac{-\sqrt{1-9u^2}}{-3u} = \frac{\sqrt{1-9u^2}}{3u}$$

$$\sec\theta = \frac{1}{\cos\theta} = \frac{1}{-\sqrt{1-9u^2}} = -\frac{\sqrt{1-9u^2}}{1 - 9u^2}$$

$$\csc\theta = \frac{1}{\sin\theta} = \frac{1}{-3u} = -\frac{1}{3u}$$

13. Substituting $\cos\theta = \dfrac{u}{\sqrt{3}}$ into $\sin^2\theta = 1 - \cos^2\theta$ yields:

$$\sin^2\theta = 1 - \left(\frac{u}{\sqrt{3}}\right)^2 = 1 - \frac{u^2}{3} = \frac{3-u^2}{3}$$

Since the terminal side of θ lies in quadrant 1, $\sin\theta > 0$ and thus:

$$\sin\theta = \sqrt{\frac{3-u^2}{3}} = \frac{\sqrt{3-u^2}}{\sqrt{3}} = \frac{\sqrt{9-3u^2}}{3}$$

Now compute the remaining four trigonometric values:

$$\tan\theta = \frac{\sin\theta}{\cos\theta} = \frac{\frac{\sqrt{9-3u^2}}{3}}{\frac{u}{\sqrt{3}}} = \frac{\sqrt{9-3u^2}}{u\sqrt{3}} = \frac{\sqrt{3-u^2}}{u}$$

$$\cot\theta = \frac{\cos\theta}{\sin\theta} = \frac{\frac{u}{\sqrt{3}}}{\frac{\sqrt{9-3u^2}}{3}} = \frac{u\sqrt{3}}{\sqrt{9-3u^2}} = \frac{u}{\sqrt{3-u^2}} = \frac{u\sqrt{3-u^2}}{3-u^2}$$

$$\sec\theta = \frac{1}{\cos\theta} = \frac{1}{\frac{u}{\sqrt{3}}} = \frac{\sqrt{3}}{u}$$

$$\csc\theta = \frac{1}{\sin\theta} = \frac{1}{\frac{\sqrt{9-3u^2}}{3}} = \frac{3}{\sqrt{9-3u^2}} = \frac{3\sqrt{9-3u^2}}{9-3u^2} = \frac{\sqrt{9-3u^2}}{3-u^2}$$

15. Using the identities for $\sec\theta$ and $\csc\theta$, we have:

$$\sin\theta\cos\theta\sec\theta\csc\theta = \frac{\sin\theta\cos\theta}{\sin\theta\cos\theta} = 1$$

17. Using the identities for $\sec\theta$ and $\tan\theta$, we have:

$$\frac{\sin\theta\sec\theta}{\tan\theta} = \frac{\sin\theta \cdot \frac{1}{\cos\theta}}{\frac{\sin\theta}{\cos\theta}} = \frac{\frac{\sin\theta}{\cos\theta}}{\frac{\sin\theta}{\cos\theta}} = 1$$

19. Working from the right-hand side and using the identities for $\sec x$ and $\tan x$, we have:

$$\sec x - 5\tan x = \frac{1}{\cos x} - \frac{5\sin x}{\cos x} = \frac{1 - 5\sin x}{\cos x}$$

21. Using the identity for $\sec A$, we have:

$$\cos A(\sec A - \cos A) = \cos A\left(\frac{1}{\cos A} - \cos A\right) = 1 - \cos^2 A = \sin^2 A$$

23. Multiplying out parentheses and using the identities for $\sec\theta$ and $\tan\theta$, we have:

$$(1-\sin\theta)(\sec\theta+\tan\theta)=\sec\theta+\tan\theta-\sin\theta\sec\theta-\sin\theta\tan\theta$$

$$=\frac{1}{\cos\theta}+\frac{\sin\theta}{\cos\theta}-\sin\theta\cdot\frac{1}{\cos\theta}-\sin\theta\cdot\frac{\sin\theta}{\cos\theta}$$

$$=\frac{1+\sin\theta-\sin\theta-\sin^2\theta}{\cos\theta}$$

$$=\frac{1-\sin^2\theta}{\cos\theta}$$

$$=\frac{\cos^2\theta}{\cos\theta}$$

$$=\cos\theta$$

25. Multiplying out parentheses and using the identities for $\sec\alpha$ and $\tan\alpha$, we have:

$$(\sec\alpha-\tan\alpha)^2=\left(\frac{1}{\cos\alpha}-\frac{\sin\alpha}{\cos\alpha}\right)^2$$

$$=\frac{(1-\sin\alpha)^2}{\cos^2\alpha}$$

$$=\frac{(1-\sin\alpha)^2}{1-\sin^2\alpha}$$

$$=\frac{(1-\sin\alpha)^2}{(1+\sin\alpha)(1-\sin\alpha)}$$

$$=\frac{1-\sin\alpha}{1+\sin\alpha}$$

27. Working from the right-hand side and using identities for $\cot A$ and $\tan A$, we have:

$$\frac{\sin A}{1-\cot A}-\frac{\cos A}{\tan A-1}=\frac{\sin A}{1-\frac{\cos A}{\sin A}}-\frac{\cos A}{\frac{\sin A}{\cos A}-1}$$

$$=\frac{\sin^2 A}{\sin A-\cos A}-\frac{\cos^2 A}{\sin A-\cos A}$$

$$=\frac{(\sin A-\cos A)(\sin A+\cos A)}{\sin A-\cos A}$$

$$=\sin A+\cos A$$

29. Working from the left-hand side and using identities for csc θ and sec θ, we have:

$$
\begin{aligned}
\csc^2\theta + \sec^2\theta &= \frac{1}{\sin^2\theta} + \frac{1}{\cos^2\theta} \\
&= \frac{\cos^2\theta + \sin^2\theta}{\sin^2\theta\cos^2\theta} \\
&= \frac{1}{\sin^2\theta\cos^2\theta} \\
&= \frac{1}{\sin^2\theta} \cdot \frac{1}{\cos^2\theta} \\
&= \csc^2\theta\sec^2\theta
\end{aligned}
$$

31. Using the identity for $\tan A$ and $\sin^2 A = 1 - \cos^2 A$, we have:

$$
\sin A \tan A = \sin A \cdot \frac{\sin A}{\cos A} = \frac{\sin^2 A}{\cos A} = \frac{1 - \cos^2 A}{\cos A}
$$

33. Working from the right-hand side, we have:

$$
\begin{aligned}
-\cot^4 A + \csc^4 A &= \frac{-\cos^4 A}{\sin^4 A} + \frac{1}{\sin^4 A} \\
&= \frac{\left(1 - \cos^2 A\right)\left(1 + \cos^2 A\right)}{\sin^4 A} \\
&= \frac{\sin^2 A\left(1 + \cos^2 A\right)}{\sin^4 A} \\
&= \frac{1 + \cos^2 A}{\sin^2 A} \\
&= \frac{1}{\sin^2 A} + \frac{\cos^2 A}{\sin^2 A} \\
&= \csc^2 A + \cot^2 A
\end{aligned}
$$

35. Working from the left-hand side, we have:

$$
\begin{aligned}
\frac{\sin A - \cos A}{\sin A} &+ \frac{\cos A - \sin A}{\cos A} \\
&= \frac{\cos A(\sin A - \cos A) + \sin A(\cos A - \sin A)}{\sin A\cos A} \\
&= \frac{\cos A\sin A - \cos^2 A + \sin A\cos A - \sin^2 A}{\sin A\cos A} \\
&= \frac{2\cos A\sin A - \left(\cos^2 A + \sin^2 A\right)}{\sin A\cos A} \\
&= 2 - \frac{1}{\sin A\cos A} \\
&= 2 - \sec A\csc A
\end{aligned}
$$

37. **(a)** We start by testing the value $\alpha = 30°$. Since $\cos 30° = \frac{\sqrt{3}}{2}$ and $\csc 30° = 2$, the equation states that:

$$\frac{2^2 - 1}{2^2} = \frac{\sqrt{3}}{2}$$

$$\frac{3}{4} = \frac{\sqrt{3}}{2}$$

Since the left-hand side is the square of the right-hand side, this is not an identity.

(b) Proceeding as in (a), we test the value $\alpha = 30°$. Since $\sec 30° = \frac{2}{\sqrt{3}}$ and $\csc 30° = 2$, the equation states that:

$$\left[\left(\tfrac{2}{\sqrt{3}}\right)^2 - 1\right]\left[(2)^2 - 1\right] = 1$$

$$\left[\tfrac{4}{3} - 1\right][4 - 1] = 1$$

$$\tfrac{1}{3} \cdot 3 = 1$$

Since this is true, we proceed by proving the identity:

$$\left(\sec^2 \alpha - 1\right)\left(\csc^2 \alpha - 1\right) = \left(\frac{1}{\cos^2 \alpha} - 1\right)\left(\frac{1}{\sin^2 \alpha} - 1\right)$$

$$= \frac{1 - \cos^2 \alpha}{\cos^2 \alpha} \cdot \frac{1 - \sin^2 \alpha}{\sin^2 \alpha}$$

$$= \frac{\sin^2 \alpha}{\cos^2 \alpha} \cdot \frac{\cos^2 \alpha}{\sin^2 \alpha}$$

$$= 1$$

39. **(a)** Starting with the left-hand side, we multiply the numerator and denominator by $1 + \cos \theta$ to obtain:

$$\frac{\sin \theta}{1 - \cos \theta} \cdot \frac{1 + \cos \theta}{1 + \cos \theta} = \frac{\sin \theta(1 + \cos \theta)}{1 - \cos^2 \theta} = \frac{\sin \theta(1 + \cos \theta)}{\sin^2 \theta} = \frac{1 + \cos \theta}{\sin \theta}$$

(b) Starting with the left-hand side, we multiply the numerator and denominator by $\sin \theta$ to obtain:

$$\frac{\sin \theta}{1 - \cos \theta} \cdot \frac{\sin \theta}{\sin \theta} = \frac{\sin^2 \theta}{\sin \theta(1 - \cos \theta)}$$

$$= \frac{1 - \cos^2 \theta}{\sin \theta(1 - \cos \theta)}$$

$$= \frac{(1 + \cos \theta)(1 - \cos \theta)}{\sin \theta(1 - \cos \theta)}$$

$$= \frac{1 + \cos \theta}{\sin \theta}$$

41. Starting with the left-hand side, we use the identities for $\sec\theta$ and $\csc\theta$, then multiply the resulting fraction by $\sin\theta$ to obtain:

$$\frac{\sec\theta - \csc\theta}{\sec\theta + \csc\theta} = \frac{\frac{1}{\cos\theta} - \frac{1}{\sin\theta}}{\frac{1}{\cos\theta} + \frac{1}{\sin\theta}} = \frac{\frac{\sin\theta}{\cos\theta} - \frac{\sin\theta}{\sin\theta}}{\frac{\sin\theta}{\cos\theta} + \frac{\sin\theta}{\sin\theta}} = \frac{\tan\theta - 1}{\tan\theta + 1}$$

43. Starting with the left-hand side, we use the identity for $\cot\theta$ to obtain:

$$\sin^2\theta\left(1 + n\cot^2\theta\right) = \sin^2\theta\left(1 + n\cdot\frac{\cos^2\theta}{\sin^2\theta}\right)$$

$$= \sin^2\theta + n\cos^2\theta$$

$$= \cos^2\theta\left(\frac{\sin^2\theta}{\cos^2\theta} + n\right)$$

$$= \cos^2\theta\left(n + \tan^2\theta\right)$$

45. **(a)** Using the difference of cubes formula $A^3 - B^3 = (A - B)\left(A^2 + AB + B^2\right)$, we obtain:

$$\cos^3\theta - \sin^3\theta = (\cos\theta - \sin\theta)\left(\cos^2\theta + \cos\theta\sin\theta + \sin^2\theta\right)$$

$$= (\cos\theta - \sin\theta)(1 + \cos\theta\sin\theta)$$

(b) Starting with the left-hand side, we use the identities for $\cot\phi$, $\tan\phi$, $\sec\phi$ and $\csc\phi$ to obtain:

$$\frac{\cos\phi\cot\phi - \sin\phi\tan\phi}{\csc\phi - \sec\phi} = \frac{\cos\phi\cdot\frac{\cos\phi}{\sin\phi} - \sin\phi\cdot\frac{\sin\phi}{\cos\phi}}{\frac{1}{\sin\phi} - \frac{1}{\cos\phi}}$$

$$= \frac{\frac{\cos^2\phi}{\sin\phi} - \frac{\sin^2\phi}{\cos\phi}}{\frac{\cos\phi - \sin\phi}{\sin\phi\cos\phi}}$$

$$= \frac{\cos^3\phi - \sin^3\phi}{\cos\phi - \sin\phi}$$

$$= \frac{(\cos\phi - \sin\phi)(1 + \cos\phi\sin\phi)}{\cos\phi - \sin\phi}$$

$$= 1 + \sin\phi\cos\phi$$

47. Since $\tan\alpha\tan\beta = 1$, then $\tan\alpha = \dfrac{1}{\tan\beta}$. Using the identities from Exercise 46, we have:

$$\sec^2\alpha = 1 + \tan^2\alpha = 1 + \frac{1}{\tan^2\beta} = 1 + \cot^2\beta = \csc^2\beta$$

Since α and β are acute angles, then $\sec\alpha > 0$ and $\csc\beta > 0$, so $\sec^2\alpha = \csc^2\beta$ implies $\sec\alpha = \csc\beta$.

49. Using the identity $\cos^2 \theta = 1 - \sin^2 \theta$, we have:

$$\cos^2 \theta = 1 - \frac{(p-q)^2}{(p+q)^2}$$

$$= 1 - \frac{p^2 - 2pq + q^2}{p^2 + 2pq + q^2}$$

$$= \frac{p^2 + 2pq + q^2 - p^2 + 2pq - q^2}{p^2 + 2pq + q^2}$$

$$= \frac{4pq}{(p+q)^2}$$

Since the terminal side of θ lies in quadrant 2, then $\cos \theta < 0$ and thus:

$$\cos \theta = -\sqrt{\frac{4pq}{(p+q)^2}} = -\frac{2\sqrt{pq}}{p+q} \quad \text{(since } p > 0, q > 0\text{)}$$

Now we find $\tan \theta$:

$$\tan \theta = \frac{\sin \theta}{\cos \theta} = \frac{\dfrac{p-q}{p+q}}{-\dfrac{2\sqrt{pq}}{p+q}} = \frac{p-q}{-2\sqrt{pq}} = \frac{q-p}{2\sqrt{qp}}$$

Chapter Six Review Exercises

1. Since the terminal side of $135°$ lies in quadrant 2, then $\sin 135° > 0$. Since the reference angle for $135°$ is $45°$, then:

$$\sin 135° = \sin 45° = \tfrac{\sqrt{2}}{2}$$

3. Since the terminal side of $-240°$ lies in quadrant 2, then $\tan(-240°) < 0$. Since the reference angle for $-240°$ is $60°$, then:

$$\tan(-240°) = -\tan 60° = -\sqrt{3}$$

5. Since the terminal side of $210°$ lies in quadrant 3, then $\csc(210°) < 0$. Since the reference angle for $210°$ is $30°$, then:

$$\csc 210° = -\csc 30° = -2$$

7. Since $270°$ corresponds to the point $(0, -1)$ on the unit circle, then $\sin 270° = -1$, which is the y-coordinate.

9. Since the terminal side of $-315°$ lies in quadrant 1, then $\cos(-315°) > 0$. Since the reference angle for $-315°$ is $45°$, then:

$$\cos(-315°) = \cos 45° = \tfrac{\sqrt{2}}{2}$$

11. Since $1800°$ corresponds to the point $(1,0)$ on the unit circle, then $\cos 1800° = 1$, which is the x-coordinate.

13. Since the terminal side of $240°$ lies in quadrant 3, then $\csc 240° < 0$. Since the reference angle for $240°$ is $60°$, then:
$$\csc 240° = -\csc 60° = -\frac{2\sqrt{3}}{3}$$

15. Since the terminal side of $780°$ lies in quadrant 1, then $\sec 780° > 0$. Since the reference angle for $780°$ is $60°$, then:
$$\sec 780° = \sec 60° = 2$$

17. Since $\sin^2 17° + \cos^2 17° = 1$ while $\ln e^{\pi} = \pi$, we have:
$$\ln\left(\sin^2 17° + \cos^2 17°\right) - \cos\left(\ln e^{\pi}\right) = \ln(1) - \cos(\pi) = 0 - (-1) = 1$$

19. We draw the triangle:

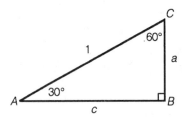

Now compute a and c:
$$\sin 30° = \frac{a}{1}, \text{ so } a = \sin 30° = \frac{1}{2}$$
$$\cos 30° = \frac{c}{1}, \text{ so } c = \cos 30° = \frac{\sqrt{3}}{2}$$

21. We draw a triangle:

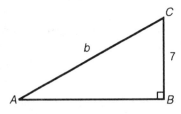

Now compute b:
$$\sin A = \frac{7}{b}, \text{ so } b = \frac{7}{\sin A} = \frac{7}{\frac{2}{5}} = \frac{35}{2}$$

23. We draw a triangle:

Using the Pythagorean theorem, we have:
$$3^2 + b^2 = 8^2$$
$$9 + b^2 = 64$$
$$b^2 = 55$$
$$b = \sqrt{55}$$
From the triangle, we have:
$$\sin A = \frac{\sqrt{55}}{8} \text{ and } \cot A = \frac{3}{\sqrt{55}} = \frac{3\sqrt{55}}{55}$$

25. We draw the triangle:

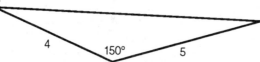

Now compute the area:
$$\tfrac{1}{2} ab \sin \theta = \tfrac{1}{2}(5)(4)\sin 150° = 10 \cdot \tfrac{1}{2} = 5 \text{ square units}$$

27. We draw a triangle:

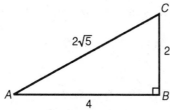

Using the Pythagorean theorem, we have:
$$4^2 + 2^2 = c^2$$
$$16 + 4 = c^2$$
$$c^2 = 20$$
$$c = 2\sqrt{5}$$
From the triangle, we have:
$$\sin A = \frac{2}{2\sqrt{5}} = \frac{1}{\sqrt{5}} = \frac{\sqrt{5}}{5} \text{ and } \cos B = 0$$
Therefore:
$$\sin^2 A + \cos^2 B = \left(\frac{\sqrt{5}}{5}\right)^2 + (0)^2 = \frac{5}{25} + 0 = \frac{1}{5}$$

29. We draw a triangle:

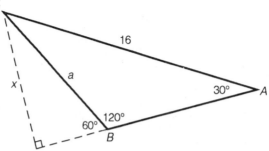

We first find x:

$$\sin 30° = \frac{x}{16}, \text{ so } x = 16\sin 30° = \tfrac{1}{2}(16) = 8$$

Now, using the smaller triangle:

$$\sin 60° = \frac{8}{a}, \text{ so } a = \frac{8}{\sin 60°} = \frac{8}{\frac{\sqrt{3}}{2}} = \frac{16}{\sqrt{3}} = \frac{16\sqrt{3}}{3}$$

31. We draw a triangle:

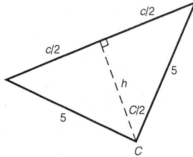

Notice that the bisector drawn for C bisects the opposite side (of length c) into two pieces of length $\frac{c}{2}$. This is true since $a = b = 5$. Using one of the smaller triangles, we have:

$$\sin\left(\tfrac{1}{2}C\right) = \frac{\frac{c}{2}}{5}$$

$$\frac{9}{10} = \frac{c}{10}$$

$$9 = c$$

To find the height h of the triangle, we use the Pythagorean theorem:

$$\left(\tfrac{c}{2}\right)^2 + h^2 = 5^2$$

$$\left(\tfrac{9}{2}\right)^2 + h^2 = 25$$

$$\tfrac{81}{4} + h^2 = 25$$

$$h^2 = \tfrac{19}{4}$$

$$h = \tfrac{\sqrt{19}}{2}$$

So the area of $\triangle ABC$ is given by:

$$\tfrac{1}{2}(\text{base})(\text{height}) = \tfrac{1}{2}(9)\left(\tfrac{\sqrt{19}}{2}\right) = \tfrac{9\sqrt{19}}{4}$$

33. Letting z be the side opposite the $90° - \beta$ angle (adjacent to β), we have:

$$\cot\alpha = \frac{x+z}{y} \quad \text{and} \quad \cot\beta = \frac{z}{y}$$

Thus:

$$\cot\alpha - \cot\beta = \frac{x+z}{y} - \frac{z}{y}$$

$$\cot\alpha - \cot\beta = \frac{x}{y}$$

$$y(\cot\alpha - \cot\beta) = x$$

$$y = \frac{x}{\cot\alpha - \cot\beta}$$

35. Since $\sin^2\theta + \cos^2\theta = 1$, we have:

$$\left(\frac{2p^2q^2}{p^4+q^4}\right)^2 + \cos^2\theta = 1$$

$$\frac{4p^4q^4}{\left(p^4+q^4\right)^2} + \cos^2\theta = 1$$

$$\cos^2\theta = \frac{\left(p^4+q^4\right)^2 - 4p^4q^4}{\left(p^4+q^4\right)^2}$$

$$\cos^2\theta = \frac{p^8 - 2p^4q^4 + q^8}{\left(p^4+q^4\right)^2}$$

$$\cos^2\theta = \frac{\left(p^4-q^4\right)^2}{\left(p^4+q^4\right)^2}$$

$$\cos\theta = \frac{p^4-q^4}{p^4+q^4}$$

Now find $\tan\theta$:

$$\tan\theta = \frac{\sin\theta}{\cos\theta} = \frac{2p^2q^2}{p^4-q^4}$$

37. Using the identities for $\sec A$ and $\csc A$, we have:

$$\frac{\sin A + \cos A}{\sec A + \csc A} = \frac{\sin A + \cos A}{\frac{1}{\cos A} + \frac{1}{\sin A}} = \frac{\sin A \cos A(\sin A + \cos A)}{\sin A + \cos A} = \sin A \cos A$$

39. Using the identities for $\sec A$, $\tan A$ and $\cot A$, we have:

$$\frac{\sin A \sec A}{\tan A + \cot A} = \frac{\frac{\sin A}{\cos A}}{\frac{\sin A}{\cos A} + \frac{\cos A}{\sin A}} = \frac{\sin^2 A}{\sin^2 A + \cos^2 A} = \sin^2 A$$

41. Using the identities for $\tan A$ and $\cot A$, we have:

$$\frac{\cos A}{1 - \tan A} + \frac{\sin A}{1 - \cot A} = \frac{\cos A}{1 - \frac{\sin A}{\cos A}} + \frac{\sin A}{1 - \frac{\cos A}{\sin A}}$$

$$= \frac{\cos^2 A}{\cos A - \sin A} + \frac{\sin^2 A}{\sin A - \cos A}$$

$$= \frac{\cos^2 A}{\cos A - \sin A} - \frac{\sin^2 A}{\cos A - \sin A}$$

$$= \frac{\cos^2 A - \sin^2 A}{\cos A - \sin A}$$

$$= \frac{(\cos A + \sin A)(\cos A - \sin A)}{\cos A - \sin A}$$

$$= \cos A + \sin A$$

43. Using the identities for $\sec A$ and $\csc A$, we have:

$$(\sec A + \csc A)^{-1}\left[(\sec A)^{-1} + (\csc A)^{-1}\right] = \frac{1}{\sec A + \csc A} \cdot \left[\frac{1}{\sec A} + \frac{1}{\csc A}\right]$$

$$= \frac{1}{\sec A + \csc A} \cdot \frac{\csc A + \sec A}{\sec A \csc A}$$

$$= \frac{1}{\sec A \csc A}$$

$$= \frac{1}{\frac{1}{\cos A} \cdot \frac{1}{\sin A}}$$

$$= \sin A \cos A$$

45. Simplifying the complex fraction, we have:

$$\frac{\frac{\sin A + \cos A}{\sin A - \cos A} - \frac{\sin A - \cos A}{\sin A + \cos A}}{\frac{\sin A + \cos A}{\sin A - \cos A} + \frac{\sin A - \cos A}{\sin A + \cos A}} = \frac{(\sin A + \cos A)^2 - (\sin A - \cos A)^2}{(\sin A + \cos A)^2 + (\sin A - \cos A)^2}$$

$$= \frac{1 + 2\sin A \cos A - 1 + 2\sin A \cos A}{1 + 2\sin A \cos A + 1 - 2\sin A \cos A}$$

$$= \frac{4\sin A \cos A}{2}$$

$$= 2\sin A \cos A$$

47. The x-coordinate of P is approximately 0.9848, so $\cos 10° \approx 0.9848$.

49. The y-coordinate of P is approximately 0.1736, so $\sin 10° \approx 0.1736$.

51. We draw a triangle:

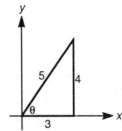

Using the Pythagorean theorem, we have:
$$3^2 + b^2 = 5^2$$
$$9 + b^2 = 25$$
$$b^2 = 16$$
$$b = 4$$
Since $0° < \theta < 90°$, both $\sin \theta$ and $\tan \theta$ are positive, so:
$$\sin \theta = \tfrac{4}{5} \text{ and } \tan \theta = \tfrac{4}{3}$$

53. We draw a triangle:

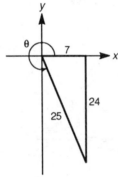

Using the Pythagorean theorem, we have:
$$7^2 + b^2 = 25^2$$
$$49 + b^2 = 625$$
$$b^2 = 576$$
$$b = 24$$
Since $270° < \theta < 360°$, $\tan \theta$ is negative, so:
$$\tan \theta = -\tfrac{24}{7}$$

55. We draw a triangle:

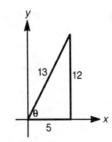

Using the Pythagorean theorem, we have:
$$a^2 + 12^2 = 13^2$$
$$a^2 + 144 = 169$$
$$a^2 = 25$$
$$a = 5$$
Since $0° < \theta < 90°$, $\cot \theta$ is positive, so:
$$\cot \theta = \tfrac{5}{12}$$

57. We draw a triangle:

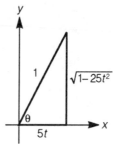

Using the Pythagorean theorem, we have:
$$(5t)^2 + b^2 = 1^2$$
$$25t^2 + b^2 = 1$$
$$b^2 = 1 - 25t^2$$
$$b = \sqrt{1 - 25t^2}$$
Since $0° < \theta < 90°$, $\tan(90° - \theta)$ is positive, so:
$$\tan(90° - \theta) = \cot \theta = \frac{5t}{\sqrt{1 - 25t^2}} = \frac{5t\sqrt{1 - 25t^2}}{1 - 25t^2}$$

59. We draw a triangle:

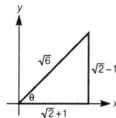

Using the Pythagorean theorem, we have:
$$\left(\sqrt{2} + 1\right)^2 + \left(\sqrt{2} - 1\right)^2 = c^2$$
$$2 + 2\sqrt{2} + 1 + 2 - 2\sqrt{2} + 1 = c^2$$
$$6 = c^2$$
$$\sqrt{6} = c$$
Since $0° < \theta < 90°$, $\sin \theta$ is positive, so:
$$\sin \theta = \frac{\sqrt{2} - 1}{\sqrt{6}} = \frac{2\sqrt{3} - \sqrt{6}}{6}$$

61. Using the identities for $\tan \theta$ and $\cot \theta$, we have:

$$\tan \theta + \cot \theta = 2$$

$$\frac{\sin \theta}{\cos \theta} + \frac{\cos \theta}{\sin \theta} = 2$$

$$\frac{\sin^2 \theta + \cos^2 \theta}{\sin \theta \cos \theta} = 2$$

$$1 = 2 \sin \theta \cos \theta$$

Now $(\sin \theta + \cos \theta)^2 = \sin^2 \theta + 2 \sin \theta \cos \theta + \cos^2 \theta = 1 + 1 = 2$, so $\sin \theta + \cos \theta = \sqrt{2}$ since $0° < \theta < 90°$.

63. We draw the figure:

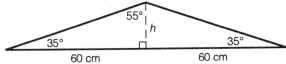

Note that the altitude of the triangle must bisect the upper vertex angle, which is $110°$ $(180° - 35° - 35°)$. Now we find the height h:

$$\tan 35° = \frac{h}{60}, \text{ so } h = 60 \tan 35°$$

Thus the area is given by:

$$\tfrac{1}{2}(\text{base})(\text{height}) = \tfrac{1}{2}(120)(60 \tan 35°) = 3600 \tan 35° \approx 2521 \text{ cm}^2$$

65. We draw the figure:

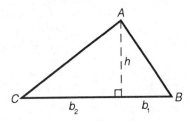

Now:

$$\cot B = \frac{b_1}{h}, \text{ so } b_1 = h \cot B$$

$$\cot C = \frac{b_2}{h}, \text{ so } b_2 = h \cot C$$

So we have:

$$b_1 + b_2 = a$$

$$h \cot B + h \cot C = a$$

$$h(\cot B + \cot C) = a$$

$$h = \frac{a}{\cot B + \cot C}$$

67. We draw the figure:

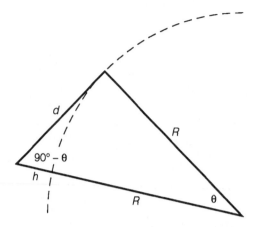

We have:
$$d^2 + R^2 = (R+h)^2$$
$$d^2 = h^2 + 2Rh$$
$$d = \sqrt{2Rh + h^2}$$

Therefore:
$$\cot\theta = \frac{R}{d} = \frac{R}{\sqrt{2Rh + h^2}}$$

69. Using the identities $\sin(90° - \theta) = \cos\theta$ and $\tan(90° - \theta) = \cot\theta$, we have:
$$\sin^2(90° - \theta)\csc\theta - \tan^2(90° - \theta)\sin\theta = \cos^2\theta \cdot \frac{1}{\sin\theta} - \cot^2\theta\sin\theta$$
$$= \frac{\cos^2\theta}{\sin\theta} - \frac{\cos^2\theta}{\sin^2\theta} \cdot \sin\theta$$
$$= \frac{\cos^2\theta}{\sin\theta} - \frac{\cos^2\theta}{\sin\theta}$$
$$= 0$$

71. Working from the right-hand side and using the identity for $\csc A$, we have:
$$\csc A - \sin A = \frac{1}{\sin A} - \sin A = \frac{1 - \sin^2 A}{\sin A} = \frac{\cos^2 A}{\sin A} = \frac{\cos A}{\sin A} \cdot \cos A = \cos A \cot A$$

73. Using the identity for $\cot A$, we have:
$$\frac{\cot A - 1}{\cot A + 1} = \frac{\frac{\cos A}{\sin A} - 1}{\frac{\cos A}{\sin A} + 1} = \frac{\cos A - \sin A}{\cos A + \sin A}$$

75. Simplifying the left-hand side, we have:
$$\frac{1}{1 - \cos A} + \frac{1}{1 + \cos A} = \frac{1 + \cos A + 1 - \cos A}{(1 - \cos A)(1 + \cos A)} = \frac{2}{1 - \cos^2 A} = \frac{2}{\sin^2 A}$$

Simplifying the right-hand side using the identity for $\cot A$, we have:

$$2 + 2\cot^2 A = 2 + 2 \cdot \frac{\cos^2 A}{\sin^2 A} = \frac{2\sin^2 A + 2\cos^2 A}{\sin^2 A} = \frac{2(\sin^2 A + \cos^2 A)}{\sin^2 A} = \frac{2}{\sin^2 A}$$

Since both sides of the equality simplify to the same quantity, the original equality is an identity.

77. First simplifying two of the fractions using the identities for $\sec A$ and $\csc A$, we have:

$$\frac{1}{1 + \sec^2 A} = \frac{1}{1 + \frac{1}{\cos^2 A}} = \frac{\cos^2 A}{\cos^2 A + 1}$$

$$\frac{1}{1 + \csc^2 A} = \frac{1}{1 + \frac{1}{\sin^2 A}} = \frac{\sin^2 A}{\sin^2 A + 1}$$

Replacing the original fractions with these simplified fractions, the left-hand side becomes:

$$\frac{1}{1 + \sin^2 A} + \frac{1}{1 + \cos^2 A} + \frac{\cos^2 A}{1 + \cos^2 A} + \frac{\sin^2 A}{1 + \sin^2 A} = \frac{1 + \sin^2 A}{1 + \sin^2 A} + \frac{1 + \cos^2 A}{1 + \cos^2 A} = 2$$

79. (a)   Since $OP = 1$, then $PN = \sin\theta$.

(b)   Since $OP = 1$, then $ON = \cos\theta$.

(c)   Since $\cos\theta = \dfrac{PN}{PT}$ (note $\theta = \angle TPN$), then $\cos\theta = \dfrac{\sin\theta}{PT}$. Hence $PT = \tan\theta$.

(d)   Here $\cos\theta = \dfrac{1}{OT}$ (using $\triangle OPT$), so $OT = \dfrac{1}{\cos\theta} = \sec\theta$.

(e)   We simplify $NA = 1 - ON = 1 - \cos\theta$.

(f)   Since $NT = OT - ON$, we have:

$$\begin{aligned}
NT &= \sec\theta - \cos\theta \\
&= \frac{1}{\cos\theta} - \cos\theta \\
&= \frac{1 - \cos^2\theta}{\cos\theta} \\
&= \frac{\sin^2\theta}{\cos\theta} \\
&= \sin\theta \cdot \frac{\sin\theta}{\cos\theta} \\
&= \sin\theta\tan\theta
\end{aligned}$$

81. We re-draw the figure (note the labels):

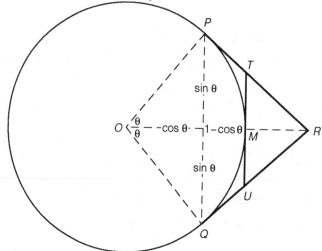

Using $\triangle QOR$:

$$\cos \theta = \frac{1}{1 + MR}$$

$$1 + MR = \frac{1}{\cos \theta}$$

$$MR = \frac{1}{\cos \theta} - 1$$

Since $\triangle RTU$ and $\triangle RPQ$ are similar, then the corresponding sides are proportional:

$$\frac{TU}{PQ} = \frac{MR}{1 - \cos \theta + MR}$$

But $PQ = 2 \sin \theta$ and $MR = \dfrac{1}{\cos \theta} - 1$, so we have:

$$\frac{TU}{2 \sin \theta} = \frac{\frac{1}{\cos \theta} - 1}{1 - \cos \theta + \frac{1}{\cos \theta} - 1} = \frac{1 - \cos \theta}{1 - \cos^2 \theta} = \frac{1 - \cos \theta}{\sin^2 \theta}$$

$$TU = \frac{2(1 - \cos \theta)}{\sin \theta}$$

Thus the area is given by:

$$A = \frac{1}{2}(TU)(MR)$$

$$= \frac{1}{2} \cdot \frac{2(1 - \cos \theta)}{\sin \theta} \cdot \left(\frac{1}{\cos \theta} - 1 \right)$$

$$= \frac{1 - \cos \theta}{\sin \theta} \cdot \frac{1 - \cos \theta}{\cos \theta}$$

$$= \frac{(1 - \cos \theta)^2}{\sin \theta \cos \theta}$$

83. (a) Using $\triangle ODC$, $OD = \cos\theta$ and $DC = \sin\theta$ (recall that $OC = 1$).

Thus, using $\triangle ADC$, $\tan\dfrac{\theta}{2} = \dfrac{CD}{AD} = \dfrac{\sin\theta}{1+\cos\theta}$, since $AO = 1$.

(b) We compute $\tan 15°$ using the formula:

$$\tan 15° = \frac{\sin 30°}{1+\cos 30°} = \frac{\frac{1}{2}}{1+\frac{\sqrt{3}}{2}} = \frac{1}{2+\sqrt{3}} \cdot \frac{2-\sqrt{3}}{2-\sqrt{3}} = \frac{2-\sqrt{3}}{4-3} = 2-\sqrt{3}$$

We compute $\tan 22.5°$ using the formula:

$$\tan 22.5° = \frac{\sin 45°}{1+\cos 45°} = \frac{\frac{\sqrt{2}}{2}}{1+\frac{\sqrt{2}}{2}} = \frac{\sqrt{2}}{2+\sqrt{2}} \cdot \frac{2-\sqrt{2}}{2-\sqrt{2}} = \frac{2\sqrt{2}-2}{4-2} = \sqrt{2}-1$$

Chapter Six Test

1. (a) Since the terminal side of $30°$ lies in the first quadrant, $\tan 30° > 0$, so $\tan 30° = \frac{\sqrt{3}}{3}$.

(b) Since the terminal side of $45°$ lies in the first quadrant, $\sec 45° > 0$, so $\sec 45° = \sqrt{2}$.

(c) Since $\sin^2\theta + \cos^2\theta = 1$ for all angles θ, then $\sin^2 25° + \cos^2 25° = 1$.

(d) Since $\sin 53° = \cos(90°-53°) = \cos 37°$, then:
$$\sin 53° - \cos 37° = \cos 37° - \cos 37° = 0$$

2. (a) Since $\cos\theta$ is negative, then $\sec\theta$ is negative.

(b) Since $\sin\theta$ is negative, then $\csc\theta$ is negative.

(c) Since both $\sin\theta$ and $\cos\theta$ are negative, then $\cot\theta$ is positive.

3. (a) Since $-270°$ corresponds to the point $(0,1)$ on the unit circle, then $\sin(-270°) = 1$.

(b) Since $180°$ corresponds to the point $(-1,0)$ on the unit circle, then $\cos 180° = -1$.

(c) Since $720°$ corresponds to the point $(1,0)$ on the unit circle, then:
$$\tan 720° = \frac{y}{x} = \frac{0}{1} = 0$$

4. Factoring as a trinomial, we have:
$$2\cot^2\theta + 11\cot\theta + 12 = (2\cot\theta + 3)(\cot\theta + 4)$$

5. Using the formula for area, we have:
$$\tfrac{1}{2}ab\sin\theta = \tfrac{1}{2}(8)(9)\sin 150° = 36 \cdot \tfrac{1}{2} = 18 \text{ cm}^2$$

6. (a) Since the terminal side of $-225°$ lies in quadrant 2, $\sin(-225°) > 0$. Using a reference angle of $45°$, we have:
$$\sin(-225°) = \sin 45° = \frac{\sqrt{2}}{2}$$

(b) Since the terminal side of 330° lies in quadrant 4, tan 330° < 0. Using a reference angle of 30°, we have:
$$\tan 330° = -\tan 30° = -\frac{\sqrt{3}}{3}$$

(c) Since the terminal side of 120° lies in quadrant 2, sec 120° < 0. Using a reference angle of 60°, we have:
$$\sec 120° = -\sec 60° = -2$$

7. We draw a triangle:

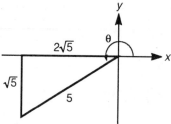

Using the Pythagorean theorem, we have:
$$a^2 + \left(\sqrt{5}\right)^2 = 5^2$$
$$a^2 + 5 = 25$$
$$a^2 = 20$$
$$a = 2\sqrt{5}$$

Since $180° < \theta < 270°$, $\cos \theta < 0$ and $\tan \theta > 0$, so:
$$\cos \theta = -\frac{2\sqrt{5}}{5} \quad \text{and} \quad \tan \theta = \frac{\sqrt{5}}{2\sqrt{5}} = \frac{1}{2}$$

8. We draw a triangle:

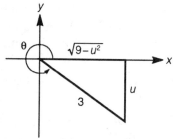

Using the Pythagorean theorem, we have:
$$a^2 + u^2 = 3^2$$
$$a^2 + u^2 = 9$$
$$a^2 = 9 - u^2$$
$$a = \sqrt{9 - u^2}$$

Since $270° < \theta < 360°$, $\cos \theta > 0$ and $\cot \theta < 0$, so:
$$\cos \theta = \frac{\sqrt{9 - u^2}}{3} \quad \text{and} \quad \cot \theta = -\frac{\sqrt{9 - u^2}}{u}$$

9.   Multiplying the numerator and denominator by $\cos\theta$, we have:

$$\frac{\frac{\cos\theta+1}{\cos\theta}+1}{\frac{\cos\theta-1}{\cos\theta}-1}=\frac{\cos\theta+1+\cos\theta}{\cos\theta-1-\cos\theta}=\frac{2\cos\theta+1}{-1}=-2\cos\theta-1$$

10.   The central angle of a triangle drawn from the center to two adjacent vertices is given by $9\theta=360°$, so $\theta=40°$. The area of this triangle is therefore:

$$\tfrac{1}{2}ab\sin\theta=\tfrac{1}{2}(3)(3)\sin40°=\tfrac{9}{2}\sin40°$$

Since there are nine such triangles which form the polygon, its area is:

$$9\!\left(\tfrac{9}{2}\sin40°\right)=\tfrac{81}{2}\sin40°\approx26.033\text{ m}^2$$

11.   Using the identities for $\cot\theta$, $\tan\theta$, $\csc\theta$ and $\sec\theta$, we have:

$$\frac{\cot^2\theta}{\csc^2\theta}+\frac{\tan^2\theta}{\sec^2\theta}=\frac{\frac{\cos^2\theta}{\sin^2\theta}}{\frac{1}{\sin^2\theta}}+\frac{\frac{\sin^2\theta}{\cos^2\theta}}{\frac{1}{\cos^2\theta}}=\cos^2\theta+\sin^2\theta=1$$

12.   We will find BD and BC. Using the smaller right triangle, we have:

$$\tan25°=\frac{BD}{50},\text{ so }BD=50\tan25°$$

Using the larger right triangle, we have:

$$\tan55°=\frac{BC}{50},\text{ so }BC=50\tan55°$$

Since $CD=BC-BD$, then:

$$CD=50\tan55°-50\tan25°=50\left(\tan55°-\tan25°\right)$$

13.   (a)   Since $85°$ has a much larger y-coordinate than $5°$, $\sin85°$ is larger.

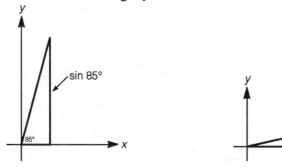

(b)   Since the x-coordinate is much larger at $5°$ than the y-coordinate, $\cos5°$ is larger.

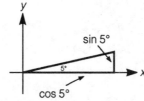

(c) Since the terminal side for 175° lies in quadrant 2, $\tan 175° < 0$. Since the terminal side for 185° lies in quadrant 3, $\tan 185° > 0$. Thus $\tan 185°$ is larger. Note that a sketch isn't necessary to make this comparison.

14. (a) To find AD, we use $\triangle ADC$:
$$\cos 20° = \frac{AD}{2.75}, \text{ so } AD = 2.75\cos 20°$$
To find CD, we use $\triangle ADC$:
$$\sin 20° = \frac{CD}{2.75}, \text{ so } CD = 2.75\sin 20°$$
To find DB, note that $DB = AB - AD$, so:
$$DB = 3.25 - 2.75\cos 20°$$

(b) Using the Pythagorean theorem in $\triangle CDB$, we have:
$$(CB)^2 = (DB)^2 + (CD)^2$$
$$(CB)^2 = (3.25 - 2.75\cos 20°)^2 + (2.75\sin 20°)^2$$
$$(CB)^2 = 10.5625 - 17.875\cos 20° + 7.5625\cos^2 20° + 7.5625\sin^2 20°$$
$$(CB)^2 = 10.5625 - 17.875\cos 20° + 7.5625(\cos^2 20° + \sin^2 20°)$$
$$(CB)^2 = 10.5625 - 17.875\cos 20° + 7.5625$$
$$(CB)^2 = 18.125 - 17.875\cos 20°$$
$$CB = \sqrt{18.125 - 17.875\cos 20°} \approx 1.15 \text{ cm}$$

Chapter Seven
Trigonometric Functions of Real Numbers

7.1 Radian Measure

1. (a) Multiplying by the conversion factor $\frac{\pi}{180°}$, we have:

   $$45° \cdot \frac{\pi}{180°} = \frac{\pi}{4} \approx 0.79 \text{ radians}$$

   (b) Multiplying by the conversion factor $\frac{\pi}{180°}$, we have:

   $$90° \cdot \frac{\pi}{180°} = \frac{\pi}{2} \approx 1.57 \text{ radians}$$

   (c) Multiplying by the conversion factor $\frac{\pi}{180°}$, we have:

   $$135° \cdot \frac{\pi}{180°} = \frac{3\pi}{4} \approx 2.36 \text{ radians}$$

3. (a) Multiplying by the conversion factor $\frac{\pi}{180°}$, we have:

   $$0° \cdot \frac{\pi}{180°} = 0 \text{ radians}$$

   (b) Multiplying by the conversion factor $\frac{\pi}{180°}$, we have:

   $$360° \cdot \frac{\pi}{180°} = 2\pi \approx 6.28 \text{ radians}$$

   (c) Multiplying by the conversion factor $\frac{\pi}{180°}$, we have:

   $$450° \cdot \frac{\pi}{180°} = \frac{5\pi}{2} \approx 7.85 \text{ radians}$$

5.   (a)   Multiplying by the conversion factor $\frac{180°}{\pi}$, we have:

$$\frac{\pi}{12} \cdot \frac{180°}{\pi} = 15°$$

(b)   Multiplying by the conversion factor $\frac{180°}{\pi}$, we have:

$$\frac{\pi}{6} \cdot \frac{180°}{\pi} = 30°$$

(c)   Multiplying by the conversion factor $\frac{180°}{\pi}$, we have:

$$\frac{\pi}{4} \cdot \frac{180°}{\pi} = 45°$$

7.   (a)   Multiplying by the conversion factor $\frac{180°}{\pi}$, we have:

$$\frac{\pi}{3} \cdot \frac{180°}{\pi} = 60°$$

(b)   Multiplying by the conversion factor $\frac{180°}{\pi}$, we have:

$$\frac{5\pi}{3} \cdot \frac{180°}{\pi} = 300°$$

(c)   Multiplying by the conversion factor $\frac{180°}{\pi}$, we have:

$$4\pi \cdot \frac{180°}{\pi} = 720°$$

9.   (a)   Multiplying by the conversion factor $\frac{180°}{\pi}$, we have:

$$2 \cdot \frac{180°}{\pi} = \frac{360°}{\pi} \approx 114.59°$$

(b)   Multiplying by the conversion factor $\frac{180°}{\pi}$, we have:

$$3 \cdot \frac{180°}{\pi} = \frac{540°}{\pi} \approx 171.89°$$

(c)   Multiplying by the conversion factor $\frac{180°}{\pi}$, we have:

$$\pi^2 \cdot \frac{180°}{\pi} = 180\pi° \approx 565.49°$$

11. Since a right angle has radian measure of $\frac{\pi}{2}$, which is larger than $\frac{3}{2}$ (since π is larger than 3), this angle is smaller than a right angle.

13. Multiplying by the conversion factor $\frac{\pi}{180°}$, we have:

$$30° = 30° \cdot \frac{\pi}{180°} = \frac{\pi}{6} \text{ radians}$$
$$45° = 45° \cdot \frac{\pi}{180°} = \frac{\pi}{4} \text{ radians}$$
$$60° = 60° \cdot \frac{\pi}{180°} = \frac{\pi}{3} \text{ radians}$$
$$120° = 120° \cdot \frac{\pi}{180°} = \frac{2\pi}{3} \text{ radians}$$
$$135° = 135° \cdot \frac{\pi}{180°} = \frac{3\pi}{4} \text{ radians}$$
$$150° = 150° \cdot \frac{\pi}{180°} = \frac{5\pi}{6} \text{ radians}$$

15. Using the formula $\theta = \frac{s}{r}$, where $s = 5$ cm and $r = 2$ cm, we have:

$$\theta = \frac{s}{r} = \frac{5 \text{ cm}}{2 \text{ cm}} = 2.5 \text{ radians}$$

17. Using the formula $\theta = \frac{s}{r}$ where $s = 200$ cm $= 2$ m and $r = 1$ m, we have:

$$\theta = \frac{s}{r} = \frac{2 \text{ m}}{1 \text{ m}} = 2 \text{ radians}$$

19. From the figure we have $0.5 < \cos 1 < 0.6$ and $0.8 < \sin 1 < 0.9$. Using a calculator, we have $\cos 1 \approx 0.54$ and $\sin 1 \approx 0.84$.

21. From the figure we have $0.5 < \cos(-1) < 0.6$ and $-0.9 < \sin(-1) < -0.8$. Using a calculator, we have $\cos(-1) \approx 0.54$ and $\sin(-1) \approx -0.84$.

23. From the figure we have $-0.7 < \cos 4 < -0.6$ and $-0.8 < \sin 4 < -0.7$. Using a calculator, we have $\cos 4 \approx -0.65$ and $\sin 4 \approx -0.76$.

25. From the figure, we have $-0.7 < \cos(-4) < -0.6$ and $0.7 < \sin(-4) < 0.8$. Using a calculator, we have $\cos(-4) \approx -0.65$ and $\sin(-4) \approx 0.76$.

27. Since $\sin(1 + 2\pi) = \sin 1$, we can use the figure to obtain $0.8 < \sin(1 + 2\pi) < 0.9$. Using a calculator, we have $\sin(1 + 2\pi) \approx 0.84$.

29. Using the formula $s = r\theta$ where $r = 3$ ft and $\theta = \frac{4\pi}{3}$, we have:

$$s = r\theta = 3 \cdot \frac{4\pi}{3} = 4\pi \text{ ft}$$

31. We must first convert $45°$ to radian measure:

$$45° = 45° \cdot \frac{\pi}{180°} = \frac{\pi}{4} \text{ radians}$$

Now using the formula $s = r\theta$ where $r = 2$ cm and $\theta = \frac{\pi}{4}$, we have:

$$s = r\theta = 2 \cdot \frac{\pi}{4} = \frac{\pi}{2} \text{ cm}$$

33. (a) The area is given by:
$$A = \tfrac{1}{2}r^2\theta = \tfrac{1}{2}(6)^2\left(\tfrac{2\pi}{3}\right) = 12\pi \text{ cm}^2 \approx 37.70 \text{ cm}^2$$

(b) We first convert θ to radian measure:
$$\theta = 80° \cdot \tfrac{\pi}{180°} = \tfrac{4\pi}{9} \text{ radians}$$
The area is given by:
$$A = \tfrac{1}{2}r^2\theta = \tfrac{1}{2}(5)^2\left(\tfrac{4\pi}{9}\right) = \tfrac{50\pi}{9} \text{ m}^2 \approx 17.45 \text{ m}^2$$

(c) The area is given by:
$$A = \tfrac{1}{2}r^2\theta = \tfrac{1}{2}(24)^2\left(\tfrac{\pi}{20}\right) = \tfrac{72\pi}{5} \text{ m}^2 \approx 45.24 \text{ m}^2$$

(d) We first convert θ to radian measure:
$$\theta = 144° \cdot \tfrac{\pi}{180°} = \tfrac{4\pi}{5} \text{ radians}$$
The area is given by:
$$A = \tfrac{1}{2}r^2\theta = \tfrac{1}{2}(1.8)^2\left(\tfrac{4\pi}{5}\right) = \tfrac{12.96\pi}{10} = 1.296\pi \text{ cm}^2 \approx 4.07 \text{ cm}^2$$

35. We have $r = 1$ cm and $A = \tfrac{\pi}{5}$ cm^2. Substituting into $A = \tfrac{1}{2}r^2\theta$, we have:
$$\tfrac{\pi}{5} = \tfrac{1}{2}(1)^2\theta$$
$$\tfrac{\pi}{5} = \tfrac{\theta}{2}$$
$$\theta = \tfrac{2\pi}{5} \text{ radians}$$

37. (a) Each revolution of the wheel is 2π radians, so in 6 revolutions there are $\theta = 6(2\pi) = 12\pi$ radians. Consequently, we have:
$$\omega = \tfrac{\theta}{t} = \tfrac{12\pi \text{ radians}}{1 \text{ sec}} = 12\pi \, \tfrac{\text{radians}}{\text{sec}}$$

(b) Using the formula $s = r\theta$, where $r = 12$ cm and $\theta = 12\pi$ radians, we have $s = (12 \text{ cm})(12\pi \text{ radians}) = 144\pi$ cm. The linear speed, therefore, is:
$$v = \tfrac{d}{t} = \tfrac{144\pi \text{ cm}}{1 \text{ sec}} = 144\pi \, \tfrac{\text{cm}}{\text{sec}}$$

(c) Using the formula $s = r\theta$, where $r = 6$ cm and $\theta = 12\pi$ radians, we have $s = (6 \text{ cm})(12\pi \text{ radians}) = 72\pi$ cm. Thus, we have:
$$v = \tfrac{d}{t} = \tfrac{72\pi \text{ cm}}{1 \text{ sec}} = 72\pi \, \tfrac{\text{cm}}{\text{sec}}$$

39. (a) In 1 second, the wheel has rotated $1080° \cdot \tfrac{\pi \text{ radians}}{180°} = 6\pi$ radians.
Consequently, we have:
$$\omega = \tfrac{\theta}{t} = \tfrac{6\pi \text{ radians}}{1 \text{ sec}} = 6\pi \, \tfrac{\text{radians}}{\text{sec}}$$

(b)   Using $\theta = 6\pi$ radians and $r = 25$ cm, then:
$$s = r\theta = (25 \text{ cm})(6\pi \text{ radians}) = 150\pi \text{ cm}$$
Thus, we have:
$$v = \frac{150\pi \text{ cm}}{1 \sec} = 150\pi \, \frac{\text{cm}}{\sec}$$

(c)   Using $\theta = 6\pi$ radians and $r = \frac{25}{2}$ cm, then:
$$s = r\theta = \left(\frac{25}{2} \text{ cm}\right)(6\pi \text{ radians}) = 75\pi \text{ cm}$$
Thus, we have:
$$v = \frac{75\pi \text{ cm}}{1 \sec} = 75\pi \, \frac{\text{cm}}{\sec}$$

41. (a)   In 1 minute, the wheel has rotated 500 rev. Since each revolution is equal to 2π radians, we have:
$$\theta = (500)(2\pi) = 1000\pi \text{ radians}$$
Consequently, we have:
$$\omega = \frac{\theta}{t} = \frac{1000\pi \text{ radians}}{60 \sec} = \frac{50\pi}{3} \, \frac{\text{radians}}{\sec}$$

(b)   Using $\theta = 1000\pi$ radians and $r = 45$ cm, then:
$$s = r\theta = (45 \text{ cm})(1000\pi \text{ radians}) = 45000\pi \text{ cm}$$
Thus, we have:
$$v = \frac{45000\pi \text{ cm}}{60 \sec} = 750\pi \, \frac{\text{cm}}{\sec}$$

(c)   Using $\theta = 1000\pi$ radians and $r = \frac{45}{2}$ cm, then:
$$s = r\theta = \left(\frac{45}{2} \text{ cm}\right)(1000\pi \text{ radians}) = 22500\pi \text{ cm}$$
Thus, we have:
$$v = \frac{22500\pi \text{ cm}}{60 \sec} = 375\pi \, \frac{\text{cm}}{\sec}$$

43.   Since the triangle is equilateral with sides 1 ft, the circle formed will have a radius of 1 ft and arc length of 1 ft, so:
$$\theta = \frac{s}{r} = \frac{1}{1} = 1 \text{ radian}$$
Converting to degrees, the angle at A will have a measure of:
$$1 \cdot \frac{180°}{\pi} = \frac{180°}{\pi} \approx 57.30°$$

45. (a)   The angular speed of the larger wheel is given by:
$$\omega = 100 \, \frac{\text{rev}}{\text{min}} \cdot 2\pi \, \frac{\text{rad}}{\text{rev}} = 200\pi \, \frac{\text{rad}}{\text{min}}$$

(b)   In 1 minute, a point on the larger wheel has traveled:
$$s = R\theta = (10 \text{ cm})(200\pi \text{ rad}) = 2000\pi \text{ cm}$$

So the linear speed is given by:
$$v = \frac{2000\pi\,\text{cm}}{1\,\text{min}} = 2000\pi\,\tfrac{\text{cm}}{\text{min}}$$

(c) Since the linear speed on the smaller wheel must also be $2000\pi\,\tfrac{\text{cm}}{\text{min}}$, and the radius of the smaller wheel is $r = 6$ cm, its angular speed is:
$$\omega = \tfrac{v}{r} = \frac{2000\pi\,\text{cm/min}}{6\,\text{cm}} = \tfrac{1000}{3}\pi\,\tfrac{\text{radians}}{\text{min}}$$

(d) In rpm, the angular speed of the smaller wheel is:
$$\frac{1\,\text{rev}}{2\pi\,\text{rad}} \cdot \tfrac{1000}{3}\pi\,\tfrac{\text{rad}}{\text{min}} = \tfrac{500}{3}\,\text{rpm}$$

47. We first convert θ to radian measure:
$$\theta = 71°23' = 71\tfrac{23}{60}° \cdot \tfrac{\pi}{180°} = \tfrac{4283\pi}{10800}$$
Now compute the arc length:
$$s = r\theta = 3960\,\text{mi} \cdot \tfrac{4283\pi}{10800} = \tfrac{47113\pi}{30}\,\text{mi} \approx 4930\,\text{mi}$$

49. We first convert θ to radian measure:
$$\theta = 21°19' = 21\tfrac{19}{60}° \cdot \tfrac{\pi}{180°} = \tfrac{1279\pi}{10800}$$
Now compute the arc length:
$$s = r\theta = 3960\,\text{mi} \cdot \tfrac{1279\pi}{10800} = \tfrac{14069\pi}{30}\,\text{mi} \approx 1470\,\text{mi}$$

51. We first convert θ to radian measure:
$$\theta = 38°54' = 38\tfrac{54}{60}° \cdot \tfrac{\pi}{180°} = \tfrac{2334\pi}{10800}$$
Now compute the arc length:
$$s = r\theta = 3960\,\text{mi} \cdot \tfrac{2334\pi}{10800} = \tfrac{8558\pi}{10}\,\text{mi} \approx 2690\,\text{mi}$$

53. (a) We first find the arc length $s = r(2\theta) = 2r\theta$. Since the perimeter is 12 cm, we have:
$$r + r + 2r\theta = 12$$
$$2r + 2r\theta = 12$$
$$r + r\theta = 6$$
$$r(1 + \theta) = 6$$
$$r = \tfrac{6}{1+\theta}$$

(b) The area is given by:
$$A = \tfrac{1}{2}r^2(2\theta) = \tfrac{1}{2}\left(\tfrac{6}{1+\theta}\right)^2(2\theta) = \tfrac{36\theta}{(1+\theta)^2}$$

7.2 Trigonometric Functions of Real Numbers

1. We complete the table:

θ	$\sin\theta$	$\cos\theta$	$\tan\theta$	$\csc\theta$	$\sec\theta$	$\cot\theta$
0	0	1	0	undef.	1	undef.
$\pi/6$	$1/2$	$\sqrt{3}/2$	$\sqrt{3}/3$	2	$2\sqrt{3}/3$	$\sqrt{3}$
$\pi/4$	$\sqrt{2}/2$	$\sqrt{2}/2$	1	$\sqrt{2}$	$\sqrt{2}$	1
$\pi/3$	$\sqrt{3}/2$	$1/2$	$\sqrt{3}$	$2\sqrt{3}/3$	2	$\sqrt{3}/3$
$\pi/2$	1	0	undef.	1	undef.	0
$2\pi/3$	$\sqrt{3}/2$	$-1/2$	$-\sqrt{3}$	$2\sqrt{3}/3$	-2	$-\sqrt{3}/3$
$3\pi/4$	$\sqrt{2}/2$	$-\sqrt{2}/2$	-1	$\sqrt{2}$	$-\sqrt{2}$	-1
$5\pi/6$	$1/2$	$-\sqrt{3}/2$	$-\sqrt{3}/3$	2	$-2\sqrt{3}/3$	$-\sqrt{3}$
π	0	-1	0	undef.	-1	undef.

3. (a) Checking the identity $\sin^2 t + \cos^2 t = 1$, we have:

$$\sin^2 \tfrac{\pi}{3} + \cos^2 \tfrac{\pi}{3} = \left(\tfrac{\sqrt{3}}{2}\right)^2 + \left(\tfrac{1}{2}\right)^2 = \tfrac{3}{4} + \tfrac{1}{4} = 1$$

(b) Checking the identity $\sin^2 t + \cos^2 t = 1$, we have:

$$\sin^2 \tfrac{5\pi}{4} + \cos^2 \tfrac{5\pi}{4} = \left(-\tfrac{\sqrt{2}}{2}\right)^2 + \left(-\tfrac{\sqrt{2}}{2}\right)^2 = \tfrac{2}{4} + \tfrac{2}{4} = 1$$

(c) Checking the identity $\sin^2 t + \cos^2 t = 1$ and using a calculator, we have:

$$\sin^2(-53) + \cos^2(-53) \approx (-0.3959)^2 + (-0.9183)^2 \approx 0.1568 + 0.8432 = 1$$

5. (a) Checking the identity $\cot^2 t + 1 = \csc^2 t$, we have:

$$\cot^2\left(-\tfrac{\pi}{6}\right) + 1 = \left(-\sqrt{3}\right)^2 + 1 = 3 + 1 = 4$$
$$\csc^2\left(-\tfrac{\pi}{6}\right) = (-2)^2 = 4$$

(b) Checking the identity $\cot^2 t + 1 = \csc^2 t$, we have:

$$\cot^2\left(\tfrac{7\pi}{4}\right) + 1 = (-1)^2 + 1 = 1 + 1 = 2$$
$$\csc^2\left(\tfrac{7\pi}{4}\right) = \left(-\sqrt{2}\right)^2 = 2$$

(c) Checking the identity $\cot^2 t + 1 = \csc^2 t$ and using a calculator, we have:

$$\cot^2(0.12) + 1 \approx 68.7787 + 1 = 69.7787$$
$$\csc^2(0.12) \approx 69.7787$$

7.   (a)   Checking the identity $\sin(-t) = -\sin t$, we have:
$$\sin\left(-\tfrac{3\pi}{2}\right) = 1$$
$$-\sin\tfrac{3\pi}{2} = -(-1) = 1$$

(b)   Checking the identity $\sin(-t) = -\sin t$, we have:
$$\sin\tfrac{5\pi}{6} = \tfrac{1}{2}$$
$$-\sin\left(-\tfrac{5\pi}{6}\right) = -\left(-\tfrac{1}{2}\right) = \tfrac{1}{2}$$

(c)   Checking the identity $\sin(-t) = -\sin t$, and using a calculator, we have:
$$\sin(-13.24) \approx -0.6238$$
$$-\sin 13.24 \approx -(0.6238) \doteq -0.6238$$

9.   (a)   Checking the identity $\sin(t + 2\pi) = \sin t$, we have:
$$\sin\left(\tfrac{5\pi}{3} + 2\pi\right) = \sin\left(\tfrac{11\pi}{3}\right) = -\tfrac{\sqrt{3}}{2}$$
$$\sin\tfrac{5\pi}{3} = -\tfrac{\sqrt{3}}{2}$$

(b)   Checking the identity $\sin(t + 2\pi) = \sin t$, we have:
$$\sin\left(-\tfrac{3\pi}{2} + 2\pi\right) = \sin\tfrac{\pi}{2} = 1$$
$$\sin\left(-\tfrac{3\pi}{2}\right) = 1$$

(c)   Checking the identity $\sin(t + 2\pi) = \sin t$ and using a calculator, we have:
$$\sin(\sqrt{19} + 2\pi) \approx \sin(10.6421) \approx -0.9382$$
$$\sin(\sqrt{19}) \approx \sin(4.3589) \approx -0.9382$$

11.   Evaluating each side of the "equality" when $t = \tfrac{\pi}{6}$, we have:
$$\cos\left(2 \cdot \tfrac{\pi}{6}\right) = \cos\tfrac{\pi}{3} = \tfrac{1}{2}$$
$$2\cos\tfrac{\pi}{6} = 2 \cdot \tfrac{\sqrt{3}}{2} = \sqrt{3}$$
Since these results are unequal, $\cos 2t = 2\cos t$ is not an identity.

13.   Using $\sin^2 t + \cos^2 t = 1$, we have:
$$\cos t = \pm\sqrt{1 - \sin^2 t} = \pm\sqrt{1 - \tfrac{9}{25}} = \pm\tfrac{4}{5}$$

Since $\pi < t < \tfrac{3\pi}{2}$, $\cos t = -\tfrac{4}{5}$ and thus:
$$\tan t = \frac{\sin t}{\cos t} = \frac{-\tfrac{3}{5}}{-\tfrac{4}{5}} = \tfrac{3}{4}$$

15. Since $\frac{\pi}{2} < t < \pi$, $\cos t < 0$ and thus:

$$\cos t = -\sqrt{1 - \sin^2 t} = -\sqrt{1 - \frac{3}{16}} = -\sqrt{\frac{13}{16}} = -\frac{\sqrt{13}}{4}$$

Thus:

$$\tan t = \frac{\sin t}{\cos t} = \frac{\frac{\sqrt{3}}{4}}{-\frac{\sqrt{13}}{4}} = -\frac{\sqrt{3}}{\sqrt{13}} = -\frac{\sqrt{39}}{13}$$

17. Using the identity $1 + \tan^2 \alpha = \sec^2 \alpha$, we have:

$$1 + \left(\frac{12}{5}\right)^2 = \sec^2 \alpha$$
$$1 + \frac{144}{25} = \sec^2 \alpha$$
$$\frac{169}{25} = \sec^2 \alpha$$
$$\pm\frac{13}{5} = \sec \alpha$$

Since $\cos \alpha > 0$, $\sec \alpha > 0$ and thus we pick $\sec \alpha = \frac{13}{5}$. Thus $\cos \alpha = \frac{5}{13}$, and we use $\sin^2 \alpha + \cos^2 \alpha = 1$ to get:

$$\sin^2 \alpha = 1 - \left(\frac{5}{13}\right)^2 = \frac{144}{169}, \text{ thus } \sin \alpha = \pm\frac{12}{13}$$

We pick the positive value since, if both the tangent and cosine are positive, so is the sine. Thus $\sin \alpha = \frac{12}{13}$.

19. For $0 < \theta < \frac{\pi}{2}$, we have:

$$\sqrt{9 - x^2} = \sqrt{9 - (3\sin \theta)^2} = \sqrt{9(1 - \sin^2 \theta)} = \sqrt{9\cos^2 \theta} = 3\cos \theta$$

We choose the positive root since $\cos \theta > 0$ for $0 < \theta < \frac{\pi}{2}$.

21. Since $0 < \theta < \frac{\pi}{2}$, $\tan \theta > 0$. Thus we have:

$$\frac{1}{\left(u^2 - 25\right)^{3/2}} = \frac{1}{\left(25\sec^2 \theta - 25\right)^{3/2}} = \frac{1}{125\tan^3 \theta} = \frac{\cot^3 \theta}{125}$$

23. Since $\sec \theta > 0$, we have:

$$\frac{1}{\sqrt{u^2 + 7}} = \frac{1}{\sqrt{7\tan^2 \theta + 7}} = \frac{1}{\sqrt{7}\sec \theta} = \frac{\cos \theta}{\sqrt{7}} = \frac{\sqrt{7}\cos \theta}{7}$$

25. (a) Since $\sin(-t) = -\sin t$, we have:

$$\sin(-t) = -\sin t = -\frac{2}{3}$$

(b) Since $\sin(-\phi) = -\sin \phi$, we have:

$$\sin(-\phi) = -\sin \phi = -\left(-\frac{1}{4}\right) = \frac{1}{4}$$

(c)   Since $\cos(-\alpha) = \cos\alpha$, we have:
$$\cos(-\alpha) = \cos\alpha = \tfrac{1}{5}$$

(d)   Since $\cos(-s) = \cos s$, we have:
$$\cos(-s) = \cos s = -\tfrac{1}{5}$$

27.   (a)   Since $\cos t = -\tfrac{1}{3}$ and $\tfrac{\pi}{2} < t < \pi$, we have $\sin t > 0$ and thus:
$$\sin t = \sqrt{1 - \tfrac{1}{9}} = \sqrt{\tfrac{8}{9}} = \tfrac{2\sqrt{2}}{3}$$
Therefore we have the values:
$$\sin(-t) = -\sin t = -\tfrac{2\sqrt{2}}{3}$$
$$\cos(-t) = \cos t = -\tfrac{1}{3}$$
Thus:
$$\sin(-t) + \cos(-t) = -\tfrac{2\sqrt{2}}{3} - \tfrac{1}{3} = -\tfrac{1+2\sqrt{2}}{3}$$

(b)   Note that $\sin^2(-t) + \cos^2(-t) = 1$, regardless of the value of t.

29.   (a)   Using the identity $\cos(\theta + 2\pi k) = \cos\theta$, we have:
$$\cos\left(\tfrac{\pi}{4} + 2\pi\right) = \cos\tfrac{\pi}{4} = \tfrac{\sqrt{2}}{2}$$

(b)   Using the identity $\sin(\theta + 2\pi k) = \sin\theta$, we have:
$$\sin\left(\tfrac{\pi}{3} + 2\pi\right) = \sin\tfrac{\pi}{3} = \tfrac{\sqrt{3}}{2}$$

(c)   Using the identity $\sin(\theta + 2\pi k) = \sin\theta$, we have:
$$\sin\left(\tfrac{\pi}{2} - 6\pi\right) = \sin\tfrac{\pi}{2} = 1$$

31.   Using the Pythagorean identities, we have:
$$\frac{\sin^2 t + \cos^2 t}{\tan^2 t + 1} = \frac{1}{\sec^2 t} = \cos^2 t$$

33.   Using the Pythagorean identities, we have:
$$\frac{\sec^2\theta - \tan^2\theta}{1 + \cot^2\theta} = \frac{\tan^2\theta + 1 - \tan^2\theta}{\csc^2\theta} = \frac{1}{\csc^2\theta} = \sin^2\theta$$

35.   Working from the right-hand side and using the identity for $\cot t$, we have:
$$\sin t + \cot t \cos t = \sin t + \frac{\cos t}{\sin t}(\cos t) = \frac{\sin^2 t}{\sin t} + \frac{\cos^2 t}{\sin t} = \frac{\sin^2 t + \cos^2 t}{\sin t} = \frac{1}{\sin t} = \csc t$$

37. Combining fractions on the left-hand side, we have:

$$\frac{1}{1+\sec s}+\frac{1}{1-\sec s}=\frac{(1-\sec s)+(1+\sec s)}{(1-\sec s)(1+\sec s)}=\frac{2}{1-\sec^2 s}=\frac{-2}{\tan^2 s}=-2\cot^2 s$$

39. Denoting $\sin\theta$ by S and $\cos\theta$ by C, we have:

$$\frac{\cot\theta}{1-\tan\theta}+\frac{\tan\theta}{1-\cot\theta}=\frac{\frac{C}{S}}{1-\frac{S}{C}}+\frac{\frac{S}{C}}{1-\frac{C}{S}}$$

$$=\frac{C^2}{S(C-S)}-\frac{S^2}{C(C-S)}$$

$$=\frac{C^3-S^3}{SC(C-S)}$$

$$=\frac{(C-S)(C^2+SC+S^2)}{SC(C-S)}$$

$$=\frac{1+SC}{SC}$$

$$=\frac{1}{SC}+1$$

Now we replace C with $\cos\theta$ and S with $\sin\theta$:

$$\frac{1}{SC}+1=\frac{\cos^2\theta+\sin^2\theta}{\sin\theta\cos\theta}+1=\frac{\cos\theta}{\sin\theta}+\frac{\sin\theta}{\cos\theta}+1=\cot\theta+\tan\theta+1$$

41. The expression on the left-hand side becomes:

$$\tan\theta-(\tan\theta\cot\theta)(\cot\theta)+\cot\theta-(\cot\theta\tan\theta)(\tan\theta)$$

This multiplies out to become:

$$\tan\theta-\cot\theta+\cot\theta-\tan\theta=0$$

43. If $\sec t=\frac{13}{5}$, then $\cos t=\frac{5}{13}$. Given $\frac{3\pi}{2}<t<2\pi$, then $\sin t<0$ (fourth quadrant), thus:

$$\sin t=-\sqrt{1-\cos^2 t}=-\sqrt{1-\frac{25}{169}}=-\sqrt{\frac{144}{169}}=-\frac{12}{13}$$

So we have:

$$\frac{2\sin t-3\cos t}{4\sin t-9\cos t}=\frac{2\left(-\frac{12}{13}\right)-3\left(\frac{5}{13}\right)}{4\left(-\frac{12}{13}\right)-9\left(\frac{5}{13}\right)}=\frac{-24-15}{-48-45}=\frac{-39}{-93}=\frac{13}{31}$$

45. Assuming that the radius of the circle is 1, the coordinates of the point labeled (x,y) are $(\cos t,\sin t)$, and the coordinates of the point labeled $(-x,-y)$ are $(\cos(t+\pi),\sin(t+\pi))$. So we have $y=\sin t$ and $-y=\sin(t+\pi)$, from which it follows that $\sin(t+\pi)=-\sin t$. Similarly, $-x=\cos(t+\pi)$ and $x=\cos t$, from which it follows that $\cos(t+\pi)=-\cos t$. Since $t-\pi$ results in the same intersection point with the unit circle as $t+\pi$, identities (ii) and (iv) follow in a similar manner.

47. (a) When $t = \frac{\pi}{6}$, we have:

$$2\sin^2\tfrac{\pi}{6} - \sin\tfrac{\pi}{6} = 2\left(\tfrac{1}{2}\right)^2 - \tfrac{1}{2} = \tfrac{1}{2} - \tfrac{1}{2} = 0$$

$$2\sin\tfrac{\pi}{6}\cos\tfrac{\pi}{6} - \cos\tfrac{\pi}{6} = 2 \bullet \tfrac{1}{2} \bullet \tfrac{\sqrt{3}}{2} - \tfrac{\sqrt{3}}{2} = \tfrac{\sqrt{3}}{2} - \tfrac{\sqrt{3}}{2} = 0$$

(b) When $t = \frac{\pi}{4}$, we have:

$$2\sin^2\tfrac{\pi}{4} - \sin\tfrac{\pi}{4} = 2\left(\tfrac{\sqrt{2}}{2}\right)^2 - \tfrac{\sqrt{2}}{2} = 1 - \tfrac{\sqrt{2}}{2} = \tfrac{2-\sqrt{2}}{2}$$

$$2\sin\tfrac{\pi}{4}\cos\tfrac{\pi}{4} - \cos\tfrac{\pi}{4} = 2 \bullet \tfrac{\sqrt{2}}{2} \bullet \tfrac{\sqrt{2}}{2} - \tfrac{\sqrt{2}}{2} = 1 - \tfrac{\sqrt{2}}{2} = \tfrac{2-\sqrt{2}}{2}$$

(c) No. Using the value $t = 0$, we have:

$$2\sin^2 0 - \sin 0 = 2(0)^2 - 0 = 0 - 0 = 0$$

$$2\sin 0\cos 0 - \cos 0 = 2(0)(1) - 1 = 0 - 1 = -1$$

Since the two sides of the equation are not equal, the given equation is not an identity.

49. We complete the table:

θ	$1 - \frac{\theta^2}{2}$	$\cos\theta$
0.1	0.995	0.995004...
0.2	0.980	0.980066...
0.3	0.955	0.955336...

7.3 Graphs of the Sine and Cosine Functions

1. A cycle is completed every 2 units, so the period is 2. The curve has high and low points of 1 and −1, respectively, so the amplitude is 1.

3. A cycle is completed every 4 units, so the period is 4. The curve has high and low points of 6 and −6, respectively, so the amplitude is 6.

5. A cycle is completed every 4 units, so the period is 4. The curve has high and low points of 6 and 2, respectively, so the amplitude is $\frac{6-2}{2} = 2$.

7. A cycle is completed every 6 units, so the period is 6. The curve has high and low points of −3 and −6, respectively, so the amplitude is $\frac{-3-(-6)}{2} = \frac{3}{2}$.

9. The coordinates of point C are $\left(-\frac{7\pi}{2}, 1\right) \approx (-10.996, 1)$.

11. The coordinates of point G are $\left(\frac{5\pi}{2}, 1\right) \approx (7.854, 1)$.

13. The coordinates of point B are $(-4\pi,0) \approx (-12.566,0)$.

15. The coordinates of point D are $(-3\pi,0) \approx (-9.425,0)$.

17. The coordinates of point E are $(-\pi,0) \approx (-3.142,0)$.

19. Referring to the graph of $y = \sin x$, we see that $y = \sin x$ is increasing on the interval $\frac{3\pi}{2} < x < 2\pi$.

21. Referring to the graph of $y = \sin x$, we see that $y = \sin x$ is decreasing on the interval $\frac{5\pi}{2} < x < \frac{7\pi}{2}$.

23. We complete the table:

x	$\sin x$	$x - \sin x$
0.453	0.43766	0.01534
0.253	0.25031	0.00269
0.0253	0.02530	0.00000
0.00253	0.00253	0.00000

25. The coordinates of point J are $\left(\frac{9\pi}{2},0\right) \approx (14.137,0)$.

27. The coordinates of point A are $(-4\pi,1) \approx (-12.566,1)$.

29. The coordinates of point E are $\left(\frac{\pi}{2},0\right) \approx (1.571,0)$.

31. The coordinates of point I are $(4\pi,1) \approx (12.566,1)$.

33. The coordinates of point B are $\left(-\frac{5\pi}{2},0\right) \approx (-7.854,0)$.

35. Referring to the graph of $y = \cos x$, we see that $y = \cos x$ is decreasing on the interval $0 < x < \pi$.

37. Referring to the graph of $y = \cos x$, we see that $y = \cos x$ is increasing on the interval $-\frac{\pi}{2} < x < 0$.

39. (a) Since C lies on the unit circle, its coordinates are $C(\cos\theta, \sin\theta)$.
     (b) Since $\angle AOB = \theta = \angle COD$ and $AO = CO$, the two triangles are congruent.
     (c) Since $AB = CD$ and $OB = OD$, the coordinates of A are $A(-\sin\theta, \cos\theta)$.
     (d) Matching up x- and y-coordinates, we have:
     $$\cos\left(\theta + \tfrac{\pi}{2}\right) = -\sin\theta \text{ and } \sin\left(\theta + \tfrac{\pi}{2}\right) = \cos\theta$$

7.4 Graphs of $y = A \sin(Bx - C)$ and $y = A \cos(Bx - C)$

1. (a)   The amplitude is 2, the period is 2π, and the x-intercepts are 0, π, 2π. The function is increasing on the intervals $\left(0, \frac{\pi}{2}\right)$ and $\left(\frac{3\pi}{2}, 2\pi\right)$.

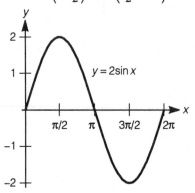

   (b)   The amplitude is 1 and the period is $\frac{2\pi}{2} = \pi$. The x-intercepts are 0, $\frac{\pi}{2}$, π, and the function is increasing on the interval $\left(\frac{\pi}{4}, \frac{3\pi}{4}\right)$. Notice that the graph is a reflection of $y = \sin 2x$ across the x-axis.

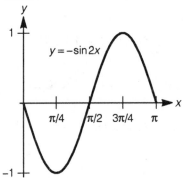

3. (a)   The amplitude is 1 and the period is $\frac{2\pi}{2} = \pi$. The x-intercepts are $\frac{\pi}{4}$, $\frac{3\pi}{4}$, and the function is increasing on the interval $\left(\frac{\pi}{2}, \pi\right)$.

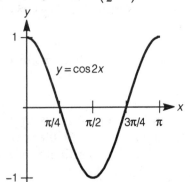

(b) The amplitude is 2 and the period is $\frac{2\pi}{2} = \pi$. The x-intercepts are $\frac{\pi}{4}, \frac{3\pi}{4}$, and the function is increasing on the interval $\left(\frac{\pi}{2}, \pi\right)$.

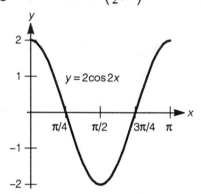

5. (a) The amplitude is 3 and the period is $\frac{2\pi}{\pi/2} = 4$. The x-intercepts are 0, 2, 4, and the function is increasing on the intervals $(0, 1)$ and $(3, 4)$.

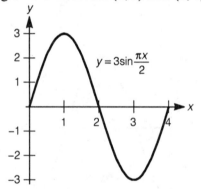

(b) The amplitude is 3 and the period is $\frac{2\pi}{\pi/2} = 4$. The x-intercepts are 0, 2, 4, and the function is increasing on the interval $(1, 3)$. Notice that the graph is a reflection of $y = 3\sin\frac{\pi x}{2}$ across the x-axis.

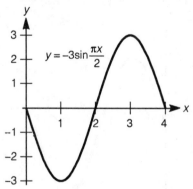

7.   (a)   The amplitude is 1 and the period is $\frac{2\pi}{2\pi} = 1$. The x-intercepts are $\frac{1}{4}, \frac{3}{4}$, and the function is increasing on the interval $\left(\frac{1}{2}, 1\right)$.

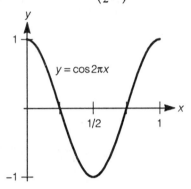

(b)   The amplitude is 4 and the period is $\frac{2\pi}{2\pi} = 1$. The x-intercepts are $\frac{1}{4}, \frac{3}{4}$, and the function is increasing on the interval $\left(0, \frac{1}{2}\right)$. Notice that the graph is a reflection of $y = 4\cos 2\pi x$ across the x-axis.

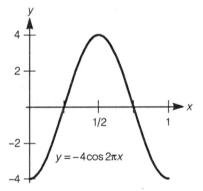

9.   The amplitude is 1 and the period is $\frac{2\pi}{2} = \pi$. The x-intercept is $\frac{3\pi}{4}$, and the function is increasing on the intervals $\left(0, \frac{\pi}{4}\right)$ and $\left(\frac{3\pi}{4}, \pi\right)$. Notice that the graph is a displacement of $y = \sin 2x$ up 1 unit.

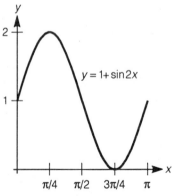

11. The amplitude is 1 and the period is $\frac{2\pi}{\pi/3} = 6$. The x-intercepts are 0, 6, and the function is increasing on the interval $(0, 3)$. Notice that the graph is a reflection of $y = \cos\frac{\pi x}{3}$ across the x-axis, then a displacement up 1 unit.

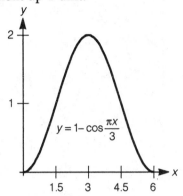

13. The amplitude is 1, the period is 2π, and the phase shift is $\frac{\pi}{6}$. The x-intercepts are $\frac{\pi}{6}, \frac{7\pi}{6}, \frac{13\pi}{6}$, the high point is $\left(\frac{2\pi}{3}, 1\right)$ and the low point is $\left(\frac{5\pi}{3}, -1\right)$.

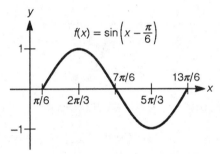

15. The amplitude is 1, the period is 2π, and the phase shift is $-\frac{\pi}{4}$. The x-intercepts are $\frac{\pi}{4}, \frac{5\pi}{4}$, the high point is $\left(\frac{3\pi}{4}, 1\right)$, and the low points are $\left(-\frac{\pi}{4}, -1\right)$ and $\left(\frac{7\pi}{4}, -1\right)$. Notice that the graph is a reflection of $y = \cos\left(x + \frac{\pi}{4}\right)$ across the x-axis.

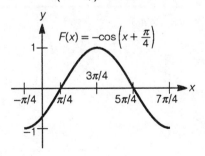

17. The amplitude is 1, the period is $\frac{2\pi}{2} = \pi$, and the phase shift is $\frac{\pi/2}{2} = \frac{\pi}{4}$. The x-intercepts are $\frac{\pi}{4}, \frac{3\pi}{4}, \frac{5\pi}{4}$, the high point is $\left(\frac{\pi}{2}, 1\right)$, and the low point is $(\pi, -1)$.

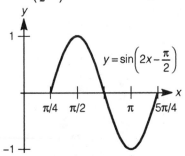

19. The amplitude is 1, the period is $\frac{2\pi}{2} = \pi$, and the phase shift is $\frac{\pi}{2}$. The x-intercepts are $\frac{3\pi}{4}, \frac{5\pi}{4}$, the high points are $\left(\frac{\pi}{2}, 1\right)$ and $\left(\frac{3\pi}{2}, 1\right)$, and the low point is $(\pi, -1)$.

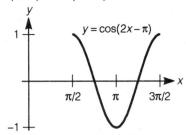

21. The amplitude is 3, the period is $\frac{2\pi}{1/2} = 4\pi$, and the phase shift is $\frac{-\pi/6}{1/2} = -\frac{\pi}{3}$. The x-intercepts are $-\frac{\pi}{3}, \frac{5\pi}{3}, \frac{11\pi}{3}$, the high point is $\left(\frac{2\pi}{3}, 3\right)$, and the low point is $\left(\frac{8\pi}{3}, -3\right)$.

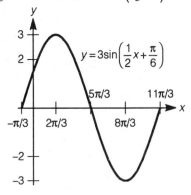

23. The amplitude is 4, the period is $\frac{2\pi}{3}$, and the phase shift is $\frac{\pi/4}{3} = \frac{\pi}{12}$. The x-intercepts are $\frac{\pi}{4}, \frac{7\pi}{12}$, the high points are $\left(\frac{\pi}{12}, 4\right)$ and $\left(\frac{3\pi}{4}, 4\right)$, and the low point is $\left(\frac{5\pi}{12}, -4\right)$.

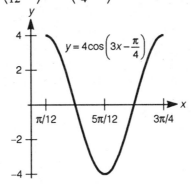

25. The amplitude is $\frac{1}{2}$, the period is $\frac{2\pi}{\pi/2} = 4$, and the phase shift is $\frac{\pi^2}{\pi/2} = 2\pi$. The x-intercepts are 2π, $2\pi + 2$, $2\pi + 4$, the high point is $\left(2\pi + 1, \frac{1}{2}\right)$, and the low point is $\left(2\pi + 3, -\frac{1}{2}\right)$.

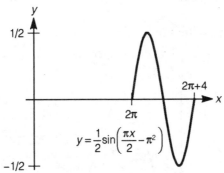

27. The amplitude is 1, the period is $\frac{2\pi}{2} = \pi$, and the phase shift is $\frac{\pi/3}{2} = \frac{\pi}{6}$. The x-intercepts are $\frac{\pi}{6}$ and $\frac{7\pi}{6}$, the high point is $\left(\frac{2\pi}{3}, 2\right)$, and the low points are $\left(\frac{\pi}{6}, 0\right)$ and $\left(\frac{7\pi}{6}, 0\right)$. Notice that the graph is a reflection of $y = \cos\left(2x - \frac{\pi}{3}\right)$ across the x-axis, then a displacement up 1 unit.

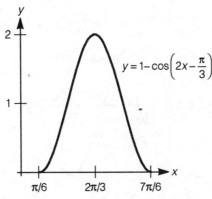

29. This is a sine function where the amplitude is 2, so $A = 2$. Since the period is 4π, we have:

$$\frac{2\pi}{B} = 4\pi$$
$$2\pi = 4\pi B$$
$$\frac{1}{2} = B$$

31. This is a sine function (reflected across the x-axis) where the amplitude is 3, so $A = -3$. Since the period is 2, we have:

$$\frac{2\pi}{B} = 2$$
$$2\pi = 2B$$
$$\pi = B$$

33. This is a cosine function (reflected across the x-axis) where the amplitude is 4, so $A = -4$. Since the period is 10π, we have:

$$\frac{2\pi}{B} = 10\pi$$
$$2\pi = 10\pi B$$
$$\frac{1}{5} = B$$

35. We graph $y = \frac{1}{2} - \frac{1}{2}\cos 2x$, which has an amplitude of $\frac{1}{2}$ and a period of $\frac{2\pi}{2} = \pi$. The graph will be a reflection of $y = \frac{1}{2}\cos 2x$ across the x-axis, then a displacement up $\frac{1}{2}$ unit.

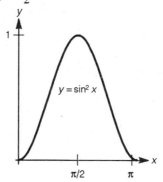

37. We graph $y = \frac{1}{2}\sin 2x$, which has an amplitude of $\frac{1}{2}$ and a period of $\frac{2\pi}{2} = \pi$.

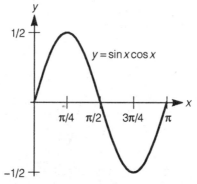

39. We know that the curve $y = \cos x$ begins a period when $x = 0$ and completes that period when $x = 2\pi$. Thus $y = A \cos Bx$ will begin a period when $Bx = 0$ and complete that period when $Bx = 2\pi$, which yields $x = 0$ and $x = \frac{2\pi}{B}$. Thus the period of $y = A \cos Bx$ is $\frac{2\pi}{B}$.

Optional TI-81 Graphing Calculator Exercises for Sections 7.3 and 7.4

1. (a) Using the indicated settings, the graph should appear as:

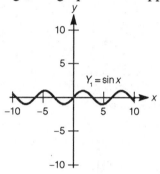

(b) Using the indicated settings, the graph should appear as:

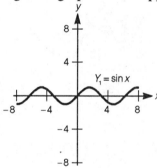

(c) The x-coordinate for the intercept between 0 and 4 is π. The x-coordinate for the intercept between 4 and 8 is 2π.

(d) The x-coordinate for the intercept between 0 and 4 is $\frac{\pi}{2}$. Using the same RANGE settings as in (b), the graph should appear as:

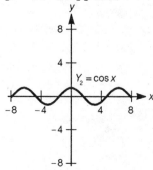

(e)   Using ZOOM-7, the graph appears as:

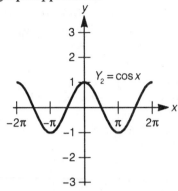

(f)   The graph of $Y_1 = \sin x$ would have to be shifted $\frac{\pi}{2}$ to the left to coincide with the graph of $Y_2 = \cos x$. Using the same settings as in (e), the graphs of both functions appear as:

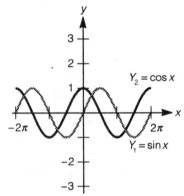

3.   (a)   Using ZOOM-7, the graphs of all three functions appear as:

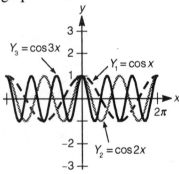

(b)   The amplitude for all three functions is 1. The period of $Y_1 = \cos x$ is 2π, the period of $Y_2 = \cos 2x$ is $\frac{2\pi}{2} = \pi$, and the period of $Y_3 = \cos 3x$ is $\frac{2\pi}{3}$. Using the indicated RANGE settings, the graphs of all three functions appear as:

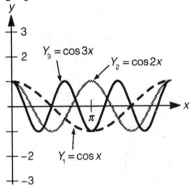

5.   (a)   The amplitude is 2.5, the period is $\frac{2\pi}{3\pi} = \frac{2}{3}$, and the phase shift is $-\frac{4}{3\pi}$.

(b)   We use $X_{min} = -\frac{4}{3\pi} \approx -0.4244$, $X_{max} = -\frac{4}{3\pi} + \frac{4}{3} \approx 0.9089$, $Y_{min} = -2.5$ and $Y_{max} = 2.5$ to obtain the graph:

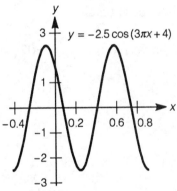

(c)   The highest points occur at $\left(\frac{1}{3} - \frac{4}{3\pi}, 2.5\right) \approx (-0.09, 2.5)$ and $\left(1 - \frac{4}{3\pi}, 2.5\right) \approx (0.58, 2.5)$, and the lowest points occur at $\left(-\frac{4}{3\pi}, -2.5\right) \approx (-0.42, -2.5)$, $\left(\frac{2}{3} - \frac{4}{3\pi}, -2.5\right) \approx (0.24, -2.5)$, and $\left(\frac{4}{3} - \frac{4}{3\pi}, -2.5\right) \approx (0.91, -2.5)$.

(d)   Tracing the curve verifies these results.

7. (a) The amplitude is 1, the period is $\frac{2\pi}{0.5} = 4\pi$, and the phase shift is $\frac{-0.75}{0.5} = -1.5$.

    (b) We use $X_{min} = -1.5$, $X_{max} = -1.5 + 8\pi \approx 23.63$, $Y_{min} = -1$ and $Y_{max} = 1$ to obtain the graph:

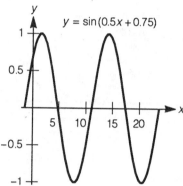

    (c) The highest points occur at $(-1.5 + \pi, 1) \approx (1.64, 1)$ and $(-1.5 + 5\pi, 1) \approx (14.2, 1)$, and the lowest points occur at $(-1.5 + 3\pi, -1) \approx (7.92, -1)$ and $(-1.5 + 7\pi, -1) \approx (20.49, -1)$.

    (d) Tracing the curve verifies these results.

9. (a) The amplitude is 0.02, the period is $\frac{2\pi}{100\pi} = 0.02$, and the phase shift is $-\frac{4\pi}{100\pi} = -0.04$.

    (b) We use $X_{min} = -0.04$, $X_{max} = 0$, $Y_{min} = -0.02$ and $Y_{max} = 0.02$ to obtain the graph:

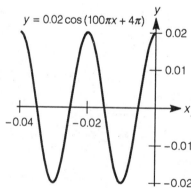

    (c) The highest points occur at $(-0.04, 0.02)$, $(-0.02, 0.02)$ and $(0, 0.02)$, and the lowest points occur at $(-0.03, -0.02)$ and $(-0.01, -0.02)$.

    (d) Tracing the curve verifies these results.

11. **(b)**   The graph of $Y_1 = \sin x$ with the indicated settings appears as:

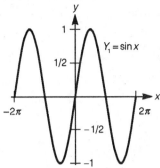

The graph of $Y_2 = \sin(\sin x)$ with ZOOM-7 settings appears as:

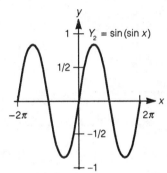

Although similar, the two graphs are not identical. The second graph appears to have a smaller amplitude as illustrated in part (c).

(c)   We graph the two functions together:

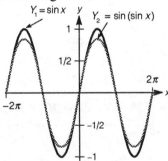

(d)   The coordinates of the highest point are approximately $(1.58, 0.84)$. The y-coordinate gives the amplitude.

(e)   Since $\sin x$ is largest when $x = \frac{\pi}{2}$ (then $\sin x = 1$), and $\sin x$ is increasing for $0 < x < \frac{\pi}{2}$, $\sin(\sin x)$ should have its largest value when $x = \frac{\pi}{2}$. Its value then is $\sin\left(\sin \frac{\pi}{2}\right) = \sin 1$.

(f)   Since $\sin 1 \approx 0.84$, our results from (d) and (e) are confirmed.

7.5 Graphs of the Tangent and the Reciprocal Functions

1. (a) The x-intercept is $-\frac{\pi}{4}$, the y-intercept is 1, and the asymptotes are $x = -\frac{3\pi}{4}$ and $x = \frac{\pi}{4}$. Notice that this graph is $y = \tan x$ displaced $\frac{\pi}{4}$ units to the left.

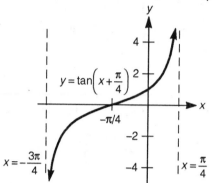

(b) The x-intercept is $-\frac{\pi}{4}$, the y-intercept is -1, and the asymptotes are $x = -\frac{3\pi}{4}$ and $x = \frac{\pi}{4}$. Notice that this graph is $y = \tan\left(x + \frac{\pi}{4}\right)$ reflected across the x-axis.

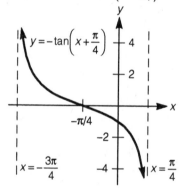

3. (a) The x- and y-intercepts are both 0, and the asymptotes are $x = -\frac{3\pi}{2}$ and $x = \frac{3\pi}{2}$. Notice that the period is $\frac{\pi}{1/3} = 3\pi$.

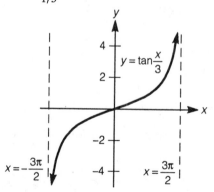

(b)   The x- and y-intercepts are both 0, and the asymptotes are $x = -\frac{3\pi}{2}$ and $x = \frac{3\pi}{2}$. Notice that this graph is $y = \tan\frac{x}{3}$ reflected across the x-axis.

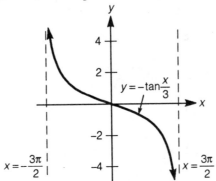

5.   The x- and y-intercepts are both 0, and the asymptotes are $x = -1$ and $x = 1$. Notice that the period is $\frac{\pi}{\pi/2} = 2$.

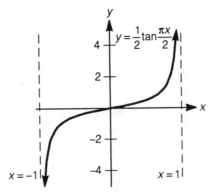

7.   The x-intercept is 1, there is no y-intercept, and the asymptotes are $x = 0$ and $x = 2$. Notice that the period is $\frac{\pi}{\pi/2} = 2$.

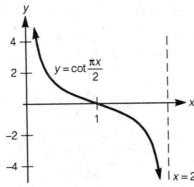

9.   The x-intercept is $\frac{3\pi}{4}$, the y-intercept is 1, and the asymptotes are $x = \frac{\pi}{4}$ and $x = \frac{5\pi}{4}$.
     Notice that this graph is $y = \cot x$ displaced $\frac{\pi}{4}$ units to the right, then reflected across the x-axis.

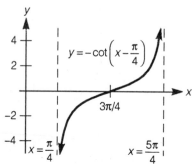

11. The x-intercept is $\frac{\pi}{4}$, there is no y-intercept, and the asymptotes are $x = 0$ and $x = \frac{\pi}{2}$.
     Notice that the period is $\frac{\pi}{2}$.

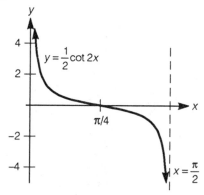

13. There is no x-intercept, the y-intercept is $-\sqrt{2}$, and the asymptotes are $x = -\frac{3\pi}{4}$, $x = \frac{\pi}{4}$,
     and $x = \frac{5\pi}{4}$. Notice that this graph is $y = \csc x$ displaced $\frac{\pi}{4}$ units to the right.

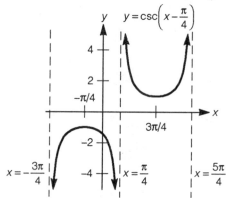

15. There are no x- or y-intercepts, and the asymptotes are $x = -2\pi$, $x = 0$ and $x = 2\pi$. Notice that the period is $\frac{2\pi}{1/2} = 4\pi$, and that this graph is $y = \csc\frac{x}{2}$ reflected across the x-axis.

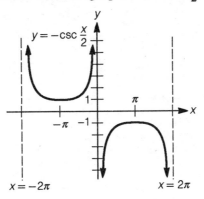

17. There are no x- or y-intercepts, and the asymptotes are $x = -1$, $x = 0$, and $x = 1$. Notice that the period is $\frac{2\pi}{\pi} = 2$.

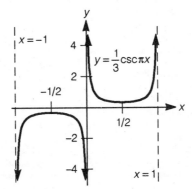

19. There is no x-intercept, the y-intercept is -1, and the asymptotes are $x = -\frac{\pi}{2}$, $x = \frac{\pi}{2}$, and $x = \frac{3\pi}{2}$. Notice that this graph is $y = \sec x$ reflected across the x-axis.

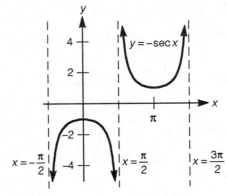

21. There is no x-intercept, the y-intercept is -1, and the asymptotes are $x = \frac{\pi}{2}$, $x = \frac{3\pi}{2}$, and $x = \frac{5\pi}{2}$. Notice that this graph is $y = \sec x$ displaced π units to the right.

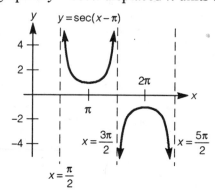

23. There is no x-intercept, the y-intercept is 3, and the asymptotes are $x = -1$, $x = 1$, and $x = 3$. Notice that the period is $\frac{2\pi}{\pi/2} = 4$.

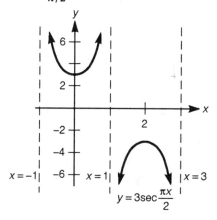

25. (a)   The x-intercepts are $-\frac{11}{18}, -\frac{5}{18}, \frac{1}{18}, \frac{7}{18}$ and $\frac{13}{18}$, and the y-intercept is -1, and there are no asymptotes. Notice that the period is $\frac{2\pi}{3\pi} = \frac{2}{3}$ and the phase shift is $\frac{\pi/6}{3\pi} = \frac{1}{18}$.

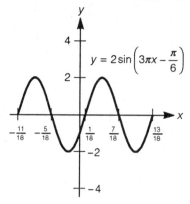

(b) There are no x-intercepts, the y-intercept is -4, and the asymptotes are $x = -\frac{11}{18}$, $x = -\frac{5}{18}$, $x = \frac{1}{18}$, $x = \frac{7}{18}$ and $x = \frac{13}{18}$. Notice that the period is $\frac{2\pi}{3\pi} = \frac{2}{3}$ and the phase shift is $\frac{\pi/6}{3\pi} = \frac{1}{18}$.

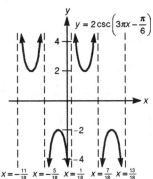

27. (a) The x-intercepts are $-\frac{5}{8}$, $-\frac{1}{8}$, $\frac{3}{8}$, and $\frac{7}{8}$, and the y-intercept is $-\frac{3\sqrt{2}}{2} \approx -2.12$, and there are no asymptotes. Notice that the period is $\frac{2\pi}{2\pi} = 1$, the phase shift is $\frac{\pi/4}{2\pi} = \frac{1}{8}$, and that this graph is $y = 3\cos\left(2\pi x - \frac{\pi}{4}\right)$ reflected across the x-axis.

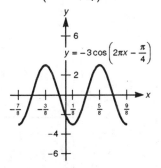

(b) There are no x-intercepts, the y-intercept is $-3\sqrt{2}$, and the asymptotes are $x = -\frac{5}{8}$, $x = -\frac{1}{8}$, $x = \frac{3}{8}$ and $x = \frac{7}{8}$. Notice that the period is $\frac{2\pi}{2\pi} = 1$, the phase shift is $\frac{\pi/4}{2\pi} = \frac{1}{8}$, and that this graph is $y = 3\sec\left(2\pi x - \frac{\pi}{4}\right)$ reflected across the x-axis.

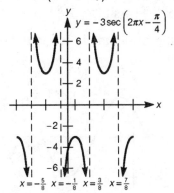

29. (a)   We first find the composition:
$$(f \circ h)(x) = f\left(\pi x - \tfrac{\pi}{6}\right) = \sin\left(\pi x - \tfrac{\pi}{6}\right)$$
The period is $\frac{2\pi}{\pi} = 2$, the amplitude is 1, and the phase shift is $\frac{\pi/6}{\pi} = \frac{1}{6}$.

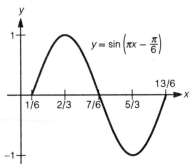

(b)   We first find the composition:
$$(g \circ h)(x) = g\left(\pi x - \tfrac{\pi}{6}\right) = \csc\left(\pi x - \tfrac{\pi}{6}\right)$$
The period is $\frac{2\pi}{\pi} = 2$, the phase shift is $\frac{\pi/6}{\pi} = \frac{1}{6}$, and the asymptotes are $x = \frac{1}{6}$, $x = \frac{7}{6}$ and $x = \frac{11}{6}$.

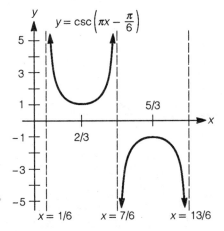

31. (a)   We first find the composition:
$$(f \circ H)(x) = f\left(\pi x + \tfrac{\pi}{4}\right) = \sin\left(\pi x + \tfrac{\pi}{4}\right)$$
The period is $\frac{2\pi}{\pi} = 2$, the amplitude is 1, and the phase shift is $\frac{-\pi/4}{\pi} = -\frac{1}{4}$.

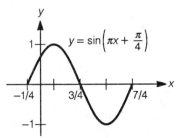

(b)   We first find the composition:
$$(g \circ H)(x) = g\left(\pi x + \tfrac{\pi}{4}\right) = \csc\left(\pi x + \tfrac{\pi}{4}\right)$$
The period is $\frac{2\pi}{\pi} = 2$, the phase shift is $\frac{-\pi/4}{\pi} = -\tfrac{1}{4}$, and the asymptotes are $x = -\tfrac{1}{4}$, $x = \tfrac{3}{4}$ and $x = \tfrac{7}{4}$.

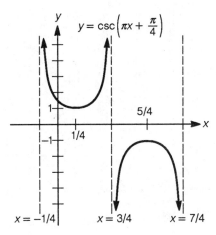

33.   We first find the composition:
$$(A \circ T)(x) = A(\tan x) = |\tan x|$$
Now draw the graph, noting that values of x where $\tan x < 0$ will be reflected across the x-axis.

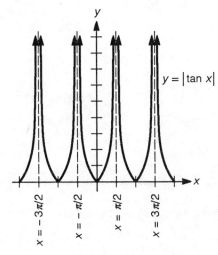

35. We first find the composition:
$$(A \circ f)(x) = A(\csc x) = |\csc x|$$
Now draw the graph, noting that values of x where $\csc x < 0$ will be reflected across the x-axis.

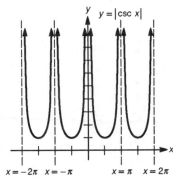

$x = -2\pi$ $x = -\pi$ $x = \pi$ $x = 2\pi$

37. (a) Since P and Q are both points on the unit circle, the coordinates are $P(\cos s, \sin s)$ and $Q\left(\cos\left(s - \frac{\pi}{2}\right), \sin\left(s - \frac{\pi}{2}\right)\right)$.

(b) Since $\triangle OAP$ is congruent to $\triangle OBQ$ (labeling the third vertex B), $OA = OB$ and $AP = BQ$. Because the y-coordinate at Q is negative, we have concluded what was required.

(c) Restating part (b), we have shown that:
$$\cos\left(s - \frac{\pi}{2}\right) = \sin s \quad \text{and} \quad \sin\left(s - \frac{\pi}{2}\right) = -\cos s$$

(d) We compute $\cot s$:
$$\cot s = \frac{\cos s}{\sin s} = \frac{-\sin\left(s - \frac{\pi}{2}\right)}{\cos\left(s - \frac{\pi}{2}\right)} = -\tan\left(s - \frac{\pi}{2}\right)$$

Optional TI-81 Graphing Calculator Exercises for Section 7.5

1. Using the indicated settings, the graph appears as:

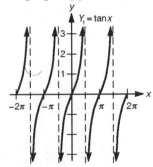

3. Using the same settings as in Exercise 1, notice that the two graphs are identical, verifying the identity $\cot x = -\tan\left(x - \frac{\pi}{2}\right)$.

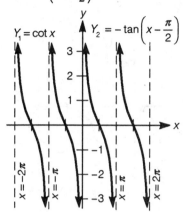

5. (a) Since the period is $\frac{\pi}{1/4} = 4\pi$, we adjust $X_{min} = -2\pi \approx -6.28$ and $X_{max} = 6\pi \approx 18.85$ to obtain the graph:

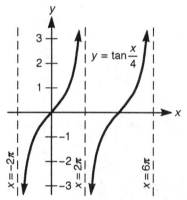

    (b) Since the period is $\frac{\pi}{4}$, we adjust $X_{min} = -\frac{\pi}{8} \approx -0.39$ and $X_{max} = \frac{3\pi}{8} \approx 1.18$ to obtain the graph:

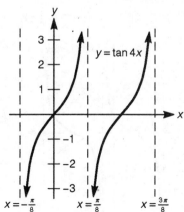

7. (a) Since the period is $\frac{\pi}{\pi} = 1$, we use the settings $X_{min} = -0.5$ and $X_{max} = 1.5$ to obtain the graph:

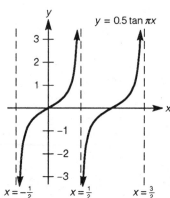

(b) The period is the same as in (a), and the phase shift is $\frac{-\pi/3}{\pi} = -\frac{1}{3}$. We use the settings $X_{min} = -\frac{1}{2} - \frac{1}{3} = -\frac{5}{6} \approx -0.83$ and $X_{max} = \frac{3}{2} - \frac{1}{3} = \frac{7}{6} \approx 1.17$ to obtain the graph:

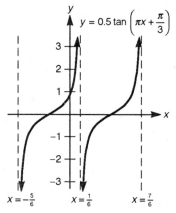

(c) The period is the same as in (a), and the phase shift is $-\frac{1}{\pi}$. We use the settings $X_{min} = -\frac{1}{2} - \frac{1}{\pi} \approx -0.82$ and $X_{max} = \frac{3}{2} - \frac{1}{\pi} \approx 1.18$ to obtain the graph:

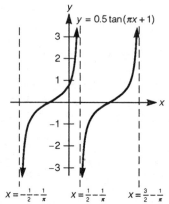

9. (a) Since the period is $\frac{\pi}{1/2} = 2\pi$, we use the settings $X_{min} = -\pi \approx -3.14$ and $X_{max} = 3\pi \approx 9.42$ to obtain the graph:

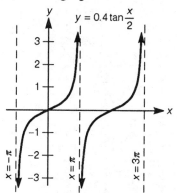

(b) Since the period is $\frac{\pi}{1/3} = 3\pi$, we use the settings $X_{min} = \frac{-3\pi}{2} \approx -4.71$ and $X_{max} = \frac{9\pi}{2} \approx 14.14$ to obtain the graph:

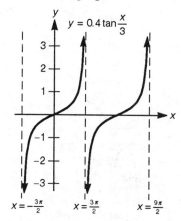

(c) Since the period is $\frac{\pi}{1/5} = 5\pi$, we use the settings $X_{min} = \frac{-5\pi}{2} \approx -7.85$ and $X_{max} = \frac{15\pi}{2} \approx 23.56$ to obtain the graph:

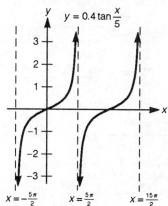

11. It appears that $\sin x = \csc x$ at odd multiples of $\frac{\pi}{2}$, such as $-\frac{3\pi}{2}$, $-\frac{\pi}{2}$, $\frac{\pi}{2}$, and $\frac{3\pi}{2}$. Since $\sin x$ and $\csc x$ have the same sign, there are no points in which $\sin x = -\csc x$. Using the indicated settings, the graphs appear as:

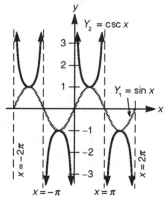

13. Since the period is $\frac{2\pi}{1/2} = 4\pi$, we use the settings $X_{min} = -4\pi \approx -12.57$, $X_{max} = 4\pi \approx 12.57$, $Y_{min} = -2$ and $Y_{max} = 2$ to obtain the graph:

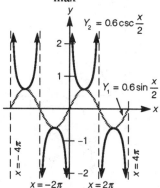

15. Since the period is $\frac{2\pi}{2} = \pi$ and the phase shift is $\frac{\pi/6}{2} = \frac{\pi}{12}$, we use the settings $X_{min} = -\frac{\pi}{4} + \frac{\pi}{12} = -\frac{\pi}{6} \approx -0.52$, $X_{max} = \frac{7\pi}{4} + \frac{\pi}{12} = \frac{11\pi}{6} \approx 5.76$, $Y_{min} = -6$ and $Y_{max} = 6$ to obtain the graph:

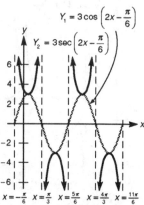

17. The two graphs are identical. This demonstrates that $\tan^2 x = \sec^2 x - 1$ is a trigonometric identity. Using the ZOOM-7 setting we obtain the graph:

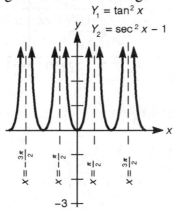

$Y_1 = \tan^2 x$

$Y_2 = \sec^2 x - 1$

19. (a) Using the ZOOM-7 setting we obtain the graph:

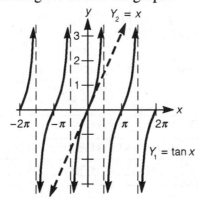

$Y_2 = x$

$Y_1 = \tan x$

(b) Using the settings $X_{min} = -1.57$ and $X_{max} = 1.57$, the graph appears as:

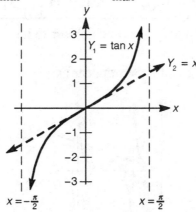

$Y_1 = \tan x$

$Y_2 = x$

$x = -\frac{\pi}{2}$ $x = \frac{\pi}{2}$

We complete the table:

x	0.000123	0.01	0.05	0.1
$\tan x$	0.000123	0.010000	0.050042	0.100335
x	0.2	0.3	0.4	0.5
$\tan x$	0.202710	0.309336	0.422793	0.546302

(c) The graphs of $Y_1 = \tan x$ and $Y_3 = x + \frac{x^3}{3}$ appear as:

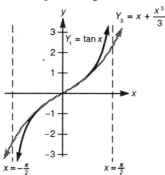

Notice how similar the two curves are near $x = 0$.

(d) We complete the table:

x	0.000123	0.01	0.05	0.1
$\tan x$	0.000123	0.010000	0.050042	0.100335
$x + \frac{x^3}{3}$	0.000123	0.010000	0.050042	0.100333
x	0.2	0.3	0.4	0.5
$\tan x$	0.202710	0.309336	0.422793	0.546302
$x + \frac{x^3}{3}$	0.202667	0.309	0.421333	0.541667

Note that the values of $x + \frac{x^3}{3}$ are much closer to $\tan x$ than are those for x.

21. (a) The graphs of $Y_1 = \tan x$ and $Y_2 = x$ appear as:

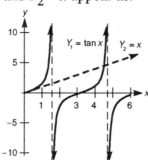

The two graphs appear to intersect at approximately $x \approx 4.5$.

(b) The intersection point appears to occur at approximately $x \approx 4.493$.

7.6 The Addition Formulas for Sine and Cosine

1. Using the identity $\sin(s+t) = \sin s \cos t + \cos s \sin t$ where $s = \theta$ and $t = 2\theta$, we have:
    $$\sin\theta\cos 2\theta + \cos\theta\sin 2\theta = \sin(\theta + 2\theta) = \sin 3\theta$$

3. Using the identity $\sin(s-t) = \sin s \cos t - \cos s \sin t$ where $s = 3\theta$ and $t = \theta$, we have:
    $$\sin 3\theta\cos\theta - \cos 3\theta\sin\theta = \sin(3\theta - \theta) = \sin 2\theta$$

5. Using the identity $\cos(s+t) = \cos s \cos t - \sin s \sin t$ where $s = 2u$ and $t = 3u$, we have:
    $$\cos 2u\cos 3u - \sin 2u\sin 3u = \cos(2u + 3u) = \cos 5u$$

7. Using the identity $\cos(s-t) = \cos s \cos t + \sin s \sin t$ where $s = \frac{2\pi}{9}$ and $t = \frac{\pi}{18}$, we have:
    $$\cos\frac{2\pi}{9}\cos\frac{\pi}{18} + \sin\frac{2\pi}{9}\sin\frac{\pi}{18} = \cos\left(\frac{2\pi}{9} - \frac{\pi}{18}\right) = \cos\frac{3\pi}{18} = \cos\frac{\pi}{6} = \frac{\sqrt{3}}{2}$$

9. Using the identity $\sin(s-t) = \sin s \cos t - \cos s \sin t$ where $s = A + B$ and $t = A$, we have:
    $$\sin(A+B)\cos A - \cos(A+B)\sin A = \sin(A+B-A) = \sin B$$

11. Using the identity $\sin(s-t) = \sin s \cos t - \cos s \sin t$ with $s = \theta$ and $t = \frac{3\pi}{2}$, we have:
    $$\sin\left(\theta - \frac{3\pi}{2}\right) = \sin\theta\cos\frac{3\pi}{2} - \cos\theta\sin\frac{3\pi}{2} = \sin\theta \cdot 0 - \cos\theta\cdot(-1) = \cos\theta$$

13. Using the identity $\cos(s+t) = \cos s \cos t - \sin s \sin t$ with $s = \theta$ and $t = \pi$, we have:
    $$\cos(\theta + \pi) = \cos\theta\cos\pi - \sin\theta\sin\pi = \cos\theta\cdot(-1) - \sin\theta\cdot 0 = -\cos\theta$$

15. Using the identity $\sin(s+t) = \sin s \cos t + \cos s \sin t$ with $s = t$ and $t = 2\pi$, we have:
    $$\sin(t + 2\pi) = \sin t\cos 2\pi + \cos t\sin 2\pi = \sin t \cdot 1 + \cos t \cdot 0 = \sin t$$
    Notice that this verifies the formula $\sin(t + 2\pi) = \sin t$.

17. Using the identity $\cos(s+t) = \cos s \cos t - \sin s \sin t$, we have:
    $$\begin{aligned}
    \cos 75° &= \cos(45° + 30°) \\
    &= \cos 45°\cos 30° - \sin 45°\sin 30° \\
    &= \frac{\sqrt{2}}{2}\cdot\frac{\sqrt{3}}{2} - \frac{\sqrt{2}}{2}\cdot\frac{1}{2} \\
    &= \frac{\sqrt{6} - \sqrt{2}}{4}
    \end{aligned}$$

19. Using the identity $\sin(s+t) = \sin s \cos t + \cos s \sin t$, we have:
    $$\sin\frac{7\pi}{12} = \sin\left(\frac{\pi}{3} + \frac{\pi}{4}\right) = \sin\frac{\pi}{3}\cos\frac{\pi}{4} + \sin\frac{\pi}{4}\cos\frac{\pi}{3} = \frac{\sqrt{3}}{2}\cdot\frac{\sqrt{2}}{2} + \frac{\sqrt{2}}{2}\cdot\frac{1}{2} = \frac{\sqrt{6}+\sqrt{2}}{4}$$

21. Using the identities $\sin(s+t) = \sin s \cos t + \cos s \sin t$ and
$\sin(s-t) = \sin s \cos t - \cos s \sin t$, we have:
$$\sin\left(\tfrac{\pi}{4}+s\right) - \sin\left(\tfrac{\pi}{4}-s\right) = \left(\sin\tfrac{\pi}{4}\cos s + \cos\tfrac{\pi}{4}\sin s\right) - \left(\sin\tfrac{\pi}{4}\cos s - \cos\tfrac{\pi}{4}\sin s\right)$$
$$= \tfrac{\sqrt{2}}{2}\cos s + \tfrac{\sqrt{2}}{2}\sin s - \tfrac{\sqrt{2}}{2}\cos s + \tfrac{\sqrt{2}}{2}\sin s$$
$$= \sqrt{2}\sin s$$

23. Using the identities $\cos(s+t) = \cos s \cos t - \sin s \sin t$ and
$\cos(s-t) = \cos s \cos t + \sin s \sin t$, we have:
$$\cos\left(\tfrac{\pi}{3}-\theta\right) - \cos\left(\tfrac{\pi}{3}+\theta\right) = \left(\cos\tfrac{\pi}{3}\cos\theta + \sin\tfrac{\pi}{3}\sin\theta\right) - \left(\cos\tfrac{\pi}{3}\cos\theta - \sin\tfrac{\pi}{3}\sin\theta\right)$$
$$= \tfrac{1}{2}\cos\theta + \tfrac{\sqrt{3}}{2}\sin\theta - \tfrac{1}{2}\cos\theta + \tfrac{\sqrt{3}}{2}\sin\theta$$
$$= \sqrt{3}\sin\theta$$

25. We must first compute $\cos\alpha$ and $\sin\beta$. Since $\tfrac{\pi}{2} < \alpha < \pi$, $\cos\alpha < 0$ and thus:
$$\cos\alpha = -\sqrt{1-\sin^2\alpha} = -\sqrt{1-\left(\tfrac{12}{13}\right)^2} = -\sqrt{1-\tfrac{144}{169}} = -\sqrt{\tfrac{25}{169}} = -\tfrac{5}{13}$$
Since $\pi < \beta < \tfrac{3\pi}{2}$, $\sin\beta < 0$ and thus:
$$\sin\beta = -\sqrt{1-\cos^2\beta} = -\sqrt{1-\left(-\tfrac{3}{5}\right)^2} = -\sqrt{1-\tfrac{9}{25}} = -\sqrt{\tfrac{16}{25}} = -\tfrac{4}{5}$$

(a) Using the identity for $\sin(\alpha+\beta)$, we have:
$$\sin(\alpha+\beta) = \sin\alpha\cos\beta + \cos\alpha\sin\beta$$
$$= \left(\tfrac{12}{13}\right)\cdot\left(-\tfrac{3}{5}\right) + \left(-\tfrac{5}{13}\right)\cdot\left(-\tfrac{4}{5}\right)$$
$$= -\tfrac{36}{65} + \tfrac{20}{65}$$
$$= -\tfrac{16}{65}$$

(b) Using the identity for $\cos(\alpha+\beta)$, we have:
$$\cos(\alpha+\beta) = \cos\alpha\cos\beta - \sin\alpha\sin\beta$$
$$= \left(-\tfrac{5}{13}\right)\cdot\left(-\tfrac{3}{5}\right) - \left(\tfrac{12}{13}\right)\cdot\left(-\tfrac{4}{5}\right)$$
$$= \tfrac{15}{65} + \tfrac{48}{65}$$
$$= \tfrac{63}{65}$$

27. We use the value for $\sin\beta$ computed in Exercise 25. Since $-2\pi < \theta < -\tfrac{3\pi}{2}$, $\sin\theta > 0$ and thus:
$$\sin\theta = \sqrt{1-\cos^2\theta} = \sqrt{1-\left(\tfrac{7}{25}\right)^2} = \sqrt{1-\tfrac{49}{625}} = \sqrt{\tfrac{576}{625}} = \tfrac{24}{25}$$

(a) Using the identity for $\sin(\theta - \beta)$, we have:

$$\sin(\theta - \beta) = \sin\theta\cos\beta - \cos\theta\sin\beta$$
$$= \left(\tfrac{24}{25}\right)\cdot\left(-\tfrac{3}{5}\right) - \left(\tfrac{7}{25}\right)\cdot\left(-\tfrac{4}{5}\right)$$
$$= -\tfrac{72}{125} + \tfrac{28}{125}$$
$$= -\tfrac{44}{125}$$

(b) Using the identity for $\sin(\theta + \beta)$, we have:

$$\sin(\theta + \beta) = \sin\theta\cos\beta + \cos\theta\sin\beta$$
$$= \left(\tfrac{24}{25}\right)\cdot\left(-\tfrac{3}{5}\right) + \left(\tfrac{7}{25}\right)\cdot\left(-\tfrac{4}{5}\right)$$
$$= -\tfrac{72}{125} - \tfrac{28}{125}$$
$$= -\tfrac{100}{125}$$
$$= -\tfrac{4}{5}$$

29. (a) Since $0 < \theta < \tfrac{\pi}{2}$, $\cos\theta > 0$ and thus:

$$\cos\theta = \sqrt{1 - \sin^2\theta} = \sqrt{1 - \left(\tfrac{1}{5}\right)^2} = \sqrt{1 - \tfrac{1}{25}} = \sqrt{\tfrac{24}{25}} = \tfrac{2\sqrt{6}}{5}$$

(b) Since $\sin 2\theta = \sin(\theta + \theta)$, we use the addition formula for $\sin(s + t)$ to obtain:

$$\sin 2\theta = \sin\theta\cos\theta + \sin\theta\cos\theta = 2\sin\theta\cos\theta = 2\cdot\tfrac{1}{5}\cdot\tfrac{2\sqrt{6}}{5} = \tfrac{4\sqrt{6}}{25}$$

31. We first find $\sin\theta$, $\cos\theta$, $\sin\beta$ and $\cos\beta$. Since $\tfrac{\pi}{2} < \theta < \pi$, then $\sec\theta < 0$, so using the identity $\sec^2\theta = 1 + \tan^2\theta$ we have:

$$\sec\theta = -\sqrt{1 + \tan^2\theta} = -\sqrt{1 + \left(-\tfrac{2}{3}\right)^2} = -\sqrt{1 + \tfrac{4}{9}} = -\sqrt{\tfrac{13}{9}} = -\tfrac{\sqrt{13}}{3}$$

Thus $\cos\theta = -\tfrac{3}{\sqrt{13}} = -\tfrac{3\sqrt{13}}{13}$. Since $\tfrac{\pi}{2} < \theta < \pi$, then $\sin\theta > 0$ and thus:

$$\sin\theta = \sqrt{1 - \cos^2\theta} = \sqrt{1 - \left(-\tfrac{3\sqrt{13}}{13}\right)^2} = \sqrt{1 - \tfrac{9}{13}} = \sqrt{\tfrac{4}{13}} = \tfrac{2}{\sqrt{13}} = \tfrac{2\sqrt{13}}{13}$$

Since $\csc\beta = 2$, $\sin\beta = \tfrac{1}{2}$. Since $0 < \beta < \tfrac{\pi}{2}$, $\cos\beta > 0$ and thus:

$$\cos\beta = \sqrt{1 - \sin^2\beta} = \sqrt{1 - \left(\tfrac{1}{2}\right)^2} = \sqrt{1 - \tfrac{1}{4}} = \sqrt{\tfrac{3}{4}} = \tfrac{\sqrt{3}}{2}$$

Now using the addition formula for $\sin(s + t)$, we have:

$$\sin(\theta + \beta) = \sin\theta\cos\beta + \cos\theta\sin\beta$$
$$= \tfrac{2\sqrt{13}}{13}\cdot\tfrac{\sqrt{3}}{2} - \tfrac{3\sqrt{13}}{13}\cdot\tfrac{1}{2}$$
$$= \tfrac{2\sqrt{39}}{26} - \tfrac{3\sqrt{13}}{26}$$
$$= \tfrac{2\sqrt{39} - 3\sqrt{13}}{26}$$

Finally using the addition formula for $\cos(s - t)$, we have:

$$\cos(\beta - \theta) = \cos\beta\cos\theta + \sin\beta\sin\theta$$
$$= \frac{\sqrt{3}}{2} \cdot \left(-\frac{3\sqrt{13}}{13}\right) + \frac{1}{2} \cdot \frac{2\sqrt{13}}{13}$$
$$= -\frac{3\sqrt{39}}{26} + \frac{2\sqrt{13}}{26}$$
$$= \frac{2\sqrt{13} - 3\sqrt{39}}{26}$$

33. Using the addition formula for $\sin(s + t)$, we have:

$$\sin\left(t + \tfrac{\pi}{4}\right) = \sin t\cos\tfrac{\pi}{4} + \cos t\sin\tfrac{\pi}{4} = \tfrac{1}{\sqrt{2}}\sin t + \tfrac{1}{\sqrt{2}}\cos t = \frac{\sin t + \cos t}{\sqrt{2}}$$

35. Using the results from Exercises 33 and 34, we have:

$$\sin\left(t + \tfrac{\pi}{4}\right) + \cos\left(t + \tfrac{\pi}{4}\right) = \frac{\sin t + \cos t}{\sqrt{2}} + \frac{\cos t - \sin t}{\sqrt{2}} = \frac{2\cos t}{\sqrt{2}} = \sqrt{2}\cos t$$

37. (a) Using the addition formula for $\sin(s + t)$, we have:

$$\sin\left(x + \tfrac{\pi}{2}\right) = \sin x\cos\tfrac{\pi}{2} + \cos x\sin\tfrac{\pi}{2} = \sin x \cdot 0 + \cos x \cdot 1 = \cos x$$

   (b) Since $\cos x = \sin\left(x + \tfrac{\pi}{2}\right)$, the graph of $y = \cos x$ is the graph of $y = \sin x$ shifted $\tfrac{\pi}{2}$ units to the left.

39. Using the addition formula for $\sin(s + t)$, we have:

$$\frac{\sin(s + t)}{\cos s\cos t} = \frac{\sin s\cos t + \cos s\sin t}{\cos s\cos t} = \frac{\sin s}{\cos s} + \frac{\sin t}{\cos t} = \tan s + \tan t$$

41. Using the addition formulas for $\cos(s - t)$ and $\cos(s + t)$, we have:

$$\cos(A - B) - \cos(A + B) = (\cos A\cos B + \sin A\sin B) - (\cos A\cos B - \sin A\sin B)$$
$$= \cos A\cos B + \sin A\sin B - \cos A\cos B + \sin A\sin B$$
$$= 2\sin A\sin B$$

43. Using the addition formulas for $\cos(s + t)$ and $\cos(s - t)$, as well as the identities $\sin^2 A = 1 - \cos^2 A$ and $\sin^2 B = 1 - \cos^2 B$, we have:

$$\cos(A + B)\cos(A - B) = (\cos A\cos B - \sin A\sin B)(\cos A\cos B + \sin A\sin B)$$
$$= (\cos A\cos B)^2 - (\sin A\sin B)^2$$
$$= \cos^2 A\cos^2 B - \sin^2 A\sin^2 B$$
$$= \cos^2 A\left(1 - \sin^2 B\right) - \left(1 - \cos^2 A\right)\sin^2 B$$
$$= \cos^2 A - \cos^2 A\sin^2 B - \sin^2 B + \cos^2 A\sin^2 B$$
$$= \cos^2 A - \sin^2 B$$

45. Using the addition formulas for $\cos(s+t)$ and $\sin(s+t)$, as well as the identity $\cos^2\beta + \sin^2\beta = 1$, we have:

$$\cos(\alpha+\beta)\cos\beta + \sin(\alpha+\beta)\sin\beta$$
$$= (\cos\alpha\cos\beta - \sin\alpha\sin\beta)\cos\beta + (\sin\alpha\cos\beta + \cos\alpha\sin\beta)\sin\beta$$
$$= \cos\alpha\cos^2\beta - \sin\alpha\sin\beta\cos\beta + \sin\alpha\sin\beta\cos\beta + \cos\alpha\sin^2\beta$$
$$= \cos\alpha\cos^2\beta + \cos\alpha\sin^2\beta$$
$$= \cos\alpha(\cos^2\beta + \sin^2\beta)$$
$$= \cos\alpha$$

47. (a) We complete the table:

t	1	2	3	4
$f(t)$	1.5	1.5	1.5	1.5

(b) Conjecture: $f(t) = 1.5$

To prove this, we first simplify the expressions:
$$\cos\left(t+\tfrac{2\pi}{3}\right) = \cos t\cos\tfrac{2\pi}{3} - \sin t\sin\tfrac{2\pi}{3} = -\tfrac{1}{2}\cos t - \tfrac{\sqrt{3}}{2}\sin t$$
$$\cos\left(t-\tfrac{2\pi}{3}\right) = \cos t\cos\tfrac{2\pi}{3} + \sin t\sin\tfrac{2\pi}{3} = -\tfrac{1}{2}\cos t + \tfrac{\sqrt{3}}{2}\sin t$$

Therefore:
$$\cos^2\left(t+\tfrac{2\pi}{3}\right) = \left(-\tfrac{1}{2}\cos t - \tfrac{\sqrt{3}}{2}\sin t\right)^2 = \tfrac{1}{4}\cos^2 t + \tfrac{\sqrt{3}}{2}\sin t\cos t + \tfrac{3}{4}\sin^2 t$$
$$\cos^2\left(t-\tfrac{2\pi}{3}\right) = \left(-\tfrac{1}{2}\cos t + \tfrac{\sqrt{3}}{2}\sin t\right)^2 = \tfrac{1}{4}\cos^2 t - \tfrac{\sqrt{3}}{2}\sin t\cos t + \tfrac{3}{4}\sin^2 t$$

Thus:
$$f(t) = \cos^2 t + \left(\tfrac{1}{4}\cos^2 t + \tfrac{\sqrt{3}}{2}\sin t\cos t + \tfrac{3}{4}\sin^2 t\right)$$
$$\qquad + \left(\tfrac{1}{4}\cos^2 t - \tfrac{\sqrt{3}}{2}\sin t\cos t + \tfrac{3}{4}\sin^2 t\right)$$
$$= \tfrac{3}{2}\cos^2 t + \tfrac{3}{2}\sin^2 t$$
$$= \tfrac{3}{2}(\cos^2 t + \sin^2 t)$$
$$= \tfrac{3}{2}$$

49. To prove this identity, we will show an alternate identity:

$$1 - \cos^2 C - \cos^2 A - \cos^2 B - 2\cos A \cos B \cos C$$
$$= \sin^2 C - \cos^2 A - \cos^2 B - 2\cos A \cos B \cos[\pi - (A+B)]$$
$$= \sin^2[\pi - (A+B)] - \cos^2 A - \cos^2 B + 2\cos A \cos B \cos(A+B)$$
$$= \sin^2(A+B) - \cos^2 A - \cos^2 B + 2\cos A \cos B(\cos A \cos B - \sin A \sin B)$$
$$= (\sin A \cos B + \sin B \cos A)^2 - \cos^2 A - \cos^2 B + 2\cos^2 A \cos^2 B$$
$$\qquad -2\cos A \cos B \sin A \sin B$$
$$= \sin^2 A \cos^2 B + \sin^2 B \cos^2 A - \cos^2 A - \cos^2 B + 2\cos^2 A \cos^2 B$$
$$= \cos^2 B(\sin^2 A - 1) + \cos^2 A(\sin^2 B - 1) + 2\cos^2 A \cos^2 B$$
$$= -\cos^2 B \cos^2 A - \cos^2 A \cos^2 B + 2\cos^2 A \cos^2 B$$
$$= 0$$

Hence $\cos^2 A + \cos^2 B + \cos^2 C + 2\cos A \cos B \cos C = 1$.

51. Since $a^2 + b^2 = 1$ and $c^2 + d^2 = 1$, there is some angle θ for which $a = \cos\theta$ and $b = \sin\theta$ and some angle φ for which $c = \cos\varphi$ and $d = \sin\varphi$. Thus:

$$|ac + bd| = |\cos\theta\cos\varphi + \sin\theta\sin\varphi| = |\cos(\theta - \varphi)| \le 1$$

53. Let $A = \frac{\pi}{3} - t$ and $B = \frac{\pi}{3} + t$, then we have:

$$\sin A \cos B + \cos A \sin B = \sin(A+B)$$

But $A + B = \frac{2\pi}{3}$, so $\sin(A+B) = \sin\frac{2\pi}{3} = \frac{\sqrt{3}}{2}$.

55. (a) Using $\triangle ABH$, $\cos(\alpha + \beta) = \frac{AB}{1}$, so $\cos(\alpha + \beta) = AB$.

   (b) Using $\triangle ACF$, $\cos\alpha = \frac{AC}{AF} = \frac{AC}{\cos\beta}$ from Exercise 54(e), so $AC = \cos\alpha\cos\beta$.

   (c) Using $\triangle EFH$, $\sin(\angle EHF) = \frac{EF}{HF}$. But $\angle EHF = \alpha$ from Exercise 54(c), and $HF = \sin\beta$ from Exercise 54(b), so $\sin\alpha = \frac{EF}{\sin\beta}$, and thus $EF = \sin\alpha\sin\beta$.

   (d) From part (a), $\cos(\alpha + \beta) = AB = AC - BC$. But $AC = \cos\alpha\cos\beta$ from part (b), and $BC = EF = \sin\alpha\sin\beta$ from part (c), so $\cos(\alpha + \beta) = \cos\alpha\cos\beta - \sin\alpha\sin\beta$.

57. (a) Using a calculator, we have:

$$\tan A + \tan B + \tan C = \tan 20° + \tan 50° + \tan 110° \approx -1.1918$$
$$\tan A \tan B \tan C = \tan 20° \tan 50° \tan 110° \approx -1.1918$$

It appears that the two values are equal.

(b) Using a calculator, we have:

$$\tan\alpha + \tan\beta + \tan\gamma = \tan\tfrac{\pi}{10} + \tan\tfrac{3\pi}{10} + \tan\tfrac{3\pi}{5} \approx -1.3764$$

$$\tan\alpha \tan\beta \tan\gamma = \tan\tfrac{\pi}{10} \cdot \tan\tfrac{3\pi}{10} \cdot \tan\tfrac{3\pi}{5} \approx -1.3764$$

It appears that the two values are equal.

(c) Since $A + B + C = \pi$, $\tan(A+B) = \tan(\pi - C) = -\tan C$. From Exercise 56 we have

$$\tan(A+B) = \frac{\tan A + \tan B}{1 - \tan A \tan B}, \text{ so:}$$

$$\frac{\tan A + \tan B}{1 - \tan A \tan B} = -\tan C$$

$$\tan A + \tan B = -\tan C + \tan A \tan B \tan C$$

$$\tan A + \tan B + \tan C = \tan A \tan B \tan C$$

7.7 Further Identities

1. (a) Using the identity for $\tan(s+t)$, we have:

$$\tan(t+s) = \frac{\tan t + \tan s}{1 - \tan t \tan s} = \frac{\tfrac{3}{4} + \tfrac{7}{24}}{1 - \tfrac{3}{4} \cdot \tfrac{7}{24}} = \frac{\tfrac{25}{24}}{1 - \tfrac{7}{32}} = \frac{\tfrac{25}{24}}{\tfrac{25}{32}} = \frac{4}{3}$$

(b) Using the identity for $\tan(s-t)$, we have:

$$\tan(t-s) = \frac{\tan t - \tan s}{1 + \tan t \tan s} = \frac{\tfrac{3}{4} - \tfrac{7}{24}}{1 + \tfrac{3}{4} \cdot \tfrac{7}{24}} = \frac{\tfrac{11}{24}}{1 + \tfrac{7}{32}} = \frac{\tfrac{11}{24}}{\tfrac{39}{32}} = \frac{44}{117}$$

3. (a) Using the identity for $\tan(s+t)$ and rationalizing denominators, we have:

$$\tan\left(x + \tfrac{3\pi}{4}\right) = \frac{\tan x + \tan\tfrac{3\pi}{4}}{1 - \tan x \tan\tfrac{3\pi}{4}}$$

$$= \frac{\sqrt{2} + (-1)}{1 - (\sqrt{2})(-1)}$$

$$= \frac{\sqrt{2} - 1}{1 + \sqrt{2}} \cdot \frac{1 - \sqrt{2}}{1 - \sqrt{2}}$$

$$= \frac{-3 + 2\sqrt{2}}{1 - 2}$$

$$= 3 - 2\sqrt{2}$$

(b)   Using the identity for tan $(s - t)$ and rationalizing denominators, we have:

$$\tan\left(x - \tfrac{5\pi}{4}\right) = \frac{\tan x - \tan\frac{5\pi}{4}}{1 + \tan x \tan\frac{5\pi}{4}}$$

$$= \frac{\sqrt{2} - 1}{1 + (\sqrt{2})(1)}$$

$$= \frac{\sqrt{2} - 1}{1 + \sqrt{2}} \cdot \frac{1 - \sqrt{2}}{1 - \sqrt{2}}$$

$$= \frac{-3 + 2\sqrt{2}}{1 - 2}$$

$$= 3 - 2\sqrt{2}$$

5.   Since $0° < \phi < 90°$, $\sin\phi > 0$ and thus:

$$\sin\phi = \sqrt{1 - \cos^2\phi} = \sqrt{1 - \left(\tfrac{7}{25}\right)^2} = \sqrt{1 - \tfrac{49}{625}} = \sqrt{\tfrac{576}{625}} = \tfrac{24}{25}$$

(a)   Using the double-angle identity for $\sin 2\phi$, we have:

$$\sin 2\phi = 2\sin\phi\cos\phi = 2 \cdot \tfrac{24}{25} \cdot \tfrac{7}{25} = \tfrac{336}{625}$$

(b)   Using the double-angle identity for $\cos 2\phi$, we have:

$$\cos 2\phi = \cos^2\phi - \sin^2\phi = \left(\tfrac{7}{25}\right)^2 - \left(\tfrac{24}{25}\right)^2 = \tfrac{49}{625} - \tfrac{576}{625} = -\tfrac{527}{625}$$

(c)   Using parts (a) and (b), we have:

$$\tan 2\phi = \frac{\sin 2\phi}{\cos 2\phi} = \frac{\frac{336}{625}}{-\frac{527}{625}} = -\tfrac{336}{527}$$

7.   Since $\frac{3\pi}{2} < u < 2\pi$, $\sec u > 0$ and thus:

$$\sec u = \sqrt{1 + \tan^2 u} = \sqrt{1 + (-4)^2} = \sqrt{1 + 16} = \sqrt{17}$$

Thus $\cos u = \frac{1}{\sqrt{17}}$. Since $\frac{3\pi}{2} < u < 2\pi$, $\sin u < 0$ and thus:

$$\sin u = -\sqrt{1 - \cos^2 u} = -\sqrt{1 - \tfrac{1}{17}} = -\sqrt{\tfrac{16}{17}} = -\tfrac{4}{\sqrt{17}}$$

(a)   Using the double-angle identity for $\sin 2u$, we have:

$$\sin 2u = 2\sin u\cos u = 2 \cdot \left(-\tfrac{4}{\sqrt{17}}\right) \cdot \left(\tfrac{1}{\sqrt{17}}\right) = -\tfrac{8}{17}$$

(b)   Using the double-angle identity for $\cos 2u$, we have:

$$\cos 2u = \cos^2 u - \sin^2 u = \left(\tfrac{1}{\sqrt{17}}\right)^2 - \left(-\tfrac{4}{\sqrt{17}}\right)^2 = \tfrac{1}{17} - \tfrac{16}{17} = -\tfrac{15}{17}$$

(c)   Using parts (a) and (b), we have:

$$\tan 2u = \frac{\sin 2u}{\cos 2u} = \frac{-\frac{8}{17}}{-\frac{15}{17}} = \tfrac{8}{15}$$

9. Since $0° < \alpha < 90°$, $\cos \alpha > 0$ and thus:

$$\cos \alpha = \sqrt{1 - \sin^2 \alpha} = \sqrt{1 - \left(\frac{\sqrt{3}}{2}\right)^2} = \sqrt{1 - \frac{3}{4}} = \sqrt{\frac{1}{4}} = \frac{1}{2}$$

(a) Since $0° < \alpha < 90°$, $0° < \frac{\alpha}{2} < 45°$ and thus $\sin \frac{\alpha}{2} > 0$. Therefore:

$$\sin \frac{\alpha}{2} = \sqrt{\frac{1 - \cos \alpha}{2}} = \sqrt{\frac{1 - \frac{1}{2}}{2}} = \sqrt{\frac{1}{4}} = \frac{1}{2}$$

(b) Since $0° < \alpha < 90°$, $0° < \frac{\alpha}{2} < 45°$ and thus $\cos \frac{\alpha}{2} > 0$. Therefore:

$$\cos \frac{\alpha}{2} = \sqrt{\frac{1 + \cos \alpha}{2}} = \sqrt{\frac{1 + \frac{1}{2}}{2}} = \sqrt{\frac{3}{4}} = \frac{\sqrt{3}}{2}$$

(c) Using parts (a) and (b), we have:

$$\tan \frac{\alpha}{2} = \frac{\sin \frac{\alpha}{2}}{\cos \frac{\alpha}{2}} = \frac{\frac{1}{2}}{\frac{\sqrt{3}}{2}} = \frac{1}{\sqrt{3}} = \frac{\sqrt{3}}{3}$$

Notice that an alternate solution is to spot that since $\sin \alpha = \frac{\sqrt{3}}{2}$ and $0° < \alpha < 90°$, then $\alpha = 60°$ and thus $\frac{\alpha}{2} = 30°$. Thus $\sin \frac{\alpha}{2} = \frac{1}{2}$, $\cos \frac{\alpha}{2} = \frac{\sqrt{3}}{2}$ and $\tan \frac{\alpha}{2} = \frac{\sqrt{3}}{3}$.

11. (a) Since $\frac{\pi}{2} < \theta < \pi$, $\frac{\pi}{4} < \frac{\theta}{2} < \frac{\pi}{2}$ and thus $\sin \frac{\theta}{2} > 0$. Therefore:

$$\sin \frac{\theta}{2} = \sqrt{\frac{1 - \cos \theta}{2}} = \sqrt{\frac{1 + \frac{7}{9}}{2}} = \sqrt{\frac{8}{9}} = \frac{2\sqrt{2}}{3}$$

(b) Since $\frac{\pi}{2} < \theta < \pi$, $\frac{\pi}{4} < \frac{\theta}{2} < \frac{\pi}{2}$ and thus $\cos \frac{\theta}{2} > 0$. Therefore:

$$\cos \frac{\theta}{2} = \sqrt{\frac{1 + \cos \theta}{2}} = \sqrt{\frac{1 - \frac{7}{9}}{2}} = \sqrt{\frac{1}{9}} = \frac{1}{3}$$

(c) Using parts (a) and (b), we have:

$$\tan \frac{\theta}{2} = \frac{\sin \frac{\theta}{2}}{\cos \frac{\theta}{2}} = \frac{\frac{2\sqrt{2}}{3}}{\frac{1}{3}} = 2\sqrt{2}$$

13. Since $\frac{\pi}{2} < \theta < \pi$, $\cos \theta < 0$ and thus:

$$\cos \theta = -\sqrt{1 - \sin^2 \theta} = -\sqrt{1 - \left(\frac{3}{4}\right)^2} = -\sqrt{1 - \frac{9}{16}} = -\sqrt{\frac{7}{16}} = -\frac{\sqrt{7}}{4}$$

(a) Using the double-angle identity for $\sin 2\theta$, we have:

$$\sin 2\theta = 2 \sin \theta \cos \theta = 2 \cdot \left(\frac{3}{4}\right) \cdot \left(-\frac{\sqrt{7}}{4}\right) = -\frac{3\sqrt{7}}{8}$$

(b)　Using the double-angle identity for $\cos 2\theta$, we have:

$$\cos 2\theta = \cos^2 \theta - \sin^2 \theta = \left(-\tfrac{\sqrt{7}}{4}\right)^2 - \left(\tfrac{3}{4}\right)^2 = \tfrac{7}{16} - \tfrac{9}{16} = -\tfrac{1}{8}$$

(c)　Since $\tfrac{\pi}{4} < \tfrac{\theta}{2} < \tfrac{\pi}{2}$, $\sin \tfrac{\theta}{2} > 0$ and thus:

$$\sin \tfrac{\theta}{2} = \sqrt{\frac{1 - \cos \theta}{2}} = \sqrt{\frac{1 - \left(-\tfrac{\sqrt{7}}{4}\right)}{2}} = \sqrt{\frac{4 + \sqrt{7}}{8}} = \frac{\sqrt{8 + 2\sqrt{7}}}{4}$$

(d)　Since $\tfrac{\pi}{4} < \tfrac{\theta}{2} < \tfrac{\pi}{2}$, $\cos \tfrac{\theta}{2} > 0$ and thus:

$$\cos \tfrac{\theta}{2} = \sqrt{\frac{1 + \cos \theta}{2}} = \sqrt{\frac{1 - \tfrac{\sqrt{7}}{4}}{2}} = \sqrt{\frac{4 - \sqrt{7}}{8}} = \frac{\sqrt{8 - 2\sqrt{7}}}{4}$$

15.　Since $180° < \theta < 270°$, $\sin \theta < 0$ and thus:

$$\sin \theta = -\sqrt{1 - \cos^2 \theta} = -\sqrt{1 - \left(-\tfrac{1}{3}\right)^2} = -\sqrt{1 - \tfrac{1}{9}} = -\sqrt{\tfrac{8}{9}} = -\frac{2\sqrt{2}}{3}$$

(a)　Using the double-angle identity for $\sin 2\theta$, we have:

$$\sin 2\theta = 2 \sin \theta \cos \theta = 2 \cdot \left(-\tfrac{2\sqrt{2}}{3}\right) \cdot \left(-\tfrac{1}{3}\right) = \tfrac{4\sqrt{2}}{9}$$

(b)　Using the double-angle identity for $\cos 2\theta$, we have:

$$\cos 2\theta = \cos^2 \theta - \sin^2 \theta = \left(-\tfrac{1}{3}\right)^2 - \left(-\tfrac{2\sqrt{2}}{3}\right)^2 = \tfrac{1}{9} - \tfrac{8}{9} = -\tfrac{7}{9}$$

(c)　Since $90° < \tfrac{\theta}{2} < 135°$, $\sin \tfrac{\theta}{2} > 0$ and thus:

$$\sin \tfrac{\theta}{2} = \sqrt{\frac{1 - \cos \theta}{2}} = \sqrt{\frac{1 + \tfrac{1}{3}}{2}} = \sqrt{\tfrac{2}{3}} = \frac{\sqrt{6}}{3}$$

(d)　Since $90° < \tfrac{\theta}{2} < 135°$, $\cos \tfrac{\theta}{2} < 0$ and thus:

$$\cos \tfrac{\theta}{2} = -\sqrt{\frac{1 + \cos \theta}{2}} = -\sqrt{\frac{1 - \tfrac{1}{3}}{2}} = -\sqrt{\tfrac{1}{3}} = -\frac{\sqrt{3}}{3}$$

17.　(a)　Since $0 < \tfrac{\pi}{12} < \tfrac{\pi}{2}$, $\sin \tfrac{\pi}{12} > 0$. Using the half-angle formula for $\sin \tfrac{1}{2} x$, we have:

$$\sin \tfrac{\pi}{12} = \sqrt{\frac{1 - \cos \tfrac{\pi}{6}}{2}} = \sqrt{\frac{1 - \tfrac{\sqrt{3}}{2}}{2}} = \sqrt{\frac{2 - \sqrt{3}}{4}} = \frac{\sqrt{2 - \sqrt{3}}}{2}$$

(b)　Since $0 < \tfrac{\pi}{12} < \tfrac{\pi}{2}$, $\cos \tfrac{\pi}{12} > 0$. Using the half-angle formula for $\cos \tfrac{1}{2} x$, we have:

$$\cos \tfrac{\pi}{12} = \sqrt{\frac{1 + \cos \tfrac{\pi}{6}}{2}} = \sqrt{\frac{1 + \tfrac{\sqrt{3}}{2}}{2}} = \frac{\sqrt{2 + \sqrt{3}}}{2}$$

(c) Using the half-angle formula for $\tan\frac{1}{2}x$, we have:

$$\tan\frac{\pi}{12} = \frac{\sin\frac{\pi}{6}}{1+\cos\frac{\pi}{6}} = \frac{\frac{1}{2}}{1+\frac{\sqrt{3}}{2}} = \frac{1}{2+\sqrt{3}}\cdot\frac{2-\sqrt{3}}{2-\sqrt{3}} = \frac{2-\sqrt{3}}{4-3} = 2-\sqrt{3}$$

Note: We could also use parts (a) and (b), as follows:

$$\tan\frac{\pi}{12} = \frac{\sin\frac{\pi}{12}}{\cos\frac{\pi}{12}}$$
$$= \frac{\frac{\sqrt{2-\sqrt{3}}}{2}}{\frac{\sqrt{2+\sqrt{3}}}{2}}$$
$$= \frac{\sqrt{2-\sqrt{3}}}{\sqrt{2+\sqrt{3}}}\cdot\frac{\sqrt{2+\sqrt{3}}}{\sqrt{2+\sqrt{3}}}$$
$$= \frac{\sqrt{4-3}}{2+\sqrt{3}}\cdot\frac{2-\sqrt{3}}{2-\sqrt{3}}$$
$$= \frac{2-\sqrt{3}}{4-3}$$
$$= 2-\sqrt{3}$$

19. (a) Since $90° < 105° < 180°$, $\sin 105° > 0$. Using the half-angle formula for $\sin\frac{1}{2}x$, we have:

$$\sin 105° = \sqrt{\frac{1-\cos 210°}{2}} = \sqrt{\frac{1+\frac{\sqrt{3}}{2}}{2}} = \sqrt{\frac{2+\sqrt{3}}{4}} = \frac{\sqrt{2+\sqrt{3}}}{2}$$

(b) Since $90° < 105° < 180°$, $\cos 105° < 0$. Using the half-angle formula for $\cos\frac{1}{2}x$, we have:

$$\cos 105° = -\sqrt{\frac{1+\cos 210°}{2}} = -\sqrt{\frac{1-\frac{\sqrt{3}}{2}}{2}} = -\sqrt{\frac{2-\sqrt{3}}{4}} = -\frac{\sqrt{2-\sqrt{3}}}{2}$$

(c) Using the half-angle formula for $\tan\frac{1}{2}x$, we have:

$$\tan 105° = \frac{\sin 210°}{1+\cos 210°} = \frac{-\frac{1}{2}}{1-\frac{\sqrt{3}}{2}} = -\frac{1}{2-\sqrt{3}}\cdot\frac{2+\sqrt{3}}{2+\sqrt{3}} = -\frac{2+\sqrt{3}}{4-3} = -2-\sqrt{3}$$

21. From the first triangle we have $\sin\theta = \frac{3}{5}$, $\cos\theta = \frac{4}{5}$ and $\tan\theta = \frac{3}{4}$.

(a) By the double-angle identity for $\sin 2\theta$, we have:
$$\sin 2\theta = 2\sin\theta\cos\theta = 2\cdot\frac{3}{5}\cdot\frac{4}{5} = \frac{24}{25}$$

(b) By the double-angle identity for $\cos 2\theta$, we have:
$$\cos 2\theta = \cos^2\theta - \sin^2\theta = \left(\frac{4}{5}\right)^2 - \left(\frac{3}{5}\right)^2 = \frac{16}{25} - \frac{9}{25} = \frac{7}{25}$$

(c)   By the double-angle identity for $\tan 2\theta$, we have:

$$\tan 2\theta = \frac{2\tan\theta}{1-\tan^2\theta} = \frac{2\cdot\frac{3}{4}}{1-\left(\frac{3}{4}\right)^2} = \frac{\frac{3}{2}}{\frac{7}{16}} = \frac{24}{7}$$

Note: An easier approach, after doing (a) and (b), would be to say:

$$\tan 2\theta = \frac{\sin 2\theta}{\cos 2\theta} = \frac{\frac{24}{25}}{\frac{7}{25}} = \frac{24}{7}$$

23.   From the first triangle we have $\sin\beta = \frac{4}{5}$ and $\cos\beta = \frac{3}{5}$.

(a)   By the double-angle identity for $\sin 2\beta$, we have:

$$\sin 2\beta = 2\sin\beta\cos\beta = 2\cdot\frac{4}{5}\cdot\frac{3}{5} = \frac{24}{25}$$

(b)   By the double-angle identity for $\cos 2\beta$, we have:

$$\cos 2\beta = \cos^2\beta - \sin^2\beta = \left(\frac{3}{5}\right)^2 - \left(\frac{4}{5}\right)^2 = \frac{9}{25} - \frac{16}{25} = -\frac{7}{25}$$

(c)   Using the identity for $\tan x$, we have:

$$\tan 2\beta = \frac{\sin 2\beta}{\cos 2\beta} = \frac{\frac{24}{25}}{-\frac{7}{25}} = -\frac{24}{7}$$

Note: We could also have used the double-angle formula for $\tan 2\beta$.

25.   From the first triangle we have $\sin\theta = \frac{3}{5}$ and $\cos\theta = \frac{4}{5}$.

(a)   Since $\sin\frac{\theta}{2} > 0$, we use the half-angle formula for $\sin\frac{\theta}{2}$ to obtain:

$$\sin\frac{\theta}{2} = \sqrt{\frac{1-\cos\theta}{2}} = \sqrt{\frac{1-\frac{4}{5}}{2}} = \sqrt{\frac{1}{10}} = \frac{\sqrt{10}}{10}$$

(b)   Since $\cos\frac{\theta}{2} > 0$, we use the half-angle formula for $\cos\frac{\theta}{2}$ to obtain:

$$\cos\frac{\theta}{2} = \sqrt{\frac{1+\cos\theta}{2}} = \sqrt{\frac{1+\frac{4}{5}}{2}} = \sqrt{\frac{9}{10}} = \frac{3\sqrt{10}}{10}$$

(c)   Using the half-angle formula for $\tan\frac{\theta}{2}$, we obtain:

$$\tan\frac{\theta}{2} = \frac{\sin\theta}{1+\cos\theta} = \frac{\frac{3}{5}}{1+\frac{4}{5}} = \frac{3}{9} = \frac{1}{3}$$

Note: We could also have computed this directly after parts (a) and (b) as:

$$\tan\frac{\theta}{2} = \frac{\sin\frac{\theta}{2}}{\cos\frac{\theta}{2}} = \frac{\frac{\sqrt{10}}{10}}{\frac{3\sqrt{10}}{10}} = \frac{1}{3}$$

27. From the first triangle we have $\cos \beta = \frac{3}{5}$.

    (a) Since $\sin \frac{\beta}{2} > 0$, we use the half-angle formula for $\sin \frac{\beta}{2}$ to obtain:

$$\sin \frac{\beta}{2} = \sqrt{\frac{1 - \cos \beta}{2}} = \sqrt{\frac{1 - \frac{3}{5}}{2}} = \sqrt{\frac{1}{5}} = \frac{\sqrt{5}}{5}$$

    (b) Since $\cos \frac{\beta}{2} > 0$, we use the half-angle formula for $\cos \frac{\beta}{2}$ to obtain:

$$\cos \frac{\beta}{2} = \sqrt{\frac{1 + \cos \beta}{2}} = \sqrt{\frac{1 + \frac{3}{5}}{2}} = \sqrt{\frac{4}{5}} = \frac{2\sqrt{5}}{5}$$

    (c) Using the identity for $\tan x$ and parts (a) and (b), we have:

$$\tan \frac{\beta}{2} = \frac{\sin \frac{\beta}{2}}{\cos \frac{\beta}{2}} = \frac{\frac{\sqrt{5}}{5}}{\frac{2\sqrt{5}}{5}} = \frac{1}{2}$$

29. Since $0 < \theta < \frac{\pi}{2}$, $\cos \theta > 0$. Since $\sin \theta = \frac{x}{5}$, we have:

$$\cos \theta = \sqrt{1 - \sin^2 \theta} = \sqrt{1 - \left(\frac{x}{5}\right)^2} = \sqrt{1 - \frac{x^2}{25}} = \frac{\sqrt{25 - x^2}}{5}$$

We now apply the double angle formulas:

$$\sin 2\theta = 2\sin \theta \cos \theta = 2 \cdot \frac{x}{5} \cdot \frac{\sqrt{25 - x^2}}{5} = \frac{2x\sqrt{25 - x^2}}{25}$$

$$\cos 2\theta = \cos^2 \theta - \sin^2 \theta = \left(\frac{\sqrt{25 - x^2}}{5}\right)^2 - \left(\frac{x}{5}\right)^2 = \frac{25 - x^2}{25} - \frac{x^2}{25} = \frac{25 - 2x^2}{25}$$

31. Since $0 < \theta < \frac{\pi}{2}$, $\cos \theta > 0$. Since $\sin \theta = \frac{x-1}{2}$, we have:

$$\cos \theta = \sqrt{1 - \sin^2 \theta} = \sqrt{1 - \left(\frac{x-1}{2}\right)^2} = \sqrt{1 - \frac{x^2 - 2x + 1}{4}} = \frac{\sqrt{3 + 2x - x^2}}{2}$$

We now apply the double-angle formulas:

$$\sin 2\theta = 2\sin \theta \cos \theta = 2 \cdot \frac{x-1}{2} \cdot \frac{\sqrt{3 + 2x - x^2}}{2} = \frac{(x-1)\sqrt{3 + 2x - x^2}}{2}$$

$$\cos 2\theta = \cos^2 \theta - \sin^2 \theta$$

$$= \left(\frac{\sqrt{3 + 2x - x^2}}{2}\right)^2 - \left(\frac{x-1}{2}\right)^2$$

$$= \frac{3 + 2x - x^2}{4} - \frac{x^2 - 2x + 1}{4}$$

$$= \frac{2 + 4x - 2x^2}{4}$$

$$= \frac{1 + 2x - x^2}{2}$$

33. Using the identity $\sin^2 \theta = \frac{1-\cos 2\theta}{2}$, we have:

$$\sin^4 \theta = \left(\sin^2 \theta\right)^2$$

$$= \left(\frac{1-\cos 2\theta}{2}\right)^2$$

$$= \frac{1-2\cos 2\theta+(\cos 2\theta)^2}{4}$$

$$= \frac{1-2\cos 2\theta+\frac{1+\cos 4\theta}{2}}{4}$$

$$= \frac{2-4\cos 2\theta+1+\cos 4\theta}{8}$$

$$= \frac{3-4\cos 2\theta+\cos 4\theta}{8}$$

35. Using the identities $\sin^2 \frac{\theta}{2} = \frac{1-\cos \theta}{2}$ and $\cos^2 \theta = \frac{1+\cos 2\theta}{2}$, we have:

$$\sin^4 \frac{\theta}{2} = \left(\sin^2 \frac{\theta}{2}\right)^2$$

$$= \left(\frac{1-\cos \theta}{2}\right)^2$$

$$= \frac{1-2\cos \theta+\cos^2 \theta}{4}$$

$$= \frac{1-2\cos \theta+\frac{1+\cos 2\theta}{2}}{4}$$

$$= \frac{2-4\cos \theta+1+\cos 2\theta}{8}$$

$$= \frac{3-4\cos \theta+\cos 2\theta}{8}$$

37. (a) Replacing 2θ with $\theta+\theta$ and using the addition formula for $\cos(s+t)$, we have:

$$\cos 2\theta = \cos(\theta+\theta) = \cos \theta \cos \theta - \sin \theta \sin \theta = \cos^2 \theta - \sin^2 \theta$$

(b) Replacing 2θ with $\theta+\theta$ and using the addition formula for $\tan(s+t)$, we have:

$$\tan 2\theta = \tan(\theta+\theta) = \frac{\tan \theta + \tan \theta}{1-\tan \theta \tan \theta} = \frac{2\tan \theta}{1-\tan^2 \theta}$$

39. Working from the right-hand side, we have:

$$\frac{1-\tan^2 s}{1+\tan^2 s} = \frac{1-\frac{\sin^2 s}{\cos^2 s}}{1+\frac{\sin^2 s}{\cos^2 s}} = \frac{\cos^2 s - \sin^2 s}{\cos^2 s + \sin^2 s} = \frac{\cos 2s}{1} = \cos 2s$$

41. Writing $\theta = 2 \cdot \frac{\theta}{2}$, we apply the double-angle formula:
$$\cos\theta = \cos\left(2 \cdot \frac{\theta}{2}\right) = \cos^2\frac{\theta}{2} - \sin^2\frac{\theta}{2} = \cos^2\frac{\theta}{2} - \left(1 - \cos^2\frac{\theta}{2}\right) = 2\cos^2\frac{\theta}{2} - 1$$

43. Using the identities $\sin^2\theta = \frac{1-\cos 2\theta}{2}$ and $\cos^2 2\theta = \frac{1+\cos 4\theta}{2}$, we have:
$$\sin^4\theta = \left(\sin^2\theta\right)^2$$
$$= \left(\frac{1-\cos 2\theta}{2}\right)^2$$
$$= \frac{1 - 2\cos 2\theta + \cos^2 2\theta}{4}$$
$$= \frac{1 - 2\cos 2\theta + \frac{1+\cos 4\theta}{2}}{4}$$
$$= \frac{2 - 4\cos 2\theta + 1 + \cos 4\theta}{8}$$
$$= \frac{3 - 4\cos 2\theta + \cos 4\theta}{8}$$

45. Working from the right-hand side, we have:
$$\frac{2\tan\theta}{1+\tan^2\theta} = \frac{2 \cdot \frac{\sin\theta}{\cos\theta}}{1 + \frac{\sin^2\theta}{\cos^2\theta}} = \frac{2\sin\theta\cos\theta}{\cos^2\theta + \sin^2\theta} = \frac{\sin 2\theta}{1} = \sin 2\theta$$

47. Working from the right-hand side, we have:
$$2\sin^3\theta\cos\theta + 2\sin\theta\cos^3\theta = (2\sin\theta\cos\theta)\left(\sin^2\theta + \cos^2\theta\right) = \sin 2\theta \cdot 1 = \sin 2\theta$$

49. Working from the left-hand side, we have:
$$\frac{1+\tan\frac{\theta}{2}}{1-\tan\frac{\theta}{2}} = \frac{1 + \frac{\sin\frac{\theta}{2}}{\cos\frac{\theta}{2}}}{1 - \frac{\sin\frac{\theta}{2}}{\cos\frac{\theta}{2}}}$$
$$= \frac{\cos\frac{\theta}{2} + \sin\frac{\theta}{2}}{\cos\frac{\theta}{2} - \sin\frac{\theta}{2}}$$
$$= \frac{\left(\cos\frac{\theta}{2} + \sin\frac{\theta}{2}\right)^2}{\cos^2\frac{\theta}{2} - \sin^2\frac{\theta}{2}}$$
$$= \frac{1 + 2\cos\frac{\theta}{2}\sin\frac{\theta}{2}}{\cos\theta}$$
$$= \frac{1 + \sin\theta}{\cos\theta}$$
$$= \tan\theta + \sec\theta$$

51. Working from the left-hand side and using the addition formula for sin $(s - t)$, we have:

$$2\sin^2(45° - \theta) = 2(\sin 45° \cos\theta - \cos 45° \sin\theta)^2$$

$$= 2\left(\frac{\cos\theta - \sin\theta}{\sqrt{2}}\right)^2$$

$$= 2 \cdot \frac{\cos^2\theta - 2\sin\theta\cos\theta + \sin^2\theta}{2}$$

$$= 1 - 2\cos\theta\sin\theta$$

$$= 1 - \sin 2\theta$$

53. Simplifying the left-hand side, we have:

$$1 + \tan\theta\tan 2\theta = 1 + \tan\theta \cdot \frac{2\tan\theta}{1 - \tan^2\theta}$$

$$= 1 + \frac{2\tan^2\theta}{1 - \tan^2\theta}$$

$$= \frac{1 - \tan^2\theta + 2\tan^2\theta}{1 - \tan^2\theta}$$

$$= \frac{1 + \tan^2\theta}{1 - \tan^2\theta}$$

Simplifying the right-hand side, we have:

$$\tan 2\theta\cot\theta - 1 = \frac{2\tan\theta}{1 - \tan^2\theta} \cdot \frac{1}{\tan\theta} - 1 = \frac{2}{1 - \tan^2\theta} - 1 = \frac{2 - 1 + \tan^2\theta}{1 - \tan^2\theta} = \frac{1 + \tan^2\theta}{1 - \tan^2\theta}$$

Since both sides simplify to the same quantity, the original equation is an identity.

55. Writing $\alpha + 2\beta = (\alpha + \beta) + \beta$, we apply the addition formula for sin $(s + t)$ to obtain:

$$\sin(\alpha + 2\beta) = \sin\left[(\alpha + \beta) + \beta\right]$$

$$= \sin(\alpha + \beta)\cos\beta + \sin\beta\cos(\alpha + \beta)$$

$$= \left[\sin\alpha\cos\beta + \sin\beta\cos\alpha\right]\cos\beta + \sin\beta \cdot 0$$

$$= \sin\alpha\cos^2\beta + \cos\alpha\sin\beta\cos\beta$$

$$= (\sin\alpha)(1 - \sin^2\beta) + \cos\alpha\sin\beta\cos\beta$$

$$= \sin\alpha + \sin\beta(\cos\alpha\cos\beta - \sin\alpha\sin\beta)$$

$$= \sin\alpha + \sin\beta\cos(\alpha + \beta)$$

$$= \sin\alpha + \sin\beta \cdot 0$$

$$= \sin\alpha$$

57. Since $0 < \theta < \frac{\pi}{2}$, $\cos\theta > 0$ and thus (since $a > 0$, $b > 0$):

$$\cos\theta = \sqrt{1 - \sin^2\theta}$$

$$= \sqrt{1 - \left(\frac{a^2 - b^2}{a^2 + b^2}\right)^2}$$

$$= \sqrt{\frac{(a^2 + b^2)^2 - (a^2 - b^2)^2}{(a^2 + b^2)^2}}$$

$$= \sqrt{\frac{4a^2b^2}{(a^2 + b^2)^2}}$$

$$= \frac{2ab}{a^2 + b^2}$$

Now using the half-angle formula for $\tan \frac{\theta}{2}$, we have:

$$\tan \frac{\theta}{2} = \frac{\sin \theta}{1+\cos \theta} = \frac{a^2 - b^2}{a^2 + b^2 + 2ab} = \frac{a^2 - b^2}{(a+b)^2} = \frac{a-b}{a+b}$$

59. (a) Using $\triangle ODC$ and the fact that $OC = 1$, we have $OD = \cos \theta$ and $DC = \sin \theta$. Now using $\triangle ADC$ and the fact that $AO = 1$, we have:

$$\tan \frac{\theta}{2} = \frac{CD}{AO + OD} = \frac{\sin \theta}{1 + \cos \theta}$$

(b) Using the formula from (a) and rationalizing denominators, we have:

$$\tan 15° = \frac{\sin 30°}{1 + \cos 30°} = \frac{\frac{1}{2}}{1 + \frac{\sqrt{3}}{2}} = \frac{1}{2 + \sqrt{3}} \cdot \frac{2 - \sqrt{3}}{2 - \sqrt{3}} = \frac{2 - \sqrt{3}}{4 - 3} = 2 - \sqrt{3}$$

Using the formula from (a) and rationalizing denominators, we have:

$$\tan \frac{\pi}{8} = \frac{\sin \frac{\pi}{4}}{1 + \cos \frac{\pi}{4}} = \frac{\frac{\sqrt{2}}{2}}{1 + \frac{\sqrt{2}}{2}} = \frac{\sqrt{2}}{2 + \sqrt{2}} \cdot \frac{2 - \sqrt{2}}{2 - \sqrt{2}} = \frac{2\sqrt{2} - 2}{4 - 2} = \sqrt{2} - 1$$

61. (a) Following the suggestion and applying the double-angle formulas for $\sin 4\theta$ and $\sin 2\theta$, we have:

$$\frac{\sin 4\theta}{4\sin \theta} = \frac{2\sin 2\theta \cos 2\theta}{4\sin \theta} = \frac{4\sin \theta \cos \theta \cos 2\theta}{4\sin \theta} = \cos \theta \cos 2\theta$$

(b) Following the suggestion and applying the double-angle formulas for $\sin 8\theta$, $\sin 4\theta$ and $\sin 2\theta$, we have:

$$\begin{aligned}
\frac{\sin 8\theta}{8\sin \theta} &= \frac{2\sin 4\theta \cos 4\theta}{8\sin \theta} \\
&= \frac{4\sin 2\theta \cos 2\theta \cos 4\theta}{8\sin \theta} \\
&= \frac{8\sin \theta \cos \theta \cos 2\theta \cos 4\theta}{8\sin \theta} \\
&= \cos \theta \cos 2\theta \cos 4\theta
\end{aligned}$$

(c) Following the suggestion and applying the double-angle formulas for $\sin 16\theta$, $\sin 8\theta$, $\sin 4\theta$ and $\sin 2\theta$, we have:

$$\begin{aligned}
\frac{\sin 16\theta}{16\sin \theta} &= \frac{2\sin 8\theta \cos 8\theta}{16\sin \theta} \\
&= \frac{4\sin 4\theta \cos 4\theta \cos 8\theta}{16\sin \theta} \\
&= \frac{8\sin 2\theta \cos 2\theta \cos 4\theta \cos 8\theta}{16\sin \theta} \\
&= \frac{16\sin \theta \cos \theta \cos 2\theta \cos 4\theta \cos 8\theta}{16\sin \theta} \\
&= \cos \theta \cos 2\theta \cos 4\theta \cos 8\theta
\end{aligned}$$

63. (a) Using the product-to-sum identity for $\sin A \cos B$, we have:
$$4\sin 6\theta \cos 2\theta = 4 \cdot \tfrac{1}{2}[\sin(6\theta + 2\theta) + \sin(6\theta - 2\theta)] = 2\sin 8\theta + 2\sin 4\theta$$

(b) Using the product-to-sum identity for $\sin A \sin B$, we have:
$$2\sin A \sin 3A = 2 \cdot \tfrac{1}{2}[\cos(A - 3A) - \cos(A + 3A)] = \cos 2A - \cos 4A$$

(c) Using the product-to-sum identity for $\cos A \cos B$, we have:
$$\cos 4\theta \cos 2\theta = \tfrac{1}{2}[\cos(4\theta + 2\theta) + \cos(4\theta - 2\theta)] = \tfrac{1}{2}\cos 6\theta + \tfrac{1}{2}\cos 2\theta$$

65. (a) Using the product-to-sum identity, we have:
$$\sin(A + B)\sin(A - B) = \frac{\cos[(A+B) - (A-B)] - \cos[(A+B) + (A-B)]}{2}$$
$$= \frac{\cos 2B - \cos 2A}{2}$$
$$= \frac{1 - 2\sin^2 B - (1 - 2\sin^2 A)}{2}$$
$$= \frac{2\sin^2 A - 2\sin^2 B}{2}$$
$$= \sin^2 A - \sin^2 B$$

(b) Using the product-to-sum identity, we have:
$$\cos(A + B)\cos(A - B) = \frac{\cos[(A+B) + (A-B)] + \cos[(A+B) - (A-B)]}{2}$$
$$= \frac{\cos 2A + \cos 2B}{2}$$
$$= \frac{2\cos^2 A - 1 + 1 - 2\sin^2 B}{2}$$
$$= \frac{2\cos^2 A - 2\sin^2 B}{2}$$
$$= \cos^2 A - \sin^2 B$$

67. (a) Letting $\alpha = A + B$ and $\beta = B - A$, then $B = \frac{\alpha+\beta}{2}$ and $A = \frac{\alpha-\beta}{2}$. Now using the product-to-sum identity for $\sin A \cos B$, we have:
$$2\cos\frac{\alpha+\beta}{2}\sin\frac{\alpha-\beta}{2} = \sin\alpha - \sin\beta$$

(b) Letting $\alpha = A + B$ and $\beta = A - B$, then $A = \frac{\alpha+\beta}{2}$ and $B = \frac{\alpha-\beta}{2}$. Now using the product-to-sum identity for $\cos A \cos B$, we have:
$$2\cos\frac{\alpha+\beta}{2}\cos\frac{\alpha-\beta}{2} = \cos\alpha + \cos\beta$$

(c)   Letting $\beta = A + B$ and $\alpha = B - A$, then $A = \frac{\beta-\alpha}{2}$ and $B = \frac{\alpha+\beta}{2}$. Now using the product-to-sum identity for $\sin A \sin B$, we have:

$$-2\sin\frac{\alpha-\beta}{2}\sin\frac{\alpha+\beta}{2} = \cos\alpha - \cos\beta$$

69.   Using the sum-to-product identities for $\sin\alpha + \sin\beta$ and $\cos\alpha + \cos\beta$, we have:

$$\frac{\sin\theta + \sin 3\theta}{\cos\theta + \cos 3\theta} = \frac{2\sin\left(\frac{\theta+3\theta}{2}\right)\cos\left(\frac{\theta-3\theta}{2}\right)}{2\cos\left(\frac{\theta+3\theta}{2}\right)\cos\left(\frac{\theta-3\theta}{2}\right)} = \frac{\sin 2\theta\,\cos(-\theta)}{\cos 2\theta\,\cos(-\theta)} = \tan 2\theta$$

71.   Using the sum-to-product identities for $\sin\alpha + \sin\beta$ and $\cos\alpha + \cos\beta$, we have:

$$\frac{\sin 2x + \sin 2y}{\cos 2x + \cos 2y} = \frac{2\sin\left(\frac{2x+2y}{2}\right)\cos\left(\frac{2x-2y}{2}\right)}{2\cos\left(\frac{2x+2y}{2}\right)\cos\left(\frac{2x-2y}{2}\right)} = \frac{\sin(x+y)}{\cos(x+y)} = \tan(x+y)$$

73.   (a)   Using a calculator, we have $\cos 72° + \cos 144° = -0.5$.

(b)   Using the observation and the addition formulas for cosine, we have:
$$\cos 72° + \cos 144°$$
$$= \cos(108° - 36°) + \cos(108° + 36°)$$
$$= (\cos 108°\cos 36° + \sin 108°\sin 36°) + (\cos 108°\cos 36° - \sin 108°\sin 36°)$$
$$= 2\cos 108°\cos 36°$$

(c)   Since $108° = 180° - 72°$, $\cos 108° = -\cos 72°$. Similarly, since $36° = 180° - 144°$, $\cos 36° = -\cos 144°$. Therefore:
$$\cos 108°\cos 36° = (-\cos 72°)(-\cos 144°) = \cos 72°\cos 144°$$

(d)   Using the above relationships, we have:
$$\cos 72° + \cos 144° = 2\cos 108°\cos 36° = 2\cos 72°\cos 144° = 2\left(-\tfrac{1}{4}\right) = -\tfrac{1}{2}$$

75.   The coordinates of points A_1, A_2, and A_3 are:

$$A_1(1,0)$$
$$A_2(\cos 120°, \sin 120°) = A_2\left(-\tfrac{1}{2}, \tfrac{\sqrt{3}}{2}\right)$$
$$A_3(\cos 240°, \sin 240°) = A_3\left(-\tfrac{1}{2}, -\tfrac{\sqrt{3}}{2}\right)$$

Using the distance formula from the point $P(x,0)$, we have:

$$PA_1 = \sqrt{(x-1)^2 + (0-0)^2} = \sqrt{(x-1)^2} = |x-1| = 1-x, \text{ since } x < 1$$

$$PA_2 = \sqrt{\left(x+\tfrac{1}{2}\right)^2 + \left(0 - \tfrac{\sqrt{3}}{2}\right)^2} = \sqrt{x^2 + x + \tfrac{1}{4} + \tfrac{3}{4}} = \sqrt{x^2 + x + 1}$$

$$PA_3 = \sqrt{\left(x+\tfrac{1}{2}\right)^2 + \left(0 + \tfrac{\sqrt{3}}{2}\right)^2} = \sqrt{x^2 + x + \tfrac{1}{4} + \tfrac{3}{4}} = \sqrt{x^2 + x + 1}$$

We can now compute the product:

$$(PA_1)(PA_2)(PA_3) = (1-x)\sqrt{x^2+x+1}\sqrt{x^2+x+1} = (1-x)(1+x+x^2) = 1-x^3$$

7.8 Trigonometric Equations

1.   For $\theta = \frac{\pi}{2}$, $2\cos^2\theta - 3\cos\theta = 2(0)^2 - 3(0) = 0$, so $\theta = \frac{\pi}{2}$ is a solution.

3.   For $x = \frac{3\pi}{4}$, $\tan^2 x - 3\tan x + 2 = (-1)^2 + 3 + 2 = 6$, so $x = \frac{3\pi}{4}$ is not a solution.

5.   Since $\sin\theta = \frac{\sqrt{3}}{2}$, $\theta = \frac{\pi}{3}$ and $\theta = \frac{2\pi}{3}$ are the primary solutions. All solutions will be of the form $\theta = \frac{\pi}{3} + 2\pi k$ or $\theta = \frac{2\pi}{3} + 2\pi k$, where k is any integer.

7.   Since $\sin\theta = -\frac{1}{2}$, $\theta = \frac{7\pi}{6}$ and $\theta = \frac{11\pi}{6}$ are the primary solutions. All solutions will be of the form $\theta = \frac{7\pi}{6} + 2\pi k$ or $\theta = \frac{11\pi}{6} + 2\pi k$, where k is any integer.

9.   Since $\cos\theta = -1$, $\theta = \pi$ is the primary solution. All solutions will be of the form $\theta = \pi + 2\pi k$, where k is any integer.

11.   Since $\tan\theta = \sqrt{3}$, $\theta = \frac{\pi}{3}$ is the primary solution. All solutions will be of the form $\theta = \frac{\pi}{3} + \pi k$, where k is any integer.

13.   Since $\tan x = 0$, $x = 0$ is the primary solution. All solutions will be of the form $x = 0 + \pi k = \pi k$, where k is any integer.

15.   Since $2\cos^2\theta + \cos\theta = \cos\theta(2\cos\theta + 1) = 0$, the primary solutions are the solutions of $\cos\theta = 0$ or $\cos\theta = -\frac{1}{2}$, which are $\theta = \frac{\pi}{2}$, $\theta = \frac{3\pi}{2}$, $\theta = \frac{2\pi}{3}$ or $\theta = \frac{4\pi}{3}$. All solutions will be of the form $\theta = \frac{\pi}{2} + \pi k$, $\theta = \frac{2\pi}{3} + 2\pi k$ or $\theta = \frac{4\pi}{3} + 2\pi k$, where k is any integer.

17.   Since $\cos^2 t\sin t - \sin t = \sin t(\cos^2 t - 1) = 0$, $\sin t = 0$ or $\cos t = \pm 1$. Therefore primary solutions are $t = 0$ or $t = \pi$. All solutions are of the form $t = \pi k$, where k is any integer.

19.   Using the identity $\cos^2 x = 1 - \sin^2 x$, we have:
$$2\cos^2 x - \sin x - 1 = 2(1 - \sin^2 x) - \sin x - 1$$
$$= -2\sin^2 x - \sin x + 1$$
$$= (-2\sin x + 1)(\sin x + 1)$$

So $\sin x = \frac{1}{2}$ or $\sin x = -1$. Thus the primary solutions are $x = \frac{\pi}{6}$, $x = \frac{5\pi}{6}$ or $x = \frac{3\pi}{2}$. All solutions are of the form $x = \frac{\pi}{6} + 2\pi k$, $x = \frac{5\pi}{6} + 2\pi k$ or $x = \frac{3\pi}{2} + 2\pi k$, where k is any integer.

21. Since $\sqrt{3}\sin t - \sqrt{1 + \sin^2 t} = 0$ is equivalent to $3\sin^2 t = 1 + \sin^2 t$ by squaring each side, $2\sin^2 t = 1$ and thus $\sin^2 t = \frac{1}{2}$ and $\sin t = \pm\frac{\sqrt{2}}{2}$. This would have primary solutions of $\frac{\pi}{4}, \frac{3\pi}{4}, \frac{5\pi}{4}$ or $\frac{7\pi}{4}$, but $\frac{5\pi}{4}$ and $\frac{7\pi}{4}$ do not work in the original equation. So the primary solutions are $t = \frac{\pi}{4}$ or $t = \frac{3\pi}{4}$. All solutions are of the form $t = \frac{\pi}{4} + 2\pi k$ or $t = \frac{3\pi}{4} + 2\pi k$, where k is any integer.

23. Since $\cos 3\theta = 1$, $3\theta = 2\pi k$ for any integer k. So $\theta = \frac{2\pi k}{3}$. Thus the values of θ in the interval $[0°, 360°)$ are $0°$, $120°$ and $240°$.

25. Since $\sin 3\theta = -\frac{\sqrt{2}}{2}$, $3\theta = \frac{5\pi}{4} + 2\pi k$ or $\frac{7\pi}{4} + 2\pi k$. Thus $\theta = \frac{5\pi}{12} + \frac{2\pi k}{3}$ or $\frac{7\pi}{12} + \frac{2\pi k}{3}$. So the primary solutions are $75°$, $105°$, $195°$, $225°$, $315°$ or $345°$.

27. Using the hint, we have:
$$\sin\theta = \cos\frac{\theta}{2}$$
$$2\sin\frac{\theta}{2}\cos\frac{\theta}{2} = \cos\frac{\theta}{2}$$
$$\left(\cos\frac{\theta}{2}\right)\left(2\sin\frac{\theta}{2} - 1\right) = 0$$
Thus $\cos\frac{\theta}{2} = 0$ or $\sin\frac{\theta}{2} = \frac{1}{2}$. Now $\sin\frac{\theta}{2} = \frac{1}{2}$ when $\frac{\theta}{2} = 30°$ or $\frac{\theta}{2} = 150°$, and therefore when $\theta = 60°$ or $300°$. When $\cos\frac{\theta}{2} = 0$ we have $\frac{\theta}{2} = 90° + 360°k$ or $270° + 360°k$, so $\theta = 180°$. Combining we have $\theta = 60°$, $180°$ or $300°$.

29. Dividing each side by $\cos 2\theta$ results in $\tan 2\theta = \sqrt{3}$, thus $2\theta = 60° + 180°k$ and so $\theta = 30° + 90°k$. So the solutions in the interval $[0°, 360°)$ are $\theta = 30°$, $120°$, $210°$ or $300°$.

31. Since $\sin\theta = \frac{1}{4}$, $\theta \approx 14.5°$ or $165.5°$.

33. Since $2\tan\theta = -4$, $\tan\theta = -2$ and thus $\theta \approx 116.6° + 180°k$. So the solutions in the interval $[0°, 360°)$ are $\theta \approx 116.6°$ or $296.6°$.

35. Using the quadratic formula, we have:
$$\cos x = \frac{1 \pm \sqrt{(-1)^2 - 4(-1)}}{2(1)} = \frac{1 \pm \sqrt{1 + 4}}{2} = \frac{1 \pm \sqrt{5}}{2}$$
Since $\frac{1+\sqrt{5}}{2} > 1$, the only solutions are those for which $\cos x = \frac{1-\sqrt{5}}{2} \approx -0.62$, and thus $x \approx 128.2°$ or $231.8°$.

37. Since $\cos x = 0.184$, $x \approx 1.39$ or $x \approx 4.90$.

39. Since $\sin x = \frac{1}{\sqrt{5}}$, $x \approx 0.46$ or $x \approx 2.68$.

41. Since $\tan x = 6$, $x \approx 1.41$ or $x \approx 4.55$.

43. Dividing through by $\cos t$ results in $\tan t = 5$, thus $t \approx 1.37$ or $t \approx 4.51$.

45. Since $\sec t = 2.24$, $\cos t = \frac{1}{2.24} \approx 0.45$. Thus $t \approx 1.11$ or $t \approx 5.18$.

47. Following the hint, we use the addition formula and double-angle formula for tangent to obtain:
$$\tan 3x = \tan(2x + x) = \frac{\tan 2x + \tan x}{1 - \tan x \tan 2x} = \frac{\frac{2\tan x}{1-\tan^2 x} + \tan x}{1 - \tan x \frac{2\tan x}{1-\tan^2 x}} = \frac{3\tan x - \tan^3 x}{1 - 3\tan^2 x}$$

So the left-hand side of the equation becomes:
$$\tan 3x - \tan x = \frac{3\tan x - \tan^3 x}{1 - 3\tan^2 x} - \tan x$$
$$= \frac{3\tan x - \tan^3 x - \tan x + 3\tan^3 x}{1 - 3\tan^2 x}$$
$$= \frac{2\tan x + 2\tan^3 x}{1 - 3\tan^2 x}$$
$$= \frac{2\tan x(1 + \tan^2 x)}{1 - 3\tan^2 x}$$

Since $1 + \tan^2 x \neq 0$, we must have $\tan x = 0$ and thus $x = 0$ or $x = \pi$.

49. By the half-angle formula for cosine, we have:
$$\cos\tfrac{x}{2} = \pm\sqrt{\frac{1 + \cos x}{2}}$$

Writing the original equation, then squaring, we have:
$$\pm\sqrt{\frac{1 + \cos x}{2}} = 1 + \cos x$$
$$\frac{1 + \cos x}{2} = 1 + 2\cos x + \cos^2 x$$
$$1 + \cos x = 2 + 4\cos x + 2\cos^2 x$$
$$0 = 2\cos^2 x + 3\cos x + 1$$
$$0 = (2\cos x + 1)(\cos x + 1)$$
$$\cos x = -\tfrac{1}{2} \text{ or } \cos x = -1$$

These equations have solutions of $x = \frac{2\pi}{3}, \frac{4\pi}{3}$ or π. Upon checking we find that $x = \frac{4\pi}{3}$ is not a solution, and thus the solutions are $x = \frac{2\pi}{3}$ or $x = \pi$.

51. Using the identity for $\sec 4\theta$ and the double-angle identity for sine, we have:
$$\sec 4\theta + 2\sin 4\theta = 0$$
$$\frac{1}{\cos 4\theta} + 2\sin 4\theta = 0$$
$$1 + 2\sin 4\theta \cos 4\theta = 0$$
$$\sin 8\theta = -1$$
Thus $8\theta = \frac{3\pi}{2} + 2\pi k$, and so $\theta = \frac{3\pi}{16} + \frac{k\pi}{4}$, where k is any integer.

53. Following the hint, we have:
$$4\sin\theta - 3\cos\theta = 2$$
$$4\sin\theta = 3\cos\theta + 2$$
$$16\sin^2\theta = 9\cos^2\theta + 12\cos\theta + 4$$
$$16(1 - \cos^2\theta) = 9\cos^2\theta + 12\cos\theta + 4$$
$$16 - 16\cos^2\theta = 9\cos^2\theta + 12\cos\theta + 4$$
$$25\cos^2\theta + 12\cos\theta - 12 = 0$$
This will not factor, so we use the quadratic formula:
$$\cos\theta = \frac{-12 \pm \sqrt{(12)^2 - 4(25)(-12)}}{2(25)} = \frac{-12 \pm \sqrt{1344}}{50} = \frac{-12 \pm 8\sqrt{21}}{50} = \frac{-6 \pm 4\sqrt{21}}{25}$$
So $\cos\theta = 0.4932$ or $\cos\theta = -0.9732$, and thus $\theta = 60.45°$ (the other solution is not in the required interval).

55. (a) Squaring each side, we have:
$$\sin^2 x \cos^2 x = 1$$
$$\sin^2 x (1 - \sin^2 x) = 1$$
$$\sin^2 x - \sin^4 x = 1$$
$$\sin^4 x - \sin^2 x + 1 = 0$$

(b) Using the quadratic formula:
$$\sin^2 x = \frac{1 \pm \sqrt{(-1)^2 - 4(1)(1)}}{2(1)} = \frac{1 \pm \sqrt{1 - 4}}{2} = \frac{1 \pm \sqrt{-3}}{2}$$
Thus the original equation has no real-number solutions.

57. (a) Since $\cos x = 0.412$, $x \approx 1.146$ or $x \approx -1.146$. We must now add multiples of 2π on to these values until we reach an x-value greater than 1000. Note that $\frac{1000}{2\pi} \approx 159$ so we check $x = 1.146 + 159(2\pi)$ as a starting point. The first such value is at $x \approx 1000.173$.

(b) Since $\cos x = -0.412$, $x \approx 1.995$ or $x \approx -1.995$. Again, we add multiples of 2π on to these values until we reach an x-value greater than 1000. See the note from part (a). The first such value is at $x \approx 1001.022$.

Optional TI-81 Graphing Calculator Exercises for Section 7.8

1. (a) The graphs of $Y_1 = \cos x$ and $Y_2 = 0.351$ with the indicated settings appear as:

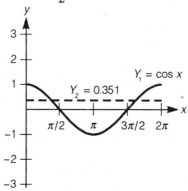

    (b) The graphs of $Y_1 = \cos x$ and $Y_2 = 0.351$ with the indicated settings appear as:

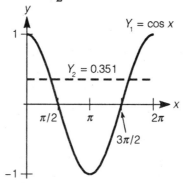

    (c) The approximation $x \approx 1.19$ is confirmed.

    (d) The approximation $x \approx 1.215$ is confirmed.

    (e) The approximation $x \approx 1.212$ is confirmed.

Note: For Exercises 3-9 be sure to set your calculator to the radian mode and use the settings from Exercise 1.

3. Using the trace and zoom technique, the two solutions are $x \approx 0.90$ and $x \approx 5.38$.

5. Using the trace and zoom technique, the two solutions are $x \approx 0.43$ and $x \approx 2.71$.

7. (a) Using the trace and zoom technique, there are no solutions to this equation.

    (b) Using the trace and zoom technique, the four solutions are $x \approx 1.93$, $x \approx 2.28$, $x \approx 5.07$ and $x \approx 5.42$.

9.   Using the trace and zoom technique, the three solutions are $x = 0$, $x \approx 0.69$ and $x \approx 4.26$.

Note: For Exercises 11-15 be sure to set your calculator to the degree mode and use the settings from Exercise 2.

11.   Using the trace and zoom technique, the two solutions are $x \approx 67.5°$ and $x \approx 247.5°$.

13.   Using the trace and zoom technique, the two solutions are $x \approx 120.8°$ and $x \approx 329.2°$.

15.   Using the trace and zoom technique, the three solutions are $x = 0°$, $x \approx 1.0°$ and $x \approx 193.9°$.

7.9 The Inverse Trigonometric Functions

1.   We are asked to find the number x in the interval $\left[-\frac{\pi}{2},\frac{\pi}{2}\right]$ such that $\sin x = \frac{\sqrt{3}}{2}$. Since $x = \frac{\pi}{3}$ is that number, $\sin^{-1}\left(\frac{\sqrt{3}}{2}\right) = \frac{\pi}{3}$.

3.   We are asked to find the number x in the interval $\left(-\frac{\pi}{2},\frac{\pi}{2}\right)$ such that $\tan x = \sqrt{3}$. Since $x = \frac{\pi}{3}$ is that number, $\tan^{-1}\sqrt{3} = \frac{\pi}{3}$.

5.   We are asked to find the number x in the interval $\left(-\frac{\pi}{2},\frac{\pi}{2}\right)$ such that $\tan x = -\frac{1}{\sqrt{3}}$. Since $x = -\frac{\pi}{6}$ is that number, $\arctan\left(-\frac{1}{\sqrt{3}}\right) = -\frac{\pi}{6}$.

7.   We are asked to find the number x in the interval $\left(-\frac{\pi}{2},\frac{\pi}{2}\right)$ such that $\tan x = 1$. Since $x = \frac{\pi}{4}$ is that number, $\tan^{-1}1 = \frac{\pi}{4}$.

9.   We are asked to find the number x in the interval $[0, \pi]$ such that $\cos x = 2\pi$. Since $\cos x \leq 1$ for all x, this value for x does not exist, thus $\cos^{-1}2\pi$ is undefined.

11.   If $x = \sin^{-1}\left(\frac{1}{4}\right)$, then $\sin x = \frac{1}{4}$. Thus $\sin\left[\sin^{-1}\frac{1}{4}\right] = \frac{1}{4}$.

13.   If $x = \cos^{-1}\left(\frac{3}{4}\right)$, then $\cos x = \frac{3}{4}$. Thus $\cos\left[\cos^{-1}\frac{3}{4}\right] = \frac{3}{4}$.

15.   We are asked to find the number x in the interval $\left(-\frac{\pi}{2},\frac{\pi}{2}\right)$ such that $\tan x = \tan\left(-\frac{\pi}{7}\right)$. Since $x = -\frac{\pi}{7}$ is that number, $\arctan\left[\tan\left(-\frac{\pi}{7}\right)\right] = -\frac{\pi}{7}$.

17. Since $\sin \frac{\pi}{2} = 1$, we must find arcsin 1. We are asked to find the number x in the interval $\left[-\frac{\pi}{2}, \frac{\pi}{2}\right]$ such that $\sin x = 1$. Since $x = \frac{\pi}{2}$ is that number, $\arcsin\left[\sin \frac{\pi}{2}\right] = \frac{\pi}{2}$.

19. Since $\cos 2\pi = 1$, we must find arccos 1. We are asked to find the number x in the interval $[0, \pi]$ such that $\cos x = 1$. Since $x = 0$ is that number, $\arccos(\cos 2\pi) = 0$.

21. If $x = \sin^{-1}\left(\frac{4}{5}\right)$, then $\sin x = \frac{4}{5}$ and thus $\cos x = \frac{3}{5}$. Therefore:

$$\tan\left[\sin^{-1}\left(\frac{4}{5}\right)\right] = \tan x = \frac{\sin x}{\cos x} = \frac{\frac{4}{5}}{\frac{3}{5}} = \frac{4}{3}$$

23. Let $x = \tan^{-1} 1$, so $\tan x = 1$ and thus $x = \frac{\pi}{4}$. Therefore:

$$\sin(\tan^{-1} 1) = \sin \frac{\pi}{4} = \frac{\sqrt{2}}{2}$$

25. If $x = \arccos \frac{5}{13}$, then $\cos x = \frac{5}{13}$ and thus $\sin x = \frac{12}{13}$. Therefore:

$$\tan\left(\arccos \frac{5}{13}\right) = \tan x = \frac{\sin x}{\cos x} = \frac{\frac{12}{13}}{\frac{5}{13}} = \frac{12}{5}$$

27. Since $\arctan \sqrt{3} = \frac{\pi}{3}$:

$$\cos(\arctan \sqrt{3}) = \cos\left(\frac{\pi}{3}\right) = \frac{1}{2}$$

29. If $x = \arccos\left(-\frac{1}{3}\right)$, then $\cos x = -\frac{1}{3}$ and since x is in the interval $[0, \pi]$, x must lie in the second quadrant, so:

$$\sin x = \sqrt{1 - \cos^2 x} = \sqrt{1 - \left(-\frac{1}{3}\right)^2} = \sqrt{1 - \frac{1}{9}} = \sqrt{\frac{8}{9}} = \frac{2\sqrt{2}}{3}$$

Therefore:

$$\sin\left[\arccos\left(-\frac{1}{3}\right)\right] = \sin x = \frac{2\sqrt{2}}{3}$$

31. (a) $\sin^{-1}\left(\frac{3}{4}\right) \approx 0.84$ radians or $48.59°$

     (b) $\cos^{-1}\left(\frac{2}{3}\right) \approx 0.84$ radians or $48.18°$

     (c) $\tan^{-1} \pi \approx 1.26$ radians or $72.34°$

     (d) $\tan^{-1}(\tan^{-1} \pi) \approx \tan^{-1} 1.26 \approx 0.90$ radians or $51.57°$

33. We first compute $\cos^{-1}\left(\frac{\sqrt{2}}{2}\right)$ and $\sin^{-1}(-1)$:

$$\cos^{-1}\left(\frac{\sqrt{2}}{2}\right) = \frac{\pi}{4} \text{ since } \cos \frac{\pi}{4} = \frac{\sqrt{2}}{2} \text{ and } \frac{\pi}{4} \text{ is in the interval } [0, \pi]$$

$$\sin^{-1}(-1) = -\frac{\pi}{2} \text{ since } \sin\left(-\frac{\pi}{2}\right) = -1 \text{ and } -\frac{\pi}{2} \text{ is in the interval } \left[-\frac{\pi}{2}, \frac{\pi}{2}\right]$$

Therefore we have:

$$\sec\left[\cos^{-1}\left(\tfrac{\sqrt{2}}{2}\right)+\sin^{-1}(-1)\right]=\sec\left(\tfrac{\pi}{4}-\tfrac{\pi}{2}\right)=\sec\left(-\tfrac{\pi}{4}\right)=\frac{1}{\cos\left(-\tfrac{\pi}{4}\right)}=\frac{1}{\tfrac{\sqrt{2}}{2}}=\sqrt{2}$$

35. Let $u=\sin^{-1}x$, so $\sin u=x$ and u is in the interval $-\tfrac{\pi}{2}\le u\le\tfrac{\pi}{2}$. Since $\cos u\ge 0$, we have:

$$\cos u=\sqrt{1-\sin^2 u}=\sqrt{1-x^2}$$

Therefore:

$$\cos\left(\sin^{-1}x\right)=\cos u=\sqrt{1-x^2}$$

37. Since $\sin 2\theta=2\sin\theta\cos\theta$ by the double-angle identity for sine, we must find $\cos\theta$. Since $0<\theta<\tfrac{\pi}{2}$, $\cos\theta>0$ and thus:

$$\cos\theta=\sqrt{1-\sin^2\theta}=\sqrt{1-\left(\frac{3x}{2}\right)^2}=\sqrt{1-\frac{9x^2}{4}}=\frac{\sqrt{4-9x^2}}{2}$$

Since $\sin\theta=\tfrac{3x}{2}$, $\theta=\sin^{-1}\left(\tfrac{3x}{2}\right)$, and therefore:

$$\tfrac{\theta}{4}-\sin 2\theta=\tfrac{1}{4}\theta-2\sin\theta\cos\theta$$

$$=\tfrac{1}{4}\sin^{-1}\left(\tfrac{3x}{2}\right)-2\cdot\tfrac{3x}{2}\cdot\frac{\sqrt{4-9x^2}}{2}$$

$$=\tfrac{1}{4}\sin^{-1}\left(\tfrac{3x}{2}\right)-\frac{3x\sqrt{4-9x^2}}{2}$$

39. Given $\tan\theta=\tfrac{x-1}{2}$, we can construct the triangle:

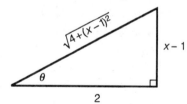

Therefore:

$$\theta-\cos\theta=\tan^{-1}\left(\tfrac{x-1}{2}\right)-\frac{2}{\sqrt{4+(x-1)^2}}=\tan^{-1}\left(\tfrac{x-1}{2}\right)-\frac{2}{\sqrt{5-2x+x^2}}$$

41. Let $\theta=\tan^{-1}4$, so $\tan\theta=4$. We draw the triangle:

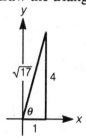

Using the double-angle formula for $\sin 2\theta$, we have:

$$\sin\left(2\tan^{-1}4\right) = \sin 2\theta = 2\sin\theta\cos\theta = 2 \cdot \frac{4}{\sqrt{17}} \cdot \frac{1}{\sqrt{17}} = \frac{8}{17}$$

43. Let $s = \arccos\frac{3}{5}$, so $\cos s = \frac{3}{5}$. Drawing a triangle:

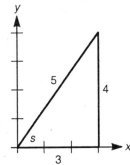

Let $t = \arctan\frac{7}{13}$, so $\tan t = \frac{7}{13}$. Drawing a triangle:

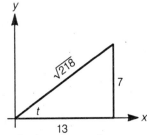

Using the addition formula for $\sin(s - t)$, we have:

$$\begin{aligned}
\sin\left(\arccos\frac{3}{5} - \arctan\frac{7}{13}\right) &= \sin(s - t)\\
&= \sin s \cos t - \cos s \sin t\\
&= \frac{4}{5} \cdot \frac{13}{\sqrt{218}} - \frac{3}{5} \cdot \frac{7}{\sqrt{218}}\\
&= \frac{31}{5\sqrt{218}}\\
&= \frac{31\sqrt{218}}{1090}
\end{aligned}$$

45. (a)   Since $\alpha = \sin^{-1}x$, $-\frac{\pi}{2} \le \alpha \le \frac{\pi}{2}$ and since $\beta = \cos^{-1}x$, then $0 \le \beta \le \pi$. Thus:

$$-\frac{\pi}{2} + 0 \le \alpha + \beta \le \frac{\pi}{2} + \pi$$
$$-\frac{\pi}{2} \le \alpha + \beta \le \frac{3\pi}{2}$$

   (b)   Since $\alpha = \sin^{-1}x$, $\sin\alpha = x$ and thus:

$$\cos\alpha = \sqrt{1 - \sin^2\alpha} = \sqrt{1 - x^2}$$

Similarly, since $\beta = \cos^{-1}x$, $\cos\beta = x$ and thus:

$$\sin\beta = \sqrt{1 - \cos^2\beta} = \sqrt{1 - x^2}$$

Using the addition formula for $\sin(\alpha + \beta)$, we have:
$$\sin(\alpha + \beta) = \sin\alpha\cos\beta + \cos\alpha\sin\beta$$
$$= x \cdot x + \sqrt{1-x^2} \cdot \sqrt{1-x^2}$$
$$= x^2 + 1 - x^2$$
$$= 1$$
But if $-\frac{\pi}{2} \le \alpha + \beta \le \frac{3\pi}{2}$ (from part (a)) and $\sin(\alpha + \beta) = 1$, then $\alpha + \beta = \frac{\pi}{2}$.

47. We take the sine of each side to get:
$$\sin\left(\sin^{-1}(3t-2)\right) = \sin\left(\sin^{-1}t - \cos^{-1}t\right)$$
$$3t - 2 = \sin\left(\sin^{-1}t - \cos^{-1}t\right)$$
Let $u = \sin^{-1}t$, so $\sin u = t$ and $\cos u = \sqrt{1-t^2}$. Also, let $v = \cos^{-1}t$, so $\cos v = t$ and $\sin v = \sqrt{1-t^2}$. Thus:
$$\sin\left(\sin^{-1}t - \cos^{-1}t\right) = \sin(u - v)$$
$$= \sin u \cos v - \cos u \sin v$$
$$= t \cdot t - \sqrt{1-t^2}\sqrt{1-t^2}$$
$$= t^2 - 1 + t^2$$
$$= 2t^2 - 1$$
Thus, we have the equation:
$$3t - 2 = 2t^2 - 1$$
$$0 = 2t^2 - 3t + 1$$
$$0 = (2t - 1)(t - 1)$$
$$t = \tfrac{1}{2}, 1$$

49. Let $s = \arctan x$ and $t = \arctan y$, so $\tan s = x$ and $\tan t = y$. Using the addition formula for $\tan(s + t)$, we have:
$$\tan(s + t) = \frac{\tan s + \tan t}{1 - \tan s \tan t} = \frac{x + y}{1 - xy}$$
Then:
$$\arctan(\tan(s + t)) = \arctan\left(\tfrac{x+y}{1-xy}\right)$$
So $s + t = \arctan x + \arctan y = \arctan\left(\tfrac{x+y}{1-xy}\right)$.

51. Following the hint, we take the tangent of each side to obtain:
$$\tan\left(2\tan^{-1}x\right) = \tan\left[\tan^{-1}\left(\tfrac{1}{4x}\right)\right]$$
$$\tan\left(2\tan^{-1}x\right) = \tfrac{1}{4x}$$

Let $\alpha = \tan^{-1} x$, so using the double-angle formula for $\tan 2\alpha$:

$$\frac{2 \tan \alpha}{1 - \tan^2 \alpha} = \frac{1}{4x}$$

$$\frac{2x}{1 - x^2} = \frac{1}{4x}$$

$$8x^2 = 1 - x^2$$

$$9x^2 = 1$$

$$x^2 = \tfrac{1}{9}$$

$$x = \pm \tfrac{1}{3}$$

53. We take the tangent of each side of the equation to obtain:

$$\tan\left(2 \tan^{-1} \sqrt{t - t^2}\right) = \tan\left(\tan^{-1} t + \tan^{-1}(1 - t)\right)$$

Let $\alpha = \tan^{-1} \sqrt{t - t^2}$, so $\tan \alpha = \sqrt{t - t^2}$. Let $\beta = \tan^{-1} t$, so $\tan \beta = t$. Let $\gamma = \tan^{-1}(1 - t)$, so $\tan \gamma = 1 - t$. Now simplify each side of the equation, using the double-angle and sum-angle identities for tangent:

$$\tan\left(2 \tan^{-1} \sqrt{t - t^2}\right) = \tan 2\alpha = \frac{2 \tan \alpha}{1 - \tan^2 \alpha} = \frac{2\sqrt{t - t^2}}{1 - \left(t - t^2\right)} = \frac{2\sqrt{t - t^2}}{1 - t + t^2}$$

$$\tan\left(\tan^{-1} + \tan^{-1}(1 - t)\right) = \tan(\beta + \gamma) = \frac{\tan \beta + \tan \gamma}{1 - \tan \beta \tan \gamma} = \frac{t + 1 - t}{1 - t(1 - t)} = \frac{1}{1 - t + t^2}$$

So, our original equation becomes:

$$\frac{2\sqrt{t - t^2}}{1 - t + t^2} = \frac{1}{1 - t + t^2}$$

$$2\sqrt{t - t^2} = 1$$

Squaring each side, we have:

$$4\left(t - t^2\right) = 1$$

$$4t - 4t^2 = 1$$

$$0 = 4t^2 - 4t + 1$$

$$0 = (2t - 1)^2$$

$$t = \tfrac{1}{2}$$

55. (a) We graph $y = \sec x$ on the domain $\left[0, \frac{\pi}{2}\right) \cup \left[\pi, \frac{3\pi}{2}\right)$, noting that it is one-to-one:

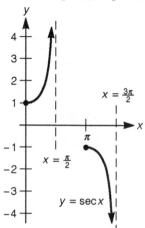

(b) We graph $y = \sec^{-1}x$:

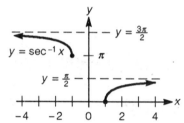

(c) Let $x = \sec^{-1}\left(\frac{2}{\sqrt{3}}\right)$, so $\sec x = \frac{2}{\sqrt{3}}$. Then $\cos x = \frac{\sqrt{3}}{2}$, so $x = \frac{\pi}{6}$. Let $y = \sec^{-1}\left(-\frac{2}{\sqrt{3}}\right)$, so $\sec y = -\frac{2}{\sqrt{3}}$. Then $\cos y = -\frac{\sqrt{3}}{2}$, so $y = \frac{7\pi}{6}$. Thus $\sec^{-1}\left(\frac{2}{\sqrt{3}}\right) = \frac{\pi}{6}$ and $\sec^{-1}\left(-\frac{2}{\sqrt{3}}\right) = \frac{7\pi}{6}$.

(d) Let $x = \sec^{-1}\left(\sqrt{2}\right)$, so $\sec x = \sqrt{2}$. Then $\cos x = \frac{1}{\sqrt{2}}$, so $x = \frac{\pi}{4}$. Let $y = \sec^{-1}\left(-\sqrt{2}\right)$, so $\sec y = -\sqrt{2}$. Then $\cos y = -\frac{1}{\sqrt{2}}$, so $y = \frac{5\pi}{4}$. Thus $\sec^{-1}\left(\sqrt{2}\right) = \frac{\pi}{4}$ and $\sec^{-1}\left(-\sqrt{2}\right) = \frac{5\pi}{4}$.

(e) Since 2 is in the domain of $y = \sec^{-1}x$, $\sec\left(\sec^{-1}2\right) = 2$. Since 0 is in the domain of the restricted secant function, $\sec^{-1}(\sec 0) = 0$.

57. (a) If A corresponds to $A\,(0,0)$, then the coordinates of B and C are $B\,(3,-2)$ and $C\,(5,1)$. Following the hint, we compute the slopes:

$$m_{\overline{AB}} = \frac{-2-0}{3-0} = -\frac{2}{3}$$
$$m_{\overline{BC}} = \frac{1-(-2)}{5-3} = \frac{3}{2}$$

Since the product of these slopes is $\left(-\frac{2}{3}\right)\left(\frac{3}{2}\right) = -1$, $\overline{AB}$ is perpendicular to $\overline{BC}$, and thus $\angle ABC = \frac{\pi}{2}$.

(b) We compute each distance using the coordinates specified in part (a) and the distance formula:

$$AB = \sqrt{(3-0)^2 + (-2-0)^2} = \sqrt{9+4} = \sqrt{13}$$
$$BC = \sqrt{(5-3)^2 + (1+2)^2} = \sqrt{4+9} = \sqrt{13}$$

Thus $AB = BC$. Since $AB = BC$ and $\angle ABC = \frac{\pi}{2}$, $\triangle ABC$ is an isosceles triangle and thus $\angle BAC = \angle ACB = \frac{\pi}{4}$.

(c) The relationships $\alpha = \tan^{-1}\left(\frac{2}{3}\right)$ and $\beta = \tan^{-1}\left(\frac{1}{5}\right)$ follow directly from the right triangles that contain α and β as acute angles. Since $\alpha + \beta = \frac{\pi}{4}$, $\tan^{-1}\left(\frac{2}{3}\right) + \tan^{-1}\left(\frac{1}{5}\right) = \frac{\pi}{4}$.

Optional TI-81 Graphing Calculator Exercises for Section 7.9

1. (a) The graphs of $Y_1 = x$ and $Y_2 = \sin^{-1}(\sin x)$ are the same using the indicated settings.

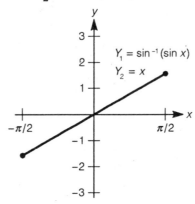

(b) The results demonstrate that $\sin^{-1}(\sin x) = x$, and thus $\sin^{-1} x$ is the inverse function of $\sin x$ on the interval $-\frac{\pi}{2} \le x \le \frac{\pi}{2}$.

3. (a) The graph of $Y_1 = \sin^{-1} x + \cos^{-1} x$ appears to be constant on the interval $-1 \le x \le 1$.

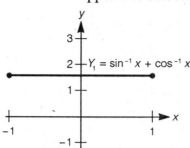

(b) The y-intercept is $\frac{\pi}{2}$. This demonstrates the identity $\sin^{-1} x + \cos^{-1} x = \frac{\pi}{2}$.

5. (a) Using the Zoom-7 settings, the graphs of $Y_1 = \tan(\tan^{-1} x)$ and $Y_2 = x$ are identical. This demonstrates that $\tan(\tan^{-1} x) = x$, at least for the values of x that are graphed.

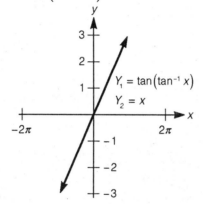

(b) The graphs of $Y_1 = \tan^{-1}(\tan x)$ and $Y_2 = x$ are identical. This demonstrates that $\tan^{-1}(\tan x) = x$ for $-\frac{\pi}{2} < x < \frac{\pi}{2}$.

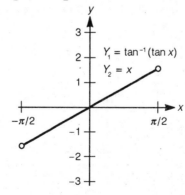

(c) It was necessary to change the range settings since $\tan x$ is one-to-one on the interval $-\frac{\pi}{2} < x < \frac{\pi}{2}$. Using the interval $-2\pi < x < 2\pi$ will not produce the same result (try it!).

7.   Using the trace and zoom technique, the single solution is $x \approx 0.79$.

9.   Using the trace and zoom technique, the single solution is $x \approx 0.80$.

11.   Using the trace and zoom technique, there are two solutions of $x = 0$ and $x \approx 0.80$.

13.   (a)   Using the trace and zoom technique, there are two solutions of $x = 0$ and $x \approx 0.98$.
        (b)   Using the trace and zoom technique, the single solution is $x = 0$.

15.   (a)   Using the trace and zoom technique, the single solution is $x \approx 0.54$.
        (b)   Using the trace and zoom technique, the single solution is $x \approx 0.50$.

Chapter Seven Review Exercises

1.   (a)   Since the terminal side of $\frac{5\pi}{3}$ lies in the fourth quadrant, $\sin\frac{5\pi}{3} < 0$. Using a reference angle of $\frac{\pi}{3}$, we have:
$$\sin\frac{5\pi}{3} = -\sin\frac{\pi}{3} = -\frac{\sqrt{3}}{2}$$

(b)   Since the terminal side of $\frac{11\pi}{6}$ lies in the fourth quadrant, $\cot\frac{11\pi}{6} < 0$. Using a reference angle of $\frac{\pi}{6}$, we have:
$$\cot\frac{11\pi}{6} = -\cot\frac{\pi}{6} = -\frac{\cos\frac{\pi}{6}}{\sin\frac{\pi}{6}} = -\frac{\frac{\sqrt{3}}{2}}{\frac{1}{2}} = -\sqrt{3}$$

3.   Using the identities $\cos(-t) = \cos t$ and $\sin(-t) = -\sin t$, we have:
$$\cos t - \cos(-t) + \sin t - \sin(-t) = \cos t - \cos t + \sin t - (-\sin t) = 2\sin t$$

5.   The period is $\frac{2\pi}{2\pi} = 1$ and the phase shift is $\frac{3}{2\pi}$. The asymptotes will occur at $x = -\frac{1}{4} + \frac{3}{2\pi}$, $x = \frac{1}{4} + \frac{3}{2\pi}$ and $x = \frac{3}{4} + \frac{3}{2\pi}$.

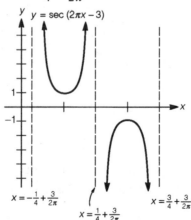

7. The amplitude is 1, the period is $\frac{2\pi}{2} = \pi$, and the phase shift is $\frac{\pi}{2}$. Notice that this graph is $y = \sin(2x - \pi)$ reflected across the x-axis.

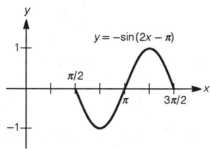

9. Since the terminal side of π radians intersects the unit circle at the point $(-1,0)$, $\cos \pi = -1$.

11. Since the terminal side of $\frac{2\pi}{3}$ radians lies in the second quadrant, $\csc \frac{2\pi}{3} > 0$. Using a reference angle of $\frac{\pi}{3}$ radians, we have:

$$\csc \frac{2\pi}{3} = \csc \frac{\pi}{3} = \frac{1}{\sin \frac{\pi}{3}} = \frac{2}{\sqrt{3}} = \frac{2\sqrt{3}}{3}$$

13. Since the terminal side of $\frac{11\pi}{6}$ radians lies in the fourth quadrant, $\tan \frac{11\pi}{6} < 0$. Using a reference angle of $\frac{\pi}{6}$ radians, we have:

$$\tan \frac{11\pi}{6} = -\tan \frac{\pi}{6} = -\frac{\sin \frac{\pi}{6}}{\cos \frac{\pi}{6}} = -\frac{\frac{1}{2}}{\frac{\sqrt{3}}{2}} = -\frac{1}{\sqrt{3}} = -\frac{\sqrt{3}}{3}$$

15. Since the terminal side of $\frac{\pi}{6}$ radians lies in the first quadrant, $\sin \frac{\pi}{6} > 0$. Thus $\sin \frac{\pi}{6} = \frac{1}{2}$.

17. Since the terminal side of $\frac{5\pi}{4}$ radians lies in the third quadrant, $\cot \frac{5\pi}{4} > 0$. Using a reference angle of $\frac{\pi}{4}$ radians, we have:

$$\cot \frac{5\pi}{4} = \cot \frac{\pi}{4} = \frac{\cos \frac{\pi}{4}}{\sin \frac{\pi}{4}} = \frac{\frac{\sqrt{2}}{2}}{\frac{\sqrt{2}}{2}} = 1$$

19. Since the terminal side of $-\frac{5\pi}{6}$ radians lies in the third quadrant, $\csc\left(-\frac{5\pi}{6}\right) < 0$. Using a reference angle of $\frac{\pi}{6}$ radians, we have:

$$\csc\left(-\frac{5\pi}{6}\right) = -\csc \frac{\pi}{6} = -\frac{1}{\sin \frac{\pi}{6}} = -\frac{1}{\frac{1}{2}} = -2$$

21. Using a calculator, we find $\sin 1 \approx 0.841$.

23. Since $\frac{3\pi}{2}$ radians intersects the unit circle at the point $(0,-1)$, $\sin\frac{3\pi}{2} = -1$. A calculator also verifies this value.

25. Using a calculator, we find $\sin(\sin 0.0123) \approx 0.0123$.

27. Since $\sin 1776 \approx -0.842$ and $\cos 1776 \approx -0.540$, we have:
$$\sin^2 1776 + \cos^2 1776 \approx (-0.842)^2 + (-0.540)^2 = 1$$

29. Since $\cos(0.25) \approx 0.969$ and $\sin(0.25) \approx 0.247$:
$$\cos^2(0.25) - \sin^2(0.25) \approx (0.969)^2 - (0.247)^2 \approx 0.878$$
Since $\cos(0.5) \approx 0.878$, the equality is verified.

31. Since $0 < \theta < \frac{\pi}{2}$, $\cos\theta > 0$ and thus:
$$\sqrt{25 - x^2} = \sqrt{25 - 25\sin^2\theta} = 5\sqrt{1 - \sin^2\theta} = 5\sqrt{\cos^2\theta} = 5\cos\theta$$

33. Since $0 < \theta < \frac{\pi}{2}$, $\tan\theta > 0$ and thus:
$$\left(x^2 - 100\right)^{1/2} = \left(100\sec^2\theta - 100\right)^{1/2} = 10\left(\sec^2\theta - 1\right)^{1/2} = 10\left(\tan^2\theta\right)^{1/2} = 10\tan\theta$$

35. Since $0 < \theta < \frac{\pi}{2}$, $\sec\theta > 0$ and thus:
$$\left(x^2 + 5\right)^{-1/2} = \left(5\tan^2\theta + 5\right)^{-1/2} = \frac{\sqrt{5}}{5}\left(\sec^2\theta\right)^{-1/2} = \frac{\sqrt{5}}{5}\cos\theta$$

37. Since $\sin\theta$ is negative:
$$\sin\theta = -\sqrt{1 - \cos^2\theta} = -\sqrt{1 - \left(\frac{8}{17}\right)^2} = -\sqrt{1 - \frac{64}{289}} = -\sqrt{\frac{225}{289}} = -\frac{15}{17}$$
Thus:
$$\tan\theta = \frac{\sin\theta}{\cos\theta} = \frac{-\frac{15}{17}}{\frac{8}{17}} = -\frac{15}{8}$$

39. Since $\angle BPA = \theta$, $\angle APC = \pi - \theta$, and the area of the sector formed by $\angle APC$ is:
$$\tfrac{1}{2}r^2\theta = \tfrac{1}{2}(\sqrt{2})^2(\pi - \theta) = \pi - \theta$$
Now using the area formula for $\triangle APC$:
$$\text{Area}_{\triangle APC} = \tfrac{1}{2}ab\sin(\pi - \theta) = \tfrac{1}{2}(\sqrt{2})(\sqrt{2})\sin\theta = \sin\theta$$
Thus, the area of the shaded region is given by:
$$A(\theta) = \pi - \theta - \sin\theta$$

41. Using the hint, we have two congruent shaded regions each with $r = 1$ cm and $\theta = \frac{\pi}{2}$, and thus the total area is:
$$2 \cdot \tfrac{1}{2}\left(\tfrac{\pi}{2} - 1\right) \text{cm}^2 = \tfrac{\pi - 2}{2} \text{cm}^2$$

43. This is a sine function where the amplitude is 4, so $A = 4$. Since the period is 2π, we have:

$$\frac{2\pi}{B} = 2\pi$$
$$2\pi = 2\pi B$$
$$1 = B$$

So the equation is $y = 4 \sin x$.

45. This is a cosine function (reflected across the x-axis) where the amplitude is 2, so $A = -2$. Since the period is $\frac{\pi}{2}$, we have:

$$\frac{2\pi}{B} = \frac{\pi}{2}$$
$$4\pi = \pi B$$
$$4 = B$$

So the equation is $y = -2 \cos 4x$.

47. The x-intercepts are $\frac{\pi}{8}$ and $\frac{3\pi}{8}$, the high point is $\left(\frac{\pi}{4}, 3\right)$, and the low points are $(0, -3)$ and $\left(\frac{\pi}{2}, -3\right)$. Notice that the period is $\frac{2\pi}{4} = \frac{\pi}{2}$, and that this graph is $y = 3\cos 4x$ reflected across the x-axis.

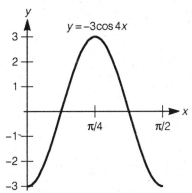

49. The x-intercepts are $\frac{1}{2}$, $\frac{5}{2}$ and $\frac{9}{2}$, the high point is $\left(\frac{3}{2}, 2\right)$, and the low point is $\left(\frac{7}{2}, -2\right)$. Notice that the period is $\frac{2\pi}{\pi/2} = 4$, and the phase shift is $\frac{\pi/4}{\pi/2} = \frac{1}{2}$.

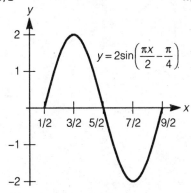

51. The x-intercepts are $\frac{5}{2}$ and $\frac{11}{2}$, the high points are $(1, 3)$ and $(7, 3)$, and the low point is $(4, -3)$. Notice that the period is $\frac{2\pi}{\pi/6} = 6$, and the phase shift is $\frac{\pi/3}{\pi/3} = 1$.

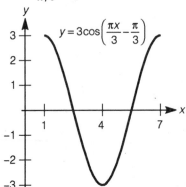

53. **(a)** Notice that the period is $\frac{\pi}{\pi/4} = 4$, and the asymptotes occur at $x = 2$ and $x = -2$.

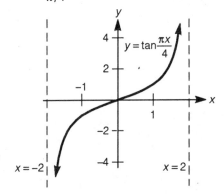

(b) Notice that the period is $\frac{\pi}{\pi/4} = 4$, and the asymptotes occur at $x = 0$ and $x = 4$.

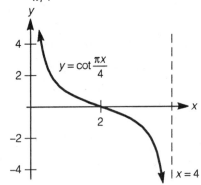

55. (a)   Notice that the period is $\frac{2\pi}{1/4} = 8\pi$, and the asymptotes occur at $x = -2\pi$, $x = 2\pi$ and $x = 6\pi$.

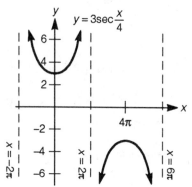

  (b)   Notice that the period is $\frac{2\pi}{1/4} = 8\pi$, and the asymptotes occur at $x = -4\pi$, $x = 0$ and $x = 4\pi$.

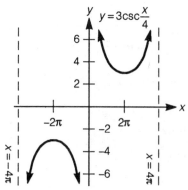

57.   Using the addition formula for sin $(s + t)$, we have:

$$\sin\left(x + \tfrac{3\pi}{2}\right) = \sin x \cos\tfrac{3\pi}{2} + \cos x \sin\tfrac{3\pi}{2} = \sin x \cdot 0 + \cos x \cdot (-1) = -\cos x$$

59.   Using the addition formula for cos $(s - t)$, we have:

$$\cos(\pi - x) = \cos \pi \cos x + \sin \pi \sin x = -1 \cdot \cos x + 0 \cdot \sin x = -\cos x$$

61.   Using the addition formula for cos $(s - t)$, we have:

$$\cos 175° \cos 25° + \sin 175° \sin 25° = \cos(175° - 25°) = \cos 150° = -\tfrac{\sqrt{3}}{2}$$

63.   Using the addition formulas for cos $(s - t)$ and cos $(s + t)$, we have:

$$\cos\left(x - \tfrac{2\pi}{3}\right) = \cos x \cos\tfrac{2\pi}{3} + \sin x \sin\tfrac{2\pi}{3}$$
$$= \cos x \cdot \left(-\tfrac{1}{2}\right) + \sin x \cdot \tfrac{\sqrt{3}}{2}$$
$$= -\tfrac{1}{2}\cos x + \tfrac{\sqrt{3}}{2}\sin x$$

$$\cos\left(x + \tfrac{2\pi}{3}\right) = \cos x \cos \tfrac{2\pi}{3} - \sin x \sin \tfrac{2\pi}{3}$$
$$= \cos x \cdot \left(-\tfrac{1}{2}\right) - \sin x \cdot \tfrac{\sqrt{3}}{2}$$
$$= -\tfrac{1}{2}\cos x - \tfrac{\sqrt{3}}{2}\sin x$$

Therefore:

$$\cos\left(x - \tfrac{2\pi}{3}\right) - \cos\left(x + \tfrac{2\pi}{3}\right) = \left(-\tfrac{1}{2}\cos x + \tfrac{\sqrt{3}}{2}\sin x\right) - \left(-\tfrac{1}{2}\cos x - \tfrac{\sqrt{3}}{2}\sin x\right)$$
$$= -\tfrac{1}{2}\cos x + \tfrac{\sqrt{3}}{2}\sin x + \tfrac{1}{2}\cos x + \tfrac{\sqrt{3}}{2}\sin x$$
$$= \sqrt{3}\sin x$$

65. Using the addition formulas for $\tan(s+t)$ and $\tan(s-t)$, we have:

$$\tan(x + 45°) = \frac{\tan x + \tan 45°}{1 - \tan x \tan 45°} = \frac{\tan x + 1}{1 - \tan x}$$
$$\tan(x - 45°) = \frac{\tan x - \tan 45°}{1 + \tan x \tan 45°} = \frac{\tan x - 1}{1 + \tan x}$$

Multiplying, we have:

$$\tan(x + 45°)\tan(x - 45°) = \frac{\tan x + 1}{1 - \tan x} \cdot \frac{\tan x - 1}{1 + \tan x} = \frac{\tan^2 x - 1}{1 - \tan^2 x} = -1$$

67. (a) Using the addition formula for $\tan(s-t)$, we have:

$$\tan \tfrac{\pi}{12} = \tan\left(\tfrac{\pi}{4} - \tfrac{\pi}{6}\right)$$
$$= \frac{\tan \tfrac{\pi}{4} - \tan \tfrac{\pi}{6}}{1 + \tan \tfrac{\pi}{4} \tan \tfrac{\pi}{6}}$$
$$= \frac{1 - \tfrac{\sqrt{3}}{3}}{1 + 1 \cdot \tfrac{\sqrt{3}}{3}}$$
$$= \frac{3 - \sqrt{3}}{3 + \sqrt{3}} \cdot \frac{3 - \sqrt{3}}{3 - \sqrt{3}}$$
$$= \frac{9 - 6\sqrt{3} + 3}{9 - 3}$$
$$= \frac{12 - 6\sqrt{3}}{6}$$
$$= 2 - \sqrt{3}$$

(b) Using the identity for $\cot x$, we have:

$$\cot \tfrac{\pi}{12} = \frac{1}{\tan \tfrac{\pi}{12}} = \frac{1}{2 - \sqrt{3}} \cdot \frac{2 + \sqrt{3}}{2 + \sqrt{3}} = \frac{2 + \sqrt{3}}{4 - 3} = 2 + \sqrt{3}$$

69. Using the identity for $\cot x$ and the addition formula for $\tan(s+t)$, we have:

$$\cot(x + y) = \frac{1}{\tan(x + y)} = \frac{1 - \tan x \tan y}{\tan x + \tan y} \cdot \frac{\cot x \cot y}{\cot x \cot y} = \frac{\cot x \cot y - 1}{\cot y + \cot x}$$

71. Working from the right-hand side, we have:

$$\frac{2\tan x}{1+\tan^2 x} = \frac{\frac{2\sin x}{\cos x}}{1+\frac{\sin^2 x}{\cos^2 x}} = \frac{2\sin x\cos x}{\cos^2 x+\sin^2 x} = \frac{\sin 2x}{1} = \sin 2x$$

73. Using the addition formulas for $\sin(x+y)$ and $\sin(x-y)$, we have:

$$\frac{\sin(x+y)\sin(x-y)}{\cos^2 x\cos^2 y} = \frac{(\sin x\cos y+\cos x\sin y)(\sin x\cos y-\cos x\sin y)}{\cos^2 x\cos^2 y}$$

$$= \frac{\sin^2 x\cos^2 y-\cos^2 x\sin^2 y}{\cos^2 x\cos^2 y}$$

$$= \frac{\sin^2 x}{\cos^2 x}-\frac{\sin^2 y}{\cos^2 y}$$

$$= \tan^2 x-\tan^2 y$$

75. Using the half-angle formula for $\tan\frac{x}{2}$, we have:

$$\sin x\left(\tan\frac{x}{2}+\cot\frac{x}{2}\right) = \sin x\left[\frac{\sin x}{1+\cos x}+\frac{1+\cos x}{\sin x}\right]$$

$$= \sin x\left[\frac{\sin^2 x+1+2\cos x+\cos^2 x}{(\sin x)(1+\cos x)}\right]$$

$$= \sin x\left[\frac{2+2\cos x}{\sin x(1+\cos x)}\right]$$

$$= \frac{2(1+\cos x)}{1+\cos x}$$

$$= 2$$

77. Using the addition formula for $\tan(s-t)$, we have:

$$\tan\left(\frac{\pi}{4}-x\right) = \frac{\tan\frac{\pi}{4}-\tan x}{1+\tan x\tan\frac{\pi}{4}} = \frac{1-\tan x}{1+\tan x}$$

Now using the result of Exercise 76, we have:

$$\tan\left(\frac{\pi}{4}+x\right)-\tan\left(\frac{\pi}{4}-x\right) = \frac{1+\tan x}{1-\tan x}-\frac{1-\tan x}{1+\tan x} = \frac{4\tan x}{1-\tan^2 x} = 2\tan 2x$$

79. Using the double-angle formula for $\sin 2\theta$, we have:

$$2\sin\left(\frac{\pi}{4}-\frac{x}{2}\right)\cos\left(\frac{\pi}{4}-\frac{x}{2}\right) = \sin\left[2\cdot\left(\frac{\pi}{4}-\frac{x}{2}\right)\right] = \sin\left(\frac{\pi}{2}-x\right) = \cos x$$

81. We first simplify $\tan\left(\frac{\pi}{4}-t\right)$ using the addition formula for $\tan(s-t)$:

$$\tan\left(\frac{\pi}{4}-t\right) = \frac{\tan\frac{\pi}{4}-\tan t}{1+\tan\frac{\pi}{4}\tan t} = \frac{1-\tan t}{1+\tan t}$$

Therefore:

$$\frac{1 - \tan\left(\frac{\pi}{4} - t\right)}{1 + \tan\left(\frac{\pi}{4} - t\right)} = \frac{1 - \frac{1 - \tan t}{1 + \tan t}}{1 + \frac{1 - \tan t}{1 + \tan t}} = \frac{1 + \tan t - 1 + \tan t}{1 + \tan t + 1 - \tan t} = \frac{2\tan t}{2} = \tan t$$

83. Using the addition formula for $\tan(s + t)$, we have:

$$\frac{\tan(\alpha - \beta) + \tan\beta}{1 - \tan(\alpha - \beta)\tan\beta} = \tan\left[(\alpha - \beta) + \beta\right] = \tan\alpha$$

85. Using the addition formula for $\tan(s + t)$ and the double-angle formula for $\tan 2\theta$, we have:

$$\begin{aligned}
\tan 3\theta &= \tan(\theta + 2\theta) \\
&= \frac{\tan\theta + \tan 2\theta}{1 - \tan\theta\tan 2\theta} \\
&= \frac{\tan\theta + \frac{2\tan\theta}{1 - \tan^2\theta}}{1 - \tan\theta \cdot \frac{2\tan\theta}{1 - \tan^2\theta}} \\
&= \frac{\tan\theta\left(1 - \tan^2\theta\right) + 2\tan\theta}{1 - \tan^2\theta - 2\tan^2\theta} \\
&= \frac{3\tan\theta - \tan^3\theta}{1 - 3\tan^2\theta} \\
&= \frac{3t - t^3}{1 - 3t^2}, \text{ where } t = \tan\theta
\end{aligned}$$

87. Working from the right-hand side, we have:

$$\begin{aligned}
\frac{\cos x + \sin x}{\cos x - \sin x} &= \frac{\cos x + \sin x}{\cos x - \sin x} \cdot \frac{\cos x + \sin x}{\cos x + \sin x} \\
&= \frac{\cos^2 x + 2\sin x\cos x + \sin^2 x}{\cos^2 x - \sin^2 x} \\
&= \frac{1 + \sin 2x}{\cos 2x} \\
&= \frac{1}{\cos 2x} + \frac{\sin 2x}{\cos 2x} \\
&= \tan 2x + \sec 2x
\end{aligned}$$

89. Using the double-angle formula for $\sin 2x$, we have:

$$\begin{aligned}
2\sin x + \sin 2x &= 2\sin x + 2\sin x\cos x \\
&= 2\sin x(1 + \cos x) \\
&= 2\sin x(1 + \cos x) \cdot \frac{1 - \cos x}{1 - \cos x} \\
&= \frac{2\sin x\left(1 - \cos^2 x\right)}{1 - \cos x} \\
&= \frac{2\sin^3 x}{1 - \cos x}
\end{aligned}$$

91. Working from the right-hand side, we have:

$$\frac{1-\cos x + \sin x}{1+\cos x + \sin x} = \frac{1+\sin x - \cos x}{1+\sin x + \cos x} \cdot \frac{1+\sin x + \cos x}{1+\sin x + \cos x}$$

$$= \frac{(1+\sin x)^2 - \cos^2 x}{\left[(1+\sin x) + \cos x\right]^2}$$

$$= \frac{1+2\sin x + \sin^2 x - \cos^2 x}{\left(1+2\sin x + \sin^2 x\right) + 2\cos x(1+\sin x) + \cos^2 x}$$

$$= \frac{2\sin x + 2\sin^2 x}{2+2\sin x + 2\cos x(1+\sin x)}$$

$$= \frac{2\sin x(1+\sin x)}{2(1+\sin x) + 2\cos x(1+\sin x)}$$

$$= \frac{2\sin x(1+\sin x)}{2(1+\sin x)(1+\cos x)}$$

$$= \frac{\sin x}{1+\cos x}$$

$$= \tan\tfrac{x}{2}$$

93. Using the addition formula for $\sin(s-t)$, we have:

$$\sin(x+y)\cos y - \cos(x+y)\sin y = \sin\left[(x+y)-y\right] = \sin x$$

95. Working from the left-hand side, we have:

$$\frac{1-\tan^2\tfrac{x}{2}}{1+\tan^2\tfrac{x}{2}} = \frac{1-\dfrac{\sin^2\tfrac{x}{2}}{\cos^2\tfrac{x}{2}}}{1+\dfrac{\sin^2\tfrac{x}{2}}{\cos^2\tfrac{x}{2}}} = \frac{\cos^2\tfrac{x}{2}-\sin^2\tfrac{x}{2}}{\cos^2\tfrac{x}{2}+\sin^2\tfrac{x}{2}} = \cos^2\tfrac{x}{2}-\sin^2\tfrac{x}{2} = \cos\left(2\cdot\tfrac{x}{2}\right) = \cos x$$

97. Using the double-angle formulas for $\sin 2\theta$ and $\cos 2\theta$, we have:

$$\sin 4x = 2\sin 2x\cos 2x$$

$$= 2(2\sin x\cos x)(\cos^2 x - \sin^2 x)$$

$$= 4\sin x\cos x(1-2\sin^2 x)$$

$$= 4\sin x\cos x - 8\sin^3 x\cos x$$

99. Using the addition formula for $\sin(s + t)$, we have:
$$\sin 5x = \sin(4x + x) = \sin 4x \cos x + \cos 4x \sin x$$
Simplifying each of these products using the double-angle formulas for $\sin 2\theta$ and $\cos 2\theta$, we have:
$$\sin 4x \cos x = 2 \sin 2x \cos 2x \cos x$$
$$= 2(2 \sin x \cos x)(\cos^2 x - \sin^2 x)\cos x$$
$$= 4 \sin x \cos^2 x(1 - 2\sin^2 x)$$
$$= 4 \sin x(1 - \sin^2 x)(1 - 2\sin^2 x)$$
$$= 4 \sin x - 12 \sin^3 x + 8\sin^5 x$$
$$\cos 4x \sin x = (\cos^2 2x - \sin^2 2x)(\sin x)$$
$$= \left[(\cos^2 x - \sin^2 x)^2 - (2\sin x \cos x)^2\right]\sin x$$
$$= \left[(1 - 2\sin^2 x)^2 - 4\sin^2 x \cos^2 x\right]\sin x$$
$$= \left[1 - 4\sin^2 x + 4\sin^4 x - 4\sin^2 x(1 - \sin^2 x)\right]\sin x$$
$$= \left[1 - 8\sin^2 x + 8\sin^4 x\right]\sin x$$
$$= \sin x - 8\sin^3 x + 8\sin^5 x$$
Therefore, we have:
$$\sin 5x = \sin 4x \cos x + \cos 4x \sin x$$
$$= 4\sin x - 12\sin^3 x + 8\sin^5 x + \sin x - 8\sin^3 x + 8\sin^5 x$$
$$= 16\sin^5 x - 20\sin^3 x + 5\sin x$$

101. The principal solution is $x = \tan^{-1} 4.26 \approx 1.34$. Since $\tan x$ is also positive in the third quadrant, the other solution in the interval $[0, 2\pi]$ is $1.34 + \pi \approx 4.48$.

103. Since $\csc x = 2.24$, $\sin x = \frac{1}{2.24} \approx 0.45$. The principal solution is $x = \sin^{-1} 0.45 \approx 0.46$. Since $\sin x$ is also positive in the second quadrant, the other solution in the interval $[0, 2\pi)$ is $\pi - 0.46 \approx 2.68$.

105. We have $\tan^2 x - 3 = 0$, so $\tan^2 x = 3$ and $\tan x = \pm\sqrt{3}$. If $\tan x = \sqrt{3}$ then $x = \frac{\pi}{3}$ or $\frac{4\pi}{3}$ while if $\tan x = -\sqrt{3}$ then $x = \frac{2\pi}{3}$ or $\frac{5\pi}{3}$. So the solutions in the interval $[0, 2\pi)$ are $x = \frac{\pi}{3}, \frac{2\pi}{3}, \frac{4\pi}{3}$ or $\frac{5\pi}{3}$.

107. Using the double-angle formula for $\cos 2x$, we have:
$$\sin x - (\cos^2 x - \sin^2 x) + 1 = 0$$
$$\sin x - (1 - 2\sin^2 x) + 1 = 0$$
$$2\sin^2 x + \sin x = 0$$
$$\sin x(2\sin x + 1) = 0$$
So $\sin x = 0$ or $\sin x = -\frac{1}{2}$. Thus the solutions are $x = 0, \pi, \frac{7\pi}{6}$ or $\frac{11\pi}{6}$.

109. Solving the equation, we have:
$$3\csc x - 4\sin x = 0$$
$$\frac{3}{\sin x} = 4\sin x$$
$$\sin^2 x = \tfrac{3}{4}$$
$$\sin x = \pm\tfrac{\sqrt{3}}{2}$$
So the solutions are $x = \frac{\pi}{3}, \frac{2\pi}{3}, \frac{4\pi}{3}$ or $\frac{5\pi}{3}$.

111. Factoring, we have:
$$2\sin^4 x - 3\sin^2 x + 1 = 0$$
$$(2\sin^2 x - 1)(\sin^2 x - 1) = 0$$
$$\sin^2 x = \tfrac{1}{2} \qquad \text{or} \qquad \sin^2 x = 1$$
$$\sin x = \pm\tfrac{\sqrt{2}}{2} \qquad \text{or} \qquad \sin x = \pm 1$$
So the solutions are $x = \frac{\pi}{4}, \frac{\pi}{2}, \frac{3\pi}{4}, \frac{5\pi}{4}, \frac{3\pi}{2}$ or $\frac{7\pi}{4}$.

113. We are asked to find the number x in $[0, \pi]$ such that $\cos x = -\frac{\sqrt{2}}{2}$. Since $x = \frac{3\pi}{4}$ is that number, $\cos^{-1}\left(-\frac{\sqrt{2}}{2}\right) = \frac{3\pi}{4}$.

115. We are asked to find the number x in $\left[-\frac{\pi}{2}, \frac{\pi}{2}\right]$ such that $\sin x = 0$. Since $x = 0$ is that number, $\sin^{-1} 0 = 0$.

117. Let $\theta = \arccos\left(-\frac{1}{2}\right)$, so $\cos\theta = -\frac{1}{2}$ and θ is in the interval $[0, \pi]$, thus $\theta = \frac{2\pi}{3}$. Therefore:
$$\sin\left[\arccos\left(-\tfrac{1}{2}\right)\right] = \sin\tfrac{2\pi}{3} = \tfrac{\sqrt{3}}{2}$$

119. Let $\theta = \cos^{-1}\left(\frac{1}{2}\right)$, so $\cos\theta = \frac{1}{2}$ and θ is in the interval $[0, \pi]$, thus $\theta = \frac{\pi}{3}$. Therefore:
$$\cot\left[\cos^{-1}\left(\tfrac{1}{2}\right)\right] = \cot\tfrac{\pi}{3} = \frac{\cos\frac{\pi}{3}}{\sin\frac{\pi}{3}} = \frac{\frac{1}{2}}{\frac{\sqrt{3}}{2}} = \frac{1}{\sqrt{3}} = \frac{\sqrt{3}}{3}$$

121. Let $\theta = \arccos\frac{3}{5}$, so $\cos\theta = \frac{3}{5}$ and θ is in the interval $[0, \pi]$. We draw the triangle:

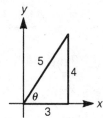

Using the addition formula for $\sin(s + t)$, we have:

$$\sin\left(\tfrac{3\pi}{2} + \theta\right) = \sin\tfrac{3\pi}{2}\cos\theta + \cos\tfrac{3\pi}{2}\sin\theta = -1 \cdot \tfrac{3}{5} + 0 \cdot \tfrac{4}{5} = -\tfrac{3}{5}$$

123. (a) We re-draw the figure, adding additional labels:

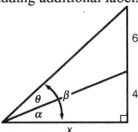

From the smaller right triangle, note that $\tan\alpha = \tfrac{4}{x}$, so $\alpha = \tan^{-1}\left(\tfrac{4}{x}\right)$. From the larger right triangle, note that $\tan\beta = \tfrac{10}{x}$, so $\beta = \tan^{-1}\left(\tfrac{10}{x}\right)$. Therefore:

$$\theta = \beta - \alpha = \tan^{-1}\left(\tfrac{10}{x}\right) - \tan^{-1}\left(\tfrac{4}{x}\right)$$

(b) We complete the table:

x	0.1	1	2	3	10	100
θ	0.01	0.15	0.27	0.35	0.40	0.06

From the table $x = 10$ yields the largest value for θ.

(c) Using a calculator $2\sqrt{10} \approx 6.32$, which is consistent with the graph. The corresponding value of θ is:

$$\theta = \tan^{-1}\left(\tfrac{10}{2\sqrt{10}}\right) - \tan^{-1}\left(\tfrac{4}{2\sqrt{10}}\right) = \tan^{-1}\left(\tfrac{\sqrt{10}}{2}\right) - \tan^{-1}\left(\tfrac{\sqrt{10}}{5}\right) \approx 0.44$$

The coordinates of the highest point on the graph are approximately $(6.32, 0.44)$.

(d) Converting to degree measure, we have:

$$0.44 \cdot \tfrac{180°}{\pi} \approx 25.4°$$

125. Let $\theta = \sin^{-1}x$, so $\sin\theta = x$. We draw the triangle:

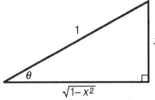

Since $\tan\theta = \tfrac{x}{\sqrt{1-x^2}}$, $\theta = \tan^{-1}\left(\tfrac{x}{\sqrt{1-x^2}}\right)$. Thus $\sin^{-1}x = \tan^{-1}\left(\tfrac{x}{\sqrt{1-x^2}}\right)$.

127. Let $\theta = \cos^{-1} x$, so $\cos \theta = x$. We draw the triangle:

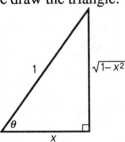

Using the double-angle formula for $\cos 2\theta$, we have:

$$\cos(2\cos^{-1} x) = \cos 2\theta = \cos^2 \theta - \sin^2 \theta = x^2 - \left(\sqrt{1-x^2}\right)^2 = x^2 - 1 + x^2 = 2x^2 - 1$$

129. (a) Using a calculator we find $\cos 20° \cos 40° \cos 60° \cos 80° = 0.0625$.

(b) Since $\cos 60° = \frac{1}{2}$, we have:

$$\cos 20° \cos 40° \cos 60° \cos 80° = \tfrac{1}{2}\cos 20° \cos 40° \cos 80°$$

Using the product-to-sum formula for $\cos A \cos B$, we have:

$$\cos 20° \cos 40° = \tfrac{1}{2}[\cos 60° + \cos(-20°)] = \tfrac{1}{2}\left[\tfrac{1}{2} + \cos 20°\right] = \tfrac{1}{4} + \tfrac{1}{2}\cos 20°$$

Therefore:

$$\cos 20° \cos 40° \cos 60° \cos 80° = \tfrac{1}{2}\cos 20° \cos 40° \cos 80°$$
$$= \tfrac{1}{2}\left(\tfrac{1}{4} + \tfrac{1}{2}\cos 20°\right)\cos 80°$$
$$= \tfrac{1}{8}\cos 80° + \tfrac{1}{4}\cos 20° \cos 80°$$

Using the product-to-sum formula for $\cos A \cos B$, and the identity $\cos \theta = -\cos(180° - \theta)$, we have:

$$\cos 20° \cos 80° = \tfrac{1}{2}[\cos 100° + \cos(-60°)] = \tfrac{1}{2}\left[-\cos 80° + \tfrac{1}{2}\right] = -\tfrac{1}{2}\cos 80° + \tfrac{1}{4}$$

Therefore:

$$\cos 20° \cos 40° \cos 60° \cos 80° = \tfrac{1}{8}\cos 80° + \tfrac{1}{4}\cos 20° \cos 80°$$
$$= \tfrac{1}{8}\cos 80° + \tfrac{1}{4}\left(-\tfrac{1}{2}\cos 80° + \tfrac{1}{4}\right)$$
$$= \tfrac{1}{8}\cos 80° - \tfrac{1}{8}\cos 80° + \tfrac{1}{16}$$
$$= \tfrac{1}{16}$$
$$= 0.0625$$

Chapter Seven Test

1. (a) Since the terminal side of $\frac{4\pi}{3}$ radians lies in the third quadrant, $\cos \frac{4\pi}{3} < 0$. Using a reference angle of $\frac{\pi}{3}$ radians, we have:

$$\cos \tfrac{4\pi}{3} = -\cos \tfrac{\pi}{3} = -\tfrac{1}{2}$$

(b)   Since the terminal side of $-\frac{5\pi}{6}$ radians lies in the third quadrant, $\csc\left(-\frac{5\pi}{6}\right) < 0$.

Using a reference angle of $\frac{\pi}{6}$ radians, we have:

$$\csc\left(-\frac{5\pi}{6}\right) = -\csc\frac{\pi}{6} = -\frac{1}{\sin\frac{\pi}{6}} = -\frac{1}{\frac{1}{2}} = -2$$

(c)   Since $\sin^2 t + \cos^2 t = 1$ for all values of t:

$$\sin^2\frac{3\pi}{4} + \cos^2\frac{3\pi}{4} = 1$$

2.   Substituting $t = 4\sin u$ and noting that $\sin u > 0$ when $0 < u < \frac{\pi}{2}$, we have:

$$\frac{1}{\sqrt{16 - t^2}} = \frac{1}{\sqrt{16 - 16\cos^2 u}} = \frac{1}{\sqrt{16\left(1 - \cos^2 u\right)}} = \frac{1}{\sqrt{16\sin^2 u}} = \frac{1}{4\sin u} = \frac{1}{4}\csc u$$

3.   Noting that the period is $\frac{2\pi}{4\pi} = \frac{1}{2}$, and the phase shift is $\frac{1}{4\pi}$, then the asymptotes are $x = -\frac{1}{8} + \frac{1}{4\pi}$, $x = \frac{1}{8} + \frac{1}{4\pi}$, and $x = \frac{3}{8} + \frac{1}{4\pi}$.

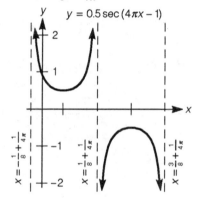

4.   The amplitude is 1, the period is $\frac{2\pi}{3}$, and the phase shift is $\frac{\pi/4}{3} = \frac{\pi}{12}$. Notice that this graph is $y = \sin\left(3x - \frac{\pi}{4}\right)$ reflected across the x-axis.

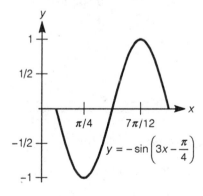

5. Notice that the period is $\frac{\pi}{\pi/4} = 4$, and that the asymptote occurs at $x = 2$.

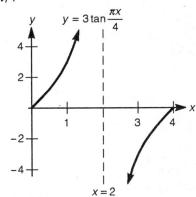

6. (a) Multiplying by the conversion factor $\frac{\pi}{180°}$, we have:
        $$175° \cdot \frac{\pi}{180°} = \frac{35}{36}\pi \text{ radians}$$

    (b) Multiplying by the conversion factor $\frac{180°}{\pi}$, we have:
        $$5 \cdot \frac{180°}{\pi} = \frac{900°}{\pi}$$

7. (a) First converting $\theta = \angle BAC$ to radians, we have:
        $$\theta = 75° \cdot \frac{\pi}{180°} = \frac{5\pi}{12} \text{ radians}$$
        Thus the arc length is given by:
        $$S = r\theta = 5 \text{ cm} \cdot \frac{5\pi}{12} = \frac{25\pi}{12} \text{ cm}$$

    (b) The area of the sector is given by:
        $$A = \tfrac{1}{2}r^2\theta = \tfrac{1}{2}(5)^2 \cdot \frac{5\pi}{12} = \frac{125\pi}{24} \text{ cm}^2$$

8. (a) Converting the angular speed to radians, we have:
        $$\omega = 25 \tfrac{\text{rev}}{\text{sec}} \cdot 2\pi \tfrac{\text{rad}}{\text{rev}} = 50\pi \tfrac{\text{rad}}{\text{sec}}$$

    (b) In 1 second, the wheel has rotated 50π radians, so the total distance traveled by a point 5 cm from the center is:
        $$d = (5 \text{ cm})(50\pi \text{ rad}) = 250\pi \text{ cm}$$
        Thus the linear velocity is given by:
        $$v = \frac{d}{t} = \frac{250\pi \text{ cm}}{1 \text{ sec}} = 250\pi \tfrac{\text{cm}}{\text{sec}}$$

9. We will simplify each side of the equation, using the identities $\sin(-\theta) = -\sin\theta$ and $\cos(-\theta) = \cos\theta$:

$$\frac{\cot\theta}{1+\tan(-\theta)} + \frac{\tan\theta}{1+\cot(-\theta)} = \frac{\frac{\cos\theta}{\sin\theta}}{1+\frac{\sin(-\theta)}{\cos(-\theta)}} + \frac{\frac{\sin\theta}{\cos\theta}}{1+\frac{\cos(-\theta)}{\sin(-\theta)}}$$

$$= \frac{\frac{\cos\theta}{\sin\theta}}{1-\frac{\sin\theta}{\cos\theta}} + \frac{\frac{\sin\theta}{\cos\theta}}{1-\frac{\cos\theta}{\sin\theta}}$$

$$= \frac{\cos^2\theta}{\sin\theta\cos\theta-\sin^2\theta} + \frac{\sin^2\theta}{\sin\theta\cos\theta-\cos^2\theta}$$

$$= \frac{\cos^2\theta}{\sin\theta(\cos\theta-\sin\theta)} + \frac{\sin^2\theta}{\cos\theta(\sin\theta-\cos\theta)}$$

$$= \frac{-\cos^3\theta}{\sin\theta\cos\theta(\sin\theta-\cos\theta)} + \frac{\sin^3\theta}{\sin\theta\cos\theta(\sin\theta-\cos\theta)}$$

$$= \frac{\sin^3\theta-\cos^3\theta}{\sin\theta\cos\theta(\sin\theta-\cos\theta)}$$

$$= \frac{(\sin\theta-\cos\theta)(\sin^2\theta+\sin\theta\cos\theta+\cos^2\theta)}{\sin\theta\cos\theta(\sin\theta-\cos\theta)}$$

$$= \frac{1+\sin\theta\cos\theta}{\sin\theta\cos\theta}$$

$$\cot\theta+\tan\theta+1 = \frac{\cos\theta}{\sin\theta} + \frac{\sin\theta}{\cos\theta}+1 = \frac{\cos^2\theta+\sin^2\theta+\sin\theta\cos\theta}{\sin\theta\cos\theta} = \frac{1+\sin\theta\cos\theta}{\sin\theta\cos\theta}$$

Since both sides of the equation simplify to the same quantity, the equation is an identity.

10. Using the addition formula for $\sin(s+t)$, we have:

$$\sin\left(\theta+\tfrac{3\pi}{2}\right) = \sin\theta\cos\tfrac{3\pi}{2} + \cos\theta\sin\tfrac{3\pi}{2} = \sin\theta\cdot 0 + \cos\theta\cdot(-1) = -\cos\theta$$

11. Since $\frac{3\pi}{2} < t < 2\pi$, we draw the triangle:

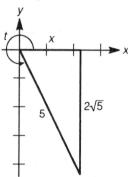

By the Pythagorean theorem:
$$x^2 + (2\sqrt{5})^2 = (5)^2$$
$$x^2 + 20 = 25$$
$$x^2 = 5$$
$$x = \sqrt{5}$$

So $\cos t = \frac{\sqrt{5}}{5}$, now use the double-angle formula for $\cos 2t$:
$$\cos 2t = \cos^2 t - \sin^2 t = \left(\frac{\sqrt{5}}{5}\right)^2 - \left(-\frac{2\sqrt{5}}{5}\right)^2 = \frac{1}{5} - \frac{4}{5} = -\frac{3}{5}$$

12. Dividing each side of the equation by $\cos x$ yields $\tan x = 3$, which has a principal solution of $x = \tan^{-1} 3 \approx 1.25$. Since $\tan x$ is also positive in the third quadrant, the other solution in the interval $(0, 2\pi)$ is $1.25 + \pi \approx 4.39$.

13. Factoring, we have:
$$2\sin^2 x + 7\sin x + 3 = 0$$
$$(2\sin x + 1)(\sin x + 3) = 0$$
$$\sin x = -\frac{1}{2} \quad \text{or} \quad \sin x = -3 \ \text{(impossible)}$$
So $x = \frac{7\pi}{6}, \frac{11\pi}{6}$.

14. Since $\cos \alpha = \frac{2}{\sqrt{5}}$ and $\frac{3\pi}{2} < \alpha < 2\pi$ (fourth quadrant):
$$\sin \alpha = -\sqrt{1 - \cos^2 \alpha} = -\sqrt{1 - \frac{4}{5}} = -\frac{1}{\sqrt{5}}$$
Similarly, since $\sin \beta = \frac{4}{5}$ and $\frac{\pi}{2} < \beta < \pi$ (second quadrant):
$$\cos \beta = -\sqrt{1 - \sin^2 \beta} = -\sqrt{1 - \frac{16}{25}} = -\sqrt{\frac{9}{25}} = -\frac{3}{5}$$
Using the addition formula for $\sin(\beta - \alpha)$, we have:
$$\sin(\beta - \alpha) = \sin \beta \cos \alpha - \cos \beta \sin \alpha = \frac{4}{5} \cdot \frac{2}{\sqrt{5}} - \left(-\frac{3}{5}\right)\left(-\frac{1}{\sqrt{5}}\right) = \frac{8}{5\sqrt{5}} - \frac{3}{5\sqrt{5}} = \frac{5}{5\sqrt{5}} = \frac{\sqrt{5}}{5}$$

15. First, we simplify the left-hand side using the addition formula for $\sin(\alpha + \beta)$:
$$\sin(x + 30°) = \sin x \cos 30° + \cos x \sin 30° = \frac{\sqrt{3}}{2} \sin x + \frac{1}{2} \cos x$$
Thus, we have the equation:
$$\frac{\sqrt{3}}{2} \sin x + \frac{1}{2} \cos x = \sqrt{3} \sin x$$
$$\sqrt{3} \sin x + \cos x = 2\sqrt{3} \sin x$$
$$\cos x = \sqrt{3} \sin x$$
$$\tan x = \frac{1}{\sqrt{3}}$$
$$x = 30°$$

16. If $\csc \theta = -3$, then $\sin \theta = -\frac{1}{3}$. Since $\pi < \theta < \frac{3\pi}{2}$ (third quadrant), we have:
$$\cos \theta = -\sqrt{1 - \sin^2 \theta} = -\sqrt{1 - \frac{1}{9}} = -\sqrt{\frac{8}{9}} = -\frac{2\sqrt{2}}{3}$$

Now $\frac{\pi}{2} < \frac{\theta}{2} < \frac{3\pi}{4}$ (second quadrant), so using the half-angle formula for $\sin\frac{\theta}{2}$ we have:

$$\sin\frac{\theta}{2} = \sqrt{\frac{1-\cos\theta}{2}} = \sqrt{\frac{1+\frac{2\sqrt{2}}{3}}{2}} = \sqrt{\frac{3+2\sqrt{2}}{6}} = \frac{\sqrt{18+12\sqrt{2}}}{6}$$

17. (a)   Since $\sin^{-1}(\sin x) = x$ for every x in the interval $\left[-\frac{\pi}{2},\frac{\pi}{2}\right]$, $\sin^{-1}\left(\sin\frac{\pi}{10}\right) = \frac{\pi}{10}$.

(b)   Since $\sin 2\pi = 0$, we are asked to find $x = \sin^{-1}0$. Then $\sin x = 0$ and x is in the interval $\left[-\frac{\pi}{2},\frac{\pi}{2}\right]$, thus $x = 0$. So $\sin^{-1}(\sin 2\pi) = 0$. Notice that we cannot use the identity $\sin^{-1}(\sin x) = x$, since 2π is not in the interval $\left[-\frac{\pi}{2},\frac{\pi}{2}\right]$.

18. Let $\theta = \arcsin\frac{3}{4}$, so $\sin\theta = \frac{3}{4}$ and θ is in the interval $\left[-\frac{\pi}{2},\frac{\pi}{2}\right]$. We draw the triangle:

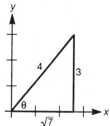

Therefore:
$$\cos\left(\arcsin\frac{3}{4}\right) = \cos\theta = \frac{\sqrt{7}}{4}$$

19. Using the addition formula for $\tan(s+t)$, we have:
$$\tan\left(\frac{\pi}{4}+\frac{\theta}{2}\right) = \frac{\tan\frac{\pi}{4}+\tan\frac{\theta}{2}}{1-\tan\frac{\pi}{4}\tan\frac{\theta}{2}} = \frac{1+\tan\frac{\theta}{2}}{1-\tan\frac{\theta}{2}}$$
Now using the half-angle formula for $\tan\frac{\theta}{2}$, we have:
$$\tan\left(\frac{\pi}{4}+\frac{\theta}{2}\right) = \frac{1+\tan\frac{\theta}{2}}{1-\tan\frac{\theta}{2}} = \frac{1+\frac{\sin\theta}{1+\cos\theta}}{1-\frac{\sin\theta}{1+\cos\theta}} = \frac{1+\cos\theta+\sin\theta}{1+\cos\theta-\sin\theta}$$

20. Let $\theta = \arctan\sqrt{x^2-1}$, so $\tan\theta = \sqrt{x^2-1}$. We draw the triangle:

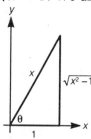

Therefore:
$$\sec\left[\arctan\sqrt{x^2-1}\right] = \sec\theta = \frac{x}{1} = x$$

Chapter Eight
Additional Topics in Trigonometry

8.1 The Law of Sines and the Law of Cosines

1. We draw the triangle:

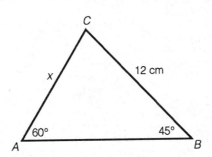

Using the law of sines:
$$\frac{\sin 45°}{x} = \frac{\sin 60°}{12}$$

Thus:

$$x = 12 \cdot \frac{\sin 45°}{\sin 60°} = 12 \cdot \frac{\sqrt{2}}{2} \cdot \frac{2}{\sqrt{3}} = \frac{12\sqrt{2}}{\sqrt{3}} = 4\sqrt{6} \text{ cm}$$

3.   We draw the triangle:

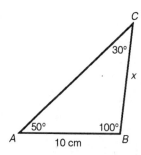

Using the law of sines:
$$\frac{\sin 50°}{x} = \frac{\sin 30°}{10}$$
Thus:
$$x = \frac{10\sin 50°}{\sin 30°} = \frac{10\sin 50°}{\frac{1}{2}} = 20\sin 50° \text{ cm}$$

5.   We draw a triangle:

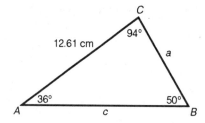

Using the law of sines:
$$\frac{\sin 36°}{a} = \frac{\sin 50°}{12.61}$$
Thus:
$$a = \frac{12.61\sin 36°}{\sin 50°} \approx 9.7 \text{ cm}$$
Using the law of sines:
$$\frac{\sin 94°}{c} = \frac{\sin 50°}{12.61}$$
Thus:
$$c = \frac{12.61\sin 94°}{\sin 50°} \approx 16.4 \text{ cm}$$

7.   We draw a triangle:

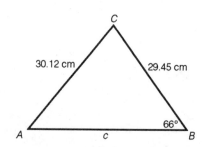

Using the law of sines:
$$\frac{\sin A}{29.45\text{ cm}} = \frac{\sin 66°}{30.12\text{ cm}}$$
Thus:
$$\sin A = \frac{29.45\sin 66°}{30.12} \approx 0.8932, \text{ so } A \approx 63.3°$$
Then $C = 180° - 66° - 63.3° \approx 50.7°$. Using the law of sines:
$$\frac{\sin 50.7°}{c} = \frac{\sin 66°}{30.12}$$
Thus:
$$c = \frac{30.12\sin 50.7°}{\sin 66°} \approx 25.5\text{ cm}$$

9. (a) Since $\sin B = \frac{\sqrt{2}}{2}$, $\angle B = 45°$ or $\angle B = 135°$.

(b) Since $\cos E = \frac{\sqrt{2}}{2}$, $\angle E = 45°$. Note that $\angle E \neq 135°$.

(c) Since $\sin H = \frac{1}{4}$, $\angle H \approx 14.5°$ or $\angle H \approx 165.5°$.

(d) Since $\cos K = -\frac{2}{3}$, $\angle K \approx 131.8°$.

11. (a) Using the law of sines, we have:
$$\frac{\sin 23.1°}{2.0} = \frac{\sin B}{6.0}$$
Thus:
$$\sin B = \frac{6.0\sin 23.1°}{2.0} \approx 1.18$$
But $\sin B \leq 1$, so no such triangle exists.

(b) We first find $\angle B$:
$$\frac{\sin 23.1°}{2.0} = \frac{\sin B}{3.0}$$
Thus:
$$\sin B = \frac{3.0\sin 23.1°}{2.0} \approx 0.5885, \text{ so } B \approx 36.05° \text{ or } B \approx 143.95°$$
Since $\angle B$ is obtuse, $\angle B \approx 143.95°$. Therefore
$\angle C = 180° - 23.1° - 143.95° \approx 12.95°$, so by the law of sines:
$$\frac{\sin 12.95°}{c} = \frac{\sin 23.1°}{2.0}$$
Thus:
$$c = \frac{2.0\sin 12.95°}{\sin 23.1°} \approx 1.1\text{ feet}$$

13. **(a)** Using the law of sines, we have:
$$\frac{\sin A}{\sqrt{2}} = \frac{\sin 30^\circ}{1}$$
Thus:
$$\sin A = \sqrt{2} \sin 30^\circ = \sqrt{2} \cdot \tfrac{1}{2} = \tfrac{\sqrt{2}}{2}$$
Therefore $\angle A = 45^\circ$ or $\angle A = 135^\circ$.

(b) If $\angle A = 45^\circ$, then $\angle C = 180^\circ - 45^\circ - 30^\circ = 105^\circ$. Using the law of sines, we have:
$$\frac{\sin 105^\circ}{c} = \frac{\sin 30^\circ}{1}$$
Thus:
$$c = \frac{\sin 105^\circ}{\sin 30^\circ} \approx 1.93$$

(c) If $\angle A = 135^\circ$, then $\angle C = 180^\circ - 135^\circ - 30^\circ = 15^\circ$. Using the law of sines, we have:
$$\frac{\sin 15^\circ}{c} = \frac{\sin 30^\circ}{1}$$
$$c = \frac{\sin 15^\circ}{\sin 30^\circ} \approx 0.52$$

(d) We find the area of each triangle:
$$A = \tfrac{1}{2} ab \sin C = \tfrac{1}{2} \cdot \sqrt{2} \cdot 1 \sin 105^\circ \approx 0.68$$
$$A = \tfrac{1}{2} ab \sin C = \tfrac{1}{2} \cdot \sqrt{2} \sin 15^\circ \approx 0.18$$

15. Using the law of sines, we have:
$$\frac{\sin 20^\circ}{2} = \frac{\sin 100^\circ}{a} = \frac{\sin 50^\circ}{b}$$
$$\frac{\sin 70^\circ}{c} = \frac{\sin 95^\circ}{b} = \frac{\sin 15^\circ}{d}$$
Hence:
$$a = \frac{2\sin 110^\circ}{\sin 20^\circ} = \frac{2\sin 70^\circ}{\sin 20^\circ} \text{ cm}$$
$$b = \frac{2\sin 50^\circ}{\sin 20^\circ} \text{ cm}$$
$$c = \frac{b\sin 70^\circ}{\sin 95^\circ} = \frac{2\sin 50^\circ \sin 70^\circ}{\sin 20^\circ \sin 85^\circ} \text{ cm}$$
$$d = \frac{b\sin 15^\circ}{\sin 95^\circ} = \frac{2\sin 50^\circ \sin 15^\circ}{\sin 20^\circ \sin 85^\circ} \text{ cm}$$

17. We sketch the triangle:

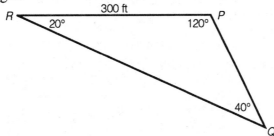

Applying the law of sines, we have:
$$\frac{\sin 40°}{300} = \frac{\sin 20°}{PQ}$$

So $PQ = \dfrac{300\sin 20°}{\sin 40°} \approx 160$ ft.

19. (a) Using the law of cosines:
$$x^2 = 5^2 + 8^2 - 2(5)(8)\cos 60° = 25 + 64 - 80 \cdot \tfrac{1}{2} = 49$$
So $x = \sqrt{49} = 7$ cm.

(b) Using the law of cosines:
$$x^2 = 5^2 + 8^2 - 2(5)(8)\cos 120° = 25 + 64 - 80 \cdot \left(-\tfrac{1}{2}\right) = 129$$
So $x = \sqrt{129}$ cm.

21. (a) Using the law of cosines:
$$\begin{aligned}
x^2 &= (7.3)^2 + (11.5)^2 - 2(7.3)(11.5)\cos 40° \\
&= 53.29 + 132.25 - 167.9\cos 40° \\
&= 185.54 - 167.9\cos 40° \\
&\approx 56.92
\end{aligned}$$
So $x \approx \sqrt{56.92} \approx 7.5$ cm.

(b) Using the law of cosines:
$$\begin{aligned}
x^2 &= (7.3)^2 + (11.5)^2 - 2(7.3)(11.5)\cos 140° \\
&= 53.29 + 132.25 - 167.9\cos 140° \\
&= 185.54 - 167.9\cos 140° \\
&\approx 314.16
\end{aligned}$$
So $x \approx \sqrt{314.16} \approx 17.7$ cm.

23. This is incorrect because x is not the side opposite the $130°$ angle. The correct equation is:
$$6^2 = x^2 + 3^2 - 2(x)(3)\cos 130°$$

25. Using $a = 6$, $b = 7$, $c = 10$ and the law of cosines, we have:
$$6^2 = 7^2 + 10^2 - 2(7)(10)\cos A, \text{ so } \cos A = \tfrac{113}{140}$$
$$7^2 = 6^2 + 10^2 - 2(6)(10)\cos B, \text{ so } \cos B = \tfrac{87}{120} = \tfrac{29}{40}$$
$$10^2 = 6^2 + 7^2 - 2(6)(7)\cos C, \text{ so } \cos C = -\tfrac{15}{84} = -\tfrac{5}{28}$$

27. We use the law of cosines to find angle A:
$$7^2 = 8^2 + 13^2 - 2(8)(13)\cos A$$
$$49 = 64 + 169 - 208\cos A$$
$$-184 = -208\cos A$$
$$\cos A = \tfrac{184}{208}$$
$$A \approx 27.8°$$
Now use the law of cosines to find angle B:
$$8^2 = 7^2 + 13^2 - 2(7)(13)\cos B$$
$$64 = 49 + 169 - 182\cos B$$
$$-154 = -182\cos B$$
$$\cos B = \tfrac{154}{182}$$
$$B \approx 32.2°$$
Since $A + B + C = 180°$, we have:
$$C = 180° - A - B \approx 180° - 27.8° - 32.2° \approx 120°$$

29. We use the law of cosines to find angle A:
$$\left(\tfrac{2}{\sqrt{3}}\right)^2 = \left(\tfrac{2}{\sqrt{3}}\right)^2 + 2^2 - 2 \cdot \tfrac{2}{\sqrt{3}} \cdot 2\cos A$$
$$\tfrac{4}{3} = \tfrac{4}{3} + 4 - \tfrac{8}{\sqrt{3}}\cos A$$
$$-4 = -\tfrac{8}{\sqrt{3}}\cos A$$
$$\cos A = \tfrac{\sqrt{3}}{2}$$
$$A = 30°$$
Since $a = b$, $B = 30°$. Since $A + B + C = 180°$, we have:
$$C = 180° - A - B = 180° - 30° - 30° = 120°$$

31. We first draw a figure, noting that the central angle is $\frac{360°}{5} = 72°$:

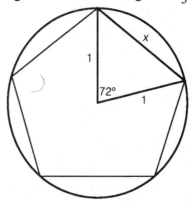

We apply the law of cosines to find x:
$$x^2 = 1^2 + 1^2 - 2(1)(1)\cos 72° = 2 - 2\cos 72° \approx 1.382$$
So $x \approx \sqrt{1.382} \approx 1.18$, and thus the perimeter is $5(1.18) \approx 5.9$ units.

33. (a) Using the law of cosines, we have:
$$\begin{aligned} a^2 &= (6.1)^2 + (3.2)^2 - 2(6.1)(3.2)\cos 40° \\ &= 37.21 + 10.24 - 39.04\cos 40° \\ &= 47.45 - 39.04\cos 40° \\ &\approx 17.54 \end{aligned}$$
So $a \approx \sqrt{17.54} \approx 4.2$ cm.

(b) Using the law of sines, we have:
$$\frac{\sin C}{3.2} = \frac{\sin 40°}{4.2}$$
$$\sin C = \frac{3.2 \sin 40°}{4.2}$$
$$\sin C \approx 0.49$$
$$C \approx 29.3°$$

(c) Since $A + B + C = 180°$, we have:
$$B = 180° - A - C \approx 180° - 40° - 29.3° \approx 110.7°$$

35. We first draw a figure indicating the relationship between Town A, Town B and Town C, where Town A is centered at the origin:

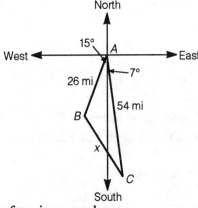

Using $\triangle ABC$ and the law of cosines, we have:
$$\begin{aligned} x^2 &= 26^2 + 54^2 - 2(26)(54)\cos 22° \\ &= 676 + 2916 - 2808\cos 22° \\ &= 3592 - 2808\cos 22° \\ &\approx 988.5 \end{aligned}$$
So $x \approx \sqrt{988.5} \approx 31$. The distance between Towns B and C is approximately 31 miles.

37. Using P to denote the plane, we have the following figure:

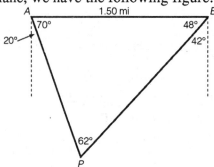

We use the law of sines to find AP:
$$\frac{AP}{\sin 48°} = \frac{1.5}{\sin 62°}$$
$$AP = \frac{1.5\sin 48°}{\sin 62°} \approx 1.26 \text{ mi}$$
We use the law of sines to find BP:
$$\frac{BP}{\sin 70°} = \frac{1.5}{\sin 62°}$$
$$BP = \frac{1.5\sin 70°}{\sin 62°} \approx 1.60 \text{ mi}$$
So the distance from the plane to lighthouse A is 1.26 miles and to lighthouse B is 1.60 miles.

39. Using the law of cosines, we have:
$$\begin{aligned}
D^2 &= d^2 + d^2 - 2(d)(d)\cos\tfrac{32°}{60} \\
&= 2d^2 - 2d^2\cos\tfrac{32°}{60} \\
&= 2d^2\left(1 - \cos\tfrac{32°}{60}\right) \\
&= 2(92{,}690{,}000)^2\left(1 - \cos\tfrac{32°}{60}\right) \\
&\approx 744{,}414{,}483{,}000
\end{aligned}$$
So $D \approx \sqrt{744{,}414{,}483{,}000} \approx 860{,}000$ miles.
Note: That is about 100 times the diameter of Earth!

41. (a) Using the law of cosines, we have:
$$\left(m^2 + n^2 + mn\right)^2 = \left(2mn + n^2\right)^2 + \left(m^2 - n^2\right)^2 - 2\left(2mn + n^2\right)\left(m^2 - n^2\right)\cos C$$
After carrying out the indicated squaring operations, and then combining like terms, the equation becomes:
$$2m^3 n - 2mn^3 + m^2 n^2 - n^4 = -2\left(2mn + n^2\right)\left(m^2 - n^2\right)\cos C$$
$$2mn\left(m^2 - n^2\right) + n^2\left(m^2 - n^2\right) = -2\left(2mn + n^2\right)\left(m^2 - n^2\right)\cos C$$
$$\left(m^2 - n^2\right)\left(2mn + n^2\right) = -2\left(2mn + n^2\right)\left(m^2 - n^2\right)\cos C$$
Therefore $\cos C = -\tfrac{1}{2}$, and consequently $C = 120°$.

(b)   Let $m = 2$ and $n = 1$. Then by means of the expressions in part (a), we obtain $a = 5$, $b = 3$, $c = 7$.

43.   We draw the figure:

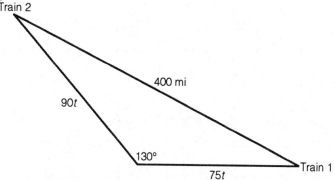

If t is the time of travel (in hours), then the trains have traveled $75t$ mi and $90t$ mi, as indicated in the figure. We use the law of cosines:

$$(400)^2 = (90t)^2 + (75t)^2 - 2(90t)(75t)\cos 130°$$
$$160000 = 8100t^2 + 5625t^2 - 13500\cos 130° t^2$$
$$t^2 = \frac{160000}{13725 - 13500\cos 130°} \approx 7.142$$
$$t \approx 2.672 \text{ hr} \approx 160 \text{ minutes}$$

The trains are 400 mi apart after 160 minutes, which occurs at 2:40 P.M.

45.   Following the hint that is given, we obtain the two equations:

$$\lambda^2 = a^2 + d^2 - 2ad\cos(180° - \theta) \qquad (1)$$
$$\lambda^2 = b^2 + c^2 - 2bc\cos\theta \qquad (2)$$

Equation (1) can be rewritten:

$$\lambda^2 = a^2 + d^2 + 2ad\cos\theta$$

Upon solving this last equation for $\cos\theta$, we obtain:

$$\cos\theta = \frac{\lambda^2 - a^2 - d^2}{2ad}$$

Now we use this expression for $\cos\theta$ in equation (2) to obtain:

$$\lambda^2 = b^2 + c^2 - 2bc\left(\frac{\lambda^2 - a^2 - d^2}{2ad}\right)$$
$$\lambda^2 ad = b^2 ad + c^2 ad - bc\lambda^2 + a^2 bc + bcd^2$$
$$\lambda^2(ad + bc) = b^2 ad + c^2 ad + a^2 bc + d^2 bc$$
$$\lambda^2(ad + bc) = (c^2 ad + a^2 bc) + (b^2 ad + d^2 bc)$$
$$\lambda^2(ad + bc) = ac(cd + ab) + bd(ab + cd)$$
$$\lambda^2(ad + bc) = (ab + cd)(ac + bd)$$
$$\lambda^2 = \frac{(ab + cd)(ac + bd)}{ad + bc}$$

47. Using the hint, we have:
$$a^4 + b^4 + c^4 = 2(a^2 + b^2)c^2$$
$$c^4 - 2(a^2 + b^2)c^2 + (a^4 + b^4) = 0$$
Using the quadratic formula, we have:
$$c^2 = \frac{2(a^2 + b^2) \pm \sqrt{4(a^2 + b^2)^2 - 4(a^4 + b^4)}}{2}$$
$$= a^2 + b^2 \pm \sqrt{a^4 + 2a^2b^2 + b^4 - a^4 - b^4}$$
$$= a^2 + b^2 \pm \sqrt{2a^2b^2}$$
$$= a^2 + b^2 \pm ab\sqrt{2}$$
By the law of cosines, we have:
$$c^2 = a^2 + b^2 - 2ab\cos C$$
Setting these expressions equal:
$$a^2 + b^2 - 2ab\cos C = a^2 + b^2 \pm ab\sqrt{2}$$
$$-2ab\cos C = \pm ab\sqrt{2}$$
$$\cos C = \pm \tfrac{\sqrt{2}}{2}$$
$$C = 45° \text{ or } 135°$$

49. (a) Simplifying each of the fractions:
$$\frac{\sin A}{a} = \frac{\frac{\sqrt{T}}{2bc}}{a} = \frac{\sqrt{T}}{2abc}$$
$$\frac{\sin B}{b} = \frac{\frac{\sqrt{T}}{2ac}}{b} = \frac{\sqrt{T}}{2abc}$$
$$\frac{\sin C}{c} = \frac{\frac{\sqrt{T}}{2ab}}{c} = \frac{\sqrt{T}}{2abc}$$

(b) Since each of the fractions is equal to $\frac{\sqrt{T}}{2abc}$, then:
$$\frac{\sin A}{a} = \frac{\sin B}{b} = \frac{\sin C}{c}$$

51. (a) Following the hint, we drop a perpendicular from O to $\overline{AD}$, and call the intersection point P. Using $\triangle APO$, we have:
$$\cos \alpha = \frac{AP}{AO} = \frac{AP}{1} = AP$$
Since $AD = 2 \cdot AP$, $AD = 2\cos \alpha$.

(b) Following steps similar to part (a), notice that $\triangle AOE$ is an isosceles triangle. Drop a perpendicular from O to $\overline{AE}$, and call the intersection point Q. Using $\triangle AQO$, we have:
$$\cos \beta = \frac{AQ}{AO} = \frac{AQ}{1} = AQ$$
Since $AE = 2 \cdot AQ$, $AE = 2\cos \beta$.

(c) Since $\angle AFC = 60°$, $\angle AFB = 180° - 60° = 120°$. Since the angles of $\triangle AFB$ must sum to $180°$, we have:
$$\alpha + 120° + \angle B = 180°$$
$$\angle B = 60° - \alpha$$
Similarly, since the angles of $\triangle AFC$ must sum to $180°$, we have:
$$\beta + 60° + \angle C = 180°$$
$$\angle C = 120° - \beta$$

(d) Using the law of sines on $\triangle ABC$, we have:
$$\frac{\sin(\angle A)}{BC} = \frac{\sin(\angle B)}{AC} = \frac{\sin(\angle C)}{AB}$$
Since $\angle A = \alpha + \beta$, $\angle B = 60° - \alpha$ and $\angle C = 120° - \beta$, we have:
$$\frac{\sin(\alpha + \beta)}{BC} = \frac{\sin(60° - \alpha)}{AC}, \text{ so } AC = \frac{BC \cdot \sin(60° - \alpha)}{\sin(\alpha + \beta)}$$
$$\frac{\sin(\alpha + \beta)}{BC} = \frac{\sin(120° - \beta)}{AB}, \text{ so } AB = \frac{BC \cdot \sin(120° - \beta)}{\sin(\alpha + \beta)}$$

(e) Using the results in parts (a), (b) and (d), we have:
$$AD \cdot AB - AE \cdot AC$$
$$= (2\cos\alpha) \cdot \frac{BC\sin(120° - \beta)}{\sin(\alpha + \beta)} - (2\cos\beta) \cdot \frac{BC\sin(60° - \alpha)}{\sin(\alpha + \beta)}$$
$$= BC \cdot \left[\frac{2\cos\alpha\sin(120° - \beta) - 2\cos\beta\sin(60° - \alpha)}{\sin(\alpha + \beta)} \right]$$

(f) Using the addition formulas for sine, we have:
$$\sin(120° - \beta) = \sin 120°\cos\beta - \cos 120°\sin\beta = \tfrac{\sqrt{3}}{2}\cos\beta + \tfrac{1}{2}\sin\beta$$
$$\sin(60° - \alpha) = \sin 60°\cos\alpha - \cos 60°\sin\alpha = \tfrac{\sqrt{3}}{2}\cos\alpha - \tfrac{1}{2}\sin\alpha$$
Therefore:
$$2\cos\alpha\sin(120° - \beta) - 2\cos\beta\sin(60° - \alpha)$$
$$= 2\cos\alpha\left(\tfrac{\sqrt{3}}{2}\cos\beta + \tfrac{1}{2}\sin\beta\right) - 2\cos\beta\left(\tfrac{\sqrt{3}}{2}\cos\alpha - \tfrac{1}{2}\sin\alpha\right)$$
$$= \sqrt{3}\cos\alpha\cos\beta + \cos\alpha\sin\beta - \sqrt{3}\cos\alpha\cos\beta + \sin\alpha\cos\beta$$
$$= \sin\alpha\cos\beta + \cos\alpha\sin\beta$$
$$= \sin(\alpha + \beta)$$
This completes the proof of the result.

8.2 Vectors in the Plane. A Geometric Approach

1.   We graph the vector:

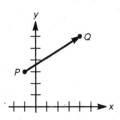

The magnitude is given by:

$$|\vec{PQ}| = \sqrt{(4-(-1))^2 + (6-3)^2} = \sqrt{25+9} = \sqrt{34}$$

3.   We graph the vector:

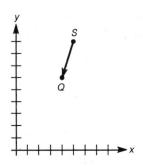

The magnitude is given by:

$$|\vec{SQ}| = \sqrt{(4-5)^2 + (6-9)^2} = \sqrt{1+9} = \sqrt{10}$$

5.   We graph the vector:

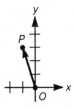

The magnitude is given by:

$$|\vec{OP}| = \sqrt{(-1-0)^2 + (3-0)^2} = \sqrt{1+9} = \sqrt{10}$$

7.   We graph the vector sum:

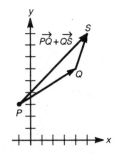

The magnitude is given by:

$$|\vec{PQ} + \vec{QS}| = |\vec{PS}| = \sqrt{(5-(-1))^2 + (9-3)^2} = \sqrt{36+36} = 6\sqrt{2}$$

9.   We graph the vector sum:

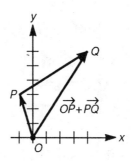

The magnitude is given by:

$$|\vec{OP} + \vec{PQ}| = |\vec{OQ}| = \sqrt{(4-0)^2 + (6-0)^2} = \sqrt{16+36} = 2\sqrt{13}$$

11.   We graph the vector sum:

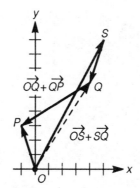

The magnitude is given by:

$$|\vec{OS} + \vec{SQ} + \vec{QP}| = |\vec{OQ} + \vec{QP}| = |\vec{OP}| = \sqrt{(-1-0)^2 + (3-0)^2} = \sqrt{1+9} = \sqrt{10}$$

13.   We graph the vector sum:

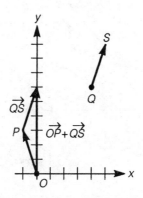

The magnitude is given by:

$$|\vec{OP} + \vec{QS}| = \sqrt{(0-0)^2 + (6-0)^2} = \sqrt{0+36} = 6$$

15. We graph the vector sum:

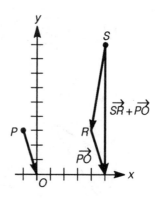

The magnitude is given by:

$$|\overrightarrow{SR} + \overrightarrow{PO}| = \sqrt{(5-5)^2 + (9-0)^2} = \sqrt{0+81} = 9$$

17. We graph the vector sum:

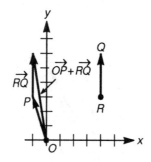

The magnitude is given by:

$$|\overrightarrow{OP} + \overrightarrow{RQ}| = \sqrt{(-1-0)^2 + (6-0)^2} = \sqrt{1+36} = \sqrt{37}$$

19. We graph the vector sum:

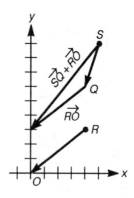

The magnitude is given by:

$$|\overrightarrow{SQ} + \overrightarrow{RO}| = \sqrt{(5-0)^2 + (9-3)^2} = \sqrt{25+36} = \sqrt{61}$$

21. We graph the vector sum:

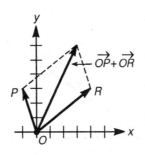

The magnitude is given by:

$$|\vec{OP} + \vec{OR}| = \sqrt{(3-0)^2 + (6-0)^2} = \sqrt{9+36} = 3\sqrt{5}$$

23. We graph the vector sum:

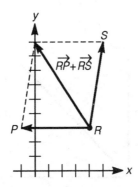

The magnitude is given by:

$$|\vec{RP} + \vec{RS}| = \sqrt{(0-4)^2 + (9-3)^2} = \sqrt{16+36} = 2\sqrt{13}$$

25. We graph the vector sum:

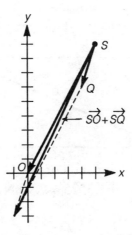

The magnitude is given by:

$$|\vec{SO} + \vec{SQ}| = \sqrt{(-1-5)^2 + (-3-9)^2} = \sqrt{36+144} = 6\sqrt{5}$$

27. We draw the figure:

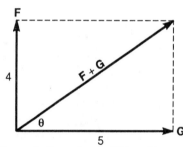

Now compute the magnitude and direction of **F** + **G**:

$$|\mathbf{F}+\mathbf{G}| = \sqrt{4^2+5^2} = \sqrt{16+25} = \sqrt{41} \text{ N}$$

$$\theta = \tan^{-1}\left(\tfrac{4}{5}\right) \approx 38.7°$$

29. We draw the figure:

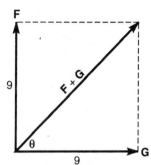

Now compute the magnitude and direction of **F** + **G**:

$$|\mathbf{F}+\mathbf{G}| = \sqrt{9^2+9^2} = 9\sqrt{2} \text{ N}$$

$$\theta = \tan^{-1}\left(\tfrac{9}{9}\right) = \tan^{-1}1 = 45°$$

31. We draw the figure:

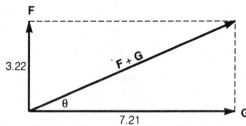

Now compute the magnitude and direction of **F** + **G**:

$$|\mathbf{F}+\mathbf{G}| = \sqrt{3.22^2+7.21^2} = \sqrt{62.3525} \approx 7.90 \text{ N}$$

$$\theta = \tan^{-1}\left(\tfrac{3.22}{7.21}\right) \approx 24.1°$$

33. We draw the parallelogram:

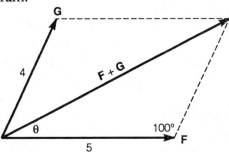

Let $d = |\mathbf{F} + \mathbf{G}|$. Then using the law of cosines:
$$d^2 = 5^2 + 4^2 - 2(5)(4)\cos 100°$$
$$d^2 = 41 - 40\cos 100°$$
$$d = \sqrt{41 - 40\cos 100°} \approx 6.92 \text{ N}$$
We find θ by the law of sines:
$$\frac{\sin \theta}{4} = \frac{\sin 100°}{d}$$
$$\sin \theta = \frac{4\sin 100°}{\sqrt{41 - 40\cos 100°}} \approx 0.5689$$
$$\theta \approx 34.67°$$

35. We draw the parallelogram:

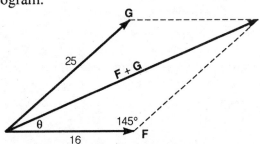

Let $d = |\mathbf{F} + \mathbf{G}|$. Then using the law of cosines:
$$d^2 = 16^2 + 25^2 - 2(16)(25)\cos 145°$$
$$d^2 = 881 - 800\cos 145°$$
$$d = \sqrt{881 - 800\cos 145°} \approx 39.20 \text{ N}$$
We find θ by the law of sines:
$$\frac{\sin \theta}{25} = \frac{\sin 145°}{d}$$
$$\sin \theta = \frac{25\sin 145°}{\sqrt{881 - 800\cos 145°}} \approx 0.3658$$
$$\theta \approx 21.46°$$

37. We draw the parallelogram:

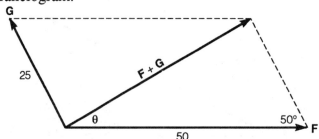

Let $d = |\mathbf{F} + \mathbf{G}|$. Then using the law of cosines:

$$d^2 = 50^2 + 25^2 - 2(50)(25)\cos 50°$$
$$d^2 = 3125 - 2500\cos 50°$$
$$d = \sqrt{3125 - 2500\cos 50°} \approx 38.96 \text{ N}$$

We find θ by the law of sines:

$$\frac{\sin \theta}{25} = \frac{\sin 50°}{d}$$
$$\sin \theta = \frac{25\sin 50°}{\sqrt{3125 - 2500\cos 50°}} \approx 0.4915$$
$$\theta \approx 29.44°$$

39. We compute the horizontal and vertical components:

$$V_x = 16\cos 30° \approx 13.86 \text{ cm/sec}$$
$$V_y = 16\sin 30° = 8 \text{ cm/sec}$$

41. We compute the horizontal and vertical components:

$$F_x = 14\cos 75° \approx 3.62 \text{ N}$$
$$F_y = 14\sin 75° \approx 13.52 \text{ N}$$

43. We compute the horizontal and vertical components:

$$V_x = 1\cos 135° \approx -0.71 \text{ cm/sec}$$
$$V_y = 1\sin 135° \approx 0.71 \text{ cm/sec}$$

45. We compute the horizontal and vertical components:

$$F_x = 1.25\cos 145° \approx -1.02 \text{ N}$$
$$F_y = 1.25\sin 145° \approx 0.72 \text{ N}$$

47. We draw the vectors:

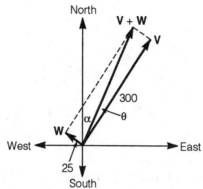

Let θ be the drift angle. Then:

$\tan\theta = \frac{25}{300} = \frac{1}{12}$

$\theta = \tan^{-1}\left(\frac{1}{12}\right) \approx 4.76°$

The ground speed is given by:

$|\mathbf{V} + \mathbf{W}| = \sqrt{25^2 + 300^2} = \sqrt{90625} \approx 301.04$ mph

Let α be the bearing. Then:

$\alpha = 30° - \theta \approx 30° - 4.76° \approx 25.24°$

49. We draw the vectors:

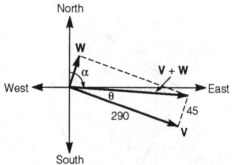

Let θ be the drift angle. Then:

$\tan\theta = \frac{45}{290} = \frac{9}{58}$

$\theta = \tan^{-1}\left(\frac{9}{58}\right) \approx 8.82°$

The ground speed is given by:

$|\mathbf{V} + \mathbf{W}| = \sqrt{290^2 + 45^2} = \sqrt{86125} \approx 293.47$ mph

Let α be the bearing. Then:

$\alpha = 100° - \theta \approx 100° - 8.82° \approx 91.18°$

51. We draw a figure:

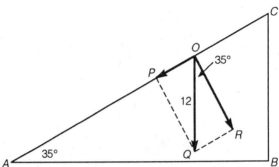

Notice that $\angle QOR = 35°$ since it is complementary to $\angle POQ$, but $\angle POQ = \angle ACB$. So the desired components are given by:

$$|\overrightarrow{OR}| = 12\cos 35° \approx 9.83 \text{ lb}$$

$$|\overrightarrow{OP}| = 12\sin 35° \approx 6.88 \text{ lb}$$

53. Using the same approach as in Exercise 51, we have:

perpendicular: $12\cos 10° \approx 11.82$ lb

parallel: $12\sin 10° \approx 2.08$ lb

55. (a)   We draw the vector sum $(\mathbf{A} + \mathbf{B}) + \mathbf{C}$:

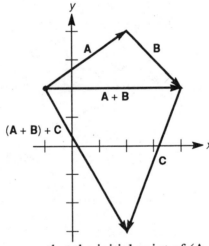

From the diagram, we see that the initial point of $(\mathbf{A} + \mathbf{B}) + \mathbf{C}$ is $(-1, 2)$ and the terminal point is $(2, -3)$.

(b)   We draw the vector sum **A** + (**B** + **C**):

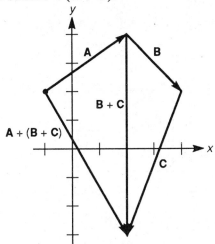

From the diagram, we see that the initial point of **A** + (**B** + **C**) is $(-1, 2)$ and the terminal point is $(2, -3)$.

8.3 Vectors in the Plane, An Algebraic Approach

1.   We compute the length of the vector:
$$|\langle 4, 3 \rangle| = \sqrt{4^2 + 3^2} = \sqrt{25} = 5$$

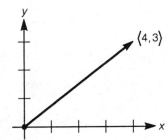

3.   We compute the length of the vector:
$$|\langle -4, 2 \rangle| = \sqrt{(-4)^2 + 2^2} = \sqrt{20} = 2\sqrt{5}$$

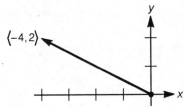

5.   We compute the length of the vector:

$$\left|\left\langle \tfrac{3}{4}, -\tfrac{1}{2} \right\rangle\right| = \sqrt{\left(\tfrac{3}{4}\right)^2 + \left(-\tfrac{1}{2}\right)^2} = \sqrt{\tfrac{9}{16} + \tfrac{1}{4}} = \tfrac{\sqrt{13}}{4}$$

7.   Subtracting components, we have:

$$\overrightarrow{PQ} = \langle 3 - 2, 7 - 3 \rangle = \langle 1, 4 \rangle$$

9.   Subtracting components, we have:

$$\overrightarrow{PQ} = \langle -3 - (-2), -2 - (-3) \rangle = \langle -3 + 2, -2 + 3 \rangle = \langle -1, 1 \rangle$$

11.   Subtracting components, we have:

$$\overrightarrow{PQ} = \langle 3 - (-5), -4 - 1 \rangle = \langle 3 + 5, -5 \rangle = \langle 8, -5 \rangle$$

13. $\mathbf{a} + \mathbf{b} = \langle 2 + 5, 3 + 4 \rangle = \langle 7, 7 \rangle$

15. $2\mathbf{a} + 4\mathbf{b} = \langle 4, 6 \rangle + \langle 20, 16 \rangle = \langle 24, 22 \rangle$

17.   Since $\mathbf{b} + \mathbf{c} = \langle 5 + 6, 4 - 1 \rangle = \langle 11, 3 \rangle$, then:

$$|\mathbf{b} + \mathbf{c}| = \sqrt{11^2 + 3^2} = \sqrt{130}$$

19.   Since $\mathbf{a} + \mathbf{c} = \langle 2 + 6, 3 - 1 \rangle = \langle 8, 2 \rangle$, then:

$$|\mathbf{a} + \mathbf{c}| = \sqrt{8^2 + 2^2} = \sqrt{68} = 2\sqrt{17}$$
$$|\mathbf{a}| = \sqrt{2^2 + 3^2} = \sqrt{13}$$
$$|\mathbf{c}| = \sqrt{6^2 + (-1)^2} = \sqrt{37}$$

So $|\mathbf{a} + \mathbf{c}| - |\mathbf{a}| - |\mathbf{c}| = 2\sqrt{17} - \sqrt{13} - \sqrt{37}$.

21.   Since $\mathbf{b} + \mathbf{c} = \langle 5 + 6, 4 - 1 \rangle = \langle 11, 3 \rangle$, then:

$$\mathbf{a} + (\mathbf{b} + \mathbf{c}) = \langle 2, 3 \rangle + \langle 11, 3 \rangle = \langle 13, 6 \rangle$$

23. $3\mathbf{a} + 4\mathbf{a} = \langle 6, 9 \rangle + \langle 8, 12 \rangle = \langle 14, 21 \rangle$

25. $\mathbf{a} - \mathbf{b} = \langle 2,3 \rangle - \langle 5,4 \rangle = \langle -3,-1 \rangle$

27. $3\mathbf{b} - 4\mathbf{d} = \langle 15,12 \rangle - \langle -8,0 \rangle = \langle 15 + 8, 12 - 0 \rangle = \langle 23,12 \rangle$

29. Since $\mathbf{b} + \mathbf{c} = \langle 11,3 \rangle$, then:
$$\mathbf{a} - (\mathbf{b} + \mathbf{c}) = \langle 2,3 \rangle - \langle 11,3 \rangle = \langle -9,0 \rangle$$

31. Since $\mathbf{c} + \mathbf{d} = \langle 4,-1 \rangle$ and $\mathbf{c} - \mathbf{d} = \langle 8,-1 \rangle$, then:
$$|\mathbf{c} + \mathbf{d}| = \sqrt{16+1} = \sqrt{17} \text{ and } |\mathbf{c} - \mathbf{d}| = \sqrt{64+1} = \sqrt{65}$$
Therefore:
$$|\mathbf{c} + \mathbf{d}|^2 - |\mathbf{c} - \mathbf{d}|^2 = 17 - 65 = -48$$

33. Separating individual components, we have:
$$\langle 3,8 \rangle = \langle 3,0 \rangle + \langle 0,8 \rangle = 3\mathbf{i} + 8\mathbf{j}$$

35. Separating individual components, we have:
$$\langle -8,-6 \rangle = \langle -8,0 \rangle + \langle 0,-6 \rangle = -8\mathbf{i} - 6\mathbf{j}$$

37. Separating individual components, we have:
$$3\langle 5,3 \rangle + 2\langle 2,7 \rangle = 3(5\mathbf{i} + 3\mathbf{j}) + 2(2\mathbf{i} + 7\mathbf{j}) = 15\mathbf{i} + 9\mathbf{j} + 4\mathbf{i} + 14\mathbf{j} = 19\mathbf{i} + 23\mathbf{j}$$

39. $\mathbf{i} + \mathbf{j} = \langle 1,1 \rangle$

41. $5\mathbf{i} - 4\mathbf{j} = \langle 5,-4 \rangle$

43. We first compute the length of the vector:
$$|\langle 4,8 \rangle| = \sqrt{4^2 + 8^2} = \sqrt{80} = 4\sqrt{5}$$
So a unit vector would be given by:
$$\tfrac{1}{4\sqrt{5}}\langle 4,8 \rangle = \left\langle \tfrac{1}{\sqrt{5}}, \tfrac{2}{\sqrt{5}} \right\rangle = \left\langle \tfrac{\sqrt{5}}{5}, \tfrac{2\sqrt{5}}{5} \right\rangle$$

45. We first compute the length of the vector:
$$|\langle 6,-3 \rangle| = \sqrt{6^2 + (-3)^2} = \sqrt{45} = 3\sqrt{5}$$
So a unit vector would be given by:
$$\tfrac{1}{3\sqrt{5}}\langle 6,-3 \rangle = \left\langle \tfrac{2}{\sqrt{5}}, -\tfrac{1}{\sqrt{5}} \right\rangle = \left\langle \tfrac{2\sqrt{5}}{5}, -\tfrac{\sqrt{5}}{5} \right\rangle$$

47. We first compute the length of the vector:
$$|8\mathbf{i} - 9\mathbf{j}| = \sqrt{8^2 + (-9)^2} = \sqrt{145}$$
So a unit vector would be given by:
$$\tfrac{1}{\sqrt{145}}(8\mathbf{i} - 9\mathbf{j}) = \tfrac{8}{\sqrt{145}}\mathbf{i} - \tfrac{9}{\sqrt{145}}\mathbf{j} = \tfrac{8\sqrt{145}}{145}\mathbf{i} - \tfrac{9\sqrt{145}}{145}\mathbf{j}$$

49. We compute the components u_1 and u_2:
$$u_1 = \cos\tfrac{\pi}{6} = \tfrac{\sqrt{3}}{2}$$
$$u_2 = \sin\tfrac{\pi}{6} = \tfrac{1}{2}$$

51. We compute the components u_1 and u_2:
$$u_1 = \cos\tfrac{2\pi}{3} = -\tfrac{1}{2}$$
$$u_2 = \sin\tfrac{2\pi}{3} = \tfrac{\sqrt{3}}{2}$$

53. We compute the components u_1 and u_2:
$$u_1 = \cos\tfrac{5\pi}{6} = -\tfrac{\sqrt{3}}{2}$$
$$u_2 = \sin\tfrac{5\pi}{6} = \tfrac{1}{2}$$

55. We verify property 1:
$$\begin{aligned}
\mathbf{u} + (\mathbf{v} + \mathbf{w}) &= \langle u_1, u_2 \rangle + \big(\langle v_1, v_2 \rangle + \langle w_1, w_2 \rangle \big) \\
&= \langle u_1, u_2 \rangle + \langle v_1 + w_1, v_2 + w_2 \rangle \\
&= \langle u_1 + v_1 + w_1, u_2 + v_2 + w_2 \rangle \\
&= \langle u_1 + v_1, u_2 + v_2 \rangle + \langle w_1, w_2 \rangle \\
&= \big(\langle u_1, u_2 \rangle + \langle v_1, v_2 \rangle \big) + \langle w_1, w_2 \rangle \\
&= (\mathbf{u} + \mathbf{v}) + \mathbf{w}
\end{aligned}$$
We verify property 2:
$$\mathbf{0} + \mathbf{v} = \langle 0, 0 \rangle + \langle v_1, v_2 \rangle = \langle 0 + v_1, 0 + v_2 \rangle = \langle v_1 + 0, v_2 + 0 \rangle = \langle v_1, v_2 \rangle + \langle 0, 0 \rangle = \mathbf{v} + \mathbf{0}$$
$$\mathbf{v} + \mathbf{0} = \langle v_1, v_2 \rangle + \langle 0, 0 \rangle = \langle v_1 + 0, v_2 + 0 \rangle = \langle v_1, v_2 \rangle = \mathbf{v}$$

57. We verify property 5:
$$\begin{aligned}
a(\mathbf{u} + \mathbf{v}) &= a\big(\langle u_1, u_2 \rangle + \langle v_1, v_2 \rangle \big) \\
&= a\langle u_1 + v_1, u_2 + v_2 \rangle \\
&= \big\langle a(u_1 + v_1), a(u_2 + v_2) \big\rangle \\
&= \langle au_1 + av_1, au_2 + av_2 \rangle \\
&= \langle au_1, au_2 \rangle + \langle av_1, av_2 \rangle \\
&= a\langle u_1, u_2 \rangle + a\langle v_1, v_2 \rangle \\
&= a\mathbf{u} + a\mathbf{v}
\end{aligned}$$
We verify property 6:
$$\begin{aligned}
(a + b)\mathbf{v} &= (a + b)\langle v_1, v_2 \rangle \\
&= \big\langle (a + b)v_1, (a + b)v_2 \big\rangle \\
&= \langle av_1 + bv_1, av_2 + bv_2 \rangle \\
&= \langle av_1, av_2 \rangle + \langle bv_1, bv_2 \rangle \\
&= a\langle v_1, v_2 \rangle + b\langle v_1, v_2 \rangle \\
&= a\mathbf{v} + b\mathbf{v}
\end{aligned}$$

59. (a) We compute the dot products:
$$\mathbf{u} \cdot \mathbf{v} = \langle -4,5 \rangle \cdot \langle 3,4 \rangle = (-4)(3) + (5)(4) = -12 + 20 = 8$$
$$\mathbf{v} \cdot \mathbf{u} = \langle 3,4 \rangle \cdot \langle -4,5 \rangle = (3)(-4) + (4)(5) = -12 + 20 = 8$$

(b) We compute the dot products:
$$\mathbf{v} \cdot \mathbf{w} = \langle 3,4 \rangle \cdot \langle 2,-5 \rangle = (3)(2) + (4)(-5) = 6 - 20 = -14$$
$$\mathbf{w} \cdot \mathbf{v} = \langle 2,-5 \rangle \cdot \langle 3,4 \rangle = (2)(3) + (-5)(4) = 6 - 20 = -14$$

(c) Let $\mathbf{A} = \langle x_1, y_1 \rangle$ and $\mathbf{B} = \langle x_2, y_2 \rangle$. We compute each dot product:
$$\mathbf{A} \cdot \mathbf{B} = \langle x_1, y_1 \rangle \cdot \langle x_2, y_2 \rangle = x_1 x_2 + y_1 y_2$$
$$\mathbf{B} \cdot \mathbf{A} = \langle x_2, y_2 \rangle \cdot \langle x_1, y_1 \rangle = x_2 x_1 + y_2 y_1 = x_1 x_2 + y_1 y_2$$
Thus $\mathbf{A} \cdot \mathbf{B} = \mathbf{B} \cdot \mathbf{A}$.

61. (a) We compute each quantity:
$$\mathbf{v} \cdot \mathbf{v} = \langle 3,4 \rangle \cdot \langle 3,4 \rangle = (3)(3) + (4)(4) = 9 + 16 = 25$$
$$|\mathbf{v}| = \sqrt{3^2 + 4^2} = \sqrt{9 + 16} = \sqrt{25} = 5, \text{ so } |\mathbf{v}|^2 = 25$$

(b) We compute each quantity:
$$\mathbf{w} \cdot \mathbf{w} = \langle 2,-5 \rangle \cdot \langle 2,-5 \rangle = (2)(2) + (-5)(-5) = 4 + 25 = 29$$
$$|\mathbf{w}| = \sqrt{2^2 + (-5)^2} = \sqrt{4 + 25} = \sqrt{29}, \text{ so } |\mathbf{w}|^2 = 29$$

63. We first do the computations:
$$|\mathbf{A}| = \sqrt{16 + 1} = \sqrt{17}$$
$$|\mathbf{B}| = \sqrt{4 + 36} = \sqrt{40} = 2\sqrt{10}$$
$$\mathbf{A} \cdot \mathbf{B} = 8 + 6 = 14$$
Thus:
$$\cos\theta = \frac{14}{\sqrt{17} \cdot 2\sqrt{10}} = \frac{7}{\sqrt{170}}$$
So $\theta \approx 57.53°$ or $\theta \approx 1.00$ radian.

65. We first do the computations:
$$|\mathbf{A}| = \sqrt{25 + 36} = \sqrt{61}$$
$$|\mathbf{B}| = \sqrt{9 + 49} = \sqrt{58}$$
$$\mathbf{A} \cdot \mathbf{B} = -15 - 42 = -57$$
Thus:
$$\cos\theta = \frac{-57}{\sqrt{61} \cdot \sqrt{58}} = \frac{-57}{\sqrt{3538}}$$
So $\theta \approx 163.39°$ or $\theta \approx 2.85$ radians.

67. (a)   We first do the computations:
$$|\mathbf{A}| = \sqrt{64+4} = \sqrt{68} = 2\sqrt{17}$$
$$|\mathbf{B}| = \sqrt{1+9} = \sqrt{10}$$
$$\mathbf{A} \cdot \mathbf{B} = -8 - 6 = -14$$
Thus:
$$\cos\theta = \frac{-14}{2\sqrt{17} \cdot \sqrt{10}} = \frac{-7}{\sqrt{170}}$$
So $\theta \approx 122.47°$ or $\theta \approx 2.14$ radians.

(b)   Again, we do the computations:
$$|\mathbf{A}| = \sqrt{64+4} = \sqrt{68} = 2\sqrt{17}$$
$$|\mathbf{B}| = \sqrt{1+9} = \sqrt{10}$$
$$\mathbf{A} \cdot \mathbf{B} = 8 + 6 = 14$$
Thus:
$$\cos\theta = \frac{14}{2\sqrt{17} \cdot \sqrt{10}} = \frac{7}{\sqrt{170}}$$
So $\theta \approx 57.53°$ or $\theta \approx 1.00$ radian.

69. (a)   We first compute:
$$|\langle 2,5 \rangle| = \sqrt{4+25} = \sqrt{29}$$
$$|\langle -5,2 \rangle| = \sqrt{25+4} = \sqrt{29}$$
$$\langle 2,5 \rangle \cdot \langle -5,2 \rangle = -10 + 10 = 0$$
So $\cos\theta = \frac{0}{29} = 0$.

(b)   Since the angle between the vectors is 90°, the vectors must be perpendicular.

(c)   We draw the sketch:

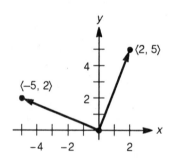

71.   Since $\cos\theta = \frac{\mathbf{A} \cdot \mathbf{B}}{|\mathbf{A}||\mathbf{B}|}$, then $\mathbf{A} \cdot \mathbf{B} = 0$ implies $\cos\theta = 0$, and thus $\theta = 90°$. So the vectors are perpendicular.

73. Call such a vector $\langle x,y \rangle$, so $x^2 + y^2 = 1$. Now we know $\langle x,y \rangle \cdot \langle -12,5 \rangle = 0$, so:

$$-12x + 5y = 0$$
$$5y = 12x$$
$$y = \tfrac{12}{5}x$$

Substituting into $x^2 + y^2 = 1$, we have:

$$x^2 + \tfrac{144}{25}x^2 = 1$$
$$\tfrac{169}{25}x^2 = 1$$
$$x^2 = \tfrac{25}{169}$$
$$x = \pm\tfrac{5}{13}$$
$$y = \pm\tfrac{12}{13}$$

So the two unit vectors are $\left\langle \tfrac{5}{13}, \tfrac{12}{13} \right\rangle$ and $\left\langle -\tfrac{5}{13}, -\tfrac{12}{13} \right\rangle$.

75. (a) We first find:

$$\mathbf{C} = \mathbf{B} - \mathbf{A} = \langle x_2, y_2 \rangle - \langle x_1, y_1 \rangle = \langle x_2 - x_1, y_2 - y_1 \rangle$$

Therefore:

$$|\mathbf{C}| = \sqrt{(x_2 - x_1)^2 + (y_2 - y_1)^2} = \sqrt{x_1^2 + y_1^2 + x_2^2 + y_2^2 - 2x_1 x_2 - 2y_1 y_2}$$

(b) Working from the left-hand side, we have:

$$|\mathbf{C}|^2 = x_1^2 + y_1^2 + x_2^2 + y_2^2 - 2x_1 x_2 - 2y_1 y_2 = |\mathbf{A}|^2 + |\mathbf{B}|^2 - 2(\mathbf{A} \cdot \mathbf{B})$$

(c) Using the suggestion given, we have:

$$|\mathbf{A}|^2 + |\mathbf{B}|^2 - 2|\mathbf{A}||\mathbf{B}|\cos\theta = |\mathbf{A}|^2 + |\mathbf{B}|^2 - 2(\mathbf{A} \cdot \mathbf{B})$$
$$-2|\mathbf{A}||\mathbf{B}|\cos\theta = -2(\mathbf{A} \cdot \mathbf{B})$$
$$|\mathbf{A}||\mathbf{B}|\cos\theta = \mathbf{A} \cdot \mathbf{B}$$
$$\cos\theta = \frac{\mathbf{A} \cdot \mathbf{B}}{|\mathbf{A}||\mathbf{B}|}$$

8.4 Parametric Equations

1. We find the x- and y-coordinates corresponding to $t = 0$:

$$x = 2 - 4(0) = 2$$
$$y = 3 - 5(0) = 3$$

The point corresponding to $t = 0$ is $(2,3)$.

3. We find the x- and y-coordinates corresponding to $t = \frac{\pi}{6}$:

$$x = 5\cos\frac{\pi}{6} = 5 \cdot \frac{\sqrt{3}}{2} = \frac{5\sqrt{3}}{2}$$
$$y = 2\sin\frac{\pi}{6} = 2 \cdot \frac{1}{2} = 1$$

The point corresponding to $t = \frac{\pi}{6}$ is $\left(\frac{5\sqrt{3}}{2}, 1\right)$.

5. We find the x- and y-coordinates corresponding to $t = \frac{\pi}{4}$:

$$x = 3\sin^3\frac{\pi}{4} = 3\left(\frac{\sqrt{2}}{2}\right)^3 = 3 \cdot \frac{\sqrt{2}}{4} = \frac{3\sqrt{2}}{4}$$
$$y = 3\cos^3\frac{\pi}{4} = 3\left(\frac{\sqrt{2}}{2}\right)^3 = 3 \cdot \frac{\sqrt{2}}{4} = \frac{3\sqrt{2}}{4}$$

The point corresponding to $t = \frac{\pi}{4}$ is $\left(\frac{3\sqrt{2}}{4}, \frac{3\sqrt{2}}{4}\right)$.

7. Solving $x = t + 1$ for t yields $t = x - 1$, now substituting:
$$y = t^2 = (x-1)^2$$
We graph the parabola which has a vertex at $(1, 0)$:

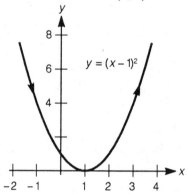

9. Solving $y = t + 1$ for t yields $t = y - 1$, now substituting:
$$x = (y-1)^2 + 1$$
$$x + 1 = (y-1)^2$$
We graph the parabola which has a vertex at $(-1, 1)$:

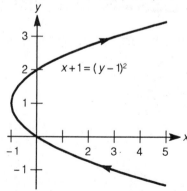

11. Multiplying the first equation by 2 and the second equation by 5 yields $2x = 10\cos t$ and $5y = 10\sin t$, so:

$$(2x)^2 + (5y)^2 = 100\cos^2 t + 100\sin^2 t$$
$$4x^2 + 25y^2 = 100$$
$$\frac{x^2}{25} + \frac{y^2}{4} = 1$$

We graph the ellipse which is centered at the origin:

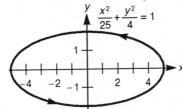

13. Multiplying the first equation by 3 and the second equation by 2 yields $3x = 12\cos 2t$ and $2y = 12\sin 2t$, so:

$$(3x)^2 + (2y)^2 = 144\cos^2 2t + 144\sin^2 2t$$
$$9x^2 + 4y^2 = 144$$
$$\frac{x^2}{16} + \frac{y^2}{36} = 1$$

We graph the ellipse which is centered at the origin:

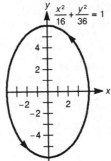

15. (a) We compute:

$$x^2 + y^2 = 4\cos^2 t + 4\sin^2 t$$
$$x^2 + y^2 = 4$$

We graph the circle which is centered at the origin:

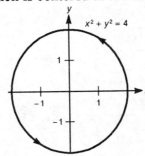

(b)   Multiplying the second equation by 2 yields $x = 4\cos t$ and $2y = 4\sin t$, so:

$$x^2 + (2y)^2 = 16\cos^2 t + 16\sin^2 t$$

$$x^2 + 4y^2 = 16$$

$$\frac{x^2}{16} + \frac{y^2}{4} = 1$$

We graph the ellipse which is centered at the origin:

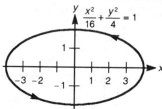

17.   (a)   When $t = 1$, we compute the x- and y-coordinates:

$$x = (100\cos 70°) \cdot 1 = 100\cos 70° \approx 34.2$$

$$y = 5 + (100\sin 70°) \cdot 1 - 16(1)^2 = 100\sin 70° - 11 \approx 83.0$$

When $t = 2$, we compute the x- and y-coordinates:

$$x = (100\cos 70°) \cdot 2 = 200\cos 70° \approx 68.4$$

$$y = 5 + (100\sin 70°) \cdot 2 - 16(2)^2 = 200\sin 70° - 59 \approx 128.9$$

When $t = 3$, we compute the x- and y-coordinates:

$$x = (100\cos 70°) \cdot 3 = 300\cos 70° \approx 102.6$$

$$y = 5 + (100\sin 70°) \cdot 3 - 16(3)^2 = 300\sin 70° - 139 \approx 142.9$$

(b)   We will find the value of t when $y = 0$:

$$5 + (100\sin 70°)t - 16t^2 = 0$$

$$16t^2 - (100\sin 70°)t - 5 = 0$$

Using the quadratic formula, we have:

$$t = \frac{100\sin 70° \pm \sqrt{10000\sin^2 70° + 320}}{32} \approx 5.93, -0.05$$

We discard the negative value and conclude that the ball is in flight for approximately 5.93 seconds. When $t \approx 5.93$, we have:

$$x \approx (100\cos 70°)(5.93) = 593\cos 70° \approx 203$$

The total horizontal distance traveled is approximately 203 feet. Note that this is consistent with the figure.

19.   We solve for t:

$$6 + (88\sin 35°)t - 16t^2 = 0$$

$$16t^2 - (88\sin 35°) - 6 = 0$$

Using the quadratic formula, we have:

$$t = \frac{88\sin 35° \pm \sqrt{7744\sin^2 35° + 384}}{32} \approx 3.27, -0.11$$

The solutions are verified.

21. Using the hint, we raise each side to the 2/3 power to obtain $x^{2/3} = \cos^2 t$ and $y^{2/3} = \sin^2 t$, so:

$$x^{2/3} + y^{2/3} = \cos^2 t + \sin^2 t = 1$$

The x-y equation for the curve is $x^{2/3} + y^{2/3} = 1$.

8.5 Introduction to Polar Coordinates

1. (a) Using $x = r\cos\theta$ and $y = r\sin\theta$, we have:

$$x = 3\cos\tfrac{2\pi}{3} = 3\left(-\tfrac{1}{2}\right) = -\tfrac{3}{2}$$

$$y = 3\sin\tfrac{2\pi}{3} = 3\left(\tfrac{\sqrt{3}}{2}\right) = \tfrac{3\sqrt{3}}{2}$$

So the rectangular coordinates are $\left(-\tfrac{3}{2}, \tfrac{3\sqrt{3}}{2}\right)$.

(b) Using $x = r\cos\theta$ and $y = r\sin\theta$, we have:

$$x = 4\cos\tfrac{11\pi}{6} = 4\left(\tfrac{\sqrt{3}}{2}\right) = 2\sqrt{3}$$

$$y = 4\sin\tfrac{11\pi}{6} = 4\left(-\tfrac{1}{2}\right) = -2$$

So the rectangular coordinates are $\left(2\sqrt{3}, -2\right)$.

(c) Using $x = r\cos\theta$ and $y = r\sin\theta$, we have:

$$x = 4\cos\left(-\tfrac{\pi}{6}\right) = 4\left(\tfrac{\sqrt{3}}{2}\right) = 2\sqrt{3}$$

$$y = 4\sin\left(-\tfrac{\pi}{6}\right) = 4\left(-\tfrac{1}{2}\right) = -2$$

So the rectangular coordinates are $\left(2\sqrt{3}, -2\right)$.

3. (a) Using $x = r\cos\theta$ and $y = r\sin\theta$, we have:

$$x = 1\cos\tfrac{\pi}{2} = 1(0) = 0$$

$$y = 1\sin\tfrac{\pi}{2} = 1(1) = 1$$

So the rectangular coordinates are $(0, 1)$.

(b) Using $x = r\cos\theta$ and $y = r\sin\theta$, we have:

$$x = 1\cos\tfrac{5\pi}{2} = 1(0) = 0$$

$$y = 1\sin\tfrac{5\pi}{2} = 1(1) = 1$$

So the rectangular coordinates are $(0, 1)$.

(c)   Using $x = r\cos\theta$ and $y = r\sin\theta$, we have (using the half-angle formulas for $\sin\theta$ and $\cos\theta$):

$$x = -1\cos\tfrac{\pi}{8} = -1\left(\sqrt{\frac{1+\frac{\sqrt{2}}{2}}{2}}\right) = -\sqrt{\frac{2+\sqrt{2}}{4}} = -\frac{\sqrt{2+\sqrt{2}}}{2}$$

$$y = -1\sin\tfrac{\pi}{8} = -1\left(\sqrt{\frac{1-\frac{\sqrt{2}}{2}}{2}}\right) = -\sqrt{\frac{2-\sqrt{2}}{4}} = -\frac{\sqrt{2-\sqrt{2}}}{2}$$

So the rectangular coordinates are $\left(-\frac{\sqrt{2+\sqrt{2}}}{2}, -\frac{\sqrt{2-\sqrt{2}}}{2}\right)$.

5.   We have:

$$r^2 = 1+1 = 2,\text{ so } r = \sqrt{2}$$
$$\theta = \tan^{-1}\left(\tfrac{-1}{-1}\right) + \pi = \tfrac{\pi}{4} + \pi = \tfrac{5\pi}{4}$$

So the polar form is $\left(\sqrt{2}, \tfrac{5\pi}{4}\right)$.

7.   If we first multiply by r, we have:

$$r^2 = 2r\cos\theta$$
$$x^2 + y^2 = 2x$$
$$x^2 - 2x + y^2 = 0$$
$$x^2 - 2x + 1 + y^2 = 1$$
$$(x-1)^2 + y^2 = 1$$

9.   Substituting for r and $\tan\theta$, we have:

$$\sqrt{x^2 + y^2} = \frac{y}{x}$$
$$x^2 + y^2 = \frac{y^2}{x^2}$$
$$x^4 + x^2 y^2 = y^2$$
$$x^4 + x^2 y^2 - y^2 = 0$$

11.   Using the double-angle formula for $\cos 2\theta$, we have:

$$r = 3\left(\cos^2\theta - \sin^2\theta\right)$$

Multiplying by r^2:

$$r^3 = 3\left(r^2\cos^2\theta - r^2\sin^2\theta\right)$$
$$\left(x^2 + y^2\right)^{3/2} = 3\left(x^2 - y^2\right)$$
$$\left(x^2 + y^2\right)^3 = 9\left(x^2 - y^2\right)^2$$

13. Multiplying each side by $2 - \sin^2\theta$ yields:

$$2r^2 - r^2\sin^2\theta = 8$$
$$2(x^2 + y^2) - y^2 = 8$$
$$2x^2 + 2y^2 - y^2 = 8$$
$$2x^2 + y^2 = 8$$
$$\frac{x^2}{4} + \frac{y^2}{8} = 1$$

15. Using the addition formula for $\cos(s - t)$, we have:

$$r\cos\left(\theta - \tfrac{\pi}{6}\right) = 2$$
$$r\left(\cos\theta\cos\tfrac{\pi}{6} + \sin\theta\sin\tfrac{\pi}{6}\right) = 2$$
$$\tfrac{\sqrt{3}}{2}r\cos\theta + \tfrac{1}{2}r\sin\theta = 2$$
$$\tfrac{\sqrt{3}}{2}x + \tfrac{1}{2}y = 2$$
$$\sqrt{3}x + y = 4$$
$$y = -\sqrt{3}x + 4$$

17. Substituting $x = r\cos\theta$ and $y = r\sin\theta$, we have:

$$3r\cos\theta - 4r\sin\theta = 2$$
$$r(3\cos\theta - 4\sin\theta) = 2$$
$$r = \frac{2}{3\cos\theta - 4\sin\theta}$$

19. Substituting $x = r\cos\theta$ and $y = r\sin\theta$, we have:

$$r^2\sin^2\theta = r^3\cos^3\theta$$
$$\sin^2\theta = r\cos^3\theta$$
$$r = \frac{\sin^2\theta}{\cos^3\theta}$$
$$r = \tan^2\theta\sec\theta$$

21. Substituting $x = r\cos\theta$ and $y = r\sin\theta$, we have:

$$2(r\cos\theta)(r\sin\theta) = 1$$
$$r^2(2\sin\theta\cos\theta) = 1$$
$$r^2\sin 2\theta = 1$$
$$r^2 = \frac{1}{\sin 2\theta}$$
$$r^2 = \csc 2\theta$$

23. Substituting $x = r \cos \theta$ and $y = r \sin \theta$, we have:

$$9r^2 \cos^2 \theta + r^2 \sin^2 \theta = 9$$
$$r^2 \left(9 \cos^2 \theta + \sin^2 \theta \right) = 9$$
$$r^2 = \frac{9}{9 \cos^2 \theta + \sin^2 \theta}.$$

25. *A:*   Since $\theta = \frac{\pi}{6}$, we have:

$$r = \frac{4}{1 + \sin \frac{\pi}{6}} = \frac{4}{1 + \frac{1}{2}} = \frac{4}{\frac{3}{2}} = \frac{8}{3}$$

B: Since $\theta = \frac{5\pi}{6}$, we have:

$$r = \frac{4}{1 + \sin \frac{5\pi}{6}} = \frac{4}{1 + \frac{1}{2}} = \frac{4}{\frac{3}{2}} = \frac{8}{3}$$

C: Since $\theta = \pi$, we have:

$$r = \frac{4}{1 + \sin \pi} = \frac{4}{1 + 0} = 4$$

D: Since $\theta = \frac{7\pi}{6}$, we have:

$$r = \frac{4}{1 + \sin \frac{7\pi}{6}} = \frac{4}{1 - \frac{1}{2}} = \frac{4}{\frac{1}{2}} = 8$$

The coordinates of the points are $A\left(\frac{8}{3}, \frac{\pi}{6}\right)$, $B\left(\frac{8}{3}, \frac{5\pi}{6}\right)$, $C(4, \pi)$, and $D\left(8, \frac{7\pi}{6}\right)$.

27. *A:*   Since $\theta = \frac{\pi}{6}$, we have:

$$r = 2 \cos \frac{\pi}{3} = 2 \cdot \frac{1}{2} = 1$$

B: Since $\theta = \frac{5\pi}{6}$, we have:

$$r = 2 \cos \frac{5\pi}{3} = 2 \cdot \frac{1}{2} = 1$$

C: Since $\theta = \frac{7\pi}{6}$, we have:

$$r = 2 \cos \frac{7\pi}{3} = 2 \cdot \frac{1}{2} = 1$$

D: Since $\theta = \frac{\pi}{2}$, we have:

$$r = 2 \cos \pi = 2 \cdot (-1) = -2$$

The coordinates of the points are $A\left(1, \frac{\pi}{6}\right)$, $B\left(1, \frac{5\pi}{6}\right)$, $C\left(1, \frac{7\pi}{6}\right)$, and $D\left(-2, \frac{\pi}{2}\right)$.

29. *A:*   Since $\theta = 0$, we have:

$$r = e^{0/6} = 1$$

B: Since $\theta = \frac{\pi}{4}$, we have:

$$r = e^{\pi/24} \approx 1.14$$

C: Since $\theta = \frac{3\pi}{4}$, we have:
$$r = e^{\pi/8} \approx 1.48$$

D: Since $\theta = \pi$, we have:
$$r = e^{\pi/6} \approx 1.69$$

E: Since $\theta = \frac{5\pi}{4}$, we have:
$$r = e^{5\pi/24} \approx 1.92$$

F: Since $\theta = \frac{3\pi}{2}$, we have:
$$r = e^{\pi/4} \approx 2.19$$

G: Since $\theta = \frac{7\pi}{4}$, we have:
$$r = e^{7\pi/24} \approx 2.50$$

H: Since $\theta = 2\pi$, we have:
$$r = e^{\pi/3} \approx 2.85$$

I: Since $\theta = \frac{9\pi}{4}$, we have:
$$r = e^{3\pi/8} \approx 3.25$$

J: Since $\theta = \frac{11\pi}{4}$, we have:
$$r = e^{11\pi/24} \approx 4.22$$

K: Since $\theta = \frac{13\pi}{4}$, we have:
$$r = e^{13\pi/24} \approx 5.48$$

The coordinates of the points are $A(1,0)$, $B\left(1.14,\frac{\pi}{4}\right)$, $C\left(1.48,\frac{3\pi}{4}\right)$, $D(1.69,\pi)$, $E\left(1.92,\frac{5\pi}{4}\right)$, $F\left(2.19,\frac{3\pi}{2}\right)$, $G\left(2.50,\frac{7\pi}{4}\right)$, $H(2.85,2\pi)$, $I\left(3.25,\frac{9\pi}{4}\right)$, $J\left(4.22,\frac{11\pi}{4}\right)$, and $K\left(5.48,\frac{13\pi}{4}\right)$.

31. We use the distance formula, taking the points $\left(r_1,\theta_1\right)$ and $\left(r_2,\theta_2\right)$ to be $\left(2,\frac{2\pi}{3}\right)$ and $\left(4,\frac{\pi}{6}\right)$:
$$d^2 = r_1^2 + r_2^2 - 2r_1r_2 \cos\left(\theta_2 - \theta_1\right)$$
$$= 2^2 + 4^2 - 2(2)(4)\cos\left(\frac{2\pi}{3} - \frac{\pi}{6}\right)$$
$$= 20 - 16\cos\frac{\pi}{2}$$
$$= 20 - 16 \cdot 0$$
$$= 20$$

So $d = \sqrt{20} = 2\sqrt{5}$.

33. We use the distance formula, taking the points $\left(r_1, \theta_1\right)$ and $\left(r_2, \theta_2\right)$ to be $\left(4, \frac{4\pi}{3}\right)$ and $(1,0)$:

$$
\begin{aligned}
d^2 &= r_1^2 + r_2^2 - 2r_1r_2 \cos\left(\theta_2 - \theta_1\right) \\
&= 4^2 + 1^2 - 2(4)(1)\cos\left(\frac{4\pi}{3} - 0\right) \\
&= 17 - 8\cos\frac{4\pi}{3} \\
&= 17 - 8 \cdot \left(-\frac{1}{2}\right) \\
&= 21
\end{aligned}
$$

So $d = \sqrt{21}$.

35. (a) Using the equation $r^2 + r_0^2 - 2rr_0 \cos\left(\theta - \theta_0\right) = a^2$, we have:

$$
\begin{aligned}
r^2 + 4^2 - 2(r)(4)\cos(\theta - 0) &= 2^2 \\
r^2 + 16 - 8r\cos\theta &= 4 \\
r^2 - 8r\cos\theta &= -12
\end{aligned}
$$

(b) Using the equation $r^2 + r_0^2 - 2rr_0 \cos\left(\theta - \theta_0\right) = a^2$, we have:

$$
\begin{aligned}
r^2 + 4^2 - 2(r)(4)\cos\left(\theta - \frac{2\pi}{3}\right) &= 2^2 \\
r^2 + 16 - 8r\cos\left(\theta - \frac{2\pi}{3}\right) &= 4 \\
r^2 - 8r\cos\left(\theta - \frac{2\pi}{3}\right) &= -12
\end{aligned}
$$

(c) Using the equation $r^2 + r_0^2 - 2rr_0 \cos\left(\theta - \theta_0\right) = a^2$, we have:

$$
\begin{aligned}
r^2 + 0^2 - 2(r)(0)\cos(\theta - 0) &= 2^2 \\
r^2 &= 4
\end{aligned}
$$

37. (a) Using the equation $r^2 + r_0^2 - 2rr_0 \cos\left(\theta - \theta_0\right) = a^2$, we have:

$$
\begin{aligned}
r^2 + 1^2 - 2(r)(1)\cos\left(\theta - \frac{3\pi}{2}\right) &= 1^2 \\
r^2 + 1 - 2r\cos\left(\theta - \frac{3\pi}{2}\right) &= 1 \\
r^2 &= 2r\cos\left(\theta - \frac{3\pi}{2}\right) \\
r &= 2\cos\left(\theta - \frac{3\pi}{2}\right)
\end{aligned}
$$

(b)   Using the equation $r^2 + r_0^2 - 2rr_0 \cos(\theta - \theta_0) = a^2$, we have:

$$r^2 + 1^2 - 2(r)(1)\cos\left(\theta - \tfrac{\pi}{4}\right) = 1^2$$
$$r^2 + 1 - 2r\cos\left(\theta - \tfrac{\pi}{4}\right) = 1$$
$$r^2 = 2r\cos\left(\theta - \tfrac{\pi}{4}\right)$$
$$r = 2\cos\left(\theta - \tfrac{\pi}{4}\right)$$

39.   (a)   Since the line is of the form $r\cos(\theta - \alpha) = d$, $\alpha = \tfrac{\pi}{6}$ and $d = 2$. So the desired perpendicualr distance is 2.

(b)   When $\theta = 0$, we have:

$$r\cos\left(0 - \tfrac{\pi}{6}\right) = 2$$
$$r\left(\tfrac{\sqrt{3}}{2}\right) = 2$$
$$r = \frac{4}{\sqrt{3}} = \frac{4\sqrt{3}}{3}$$

When $\theta = \tfrac{\pi}{2}$, we have:

$$r\cos\left(\tfrac{\pi}{2} - \tfrac{\pi}{6}\right) = 2$$
$$r \bullet \tfrac{1}{2} = 2$$
$$r = 4$$

The required polar points are $\left(\dfrac{4\sqrt{3}}{3}, 0\right)$ and $\left(4, \tfrac{\pi}{2}\right)$.

(c)   These coordinates are (d, α), which is the polar point $\left(2, \tfrac{\pi}{6}\right)$.

(d)   We sketch the line:

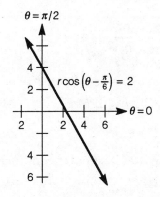

41. (a) Since the line is of the form $r\cos(\theta - \alpha) = d$, $\alpha = -\frac{2\pi}{3}$ and $d = 4$. So the desired perpendicualr distance is 4.

(b) When $\theta = 0$, we have:
$$r\cos\left(0 + \tfrac{2\pi}{3}\right) = 4$$
$$r\left(-\tfrac{1}{2}\right) = 4$$
$$r = -8$$
When $\theta = \frac{\pi}{2}$, we have:
$$r\cos\left(\tfrac{\pi}{2} + \tfrac{2\pi}{3}\right) = 4$$
$$r\cos\tfrac{7\pi}{6} = 4$$
$$r\cdot\left(-\tfrac{\sqrt{3}}{2}\right) = 4$$
$$r = -\frac{8}{\sqrt{3}} = -\frac{8\sqrt{3}}{3}$$
The required polar points are $(-8,0)$ and $\left(-\tfrac{8}{3}\sqrt{3}, \tfrac{\pi}{2}\right)$.

(c) These coordinates are (d, α), which is the polar point $\left(4, -\tfrac{2\pi}{3}\right)$.

(d) We sketch the line:

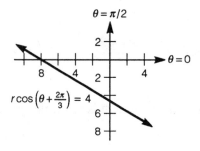

43. (a) Since $x^2 + y^2 = r^2$, we have:
$$r = \sqrt{x^2 + y^2} = \sqrt{\left(-\sqrt{3}\right)^2 + (1)^2} = \sqrt{4} = 2$$
Now using $x = r\cos\theta$ and $y = r\sin\theta$, we have:
$$2\cos\theta = -\sqrt{3}, \text{ so } \cos\theta = -\frac{\sqrt{3}}{2}$$
$$2\sin\theta = 1, \text{ so } \sin\theta = \frac{1}{2}$$
Therefore a value of θ is $\frac{5\pi}{6}$. The polar coordinates of P are $\left(2, \tfrac{5\pi}{6}\right)$.

(b) In the general equation $r\cos(\theta - \alpha) = d$, we use the values $\alpha = \frac{5\pi}{6}$ and $d = 2$ to obtain:

$$r\cos\left(\theta - \frac{5\pi}{6}\right) = 2$$

This is the polar equation for the tangent line.

(c) The x-axis corresponds to $\theta = 0$, therefore:

$$r\cos\left(0 - \frac{5\pi}{6}\right) = 2$$

$$r\left(-\frac{\sqrt{3}}{2}\right) = 2$$

$$r = -\frac{4}{\sqrt{3}} = -\frac{4\sqrt{3}}{3}$$

So the x-intercept of the line is $-\frac{4\sqrt{3}}{3}$. The y-axis corresponds to $\theta = \frac{\pi}{2}$, so:

$$r\cos\left(\frac{\pi}{2} - \frac{5\pi}{6}\right) = 2$$

$$r\cos\left(-\frac{\pi}{3}\right) = 2$$

$$r \cdot \frac{1}{2} = 2$$

$$r = 4$$

So the y-intercept of the line is 4.

45. (a) Since $r\cos\theta = x$, this is the graph of $x = 3$:

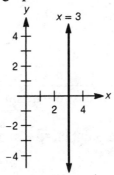

(b) Since $r\sin\theta = y$, this is the graph of $y = 3$:

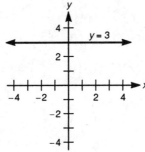

47. (a)   Since d is the perpendicular distance from the origin to L, and (d, α) is the foot of the perpendicular on L, the polar equation for L must be $r\cos(\theta - \alpha) = d$.

   (b)   Using the formula $\cos(s - t) = \cos s \cos t + \sin s \sin t$, we have:
$$r\cos(\theta - \alpha) = d$$
$$r(\cos\theta\cos\alpha + \sin\theta\sin\alpha) = d$$
$$(r\cos\theta)\cos\alpha + (r\sin\theta)\sin\alpha = d$$
$$x\cos\alpha + y\sin\alpha = d$$

49. We first establish an identity for $\sin 3\theta$:
$$\sin 3\theta = \sin(2\theta + \theta)$$
$$= \sin 2\theta\cos\theta + \cos 2\theta\sin\theta$$
$$= (2\sin\theta\cos\theta)\cos\theta + (\cos^2\theta - \sin^2\theta)\sin\theta$$
$$= 3\sin\theta\cos^2\theta - \sin^3\theta$$
Now we multiply the equation $r = a\sin 3\theta$ by r^3 and substitute:
$$r^4 = a\left((3r\sin\theta)(r^2\cos^2\theta) - r^3\sin^3\theta\right)$$
$$\left(x^2 + y^2\right)^2 = a\left(3yx^2 - y^3\right)$$
$$\left(x^2 + y^2\right)^2 = ay\left(3x^2 - y^2\right)$$

51. Substituting $x = r\cos\theta$ and $y = r\sin\theta$ results in the equation:
$$\frac{r^2\cos^2\theta}{a^2} - \frac{r^2\sin^2\theta}{b^2} = 1$$
$$\frac{r^2\left(b^2\cos^2\theta - a^2\sin^2\theta\right)}{a^2 b^2} = 1$$
$$r^2 = \frac{a^2 b^2}{b^2\cos^2\theta - a^2\sin^2\theta}$$

53. (a) A is the point $(1, 0)$, B is the point $\left(1, \frac{2\pi}{3}\right)$, and C is the point $\left(1, \frac{4\pi}{3}\right)$.

   (b)   We use the distance formula $d^2 = r_1^2 + r_2^2 - 2r_1 r_2\cos(\theta_2 - \theta_1)$ to find each distance:
$$(PA)^2 = r^2 + 1^2 - 2(r)(1)\cos(0 - 0) = r^2 + 1 - 2r$$
$$(PB)^2 = r^2 + 1^2 - 2(r)(1)\cos\left(0 - \frac{2\pi}{3}\right) = r^2 + 1 + r$$
$$(PC)^2 = r^2 + 1^2 - 2(r)(1)\cos\left(0 - \frac{4\pi}{3}\right) = r^2 + 1 + r$$

Therefore:

$$(PA)^2 (PB)^2 (PC)^2 = \left(r^2 - 2r + 1\right)\left(r^2 + r + 1\right)^2$$
$$= (r-1)^2 \left(r^2 + r + 1\right)^2$$
$$= \left(r^3 - 1\right)^2$$

So:

$$(PA)(PB)(PC) = \sqrt{\left(r^3 - 1\right)^2} = 1 - r^3, \text{ since } r < 1$$

8.6 Curves in Polar Coordinates

1.   We sketch the polar curve:

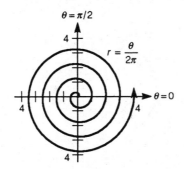

3.   We sketch the polar curve:

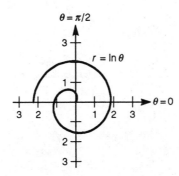

5.   We sketch the polar curve:

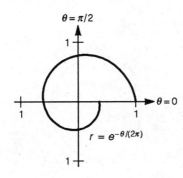

7.   We sketch the polar curve:

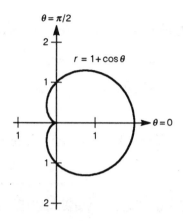

9.   We sketch the polar curve:

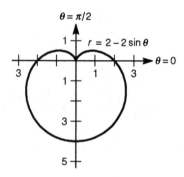

11.   We sketch the polar curve:

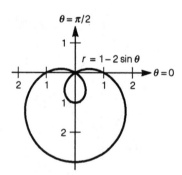

13.   We sketch the polar curve:

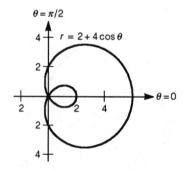

15. We sketch the polar curve:

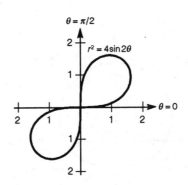

17. We sketch the polar curve:

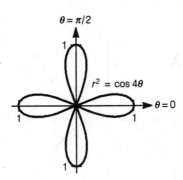

19. We sketch the polar curve:

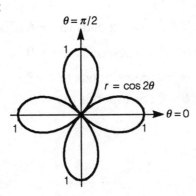

21. We sketch the polar curve:

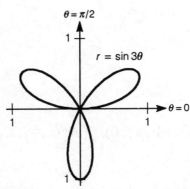

23. We sketch the polar curve:

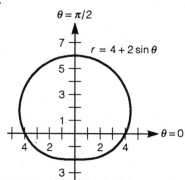

25. We sketch the polar curve:

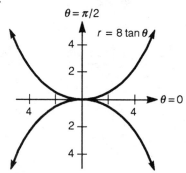

27. (a) The graph is C, since $\theta = \frac{\pi}{2}$ corresponds to $r = 6$.

(b) The graph is B, since $\theta = \frac{\pi}{2}$ corresponds to $r = 0$.

(c) The graph is D, since $\theta = \frac{\pi}{2}$ corresponds to $r = 3$.

(d) The graph is A, since $\theta = 0$ corresponds to $r = 6$.

29. (a) For A, let $\theta = 0$:

$\ln r = 0$, so $r = 1$

For B, let $\theta = \frac{\pi}{2}$:

$$\ln r = \frac{a\pi}{2}, \text{ so } r = e^{a\pi/2}$$

For C, let $\theta = \pi$:

$\ln r = a\pi$, so $r = e^{a\pi}$

For D, let $\theta = \frac{3\pi}{2}$:

$$\ln r = \frac{3a\pi}{2}, \text{ so } r = e^{3a\pi/2}$$

So the polar coordinates are $A(1, 0)$, $B\left(e^{a\pi/2}, \frac{\pi}{2}\right)$, $C\left(e^{a\pi}, \pi\right)$, and $D\left(e^{3a\pi/2}, \frac{3\pi}{2}\right)$.

Since $OA = 1, OB = e^{a\pi/2}$, $OC = e^{a\pi}$, and $OD = e^{3a\pi/2}$, the required ratios are therefore:

$$\frac{OD}{OC} = \frac{e^{3a\pi/2}}{e^{a\pi}} = e^{3a\pi/2 - a\pi} = e^{a\pi/2}$$

$$\frac{OC}{OB} = \frac{e^{a\pi}}{e^{a\pi/2}} = e^{a\pi - a\pi/2} = e^{a\pi/2}$$

$$\frac{OB}{OA} = \frac{e^{a\pi/2}}{1} = e^{a\pi/2}$$

Thus:

$$\frac{OD}{OC} = \frac{OC}{OB} = \frac{OB}{OA} = e^{a\pi/2}$$

(b)   Using the hint, the rectangular coordinates are $A(1, 0)$, $B\left(0, e^{a\pi/2}\right)$, $C\left(-e^{a\pi}, 0\right)$, and $D\left(0, -e^{3a\pi/2}\right)$. Now compute the slope of line segments $\overline{AB}$, $\overline{BC}$, and $\overline{CD}$:

$$m_{\overline{AB}} = \frac{e^{a\pi/2} - 0}{0 - 1} = -e^{a\pi/2}$$

$$m_{\overline{BC}} = \frac{0 - e^{a\pi/2}}{-e^{a\pi} - 0} = \frac{1}{e^{a\pi/2}}$$

$$m_{\overline{CD}} = \frac{-e^{3a\pi/2} - 0}{0 + e^{a\pi}} = -e^{a\pi/2}$$

Since $m_{\overline{AB}} \cdot m_{\overline{BC}} = -1$ and $m_{\overline{BC}} \cdot m_{\overline{CD}} = -1$, $\angle ABC$ and $\angle BCD$ are right angles.

31. (a)   We graph the curves. Note that the curves are identical.

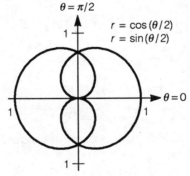

$\theta = \pi/2$

$r = \cos(\theta/2)$
$r = \sin(\theta/2)$

$\theta = 0$

(b)   We square each side of the polar curve $r = \cos\dfrac{\theta}{2}$ and apply the half-angle identity:

$$r^2 = \cos^2\frac{\theta}{2}$$

$$r^2 = \frac{1+\cos\theta}{2}$$

$$2r^2 = 1 + \cos\theta$$

$$2(x^2 + y^2) = 1 + \frac{x}{\sqrt{x^2 + y^2}}$$

$$2(x^2 + y^2)^{3/2} = (x^2 + y^2)^{1/2} + x$$

$$(x^2 + y^2)^{1/2}(2x^2 + 2y^2 - 1) = x$$

$$(x^2 + y^2)(2x^2 + 2y^2 - 1)^2 = x^2$$

$$(x^2 + y^2)(2x^2 + 2y^2 - 1)^2 - x^2 = 0$$

Now square each side of the polar curve $r = \sin\dfrac{\theta}{2}$ and apply the half-angle identity:

$$r^2 = \sin^2\frac{\theta}{2}$$

$$r^2 = \frac{1-\cos\theta}{2}$$

$$2r^2 = 1 - \cos\theta$$

$$2(x^2 + y^2) = 1 - \frac{x}{\sqrt{x^2 + y^2}}$$

$$2(x^2 + y^2)^{3/2} = (x^2 + y^2)^{1/2} - x$$

$$(x^2 + y^2)^{1/2}(2x^2 + 2y^2 - 1) = -x$$

$$(x^2 + y^2)(2x^2 + 2y^2 - 1)^2 = x^2$$

$$(x^2 + y^2)(2x^2 + 2y^2 - 1)^2 - x^2 = 0$$

So both curves convert to the same rectangular equation.

Optional TI-81 Graphing Calculator Exercises for Sections 8.4 – 8.6

1. (a)   Using the indicated settings, we obtain the graph:

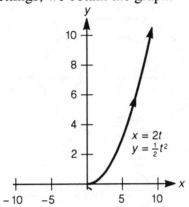

(b)   Changing $T_{min} = -6.28$, we obtain the graph:

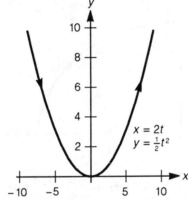

Notice the graph now appears more as a parabola.

(c)   Setting $T_{min} = 2$ and $T_{max} = 3$, we obtain the graph:

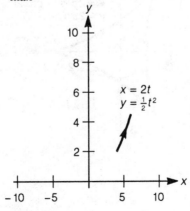

(d)   Using TRACE verifies the endpoint has coordinates $(4, 2)$ when $t = 2$.

3.   Using settings $X_{min} = -4$, $X_{max} = 4$, $Y_{min} = -4$, $Y_{max} = 4$, we obtain the graph:

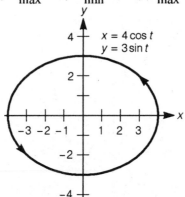

5.   Using the settings $X_{min} = 0$, $X_{max} = 4$, $Y_{min} = 0$, $Y_{max} = 3$, we obtain the graph:

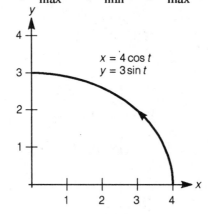

7.   Using the settings $X_{min} = -3$, $X_{max} = 3$, $Y_{min} = -3$, $Y_{max} = 3$, we obtain the graph:

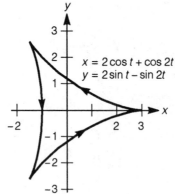

9.   Using the settings $X_{min} = -4$, $X_{max} = 4$, $Y_{min} = 0$, $Y_{max} = 3$, we obtain the graph:

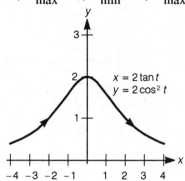

11.   (a)   Graphing the polar curve, we obtain:

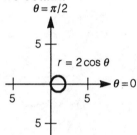

(b)   Using the indicated settings, we obtain the curve:

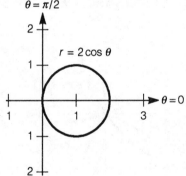

(c)   Substituting $r = \sqrt{x^2 + y^2}$ and $\cos\theta = \dfrac{x}{\sqrt{x^2 + y^2}}$, we have:

$$\sqrt{x^2 + y^2} = \frac{2x}{\sqrt{x^2 + y^2}}$$
$$x^2 + y^2 = 2x$$
$$x^2 - 2x + y^2 = 0$$
$$x^2 - 2x + 1 + y^2 = 1$$
$$(x - 1)^2 + y^2 = 1$$

This is a circle centered at $(1, 0)$, with radius $= 1$, which is consistent with our graph from part (b).

13. (a) We graph the polar curve $x = \cos 4\theta \cos \theta$, $y = \cos 4\theta \sin \theta$:

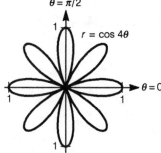

(b) We graph the polar curve $x = \sin 4\theta \cos \theta$, $y = \sin 4\theta \sin \theta$:

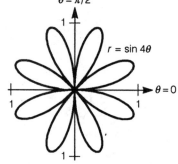

15. (a) We graph the polar curve $x = \cos 5\theta \cos \theta$, $y = \cos 5\theta \sin \theta$:

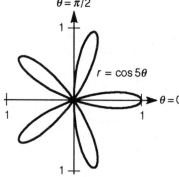

(b) We graph the polar curve $x = \sin 5\theta \cos \theta$, $y = \sin 5\theta \sin \theta$:

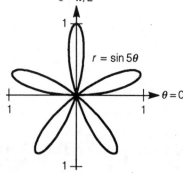

17. We graph the polar curve $x = \theta \cos \theta$, $y = \theta \sin \theta$:

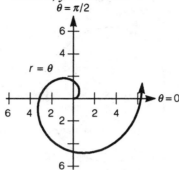

19. (a)   We graph the polar curve $x = (2\cos \theta + 1)\cos \theta$, $y = (2\cos \theta + 1)\sin \theta$:

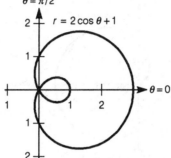

(b)   We graph the polar curve $x = (2\cos \theta - 1)\cos \theta$, $y = (2\cos \theta - 1)\sin \theta$:

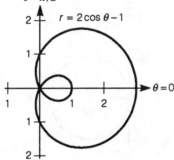

(c)   We graph the polar curve $x = (\cos \theta + 2)\cos \theta$, $y = (\cos \theta + 2)\sin \theta$:

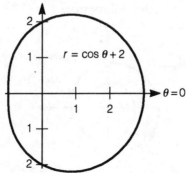

(d)   We graph the polar curve $x = (\cos\theta - 2)\cos\theta$, $y = (\cos\theta - 2)\sin\theta$:

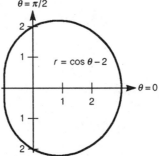

(e)   We graph the polar curve $x = (2\cos\theta + 2)\cos\theta$, $y = (2\cos\theta + 2)\sin\theta$:

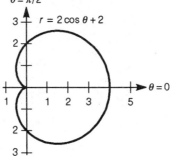

(f)   We graph the polar curve $x = \cos\theta\cos\theta$, $y = \cos\theta\sin\theta$:

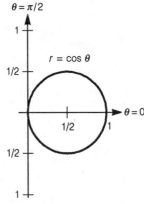

21. (a)   Using $T_{max} = 10$, we graph the polar curve $x = e^{\theta/10}\cos\theta$, $y = e^{\theta/10}\sin\theta$:

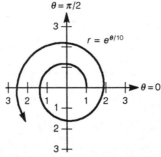

(b)   Using $T_{max} = 20$, we graph the polar curve $x = e^{\theta/10} \cos \theta$, $y = e^{\theta/10} \sin \theta$:

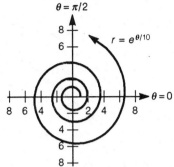

(c)   Using $T_{max} = 25$, we graph the polar curve $x = e^{\theta/10} \cos \theta$, $y = e^{\theta/10} \sin \theta$:

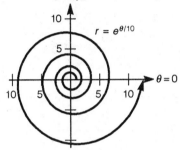

23. (a)   We graph the polar curve:

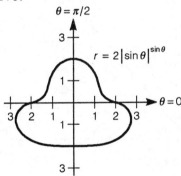

(b)   We graph the polar curve:

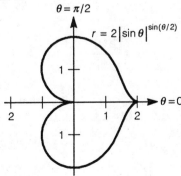

Chapter Eight Review Exercises

1.   We draw the triangle:

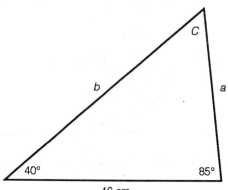

16 cm

First note that $\angle C = 180° - 40° - 85° = 55°$. Using the law of sines:

$$\frac{\sin 40°}{a} = \frac{\sin 55°}{16}, \text{ so } a = \frac{16\sin 40°}{\sin 55°} \approx 12.6 \text{ cm}$$

Using the law of sines:

$$\frac{\sin 85°}{b} = \frac{\sin 55°}{16}, \text{ so } b = \frac{16\sin 85°}{\sin 55°} \approx 19.5 \text{ cm}$$

3.   (a)   We draw the triangle:

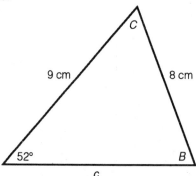

9 cm   8 cm

52°   B

c

Using the law of sines:

$$\frac{\sin B}{9} = \frac{\sin 52°}{8}, \text{ so } \sin B = \frac{9\sin 52°}{8} \approx 0.8865$$

Thus $\angle B \approx 62.4°$. Now note that $\angle C \approx 180° - 52° - 62.4° \approx 65.6°$. Using the law of sines:

$$\frac{\sin 65.6°}{c} = \frac{\sin 52°}{8}, \text{ so } c = \frac{8\sin 65.6°}{\sin 52°} \approx 9.2 \text{ cm}$$

(b)   We have $\sin B \approx 0.8865$, thus $\angle B \approx 117.6°$. Now note that $\angle C = 180° - 52° - 117.6° \approx 10.4°$. Using the law of sines:

$$\frac{\sin 10.4°}{c} = \frac{\sin 52°}{8}, \text{ so } c = \frac{8\sin 10.4°}{\sin 52°} \approx 1.8 \text{ cm}$$

5.   We draw the triangle:

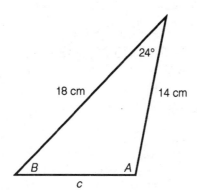

Using the law of cosines:

$c^2 = 18^2 + 14^2 - 2(18)(14)\cos 24°$, so $c \approx 7.7$ cm

Using the law of sines:

$$\frac{\sin A}{18} = \frac{\sin 24°}{7.7}, \text{ so } \sin A = \frac{18\sin 24°}{7.7} \approx 0.9486$$

Then either $\angle A \approx 71.5°$ or $\angle A \approx 108.5°$. If $\angle A \approx 71.5°$, then $\angle B \approx 180° - 24° - 71.5° \approx 84.5°$. But this is impossible since $\angle B < \angle A$. If $\angle A \approx 108.5°$, then $\angle B \approx 180° - 24° - 108.5° \approx 47.5°$.

7.   We draw the triangle:

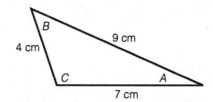

Using the law of cosines:

$9^2 = 4^2 + 7^2 - 2(4)(7)\cos C$, so $\cos C \approx -0.2857$, thus $C \approx 106.6°$

Using the law of sines:

$$\frac{\sin 106.6°}{9} = \frac{\sin B}{7}, \text{ so } \sin B = \frac{7\sin 106.6°}{9} \approx 0.7454$$

Thus $B \approx 48.2°$. Therefore $\angle A \approx 180° - 106.6° - 48.2° \approx 25.2°$.

9.   Note that $\angle AEB = 86°$. Using the law of sines, we have:

$$\frac{\sin 50°}{BE} = \frac{\sin 86°}{12}, \text{ so } BE = \frac{12\sin 50°}{\sin 86°} \approx 9.21 \text{ cm}$$

11. Using the area formula, we have:

$A = \frac{1}{2}(BC)(BE)\sin 36° = \frac{1}{2}(12 \text{ cm})(9.21 \text{ cm})(\sin 36°) \approx 32.48 \text{ cm}^2$

13. Using the area formula, we have:

$A = \frac{1}{2}(12 \text{ cm})(BD\sin 44°) = \frac{1}{2}(12 \text{ cm})(13.25 \text{ cm})(\sin 44°) \approx 55.23 \text{ cm}^2$

15. Using the law of cosines, we have:
$$(CD)^2 = (BC)^2 + (BD)^2 - 2(BC)(BD)\cos 36°$$
$$\approx (12)^2 + (13.25)^2 - 2(12)(13.25)\cos 36°$$
$$\approx 319.56 - 318\cos 36°$$
$$\approx 62.29$$
So $CD \approx \sqrt{62.29} \approx 7.89$ cm.

17. Using the law of sines, we have:
$$\frac{\sin 80°}{AC} = \frac{\sin 50°}{12}, \text{ so } AC = \frac{12\sin 80°}{\sin 50°} \approx 15.43 \text{ cm}$$

19. We re-draw the figure:

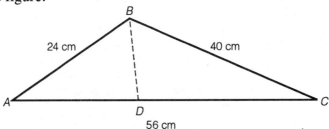

We use the law of cosines to find $\angle ABC$:
$$56^2 = 40^2 + 24^2 - 2(40)(24)\cos(\angle ABC)$$
$$-0.5 = \cos(\angle ABC)$$
$$\angle ABC = 120°$$
We now use the law of sines to find $\angle A$:
$$\frac{\sin A}{40} = \frac{\sin 120°}{56}, \text{ so } \sin A \approx 0.6186, \text{ thus } A \approx 38.21°$$
Then $\angle ADB \approx 180° - 60° - 38.21° \approx 81.79°$. We find BD by the law of sines:
$$\frac{\sin 81.79°}{24} = \frac{\sin 38.21°}{BD}, \text{ so } BD = \frac{24\sin 38.21°}{\sin 81.79°} \approx 15 \text{ cm}$$

21. We re-draw the figure:

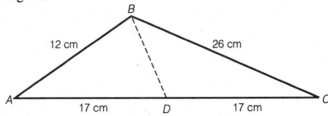

Using the law of cosines, we find $\angle A$:
$$26^2 = 12^2 + 34^2 - 2(12)(34)\cos A$$
$$\cos A \approx 0.7647$$
$$A \approx 40.12°$$

We can now find BD by using the law of cosines (on $\triangle ADB$):
$$(BD)^2 = 12^2 + 17^2 - 2(12)(17)\cos 40.12°$$
$$(BD)^2 = 121$$
$$BD = 11\,\text{cm}$$

23. (a) Using the law of cosines, we have:
$$a^2 = b^2 + c^2 - 2bc\cos A$$
$$4^2 = 5^2 + 6^2 - 2(5)(6)\cos A$$
$$16 = 25 + 36 - 60\cos A$$
$$-45 = -60\cos A$$
$$\tfrac{3}{4} = \cos A$$
Also using the law of cosines, we have:
$$c^2 = a^2 + b^2 - 2ab\cos C$$
$$6^2 = 4^2 + 5^2 - 2(4)(5)\cos C$$
$$36 = 16 + 25 - 40\cos C$$
$$-5 = -40\cos C$$
$$\tfrac{1}{8} = \cos C$$

(b) Since $\cos A = \tfrac{3}{4}$, then:
$$\sin A = \sqrt{1 - \cos^2 A} = \sqrt{1 - \tfrac{9}{16}} = \sqrt{\tfrac{7}{16}} = \tfrac{\sqrt{7}}{4}$$
Therefore:
$$\cos 2A = \cos^2 A - \sin^2 A = \left(\tfrac{3}{4}\right)^2 - \left(\tfrac{\sqrt{7}}{4}\right)^2 = \tfrac{9}{16} - \tfrac{7}{16} = \tfrac{1}{8}$$
But since $\cos C = \tfrac{1}{8}$, then $C = 2A$.

25. Since $BE = BD$, then $\triangle BDE$ is isosceles and thus its base angles are congruent. Therefore, calling $\angle BED = \theta$ we have:
$$\theta + \theta + 108° = 180°$$
$$2\theta = 72°$$
$$\theta = 36°$$
So $\angle BED = 36°$.

27. Note that from Exercise 26 $\angle ABE = 36°$. Using the law of sines on $\triangle ABE$, we have:
$$\frac{\sin \angle AEB}{16} = \frac{\sin 36°}{11}$$
$$\sin \angle AEB = \frac{16\sin 36°}{11} \approx 0.8550$$
$$\angle AEB \approx 58.76° \text{ or } 121.24°$$
Since the given figure indicates that $\angle AEB$ is an obtuse angle, we have $\angle AEB \approx 121.24°$.

29. Note that from Exercise 28 $\angle BAE \approx 22.76°$. Using the law of sines on $\triangle ABE$, we have:

$$\frac{\sin 22.76°}{BE} = \frac{\sin 36°}{11}$$

$$BE = \frac{11\sin 22.76°}{\sin 36°} \approx 7.24 \text{ m}$$

31. (a) Let O be the center of the circle, so that $OA = OB = 10$ cm. Since $\angle AOB = \frac{360°}{5} = 72°$, we can find AB using the law of cosines on $\triangle AOB$:

$$(AB)^2 = (OA)^2 + (OB)^2 - 2(OA)(OB)\cos 72°$$
$$= (10)^2 + (10)^2 - 2(10)(10)\cos 72°$$
$$= 200 - 200\cos 72°$$

So $AB = \sqrt{200 - 200\cos 72°} \approx 11.76$ cm.

(b) The interior angles of the pentagon are $\frac{180°(5-2)}{5} = 108°$, so $\angle ABC = 108°$. Using the law of cosines on $\triangle ABC$, we have (using the exact value from (a)):

$$(AC)^2 = (AB)^2 + (BC)^2 - 2(AB)(BC)\cos 108°$$
$$\approx (11.76)^2 + (11.76)^2 - 2(11.76)(11.76)\cos 108°$$
$$\approx 361.80$$

So $AC \approx \sqrt{361.80} \approx 19.02$ cm.

33. For the triangle, note the following figure:

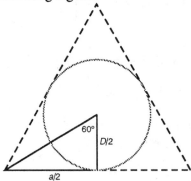

So $\tan 60° = \frac{a/2}{D/2}$, or $\sqrt{3} = \frac{a}{D}$, thus $a = D\sqrt{3}$. For the hexagon, note the following figure:

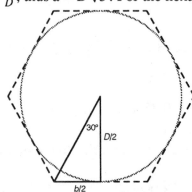

So $\tan 30° = \frac{b/2}{D/2}$, or $\frac{1}{\sqrt{3}} = \frac{b}{D}$, thus $b = \frac{D}{\sqrt{3}}$. Thus we have:

$$ab = \left(D\sqrt{3}\right)\left(\frac{D}{\sqrt{3}}\right) = D^2$$

35. The area of the triangle is given by:

$$A = \tfrac{1}{2}(\text{base})(\text{height}) = \tfrac{1}{2}(a)(b\sin 60°) = \tfrac{\sqrt{3}}{4}ab$$

Since the area is $10\sqrt{3}$ cm^2, we have:

$$\tfrac{\sqrt{3}}{4}ab = 10\sqrt{3}$$
$$ab = 40$$

We can find the third side d in terms of a and b using the law of cosines:

$$d^2 = a^2 + b^2 - 2ab\cos 60°$$
$$d^2 = a^2 + b^2 - 2(40)\left(\tfrac{1}{2}\right)$$
$$d = \sqrt{a^2 + b^2 - 40}$$

Since the perimeter of the triangle is 20 cm, we have:

$$a + b + \sqrt{a^2 + b^2 - 40} = 20$$
$$\sqrt{a^2 + b^2 - 40} = 20 - (a + b)$$
$$a^2 + b^2 - 40 = 400 - 40(a + b) + (a + b)^2$$
$$a^2 + b^2 - 40 = 400 - 40a - 40b + a^2 + 2ab + b^2$$
$$-440 = -40a - 40b + 2(40)$$
$$-520 = -40a - 40b$$
$$13 = a + b$$

So, we have the system of equations:

$$a + b = 13$$
$$ab = 40$$

Solving the first equation for b yields $b = 13 - a$, now substitute:

$$a(13 - a) = 40$$
$$13a - a^2 = 40$$
$$a^2 - 13a + 40 = 0$$
$$(a - 8)(a - 5) = 0$$
$$a = 8, 5$$

Since a is the smaller of the two numbers, we have $a = 5$ and $b = 8$.

37. We draw a figure:

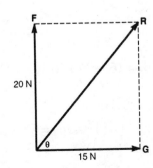

The resultant has a magnitude and direction given by:
$$|\mathbf{R}| = \sqrt{15^2 + 20^2} = \sqrt{625} = 25 \text{ N}$$
$$\theta = \tan^{-1}\left(\tfrac{20}{15}\right) \approx 53.1°$$

39. The components are given by:
$$v_x = 50\cos 35° \approx 41.0 \text{ cm/sec}$$
$$v_y = 50\sin 35° \approx 28.7 \text{ cm/sec}$$

41. We draw the figure:

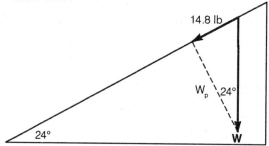

So the desired weights are given by:
$$\tan 24° = \frac{14.8}{|\mathbf{W}_p|}, \text{ so } |\mathbf{W}_p| = \frac{14.8}{\tan 24°} \approx 33.2 \text{ lb}$$
$$\sin 24° = \frac{14.8}{|\mathbf{W}|}, \text{ so } |\mathbf{W}| = \frac{14.8}{\sin 24°} \approx 36.4 \text{ lb}$$

43. We first find the lengths of $\langle 2,6 \rangle$ and $\langle -5,b \rangle$:
$$|\langle 2,6 \rangle| = \sqrt{2^2 + 6^2} = \sqrt{40}$$
$$|\langle -5,b \rangle| = \sqrt{(-5)^2 + b^2} = \sqrt{b^2 + 25}$$
We find b by solving the equation:
$$\sqrt{b^2 + 25} = \sqrt{40}$$
$$b^2 + 25 = 40$$
$$b^2 = 15$$
$$b = \pm\sqrt{15}$$

45. $\mathbf{a} + \mathbf{b} = \langle 3,5 \rangle + \langle 7,4 \rangle = \langle 10,9 \rangle$

47. We compute the quantity:
$$3\mathbf{c} + 2\mathbf{a} = 3\langle 2,-1 \rangle + 2\langle 3,5 \rangle = \langle 6,-3 \rangle + \langle 6,10 \rangle = \langle 12,7 \rangle$$

49. Since $\mathbf{b} + \mathbf{d} = \langle 7,7 \rangle$ and $\mathbf{b} - \mathbf{d} = \langle 7,1 \rangle$:
$$|\mathbf{b}+\mathbf{d}|^2 - |\mathbf{b}-\mathbf{d}|^2 = \left(7^2 + 7^2\right) - \left(7^2 + 1^2\right) = 98 - 50 = 48$$

51. $(\mathbf{a} + \mathbf{b}) + \mathbf{c} = \langle 10,9 \rangle + \langle 2,-1 \rangle = \langle 12,8 \rangle$

53. $(\mathbf{a} - \mathbf{b}) - \mathbf{c} = \langle -4,1 \rangle - \langle 2,-1 \rangle = \langle -6,2 \rangle$

55. $4\mathbf{c} + 2\mathbf{a} - 3\mathbf{b} = \langle 8,-4 \rangle + \langle 6,10 \rangle - \langle 21,12 \rangle = \langle -7,-6 \rangle$

57. $\langle 7,-6 \rangle = 7\mathbf{i} - 6\mathbf{j}$

59. We first compute the length of $\langle 6,4 \rangle$:
$$|\langle 6,4 \rangle| = \sqrt{36+16} = \sqrt{52} = 2\sqrt{13}$$
Therefore a unit vector in the same direction as $\langle 6,4 \rangle$ would be:
$$\tfrac{1}{2\sqrt{13}}\langle 6,4 \rangle = \left\langle \tfrac{3}{\sqrt{13}}, \tfrac{2}{\sqrt{13}} \right\rangle = \left\langle \tfrac{3\sqrt{13}}{13}, \tfrac{2\sqrt{13}}{13} \right\rangle$$

61. We use the distance formula, taking the points (r_1,θ_1) and (r_2,θ_2) to be $\left(3,\tfrac{\pi}{12}\right)$ and $\left(2,\tfrac{17\pi}{18}\right)$:
$$\begin{aligned}
d^2 &= r_1^2 + r_2^2 - 2r_1r_2 \cos(\theta_2 - \theta_1) \\
&= 3^2 + 2^2 - 2(3)(2)\cos\left(\tfrac{\pi}{12} - \tfrac{17\pi}{18}\right) \\
&= 13 - 12\cos\left(-\tfrac{31\pi}{36}\right) \\
&\approx 23.8757
\end{aligned}$$
So $d \approx \sqrt{23.8757} \approx 4.89$.

63. Using the equation $r^2 + r_0^2 - 2rr_0 \cos(\theta - \theta_0) = a^2$, we have:
$$\begin{aligned}
r^2 + (5)^2 - 2(r)(5)\cos\left(\theta - \tfrac{\pi}{6}\right) &= 3^2 \\
r^2 + 25 - 10r\cos\left(\theta - \tfrac{\pi}{6}\right) &= 9 \\
r^2 - 10r\cos\left(\theta - \tfrac{\pi}{6}\right) &= -16
\end{aligned}$$

65. (a) Since the line is of the form $r\cos(\theta - \alpha) = d$, $\alpha = \tfrac{\pi}{3}$ and $d = 3$. So the desired perpendicular distance is 3.

(b) When $\theta = 0$, we have:
$$\begin{aligned}
r\cos\left(0 - \tfrac{\pi}{3}\right) &= 3 \\
r\left(\tfrac{1}{2}\right) &= 3 \\
r &= 6
\end{aligned}$$

When $\theta = \frac{\pi}{2}$, we have:
$$r\cos\left(\frac{\pi}{2} - \frac{\pi}{3}\right) = 3$$
$$r\cos\frac{\pi}{6} = 3$$
$$r \cdot \frac{\sqrt{3}}{2} = 3$$
$$r = \frac{6}{\sqrt{3}} = 2\sqrt{3}$$

The required polar points are $(6,0)$ and $\left(2\sqrt{3}, \frac{\pi}{2}\right)$.

(c)   These coordinates are (d, α), which is the polar point $\left(3, \frac{\pi}{3}\right)$.

(d)   We sketch the line:

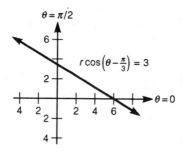

67. (a)   We plot points in polar form:

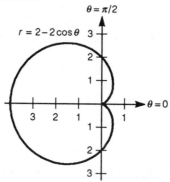

(b)   We plot points in polar form:

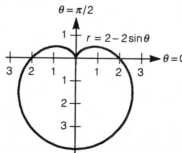

69. (a)   We plot points in polar form:

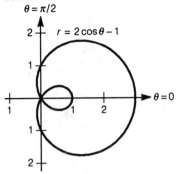

$r = 2\cos\theta - 1$

(b)   We plot points in polar form:

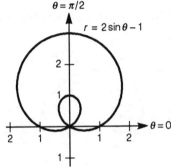

$r = 2\sin\theta - 1$

71. (a)   We plot points in polar form:

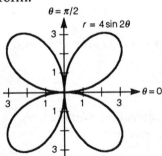

$r = 4\sin 2\theta$

(b)   We plot points in polar form:

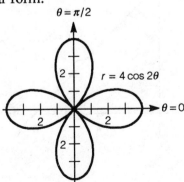

$r = 4\cos 2\theta$

73. (a)   We plot points in polar form (note that we must plot $0 \le \theta \le 4\pi$):

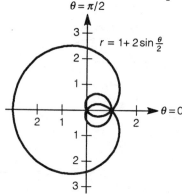

(b)   We plot points in polar form (note that we must plot $0 \le \theta \le 4\pi$):

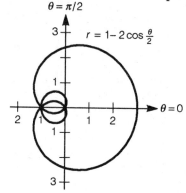

75.   Solving $y = 1 + t$ for t yields $t = y - 1$, now substituting:

$$x = 3 - 5(y - 1)$$
$$x = 3 - 5y + 5$$
$$x + 5y = 8$$

So the given parametric equations determine a line.

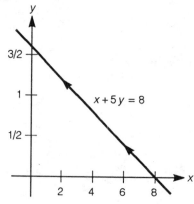

77. Multiplying the first equation by 2 yields $2x = 6 \sin t$ and $y = 6 \cos t$, so:
$$(2x)^2 + y^2 = 36 \sin^2 t + 36 \cos^2 t$$
$$4x^2 + y^2 = 36$$
$$\frac{x^2}{9} + \frac{y^2}{36} = 1$$
So the given parametric equations determine an ellipse.

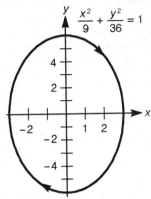

79. Multiplying the first equation by 3 and the second equation by 4 yields $3x = 12 \sec t$ and $4y = 12 \tan t$, so:
$$(3x)^2 - (4y)^2 = 144 \sec^2 t - 144 \tan^2 t$$
$$9x^2 - 16y^2 = 144$$
$$\frac{x^2}{16} - \frac{y^2}{9} = 1$$
So the given parametric equations determine a hyperbola.

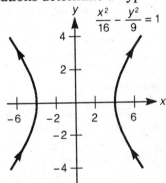

81. (a) P can be written as $\left(\dfrac{\sin \alpha}{\alpha}, \alpha \right)$, while Q can be written as $\left(\dfrac{\sin 2\alpha}{2\alpha}, 2\alpha \right)$.

(b) For point P, we have:

$$x = r\cos\theta = \frac{\sin\alpha}{\alpha}\cdot\cos\alpha = \frac{\sin\alpha\cos\alpha}{\alpha}$$

$$y = r\sin\theta = \frac{\sin\alpha}{\alpha}\cdot\sin\alpha = \frac{\sin^2\alpha}{\alpha}$$

So the rectangular coordinates are $P\left(\dfrac{\sin\alpha\cos\alpha}{\alpha}, \dfrac{\sin^2\alpha}{\alpha}\right)$. For point Q, we have:

$$x = r\cos\theta = \frac{\sin 2\alpha}{2\alpha}\cdot\cos 2\alpha = \frac{\sin 2\alpha\cos 2\alpha}{2\alpha}$$

$$y = r\sin\theta = \frac{\sin 2\alpha}{2\alpha}\cdot\sin 2\alpha = \frac{\sin^2 2\alpha}{2\alpha}$$

So the rectangular coordinates are $Q\left(\dfrac{\sin 2\alpha\cos 2\alpha}{2\alpha}, \dfrac{\sin^2 2\alpha}{2\alpha}\right)$.

(c) The slope of $\overline{OQ}$ is given by:

$$\frac{\dfrac{\sin^2 2\alpha}{2\alpha}-0}{\dfrac{\sin 2\alpha\cos 2\alpha}{2\alpha}-0} = \frac{\sin^2 2\alpha}{\sin 2\alpha\cos 2\alpha} = \frac{\sin 2\alpha}{\cos 2\alpha} = \tan 2\alpha$$

(d) The slope of $\overline{PQ}$ is given by:

$$\frac{\dfrac{\sin^2 2\alpha}{2\alpha}-\dfrac{\sin^2\alpha}{\alpha}}{\dfrac{\sin 2\alpha\cos 2\alpha}{2\alpha}-\dfrac{\sin\alpha\cos\alpha}{\alpha}} = \frac{\sin^2 2\alpha - 2\sin^2\alpha}{\sin 2\alpha\cos 2\alpha - 2\sin\alpha\cos\alpha}$$

(e) Making the substitution $\sin^2 2\alpha = 1 - \cos^2 2\alpha$ and using the double-angle identities for sine and cosine, we have:

$$\frac{\sin^2 2\alpha - 2\sin^2\alpha}{\sin 2\alpha\cos 2\alpha - 2\sin\alpha\cos\alpha} = \frac{1-\cos^2 2\alpha - 2\sin^2\alpha}{\sin 2\alpha\cos 2\alpha - \sin 2\alpha}$$

$$= \frac{\left(1-2\sin^2\alpha\right)-\cos^2 2\alpha}{\sin 2\alpha(\cos 2\alpha - 1)}$$

$$= \frac{\cos 2\alpha - \cos^2 2\alpha}{\sin 2\alpha(\cos 2\alpha - 1)}$$

$$= \frac{\cos 2\alpha(1-\cos 2\alpha)}{\sin 2\alpha(\cos 2\alpha - 1)}$$

$$= -\frac{\cos 2\alpha}{\sin 2\alpha}$$

$$= -\cot 2\alpha$$

(f)   Since the slope of $\overline{OQ}$ is $\tan 2\alpha$ and the slope of $\overline{PQ}$ is $-\cot 2\alpha$, we have:

$$m_{\overline{OQ}} \bullet m_{\overline{PQ}} = (\tan 2\alpha)(-\cot 2\alpha) = \tan 2\alpha \bullet \frac{-1}{\tan 2\alpha} = -1$$

Since the product of these slopes is –1, $\overline{OQ} \perp \overline{PQ}$.

Chapter Eight Test

1.   We draw the triangle:

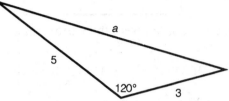

Using the law of cosines:

$$a^2 = 3^2 + 5^2 - 2(3)(5)\cos 120° = 9 + 25 - 30\left(-\tfrac{1}{2}\right) = 49$$

So $a = 7$ cm.

2.   We draw the triangle:

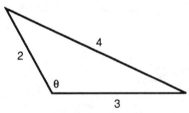

We use the law of cosines:

$$4^2 = 2^2 + 3^2 - 2(2)(3)\cos\theta$$
$$16 = 4 + 9 - 12\cos\theta$$
$$3 = -12\cos\theta$$
$$\cos\theta = -\tfrac{1}{4}$$

The angle opposite the 4 cm side must be obtuse (not acute), since its cosine is negative.

3.   We use the law of sines:

$$\frac{\sin 45°}{20\sqrt{2}} = \frac{\sin 30°}{x}$$

$$x = \frac{20\sqrt{2}\sin 30°}{\sin 45°} = \frac{20\sqrt{2} \bullet \frac{1}{2}}{\frac{\sqrt{2}}{2}} = 20 \text{ cm}$$

4.   We draw the figure:

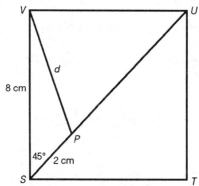

Note that the diagonals of a square bisect the vertex angles. We use the law of cosines with ΔSPV:

$$d^2 = 8^2 + 2^2 - 2(8)(2)\cos 45° = 64 + 4 - 32\left(\tfrac{\sqrt{2}}{2}\right) = 68 - 16\sqrt{2}$$
$$d = \sqrt{68 - 16\sqrt{2}} = 2\sqrt{17 - 4\sqrt{2}}$$

So $PV = 2\sqrt{17 - 4\sqrt{2}}$ cm.

5.   We use the law of cosines to find a:

$$a^2 = (5.8)^2 + (3.2)^2 - 2(5.8)(3.2)\cos 27° = 43.88 - 37.12\cos 27° \approx 10.81$$

So $a \approx \sqrt{10.81} \approx 3.3$ cm. Now we use the law of sines to find $\angle C$:

$$\frac{\sin C}{3.2} = \frac{\sin 27°}{3.3}$$
$$\sin C = \frac{3.2\sin 27°}{3.3} \approx 0.442$$
$$C \approx 26.2°$$

Finally, we find $B \approx 180° - 27° - 26.2° \approx 126.8°$.

6.   We draw the figure:

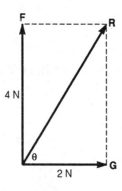

(a)   We find the magnitude of the resultant:
$$|\mathbf{R}| = \sqrt{2^2 + 4^2} = \sqrt{20} = 2\sqrt{5} \text{ N}$$

(b) $\tan \theta = \tfrac{4}{2} = 2$

7.   We draw the figure:

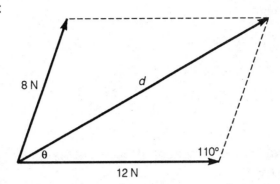

(a)   Using the law of cosines, we have:
$$d^2 = 12^2 + 8^2 - 2(12)(8)\cos 110° = 208 - 192\cos 110°$$
So $d = \sqrt{208 - 192\cos 110°} = 4\sqrt{13 - 12\cos 110°}$ N.

(b)   Using the law of sines:
$$\frac{\sin\theta}{8} = \frac{\sin 110°}{4\sqrt{13 - 12\cos 110°}}$$
$$\sin\theta = \frac{2\sin 110°}{\sqrt{13 - 12\cos 110°}}$$

8.   We draw the figure:

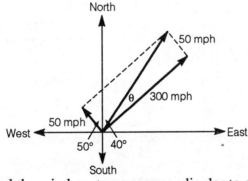

The heading vector and the wind vector are perpendicular to each other, so:
$$\text{ground speed} = \sqrt{300^2 + 50^2} = 50\sqrt{37} \text{ mph}$$
$$\tan\theta = \tfrac{50}{300} = \tfrac{1}{6}$$

9.   (a) $2\mathbf{A} + 3\mathbf{B} = 2\langle 2,4\rangle + 3\langle 3,-1\rangle = \langle 13,5\rangle$

(b)   We compute the length:
$$|2\mathbf{A} + 3\mathbf{B}| = |\langle 13,5\rangle| = \sqrt{194}$$

(c) $\mathbf{C} - \mathbf{B} = \langle 4,-4\rangle - \langle 3,-1\rangle = \langle 1,-3\rangle = \mathbf{i} - 3\mathbf{j}$

10. Subtracting components, we have:

$$\overrightarrow{PQ} = \langle -7,2 \rangle - \langle 4,5 \rangle = \langle -11,-3 \rangle$$

The required unit vector is:

$$\frac{\overrightarrow{PQ}}{|\overrightarrow{PQ}|} = \frac{1}{\sqrt{130}} \langle -11,-3 \rangle = \left\langle \frac{-11\sqrt{130}}{130}, \frac{-3\sqrt{130}}{130} \right\rangle$$

11. Using the double-angle formula for $\cos 2\theta$ and multiplying by r^2 yields:

$$r^2 = \cos^2 \theta - \sin^2 \theta$$
$$r^4 = r^2 \cos^2 \theta - r^2 \sin^2 \theta$$
$$\left(x^2 + y^2 \right)^2 = x^2 - y^2$$

12. We plot points in polar form:

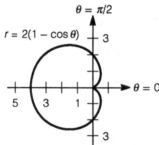

13. Multiplying the second equation by 2 yields $x = 4 \sin t$ and $2y = 4 \cos t$, so:

$$x^2 + (2y)^2 = 16 \sin^2 t + 16 \cos^2 t$$
$$x^2 + 4y^2 = 16$$
$$\frac{x^2}{16} + \frac{y^2}{4} = 1$$

So the given parametric equations determine an ellipse.

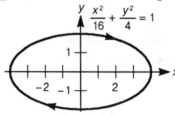

14. We use the distance formula, taking the points (r_1, θ_1) and (r_2, θ_2) to be $\left(4, \frac{10\pi}{21}\right)$ and $\left(1, \frac{\pi}{7}\right)$:

$$\begin{aligned}
d^2 &= r_1^2 + r_2^2 - 2r_1 r_2 \cos(\theta_2 - \theta_1) \\
&= 4^2 + 1^2 - 2(4)(1)\cos\left(\frac{\pi}{7} - \frac{10\pi}{21}\right) \\
&= 17 - 8\cos\left(-\frac{\pi}{3}\right) \\
&= 17 - 8 \cdot \frac{1}{2} \\
&= 13
\end{aligned}$$

So $d = \sqrt{13}$.

15. Using the equation $r^2 + r_0^2 - 2rr_0 \cos(\theta - \theta_0) = a^2$, we have:

$$\begin{aligned}
r^2 + (5)^2 - 2(r)(5)\cos\left(\theta - \frac{\pi}{2}\right) &= 2^2 \\
r^2 + 25 - 10r\cos\left(\theta - \frac{\pi}{2}\right) &= 4 \\
r^2 - 10r\cos\left(\theta - \frac{\pi}{2}\right) &= -21
\end{aligned}$$

Now substitute the polar point $\left(2, \frac{\pi}{6}\right)$ in the equation:

$$2^2 - 10(2)\cos\left(\frac{\pi}{6} - \frac{\pi}{2}\right) = 4 - 20\cos\left(-\frac{\pi}{3}\right) = 4 - 20 \cdot \frac{1}{2} = -6$$

Since we did not obtain -21, the point does not lie on the circle.

16. (a) The line will be of the form $r\cos(\theta - \alpha) = d$. Since $\alpha = \frac{5\pi}{6}$ and $d = 4$, the equation of the line is $r\cos\left(\theta - \frac{5\pi}{6}\right) = 4$.

(b) Using the addition formula for cosine, we have:

$$\begin{aligned}
r\cos\left(\theta - \frac{5\pi}{6}\right) &= 4 \\
r\left(\cos\theta\cos\frac{5\pi}{6} + \sin\theta\sin\frac{5\pi}{6}\right) &= 4 \\
r\left[\cos\theta \cdot \left(-\frac{\sqrt{3}}{2}\right) + \sin\theta \cdot \frac{1}{2}\right] &= 4 \\
-r\sqrt{3}\cos\theta + r\sin\theta &= 8 \\
-r\sqrt{3}\cos\theta + r\sin\theta - 8 &= 0
\end{aligned}$$

Chapter Nine
Systems of Equations

9.1 Systems of Two Linear Equations in Two Unknowns

1. (a) yes
    (b) no--the xy term makes it non-linear
    (c) yes
    (d) yes

3. We test $(5, 1)$ in the two equations and see if it produces true statements:
$$2(5) - 8(1) = 2$$
$$10 - 8 = 2 \qquad \text{true}$$
$$3(5) + 7(1) = 22$$
$$15 + 7 = 22 \qquad \text{true}$$

5. We test $(0, -4)$ in the two equations and see if it produces true statements:
$$\tfrac{1}{6}(0) + \tfrac{1}{2}(-4) = -2$$
$$0 - 2 = -2 \qquad \text{true}$$
$$\tfrac{2}{3}(0) + \tfrac{3}{4}(-4) = 2$$
$$0 - 3 = 2 \qquad \text{false}$$
So $(0, -4)$ is not a solution to the given system.

7.   We test $(3,-2)$ in the two equations and see if it produces true statements:
$$\tfrac{2}{7}(3) - \tfrac{1}{5}(-2) = \tfrac{44}{35}$$
$$\tfrac{6}{7} + \tfrac{2}{5} = \tfrac{44}{35}$$
$$\tfrac{30}{35} + \tfrac{14}{35} = \tfrac{44}{35} \qquad \text{true}$$
$$\tfrac{1}{3}(3) - \tfrac{5}{4}(-2) = \tfrac{7}{2}$$
$$1 + \tfrac{5}{2} = \tfrac{7}{2}$$
$$\tfrac{2}{2} + \tfrac{5}{2} = \tfrac{7}{2} \qquad \text{true}$$
So $(3,-2)$ is a solution to the given system.

9.   We solve the second equation for x:
$$x + 4y = -4$$
$$x = -4 - 4y$$
Now substitute $x = -4 - 4y$ for x in the first equation, and solve for y:
$$3x - 2y = -19$$
$$3(-4 - 4y) - 2y = -19$$
$$-12 - 12y - 2y = -19$$
$$-14y - 12 = -19$$
$$-14y = -7$$
$$y = \tfrac{1}{2}$$
We substitute this back into the second equation:
$$x = -4 - 4y = -4 - 4 \cdot \tfrac{1}{2} = -4 - 2 = -6$$
So the solution is $\left(-6, \tfrac{1}{2}\right)$.

11.   We solve the first equation for y:
$$4x + 2y = 3$$
$$2y = 3 - 4x$$
$$y = \frac{3 - 4x}{2}$$
Now substitute $y = \frac{3-4x}{2}$ for y in the second equation, and solve for x:
$$10x + 4y = 1$$
$$10x + 4\left(\frac{3 - 4x}{2}\right) = 1$$
$$10x + 2(3 - 4x) = 1$$
$$10x + 6 - 8x = 1$$
$$2x + 6 = 1$$
$$2x = -5, \text{ so } x = -\tfrac{5}{2}$$
We substitute this back into the first equation:
$$y = \frac{3 - 4x}{2} = \frac{3 - 4\left(-\tfrac{5}{2}\right)}{2} = \frac{3 + 10}{2} = \frac{13}{2}$$
The solution is $\left(-\tfrac{5}{2}, \tfrac{13}{2}\right)$.

13. We solve the second equation for y:
$$-7x + 2y = 0$$
$$2y = 7x$$
$$y = \frac{7x}{2}$$

Now substitute $y = \frac{7x}{2}$ for y in the first equation, and solve for x:
$$13x - 8y = -3$$
$$13x - 8 \cdot \frac{7x}{2} = -3$$
$$13x - 28x = -3$$
$$-15x = -3$$
$$x = \tfrac{1}{5}$$

We substitute this back into the second equation:
$$y = \frac{7x}{2} = \frac{7 \cdot \frac{1}{5}}{2} = \tfrac{7}{10}$$

The solution is $\left(\tfrac{1}{5}, \tfrac{7}{10}\right)$.

15. First we multiply each equation by 20 to clear fractions:
$$20\left(-\tfrac{2}{5}x + \tfrac{1}{4}y\right) = 20(3)$$
$$20\left(\tfrac{1}{4}x - \tfrac{2}{5}y\right) = 20(-3)$$

Therefore we have the system:
$$-8x + 5y = 60$$
$$5x - 8y = -60$$

We solve the first equation for y:
$$-8x + 5y = 60$$
$$5y = 8x + 60$$
$$y = \frac{8x + 60}{5}$$

Now substitute $y = \frac{8x+60}{5}$ for y in the second equation and solve for x:
$$5x - 8y = -60$$
$$5x - 8\left(\frac{8x + 60}{5}\right) = -60$$

Multiply by 5 to clear fractions:
$$25x - 8(8x + 60) = -300$$
$$25x - 64x - 480 = -300$$
$$-39x = 180$$
$$x = -\tfrac{180}{39} = -\tfrac{60}{13}$$

We substitute this back into the first equation:
$$y = \frac{8x + 60}{5} = \frac{8\left(-\frac{60}{13}\right) + 60}{5} = \frac{-480 + 780}{65} = \tfrac{300}{65} = \tfrac{60}{13}$$

The solution is $\left(-\tfrac{60}{13}, \tfrac{60}{13}\right)$.

17. We solve the first equation for x:

$$\sqrt{2}x - \sqrt{3}y = \sqrt{3}$$
$$\sqrt{2}x = \sqrt{3}y + \sqrt{3}$$
$$x = \frac{\sqrt{3}y + \sqrt{3}}{\sqrt{2}}$$

Now substitute $x = \frac{\sqrt{3}y+\sqrt{3}}{\sqrt{2}}$ for x in the second equation and solve for y:

$$\sqrt{3}x - \sqrt{8}y = \sqrt{2}$$
$$\sqrt{3}\left(\frac{\sqrt{3}y + \sqrt{3}}{\sqrt{2}}\right) - \sqrt{8}y = \sqrt{2}$$
$$\frac{3y + 3}{\sqrt{2}} - \sqrt{8}y = \sqrt{2}$$

Multiply by $\sqrt{2}$ to clear the fractions:

$$3y + 3 - 4y = 2$$
$$3 - y = 2$$
$$-y = -1$$
$$y = 1$$

We substitute this back into the first equation:

$$x = \frac{\sqrt{3}y + \sqrt{3}}{\sqrt{2}} = \frac{\sqrt{3} + \sqrt{3}}{\sqrt{2}} = \frac{2\sqrt{3}}{\sqrt{2}} = \frac{2\sqrt{6}}{2} = \sqrt{6}$$

The solution is $\left(\sqrt{6}, 1\right)$.

19. We multiply the second equation by 2:

$$5x + 6y = 4$$
$$4x - 6y = -6$$

Adding, we obtain:

$$9x = -2$$
$$x = -\tfrac{2}{9}$$

We substitute $x = -\tfrac{2}{9}$ into the first equation:

$$5x + 6y = 4$$
$$5\left(-\tfrac{2}{9}\right) + 6y = 4$$
$$-\tfrac{10}{9} + 6y = 4$$
$$6y = \tfrac{46}{9}$$
$$y = \tfrac{46}{54} = \tfrac{23}{27}$$

The solution is $\left(-\tfrac{2}{9}, \tfrac{23}{27}\right)$.

21. We multiply the second equation by -2:

$$4x + 13y = -5$$
$$-4x + 108y = 2$$

Adding, we obtain:

$$121y = -3$$
$$y = -\tfrac{3}{121}$$

We substitute $y = -\tfrac{3}{121}$ into the first equation:

$$4x + 13y = -5$$
$$4x + 13\left(-\tfrac{3}{121}\right) = -5$$
$$4x - \tfrac{39}{121} = -5$$
$$4x = -\tfrac{566}{121}$$
$$x = -\tfrac{283}{242}$$

The solution is $\left(-\tfrac{283}{242}, -\tfrac{3}{121}\right)$.

23. We multiply the first equation by 12 and the second equation by 70 to clear fractions:

$$12\left(\tfrac{1}{4}x - \tfrac{1}{3}y\right) = 12(4)$$
$$70\left(\tfrac{2}{7}x - \tfrac{1}{7}y\right) = 70\left(\tfrac{1}{10}\right)$$

Therefore we have the system:

$$3x - 4y = 48$$
$$20x - 10y = 7$$

We multiply the first equation by -5 and the second equation by 2:

$$-15x + 20y = -240$$
$$40x - 20y = 14$$

Adding, we obtain:

$$25x = -226$$
$$x = -\tfrac{226}{25}$$

We substitute $x = -\tfrac{226}{25}$ into the second equation:

$$20x - 10y = 7$$
$$20\left(-\tfrac{226}{25}\right) - 10y = 7$$
$$-\tfrac{904}{5} - 10y = 7$$
$$-10y = \tfrac{939}{5}$$
$$y = -\tfrac{939}{50}$$

The solution is $\left(-\tfrac{226}{25}, -\tfrac{939}{50}\right)$.

25. We multiply the second equation by -4:

$$8x + 16y = 5$$
$$-8x - 20y = -5$$

Adding, we obtain:

$$-4y = 0, \text{ so } y = 0$$

We substitute $y = 0$ into the first equation:
$$8x + 16y = 5$$
$$8x = 5$$
$$x = \tfrac{5}{8}$$
The solution is $\left(\tfrac{5}{8}, 0\right)$.

27. We divide the first equation by 5 and multiply the second equation by 10 to get:
$$25x - 8y = 9$$
$$x + y = 3$$
Multiply the second equation by 8:
$$25x - 8y = 9$$
$$8x + 8y = 24$$
Adding, we obtain:
$$33x = 33$$
$$x = 1$$
Now substitute $x = 1$ into the second equation:
$$x + y = 3$$
$$1 + y = 3$$
$$y = 2$$
The solution is $(1, 2)$.

29. The given points must satisfy the equation, so we have:
$$4 = 0^2 = b \cdot 0 + c$$
$$4 = c$$
For the other point we have:
$$14 = 2^2 + b \cdot 2 + c$$
$$10 = 2b + c$$
This system, $c = 4$ and $2b + c = 10$, is easily solved by substitution:
$$2b + c = 10$$
$$2b + 4 = 10$$
$$2b = 6$$
$$b = 3$$
$$y = x^2 + 3x + 4$$
So the equation of the parabola is $y = x^2 + 3x + 4$.

31. Again, the points satisfy the equation:
$$Ax + By = 2$$
$$A(-4) + B(5) = 2$$
$$A(7) + B(-9) = 2$$
So we have the system:
$$-4A + 5B = 2$$
$$7A - 9B = 2$$

Multiply the first equation by 7 and the second equation by 4:

$$-28A + 35B = 14$$
$$28A - 36B = 8$$
$$-B = 22$$
$$B = -22$$

Substituting for B:

$$7A - 9(-22) = 2$$
$$7A + 198 = 2$$
$$7A = -196$$
$$A = -28$$

33. First we graph the region:

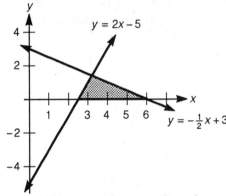

The base of the triangle is the difference between the x-intercepts, so:

$$\text{base} = 6 - \tfrac{5}{2} = \tfrac{7}{2}$$

The height is the y-coordinate of the intersection of the two lines $y = -\tfrac{1}{2}x + 3$ and $y = 2x - 5$, so:

$$-\tfrac{1}{2}x + 3 = 2x - 5$$
$$8 = \tfrac{5}{2}x$$
$$\tfrac{16}{5} = x$$

So $y = 2\left(\tfrac{16}{5}\right) - 5 = \tfrac{32}{5} - 5 = \tfrac{7}{5}$. So the area is:

$$\tfrac{1}{2}(\text{base})(\text{height}) = \tfrac{1}{2}\left(\tfrac{7}{2}\right)\left(\tfrac{7}{5}\right) = \tfrac{49}{20} \text{ square units}$$

35. Let $x =$ the amount of the 10% solution and $y =$ the amount of the 35% solution. Then $x + y = 200$ cc. Also we know that the amount of acid in each separate solution, 10% of x and 35% of y, must add to equal the acid in the mixture, 25% of $(x + y)$. This second equation is usually in need of simplifying:

$$0.10x + 0.35y = 0.25(x + y)$$
$$10x + 35y = 25(x + y)$$
$$2x + 7y = 5x + 5y$$
$$-3x + 2y = 0$$

This system is now solved by either method:
$$. \ x + \ y = 200$$
$$-3x + 2y = 0$$
Multiply the first equation by 3:
$$3x + 3y = 600$$
$$-3x + 2y = 0$$
Adding, we obtain:
$$5y = 600$$
$$y = 120$$
So $x = 80$, and we need 80 cc of the 10% solution and 120 cc of the 35% solution.

37. Let x be the amount of $5.20 coffee and y be the amount of $5.80 coffee. Then $x + y = 16$ pounds. The total value of each bean is $5.20x$ and $5.80y$, respectively, and that of the mixture is $5.50(x + y)$, so we have:
$$5.20x + 5.80y = 5.50(x + y)$$
$$52x + 58y = 55(x + y)$$
$$52x + 58y = 55x + 55y$$
$$-3x + 3y = 0$$
$$-x + y = 0$$
We solve the system:
$$x + y = 16$$
$$-x + y = 0$$
Adding, we obtain:
$$2y = 16$$
$$y = 8$$
Since $x = 16 - y$, then $x = 16 - 8 = 8$. So we mix 8 pounds of $5.20 coffee and 8 pounds of $5.80 coffee.

39. Eliminating fractions, we have:
$$bx + ay = ab$$
$$ax + by = ab$$
Therefore:
$$abx + a^2y = a^2b$$
$$-abx - b^2y = -ab^2$$
Adding, we obtain:
$$(a^2 - b^2)y = a^2b - ab^2$$
$$y = \frac{ab(a - b)}{a^2 - b^2}$$

If we factor and reduce, we have $y = \frac{ab}{a+b}$. Now substitute to solve for x:

$$\frac{x}{a} + \frac{\frac{ab}{a+b}}{b} = 1$$

$$\frac{x}{a} + \frac{a}{a+b} = 1$$

$$(a+b)x + a^2 = a(a+b)$$

$$(a+b)x = a^2 + ab - a^2$$

$$x = \frac{ab}{a+b}$$

The solution is $\left(\frac{ab}{a+b}, \frac{ab}{a+b}\right)$. Note that we cannot have $a = \pm b$ for this solution.

41. We multiply the first equation by b and the second equation by $-a$:

$$abx + a^2by = b$$
$$-abx - ab^2y = -a$$

Adding, we obtain:

$$\left(a^2b - ab^2\right)y = b - a$$

$$y = \frac{b-a}{a^2b - ab^2} = \frac{-1}{ab}$$

We substitute $y = \frac{-1}{ab}$ into the first equation:

$$ax + a^2y = 1$$

$$ax - \frac{a}{b} = 1$$

$$ax = \frac{a+b}{b}$$

$$x = \frac{a+b}{ab}$$

So the solution is $\left(\frac{a+b}{ab}, \frac{-1}{ab}\right)$. Note that we cannot have $ab = 0$, so $a \neq 0$ and $b \neq 0$.

43. Let $x = \frac{1}{s}$ and $y = \frac{1}{t}$. Then we have the system:

$$\tfrac{1}{2}x - \tfrac{1}{2}y = -10$$
$$2x + 3y = 5$$

Multiply the first equation by -4:

$$-2x + 2y = 40$$
$$2x + 3y = 5$$

Adding, we obtain:

$$5y = 45$$
$$y = 9$$

We substitute $y = 9$ into the second equation, and solve for x:
$$2x + 3y = 5$$
$$2x + 3(9) = 5$$
$$2x + 27 = 5$$
$$2x = -22$$
$$x = -11$$
Since $s = \frac{1}{x}$ and $t = \frac{1}{y}$, then $s = -\frac{1}{11}$ and $t = \frac{1}{9}$. So the solution is $\left(-\frac{1}{11}, \frac{1}{9}\right)$.

45. First we clear fractions and put into a standard form:
$$\frac{2w - 1}{3} + \frac{z + 2}{4} = 4$$
Multiplying by 12 we have the equation:
$$4(2w - 1) + 3(z + 2) = 12 \cdot 4$$
$$8w - 4 + 3z + 6 = 48$$
$$8w + 3z = 46$$
The second equation will be $w + 2z = 9$. So we have the system:
$$8w + 3z = 46$$
$$w + 2z = 9$$
Multiply the second equation by -8:
$$8w + 3z = 46$$
$$-8w - 16z = -72$$
Adding, we obtain:
$$-13z = -26$$
$$z = 2$$
$$w = 9 - 2(2) = 5$$
So the solution is $(w, z) = (5, 2)$.

47. Using substitution, we solve the first equation for x:
$$1.03x - 2.54y = 5.47$$
$$1.03x = 2.54y + 5.47$$
$$x = 2.4660y + 5.3107$$
Now substitute into the second equation:
$$3.85x + 4.29y = -1.84$$
$$3.85(2.466y + 5.3107) + 4.29y = -1.84$$
$$13.7841y = -22.2862$$
$$y \approx -1.62$$
Substituting back for x yields:
$$x = 2.4660(-1.62) + 5.3107 \approx 1.32$$
So the solution is $(1.32, -1.62)$.

49. Letting $u = \ln x$ and $v = \ln y$, we have the system:
$$2u - 5v = 11$$
$$u + v = -5$$
Using substitution, we solve the second equation for v to obtain $v = -5 - u$.

Now substitute into the first equation:
$$2u - 5(-5 - u) = 11$$
$$7u + 25 = 11$$
$$7u = -14$$
$$u = -2$$
$$v = -5 + 2 = -3$$
So $\ln x = -2$ thus $x = e^{-2}$, and $\ln y = -3$ thus $y = e^{-3}$. The solution is (e^{-2}, e^{-3}), which is approximately $(0.14, 0.05)$.

51. Letting $u = e^x$ and $v = e^y$, we have the system:
$$u - 3v = 2$$
$$3u + v = 16$$
Using substitution, we solve the first equation for u to obtain $u = 2 + 3v$.
Now substitute into the second equation:
$$3(2 + 3v) + v = 16$$
$$10v + 6 = 16$$
$$10v = 10$$
$$v = 1$$
$$u = 5$$
So $e^x = 5$ thus $x = \ln 5$, and $e^y = 1$ thus $y = 0$. The solution is $(\ln 5, 0)$, which is approximately $(1.61, 0)$.

53. Letting $u = \sqrt{x^2 - 3x}$ and $v = \sqrt{y^2 + 6y}$, we have the system:
$$4u - 3v = -4$$
$$\tfrac{1}{2}u + \tfrac{1}{2}v = 3$$
Multiplying the second equation by 6, we have the system:
$$4u - 3v = -4$$
$$3u + 3v = 18$$
Adding we obtain $7u = 14$, so $u = 2$. Substitute into the first equation:
$$4(2) - 3v = -4$$
$$8 - 3v = -4$$
$$-3v = -12$$
$$v = 4$$
So $u = 2$ and $v = 4$. Since $u = \sqrt{x^2 - 3x}$ and $v = \sqrt{y^2 + 6y}$, we have the two equations:

$$\sqrt{x^2 - 3x} = 2 \qquad\qquad \sqrt{y^2 + 6y} = 4$$
$$x^2 - 3x = 4 \qquad\qquad y^2 + 6y = 16$$
$$x^2 - 3x - 4 = 0 \qquad\qquad y^2 + 6y - 16 = 0$$
$$(x - 4)(x + 1) = 0 \qquad\qquad (y + 8)(y - 2) = 0$$
$$x = 4, -1 \qquad\qquad y = -8, 2$$

So the solutions are $(4, -8)$, $(4, 2)$, $(-1, -8)$, and $(-1, 2)$.

55. If we take tu as our number, then it is important to distinguish between its value, $10t + u$, and the sum of its digits $t + u$. Here we are told that $t + u = 14$ and that $2t = u + 1$. We solve this system:
$$t + u = 14$$
$$2t - u = 1$$
Adding, we obtain:
$$3t = 15, \text{ so } t = 5$$
Thus $u = 9$ and our original two-digit number was 59.

57. Let l and w be the length and width, respectively, of the rectangle. Since the perimeter is 34 in., then $2l + 2w = 34$, or $l + w = 17$. We are also given that $l = 2 + 2w$, so we substitute:
$$l + w = 17$$
$$(2 + 2w) + w = 17$$
$$2 + 3w = 17$$
$$3w = 15, \text{ so } w = 5$$
Thus $l = 17 - w = 17 - 5 = 12$. Thus the length is 12 in. and the width is 5 in.

59. (a)   We solve the first equation for y:
$$x + y = 100$$
$$y = 100 - x$$
Now substituting into the second equation:
$$-2x + 3y = 0$$
$$-2x + 3(100 - x) = 0$$
$$-2x + 300 - 3x = 0$$
$$300 - 5x = 0$$
$$300 = 5x, \text{ so } x = 60$$
Then $y = 100 - 60 = 40$. Thus the solution is $(60, 40)$.

(b)   We multiply the first equation by 2:
$$2x + 2y = 200$$
$$-2x + 3y = 0$$
Adding, we obtain:
$$5y = 200, \text{ so } y = 40$$
Substituting back into the first equation:
$$x + y = 100$$
$$x + 40 = 100$$
$$x = 60$$
Again, the solution is $(60, 40)$.

61. We solve each equation for x:
$$x = by - ab$$
$$x = cy - ac$$

Setting these equal:
$$by - ab = cy - ac$$
$$by - cy = ab - ac$$
$$y(b - c) = a(b - c)$$
$$y = a, \text{ as long as } b \neq c$$

Plugging into the second equation, we have:
$$x = cy - ac = ac - ac = 0$$

So the solution is $(0, a)$.

63. Let $u = \frac{1}{x}$ and $v = \frac{1}{y}$. Then we have:
$$\frac{a}{b}u + \frac{b}{a}v = a + b$$
$$bu + av = a^2 + b^2$$

Multiply the first equation by ab to clear the fractions:
$$a^2u + b^2v = a^2b + ab^2$$
$$bu + av = a^2 + b^2$$

We multiply the first equation by $-b$ and the second equation by a^2:
$$-a^2bu - b^3v = -a^2b^2 - ab^3$$
$$a^2bu + a^3v = a^4 + a^2b^2$$

Adding, we obtain:
$$\left(a^3 - b^3\right)v = a^4 - ab^3$$
$$\left(a^3 - b^3\right)v = a\left(a^3 - b^3\right)$$
$$v = a$$

Substituting $v = a$ into the second equation, we have:
$$bu + av = a^2 + b^2$$
$$bu + a^2 = a^2 + b^2$$
$$bu = b^2$$
$$u = b$$

Since $x = \frac{1}{u}$ and $y = \frac{1}{v}$, then $x = \frac{1}{b}$ and $y = \frac{1}{a}$. So the solution is $\left(\frac{1}{b}, \frac{1}{a}\right)$.

65. Since the lines are concurrent, the following two systems must possess the same solution:

I.	$7x + 5y = 4$	*II.* $x + ky = 3$
	$x + ky = 3$	$5x + y = -k$

Using either method of this section, we find that the solution of system *I* is $\left(\frac{4k-15}{7k-5}, \frac{17}{7k-5}\right)$. Also the solution of system *II* is found to be $\left(\frac{-3-k^2}{5k-1}, \frac{k+15}{5k-1}\right)$. Since the two solutions are the same, the corresponding x- and y-coordinates must be equal. Equating the y-coordinates gives us:
$$\frac{17}{7k-5} = \frac{k+15}{5k-1}$$

After clearing fractions and simplifying, this equation becomes $7k^2 + 15k - 58 = 0$. We factor to get $(7k + 29)(k - 2) = 0$ and therefore $k = -\frac{29}{7}$ or $k = 2$. These are the required values of k. If we equate the x-coordinates rather than the y-coordinates we obtain the equation $7k^3 + 15k^2 - 58k = 0$. We factor to get $k(7k + 29)(k - 2) = 0$, which has solutions $k = 0, -\frac{29}{7}$, and 2. The root $k = 0$ is extraneous, for in that case the two y-coordinates are not equal.

67. (a) We first draw the triangle:

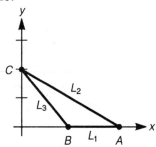

We find the equations for each side:

L_1: $y = 0$

L_2: $m = \dfrac{0-2}{2a-0} = -\dfrac{1}{a}$

Therefore:

$$y = -\frac{1}{a}x + 2$$
$$ay = -x + 2a$$
$$x + ay = 2a$$

L_3: $m = \dfrac{0-2}{2b-0} = -\dfrac{1}{b}$

Therefore:

$$y = -\frac{1}{b}x + 2$$
$$by = -x + 2b$$
$$x + by = 2b$$

(b) We first find the midpoints of each side:

L_1: mid $= \left(\dfrac{2b+2a}{2}, \dfrac{0}{2}\right) = (a+b, 0)$

L_2: mid $= \left(\dfrac{0+2a}{2}, \dfrac{2+0}{2}\right) = (a, 1)$

L_3: mid $= \left(\dfrac{0+2b}{2}, \dfrac{2+0}{2}\right) = (b, 1)$

Now we find the slopes from each vertex to the adjacent midpoint:

CL_1: $m = \dfrac{0-2}{a+b-0} = \dfrac{-2}{a+b}$

BL_2: $m = \dfrac{1-0}{a-2b} = \dfrac{1}{a-2b}$

AL_3: $m = \dfrac{1-0}{b-2a} = \dfrac{1}{b-2a}$

Finally, we use these slopes and vertices in the point-slope formula:
For CL_1, we have:

$$y - 2 = \dfrac{-2}{a+b}(x-0)$$

$$y = \dfrac{-2}{a+b}x + 2$$

$$(a+b)y = -2x + 2(a+b)$$

$$2x + (a+b)y = 2(a+b)$$

For BL_2, we have:

$$y - 0 = \dfrac{1}{a-2b}(x-2b)$$

$$y = \dfrac{1}{a-2b}x - \dfrac{2b}{a-2b}$$

$$(a-2b)y = x - 2b$$

$$-x + (a-2b)y = -2b$$

$$x + (2b-a)y = 2b$$

For AL_3, we have:

$$y - 0 = \dfrac{1}{b-2a}(x-2a)$$

$$y = \dfrac{1}{b-2a}x - \dfrac{2a}{b-2a}$$

$$(b-2a)y = x - 2a$$

$$-x + (b-2a)y = -2a$$

$$x + (2a-b)y = 2a$$

(c) We find the intersection of the three lines:

$$2x + (a+b)y = 2(a+b)$$
$$x + (2b-a)y = 2b$$
$$x + (2a-b)y = 2a$$

We will solve the second and third equations for x:

$$x = 2b - (2b-a)y$$
$$x = 2a - (2a-b)y$$

Setting these equal:
$$2b - (2b - a)y = 2a - (2a - b)y$$
$$(2a - b)y - (2b - a)y = 2a - 2b$$
$$(2a - b - 2b + a)y = 2a - 2b$$
$$(3a - 3b)y = 2a - 2b$$
$$y = \frac{2a - 2b}{3a - 3b} = \frac{2(a - b)}{3(a - b)} = \frac{2}{3}$$

We substitute $y = \frac{2}{3}$ into the second equation:
$$x = 2b - \frac{2}{3}(2b - a) = \frac{6b - 4b + 2a}{3} = \frac{2b + 2a}{3} = \frac{2(a + b)}{3}$$

We must <u>still</u> show that this point $\left(\frac{2(a+b)}{3}, \frac{2}{3}\right)$ lies on the first line:
$$2x + (a + b)y = 2(a + b)$$
$$4\left(\frac{a + b}{3}\right) + (a + b)\frac{2}{3} = 2(a + b)$$
$$6\frac{(a + b)}{3} = 2(a + b) \quad \text{true}$$

Thus the point $\left(\frac{2(a+b)}{3}, \frac{2}{3}\right)$ is the intersection of all three medians.

(d)   The altitudes will be perpendicular to the slopes of each side, which we found in (a). So:

For BL_2, we have:
$$m = a$$
$$y - 0 = a(x - 2b)$$
$$y = ax - 2ab$$
$$-ax + y = -2ab$$
$$ax - y = 2ab$$

For CL_2, we have:
$$m = \text{undefined}$$
$$x = 0$$

For AL_3, we have:
$$m = b$$
$$y - 0 = b(x - 2a)$$
$$y = bx - 2ab$$
$$-bx + y = -2ab$$
$$bx - y = 2ab$$

(e)   We find the intersection of the three lines:
$$ax - y = 2ab$$
$$bx - y = 2ab$$
$$x = 0$$

Since $x = 0$, substituting into the first equation yields:
$$-y = 2ab$$
$$y = -2ab$$
We must still show that this point $(0, -2ab)$ lies on the second line:
$$bx - y = 2ab$$
$$b(0) - (-2ab) = 2ab$$
$$2ab = 2ab \quad \text{true}$$
Thus, the point $(0, -2ab)$ is the intersection of all three altitudes.

(f) Recall that the perpendicular bisectors must be perpendicular and pass through the midpoints (which we found in (b)):

For L_1, we have:
$$\text{mid} = (a + b, 0)$$
$$m = \text{undefined}$$
$$x = a + b$$
For L_2, we have:
$$\text{mid} = (a, 1)$$
$$m = a$$
$$y - 1 = a(x - a)$$
$$y - 1 = ax - a^2$$
$$-ax + y = 1 - a^2$$
$$ax - y = a^2 - 1$$
For L_3, we have:
$$\text{mid} = (b, 1)$$
$$m = b$$
$$y - 1 = b(x - b)$$
$$y - 1 = bx - b^2$$
$$-bx + y = 1 - b^2$$
$$bx - y = b^2 - 1$$

(g) We find the intersection of the three lines:
$$ax - y = a^2 - 1$$
$$bx - y = b^2 - 1$$
$$x = a + b$$
Substituting into the first equation yields:
$$a(a + b) - y = a^2 - 1$$
$$a^2 + ab - y = a^2 - 1$$
$$-y = -ab - 1$$
$$y = ab + 1$$
We must still show that the point $(a + b, ab + 1)$ lies on the second line:
$$bx - y = b^2 - 1$$
$$b(a + b) - (ab + 1) = b^2 - 1$$
$$ab + b^2 - ab - 1 = b^2 - 1$$
$$b^2 - 1 = b^2 - 1 \quad \text{true}$$
Thus the point $(a + b, ab + 1)$ is the intersection of all three perpendicular bisectors.

(h)   We find the line that passes through $\left(\frac{2(a+b)}{3},\frac{2}{3}\right)$, $(0,-2ab)$, and $(a+b, ab+1)$. We work with the last two points:

$$m = \frac{ab+1-(-2ab)}{a+b-0} = \frac{3ab+1}{a+b}$$

Since $(0,-2ab)$ is the y-intercept, then the line should be:

$$y = \frac{3ab+1}{a+b}x - 2ab$$

We must still show that the point $\left(\frac{2(a+b)}{3},\frac{2}{3}\right)$ lies on this line:

$$\frac{2}{3} = \left(\frac{3ab+1}{a+b}\right)\left(\frac{2(a+b)}{3}\right) - 2ab$$

$$\frac{2}{3} = \frac{2(3ab+1)}{3} - 2ab$$

$$\frac{2}{3} = \frac{6ab+2-6ab}{3}$$

$$\frac{2}{3} = \frac{2}{3} \quad \text{true}$$

Thus $y = \dfrac{3ab+1}{a+b}x - 2ab$ is the Euler line.

9.2 Gaussian Elimination

1.   We are given the system:

$$2x + y + z = -9$$
$$3y - 2z = -4$$
$$8z = -8$$

Solve for z in the third equation:

$$8z = -8$$
$$z = -1$$

Substitute into the second equation:

$$3y - 2z = -4$$
$$3y - 2(-1) = -4$$
$$3y + 2 = -4$$
$$3y = -6$$
$$y = -2$$

Substitute into the first equation:

$$2x + y + z = -9$$
$$2x + (-2) + (-1) = -9$$
$$2x - 3 = 9$$
$$2x = -6$$
$$x = -3$$

So the solution is $(-3,-2,-1)$.

3.   We are given the system:
$$8x + 5y + 3z = 1$$
$$3y + 4z = 2$$
$$5z = 3$$
Solve for z in the third equation:
$$5z = 3$$
$$z = \tfrac{3}{5}$$
Substitute into the second equation:
$$3y + 4z = 2$$
$$3y + (4)\tfrac{3}{5} = 2$$
$$3y + \tfrac{12}{5} = 2$$
$$3y = -\tfrac{2}{5}$$
$$y = -\tfrac{2}{15}$$
Substitute into the first equation:
$$8x + 5y + 3z = 1$$
$$8x + 5\left(-\tfrac{2}{15}\right) + (3)\tfrac{3}{5} = 1$$
$$8x - \tfrac{2}{3} + \tfrac{9}{5} = 1$$
$$8x - \tfrac{10}{15} + \tfrac{27}{15} = 1$$
$$8x = -\tfrac{2}{15}$$
$$x = -\tfrac{1}{60}$$
So the solution is $\left(-\tfrac{1}{60}, -\tfrac{2}{15}, \tfrac{3}{5}\right)$.

5.   We are given the system:
$$-4x + 5y = 0$$
$$3y + 2z = 1$$
$$3z = -1$$
Solve for z in the third equation:
$$3z = -1$$
$$z = -\tfrac{1}{3}$$
Substitute into the second equation:
$$3y + 2z = 1$$
$$3y + 2\left(-\tfrac{1}{3}\right) = 1$$
$$3y - \tfrac{2}{3} = 1$$
$$3y = \tfrac{5}{3}$$
$$y = \tfrac{5}{9}$$

Substitute into the first equation:
$$-4x + 5y = 0$$
$$-4x + (5)\tfrac{5}{9} = 0$$
$$-4x + \tfrac{25}{9} = 0$$
$$-4x = -\tfrac{25}{9}$$
$$x = \tfrac{25}{36}$$

So the solution is $\left(\tfrac{25}{36}, \tfrac{5}{9}, -\tfrac{1}{3}\right)$.

7. We are given the system:
$$-x + 8y + 3z = 0$$
$$2z = 0$$

Solve for z in the second equation:
$$2z = 0$$
$$z = 0$$

Substitute into the first equation:
$$-x + 8y + 3z = 0$$
$$-x + 8y + 3(0) = 0$$
$$-x + 8y = 0$$
$$8y = x$$
$$y = \frac{x}{8}$$

So the solution is $\left(x, \tfrac{x}{8}, 0\right)$, where x is any real number.

9. We are given the system:
$$2x + 3y + z + w = -6$$
$$y + 3z - 4w = 23$$
$$6z - 5w = 31$$
$$-2w = 10$$

Solve for w in the fourth equation:
$$-2w = 10$$
$$w = -5$$

Substitute into the third equation:
$$6z - 5w = 31$$
$$6z - 5(-5) = 31$$
$$6z + 25 = 31$$
$$6z = 6$$
$$z = 1$$

Substitute into the second equation:
$$y + 3z - 4w = 23$$
$$y + 3(1) - 4(-5) = 23$$
$$y + 23 = 23$$
$$y = 0$$

Substitute into the first equation:
$$2x + 3y + z + w = -6$$
$$2x + 3(0) + 1 + (-5) = -6$$
$$2x - 4 = -6$$
$$2x = -2$$
$$x = -1$$
So the solution is $(-1, 0, 1, -5)$.

11. We must arrange this according to echelon form. First add -2 times the first equation to the second one, and -3 times it to the third:
$$x + y + z = 12$$
$$-3y - 3z = -25$$
$$-y - 2z = -14$$
We want to eliminate the y term from the third equation so interchange the second and third equations:
$$x + y + z = 12$$
$$-y - 2z = -14$$
$$-3y - 3z = -25$$
Add -3 times the second equation to the third:
$$x + y + z = 12$$
$$-y - 2z = -14$$
$$3z = 17$$
Solve for z in the third equation:
$$3z = 17$$
$$z = \tfrac{17}{3}$$
Substitute into the second equation:
$$-y - 2z = -14$$
$$-y - (2)\tfrac{17}{3} = -14$$
$$-y - \tfrac{34}{3} = -14$$
$$-y = -\tfrac{8}{3}$$
$$y = \tfrac{8}{3}$$
Substitute into the first equation:
$$x + y + z = 12$$
$$x + \tfrac{8}{3} + \tfrac{17}{3} = 12$$
$$x + \tfrac{25}{3} = 12$$
$$x = \tfrac{11}{3}$$
So the solution is $\left(\tfrac{11}{3}, \tfrac{8}{3}, \tfrac{17}{3}\right)$.

13. We must arrange the system in echelon form. Multiply the first equation by -2 and add it to the second equation:
$$-4x + 6y - 4z = -8$$
$$4x + 2y + 3z = 7$$

Adding, we obtain:
$$8y - z = -1$$
Multiply the first equation by $-\frac{5}{2}$ and add it to the third equation:
$$-5x + \tfrac{15}{2}y - 5z = -10$$
$$5x + 4y + 2z = 7$$
Adding, we obtain:
$$\tfrac{23}{2}y - 3z = -3$$
$$23y - 6z = -6$$
So we have the system:
$$2x - 3y + 2z = 4$$
$$8y - z = -1$$
$$23y - 6z = -6$$
Multiply the second equation by $-\frac{23}{8}$ and add it to the third equation:
$$-23y + \tfrac{23}{8}z = \tfrac{23}{8}$$
$$23y - 6z = -6$$
Adding, we obtain:
$$-\tfrac{25}{8}z = -\tfrac{25}{8}$$
So we have the system in echelon form:
$$2x - 3y + 2z = 4$$
$$8y - z = -1$$
$$-\tfrac{25}{8}z = -\tfrac{25}{8}$$
Solve for z in the third equation:
$$-\tfrac{25}{8}z = -\tfrac{25}{8}$$
$$z = 1$$
Substitute into the second equation:
$$8y - z = -1$$
$$8y - 1 = -1$$
$$8y = 0$$
$$y = 0$$
Substitute into the first equation:
$$2x - 3y + 2z = 4$$
$$2x - 3(0) + 2(1) = 4$$
$$2x + 2 = 4$$
$$2x = 2$$
$$x = 1$$
So the solution is $(1, 0, 1)$.

15. We must arrange the system in echelon form. Multiply the first equation by -2 and add it to the second equation:
$$-6x - 6y + 4z = -26$$
$$6x + 2y - 5z = 13$$
Adding, we obtain:
$$-4y - z = -13$$

Multiply the first equation by $-\frac{7}{3}$ and add it to the third equation:
$$-7x - 7y + \tfrac{14}{3}z = -\tfrac{91}{3}$$
$$7x + 5y - 3z = 26$$
Adding, we obtain:
$$-2y + \tfrac{5}{3}z = -\tfrac{13}{3}$$
$$-6y + 5z = -13$$
So we have the system:
$$3x + 3y - 2z = 13$$
$$-4y - z = -13$$
$$-6y + 5z = -13$$
Multiply the second equation by $-\frac{3}{2}$ and add it to the third equation:
$$6y + \tfrac{3}{2}z = \tfrac{39}{2}$$
$$-6y + 5z = -13$$
Adding, we obtain:
$$\tfrac{13}{2}z = \tfrac{13}{2}$$
So we have the system in echelon form:
$$3x + 3y - 2z = 13$$
$$-4y - z = -13$$
$$\tfrac{13}{2}z = \tfrac{13}{2}$$
Solve for z in the third equation:
$$\tfrac{13}{2}z = \tfrac{13}{2}$$
$$z = 1$$
Substitute into the second equation:
$$-4y - z = -13$$
$$-4y - 1 = -13$$
$$-4y = -12$$
$$y = 3$$
Substitute into the first equation:
$$3x + 3y - 2z = 13$$
$$3x + 3(3) - 2(1) = 13$$
$$3x + 7 = 13$$
$$3x = 6$$
$$x = 2$$
So the solution is $(2, 3, 1)$.

17. We must arrange the system in echelon form. Multiply the first equation by 2 and add it to the second equation:
$$2x + 2y + 2z = 2$$
$$-2x + y + z = -2$$
Adding, we obtain:
$$3y + 3z = 0$$
$$y + z = 0$$

Multiply the first equation by –3 and add it to the third equation:
$$-3x - 3y - 3z = -3$$
$$3x + 6y + 6z = 5$$
Adding, we obtain:
$$3y + 3z = 2$$
So we have the system:
$$x + y + z = 1$$
$$y + z = 0$$
$$3y + 3z = 2$$
Multiply the second equation by –3 and add it to the third equation:
$$-3y - 3z = 0$$
$$3y + 3z = 2$$
Adding, we obtain:
$$0 = 2$$
Since this equation is false, there is no solution. This system is inconsistent.

19. We first re-arrange the system as:
$$x + 3y - 2z = 2$$
$$2x - y + z = -1$$
$$-5x + 6y - 5z = 5$$
Multiply the first equation by –2 and add it to the second equation:
$$-2x - 6y + 4z = -4$$
$$2x - y + z = -1$$
Adding, we obtain:
$$-7y + 5 = -5$$
Multiply the first equation by 5 and add it to the third equation:
$$5x + 15y - 10z = 10$$
$$-5x + 6y - 5z = 5$$
Adding, we obtain:
$$21y - 15z = 15$$
$$7y - 5z = 5$$
So we have the system:
$$x + 3y - 2z = 2$$
$$-7y + 5z = -5$$
$$7y - 5z = 5$$
Adding the second and third equations yields $0 = 0$, so the system is dependent. Solving for x and y in terms of z yields the solution $\left(\frac{z+1}{-7}, \frac{5(z+1)}{7}, z\right)$, where z is any real number.

21. We first re-arrange the system as:
$$x + 3y + 2z = -1$$
$$2x - y + z = 4$$
$$7x + 5z = 11$$

Multiply the first equation by -2 and add it to the second equation:
$$-2x - 6y - 4z = 2$$
$$2x - y + z = 4$$
Adding, we obtain:
$$-7y - 3z = 6$$
Multiply the first equation by -7 and add it to the third equation:
$$-7x - 21y - 14z = 7$$
$$7x + 5z = 11$$
Adding, we obtain:
$$-21y - 9z = 18$$
$$-7y - 3z = 6$$
So we have the system:
$$x + 3y + 2z = -1$$
$$-7y - 3z = 6$$
$$-7y - 3z = 6$$
Multiply the second equation by -1 and add it to the third equation:
$$7y + 3z = -6$$
$$-7y - 3z = 6$$
Adding, we obtain:
$$0 = 0$$
So we have the system:
$$x + 3y + 2z = -1$$
$$-7y - 3z = 6$$
We solve the second equation for y:
$$-7y - 3z = 6$$
$$-7y = 3z + 6$$
$$y = \frac{-3z - 6}{7}$$
Substitute into the first equation:
$$x + 3y + 2z = -1$$
$$x + 3\left(\frac{-3z - 6}{7}\right) + 2z = -1$$
$$x + \frac{-9z - 18 + 14z}{7} = -1$$
$$x + \frac{5z - 18}{7} = -1$$
$$x = \frac{-7 - 5z + 18}{7} = \frac{-5z + 11}{7}$$
So the solution is $\left(\frac{11-5z}{7}, \frac{-3z-6}{7}, z\right)$, where z is any real number.

23. Multiply the first equation by -1 and add it to the second equation:
$$-x - y - z - w = -4$$
$$x - 2y - z - w = 3$$

Adding, we obtain:
$$-3y - 2z - 2w = -1$$
$$3y + 2z + 2w = 1$$

Multiply the first equation by -2 and add it to the third equation:
$$-2x - 2y - 2z - 2w = -8$$
$$2x - y + z - w = 2$$

Adding, we obtain:
$$-3y - z - 3w = -6$$
$$3y + z + 3w = 6$$

Multiply the first equation by -1 and add it to the fourth equation:
$$-x - y - z - w = -4$$
$$x - y + 2z - 2w = -7$$

Adding, we obtain:
$$-2y + z - 3w = -11$$
$$2y - z + 3w = 11$$

So we have the system:
$$x + y + z + w = 4$$
$$3y + 2z + 2w = 1$$
$$3y + z + 3w = 6$$
$$2y - z + 3w = 11$$

Multiply the second equation by -1 and add it to the third equation:
$$-3y - 2z - 2w = -1$$
$$3y + z + 3w = 6$$

Adding, we obtain:
$$-z + w = 5$$
$$z - w = -5$$

Multiply the second equation by $-\frac{2}{3}$ and add it to the fourth equation:
$$-2y - \frac{4}{3}z - \frac{4}{3}w = -\frac{2}{3}$$
$$2y - z + 3w = 11$$

Adding, we obtain:
$$-\frac{7}{3}z + \frac{5}{3}w = \frac{31}{3}$$
$$-7z + 5w = 31$$

So we have the system:
$$x + y + z + w = 4$$
$$3y + 2z + 2w = 1$$
$$z - w = -5$$
$$-7z + 5w = 31$$

Multiply the third equation by 7 and add it to the fourth equation:
$$7z - 7w = -35$$
$$-7z + 5w = 31$$

Adding, we obtain:
$$-2w = -4$$

So we have the system:
$$x + y + z + w = 4$$
$$3y + 2z + 2w = 1$$
$$z - w = -5$$
$$-2w = -4$$

We solve the fourth equation for w:
$$-2w = -4$$
$$w = 2$$

Substitute into the third equation:
$$z - w = -5$$
$$z - 2 = -5$$
$$z = -3$$

Substitute into the second equation:
$$3y + 2z + 2w = 1$$
$$3y + 2(-3) + 2(2) = 1$$
$$3y - 2 = 1$$
$$3y = 3$$
$$y = 1$$

Substitute into the first equation:
$$x + y + z + w = 4$$
$$x + 1 - 3 + 2 = 4$$
$$x = 4$$

So the solution is $(4, 1, -3, 2)$.

25. We first re-arrange the system as:
$$x + 4y - 3z = 1$$
$$2x + 3y + 2z = 5$$

Multiply the first equation by -2 and add it to the second equation:
$$-2x - 8y + 6z = -2$$
$$2x + 3y + 2z = 5$$

Adding, we obtain:
$$-5y + 8z = 3$$

So we have the system:
$$x + 4y - 3z = 1$$
$$-5y + 8z = 3$$

We solve the second equation for y:
$$-5y + 8z = 3$$
$$-5y = 3 - 8z$$
$$y = \frac{8z - 3}{5}$$

Substitute into the first equation:
$$x + 4y - 3z = 1$$
$$x + 4\left(\frac{8z - 3}{5}\right) - 3z = 1$$
$$x + \frac{17z - 12}{5} = 1$$
$$x = \frac{17 - 17z}{5}$$

So the solution is $\left(\frac{17 - 17z}{5}, \frac{8z - 3}{5}, z\right)$, where z is any real number.

27. Multiply the first equation by -3 and add it to the second equation:
$$-3x + 6y + 6z - 6w = 30$$
$$3x + 4y - z - 3w = 11$$
Adding, we obtain:
$$10y + 5z - 9w = 41$$
Multiply the first equation by 4 and add it to the third equation:
$$4x - 8y - 8z + 8w = -40$$
$$-4x - 3y - 3z + 8w = -21$$
Adding, we obtain:
$$-11y - 11z + 16w = -61$$
So we have the system:
$$x - 2y - 2z + 2w = -10$$
$$10y + 5z - 9w = 41$$
$$-11y - 11z + 16w = -61$$
Add the second and third equations:
$$10y + 5z - 9w = 41$$
$$-11y - 11z + 16w = -61$$
Adding, we obtain:
$$-y - 6z + 7w = -20$$
$$y + 6z - 7w = 20$$
So we have the system:
$$x - 2y - 2z + 2w = -10$$
$$y + 6z - 7w = 20$$
$$-11y - 11z + 16w = -61$$
Multiply the second equation by 11 and add it to the third equation:
$$11y + 66z - 77w = 220$$
$$-11y - 11z + 16w = -61$$
Adding, we obtain:
$$55z - 61w = 159$$
So we have the system:
$$x - 2y - 2z + 2w = -10$$
$$y + 6z - 7w = 20$$
$$55z - 61w = 159$$

Solve the third equation for z:

$$55z - 61w = 159$$
$$55z = 61w + 159$$
$$z = \frac{61w + 159}{55}$$

Substitute into the second equation:

$$y + 6z - 7w = 20$$
$$y + 6\left(\frac{61w + 159}{55}\right) - 7w = 20$$
$$y + \frac{954 - 19w}{55} = 20$$
$$y = \frac{146 + 19w}{55}$$

Substitute into the first equation:

$$x - 2y - 2z + 2w = -10$$
$$x - 2\left(\frac{146 + 19w}{55}\right) - 2\left(\frac{61w + 159}{55}\right) + 2w = -10$$
$$x + \frac{-122 - 10w}{11} = -10$$
$$x = \frac{12 + 10w}{11}$$

So the solution is $\left(\frac{12+10w}{11}, \frac{146+19w}{55}, \frac{61w+159}{55}, w\right)$, where w is any real number.

29. Solving the second equation for y yields $y = \frac{2z+3}{3}$. We substitute this into the first equation to get $x = -\frac{5z}{12}$. So the solution is $\left(-\frac{5z}{12}, \frac{2z+3}{3}, z\right)$, where z is any real number.

31. (a) Multiply by $(x-2)(x+2)$, so:

$$1 = A(x+2) + B(x-2)$$
$$1 = Ax + 2A + Bx - 2B$$
$$1 = (A+B)x + (2A - 2B)$$

Since A and B are constants (cannot contain x) we must have:

$$A + B = 0$$
$$2A - 2B = 1$$

Multiply the first equation by -2 and add it to the second equation:

$$-2A - 2B = 0$$
$$2A - 2B = 1$$

Adding, we obtain:

$$-4B = 1$$
$$B = -\tfrac{1}{4}$$

Substituting into the first equation:
$$A + B = 0$$
$$A - \tfrac{1}{4} = 0$$
$$A = \tfrac{1}{4}$$
So $A = \tfrac{1}{4}$ and $B = -\tfrac{1}{4}$.

(b)   Multiply by $(x - 2)(x + 2)$, so:
$$x = A(x + 2) + B(x - 2)$$
$$x = Ax + 2A + Bx - 2B$$
$$x = (A + B)x + (2A - 2B)$$
Since A and B are constants we have the system:
$$A + B = 1$$
$$2A - 2B = 0$$
Multiply the first equation by -2 and add it to the second equation:
$$-2A - 2B = -2$$
$$2A - 2B = 0$$
Adding, we obtain:
$$-4B = -2$$
$$b = \tfrac{1}{2}$$
Substituting into the first equation:
$$A + B = 1$$
$$A + \tfrac{1}{2} = 1$$
$$A = \tfrac{1}{2}$$
So $A = \tfrac{1}{2}$ and $B = \tfrac{1}{2}$.

33.   Multiply by $(x + 4)(x - 5)$, so:
$$3x = A(x - 5) + B(x + 4)$$
$$3x = Ax - 5A + Bx + 4B$$
$$3x = (A + B)x + (-5A + 4B)$$
Since A and B are constants we have the system:
$$A + B = 3$$
$$-5A + 4B = 0$$
Multiply the first equation by 5 and add it to the second equation:
$$5A + 5B = 15$$
$$-5A + 4B = 0$$
Adding, we obtain:
$$9B = 15$$
$$B = \tfrac{5}{3}$$
Substituting into the first equation:
$$A + B = 3$$
$$A + \tfrac{5}{3} = 3$$
$$A = \tfrac{4}{3}$$
So $A = \tfrac{4}{3}$ and $B = \tfrac{5}{3}$.

35. (a) Multiply by $(x-4)(x+4)$, so:

$$3 = A(x+4) + B(x-4)$$
$$3 = Ax + 4A + Bx - 4B$$
$$3 = (A+B)x + (4A-4B)$$

Since A and B are constants we have:

$$A + B = 0$$
$$4A - 4B = 3$$

Multiply the first equation by -4 and add it to the second equation:

$$-4A - 4B = 0$$
$$4A - 4B = 3$$

Adding, we obtain:

$$-8B = 3$$
$$B = -\tfrac{3}{8}$$

Substituting into the first equation:

$$A + B = 0$$
$$A - \tfrac{3}{8} = 0$$
$$A = \tfrac{3}{8}$$

So $A = \tfrac{3}{8}$ and $B = -\tfrac{3}{8}$.

(b) Multiply by $(x-4)(x+4)^2$, so:

$$3 = A(x+4)^2 + B(x-4)(x+4) + C(x-4)$$
$$3 = Ax^2 + 8Ax + 16A + Bx^2 - 16B + Cx - 4C$$
$$3 = (A+B)x^2 + (8A+C)x + (16A - 16B - 4C)$$

Since A, B, and C are constants we have:

$$A + B \qquad = 0$$
$$8A \qquad + C = 0$$
$$16A - 16B - 4C = 3$$

Multiply the first equation by -8 and add it to the second equation:

$$-8A - 8B \qquad = 0$$
$$8A \qquad + C = 0$$

Adding, we obtain:

$$-8B + C = 0$$

Multiply the first equation by -16 and add it to the third equation:

$$-16A - 16B \qquad = 0$$
$$16A - 16B - 4C = 3$$

Adding, we obtain:

$$-32B - 4C = 3$$

So we have the system:

$$A + B \qquad = 0$$
$$-8B + C = 0$$
$$-32B - 4C = 3$$

Multiply the second equation by -4 and add it to the third equation:

$$32B - 4C = 0$$
$$-32B - 4C = 3$$

Adding, we obtain:
$$-8C = 3$$
$$C = -\tfrac{3}{8}$$
Substitute into the second equation:
$$-8B + C = 0$$
$$-8B - \tfrac{3}{8} = 0$$
$$-8B = \tfrac{3}{8}$$
$$B = -\tfrac{3}{64}$$
Substitute into the first equation:
$$A + B = 0$$
$$A - \tfrac{3}{64} = 0$$
$$A = \tfrac{3}{64}$$
So $A = \tfrac{3}{64}$, $B = -\tfrac{3}{64}$ and $C = -\tfrac{3}{8}$.

37. Multiply by $(x + 1)(x^2 - x + 1)$, so:
$$1 = A\left(x^2 - x + 1\right) + (Bx + C)(x + 1)$$
$$1 = Ax^2 - Ax + A + Bx^2 + Cx + Bx + C$$
$$1 = (A + B)x^2 + (-A + B + C)x + (A + C)$$
Since A, B, and C are constants we have:
$$A + B \qquad = 0$$
$$-A + B + C = 0$$
$$A \qquad + C = 1$$
Add the first two equations:
$$A + B \qquad = 0$$
$$-A + B + C = 0$$
Adding, we obtain:
$$2B + C = 0$$
Multiply the first equation by -1 and add it to the third equation:
$$-A - B \qquad = 0$$
$$A \qquad + C = 1$$
Adding, we obtain:
$$-B + C = 1$$
So we have the system:
$$A + B \qquad = 0$$
$$-B + C = 1$$
$$2B + C = 0$$
Multiply the second equation by 2 and add it to the third equation:
$$-2B + 2C = 2$$
$$2B + \ C = 0$$
Adding, we obtain:
$$3C = 2$$
$$C = \tfrac{2}{3}$$

Substitute into the second equation:
$$-B + C = 1$$
$$-B + \tfrac{2}{3} = 1$$
$$-B = \tfrac{1}{3}$$
$$B = -\tfrac{1}{3}$$
Substitute into the first equation:
$$A + B = 0$$
$$A - \tfrac{1}{3} = 0$$
$$A = \tfrac{1}{3}$$
So $A = \tfrac{1}{3}$, $B = -\tfrac{1}{3}$ and $C = \tfrac{2}{3}$.

39. Multiply by $x(1-x)$, so:
$$4 = A(1-x) + B(x)$$
$$4 = (-A + B)x + A$$
Since A and B are constants we have:
$$-A + B = 0$$
$$A \quad = 4$$
Substituting yields $-4 + B = 0$, so $B = 4$. So $A = 4$ and $B = 4$.

41. Multiply by $(x^2 + x + 1)(x^2 - x + 1)$, so:
$$1 = (Ax + B)(x^2 - x + 1) + (Cx + D)(x^2 + x + 1)$$
$$1 = (A+C)x^3 + (-A+B+C+D)x^2 + (A-B+C+D)x + (B+D)$$
Since A, B, C, and D are constants we have:
$$A \qquad + C = 0$$
$$-A + B + C + D = 0$$
$$A - B + C + D = 0$$
$$B \quad + D = 1$$
Add the first and second equations:
$$A \quad + C \quad = 0$$
$$-A + B + C + D = 0$$
Adding, we obtain:
$$B + 2C + D = 0$$
Multiply the first equation by -1 and add it to the third equation:
$$-A \quad - C \quad = 0$$
$$A - B + C + D = 0$$
Adding, we obtain:
$$-B + D = 0$$
So we have the system:
$$A \quad + C \quad = 0$$
$$-B \quad + D = 0$$
$$B + 2C + D = 0$$
$$B \quad + D = 1$$

Add the second and third equations:
$$-B \qquad + D = 0$$
$$B + 2C + D = 0$$
Adding, we obtain:
$$2C + 2D = 0$$
$$C + D = 0$$
Add the second and the fourth equations:
$$-B + D = 0$$
$$B + D = 1$$
$$2D = 1$$
$$D = \tfrac{1}{2}$$
So we have the system:
$$A \quad + C \qquad = 0$$
$$-B \quad + D = 0$$
$$C + D = 0$$
$$D = \tfrac{1}{2}$$
Substituting into the third equation yields $C = -\tfrac{1}{2}$. Substitution into the second equation yields $B = \tfrac{1}{2}$. Substituting into the first equation yields $A = \tfrac{1}{2}$. So $A = \tfrac{1}{2}$, $B = \tfrac{1}{2}$, $C = -\tfrac{1}{2}$, and $D = \tfrac{1}{2}$.

43. We know its form is $(x - h)^2 + (y - k)^2 = r^2$, so:
$$(-1 - h)^2 + (-5 - k)^2 = r^2$$
$$(1 - h)^2 + (2 - k)^2 = r^2$$
$$(-2 - h)^2 + k^2 = r^2$$
Therefore:
$$1 + 2h + h^2 + 25 + 10k + k^2 = r^2$$
$$1 - 2h + h^2 + 4 - 4k + k^2 = r^2$$
$$4 + 4h + h^2 + k^2 = r^2$$
Setting the first two equations equal yields $4h + 14k = -21$ and setting the last two equations equal yields $6h + 4k = 1$. So we solve the system:
$$4h + 14k = -21$$
$$6h + 4k = 1$$
Multiplying the first equation by -3 and the second equation by 2, then adding, yields $k = -\tfrac{65}{34}$, and substituting into the second equation yields $h = \tfrac{49}{34}$. We find r by the original third equation:
$$r^2 = 4 + 4h + h^2 + k^2 = 4 + \tfrac{98}{17} + \left(\tfrac{49}{34}\right)^2 + \left(-\tfrac{65}{34}\right)^2 = \tfrac{17914}{34^2} = \tfrac{8957}{578}$$
So we have $\left(x - \tfrac{49}{34}\right)^2 + \left(y + \tfrac{65}{34}\right)^2 = \tfrac{8957}{578}$, which simplifies to:
$$17x^2 + 17y^2 - 49x + 65y - 166 = 0$$

45. Using the hint, we let $A = e^x$, $B = e^y$, and $C = e^z$. Then we have the system:

$$A + B - 2C = 2a$$
$$A + 2B - 4C = 3a$$
$$\tfrac{1}{2}A - 3B + C = -5a$$

Multiply the first equation by -1 and add it to the second equation:

$$-A - B + 2C = -2a$$
$$A + 2B - 4C = 3a$$

Adding, we obtain:

$$B - 2C = a$$

Multiply the first equation by -1 and the third equation by 2, then add the resulting equations:

$$-A - B + 2C = -2a$$
$$A - 6B + 2C = -10a$$

Adding, we obtain:

$$-7B + 4C = -12a$$

So we have the system:

$$A + B - 2C = 2a$$
$$B - 2C = a$$
$$-7B + 4C = -12a$$

Multiply the second equation by 7 and add it to the third equation:

$$7B - 14C = 7a$$
$$-7B + 4C = -12a$$

Adding, we obtain:

$$-10C = -5a$$
$$C = \tfrac{a}{2}$$

Substituting into the second equation:

$$B - 2C = a$$
$$B - 2\left(\tfrac{a}{2}\right) = a$$
$$B - a = a$$
$$B = 2a$$

Substituting into the first equation:

$$A + B - 32C = 2a$$
$$A + 2a - a = 2a$$
$$A + a = 2a$$
$$A = a$$

So $A = a$, $B = 2a$, and $C = \tfrac{a}{2}$. Since $A = e^x$, $B = e^y$, $C = e^z$, then $x = \ln a$, $y = \ln 2a$ and $z = \ln \tfrac{a}{2}$.

47. From the figure in the text, we obtain the system of three equations:

$$r_1 + r_2 = a$$
$$r_2 + r_3 = b$$
$$r_1 + r_3 = c$$

This system can be solved by repeated substitution, or by using the following technique. Add the three equations and then divide by 2 to obtain $r_1 + r_2 + r_3 = \frac{a+b+c}{2}$. Now replace $r_2 + r_3$ in this last equation by b. The result is $r_1 = \frac{a+b+c}{2} - b$. Thus $r_1 = \frac{a-b+c}{2}$. Similarly, we find that $r_2 = \frac{a+b-c}{2}$ and $r_3 = \frac{b+c-a}{2}$.

49. (a) Multiply by $(x-a)(x-b)$, so:
$$1 = A(x-b) + B(x-a)$$
$$1 = (A+B)x + (-bA - aB)$$
Since A and B are constants we have:
$$A + B = 0$$
$$-bA - aB = 1$$
Multiplying the first equation by b and adding it to the second equation yields $B = \frac{1}{b-a}$. We re-solve for A. Multiplying the first equation by a and adding it to the second equation yields $A = \frac{1}{a-b}$. So $A = \frac{1}{a-b}$ and $B = \frac{1}{b-a}$.

(b) Multiply by $(x-a)(x-b)$, so:
$$px + q = A(x-b) + B(x-a)$$
$$px + q = (A+B)x + (-bA - aB)$$
Since A and B are constants we have:
$$A + B = p$$
$$-bA - aB = q$$
Multiplying the first equation by b and adding it to the second equation yields $B = \frac{bp+q}{b-a}$. We re-solve for A. Multiplying the first equation by a and adding it to the second equation yields $A = \frac{ap+q}{a-b}$. So $A = \frac{ap+q}{a-b}$ and $B = \frac{bp+q}{b-a}$.

51. We multiply by $(x-a)(x-b)(x-c)$ to get:
$$x^2 + px + q = A(x-b)(x-c) + B(x-a)(x-c) + C(x-a)(x-b)$$
Since this equation is an identity for all x, then it must be true for $x = a$, $x = b$ and $x = c$.

$x = a$: $a^2 + pa + q = A(a-b)(a-c) + 0 + 0$
$$A = \frac{a^2 + pa + q}{(a-b)(a-c)}$$

$x = b$: $b^2 + pb + q = 0 + B(b-a)(b-c) + 0$
$$B = \frac{b^2 + pb + q}{(b-a)(b-c)}$$

$x = c$: $c^2 + pc + q = 0 + 0 + C(c-a)(c-b)$
$$C = \frac{c^2 + pc + q}{(c-a)(c-b)}$$

So $A = \dfrac{a^2 + pa + q}{(a-b)(a-c)}$, $B = \dfrac{b^2 + pb + q}{(b-a)(b-c)}$ and $C = \dfrac{c^2 + pc + q}{(c-a)(c-b)}$.

53. Let α, β and γ be the distances opposite towns A, B and C. The walking, riding and driving rates are, respectively, $\frac{1}{a}$ mi / min, $\frac{1}{b}$ mi / min and $\frac{1}{c}$ mi / min. From the statement of the problem we can obtain the following system of three equations:

$$\alpha a + \beta b + \gamma c = 60(a + c - b)$$
$$\beta a + \gamma b + \alpha c = 60(b + a - c)$$
$$\gamma a + \alpha b + \beta c = 60(c + b - a)$$

By adding all three equations, we obtain:

$$\alpha(a + b + c) + \beta(a + b + c) + \gamma(a + b + c) = 60(a + b + c)$$

Now divide through by the quantity $a + b + c$. This yields $\alpha + \beta + \gamma = 60$ miles.

9.3 Matrices

1. (a) two by three (2×3)

(b) three by two (3×2)

3. five by four (5×4)

5. The coefficient matrix is $\begin{pmatrix} 2 & 3 & 4 \\ 5 & 6 & 7 \\ 8 & 9 & 10 \end{pmatrix}$ and the augmented matrix is $\begin{pmatrix} 2 & 3 & 4 & 10 \\ 5 & 6 & 7 & 9 \\ 8 & 9 & 10 & 8 \end{pmatrix}$.

7. The coefficient matrix is $\begin{pmatrix} 1 & 0 & 1 & 1 \\ 1 & 1 & 0 & 2 \\ 0 & 1 & 1 & 1 \\ 2 & -1 & -1 & 0 \end{pmatrix}$ and the augmented matrix is

$\begin{pmatrix} 1 & 0 & 1 & 1 & -1 \\ 1 & 1 & 0 & 2 & 0 \\ 0 & 1 & 1 & 1 & 1 \\ 2 & -1 & -1 & 0 & 2 \end{pmatrix}$.

9. Form the augmented matrix:

$$\begin{pmatrix} 1 & -1 & 2 & 7 \\ 3 & 2 & -1 & -10 \\ -1 & 3 & 1 & -2 \end{pmatrix}$$

Adding -3 times row 1 to row 2 and adding row 1 to row 3 yields:

$$\begin{pmatrix} 1 & -1 & 2 & 7 \\ 0 & 5 & -7 & -31 \\ 0 & 2 & 3 & 5 \end{pmatrix}$$

Multiplying row 3 by -3 and adding to row 2 yields:

$$\begin{pmatrix} 1 & -1 & 2 & 7 \\ 0 & -1 & -16 & -46 \\ 0 & 2 & 3 & 5 \end{pmatrix}$$

Multiplying row 2 by 2 and adding to row 3 yields:

$$\begin{pmatrix} 1 & -1 & 2 & 7 \\ 0 & -1 & -16 & -46 \\ 0 & 0 & -29 & -87 \end{pmatrix}$$

So we have the system of equations:

$$x - y + 2z = 7$$
$$-y - 16z = -46$$
$$-29z = -87$$

Solve equation three for z:

$$-29z = -87$$
$$z = 3$$

Substitute into equation two:

$$-y - 48 = -46$$
$$-y = 2$$
$$y = -2$$

Substitute into equation one:

$$x + 2 + 6 = 7$$
$$x = -1$$

So the solution is $(-1, -2, 3)$.

11. Form the augmented matrix:

$$\begin{pmatrix} 1 & 0 & 1 & -2 \\ -3 & 2 & 0 & 17 \\ 1 & -1 & -1 & -9 \end{pmatrix}$$

Adding 3 times row 1 to row 2, and -1 times row 1 to row 3 yields:

$$\begin{pmatrix} 1 & 0 & 1 & -2 \\ 0 & 2 & 3 & 11 \\ 0 & -1 & -2 & -7 \end{pmatrix}$$

Switching row 2 and row 3:

$$\begin{pmatrix} 1 & 0 & 1 & -2 \\ 0 & -1 & -2 & -7 \\ 0 & 2 & 3 & 11 \end{pmatrix}$$

Multiplying row 2 by 2 and adding it to row 3 yields:

$$\begin{pmatrix} 1 & 0 & 1 & -2 \\ 0 & -1 & -2 & -7 \\ 0 & 0 & -1 & -3 \end{pmatrix}$$

So we have the system:
$$x \quad + \quad z = -2$$
$$-y - 2z = -7$$
$$- z = -3$$
Solve equation three for z:
$$-z = -3$$
$$z = 3$$
Substitute into equation two:
$$-y - 6 = -7$$
$$-y = -1$$
$$y = 1$$
Substitute into equation one:
$$x + 3 = -2$$
$$x = -5$$
So the solution is $(-5, 1, 3)$.

13. Form the augmented matrix:
$$\begin{pmatrix} 1 & 1 & 1 & -4 \\ 2 & -3 & 1 & -1 \\ 4 & 2 & -3 & 33 \end{pmatrix}$$
Adding -2 times row 1 to row 2, and -4 times row 1 to row 3 yields:
$$\begin{pmatrix} 1 & 1 & 1 & -4 \\ 0 & -5 & -1 & 7 \\ 0 & -2 & -7 & 49 \end{pmatrix}$$
Adding -3 times row 3 to row 2 (to get a 1 entry) yields:
$$\begin{pmatrix} 1 & 1 & 1 & -4 \\ 0 & 1 & 20 & -140 \\ 0 & -2 & -7 & 49 \end{pmatrix}$$
Adding 2 times row 2 to row 3 yields:
$$\begin{pmatrix} 1 & 1 & 1 & -4 \\ 0 & 1 & 20 & -140 \\ 0 & 0 & 33 & -231 \end{pmatrix}$$
So we have the system of equations:
$$x + y + \quad z = -4$$
$$y + 20z = -140$$
$$33z = -231$$
Solving the last equation for z:
$$33z = -231$$
$$z = -7$$

Substituting into the second equation:
$$y + 20z = -140$$
$$y - 140 = -140$$
$$y = 0$$
Substituting into the first equation:
$$x + y + z = -4$$
$$x + 0 - 7 = -4$$
$$x = 3$$
So the soltution is $(3, 0, -7)$.

15. Form the augmented matrix:
$$\begin{pmatrix} 3 & -2 & 6 & 0 \\ 1 & 3 & 20 & 15 \\ 10 & -11 & -10 & -9 \end{pmatrix}$$
Switching row 1 and row 2 yields:
$$\begin{pmatrix} 1 & 3 & 20 & 15 \\ 3 & -2 & 6 & 0 \\ 10 & -11 & -10 & -9 \end{pmatrix}$$
Adding -3 times row 1 to row 2, and -10 times row 1 to row 3 yields:
$$\begin{pmatrix} 1 & 3 & 20 & 15 \\ 0 & -11 & -54 & -45 \\ 0 & -41 & -210 & -159 \end{pmatrix}$$
Multiplying row 2 by -4 and adding it to row 3 yields:
$$\begin{pmatrix} 1 & 3 & 20 & 15 \\ 0 & -11 & -54 & -45 \\ 0 & 3 & 6 & 21 \end{pmatrix}$$
Dividing row 3 by 3 and then switching it with row 2 yields:
$$\begin{pmatrix} 1 & 3 & 20 & 15 \\ 0 & 1 & 2 & 7 \\ 0 & -11 & -54 & -45 \end{pmatrix}$$
Multiplying row 2 by 11 and adding it to row 3 yields:
$$\begin{pmatrix} 1 & 3 & 20 & 15 \\ 0 & 1 & 2 & 7 \\ 0 & 0 & -32 & 32 \end{pmatrix}$$
So we have the system of equations:
$$x + 3y + 20z = 15$$
$$y + 2z = 7$$
$$-32z = 32$$

Solve equation three for z:
$$-32z = 32$$
$$z = -1$$
Substitute into equation two:
$$y - 2 = 7$$
$$y = 9$$
Substitute into equation one:
$$x + 27 - 20 = 15$$
$$x + 7 = 15$$
$$x = 8$$
So the solution is $(8, 9, -1)$.

17. Form the augmented matrix:
$$\begin{pmatrix} 4 & -3 & 3 & 2 \\ 5 & 1 & -4 & 1 \\ 9 & -2 & -1 & 3 \end{pmatrix}$$
Subtracting row 1 from row 2 yields:
$$\begin{pmatrix} 4 & -3 & 3 & 2 \\ 1 & 4 & -7 & -1 \\ 9 & -2 & -1 & 3 \end{pmatrix}$$
Switching row 1 and row 2 yields:
$$\begin{pmatrix} 1 & 4 & -7 & -1 \\ 4 & -3 & 3 & 2 \\ 9 & -2 & -1 & 3 \end{pmatrix}$$
Adding -4 times row 1 to row 2, and -9 times row 1 to row 3 yields:
$$\begin{pmatrix} 1 & 4 & -7 & -1 \\ 0 & -19 & 31 & 6 \\ 0 & -38 & 62 & 12 \end{pmatrix}$$
Multiplying row 2 by -2 and adding it to row 3 yields:
$$\begin{pmatrix} 1 & 4 & -7 & -1 \\ 0 & -19 & 31 & 6 \\ 0 & 0 & 0 & 0 \end{pmatrix}$$
So we have the system of equations:
$$x + 4y - 7z = -1$$
$$-19y + 31z = 6$$
We solve equation two for y:
$$-19y = 6 - 31z$$
$$y = \frac{31z - 6}{19}$$

Substitute into equation one:

$$x + 4\left(\frac{31z - 6}{19}\right) - 7z = -1$$

$$x + \frac{-9z - 24}{19} = -1$$

$$x = \frac{9z + 5}{19}$$

So the solution is $\left(\frac{9z+5}{19}, \frac{31z-6}{19}, z\right)$, for any real number z.

19. Form the augmented matrix:

$$\begin{pmatrix} 1 & -1 & 1 & 1 & 6 \\ 1 & 1 & -1 & 1 & 4 \\ 1 & 1 & 1 & -1 & -2 \\ -1 & 1 & 1 & 1 & 0 \end{pmatrix}$$

Adding -1 times row 1 to both row 2 and row 3, and adding row 1 to row 4 yields:

$$\begin{pmatrix} 1 & -1 & 1 & 1 & 6 \\ 0 & 2 & -2 & 0 & -2 \\ 0 & 2 & 0 & -2 & -8 \\ 0 & 0 & 2 & 2 & 6 \end{pmatrix}$$

Dividing rows 2, 3, and 4 by 2 and subtracting row 2 from row 3 yields:

$$\begin{pmatrix} 1 & -1 & 1 & 1 & 6 \\ 0 & 1 & -1 & 0 & -1 \\ 0 & 0 & 1 & -1 & -3 \\ 0 & 0 & 1 & 1 & 3 \end{pmatrix}$$

Multiplying row 3 by -1 and adding it to row 4 yields:

$$\begin{pmatrix} 1 & -1 & 1 & 1 & 6 \\ 0 & 1 & -1 & 0 & -1 \\ 0 & 0 & 1 & -1 & -3 \\ 0 & 0 & 0 & 2 & 6 \end{pmatrix}$$

So we have the system of equations:

$$x - y + z + w = 6$$
$$y - z \qquad = -1$$
$$z - w = -3$$
$$2w = 6$$

Solve equation four for w:

$$2w = 6$$
$$w = 3$$

Substitute into equation three:

$$z - 3 = -3$$
$$z = 0$$

Substitute into equation two:
$$y - 0 = -1$$
$$y = -1$$
Substitute into equation one:
$$x + 1 + 0 + 3 = 6$$
$$x + 4 = 6$$
$$x = 2$$
So the solution is $(2, -1, 0, 3)$.

21. Form the augmented matrix:
$$\begin{pmatrix} 15 & 14 & 26 & 1 \\ 18 & 17 & 32 & -1 \\ 21 & 20 & 38 & 0 \end{pmatrix}$$
We reduce coefficients by subtracting row 2 from row 3, and row 1 from row 2:
$$\begin{pmatrix} 15 & 14 & 26 & 1 \\ 3 & 3 & 6 & -2 \\ 3 & 3 & 6 & 1 \end{pmatrix}$$
Subtracting row 2 from row 3 yields:
$$\begin{pmatrix} 15 & 14 & 26 & 1 \\ 3 & 3 & 6 & -2 \\ 0 & 0 & 0 & 3 \end{pmatrix}$$
But this last row corresponds to the equation $0 = 3$, which is false. So the system has no solution.

23. Adding the corresponding terms, $A + B$ will be:
$$\begin{pmatrix} 2 & 3 \\ -1 & 4 \end{pmatrix} + \begin{pmatrix} 1 & -1 \\ 3 & 0 \end{pmatrix} = \begin{pmatrix} 3 & 2 \\ 2 & 4 \end{pmatrix}$$

25. We have:
$$2A = \begin{pmatrix} 4 & 6 \\ -2 & 8 \end{pmatrix} \text{ and } 2B = \begin{pmatrix} 2 & -2 \\ 6 & 0 \end{pmatrix}$$
Therefore:
$$2A + 2B = \begin{pmatrix} 6 & 4 \\ 4 & 8 \end{pmatrix}$$

27. The multiplication is defined, since # cols in A = # rows in B:
row 1, col 1: $2(1) + 3(3) = 2 + 9 = 11$
row 1, col 2: $2(-1) + 3(0) = -2 + 0 = -2$
row 2, col 1: $-1(1) + 4(3) = -1 + 12 = 11$
row 2, col 2: $-1(-1) + 4(0) = 1 + 0 = 1$
So $AB = \begin{pmatrix} 11 & -2 \\ 11 & 1 \end{pmatrix}$.

29. The multiplication is defined, since # cols in A = # rows in C:

        row 1, col 1: $2(1) + 3(0) = 2 + 0 = 2$

        row 1, col 2: $2(0) + 3(1) = 0 + 3 = 3$

        row 2, col 1: $-1(1) + 4(0) = -1 + 0 = -1$

        row 2, col 2: $-1(0) + 4(1) = 0 + 4 = 4$

$$\text{So } AC = \begin{pmatrix} 2 & 3 \\ -1 & 4 \end{pmatrix}$$

31. This operation is not defined, since D and E do not have the same size.

33. We have:

$$2F - 3G = \begin{pmatrix} 10 & -2 \\ -8 & 0 \\ 4 & 6 \end{pmatrix} - \begin{pmatrix} 0 & 0 \\ 0 & 0 \\ 0 & 0 \end{pmatrix} = \begin{pmatrix} 10 & -2 \\ -8 & 0 \\ 4 & 6 \end{pmatrix}$$

35. The multiplication is defined, since # cols of E = # rows of D.

        row 1, col 1: $2(-1) + 1(4) = -2 + 4 = 2$

        row 1, col 2: $2(2) + 1(0) = 4 + 0 = 4$

        row 1, col 3: $2(3) + 1(5) = 6 + 5 = 11$

        row 2, col 1: $8(-1) - 1(4) = -8 - 4 = -12$

        row 2, col 2: $8(2) - 1(0) = 16 - 0 = 16$

        row 2, col 3: $8(3) - 1(5) = 24 - 5 = 19$

        row 3, col 1: $6(-1) + 5(4) = -6 + 20 = 14$

        row 3, col 2: $6(2) + 5(0) = 12 + 0 = 12$

        row 3, col 3: $6(3) + 5(5) = 18 + 25 = 43$

$$\text{So } ED = \begin{pmatrix} 2 & 4 & 11 \\ -12 & 16 & 19 \\ 14 & 12 & 43 \end{pmatrix}.$$

37. The multiplication is defined, since # cols of F = # rows of D:

        row 1, col 1: $5(-1) - 1(4) = -5 - 4 = -9$

        row 1, col 2: $5(2) - 1(0) = 10 - 0 = 10$

        row 1, col 3: $5(3) - 1(5) = 15 - 5 = 10$

        row 2, col 1: $-4(-1) + 0(4) = 4 + 0 = 4$

        row 2, col 2: $-4(2) + 0(0) = -8 + 0 = -8$

        row 2, col 3: $-4(3) + 0(5) = -12 + 0 = -12$

        row 3, col 1: $2(-1) + 3(4) = -2 + 12 = 10$

        row 3, col 2: $2(2) + 3(0) = 4 + 0 = 4$

        row 3, col 3: $2(3) + 3(5) = 6 + 15 = 21$

$$\text{So } FD = \begin{pmatrix} -9 & 10 & 10 \\ 4 & -8 & -12 \\ 10 & 4 & 21 \end{pmatrix}$$

39. The operation is not defined, since G and A are not the same size.

41. The multiplication is defined, since # cols of G = # rows of D:
$$GD = \begin{pmatrix} 0 & 0 & 0 \\ 0 & 0 & 0 \\ 0 & 0 & 0 \end{pmatrix}$$

43. We have:
$$A + (B + C) = \begin{pmatrix} 2 & 3 \\ -1 & 4 \end{pmatrix} + \begin{pmatrix} 2 & -1 \\ 3 & 1 \end{pmatrix} = \begin{pmatrix} 4 & 2 \\ 2 & 5 \end{pmatrix}$$

45. The multiplication is not defined, since # cols of $D \neq$ # rows of C.

47. The multiplication is defined, since # cols of A = # rows of A.
row 1, col 1: $2(2) + 3(-1) = 4 - 3 = 1$
row 1, col 2: $2(3) + 3(4) = 6 + 12 = 18$
row 2, col 1: $-1(2) + 4(-1) = -2 - 4 = -6$
row 2, col 2: $-1(3) + 4(4) = -3 + 16 = 13$
So $A^2 = \begin{pmatrix} 1 & 18 \\ -6 & 13 \end{pmatrix}$.

49. We have:
$$AA^2 = \begin{pmatrix} 2 & 3 \\ -1 & 4 \end{pmatrix} \begin{pmatrix} 1 & 18 \\ -6 & 13 \end{pmatrix} = \begin{pmatrix} -16 & 75 \\ -25 & 34 \end{pmatrix}$$

51. (a) We have:
$$A(B + C) = \begin{pmatrix} -1 & 3 & 4 \\ 3 & 2 & -3 \\ 9 & 1 & 6 \end{pmatrix} \begin{pmatrix} 11 & 6 & 2 \\ 2 & 1 & 6 \\ -2 & 1 & 6 \end{pmatrix} = \begin{pmatrix} -13 & 1 & 40 \\ 43 & 17 & 0 \\ 89 & 61 & 60 \end{pmatrix}$$

(b) We have:
$$AB + AC = \begin{pmatrix} -1 & 3 & 4 \\ 3 & 2 & -3 \\ 9 & 1 & 6 \end{pmatrix} \begin{pmatrix} 7 & 0 & 1 \\ 0 & 0 & 3 \\ -1 & 2 & 4 \end{pmatrix} + \begin{pmatrix} -1 & 3 & 4 \\ 3 & 2 & -3 \\ 9 & 1 & 6 \end{pmatrix} \begin{pmatrix} 4 & 6 & 1 \\ 2 & 1 & 3 \\ -1 & -1 & 2 \end{pmatrix}$$
$$= \begin{pmatrix} -11 & 8 & 24 \\ 24 & -6 & -3 \\ 57 & 12 & 36 \end{pmatrix} + \begin{pmatrix} -2 & -7 & 16 \\ 19 & 23 & 3 \\ 32 & 49 & 24 \end{pmatrix}$$
$$= \begin{pmatrix} -13 & 1 & 40 \\ 43 & 17 & 0 \\ 89 & 61 & 60 \end{pmatrix}$$

(c) We have:

$$(AB)C = \left[\begin{pmatrix} -1 & 3 & 4 \\ 3 & 2 & -3 \\ 9 & 1 & 6 \end{pmatrix} \begin{pmatrix} 7 & 0 & 1 \\ 0 & 0 & 3 \\ -1 & 2 & 4 \end{pmatrix} \right] \begin{pmatrix} 4 & 6 & 1 \\ 2 & 1 & 3 \\ -1 & -1 & 2 \end{pmatrix}$$

$$= \begin{pmatrix} -11 & 8 & 24 \\ 24 & -6 & -3 \\ 57 & 12 & 36 \end{pmatrix} \begin{pmatrix} 4 & 6 & 1 \\ 2 & 1 & 3 \\ -1 & -1 & 2 \end{pmatrix}$$

$$= \begin{pmatrix} -52 & -82 & 61 \\ 87 & 141 & 0 \\ 216 & 318 & 165 \end{pmatrix}$$

(d) We have:

$$A(BC) = \begin{pmatrix} -1 & 3 & 4 \\ 3 & 2 & -3 \\ 9 & 1 & 6 \end{pmatrix} \begin{pmatrix} 27 & 41 & 9 \\ -3 & -3 & 6 \\ -4 & -8 & 13 \end{pmatrix} = \begin{pmatrix} -52 & -82 & 61 \\ 87 & 141 & 0 \\ 216 & 318 & 165 \end{pmatrix}$$

53. (a) We compute A^2 and B^2:

$$A^2 = \begin{pmatrix} 3 & 5 \\ 7 & 9 \end{pmatrix} \begin{pmatrix} 3 & 5 \\ 7 & 9 \end{pmatrix} = \begin{pmatrix} 44 & 60 \\ 84 & 116 \end{pmatrix}$$

$$B^2 = \begin{pmatrix} 2 & 4 \\ 6 & 8 \end{pmatrix} \begin{pmatrix} 2 & 4 \\ 6 & 8 \end{pmatrix} = \begin{pmatrix} 28 & 40 \\ 60 & 88 \end{pmatrix}$$

So $A^2 - B^2 = \begin{pmatrix} 16 & 20 \\ 24 & 28 \end{pmatrix}$.

(b) We have $A - B = \begin{pmatrix} 1 & 1 \\ 1 & 1 \end{pmatrix}$ and $A + B = \begin{pmatrix} 5 & 9 \\ 13 & 17 \end{pmatrix}$, so:

$$(A - B)(A + B) = \begin{pmatrix} 1 & 1 \\ 1 & 1 \end{pmatrix} \begin{pmatrix} 5 & 9 \\ 13 & 17 \end{pmatrix} = \begin{pmatrix} 18 & 26 \\ 18 & 26 \end{pmatrix}$$

(c) We compute the product:

$$(A + B)(A - B) = \begin{pmatrix} 5 & 9 \\ 13 & 17 \end{pmatrix} \begin{pmatrix} 1 & 1 \\ 1 & 1 \end{pmatrix} = \begin{pmatrix} 14 & 14 \\ 30 & 30 \end{pmatrix}$$

(d) We compute AB and BA:

$$AB = \begin{pmatrix} 3 & 5 \\ 7 & 9 \end{pmatrix} \begin{pmatrix} 2 & 4 \\ 6 & 8 \end{pmatrix} = \begin{pmatrix} 36 & 52 \\ 68 & 100 \end{pmatrix}$$

$$BA = \begin{pmatrix} 2 & 4 \\ 6 & 8 \end{pmatrix} \begin{pmatrix} 3 & 5 \\ 7 & 9 \end{pmatrix} = \begin{pmatrix} 34 & 46 \\ 74 & 102 \end{pmatrix}$$

Therefore:

$A^2 + AB - BA - B^2 =$

$$\begin{pmatrix} 44 & 60 \\ 84 & 116 \end{pmatrix} + \begin{pmatrix} 36 & 52 \\ 68 & 100 \end{pmatrix} - \begin{pmatrix} 34 & 46 \\ 74 & 102 \end{pmatrix} - \begin{pmatrix} 28 & 40 \\ 60 & 88 \end{pmatrix} = \begin{pmatrix} 18 & 26 \\ 18 & 26 \end{pmatrix}$$

55. (a) The product AZ is given by:

$$AZ = \begin{pmatrix} 1 & 0 \\ 0 & -1 \end{pmatrix}\begin{pmatrix} x \\ y \end{pmatrix} = \begin{pmatrix} x \\ -y \end{pmatrix}$$

(b) The product BZ is given by:

$$BZ = \begin{pmatrix} -1 & 0 \\ 0 & 1 \end{pmatrix}\begin{pmatrix} x \\ y \end{pmatrix} = \begin{pmatrix} -x \\ y \end{pmatrix}$$

(c) The product AB is given by:

$$AB = \begin{pmatrix} 1 & 0 \\ 0 & -1 \end{pmatrix}\begin{pmatrix} -1 & 0 \\ 0 & 1 \end{pmatrix} = \begin{pmatrix} -1 & 0 \\ 0 & -1 \end{pmatrix}$$

So $(AB)Z = \begin{pmatrix} -1 & 0 \\ 0 & -1 \end{pmatrix}\begin{pmatrix} x \\ y \end{pmatrix} = \begin{pmatrix} -x \\ -y \end{pmatrix}$. This would represent a reflection about the origin.

57. (a) We first compute the product:

$$AB = \begin{pmatrix} 1 & 2 \\ 3 & 4 \end{pmatrix}\begin{pmatrix} 3 & -1 \\ 5 & 8 \end{pmatrix} = \begin{pmatrix} 13 & 15 \\ 29 & 29 \end{pmatrix}$$

Now compute the function values:

$f(A) = 1(4) - 2(3) = 4 - 6 = -2$
$f(B) = 3(8) - (-1)(5) = 24 + 5 = 29$
$f(AB) = 13(29) - 15(29) = -2(29) = -58$

So $f(A) \cdot f(B) = f(AB)$.

(b) We first compute the product:

$$AB = \begin{pmatrix} a & b \\ c & d \end{pmatrix}\begin{pmatrix} e & f \\ g & h \end{pmatrix} = \begin{pmatrix} ae + bg & af + bh \\ ce + dg & cf + dh \end{pmatrix}$$

Now compute the function values:

$f(A) = ad - bc$
$f(B) = eh - fg$
$f(AB) = (ae + bg)(cf + dh) - (af + bh)(ce + dg)$
$\quad\quad = (acef + bcfg + adeh + bdgh) - (acef + bceh + adfg + bdgh)$
$\quad\quad = bcfg + adeh - bceh - adfg$

So $f(A) \cdot f(B) = f(AB)$.

59. (a) Let $A = \begin{pmatrix} a & b \\ c & d \end{pmatrix}$ and $B = \begin{pmatrix} e & f \\ g & h \end{pmatrix}$. Then $A + B = \begin{pmatrix} a+e & b+f \\ c+g & d+h \end{pmatrix}$, so

$(A+B)^T = \begin{pmatrix} a+e & c+g \\ b+f & d+h \end{pmatrix}$. Now $A^T = \begin{pmatrix} a & c \\ b & d \end{pmatrix}$ and $B^T = \begin{pmatrix} e & g \\ f & h \end{pmatrix}$, so we have

$A^T + B^T = \begin{pmatrix} a+e & c+g \\ b+f & d+h \end{pmatrix}$. Thus $(A+B)^T = A^T + B^T$.

(b) Since $A^T = \begin{pmatrix} a & c \\ b & d \end{pmatrix}$, then $\left(A^T\right)^T = \begin{pmatrix} a & b \\ c & d \end{pmatrix}$. Thus $\left(A^T\right)^T = A$.

(c) Since $AB = \begin{pmatrix} ae+bg & af+bh \\ ce+dg & cf+dh \end{pmatrix}$, then $(AB)^T = \begin{pmatrix} ae+bg & ce+dg \\ af+bh & cf+dh \end{pmatrix}$. Also

$B^T A^T = \begin{pmatrix} e & g \\ f & h \end{pmatrix} \begin{pmatrix} a & c \\ b & d \end{pmatrix} = \begin{pmatrix} ae+bg & ce+dg \\ af+bh & cf+dh \end{pmatrix}$. So we see that $(AB)^T = B^T A^T$.

9.4 The Inverse of a Square Matrix

1.   We compute AI_2 and I_2A:

$$AI_2 = \begin{pmatrix} 4 & -1 \\ -5 & 2 \end{pmatrix} \begin{pmatrix} 1 & 0 \\ 0 & 1 \end{pmatrix} = \begin{pmatrix} 4 & -1 \\ -5 & 2 \end{pmatrix} = A$$

$$I_2A = \begin{pmatrix} 1 & 0 \\ 0 & 1 \end{pmatrix} \begin{pmatrix} 4 & -1 \\ -5 & 2 \end{pmatrix} = \begin{pmatrix} 4 & -1 \\ -5 & 2 \end{pmatrix} = A$$

3.   We compute CI_3 and I_3C:

$$CI_3 = \begin{pmatrix} 3 & 0 & -2 \\ 0 & 5 & 6 \\ 1 & 4 & -7 \end{pmatrix} \begin{pmatrix} 1 & 0 & 0 \\ 0 & 1 & 0 \\ 0 & 0 & 1 \end{pmatrix} = \begin{pmatrix} 3 & 0 & -2 \\ 0 & 5 & 6 \\ 1 & 4 & -7 \end{pmatrix} = C$$

$$I_3C = \begin{pmatrix} 1 & 0 & 0 \\ 0 & 1 & 0 \\ 0 & 0 & 1 \end{pmatrix} \begin{pmatrix} 3 & 0 & -2 \\ 0 & 5 & 6 \\ 1 & 4 & -7 \end{pmatrix} = \begin{pmatrix} 3 & 0 & -2 \\ 0 & 5 & 6 \\ 1 & 4 & -7 \end{pmatrix} = C$$

5.   We need to find numbers a, b, c and d such that:

$$\begin{pmatrix} 7 & 9 \\ 4 & 5 \end{pmatrix} \begin{pmatrix} a & b \\ c & d \end{pmatrix} = \begin{pmatrix} 1 & 0 \\ 0 & 1 \end{pmatrix} \text{ and } \begin{pmatrix} a & b \\ c & d \end{pmatrix} \begin{pmatrix} 7 & 9 \\ 4 & 5 \end{pmatrix} = \begin{pmatrix} 1 & 0 \\ 0 & 1 \end{pmatrix}$$

We compute the left-hand product:

$$\begin{pmatrix} 7 & 9 \\ 4 & 5 \end{pmatrix} \begin{pmatrix} a & b \\ c & d \end{pmatrix} = \begin{pmatrix} 7a+9c & 7b+9d \\ 4a+5c & 4b+5d \end{pmatrix} = \begin{pmatrix} 1 & 0 \\ 0 & 1 \end{pmatrix}$$

So we have the following systems of equations:

$$7a + 9c = 1 \qquad \text{and} \qquad 7b + 9d = 0$$
$$4a + 5c = 0 \qquad\qquad\qquad 4b + 5d = 1$$

Multiplying the first equation by 5 and the second equation by -9 yields:

$$35a + 45c = 5 \quad \text{and} \quad 35b + 45d = 0$$
$$-36a - 45c = 0 \qquad\qquad -36b - 45d = -9$$

Adding, we obtain:

$$-a = 5 \qquad\qquad\qquad -b = -9$$
$$a = -5 \qquad\qquad\qquad b = 9$$

Substituting for c and d, respectively:

$$4a + 5c = 0 \qquad\qquad 4b + 5d = 1$$
$$-20 + 5c = 0 \qquad\qquad 36 + 5d = 1$$
$$5c = 20 \qquad\qquad\qquad 5d = -35$$
$$c = 4 \qquad\qquad\qquad\quad d = -7$$

So the inverse matrix is $A^{-1} = \begin{pmatrix} -5 & 9 \\ 4 & -7 \end{pmatrix}$.

7. We need to find numbers a, b, c, and d such that:

$$\begin{pmatrix} -3 & 1 \\ 5 & 6 \end{pmatrix} \begin{pmatrix} a & b \\ c & d \end{pmatrix} = \begin{pmatrix} 1 & 0 \\ 0 & 1 \end{pmatrix}$$

We compute the left-hand product:

$$\begin{pmatrix} -3a+c & -3b+d \\ 5a+6c & 5b+6d \end{pmatrix} = \begin{pmatrix} 1 & 0 \\ 0 & 1 \end{pmatrix}$$

So we have the following systems of equations:

$$-3a + c = 1 \qquad \text{and} \qquad -3b + d = 0$$
$$5a + 6c = 0 \qquad\qquad\qquad 5b + 6d = 1$$

Solving for c and d respectively, then substituting:

$$-3a + c = 1 \qquad\qquad -3b + d = 0$$
$$c = 3a + 1 \qquad\qquad\quad d = 3b$$

Substituting:

$$5a + 6(3a+1) = 0 \qquad\qquad 5b + 6(3b) = 1$$
$$23a + 6 = 0 \qquad\qquad\qquad\quad 23b = 1$$
$$23a = -6 \qquad\qquad\qquad\qquad b = \tfrac{1}{23}$$
$$a = -\tfrac{6}{23}$$

Now substituting to find c and d:

$$c = 3a + 1 \qquad\qquad\qquad d = 3b$$
$$c = -\tfrac{18}{23} + 1 \qquad\qquad\quad d = \tfrac{3}{23}$$
$$c = \tfrac{5}{23}$$

So the inverse matrix is $A^{-1} = \begin{pmatrix} -\tfrac{6}{23} & \tfrac{1}{23} \\ \tfrac{5}{23} & \tfrac{3}{23} \end{pmatrix}$.

9. We need to find numbers a, b, c, and d such that:
$$\begin{pmatrix} -2 & 3 \\ -4 & 6 \end{pmatrix}\begin{pmatrix} a & b \\ c & d \end{pmatrix} = \begin{pmatrix} 1 & 0 \\ 0 & 1 \end{pmatrix}$$
We compute the left-hand product:
$$\begin{pmatrix} -2a+3c & -2b+3d \\ -4a+6c & -4b+6d \end{pmatrix} = \begin{pmatrix} 1 & 0 \\ 0 & 1 \end{pmatrix}$$
So we have the following systems of equations:

$$-2a+3c = 1 \qquad \text{and} \qquad -2b+3d = 0$$
$$-4a+6c = 0 \qquad\qquad\qquad -4b+6d = 1$$

Multiplying the top equation by –2 yields:

$$4a-6c = -2 \qquad\qquad\qquad 4b-6d = 0$$
$$-4a+6c = 0 \qquad\qquad\qquad -4b+6d = 1$$

Adding, we obtain:

$$0 = -2 \qquad\qquad\qquad 0 = 1$$

Since both of these equations are false, no A^{-1} exists.

11. We need to find numbers a, b, c, and d such that:
$$\begin{pmatrix} \frac{1}{3} & \frac{1}{3} \\ -\frac{1}{9} & \frac{2}{9} \end{pmatrix}\begin{pmatrix} a & b \\ c & d \end{pmatrix} = \begin{pmatrix} 1 & 0 \\ 0 & 1 \end{pmatrix}$$
We compute the left-hand product:
$$\begin{pmatrix} \frac{1}{3}a+\frac{1}{3}c & \frac{1}{3}b+\frac{1}{3}d \\ -\frac{1}{9}a+\frac{2}{9}c & -\frac{1}{9}b+\frac{2}{9}d \end{pmatrix} = \begin{pmatrix} 1 & 0 \\ 0 & 1 \end{pmatrix}$$
So we have the following systems of equations:

$$\tfrac{1}{3}a+\tfrac{1}{3}c = 1 \qquad \text{and} \qquad \tfrac{1}{3}b+\tfrac{1}{3}d = 0$$
$$-\tfrac{1}{9}a+\tfrac{2}{9}c = 0 \qquad\qquad\qquad -\tfrac{1}{9}b+\tfrac{2}{9}d = 1$$

Multiplying the first equation by 3 and the second by 9 yields:

$$a+\ c = 3 \qquad\qquad\qquad b+\ d = 0$$
$$-a+2c = 0 \qquad\qquad\qquad -b+2d = 9$$

Adding, we obtain:

$$3c = 3 \qquad\qquad\qquad 3d = 9$$
$$c = 1 \qquad\qquad\qquad d = 3$$

Now substituting to find a and b:

$$a+c = 3 \qquad\qquad\qquad b+d = 0$$
$$a+1 = 3 \qquad\qquad\qquad b+3 = 0$$
$$a = 2 \qquad\qquad\qquad b = -3$$

So the inverse is $A^{-1} = \begin{pmatrix} 2 & -3 \\ 1 & 3 \end{pmatrix}$.

13. Form the matrix:
$$\begin{pmatrix} 2 & 1 & 1 & 0 \\ 3 & 2 & 0 & 1 \end{pmatrix}$$

Multiply row 1 by -1 and add to row 2:

$$\begin{pmatrix} 2 & 1 & 1 & 0 \\ 1 & 1 & -1 & 1 \end{pmatrix}$$

Switch rows 1 and 2:

$$\begin{pmatrix} 1 & 1 & -1 & 1 \\ 2 & 1 & 1 & 0 \end{pmatrix}$$

Multiply row 1 by -2 and add to row 2:

$$\begin{pmatrix} 1 & 1 & -1 & 1 \\ 0 & -1 & 3 & -2 \end{pmatrix}$$

Add row 2 to row 1:

$$\begin{pmatrix} 1 & 0 & 2 & -1 \\ 0 & -1 & 3 & -2 \end{pmatrix}$$

Multiply row 2 by -1:

$$\begin{pmatrix} 1 & 0 & 2 & -1 \\ 0 & 1 & -3 & 2 \end{pmatrix}$$

So the inverse is $\begin{pmatrix} 2 & -1 \\ -3 & 2 \end{pmatrix}$.

15. Form the matrix:

$$\begin{pmatrix} 0 & -11 & 1 & 0 \\ 1 & 6 & 0 & 1 \end{pmatrix}$$

Switch rows 1 and 2:

$$\begin{pmatrix} 1 & 6 & 0 & 1 \\ 0 & -11 & 1 & 0 \end{pmatrix}$$

Multiply row 2 by $-\frac{1}{11}$:

$$\begin{pmatrix} 1 & 6 & 0 & 1 \\ 0 & 1 & -\frac{1}{11} & 0 \end{pmatrix}$$

Multiply row 2 by -6 and add to row 1:

$$\begin{pmatrix} 1 & 0 & \frac{6}{11} & 1 \\ 0 & 1 & -\frac{1}{11} & 0 \end{pmatrix}$$

So the inverse is $\begin{pmatrix} \frac{6}{11} & 1 \\ -\frac{1}{11} & 0 \end{pmatrix}$.

17. Form the matrix:

$$\begin{pmatrix} \frac{2}{3} & -\frac{1}{4} & 1 & 0 \\ -8 & 3 & 0 & 1 \end{pmatrix}$$

Multiply row 1 by 12:

$$\begin{pmatrix} 8 & -3 & 12 & 0 \\ -8 & 3 & 0 & 1 \end{pmatrix}$$

Add row 1 to row 2:

$$\begin{pmatrix} 8 & -3 & 12 & 0 \\ 0 & 0 & 12 & 1 \end{pmatrix}$$

So the inverse does not exist.

19. Form the matrix:

$$\begin{pmatrix} -5 & 4 & -3 & 1 & 0 & 0 \\ 10 & -7 & 6 & 0 & 1 & 0 \\ 8 & -6 & 5 & 0 & 0 & 1 \end{pmatrix}$$

Multiply row 1 by 2 and add to row 2:

$$\begin{pmatrix} -5 & 4 & -3 & 1 & 0 & 0 \\ 0 & 1 & 0 & 2 & 1 & 0 \\ 8 & -6 & 5 & 0 & 0 & 1 \end{pmatrix}$$

Add row 3 to row 1 (to reduce the numbers):

$$\begin{pmatrix} 3 & -2 & 2 & 1 & 0 & 1 \\ 0 & 1 & 0 & 2 & 1 & 0 \\ 8 & -6 & 5 & 0 & 0 & 1 \end{pmatrix}$$

Multiply row 2 by 2 and add to row 1, and also multiply row 2 by 6 and add to row 3:

$$\begin{pmatrix} 3 & 0 & 2 & 5 & 2 & 1 \\ 0 & 1 & 0 & 2 & 1 & 0 \\ 8 & 0 & 5 & 12 & 6 & 1 \end{pmatrix}$$

Multiply row 1 by $-\frac{8}{3}$ and add to row 3:

$$\begin{pmatrix} 3 & 0 & 2 & 5 & 2 & 1 \\ 0 & 1 & 0 & 2 & 1 & 0 \\ 0 & 0 & -\frac{1}{3} & -\frac{4}{3} & \frac{2}{3} & -\frac{5}{3} \end{pmatrix}$$

Multiply row 3 by -3:

$$\begin{pmatrix} 3 & 0 & 2 & 5 & 2 & 1 \\ 0 & 1 & 0 & 2 & 1 & 0 \\ 0 & 0 & 1 & 4 & -2 & 5 \end{pmatrix}$$

Multiply row 3 by -2 and add to row 1:

$$\begin{pmatrix} 3 & 0 & 0 & -3 & 6 & -9 \\ 0 & 1 & 0 & 2 & 1 & 0 \\ 0 & 0 & 1 & 4 & -2 & 5 \end{pmatrix}$$

Multiply row 1 by $\frac{1}{3}$:

$$\begin{pmatrix} 1 & 0 & 0 & -1 & 2 & -3 \\ 0 & 1 & 0 & 2 & 1 & 0 \\ 0 & 0 & 1 & 4 & -2 & 5 \end{pmatrix}$$

So the inverse is $\begin{pmatrix} -1 & 2 & -3 \\ 2 & 1 & 0 \\ 4 & -2 & 5 \end{pmatrix}$.

21. Form the matrix:

$$\begin{pmatrix} 1 & 2 & -1 & 1 & 0 & 0 \\ 0 & 3 & 0 & 0 & 1 & 0 \\ -4 & 0 & 5 & 0 & 0 & 1 \end{pmatrix}$$

Multiply row 1 by 4 and add to row 3:

$$\begin{pmatrix} 1 & 2 & -1 & 1 & 0 & 0 \\ 0 & 3 & 0 & 0 & 1 & 0 \\ 0 & 8 & 1 & 4 & 0 & 1 \end{pmatrix}$$

Multiply row 2 by $\frac{1}{3}$:

$$\begin{pmatrix} 1 & 2 & -1 & 1 & 0 & 0 \\ 0 & 1 & 0 & 0 & \frac{1}{3} & 0 \\ 0 & 8 & 1 & 4 & 0 & 1 \end{pmatrix}$$

Multiply row 2 by −2 and add to row 1, and multiply row 2 by −8 and add to row 3:

$$\begin{pmatrix} 1 & 0 & -1 & 1 & -\frac{2}{3} & 0 \\ 0 & 1 & 0 & 0 & \frac{1}{3} & 0 \\ 0 & 0 & 1 & 4 & -\frac{8}{3} & 1 \end{pmatrix}$$

Add row 3 to row 1:

$$\begin{pmatrix} 1 & 0 & 0 & 5 & -\frac{10}{3} & 1 \\ 0 & 1 & 0 & 0 & \frac{1}{3} & 0 \\ 0 & 0 & 1 & 4 & -\frac{8}{3} & 1 \end{pmatrix}$$

So the inverse is $\begin{pmatrix} 5 & -\frac{10}{3} & 1 \\ 0 & \frac{1}{3} & 0 \\ 4 & -\frac{8}{3} & 1 \end{pmatrix}$.

23. Form the matrix:

$$\begin{pmatrix} -7 & 5 & 3 & 1 & 0 & 0 \\ 3 & -2 & -2 & 0 & 1 & 0 \\ 3 & -2 & -1 & 0 & 0 & 1 \end{pmatrix}$$

Multiply row 2 by -1 and add to row 3:
$$\begin{pmatrix} -7 & 5 & 3 & 1 & 0 & 0 \\ 3 & -2 & -2 & 0 & 1 & 0 \\ 0 & 0 & 1 & 0 & -1 & 1 \end{pmatrix}$$
Multiply row 2 by 2 and add to row 1:
$$\begin{pmatrix} -1 & 1 & -1 & 1 & 2 & 0 \\ 3 & -2 & -2 & 0 & 1 & 0 \\ 0 & 0 & 1 & 0 & -1 & 1 \end{pmatrix}$$
Multiply row 1 by 3 and add to row 2:
$$\begin{pmatrix} -1 & 1 & -1 & 1 & 2 & 0 \\ 0 & 1 & -5 & 3 & 7 & 0 \\ 0 & 0 & 1 & 0 & -1 & 1 \end{pmatrix}$$
Multiply row 3 by 5 and add to row 2, and add row 3 to row 1:
$$\begin{pmatrix} -1 & 1 & 0 & 1 & 1 & 1 \\ 0 & 1 & 0 & 3 & 2 & 5 \\ 0 & 0 & 1 & 0 & -1 & 1 \end{pmatrix}$$
Multiply row 1 by -1:
$$\begin{pmatrix} 1 & -1 & 0 & -1 & -1 & -1 \\ 0 & 1 & 0 & 3 & 2 & 5 \\ 0 & 0 & 1 & 0 & -1 & 1 \end{pmatrix}$$
Add row 2 to row 1:
$$\begin{pmatrix} 1 & 0 & 0 & 2 & 1 & 4 \\ 0 & 1 & 0 & 3 & 2 & 5 \\ 0 & 0 & 1 & 0 & -1 & 1 \end{pmatrix}$$
So the inverse is $\begin{pmatrix} 2 & 1 & 4 \\ 3 & 2 & 5 \\ 0 & -1 & 1 \end{pmatrix}$.

25. Form the matrix:
$$\begin{pmatrix} 1 & 2 & 3 & 1 & 0 & 0 \\ 4 & 5 & 6 & 0 & 1 & 0 \\ 7 & 8 & 9 & 0 & 0 & 1 \end{pmatrix}$$
Multiply row 1 by -4 and add to row 2, and multiply row 1 by -7 and add to row 3:
$$\begin{pmatrix} 1 & 2 & 3 & 1 & 0 & 0 \\ 0 & -3 & -6 & -4 & 1 & 0 \\ 0 & -6 & -12 & -7 & 0 & 1 \end{pmatrix}$$

Multiply row 2 by –2 and add to row 3:

$$\begin{pmatrix} 1 & 2 & 3 & 1 & 0 & 0 \\ 0 & -3 & -6 & -4 & 1 & 0 \\ 0 & 0 & 0 & 1 & -2 & 1 \end{pmatrix}$$

So the inverse does not exist.

27. (a) Since the system can be written as $A \bullet X = B$, where $X = \begin{pmatrix} x \\ y \end{pmatrix}$ and $B = \begin{pmatrix} 5 \\ 7 \end{pmatrix}$, then

$X = A^{-1} \bullet B$. So $X = \begin{pmatrix} 11 & -8 \\ -4 & 3 \end{pmatrix}\begin{pmatrix} 5 \\ 7 \end{pmatrix} = \begin{pmatrix} -1 \\ 1 \end{pmatrix}$, thus $x = -1$ and $y = 1$.

(b) Again this system can be written as $A \bullet X = B$, where $X = \begin{pmatrix} x \\ y \end{pmatrix}$ and $B = \begin{pmatrix} -12 \\ 0 \end{pmatrix}$, then

$X = A^{-1} \bullet B$. So $X = \begin{pmatrix} 11 & -8 \\ -4 & 3 \end{pmatrix}\begin{pmatrix} -12 \\ 0 \end{pmatrix} = \begin{pmatrix} -132 \\ 48 \end{pmatrix}$, thus $x = -132$ and $y = 48$.

29. (a) Since the system can be written as $A \bullet X = B$, where $X = \begin{pmatrix} x \\ y \\ z \end{pmatrix}$ and $B = \begin{pmatrix} 28 \\ 9 \\ 22 \end{pmatrix}$, then

$X = A^{-1} \bullet B$. So $X = \begin{pmatrix} 1 & 2 & -2 \\ -1 & 3 & 0 \\ 0 & -2 & 1 \end{pmatrix}\begin{pmatrix} 28 \\ 9 \\ 22 \end{pmatrix} = \begin{pmatrix} 2 \\ -1 \\ 4 \end{pmatrix}$, thus $x = 2$, $y = -1$, $z = 4$.

(b) Again this system can be written as $A \bullet X = B$, where $X = \begin{pmatrix} x \\ y \\ z \end{pmatrix}$ and $B = \begin{pmatrix} -7 \\ -2 \\ -6 \end{pmatrix}$, then

$X = A^{-1} \bullet B$. So $X = \begin{pmatrix} 1 & 2 & -2 \\ -1 & 3 & 0 \\ 0 & -2 & 1 \end{pmatrix}\begin{pmatrix} -7 \\ -2 \\ -6 \end{pmatrix} = \begin{pmatrix} 1 \\ 1 \\ -2 \end{pmatrix}$, thus $x = 1$, $y = 1$, $z = -2$.

31. (a) We compute AA:

$$AA = \begin{pmatrix} 1 & -6 & 3 \\ 2 & -7 & 3 \\ 4 & -12 & 5 \end{pmatrix}\begin{pmatrix} 1 & -6 & 3 \\ 2 & -7 & 3 \\ 4 & -12 & 5 \end{pmatrix} = \begin{pmatrix} 1 & 0 & 0 \\ 0 & 1 & 0 \\ 0 & 0 & 1 \end{pmatrix}$$

The product is the identity matrix I_3, and thus $A^{-1} = A$.

(b) Since this can be written as $A \cdot X = B$, where $X = \begin{pmatrix} x \\ y \\ z \end{pmatrix}$ and $B = \begin{pmatrix} \frac{19}{2} \\ 11 \\ 19 \end{pmatrix}$, then

$X = A^{-1} \cdot B$. Recall from part (a) that $A^{-1} = A$, so $X = \begin{pmatrix} 1 & -6 & 3 \\ 2 & -7 & 3 \\ 4 & -12 & 5 \end{pmatrix}\begin{pmatrix} \frac{19}{2} \\ 11 \\ 19 \end{pmatrix} = \begin{pmatrix} \frac{1}{2} \\ -1 \\ 1 \end{pmatrix}$,

thus $x = \frac{1}{2}$, $y = -1$, and $z = 1$.

33. We find $\begin{pmatrix} a & b \\ c & d \end{pmatrix}$ where $\begin{pmatrix} 2 & 5 \\ 6 & 15 \end{pmatrix}\begin{pmatrix} a & b \\ c & d \end{pmatrix} = \begin{pmatrix} 1 & 0 \\ 0 & 1 \end{pmatrix}$. Carrying out the multiplication, we have:

$$\begin{pmatrix} 2a + 5c & 2b + 5d \\ 6a + 15c & 6b + 15d \end{pmatrix} = \begin{pmatrix} 1 & 0 \\ 0 & 1 \end{pmatrix}$$

So we have the following systems of equations:

$$2a + 5c = 1 \quad \text{and} \quad 2b + 5d = 0$$
$$6a + 15c = 0 \qquad\qquad\qquad 6b + 15d = 1$$

Multiplying the first equation by -3 yields:

$$-6a - 15c = -3 \qquad\qquad -6b - 15d = 0$$
$$6a + 15c = 0 \qquad\qquad\quad 6b + 15d = 1$$

Adding, we obtain:

$$0 = -3 \qquad\qquad\qquad 0 = 1$$

Neither of these systems has a solution, thus the matrix has no inverse.

35. (a) Form the matrix:

$$\begin{pmatrix} 2 & 3 & 1 & 0 \\ 4 & 5 & 0 & 1 \end{pmatrix}$$

Multiply row 1 by -2 and add to row 2:

$$\begin{pmatrix} 2 & 3 & 1 & 0 \\ 0 & -1 & -2 & 1 \end{pmatrix}$$

Multiply row 2 by 3 and add to row 1:

$$\begin{pmatrix} 2 & 0 & -5 & 3 \\ 0 & -1 & -2 & 1 \end{pmatrix}$$

Multiply row 1 by $\frac{1}{2}$ and row 2 by -1:

$$\begin{pmatrix} 1 & 0 & -\frac{5}{2} & \frac{3}{2} \\ 0 & 1 & 2 & -1 \end{pmatrix}$$

So $A^{-1} = \begin{pmatrix} -\frac{5}{2} & \frac{3}{2} \\ 2 & -1 \end{pmatrix}$ Now form the matrix:

$$\begin{pmatrix} 7 & 8 & 1 & 0 \\ 6 & 7 & 0 & 1 \end{pmatrix}$$

Subtract row 2 from row 1:

$$\begin{pmatrix} 1 & 1 & 1 & -1 \\ 6 & 7 & 0 & 1 \end{pmatrix}$$

Multiply row 1 by -6 and add to row 2:

$$\begin{pmatrix} 1 & 1 & 1 & -1 \\ 0 & 1 & -6 & 7 \end{pmatrix}$$

Subtract row 2 from row 1:

$$\begin{pmatrix} 1 & 0 & 7 & -8 \\ 0 & 1 & -6 & 7 \end{pmatrix}$$

So $B^{-1} = \begin{pmatrix} 7 & -8 \\ -6 & 7 \end{pmatrix}$.

Then $B^{-1}A^{-1} = \begin{pmatrix} 7 & -8 \\ -6 & 7 \end{pmatrix}\begin{pmatrix} -\frac{5}{2} & \frac{3}{2} \\ 2 & -1 \end{pmatrix} = \begin{pmatrix} -\frac{67}{2} & \frac{37}{2} \\ 29 & -16 \end{pmatrix}.$

(b)   We first find AB:

$$AB = \begin{pmatrix} 2 & 3 \\ 4 & 5 \end{pmatrix}\begin{pmatrix} 7 & 8 \\ 6 & 7 \end{pmatrix} = \begin{pmatrix} 32 & 37 \\ 58 & 67 \end{pmatrix}$$

Now form the matrix:

$$\begin{pmatrix} 32 & 37 & 1 & 0 \\ 58 & 67 & 0 & 1 \end{pmatrix}$$

Subtract row 1 from row 2 (to reduce numbers):

$$\begin{pmatrix} 32 & 37 & 1 & 0 \\ 26 & 30 & -1 & 1 \end{pmatrix}$$

Subtract row 2 from row 1 (to reduce numbers):

$$\begin{pmatrix} 6 & 7 & 2 & -1 \\ 26 & 30 & -1 & 1 \end{pmatrix}$$

Multiply row 1 by -4 and add to row 2:

$$\begin{pmatrix} 6 & 7 & 2 & -1 \\ 2 & 2 & -9 & 5 \end{pmatrix}$$

Multiply row 2 by -3 and add to row 1:

$$\begin{pmatrix} 0 & 1 & 29 & -16 \\ 2 & 2 & -9 & 5 \end{pmatrix}$$

Multiply row 1 by -2 and add to row 2:

$$\begin{pmatrix} 0 & 1 & 29 & -16 \\ 2 & 0 & -67 & 37 \end{pmatrix}$$

Multiply row 2 by $\frac{1}{2}$, then switch rows 1 and 2:

$$\begin{pmatrix} 1 & 0 & -\frac{67}{2} & \frac{37}{2} \\ 0 & 1 & 29 & -16 \end{pmatrix}$$

So $(AB)^{-1} = \begin{pmatrix} -\frac{67}{2} & \frac{37}{2} \\ 29 & -16 \end{pmatrix}$, which is the same as $B^{-1}A^{-1}$ from part (a).

9.5 Determinants and Cramer's Rule

1. (a) We evaluate the determinant:

$$\begin{vmatrix} 2 & -17 \\ 1 & 6 \end{vmatrix} = 2(6) - (-17)(1) = 12 + 17 = 29$$

(b) We evaluate the determinant:

$$\begin{vmatrix} 1 & 6 \\ 2 & -17 \end{vmatrix} = 1(-17) - (6)(2) = -17 - 12 = -29$$

3. (a) We evaluate the determinant:

$$\begin{vmatrix} 7 & 7 \\ 500 & 700 \end{vmatrix} = 100 \begin{vmatrix} 5 & 7 \\ 5 & 7 \end{vmatrix} = 100[5(7) - 7(5)] = 100(35 - 35) = 0$$

(b) We evaluate the determinant:

$$\begin{vmatrix} 5 & 500 \\ 7 & 700 \end{vmatrix} = 100 \begin{vmatrix} 5 & 5 \\ 7 & 7 \end{vmatrix} = 100[5(7) - 5(7)] = 100(35 - 35) = 0$$

5. We evaluate the determinant:

$$\begin{vmatrix} \cos\theta & -\sin\theta \\ \sin\theta & \cos\theta \end{vmatrix} = \cos^2\theta - (-\sin^2\theta) = \cos^2\theta + \sin^2\theta = 1$$

7. We evaluate the minor of 3:

$$\begin{vmatrix} 5 & 1 \\ 10 & -10 \end{vmatrix} = 10 \begin{vmatrix} 5 & 1 \\ 1 & -1 \end{vmatrix} = 10[5(-1) - 1(1)] = 10(-6) = -60$$

9. We evaluate the minor of -10:

$$\begin{vmatrix} -6 & 3 \\ 5 & -4 \end{vmatrix} = 3 \begin{vmatrix} -2 & 1 \\ 5 & -4 \end{vmatrix} = 3[-2(-4) - 1(5)] = 3(3) = 9$$

11. (a) We compute:

$$-6\begin{vmatrix} -4 & 1 \\ 9 & -10 \end{vmatrix} + 3\begin{vmatrix} 5 & 1 \\ 10 & -10 \end{vmatrix} + 8\begin{vmatrix} 5 & -4 \\ 10 & 9 \end{vmatrix} = -6(31) + 3(-60) + 8(85) = 314$$

(b) We compute:

$$-6\begin{vmatrix} -4 & 1 \\ 9 & -10 \end{vmatrix} - 3\begin{vmatrix} 5 & 1 \\ 10 & -10 \end{vmatrix} + 8\begin{vmatrix} 5 & -4 \\ 10 & 9 \end{vmatrix} = -6(31) - 3(-60) + 8(85) = 674$$

(c) The answer in part (b) would be the determinant.

13. (a) We expand along the second row:

$$-4\begin{vmatrix} 2 & 3 \\ 8 & 9 \end{vmatrix} + 5\begin{vmatrix} 1 & 3 \\ 7 & 9 \end{vmatrix} - 6\begin{vmatrix} 1 & 2 \\ 7 & 8 \end{vmatrix} = -4(-6) + 5(-12) - 6(-6) = 0$$

(b) We expand along the third row:

$$7\begin{vmatrix} 2 & 3 \\ 5 & 6 \end{vmatrix} - 8\begin{vmatrix} 1 & 3 \\ 4 & 6 \end{vmatrix} + 9\begin{vmatrix} 1 & 2 \\ 4 & 5 \end{vmatrix} = 7(-3) - 8(-6) + 9(-3) = 0$$

(c) We expand along the first column:

$$1\begin{vmatrix} 5 & 6 \\ 8 & 9 \end{vmatrix} - 4\begin{vmatrix} 2 & 3 \\ 8 & 9 \end{vmatrix} + 7\begin{vmatrix} 2 & 3 \\ 5 & 6 \end{vmatrix} = 1(-3) - 4(-6) + 7(-3) = 0$$

(d) We expand along the third column:

$$3\begin{vmatrix} 4 & 5 \\ 7 & 8 \end{vmatrix} - 6\begin{vmatrix} 1 & 2 \\ 7 & 8 \end{vmatrix} + 9\begin{vmatrix} 1 & 2 \\ 4 & 5 \end{vmatrix} = 3(-3) - 6(-6) + 9(-3) = 0$$

15. Factoring 5 out of the first row:

$$\begin{vmatrix} 5 & 10 & 15 \\ 1 & 2 & 3 \\ -9 & 11 & 7 \end{vmatrix} = 5\begin{vmatrix} 1 & 2 & 3 \\ 1 & 2 & 3 \\ -9 & 11 & 7 \end{vmatrix} = 5\begin{vmatrix} 0 & 0 & 0 \\ 1 & 2 & 3 \\ -9 & 11 & 7 \end{vmatrix} = 0$$

17. Expanding along the third row:

$$\begin{vmatrix} 1 & 2 & -3 \\ 4 & 5 & -9 \\ 0 & 0 & 1 \end{vmatrix} = 0\begin{vmatrix} 2 & -3 \\ 5 & -9 \end{vmatrix} - 0\begin{vmatrix} 1 & -3 \\ 4 & -9 \end{vmatrix} + 1\begin{vmatrix} 1 & 2 \\ 4 & 5 \end{vmatrix} = 1(-3) = -3$$

19. Factoring 2 out of the first row and 4 out of the second column:

$$\begin{vmatrix} -6 & -8 & 18 \\ 25 & 12 & 15 \\ -9 & 4 & 13 \end{vmatrix} = 2\begin{vmatrix} -3 & -4 & 9 \\ 25 & 12 & 15 \\ -9 & 4 & 13 \end{vmatrix} = 8\begin{vmatrix} -3 & -1 & 9 \\ 25 & 3 & 15 \\ -9 & 1 & 13 \end{vmatrix}$$

Now expanding along the first row:

$$\begin{vmatrix} -3 & -1 & 9 \\ 25 & 3 & 15 \\ -9 & 1 & 13 \end{vmatrix} = -3\begin{vmatrix} 3 & 15 \\ 1 & 13 \end{vmatrix} + 1\begin{vmatrix} 25 & 15 \\ -9 & 13 \end{vmatrix} + 9\begin{vmatrix} 25 & 3 \\ -9 & 1 \end{vmatrix} = -3(24) + 1(460) + 9(52) = 856$$

Therefore:

$$8\begin{vmatrix} -3 & -1 & 9 \\ 25 & 3 & 15 \\ -9 & 1 & 13 \end{vmatrix} = 8(856) = 6848$$

21. Factoring 16 out of the first row and 10 out of the third row:

$$\begin{vmatrix} 16 & 0 & -64 \\ -8 & 15 & -12 \\ 30 & -20 & 10 \end{vmatrix} = 16\begin{vmatrix} 1 & 0 & -4 \\ -8 & 15 & -12 \\ 30 & -20 & 10 \end{vmatrix} = 160\begin{vmatrix} 1 & 0 & -4 \\ -8 & 15 & -12 \\ 3 & -2 & 1 \end{vmatrix}$$

Now expanding along the first row:

$$\begin{vmatrix} 1 & 0 & -4 \\ -8 & 15 & -12 \\ 3 & -2 & 1 \end{vmatrix} = 1\begin{vmatrix} 15 & -12 \\ -2 & 1 \end{vmatrix} - 0\begin{vmatrix} -8 & -12 \\ 3 & 1 \end{vmatrix} - 4\begin{vmatrix} -8 & 15 \\ 3 & 2 \end{vmatrix} = 1(-9) - 4(-29) = 107$$

Therefore:

$$160\begin{vmatrix} 1 & 0 & -4 \\ -8 & 15 & -12 \\ 3 & -2 & 1 \end{vmatrix} = 160(107) = 17120$$

23. We evaluate the determinant:

$$\begin{vmatrix} 1 & x & x^2 \\ 1 & y & y^2 \\ 1 & z & z^2 \end{vmatrix} = \begin{vmatrix} 1 & x & x^2 \\ 0 & y-x & y^2-x^2 \\ 0 & z-x & z^2-x^2 \end{vmatrix}$$

$$= \begin{vmatrix} y-x & y^2-x^2 \\ z-x & z^2-x^2 \end{vmatrix}$$

$$= \begin{vmatrix} y-x & (y+x)(y-x) \\ z-x & (z+x)(z-x) \end{vmatrix}$$

$$= (y-x)(z-x)\begin{vmatrix} 1 & y+x \\ 1 & z+x \end{vmatrix}$$

$$= (y-x)(z-x)[z+x-y-x]$$

$$= (y-x)(z-x)(z-y)$$

25. Substracting row 1 from row 2 and row 1 from row 3:

$$\begin{vmatrix} 1 & 1 & 1 \\ 1 & 1+x & 1 \\ 1 & 1 & 1+y \end{vmatrix} = \begin{vmatrix} 1 & 1 & 1 \\ 0 & x & 0 \\ 0 & 0 & y \end{vmatrix} = 1\begin{vmatrix} x & 0 \\ 0 & y \end{vmatrix} = xy$$

27. Adding column 2 to column 3, we obtain:

$$\begin{vmatrix} 1 & -1 & -1 & 2 \\ 0 & 1 & 0 & 0 \\ 2 & 1 & 1 & -1 \\ -2 & 2 & 3 & 1 \end{vmatrix} = 1\begin{vmatrix} 1 & -1 & 2 \\ 2 & 1 & -1 \\ -2 & 3 & 1 \end{vmatrix}$$

Adding twice column 3 to column 1, we obtain:

$$\begin{vmatrix} 5 & -1 & 2 \\ 0 & 1 & -1 \\ 0 & 3 & 1 \end{vmatrix} = 5\begin{vmatrix} 1 & -1 \\ 3 & 1 \end{vmatrix} = 5(4) = 20$$

29. We evaluate the determinant:

$$\begin{vmatrix} 2 & 0 & 0 & 0 \\ 0 & 3 & 0 & 0 \\ 0 & 0 & 4 & 0 \\ 0 & 0 & 0 & 5 \end{vmatrix} = 2\begin{vmatrix} 3 & 0 & 0 \\ 0 & 4 & 0 \\ 0 & 0 & 5 \end{vmatrix} = 2(3)\begin{vmatrix} 4 & 0 \\ 0 & 5 \end{vmatrix} = 6(20) = 120$$

31. (a) Subtract the first column from the second and add twice the first column to the third. The result is:

$$\begin{vmatrix} -20 & 22 & -43 \\ 6 & -10 & 13 \\ -1 & 0 & 0 \end{vmatrix} = -1[22(13) - (-43)(-10)] = 144$$

(b) Multiply the second row by 2 and subtract it from the first row, and multiply the second row by 4 and subtract it from the third row. This yields:

$$\begin{vmatrix} 0 & -32 & -5 \\ 1 & 6 & 1 \\ 0 & -25 & -2 \end{vmatrix} = -1[(-32)(-2) - (-5)(-25)] = 61$$

(c) Subtract the first column from the second and add 10 times the first column to the third. This yields:

$$\begin{vmatrix} 2 & 0 & 0 \\ 1 & -5 & 16 \\ 4 & -5 & 39 \end{vmatrix} = 2[(-5)(39) - 16(-5)] = -230$$

33. We compute D:

$$D = \begin{vmatrix} 3 & 4 & -1 \\ 1 & -3 & 2 \\ 5 & 0 & -6 \end{vmatrix} = 5\begin{vmatrix} 4 & -1 \\ -3 & 2 \end{vmatrix} - 6\begin{vmatrix} 3 & 4 \\ 1 & -3 \end{vmatrix} = 5(5) - 6(-13) = 103$$

Now compute D_x:

$$D_x = \begin{vmatrix} 5 & 4 & -1 \\ 2 & -3 & 2 \\ -7 & 0 & -6 \end{vmatrix} = -7\begin{vmatrix} 4 & -1 \\ -3 & 2 \end{vmatrix} - 6\begin{vmatrix} 5 & 4 \\ 2 & -3 \end{vmatrix} = -7(5) - 6(-23) = 103$$

For D_y, adding -2 times column 1 to both column 2 and column 3 yields:

$$D_y = \begin{vmatrix} 3 & 5 & -1 \\ 1 & 2 & 2 \\ 5 & -7 & -6 \end{vmatrix} = \begin{vmatrix} 3 & -1 & -7 \\ 1 & 0 & 0 \\ 5 & -17 & -16 \end{vmatrix} = -1\begin{vmatrix} -1 & -7 \\ -17 & -16 \end{vmatrix} = -(-103) = 103$$

For D_z, adding 3 times column 1 to column 2, and -2 times column 1 to column 3 yields:

$$D_z = \begin{vmatrix} 3 & 4 & 5 \\ 1 & -3 & 2 \\ 5 & 0 & -7 \end{vmatrix} = \begin{vmatrix} 3 & 13 & -1 \\ 1 & 0 & 0 \\ 5 & 15 & -17 \end{vmatrix} = -1\begin{vmatrix} 13 & 1 \\ 15 & -17 \end{vmatrix} = -(-206) = 206$$

So $x = \frac{D_x}{D} = 1$, $y = \frac{D_y}{D} = 1$, and $z = \frac{D_z}{D} = 2$. So the solution is $(1, 1, 2)$.

35. Subtracting column 1 from both column 2 and column 3 yields:

$$D = \begin{vmatrix} 3 & 2 & -1 \\ 2 & -3 & -4 \\ 1 & 1 & 1 \end{vmatrix} = \begin{vmatrix} 3 & -1 & -4 \\ 2 & -5 & -6 \\ 1 & 0 & 0 \end{vmatrix} = 1\begin{vmatrix} -1 & -4 \\ -5 & -6 \end{vmatrix} = -14$$

Adding -5 times column 3 to column 1, and subtracting column 3 from column 2 yields:

$$D_x = \begin{vmatrix} -6 & 2 & -1 \\ -11 & -3 & -4 \\ 5 & 1 & 1 \end{vmatrix} = \begin{vmatrix} -1 & 3 & -1 \\ 9 & 1 & -4 \\ 0 & 0 & 1 \end{vmatrix} = 1\begin{vmatrix} -1 & 3 \\ 9 & 1 \end{vmatrix} = -28$$

Adding row 3 to row 1, and 4 times row 3 to row 2 yields:

$$D_y = \begin{vmatrix} 3 & -6 & -1 \\ 2 & -11 & -4 \\ 1 & 5 & 1 \end{vmatrix} = \begin{vmatrix} 4 & -1 & 0 \\ 6 & 9 & 0 \\ 1 & 5 & 1 \end{vmatrix} = 1\begin{vmatrix} 4 & -1 \\ 6 & 9 \end{vmatrix} = 42$$

Adding -2 times row 3 to row 1, and 3 times row 3 to row 2 yields:

$$D_z = \begin{vmatrix} 3 & 2 & -6 \\ 2 & 3 & -11 \\ 1 & 1 & 5 \end{vmatrix} = \begin{vmatrix} 1 & 0 & -16 \\ 5 & 0 & 4 \\ 1 & 1 & 5 \end{vmatrix} = -1\begin{vmatrix} 1 & -16 \\ 5 & 4 \end{vmatrix} = -(84) = -84$$

So $x = \frac{D_x}{D} = 2$, $y = \frac{D_y}{D} = -3$, and $z = \frac{D_z}{D} = 6$. So the solution is $(2, -3, 6)$.

37. Adding -2 times row 3 to row 1, and -3 times row 3 to row 2 yields:

$$D = \begin{vmatrix} 2 & 5 & 2 \\ 3 & -1 & -4 \\ 1 & 2 & -3 \end{vmatrix} = \begin{vmatrix} 0 & 1 & 8 \\ 0 & -7 & 5 \\ 1 & 2 & -3 \end{vmatrix} = 1 \begin{vmatrix} 1 & 8 \\ -7 & 5 \end{vmatrix} = 61$$

Now compute:

$$D_x = \begin{vmatrix} 0 & 5 & 2 \\ 0 & -1 & -4 \\ 0 & 2 & 3 \end{vmatrix} = 0$$

$$D_y = \begin{vmatrix} 2 & 0 & 2 \\ 3 & 0 & -4 \\ 1 & 0 & -3 \end{vmatrix} = 0$$

$$D_z = \begin{vmatrix} 2 & 5 & 0 \\ 3 & -1 & 0 \\ 1 & 2 & 0 \end{vmatrix} = 0$$

So $x = \frac{D_x}{D} = 0$, $y = \frac{D_y}{D} = 0$, and $z = \frac{D_z}{D} = 0$. So the solution is $(0,0,0)$.

39. Factoring 6 out of column 1 and 2 out of column 2, we have:

$$D = \begin{vmatrix} 12 & 0 & -11 \\ 6 & 6 & -4 \\ 6 & 2 & -5 \end{vmatrix} = 6 \begin{vmatrix} 2 & 0 & -11 \\ 1 & 6 & -4 \\ 1 & 2 & -5 \end{vmatrix} = 12 \begin{vmatrix} 2 & 0 & -11 \\ 1 & 3 & -4 \\ 1 & 1 & -5 \end{vmatrix}$$

Adding -3 times row 3 to row 2 yields:

$$D = 12 \begin{vmatrix} 2 & 0 & -11 \\ -2 & 0 & 11 \\ 1 & 1 & -5 \end{vmatrix} = 12(-1) \begin{vmatrix} 2 & -11 \\ -2 & 11 \end{vmatrix} = -12(0) = 0$$

So Cramer's Rule will not work. We form the augmented matrix (using equation two as row 1):

$$\begin{pmatrix} 6 & 6 & -4 & 26 \\ 12 & 0 & -11 & 13 \\ 6 & 2 & -5 & 13 \end{pmatrix}$$

Adding -2 times row 1 to row 2 and -1 times row 1 to row 3 yields:

$$\begin{pmatrix} 6 & 6 & -4 & 26 \\ 0 & -12 & -3 & -39 \\ 0 & -4 & -1 & -13 \end{pmatrix}$$

Multiply row 1 by $\frac{1}{2}$ and row 2 by $\frac{1}{3}$:

$$\begin{pmatrix} 3 & 3 & -2 & 13 \\ 0 & 4 & 1 & 13 \\ 0 & -4 & -1 & -13 \end{pmatrix}$$

Adding row 2 to row 3 yields:

$$\begin{pmatrix} 3 & 3 & -2 & 13 \\ 0 & 4 & 1 & 13 \\ 0 & 0 & 0 & 0 \end{pmatrix}$$

So we have the system:

$$3x + 3y - 2z = 13$$
$$4y + z = 13$$

Solve equation 2 for z:

$$4y + z = 13$$
$$z = 13 - 4y$$

Substitute into equation 1:

$$3x + 3y - 26 + 8y = 13$$
$$3x = 39 - 11y$$
$$x = 13 - \tfrac{11}{3}y$$

So the solution is $\left(13 - \tfrac{11}{3}y, y, 13 - 4y\right)$, for any real number y.

41. Adding row 1 to row 2, 3 times row 1 to row 3, and row 1 to row 4 yields:

$$D = \begin{vmatrix} 1 & 1 & 1 & 1 \\ 1 & -1 & 1 & -1 \\ 2 & -2 & -3 & -3 \\ 3 & 2 & 1 & -1 \end{vmatrix} = \begin{vmatrix} 1 & 1 & 1 & 1 \\ 2 & 0 & 2 & 0 \\ 5 & 1 & 0 & 0 \\ 4 & 3 & 2 & 0 \end{vmatrix} = -1\begin{vmatrix} 2 & 0 & 2 \\ 5 & 1 & 0 \\ 4 & 3 & 2 \end{vmatrix}$$

Subtracting column 3 from column 1 yields:

$$D = -1\begin{vmatrix} 0 & 0 & 2 \\ 5 & 1 & 0 \\ 2 & 3 & 2 \end{vmatrix} = -2\begin{vmatrix} 5 & 1 \\ 2 & 3 \end{vmatrix} = -2(13) = -26$$

Adding row 1 to row 2, 3 times row 1 to row 3, and row 1 to row 4 yields:

$$D_x = \begin{vmatrix} -7 & 1 & 1 & 1 \\ -11 & -1 & 1 & -1 \\ 26 & -2 & -3 & -3 \\ -9 & 2 & 1 & -1 \end{vmatrix} = \begin{vmatrix} -7 & 1 & 1 & 1 \\ -18 & 0 & 2 & 0 \\ 5 & 1 & 0 & 0 \\ -16 & 3 & 2 & 0 \end{vmatrix} = -1\begin{vmatrix} -18 & 0 & 2 \\ 5 & 1 & 0 \\ -16 & 3 & 2 \end{vmatrix}$$

Subtracting row 3 from row 1 yields:

$$D_x = -1\begin{vmatrix} -2 & -3 & 0 \\ 5 & 1 & 0 \\ -16 & 3 & 2 \end{vmatrix} = -2\begin{vmatrix} -2 & -3 \\ 5 & 1 \end{vmatrix} = -2(13) = -26$$

Adding row 1 to row 2, 3 times row 1 to row 3, and row 1 to row 4 yields:

$$D_y = \begin{vmatrix} 1 & -7 & 1 & 1 \\ 1 & -11 & 1 & -1 \\ 2 & 26 & -3 & -3 \\ 3 & -9 & 1 & -1 \end{vmatrix} = \begin{vmatrix} 1 & -7 & 1 & 1 \\ 2 & -18 & 2 & 0 \\ 5 & 5 & 0 & 0 \\ 4 & -16 & 2 & 0 \end{vmatrix} = -1 \begin{vmatrix} 2 & -18 & 2 \\ 5 & 5 & 0 \\ 4 & -16 & 2 \end{vmatrix}$$

Subtracting row 3 from row 1 yields:

$$D_y = -1 \begin{vmatrix} -2 & -2 & 0 \\ 5 & 5 & 0 \\ 4 & -16 & 2 \end{vmatrix} = -2 \begin{vmatrix} -2 & -2 \\ 5 & 5 \end{vmatrix} = -2(0) = 0$$

Adding row 1 to row 2, 3 times row 1 to row 3, and row 1 to row 4 yields:

$$D_z = \begin{vmatrix} 1 & 1 & -7 & 1 \\ 1 & -1 & -11 & -1 \\ 2 & -2 & 26 & -3 \\ 3 & 2 & -9 & -1 \end{vmatrix} = \begin{vmatrix} 1 & 1 & -7 & 1 \\ 2 & 0 & -18 & 0 \\ 5 & 1 & 5 & 0 \\ 4 & 3 & -16 & 0 \end{vmatrix} = -1 \begin{vmatrix} 2 & 0 & -18 \\ 5 & 1 & 5 \\ 4 & 3 & -16 \end{vmatrix}$$

Adding –3 times row 2 to row 3 yields:

$$D_z = -1 \begin{vmatrix} 2 & 0 & -18 \\ 5 & 1 & 5 \\ 11 & 0 & -31 \end{vmatrix} = -1 \begin{vmatrix} 2 & -18 \\ -11 & -31 \end{vmatrix} = -1(-260) = 260$$

Adding row 1 to row 2, 2 times row 1 to row 3, and –2 times row 1 to row 4 yields:

$$D_w = \begin{vmatrix} 1 & 1 & 1 & -7 \\ 1 & -1 & 1 & -11 \\ 2 & -2 & -3 & 26 \\ 3 & 2 & 1 & -9 \end{vmatrix} = \begin{vmatrix} 1 & 1 & 1 & -7 \\ 2 & 0 & 2 & -18 \\ 4 & 0 & -1 & 12 \\ 1 & 0 & -1 & 5 \end{vmatrix} = -1 \begin{vmatrix} 2 & 2 & -18 \\ 4 & -1 & 12 \\ 1 & -1 & 5 \end{vmatrix} = -2 \begin{vmatrix} 1 & 1 & -9 \\ 4 & -1 & 12 \\ 1 & -1 & 5 \end{vmatrix}$$

Adding row 1 to both row 2 and row 3 yields:

$$D_w = -2 \begin{vmatrix} 1 & 1 & -9 \\ 5 & 0 & 3 \\ 2 & 0 & -4 \end{vmatrix} = 2 \begin{vmatrix} 5 & 3 \\ 2 & -4 \end{vmatrix} = 2(-26) = -52$$

So $x = \frac{D_x}{D} = 1$, $y = \frac{D_y}{D} = 0$, $z = \frac{D_z}{D} = -10$, and $w = \frac{D_w}{D} = 2$. So the solution is $(1, 0, -10, 2)$.

43. Expanding the determinant, we have:

$$\begin{vmatrix} x-4 & 0 & 0 \\ 0 & x+4 & 0 \\ 0 & 0 & x+1 \end{vmatrix} = (x-4) \begin{vmatrix} x+4 & 0 \\ 0 & x+1 \end{vmatrix} = (x-4)(x+4)(x+1)$$

This will equal 0 when $x = 4$, $x = -4$, or $x = -1$.

45. Expanding along the first column, we have:

$$\begin{vmatrix} a & b & c \\ a & b & c \\ d & e & f \end{vmatrix} = a\begin{vmatrix} b & c \\ e & f \end{vmatrix} - a\begin{vmatrix} b & c \\ e & f \end{vmatrix} + d\begin{vmatrix} b & c \\ b & c \end{vmatrix}$$

$$= a(bf - ec) - a(bf - ec) + d(bc - bc)$$
$$= abf - aec - abf + aec + 0$$
$$= 0$$

47. Expanding the determinant:

$$\begin{vmatrix} a_1 + A_1 & b_1 & c_1 \\ a_2 + A_2 & b_2 & c_2 \\ a_3 + A_3 & b_3 & c_3 \end{vmatrix} = (a_1 + A_1)\begin{vmatrix} b_2 & c_2 \\ b_3 & c_3 \end{vmatrix} - (a_2 + A_2)\begin{vmatrix} b_1 & c_1 \\ b_3 & c_3 \end{vmatrix} + (a_3 + A_3)\begin{vmatrix} b_1 & c_1 \\ b_2 & c_2 \end{vmatrix}$$

$$= \left\{ a_1\begin{vmatrix} b_2 & c_2 \\ b_3 & c_3 \end{vmatrix} - a_2\begin{vmatrix} b_1 & c_1 \\ b_3 & c_3 \end{vmatrix} + a_3\begin{vmatrix} b_1 & c_1 \\ b_2 & c_2 \end{vmatrix} \right\}$$

$$+ \left\{ A_1\begin{vmatrix} b_2 & c_2 \\ b_3 & c_3 \end{vmatrix} - A_2\begin{vmatrix} b_1 & c_1 \\ b_3 & c_3 \end{vmatrix} + A_3\begin{vmatrix} b_1 & c_1 \\ b_2 & c_2 \end{vmatrix} \right\}$$

Now observe that the expression in the first set of braces is:

$$\begin{vmatrix} a_1 & b_1 & c_1 \\ a_2 & b_2 & c_2 \\ a_3 & b_3 & c_3 \end{vmatrix}$$

The expression in the second set of braces is:

$$\begin{vmatrix} A_1 & b_1 & c_1 \\ A_2 & b_2 & c_2 \\ A_3 & b_3 & c_3 \end{vmatrix}$$

49. Expanding the determinant on the left-hand side of the given equation along its first row, we obtain:

$$\begin{vmatrix} a_1 & b_1 & c_1 \\ a_2 & b_2 & c_2 \\ a_3 & b_3 & c_3 \end{vmatrix} = a_1\begin{vmatrix} b_2 & c_2 \\ b_3 & c_3 \end{vmatrix} - b_1\begin{vmatrix} a_2 & c_2 \\ a_3 & c_3 \end{vmatrix} + c_1\begin{vmatrix} a_2 & b_2 \\ a_3 & b_3 \end{vmatrix}$$

Next, expanding the determinant on the right-hand side of the given equation along its second row, we obtain:

$$\begin{vmatrix} a_2 & b_2 & c_2 \\ a_1 & b_1 & c_1 \\ a_3 & b_3 & c_3 \end{vmatrix} = -a_1\begin{vmatrix} b_2 & c_2 \\ b_3 & c_3 \end{vmatrix} + b_1\begin{vmatrix} a_2 & c_2 \\ a_3 & c_3 \end{vmatrix} - c_1\begin{vmatrix} a_2 & b_2 \\ a_3 & b_3 \end{vmatrix}$$

By inspection now, we observe that the two expressions for the determinants are negatives of one another.

51. Subtract the fourth row from each of the other three rows. After that, we have:

$$\begin{vmatrix} a & 0 & 0 & -d \\ 0 & b & 0 & -d \\ 0 & 0 & c & -d \\ 1 & 1 & 1 & 1+d \end{vmatrix} = abcd \begin{vmatrix} 1 & 0 & 0 & -1 \\ 0 & 1 & 0 & -1 \\ 0 & 0 & 1 & -1 \\ \frac{1}{a} & \frac{1}{b} & \frac{1}{c} & 1+\frac{1}{d} \end{vmatrix}$$

$$= abcd \begin{vmatrix} 1 & 0 & 0 & 0 \\ 0 & 1 & 0 & -1 \\ 0 & 0 & 1 & -1 \\ \frac{1}{a} & \frac{1}{b} & \frac{1}{c} & 1+\frac{1}{a}+\frac{1}{d} \end{vmatrix}$$

$$= abcd \begin{vmatrix} 1 & 0 & -1 \\ 0 & 1 & -1 \\ \frac{1}{b} & \frac{1}{c} & 1+\frac{1}{a}+\frac{1}{d} \end{vmatrix}$$

$$= abcd \begin{vmatrix} 1 & 0 & 0 \\ 0 & 1 & -1 \\ \frac{1}{b} & \frac{1}{c} & 1+\frac{1}{a}+\frac{1}{b}+\frac{1}{d} \end{vmatrix}$$

$$= abcd \begin{vmatrix} 1 & -1 \\ \frac{1}{c} & 1+\frac{1}{a}+\frac{1}{b}+\frac{1}{d} \end{vmatrix}$$

$$= abcd\left(1+\frac{1}{a}+\frac{1}{b}+\frac{1}{c}+\frac{1}{d}\right)$$

53. By expanding D along its first column, we obtain the equation:

$$a_1 D = a_1\left[a_1(b_2c_3 - b_3c_2) - a_2(b_1c_3 - b_3c_1) + a_3(b_1c_2 - b_2c_1)\right] \quad **$$

On the other hand $\begin{vmatrix} B_2 & C_2 \\ B_3 & C_3 \end{vmatrix}$ is equal to:

$$B_2C_3 - B_3C_2 = (a_1c_3 - a_3c_1)(a_1b_2 - a_2b_1) - (a_2c_1 - a_1c_2)(a_3b_1 - a_1b_3)$$

$$= a_1^2b_2c_3 - a_1a_3b_2c_1 - a_1a_2b_1c_3 + a_2a_3b_1c_1 - a_1^2b_3c_2$$
$$+ a_1a_3b_1c_2 + a_1a_2b_3c_1 - a_2a_3b_1c_1$$

$$= a_1\left[a_1b_2c_3 - a_3b_2c_1 - a_2b_1c_3 - a_1b_3c_2 + a_3b_1c_2 + a_2b_3c_1\right]$$

$$= a_1\left[a_1(b_2c_3 - b_3c_2) - a_2(b_1c_3 - b_3c_1) + a_3(b_1c_2 - b_2c_1)\right]$$

By inspection now, we see that this last expression agrees with the right-hand side of equation **. This proves that:

$$\begin{vmatrix} B_2 & C_2 \\ B_3 & C_3 \end{vmatrix} = a_1 D$$

55. (a) Expanding along the first row:
$$\begin{vmatrix} 1 & 0 & 0 \\ x & 1 & 0 \\ x & y & 1 \end{vmatrix} = 1 \begin{vmatrix} 1 & 0 \\ y & 1 \end{vmatrix} = 1(1) = 1$$

(b) Expanding along the first row:
$$\begin{vmatrix} 1 & 0 & 0 & 0 \\ x & 1 & 0 & 0 \\ x & y & 1 & 0 \\ x & y & z & 1 \end{vmatrix} = 1 \begin{vmatrix} 1 & 0 & 0 \\ y & 1 & 0 \\ y & z & 1 \end{vmatrix} = 1 \begin{vmatrix} 1 & 0 \\ z & 1 \end{vmatrix} = 1(1) = 1$$

57. Subtracting row 1 from each of row 2, row 3, and row 4 yields:
$$\begin{vmatrix} 1 & a & a & a \\ 1 & b & a & a \\ 1 & a & b & a \\ 1 & a & a & b \end{vmatrix} = \begin{vmatrix} 1 & a & a & a \\ 0 & b-a & 0 & 0 \\ 0 & 0 & b-a & 0 \\ 0 & 0 & 0 & b-a \end{vmatrix}$$
$$= 1 \begin{vmatrix} b-a & 0 & 0 \\ 0 & b-a & 0 \\ 0 & 0 & b-a \end{vmatrix}$$
$$= (b-a) \begin{vmatrix} b-a & 0 \\ 0 & b-a \end{vmatrix}$$
$$= (b-a)(b-a)^2$$
$$= (b-a)^3$$

59. We form the augmented matrix:
$$\begin{pmatrix} a & b & c & k \\ a^2 & b^2 & c^2 & k^2 \\ a^3 & b^3 & c^3 & k^3 \end{pmatrix}$$
Adding $-a$ times row 1 to row 2 and $-a^2$ times row 1 to row 3:
$$\begin{pmatrix} a & b & c & k \\ 0 & b^2 - ab & c^2 - ac & k^2 - ak \\ 0 & b^3 - a^2b & c^3 - a^2c & k^3 - a^2k \end{pmatrix}$$
Factoring:
$$\begin{pmatrix} a & b & c & k \\ 0 & b(b-a) & c(c-a) & k(k-a) \\ 0 & b(b+a)(b-a) & c(c+a)(c-a) & k(k+a)(k-a) \end{pmatrix}$$

Adding $-(b+a)$ times row 2 to row 3:

$$\begin{pmatrix} a & b & c & k \\ 0 & b(b-a) & c(c-a) & k(k-a) \\ 0 & 0 & c(c-a)(c-b) & k(k-a)(k-b) \end{pmatrix}$$

So we have the system:

$$ax + by \qquad\qquad + cz = k$$
$$b(b-a)y + c(c-a)z = k(k-a)$$
$$c(c-a)(c-b)z = k(k-a)(k-b)$$

Solving the third equation for z:

$$c(c-a)(c-b)z = k(k-a)(k-b)$$
$$z = \frac{k(k-a)(k-b)}{c(c-a)(c-b)}$$

Substitute into the second equation:

$$b(b-a)y + \frac{k(k-a)(k-b)}{c-b} = k(k-a)$$
$$b(b-a)(c-b)y = k(k-a)(c-b-k+b)$$
$$y = \frac{k(k-a)(k-c)}{b(b-a)(b-c)}$$

Substitute into the first equation:

$$ax + \frac{k(k-a)(k-c)}{(b-a)(b-c)} + \frac{k(k-a)(k-b)}{(c-a)(c-b)} = k$$
$$ax = \frac{k(k-b)(k-c)}{(b-a)(c-a)}$$
$$x = \frac{k(k-b)(k-c)}{a(a-b)(a-c)}$$

So the solution is $\left(\frac{k(k-b)(k-c)}{a(a-b)(a-c)}, \frac{k(k-a)(k-c)}{b(b-a)(b-c)}, \frac{k(k-a)(k-b)}{c(c-a)(c-b)} \right)$.

61. We evaluate the determinant by expanding along the first row:

$$\begin{vmatrix} x & y & 1 \\ -3 & -1 & 1 \\ 2 & 9 & 1 \end{vmatrix} = x(-1-9) - y(-3-2) + 1(-27+2) = -10x + 5y - 25$$

Setting this equal to 0, we have:

$$-10x + 5y - 25 = 0$$
$$5y = 10x + 25$$
$$y = 2x + 5$$

63. We re-draw the figure:

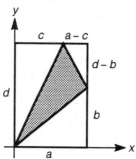

The area of the rectangle is ad, and the three triangles have areas of $\frac{1}{2}(ab)$, $\frac{1}{2}(a-c)(d-b)$, and $\frac{1}{2}(cd)$, so the area of the shaded triangle is:

$$ad - \left[\tfrac{1}{2}ab + \tfrac{1}{2}ad - \tfrac{1}{2}cd - \tfrac{1}{2}ab + \tfrac{1}{2}bc + \tfrac{1}{2}cd\right] = \tfrac{1}{2}ad - \tfrac{1}{2}bc = \tfrac{1}{2}(ad - bc) = \tfrac{1}{2}\begin{vmatrix} a & b \\ c & d \end{vmatrix}$$

65. (a) From Exercise 36 of the previous section, we found $A^{-1} = \begin{pmatrix} \frac{d}{ad-bc} & \frac{-b}{ad-bc} \\ \frac{-c}{ad-bc} & \frac{a}{ad-bc} \end{pmatrix}$. To check this, note that A times this matrix equals I_2. Since $D = ad - bc$, then

$$A^{-1} = \tfrac{1}{D}\begin{pmatrix} d & -b \\ -c & a \end{pmatrix}.$$

(b) We have $D = -6(9) - 7(1) = -54 - 7 = -61$, so the inverse is:

$$-\tfrac{1}{61}\begin{pmatrix} 9 & -7 \\ -1 & -6 \end{pmatrix} = \begin{pmatrix} -\frac{9}{61} & \frac{7}{61} \\ \frac{1}{61} & \frac{6}{61} \end{pmatrix}$$

Optional TI-81 Graphing Calculator Exercises for Sections 9.3, 9.4, and 9.5

1. (a) The product is $AB = \begin{pmatrix} -12 & 17 & -37 \\ -5 & 22 & -8 \\ 7 & -2 & 26 \end{pmatrix}$.

(b) The product is $BA = \begin{pmatrix} 19 & -7 & 6 \\ 5 & 17 & 4 \\ -5 & 16 & 0 \end{pmatrix}$.

(c) The inverse is $A^{-1} = \begin{pmatrix} -3 & 2 & -4 \\ -1 & 1 & -1 \\ 8 & -5 & 10 \end{pmatrix}$, which confirms the matrix given in the text.

(d)   The (rounded) inverse is $B^{-1} = \begin{pmatrix} -0.116 & 0.674 & -0.767 \\ 0.035 & 0.198 & -0.07 \\ -0.081 & -0.128 & 0.163 \end{pmatrix}$.

(e)   The determinant of A is 1.

3.   (a)   The determinant of A is 1.
     (b)   The determinant of B is -452.
     (c)   The determinant of AB is -452.
     (d)   This value is $1(-452) = -452$. Note this is the same answer as (c).

5.   (a)   The determinant of B^{-1} is -0.0022123894.
     (b)   This value is -0.0022123894. Note this is the same answer as (a).

9.6 Nonlinear Systems of Equations

1.   Substituting, we obtain:
$$x^2 = 3x$$
$$x^2 - 3x = 0$$
$$x(x-3) = 0$$
From which we get $x = 0$ or $x = 3$. If $x = 0$, $y = 3 \cdot 0 = 0$ and if $x = 3$, $y = 9$. Our solutions are $(0,0)$ and $(3,9)$. Note that we have found the points of intersection of a line and a parabola.

3.   Since $x^2 = 24y$, $x^2 + y^2 = 25$ becomes:
$$24y + y^2 = 25$$
$$y^2 + 24y - 25 = 0$$
$$(y+25)(y-1) = 0$$
$$y = -25 \text{ and } y = 1$$
When $y = 1$ we get $x^2 = 24$ or $x = \pm 2\sqrt{6}$, but $y = -25$ means $x^2 = -600$ which is impossible. Our two solutions are $\left(2\sqrt{6},1\right)$ and $\left(-2\sqrt{6},1\right)$.

5.   Substitute into the first equation:
$$x\left(-x^2\right) = 1$$
$$-x^3 = 1$$
$$x = -1$$
Since $y = \frac{1}{x}$, then $y = -1$. So the only solution is $(-1,-1)$.

7.   Multiply the first equation by -2:
$$-4x^2 - 2y^2 = -34$$
$$x^2 + 2y^2 = 22$$

Adding, we obtain:
$$-3x^2 = -12$$
$$x^2 = 4$$
$$x = \pm 2$$
When $x = 2$, we have:
$$2(4) + y^2 = 17$$
$$y^2 = 9$$
$$y = \pm 3$$
When $x = -2$, we have:
$$2(4) + y^2 = 17$$
$$y^2 = 9$$
$$y = \pm 3$$
So the solutions are $(2, 3)$, $(2, -3)$, $(-2, 3)$, and $(-2, -3)$.

9. Substitute into the first equation:
$$x^2 - 1 = 1 - x^2$$
$$2x^2 = 2$$
$$x^2 = 1$$
$$x = \pm 1$$
So $y = 1 - 1 = 0$ for each value of x. So the solutions are $(1, 0)$ and $(-1, 0)$.

11. Substitute into the first equation:
$$x(4x + 1) = 4$$
$$4x^2 + x - 4 = 0$$
$$x = \frac{-1 \pm \sqrt{1 + 64}}{8} = \frac{-1 \pm \sqrt{65}}{8}$$
When $x = \frac{-1+\sqrt{65}}{8}$, we have:
$$y = \frac{-1 + \sqrt{65}}{2} + 1 = \frac{1 + \sqrt{65}}{2}$$
When $x = \frac{-1-\sqrt{65}}{8}$, we have:
$$y = \frac{-1 - \sqrt{65}}{2} + 1 = \frac{1 - \sqrt{65}}{2}$$
So the solutions are $\left(\frac{-1+\sqrt{65}}{8}, \frac{1+\sqrt{65}}{2}\right)$ and $\left(\frac{-1-\sqrt{65}}{8}, \frac{1-\sqrt{65}}{2}\right)$.

13. Let $a = \frac{1}{x^2}$ and $b = \frac{1}{y^2}$, so:
$$a - 3b = 14$$
$$2a + b = 35$$
Multiply the first equation by -2:
$$-2a + 6b = -28$$
$$2a + b = 35$$

Adding, we obtain:
$$7b = 7$$
$$b = 1$$
So $a - 3 = 14$, and $a = 17$. Since $a = \frac{1}{x^2}$ and $b = \frac{1}{y^2}$, we have $x^2 = \frac{1}{17}$ and $y^2 = 1$.

So $x = \frac{\pm\sqrt{17}}{17}$ and $y = \pm 1$. So the solutions are $\left(\frac{\sqrt{17}}{17}, 1\right)$, $\left(\frac{\sqrt{17}}{17}, -1\right)$, $\left(\frac{-\sqrt{17}}{17}, 1\right)$, and $\left(\frac{-\sqrt{17}}{17}, -1\right)$.

15. Substitute into the second equation:
$$(x - 3)^2 + \left(-\sqrt{x-1}\right)^2 = 4$$
$$x^2 - 6x + 9 + x - 1 = 4$$
$$x^2 - 5x + 4 = 0$$
$$(x - 1)(x - 4) = 0$$
$$x = 1 \text{ or } x = 4$$
When $x = 1$, $y = -\sqrt{1-1} = 0$ and when $x = 4$, $y = -\sqrt{4-1} = -\sqrt{3}$. So the solutions are $(1, 0)$ and $\left(4, -\sqrt{3}\right)$.

17. Since $y = 2^{2x} - 12 = \left(2^x\right)^2 - 12$, we substitute into the second equation:
$$y = y^2 - 12$$
$$0 = y^2 - y - 12$$
$$0 = (y - 4)(y + 3)$$
$$y = 4 \text{ or } y = -3$$
When $y = 4$, we have $2^x = 4$, or $x = 2$. But $y = -3$ will not yield a solution. So the only solution is $(2, 4)$.

19. Let $u = \log_{10} x$ and $v = \log_{10} y$, so:
$$2u^2 - v^2 = -1$$
$$4u^2 - 3v^2 = -11$$
Multiply the first equation by -2:
$$-4u^2 + 2v^2 = 2$$
$$4u^2 - 3v^2 = -11$$
Adding, we obtain:
$$-v^2 = -9$$
$$v^2 = 9$$
$$v = \pm 3$$
Substitute into the first equation:
$$2u^2 - 9 = -1$$
$$2u^2 = 8$$
$$u^2 = 4$$
$$u = \pm 2$$
Since $u = \log_{10} x$, then $x = 10^{\pm 2}$. Similarly, $y = 10^{\pm 3}$. So the solutions are $(100, 1000)$, $\left(100, \frac{1}{1000}\right)$, $\left(\frac{1}{100}, 1000\right)$, and $\left(\frac{1}{100}, \frac{1}{1000}\right)$.

21. First take the logarithm of each side of the first equation:

$$\ln(2^x 3^y) = \ln 4$$
$$\ln(2^x) + \ln(3^y) = \ln 2^2$$
$$(\ln 2)x + (\ln 3)y = 2\ln 2$$

Multiply the second equation by $-\ln 2$:

$$(\ln 2)x + (\ln 3)y = 2\ln 2$$
$$(-\ln 2)x - (\ln 2)y = -5\ln 2$$

Adding, we obtain:

$$(\ln 3 - \ln 2)y = -3\ln 2$$
$$y = \frac{3\ln 2}{\ln 2 - \ln 3}$$

Substitute into the second equation:

$$x = 5 - \frac{3\ln 2}{\ln 2 - \ln 3} = \frac{2\ln 2 - 5\ln 3}{\ln 2 - \ln 3}$$

The solution is $\left(\frac{2\ln 2 - 5\ln 3}{\ln 2 - \ln 3}, \frac{3\ln 2}{\ln 2 - \ln 3}\right)$.

23. Substituting x yields:

$$y = 3x + 1 = \frac{-3 + 3\sqrt{13}}{6} + 1 = \frac{3 + 3\sqrt{13}}{6} = \frac{1 + \sqrt{13}}{2}$$

$$y = \frac{1}{x} = \frac{6}{-1 + \sqrt{13}} \cdot \frac{-1 - \sqrt{13}}{-1 - \sqrt{13}} = \frac{-6(1 + \sqrt{13})}{1 - 13} = \frac{-6(1 + \sqrt{13})}{-12} = \frac{1 + \sqrt{13}}{2}$$

They both yield the same y-value.

25. Since $ax + by = 2$, then $by = 2 - ax$. Substitute into the second equation:

$$ax(by) = 1$$
$$ax(2 - ax) = 1$$
$$2ax - a^2x^2 = 1$$
$$a^2x^2 - 2ax + 1 = 0$$
$$(ax - 1)^2 = 0$$
$$ax = 1$$
$$x = \frac{1}{a}$$

When $ax = 1$, $by = 2 - 1 = 1$, so $y = \frac{1}{b}$. So the solution is $\left(\frac{1}{a}, \frac{1}{b}\right)$.

27. Solve the second equation for y to get $y = 23 - x$. Substitute into the first equation:

$$x^3 + (23 - x)^3 = 3473$$
$$x^3 + 12167 - 1587x + 69x^2 - x^3 = 3473$$
$$69x^2 - 1587x + 8694 = 0$$
$$x^2 - 23x + 126 = 0$$
$$(x - 9)(x - 14) = 0$$
$$x = 9 \text{ or } x = 14$$

When $x = 9$, $y = 14$ and when $x = 14$, $y = 9$. So the solutions are $(9, 14)$ and $(14, 9)$.

29. First draw the rectangle:

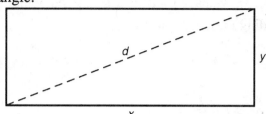

Now $x^2 + y^2 = d^2$ and $2x + 2y = 2p$. Solve the second equation for y:

$$2x + 2y = 2p$$
$$2y = 2p - 2x$$
$$y = p - x$$

Substitute into the first equation:

$$x^2 + (p - x)^2 = d^2$$
$$x^2 + p^2 - 2px + x^2 = d^2$$
$$2x^2 - 2px + p^2 - d^2 = 0$$

Using the quadratic formula, we have:

$$x = \frac{2p \pm \sqrt{4p^2 - 8(p^2 - d^2)}}{4} = \frac{2p \pm 2\sqrt{2d^2 - p^2}}{4} = \frac{p \pm \sqrt{2d^2 - p^2}}{2}$$

When $x = \dfrac{p + \sqrt{2d^2 - p^2}}{2}$, we have:

$$y = p - \frac{p + \sqrt{2d^2 - p^2}}{2} = \frac{p - \sqrt{2d^2 - p^2}}{2}$$

When $x = \dfrac{p - \sqrt{2d^2 - p^2}}{2}$, we have:

$$y = p - \frac{p - \sqrt{2d^2 - p^2}}{2} = \frac{p + \sqrt{2d^2 - p^2}}{2}$$

So the rectangle has dimenstions $\dfrac{p - \sqrt{2d^2 - p^2}}{2}$ by $\dfrac{p - \sqrt{2d^2 - p^2}}{2}$.

31. If we follow the hint we have:

$$\left(\sqrt{u + v} + \sqrt{u - v}\right)^2 = 4^2$$
$$u + v + 2\sqrt{u^2 - v^2} + u - v = 16$$

But since $u^2 - v^2 = 9$ from the first equation:

$$2u + 2\sqrt{9} = 16$$
$$2u + 6 = 16$$
$$2u = 10, \text{ so } u = 5$$

When $u = 5$, we have:

$$5^2 - v^2 = 9$$
$$v^2 = 16$$
$$v = \pm 4$$

So the solutions are $(5, 4)$ and $(5, -4)$.

33. (a)   Using the point $(2, 3)$, we have $3 = N_0 e^{2k}$. Using the point $(8, 24)$, we have $24 = N_0 e^{8k}$. Dividing these two yields:

$$\frac{24}{3} = \frac{N_0 e^{8k}}{N_0 e^{2k}}$$

$$8 = e^{6k}$$

$$\ln 8 = 6k$$

$$k = \frac{\ln 8}{6}$$

Substituting into the first equation, we have:

$$3 = N_0 e^{\frac{1}{3}\ln 8}$$

$$3 = N_0 e^{\ln(8^{1/3})}$$

$$3 = N_0 \cdot 2$$

$$N_0 = \tfrac{3}{2}$$

(b)   Using the point $\left(\frac{1}{2}, 1\right)$, we have $1 = N_0 e^{\frac{1}{2}k}$. Using the point $(4, 10)$, we have $10 = N_0 e^{4k}$. Dividing these two yields:

$$\frac{10}{1} = \frac{N_0 e^{4k}}{N_0 e^{\frac{1}{2}k}}$$

$$10 = e^{\frac{7}{2}k}$$

$$\ln 10 = \tfrac{7}{2}k$$

$$k = \tfrac{2}{7}\ln 10$$

Substituting into the first equation, we have:

$$1 = N_0 e^{\frac{1}{7}\ln 10}$$

$$1 = N_0 e^{\ln(10^{1/7})}$$

$$1 = N_0 \cdot 10^{1/7}$$

$$N_0 = 10^{-1/7}$$

35.   Let $w = x + y + z$, so:

$$xw = p^2$$

$$yw = q^2$$

$$zw = r^2$$

Adding the three equations, we obtain:

$$(x + y + z)w = p^2 + q^2 + r^2$$

$$w^2 = p^2 + q^2 + r^2$$

$$w = \pm\sqrt{p^2 + q^2 + r^2}$$

Substitute into the first equation:

$$x = \frac{p^2}{\pm\sqrt{p^2 + q^2 + r^2}}$$

Similarly $y = \frac{q^2}{\pm A}$ and $z = \frac{r^2}{\pm A}$ where $A = \sqrt{p^2 + q^2 + r^2}$. So the solutions are $\left(\frac{p^2}{A}, \frac{q^2}{A}, \frac{r^2}{A}\right)$ and $\left(\frac{-p^2}{A}, \frac{-q^2}{A}, \frac{-r^2}{A}\right)$, where $A = \sqrt{p^2 + q^2 + r^2}$.

37. The area is given by $A = \frac{1}{2}bh$, where b and h are the missing legs. We have:

$180 = \frac{1}{2}bh$, so $bh = 360$ thus $b = \frac{360}{h}$

Also from the Pythagorean theorem $b^2 + h^2 = 41^2$, so:

$$\left(\frac{360}{h}\right)^2 + h^2 = 41^2$$
$$\frac{129600}{h^2} + h^2 = 1681$$
$$129600 + h^4 = 1681h^2$$
$$h^4 - 1681h^2 + 129600 = 0$$
$$\left(h^2 - 1600\right)\left(h^2 - 81\right) = 0$$
$$h^2 = 1600 \quad \text{or} \quad h^2 = 81$$
$$h = \pm 40 \quad \text{or} \quad h = \pm 9$$

Since these must be the length of the sides, we can neglect the negative values. So $h = 40$ and $h = 9$ are solutions. Since $b = \frac{360}{h}$ we obtain $b = 9$ when $h = 40$, and $b = 40$ when $h = 9$. So the legs are 9 cm and 40 cm.

39. We have $lw = 60$ and $2l + 2w = 46$, so $l + w = 23$ and $l = 23 - w$. We substitute this into the first equation:

$$(23 - w)(w) = 60$$
$$23w - w^2 = 60$$
$$w^2 - 23w + 60 = 0$$
$$(w - 20)(w - 3) = 0$$
$$w = 20 \quad \text{or} \quad w = 3$$

When $w = 20$ we have $l = 3$, and when $w = 3$ we have $l = 20$. So the rectangle must be 3 cm by 20 cm.

41. Solve $xy = 2$ to get $y = \frac{2}{x}$. Now substitute into the first equation:

$$x^2 + \frac{4}{x^2} = 5$$
$$x^4 + 4 = 5x^2$$
$$x^4 - 5x^2 + 4 = 0$$
$$\left(x^2 - 1\right)\left(x^2 - 4\right) = 0$$
$$x^2 = 1 \quad \text{or} \quad x^2 = 4$$
$$x = \pm 1 \qquad x = \pm 2$$

When $x = 1$ we have $y = 2$, when $x = -1$ we have $y = -2$, when $x = 2$ we have $y = 1$, and when $x = -2$ we have $y = -1$. So the solutions are $(1, 2)$, $(-1, -2)$, $(2, 1)$, and $(-2, -1)$.

43. Multiply the second equation by 2:
$$2xy = 6$$
Adding to the first equation, we obtain:
$$x^2 + 2xy + y^2 = 13$$
$$(x + y)^2 = 13$$
$$x + y = \pm\sqrt{13}$$
Subtracting from the first equation, we obtain:
$$x^2 - 2xy + y^2 = 1$$
$$(x - y)^2 = 1$$
$$x - y = \pm 1$$
We solve the four systems of equations:

$x + y = \sqrt{13}$	$x + y = \sqrt{13}$	$x + y = -\sqrt{13}$	$x + y = -\sqrt{13}$
$x - y = 1$	$x - y = -1$	$x - y = 1$	$x - y = -1$
$2x = 1 + \sqrt{13}$	$2x = -1 + \sqrt{13}$	$2x = 1 - \sqrt{13}$	$2x = -1 - \sqrt{13}$
$x = \dfrac{1 + \sqrt{13}}{2}$	$x = \dfrac{-1 + \sqrt{13}}{2}$	$x = \dfrac{1 - \sqrt{13}}{2}$	$x = \dfrac{-1 - \sqrt{13}}{2}$
$y = \dfrac{-1 + \sqrt{13}}{2}$	$y = \dfrac{1 + \sqrt{13}}{2}$	$y = \dfrac{-1 - \sqrt{13}}{2}$	$y = \dfrac{1 - \sqrt{13}}{2}$

So the solutions are $\left(\frac{1+\sqrt{13}}{2}, \frac{-1+\sqrt{13}}{2}\right)$, $\left(\frac{-1+\sqrt{13}}{2}, \frac{1+\sqrt{13}}{2}\right)$, $\left(\frac{1-\sqrt{13}}{2}, \frac{-1-\sqrt{13}}{2}\right)$, and
$\left(\frac{-1-\sqrt{13}}{2}, \frac{1-\sqrt{13}}{2}\right)$.

45. Solving by factoring, we have:
$$2m^2 - 7m + 6 = 0$$
$$(2m - 3)(m - 2) = 0$$
$$m = \tfrac{3}{2} \text{ or } m = 2$$
So we have the equations:
$$x^2\left(\tfrac{9}{2} - 4\right) = 2 \qquad \text{or} \qquad x^2(6 - 4) = 2$$
$$x^2 = 4 \qquad\qquad\qquad x^2 = 1$$
$$x = \pm 2 \qquad\qquad\qquad x = \pm 1$$
$$y = \pm 3 \qquad\qquad\qquad y = \pm 2$$
So the solutions are $(2, 3)$, $(-2, -3)$, $(1, 2)$, and $(-1, -2)$.

47. Using the hint, we square the first equation:
$$x^2 y^2 + 2pqxy + p^2 q^2 = 4p^2 x^2$$
$$x^2 y^2 \qquad\quad + p^2 q^2 = 2q^2 y^2$$
Subtracting the second equation from the first:
$$2pqxy = 4p^2 x^2 - 2q^2 y^2$$
$$pqxy = 2p^2 x^2 - q^2 y^2$$
$$0 = 2p^2 x^2 - pqxy - q^2 y^2$$
$$0 = (2px + qy)(px - qy)$$

So we have the equations:

$$2px + qy = 0, \text{ so } y = -\frac{2p}{q}x$$

$$px - qy = 0, \text{ so } y = \frac{p}{q}x$$

Substituting into the first equation:

$$x\left(-\frac{2p}{q}x\right) + pq = 2px$$

$$x\left(\frac{p}{q}x\right) + pq = 2px$$

Multiplying each side by $\frac{q}{p}$:

$$-2x^2 - 2qx + q^2 = 0$$

$$x^2 - 2qx + q^2 = 0$$

The second equation factors to $(x - q)^2 = 0$, so $x = q$ and thus $y = \frac{p}{q} \cdot q = p$. Thus one solution is (q,p). This solution checks in the original (non-squared) first equation. For the first equation, we use the quadratic formula:

$$x = \frac{2q \pm \sqrt{4q^2 + 8q^2}}{-4} = \frac{2q \pm 2q\sqrt{3}}{-4} = \frac{-1 \pm \sqrt{3}}{2}q$$

$$y = -\frac{2p}{q}\left(\frac{-1 \pm \sqrt{3}}{2}\right)q = (1 \pm \sqrt{3})p$$

This results in two other solutions, $\left(\frac{-1+\sqrt{3}}{2}q, (1 - \sqrt{3})p\right)$ and $\left(\frac{-1-\sqrt{3}}{2}q, (1 + \sqrt{3})p\right)$. Both of these solutions check in the original (non-squared) first equation.

49. Taking logs in the first equation:

$$\ln(x^4) = \ln(y^6)$$

$$4\ln x = 6\ln y$$

$$2\ln x = 3\ln y$$

The second equation is:

$$\ln x - \ln y = \frac{\ln x}{\ln y}$$

$$\ln x \ln y - (\ln y)^2 = \ln x$$

Let $u = \ln x$ and $v = \ln y$, so we have the equations $2u = 3v$ and $uv - v^2 = u$. Solving the first equation for u yields $u = \frac{3v}{2}$, and substituting into the second equation yields:

$$\left(\frac{3v}{2}\right)v - v^2 = \frac{3v}{2}$$

$$3v^2 - 2v^2 = 3v$$

$$v^2 - 3v = 0$$

$$v(v - 3) = 0$$

$$v = 0 \text{ or } v = 3$$

When $v = 0$, $u = 0$ and when $v = 3$, $u = \frac{9}{2}$. Since $v = \ln y$, $v = 0$ cannot be a solution to the original second equation ($\ln y$ is the denominator). Thus $u = \ln x$ and $v = \ln y$ yields:

$\ln x = \frac{9}{2}$ so $x = e^{9/2}$

$\ln y = 3$ so $y = e^3$

So the only solution is $\left(e^{9/2}, e^3\right)$.

Optional TI-81 Graphing Calculator Exercises for Section 9.6

1. (a) Solving for y, we have $y = 2 - x$ and:

$$-3y = -2x + 9$$
$$y = \tfrac{2}{3}x - 3$$

We draw the graphs:

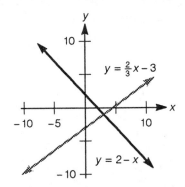

(b) The intersection point $(3, -1)$ is confirmed.

3. (a) Solving for y, we have $y = x - 1$ and $y = 2\sqrt{x + 3}$. We draw the graphs:

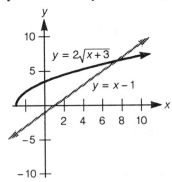

(b) Using the TRACE-ZOOM technique, the intersection point is approximately $(7.47, 6.47)$.

5. (a) We draw the graphs:

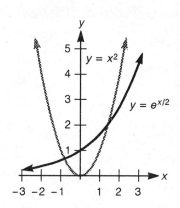

$y = x^2$
$y = e^{x/2}$

(b) Using the TRACE-ZOOM technique, the intersection points are approximately $(-0.82, 0.67)$ and $(1.43, 2.04)$

7. (a) Solving for y, we have $y = 2x - x^2$ and $y = 3x^3 - x^2 - 10x$. We draw the graphs:

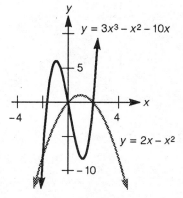

$y = 3x^3 - x^2 - 10x$
$y = 2x - x^2$

(b) Using the TRACE-ZOOM technique, the intersection points are $(-2, -8)$, $(0, 0)$, and $(2, 0)$. Note that these are exact coordinates.

9. (a) We draw the graphs:

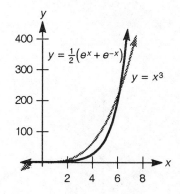

$y = \frac{1}{2}(e^x + e^{-x})$
$y = x^3$

(b) Using the TRACE-ZOOM technique, the intersection points are approximately $(1.23, 1.85)$ and $(6.14, 230.95)$.

9.7 Systems of Inequalities

1.   (a)   Substitute the pair $(1,2)$:
$$4(1) - 6(2) + 3 \geq 0$$
$$-5 \geq 0$$
Then $(1,2)$ is not a solution.

(b)   Substitute the pair $\left(0,\tfrac{1}{2}\right)$:
$$4(0) - 6\left(\tfrac{1}{2}\right) + 3 \geq 0$$
$$0 \geq 0$$
Then $\left(0,\tfrac{1}{2}\right)$ is a solution.

3.   We graph the region:

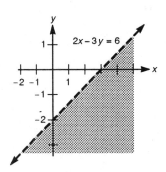

5.   We graph the region:

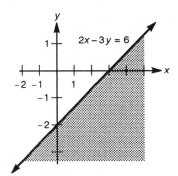

7.   We graph the region:

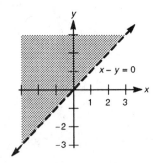

9. We graph the region:

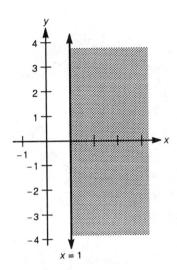

11. We graph the region:

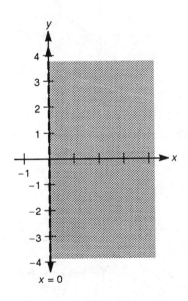

13. We graph the region:

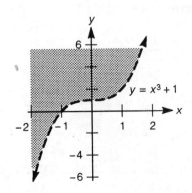

15. We graph the region:

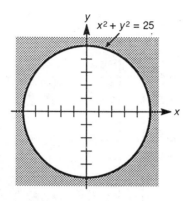

17. We graph the system of inequalities:

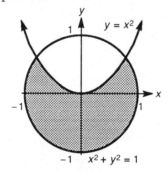

19. We graph the system of inequalities:

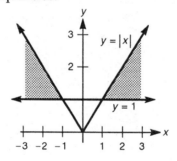

21. We graph the system of inequalities:

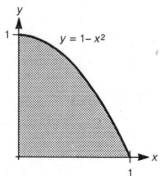

23. The region is convex and bounded. The vertices are $(0,0)$, $(7,0)$, $(3,8)$, and $(0,5)$. We graph the system of inequalities:

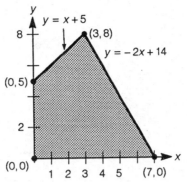

25. The region is convex and bounded. The vertices are $(0,0)$, $(0,4)$, $(3,5)$, and $(8,0)$. We graph the system of inequalities:

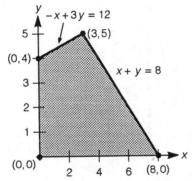

27. The region is convex but not bounded. The vertices are $(2,7)$ and $(8,5)$. We graph the system of inequalities:

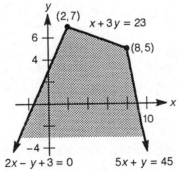

29. The region is convex but not bounded. The only vertex is $(6,0)$. We graph the system of inequalities:

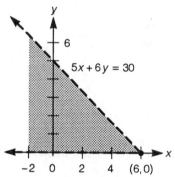

31. The region is convex and bounded. The vertices are $(0,0)$, $(0,5)$, and $(6,0)$. We graph the system of inequalities:

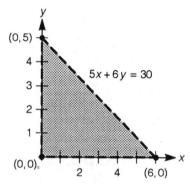

33. The region is convex and bounded. The vertices are $(5,30)$, $(10,30)$, $(20,15)$, and $(20,20)$. We graph the system of inequalities:

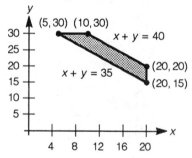

35. (a) We first find the intersection of $y = e^x$ and $y = e^{-x} + 1$:

$$e^x = e^{-x} + 1$$
$$\left(e^x\right)^2 = 1 + e^x$$
$$\left(e^x\right)^2 - e^x - 1 = 0$$

Using the quadratic formula, we have:

$$e^x = \frac{1 \pm \sqrt{1 - 4(-1)}}{2} = \frac{1 \pm \sqrt{5}}{2}$$

Since $e^x > 0$, we discard the negative root to obtain $e^x = \frac{1+\sqrt{5}}{2}$, so

$x = \ln \frac{1+\sqrt{5}}{2} \approx 0.48$.

Since $y = 1 + e^{-x}$, then $y = \frac{1+\sqrt{5}}{2} \approx 1.62$. The vertices are $(0, 1)$, $(0, 2)$ and

$\left(\ln \frac{1+\sqrt{5}}{2}, \frac{1+\sqrt{5}}{2} \right)$. We graph the inequalities:

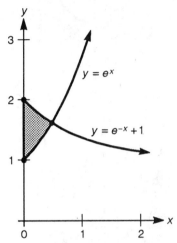

(b) Since $\sin x = \cos x$ at $x = \frac{\pi}{4}$ and $x = \frac{5\pi}{4}$, the vertices are $(0, 0)$, $(0, 1)$, $\left(\frac{\pi}{4}, \frac{\sqrt{2}}{2} \right)$,

$\left(\frac{5\pi}{4}, -\frac{\sqrt{2}}{2} \right)$, $(2\pi, 0)$, and $(2\pi, 1)$. We graph the inequalities:

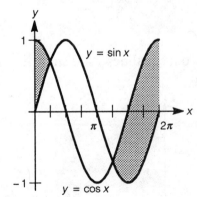

37. The domain results from the inequality $x^2 + y^2 - 1 \geq 0$, or $x^2 + y^2 \geq 1$. We sketch the graph:

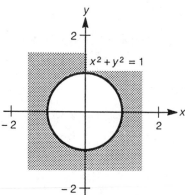

39. The domain results from the inequality $x^2 - y > 0$, so $y < x^2$. We sketch the graph:

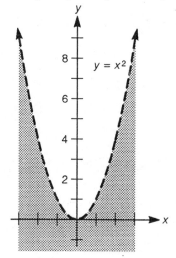

41. The domain results from two inequalities $x \geq 0$ and $y \geq 0$ (first quadrant and positive axes). We sketch the graph:

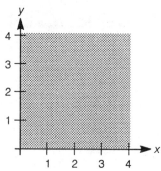

Chapter Nine Review Exercises

1. Adding the two equations yields:
$$2x = 6$$
$$x = 3$$
Substitute into equation 1:
$$3xy = -2$$
$$y = -5$$
So the solution is $(3, -5)$.

3. Multiply the first equation by -2:
$$-4x - 2y = -4$$
$$x + 2y = 7$$
Adding, we obtain:
$$-3x = 3$$
$$x = -1$$
Substitute into equation 1:
$$-2 + y = 2$$
$$y = 4$$
So the solution is $(-1, 4)$.

5. Multiply equation 1 by -5 and equation 2 by 2:
$$-35x - 10y = -45$$
$$8x + 10y = 126$$
Adding, we obtain:
$$-27x = 81$$
$$x = -3$$
Substitute into equation 1:
$$-21 + 2y = 9$$
$$2y = 30$$
$$y = 15$$
So the solution is $(-3, 15)$.

7. Multiply the first equation by 2 and the second equation by 24 to clear fractions:
$$4x - y = -16$$
$$8x + 3y = -24$$
Multiply equation 1 by -2:
$$-8x + 2y = 32$$
$$8x + 3y = -24$$
Adding, we obtain:
$$5y = 8$$
$$y = \tfrac{8}{5}$$

Substitute into equation 1:
$$4x - \tfrac{8}{5} = -16$$
$$4x = -\tfrac{72}{5}$$
$$x = -\tfrac{18}{5}$$
So the solution is $\left(-\tfrac{18}{5}, \tfrac{8}{5}\right)$.

9.   Multiply the first equation by 2:
$$6x + 10y = 2$$
$$9x - 10y = 8$$
Adding, we obtain:
$$15x = 10$$
$$x = \tfrac{2}{3}$$
Substitute into equation 1:
$$2 + 5y = 1$$
$$5y = -1$$
$$y = -\tfrac{1}{5}$$
So the solution is $\left(\tfrac{2}{3}, -\tfrac{1}{5}\right)$.

11.   Multiply the first equation by 6 and the second equation by 2 to clear fractions:
$$4x + 3y = -72$$
$$x - 2y = 4$$
Multiply the second equation by -4:
$$4x + 3y = -72$$
$$-4x + 8y = -16$$
Adding, we obtain:
$$11y = -88$$
$$y = -8$$
Substitute into equation 2:
$$x + 16 = 4$$
$$x = -12$$
So the solution is $(-12, -8)$.

13.   Let $a = \tfrac{1}{x}$ and $b = \tfrac{1}{y}$. So we have:
$$a + b = -1$$
$$2a + 5b = -14$$
Multiply the first equation by -2:
$$-2a - 2b = 2$$
$$2a + 5b = -14$$
Adding, we obtain:
$$3b = -12$$
$$b = -4$$

Substitute into equation 1:
$$a - 4 = -1$$
$$a = 3$$
Since $x = \frac{1}{a}$ and $y = \frac{1}{b}$, then $x = \frac{1}{3}$ and $y = -\frac{1}{4}$. So the solution is $\left(\frac{1}{3}, -\frac{1}{4}\right)$.

15. Multiply the second equation by $a - 1$:
$$ax + (1-a)y = 1$$
$$\left(-a^2 + 2a - 1\right)x + (a-1)y = 0$$
Adding, we obtain:
$$\left(-a^2 + 3a - 1\right)x = 1$$
$$x = \frac{-1}{a^2 - 3a + 1}$$
Substitute into equation 2:
$$\frac{a-1}{a^2 - 3a + 1} + y = 0$$
$$y = \frac{1-a}{a^2 - 3a + 1}$$
So the solution is $\left(\frac{-1}{a^2-3a+1}, \frac{1-a}{a^2-3a+1}\right)$. We must assume that $a^2 - 3a + 1 \neq 0$, or $a \neq \frac{3 \pm \sqrt{5}}{2}$.

17. Multiply the first equation by 2:
$$4x - 2y = 6a^2 - 2$$
$$x + 2y = -a^2 + 2$$
Adding, we obtain:
$$5x = 5a^2$$
$$x = a^2$$
Substitute into equation 2:
$$2y + a^2 = 2 - a^2$$
$$2y = 2 - 2a^2$$
$$y = 1 - a^2$$
So the solution is $\left(a^2, 1 - a^2\right)$.

19. Multiply the first equation by 3:
$$15x - 3y = 12a^2 - 18b^2$$
$$2x + 3y = 5a^2 + b^2$$
Adding, we obtain:
$$17x = 17a^2 - 17b^2$$
$$x = a^2 - b^2$$
Substitute into equation 1:
$$5a^2 - 5b^2 - y = 4a^2 - 6b^2$$
$$-y = -a^2 - b^2$$
$$y = a^2 + b^2$$
So the solution is $\left(a^2 - b^2, a^2 + b^2\right)$.

21. Multiply the first equation by p and the second equation by q:
$$p^2x - pqy = pq^2$$
$$q^2x + pqy = p^2q$$
Adding, we obtain:
$$(p^2 + q^2)x = pq(p + q)$$
$$x = \frac{pq(p+q)}{p^2 + q^2}$$
We re-solve the system to find y. Multiply the first equation by $-q$ and the second equation by p:
$$-pqx + q^2y = -q^3$$
$$pqx + p^2y = p^3$$
Adding we obtain:
$$(p^2 + q^2)y = p^3 - q^3$$
$$y = \frac{p^3 - q^3}{p^2 + q^2}$$
So the solution is $\left(\frac{pq(p+q)}{p^2+q^2}, \frac{p^3-q^3}{p^2+q^2}\right)$. We must assume that p and q are not both 0.

23. Let $u = \frac{1}{x}$ and $v = \frac{1}{y}$, so we have:
$$4au - 3bv = a - 7b$$
$$2a^2u - 2b^2v = 3a^2 - 5ab - 2b^2$$
Multiply the first equation by $-2b$ and the second equation by 3:
$$-8abu + 6b^2v = 14b^2 - 2ab$$
$$9a^2u - 6b^2v = 9a^2 - 15ab - 6b^2$$
Adding, we obtain:
$$(9a^2 - 8ab)u = 9a^2 - 17ab + 8b^2$$
$$u = \frac{9a^2 - 17ab + 8b^2}{9a^2 - 8ab} = \frac{(9a - 8b)(a - b)}{a(9a - 8b)} = \frac{a - b}{a}$$
Substitute into equation 1:
$$4a\left(\frac{a-b}{a}\right) - 3bv = a - 7b$$
$$4a - 4b - 3bv = a - 7b$$
$$-3bv = -3a - 3b$$
$$v = \frac{a+b}{b}$$
Since $x = \frac{1}{u}$ and $y = \frac{1}{v}$, we have $x = \frac{a}{a-b}$ and $y = \frac{b}{a+b}$. So the solution is $\left(\frac{a}{a-b}, \frac{b}{a+b}\right)$. We must assume that $ab \neq 0$ and $a \neq \pm b$.

25. Form the augmented matrix:
$$\begin{pmatrix} 1 & 1 & 1 & 9 \\ 1 & -1 & -1 & -5 \\ 2 & 1 & -2 & -1 \end{pmatrix}$$
Add -1 times row 1 to row 2 and -2 times row 1 to row 3:
$$\begin{pmatrix} 1 & 1 & 1 & 9 \\ 0 & -2 & -2 & -14 \\ 0 & -1 & -4 & -19 \end{pmatrix}$$
Switch row 2 and row 3, multiply each by -1:
$$\begin{pmatrix} 1 & 1 & 1 & 9 \\ 0 & 1 & 4 & 19 \\ 0 & 2 & 2 & 14 \end{pmatrix}$$
Add -2 times row 2 to row 3:
$$\begin{pmatrix} 1 & 1 & 1 & 9 \\ 0 & 1 & 4 & 19 \\ 0 & 0 & -6 & -24 \end{pmatrix}$$
So we have the system:
$$\begin{aligned} x + y + z &= 9 \\ y + 4z &= 19 \\ -6z &= -24 \end{aligned}$$
Solve equation 3 for z:
$$-6z = -24$$
$$z = 4$$
Substitute into equation 2:
$$y + 16 = 19$$
$$y = 3$$
Substitute into equation 1:
$$x + 3 + 4 = 9$$
$$x = 2$$
So the solution is $(2, 3, 4)$.

27. Switching equations 1 and 3, form the augmented matrix:
$$\begin{pmatrix} 1 & 1 & 1 & -3 \\ 2 & 3 & 3 & -8 \\ 4 & -4 & 1 & 4 \end{pmatrix}$$
Add -2 times row 1 to row 2 and -4 times row 1 to row 3:
$$\begin{pmatrix} 1 & 1 & 1 & -3 \\ 0 & 1 & 1 & -2 \\ 0 & -8 & -3 & 16 \end{pmatrix}$$

Add 8 times row 2 to row 3:

$$\begin{pmatrix} 1 & 1 & 1 & -3 \\ 0 & 1 & 1 & -2 \\ 0 & 0 & 5 & 0 \end{pmatrix}$$

So we have the system:

$$x + y + z = -3$$
$$y + z = -2$$
$$5z = 0$$

Solve equation 3 for z:

$$5z = 0$$
$$z = 0$$

Substitute into equation 2:

$$y + 0 = -2$$
$$y = -2$$

Substitute into equation 1:

$$x - 2 + 0 = -3$$
$$x = -1$$

So the solution is $(-1, -2, 0)$.

29. Using equation 3 as row 1, form the augmented matrix:

$$\begin{pmatrix} 1 & 1 & -2 & 4 \\ 1 & -2 & 1 & -2 \\ -2 & 1 & 1 & 1 \end{pmatrix}$$

Add -1 times row 1 to row 2 and 2 times row 1 to row 3:

$$\begin{pmatrix} 0 & 1 & -2 & 4 \\ 0 & -3 & 3 & -6 \\ 0 & 3 & -3 & 9 \end{pmatrix}$$

Multiply row 2 by $\frac{1}{3}$ and row 3 by $\frac{1}{3}$:

$$\begin{pmatrix} 1 & 1 & -2 & 4 \\ 0 & -1 & 1 & -2 \\ 0 & 1 & -1 & 3 \end{pmatrix}$$

Add row 2 to row 3:

$$\begin{pmatrix} 1 & 1 & -2 & 4 \\ 0 & -1 & 1 & -2 \\ 0 & 0 & 0 & 1 \end{pmatrix}$$

Since $0 = 1$ is false, there is no solution to the system.

31. Multiply the second equation by -2:

$$4x + 2y - 3z = 15$$
$$-4x - 2y - 6z = -6$$

Adding, we obtain:

$$-9z = 9$$
$$z = -1$$

Substitute into the original equation:

$$2x + y - 3 = 3$$
$$2x + y = 6$$
$$y = 6 - 2x$$

So the solution is $(x, 6 - 2x, -1)$, for any real number x.

33. Form the augmented matrix:

$$\begin{pmatrix} 1 & 2 & -3 & -2 \\ 2 & -1 & 1 & 1 \\ 3 & -4 & 5 & 1 \end{pmatrix}$$

Add –2 times row 1 to row 2 and –3 times row 1 to row 3:

$$\begin{pmatrix} 1 & 2 & -3 & -2 \\ 0 & -5 & 7 & 5 \\ 0 & -10 & 14 & 7 \end{pmatrix}$$

Add –2 times row 2 to row 3:

$$\begin{pmatrix} 1 & 2 & -3 & -2 \\ 0 & -5 & 7 & 5 \\ 0 & 0 & 0 & -3 \end{pmatrix}$$

Since $0 = -3$ is false, there is no solution to the system.

35. Form the augmented matrix:

$$\begin{pmatrix} 1 & 1 & 1 & a+b \\ 2 & -1 & 2 & -a+5b \\ 1 & -2 & 1 & -2a+4b \end{pmatrix}$$

Add –2 times row 1 to row 2 and –1 times row 1 to row 3:

$$\begin{pmatrix} 1 & 1 & 1 & a+b \\ 0 & -3 & 0 & -3a+3b \\ 0 & -3 & 0 & -3a+3b \end{pmatrix}$$

Add –1 times row 2 to row 3:

$$\begin{pmatrix} 1 & 1 & 1 & a+b \\ 0 & -3 & 0 & -3a+3b \\ 0 & 0 & 0 & 0 \end{pmatrix}$$

So we have the system:

$$x + y + z = a + b$$
$$-3y = -3a + 3b$$

Solve equation 2 for y:

$$-3y = -3a + 3b$$
$$y = a - b$$

Substitute into equation 1:
$$x + a - b + z = a + b$$
$$x = 2b - z$$
So the solution is $(2b - z, a - b, z)$, for any real number z.

37. Form the augmented matrix:
$$\begin{pmatrix} 1 & 1 & 1 & 1 & 8 \\ 3 & 3 & -1 & -1 & 20 \\ 4 & -1 & -1 & 2 & 18 \\ 2 & 5 & 5 & -5 & 8 \end{pmatrix}$$
Add -3 times row 1 to row 2, -4 times row 1 to row 3, and -2 times row 1 to row 4:
$$\begin{pmatrix} 1 & 1 & 1 & 1 & 8 \\ 0 & 0 & -4 & -4 & -4 \\ 0 & -5 & -5 & -2 & -14 \\ 0 & 3 & 3 & -7 & -8 \end{pmatrix}$$
Switch row 2 and row 3, and multiply row 3 by $-\frac{1}{4}$:
$$\begin{pmatrix} 1 & 1 & 1 & 1 & 8 \\ 0 & -5 & -5 & -2 & -14 \\ 0 & 0 & 1 & 1 & 1 \\ 0 & 3 & 3 & -7 & -8 \end{pmatrix}$$
Add 2 times row 4 to row 2:
$$\begin{pmatrix} 1 & 1 & 1 & 1 & 8 \\ 0 & 1 & 1 & -16 & -30 \\ 0 & 0 & 1 & 1 & 1 \\ 0 & 3 & 3 & -7 & -8 \end{pmatrix}$$
Add -3 times row 2 to row 4:
$$\begin{pmatrix} 1 & 1 & 1 & 1 & 8 \\ 0 & 1 & 1 & -16 & -30 \\ 0 & 0 & 1 & 1 & 1 \\ 0 & 0 & 0 & 41 & 82 \end{pmatrix}$$
So we have the system:
$$x + y + z + w = 8$$
$$y + z - 16w = -30$$
$$z + w = 1$$
$$41w = 82$$
Solve equation 4 for w:
$$41w = 82$$
$$w = 2$$

Substitute into equation 3:
$$z + 2 = 1$$
$$z = -1$$
Substitute into equation 2:
$$y - 1 - 32 = -30$$
$$y = 3$$
Substitute into equation 1:
$$x + 3 - 1 + 2 = 8$$
$$x = 4$$
So the solution is $(4, 3, -1, 2)$.

39. Multiply by $(x - 10)(x + 10)$, so:
$$1 = A(x + 10) + B(x - 10)$$
$$1 = (A + B)x + (10A - 10B)$$
Since A and B are constants:
$$A + B = 0$$
$$10A - 10B = 1$$
Multiply equation 1 by 10:
$$10A + 10B = 0$$
$$10A - 10B = 1$$
Adding, we obtain:
$$20A = 1$$
$$A = \tfrac{1}{20}$$
Substitute into equation 1:
$$\tfrac{1}{20} + B = 0$$
$$B = -\tfrac{1}{20}$$
So $A = \tfrac{1}{20}$ and $B = -\tfrac{1}{20}$.

41. Multiply by $(x + 1)^2$, so:
$$2x = A(x + 1) + B$$
$$2x = Ax + (A + B)$$
Since A and B are constants:
$$A = 2$$
$$A + B = 0$$
Substitute into equation 2:
$$2 + B = 0$$
$$B = -2$$
So $A = 2$ and $B = -2$.

43. Multiply by $x(x - 4)$, so:
$$5 = A(x - 4) + Bx$$
$$5 = (A + B)x - 4A$$
Since A and B are constants:
$$A + B = 0$$
$$-4A = 5$$

Solve equation 2 for A:
$$-4A = 5$$
$$A = -\tfrac{5}{4}$$
Substitute into equation 1:
$$-\tfrac{5}{4} + B = 0$$
$$B = \tfrac{5}{4}$$
So $A = -\tfrac{5}{4}$ and $B = \tfrac{5}{4}$.

45. Multiply by $(x-1)(x+3)^2$, so:
$$1 = A(x+3)^2 + B(x-1)(x+3) + C(x-1)$$
$$1 = A(x^2 + 6x + 9) + B(x^2 + 2x - 3) + C(x-1)$$
$$1 = (A+B)x^2 + (6A + 2B + C)x + (9A - 3B - C)$$
Since A, B, and C are constants:
$$A + B \qquad = 0$$
$$6A + 2B + C = 0$$
$$9A - 3B - C = 1$$
Add -6 times equation 1 to equation 2 and -9 times equation 1 to equation 3:
$$A + B \qquad = 0$$
$$-4B + C = 0$$
$$-12B - C = 1$$
Add equation 2 and 3:
$$-16B = 1$$
$$B = -\tfrac{1}{16}$$
Substitute into equation 2:
$$\tfrac{1}{4} + C = 0$$
$$C = -\tfrac{1}{4}$$
Substitute into equation 1:
$$A - \tfrac{1}{16} = 0$$
$$A = \tfrac{1}{16}$$
So $A = \tfrac{1}{16}$, $B = -\tfrac{1}{16}$, and $C = -\tfrac{1}{4}$.

47. Multiply by $(x-1)(x^2 + x + 5)$, so:
$$4x^2 + 2x + 15 = A(x^2 + x + 5) + (Bx + C)(x - 1)$$
$$4x^2 + 2x + 15 = (A+B)x^2 + (A - B + C)x + (5A - C)$$
Since A, B, and C are constants:
$$A + B \qquad = 4$$
$$A - B + C = 2$$
$$5A \qquad - C = 15$$

Add –1 times equation 1 to equation 2 and –5 times equation 1 to equation 3:

$$A + B \quad = 4$$
$$-2B + C = -2$$
$$-5B - C = -5$$

Add equation 2 and equation 3:

$$-7B = -7$$
$$B = 1$$

Substitute into equation 2:

$$-2 + C = -2$$
$$C = 0$$

Substitute into equation 1:

$$A + 1 = 4$$
$$A = 3$$

So $A = 3$, $B = 1$, and $C = 0$.

49. Multiply by $(x + 4)(x^2 - 4x + 16)$, so:

$$1 = A(x^2 - 4x + 16) + (Bx + C)(x + 4)$$
$$1 = (A + B)x^2 + (-4A + 4B + C)x + (16A + 4C)$$

Since A, B, and C are constants:

$$A + \quad B \qquad = 0$$
$$-4A + 4B + \quad C = 0$$
$$16A \qquad + 4C = 1$$

Add 4 times equation 1 to equation 2, and –16 times equation 1 to equation 3:

$$A + \quad B \qquad = 0$$
$$8B + \quad C = 0$$
$$-16B + 4C = 1$$

Add 2 times equation 2 to equation 3:

$$A + B \qquad = 0$$
$$8B + C = 0$$
$$6C = 1$$

Solve equation 3 for C:

$$6C = 1$$
$$C = \tfrac{1}{6}$$

Substitute into equation 2:

$$8B + \tfrac{1}{6} = 0$$
$$8B = -\tfrac{1}{6}$$
$$B = -\tfrac{1}{48}$$

Substitute into equation 1:

$$A - \tfrac{1}{48} = 0$$
$$A = \tfrac{1}{48}$$

So $A = \tfrac{1}{48}$, $B = -\tfrac{1}{48}$, and $C = \tfrac{1}{6}$.

51. Multiply by $(x-a)^3$, so:
$$x = A(x-a)^2 + B(x-a) + C$$
$$x = Ax^2 + (-2aA + B)x + \left(a^2A - aB + C\right)$$
Since A, B, and C are constants:
$$A = 0$$
$$-2aA + B = 1$$
$$a^2A - aB + C = 0$$
Substitute into equation 2:
$$0 + B = 1$$
$$B = 1$$
Substitute into equation 3:
$$0 - a + C = 0$$
$$C = a$$
So $A = 0$, $B = 1$, and $C = a$.

53. Multiply by $(x-a)(x-b)$, so:
$$(a-b)(a+b-x) = A(x-b) + B(x-a)$$
$$(a-b)(a+b) - (a-b)x = (A+B)x + (-bA - aB)$$
Since A and B are constants:
$$A + B = b - a$$
$$-bA - aB = a^2 - b^2$$
Multiply equation 1 by b:
$$bA + bB = b^2 - ab$$
$$-bA - aB = a^2 - b^2$$
Adding, we obtain:
$$(b-a)B = a^2 - ab$$
$$B = \frac{a(a-b)}{b-a} = -a$$
Substitute into equation 1:
$$A - a = b - a$$
$$A = b$$
So $A = b$ and $B = -a$.

55. We evaluate the determinant:
$$\begin{vmatrix} 1 & 5 \\ -6 & 4 \end{vmatrix} = 1(4) - 5(-6) = 4 + 30 = 34$$

57. Adding -2 times row 2 to row 3 yields:
$$\begin{vmatrix} 4 & 0 & 3 \\ -2 & 1 & 5 \\ 0 & 2 & -1 \end{vmatrix} = \begin{vmatrix} 4 & 0 & 3 \\ -2 & 1 & 5 \\ 4 & 0 & -11 \end{vmatrix} = 1 \cdot \begin{vmatrix} 4 & 3 \\ 4 & -11 \end{vmatrix} = -56$$

59. Subtracting row 2 from row 1 yields:

$$\begin{vmatrix} 1 & 5 & 7 \\ 1 & 5 & 7 \\ 17 & 19 & 21 \end{vmatrix} = \begin{vmatrix} 0 & 0 & 0 \\ 1 & 5 & 7 \\ 17 & 19 & 21 \end{vmatrix} = 0$$

61. We evaluate the determinant:

$$\begin{vmatrix} 1 & 0 & 0 & 0 \\ 0 & 2 & 0 & 0 \\ 0 & 0 & 3 & 0 \\ 0 & 0 & 0 & 4 \end{vmatrix} = 1\begin{vmatrix} 2 & 0 & 0 \\ 0 & 3 & 0 \\ 0 & 0 & 4 \end{vmatrix} = 2\begin{vmatrix} 3 & 0 \\ 0 & 4 \end{vmatrix} = 2(12) = 24$$

63. By expanding along column 1:

$$\begin{vmatrix} a & b & c \\ b & c & a \\ c & a & b \end{vmatrix} = a\begin{vmatrix} c & a \\ a & b \end{vmatrix} - b\begin{vmatrix} b & c \\ a & b \end{vmatrix} + c\begin{vmatrix} b & c \\ c & a \end{vmatrix}$$

$$= a(bc - a^2) - b(b^2 - ac) + c(ab - c^2)$$
$$= abc - a^3 - b^3 + abc + abc - c^3$$
$$= 3abc - a^3 - b^3 - c^3$$

65. Adding b times column 2 to column 1 yields:

$$\begin{vmatrix} a^2 + x & b & c & d \\ -b & 1 & 0 & 0 \\ -c & 0 & 1 & 0 \\ -d & 0 & 0 & 1 \end{vmatrix} = \begin{vmatrix} a^2 + b^2 + x & b & c & d \\ 0 & 1 & 0 & 0 \\ -c & 0 & 1 & 0 \\ -d & 0 & 0 & 1 \end{vmatrix} = 1\begin{vmatrix} a^2 + b^2 + x & c & d \\ -c & 1 & 0 \\ -d & 0 & 1 \end{vmatrix}$$

Adding c times column 2 to column 1 yields:

$$\begin{vmatrix} a^2 + b^2 + c^2 + x & c & d \\ 0 & 1 & 0 \\ -d & 0 & 1 \end{vmatrix} = 1\begin{vmatrix} a^2 + b^2 + c^2 + x & d \\ -d & 1 \end{vmatrix}$$

$$= a^2 + b^2 + c^2 + x - (-d^2)$$
$$= a^2 + b^2 + c^2 + d^2 + x$$

67. Substituting $(x, y) = (-2, 5)$ and $(x, y) = (2, 9)$, we obtain:

$$5 = 4a - 2b - 1$$
$$9 = 4a + 2b - 1$$

Adding, we obtain:

$$14 = 8a - 2$$
$$16 = 8a$$
$$a = 2$$

Substitute into the first equation:
$$5 = 8 - 2b - 1$$
$$-2 = -2b$$
$$b = 1$$
So $a = 2$ and $b = 1$.

69. (a)   We first graph the triangle:

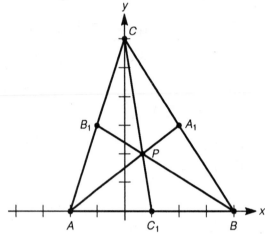

By the midpoint formula:
$$A_1 = \left(\frac{4+0}{2}, \frac{0+6}{2}\right) = (2,3)$$
$$B_1 = \left(\frac{-2+0}{2}, \frac{0+6}{2}\right) = (-1,3)$$
$$C_1 = \left(\frac{-2+4}{2}, \frac{0+0}{2}\right) = (1,0)$$
For AA_1, we have:
$$\text{slope} \;=\; \frac{3-0}{2-(-2)} = \tfrac{3}{4}$$
$$\text{point} \;=\; (-2,0)$$
$$y - 0 = \tfrac{3}{4}(x+2)$$
$$y = \tfrac{3}{4}x + \tfrac{3}{2}$$
For BB_1, we have:
$$\text{slope} \;=\; \frac{3-0}{-1-4} = -\tfrac{3}{5}$$
$$\text{point} \;=\; (4,0)$$
$$y - 0 = -\tfrac{3}{5}(x-4)$$
$$y = -\tfrac{3}{5}x + \tfrac{12}{5}$$

We solve $\frac{3}{4}x + \frac{3}{2} = -\frac{3}{5}x + \frac{12}{5}$. Multiply by 20 to clear fractions:

$$15x + 30 = -12x + 48$$
$$27x = 18$$
$$x = \frac{2}{3}$$

So $y = \frac{3}{4} \cdot \frac{2}{3} + \frac{3}{2} = 2$. Thus the point of intersection is $\left(\frac{2}{3}, 2\right)$.

(b) For CC_1, we have:

$$\text{slope} = \frac{6-0}{0-1} = -6$$
$$y = -6x + 6$$

We solve $-6x + 6 = -\frac{3}{5}x + \frac{12}{5}$.
Multiply by 5 to clear fractions:

$$-30x + 30 = -3x + 12$$
$$18 = 27x$$
$$\frac{2}{3} = x$$

So $y = -6\left(\frac{2}{3}\right) + 6 = 2$. Thus the point of intersection is $\left(\frac{2}{3}, 2\right)$.

(c) We solve $-6x + 6 = \frac{3}{4}x + \frac{3}{2}$. Multiply by 4 to clear fractions:

$$-24x + 24 = 3x + 6$$
$$18 = 27x$$
$$\frac{2}{3} = x$$

So $y = -6\left(\frac{2}{3}\right) + 6 = 2$. Thus the point of intersection is $\left(\frac{2}{3}, 2\right)$.

(d) We use the distance formula:

$$AP = \sqrt{\left(-2 - \frac{2}{3}\right)^2 + (0-2)^2} = \sqrt{\frac{64}{9} + \frac{36}{9}} = \frac{10}{3}$$
$$PA_1 = \sqrt{\left(\frac{2}{3} - 2\right)^2 + (2-3)^2} = \sqrt{\frac{16}{9} + \frac{9}{9}} = \frac{5}{3}$$

So $\dfrac{AP}{PA_1} = \dfrac{\frac{10}{3}}{\frac{5}{3}} = 2$. Again using the distance formula:

$$BP = \sqrt{\left(4 - \frac{2}{3}\right)^2 + (0-2)^2} = \sqrt{\frac{100}{9} + \frac{36}{9}} = \frac{2\sqrt{34}}{3}$$
$$PB_1 = \sqrt{\left(\frac{2}{3} + 1\right)^2 + (2-3)^2} = \sqrt{\frac{25}{9} + \frac{9}{9}} = \frac{\sqrt{34}}{3}$$

So $\dfrac{BP}{PB_1} = \dfrac{\frac{2\sqrt{34}}{3}}{\frac{\sqrt{34}}{3}} = 2$. Again using the distance formula:

$$CP = \sqrt{\left(0 - \frac{2}{3}\right)^2 + (6-2)^2} = \sqrt{\frac{4}{9} + \frac{144}{9}} = \frac{2\sqrt{37}}{3}$$
$$PC_1 = \sqrt{\left(\frac{2}{3} - 1\right)^2 + (2-0)^2} = \sqrt{\frac{1}{9} + \frac{36}{9}} = \frac{\sqrt{37}}{3}$$

So $\dfrac{CP}{PC_1} = \dfrac{\frac{2\sqrt{37}}{3}}{\frac{\sqrt{37}}{3}} = 2$. These ratios are all equal.

71. See the figure:

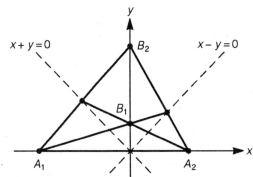

We start by finding the equation for A_2B_2:

$$m = -\frac{b_2}{a_2}$$

point: (a,a)

By the point-slope formula:

$$a_2y + b_2x = aa_2 + ab_2$$

Since $(a_2, 0)$ lies on this curve, we have $a_2b_2 = aa_2 + ab_2$. (*)

Now we find the equation for A_1B_1:

$$m = -\frac{b_1}{a_1}$$

point: (a,a)

By the point-slope formula:

$$a_1y + b_1x = aa_1 + ab_1$$

Since $(a_1, 0)$ lies on this curve, we have $a_1b_1 = aa_1 + ab_1$. (**)

Now we look at A_1B_2:

$$m = -\frac{b_2}{a_1}$$

$$b_2x + a_1y = a_1b_2$$

Also look at A_2B_1:

$$m = -\frac{b_1}{a_2}, \text{ so:}$$

$$b_1x + a_2y = a_2b_1$$

We solve the system:

$$b_2x + a_1y = a_1b_2$$
$$b_1x + a_2y = a_2b_1$$

Multiply equation 1 by $-b_1$ and equation 2 by b_2:

$$-b_1b_2x - a_1b_1y = -a_1b_1b_2$$
$$b_1b_2x + a_2b_2y = a_2b_1b_2$$

Adding, we obtain:

$$(a_2b_2 - a_1b_1)y = a_2b_1b_2 - a_1b_1b_2$$

$$y = \frac{b_1b_2(a_2 - a_1)}{a_2b_2 - a_1b_1}$$

Substituting into equation 1, we have:

$$b_2x + \frac{a_1b_1b_2(a_2 - a_1)}{a_2b_2 - a_1b_1} = a_1b_2$$

$$x = \frac{a_1a_2(b_2 - b_1)}{a_2b_2 - a_1b_1}$$

Now, we are asked to show that $x + y = 0$, so:

$$x + y = \frac{a_1a_2(b_2 - b_1) + b_1b_2(a_2 - a_1)}{a_2b_2 - a_1b_1} = \frac{a_1a_2b_2 - a_1a_2b_1 + a_2b_1b_2 - a_1b_1b_2}{a_2b_2 - a_1b_1}$$

Replacing a_1b_1 and a_2b_2 by equations (*) and (**):

$$x + y = \frac{a_1(aa_2 + ab_2) - a_2(aa_1 + ab_1) + b_1(aa_2 + ab_2) - b_2(aa_1 + ab_1)}{a_2b_2 - a_1b_1}$$

$$= \frac{a(a_1a_2 + a_1b_2 - a_1a_2 - a_2b_1 + a_2b_1 + b_1b_2 - a_1b_2 - b_1b_2)}{a_2b_2 - a_1b_1}$$

$$= 0$$

73. We find each intersection point:

$y = x - 1$	$y = x - 1$	$y = -x - 2$
$y = -x - 2$	$y = 2x + 3$	$y = 2x + 3$
$2y = -3$	$x - 1 = 2x + 3$	$-x - 2 = 2x + 3$
$y = -\frac{3}{2}$	$-4 = x$	$-5 = 3x$
		$x = -\frac{5}{3}$
		$y = -\frac{1}{3}$
$x = -\frac{1}{2}$	$y = -5$	
$\left(-\frac{1}{2}, -\frac{3}{2}\right)$	$(-4, -5)$	$\left(-\frac{5}{3}, -\frac{1}{3}\right)$

We now substitute into the equation $(x - h)^2 + (y - k)^2 = r^2$:

$$\left(-\tfrac{1}{2} - h\right)^2 + \left(-\tfrac{3}{2} - k\right)^2 = r^2$$

$$(-4 - h)^2 + (-5 - k)^2 = r^2$$

$$\left(-\tfrac{5}{3} - h\right)^2 + \left(-\tfrac{1}{3} - k\right)^2 = r^2$$

These equations, when multiplied out, become:

$$\tfrac{5}{2} + h + h^2 + 3k + k^2 = r^2$$

$$41 + 8h + h^2 + 10k + k^2 = r^2$$

$$\tfrac{26}{9} + \tfrac{10}{3}h + h^2 + \tfrac{2}{3}k + k^2 = r^2$$

Subtracting the first equation from the other two yields:

$$\tfrac{77}{2} + 7h + 7k = 0$$

$$\tfrac{7}{18} + \tfrac{7}{3}h - \tfrac{7}{3}k = 0$$

These simplify to:
$$6h + 6k = -33$$
$$6h - 6k = -1$$
Adding, we obtain:
$$12h = -34$$
$$h = -\tfrac{17}{6}$$
To find k, we subtract the two equations:
$$12k = -32$$
$$k = -\tfrac{8}{3}$$
Finally, we find r:
$$r^2 = (-4 - h)^2 + (-5 - k)^2 = \left(-\tfrac{7}{6}\right)^2 + \left(-\tfrac{7}{3}\right)^2 = \tfrac{245}{36}$$
The equation of the circle is:
$$\left(x + \tfrac{17}{6}\right)^2 + \left(y + \tfrac{8}{3}\right)^2 = \tfrac{245}{36}$$

75. $2A + 2B = \begin{pmatrix} 6 & -4 \\ 2 & 10 \end{pmatrix} + \begin{pmatrix} 4 & 2 \\ 2 & 16 \end{pmatrix} = \begin{pmatrix} 10 & -2 \\ 4 & 26 \end{pmatrix}$

77. $4B = \begin{pmatrix} 8 & 4 \\ 4 & 32 \end{pmatrix}$

79. $AB = \begin{pmatrix} 4 & -13 \\ 7 & 41 \end{pmatrix}$

81. $AB - BA = \begin{pmatrix} 7 & -13 \\ 7 & 41 \end{pmatrix} - \begin{pmatrix} 7 & 1 \\ 11 & 38 \end{pmatrix} = \begin{pmatrix} -3 & -14 \\ -4 & 3 \end{pmatrix}$

83. $C + B = \begin{pmatrix} 1 & 1 \\ 1 & 7 \end{pmatrix}$

85. $AB + AC = \begin{pmatrix} 4 & -13 \\ 7 & 41 \end{pmatrix} + \begin{pmatrix} -3 & 2 \\ -1 & -5 \end{pmatrix} = \begin{pmatrix} 1 & -11 \\ 6 & 36 \end{pmatrix}$

87. $BA + CA = \begin{pmatrix} 7 & 1 \\ 11 & 38 \end{pmatrix} + \begin{pmatrix} -3 & 2 \\ -1 & -5 \end{pmatrix} = \begin{pmatrix} 4 & 3 \\ 10 & 33 \end{pmatrix}$

89. $DE = \begin{pmatrix} -42 & 58 \\ 5 & 20 \end{pmatrix}$

91. The multiplication is undefined.

93. The multiplication is undefined.

95. $(A+B)+C = \begin{pmatrix} 5 & -1 \\ 2 & 13 \end{pmatrix} + \begin{pmatrix} -1 & 0 \\ 0 & -1 \end{pmatrix} = \begin{pmatrix} 4 & -1 \\ 2 & 12 \end{pmatrix}$

97. $(AB)C = \begin{pmatrix} 4 & -13 \\ 7 & 41 \end{pmatrix} \begin{pmatrix} -1 & 0 \\ 0 & -1 \end{pmatrix} = \begin{pmatrix} -4 & 13 \\ -7 & -41 \end{pmatrix}$

99. We compute A^2 and then A^3:

$$A^2 = \begin{pmatrix} 1 & 1 \\ 0 & 1 \end{pmatrix}\begin{pmatrix} 1 & 1 \\ 0 & 1 \end{pmatrix} = \begin{pmatrix} 1 & 2 \\ 0 & 1 \end{pmatrix}$$

$$A^3 = A \cdot A^2 = \begin{pmatrix} 1 & 1 \\ 0 & 1 \end{pmatrix}\begin{pmatrix} 1 & 2 \\ 0 & 1 \end{pmatrix} = \begin{pmatrix} 1 & 3 \\ 0 & 1 \end{pmatrix}$$

101. (a) Form the matrix:

$$\begin{pmatrix} 1 & 5 & 1 & 0 \\ 2 & 9 & 0 & 1 \end{pmatrix}$$

Multiply row 1 by –2 and add to row 2:

$$\begin{pmatrix} 1 & 5 & 1 & 0 \\ 0 & -1 & -2 & 1 \end{pmatrix}$$

Multiply row 2 by 5 and add to row 1:

$$\begin{pmatrix} 1 & 0 & -9 & 5 \\ 0 & -1 & -2 & 1 \end{pmatrix}$$

Multiply row 2 by –1:

$$\begin{pmatrix} 1 & 0 & -9 & 5 \\ 0 & 1 & 2 & -1 \end{pmatrix}$$

So the inverse is $\begin{pmatrix} -9 & 5 \\ 2 & -1 \end{pmatrix}$.

(b) Since the system is equivalent to $A \cdot X = B$, where $A = \begin{pmatrix} 1 & 5 \\ 2 & 9 \end{pmatrix}$,

$X = \begin{pmatrix} x \\ y \end{pmatrix}$, and $B = \begin{pmatrix} 3 \\ -4 \end{pmatrix}$, then:

$$X = A^{-1} \cdot B = \begin{pmatrix} -9 & 5 \\ 2 & -1 \end{pmatrix}\begin{pmatrix} 3 \\ -4 \end{pmatrix} = \begin{pmatrix} -47 \\ 10 \end{pmatrix}$$

So the solution is $(-47, 10)$.

103. (a) Form the matrix:

$$\begin{pmatrix} 1 & -2 & 3 & 1 & 0 & 0 \\ 2 & -5 & 10 & 0 & 1 & 0 \\ -1 & 2 & -2 & 0 & 0 & 1 \end{pmatrix}$$

Multiply row 1 by –2 and add to row 2, then add row 1 to row 3:

$$\begin{pmatrix} 1 & -2 & 3 & 1 & 0 & 0 \\ 0 & -1 & 4 & -2 & 1 & 0 \\ 0 & 0 & 1 & 1 & 0 & 1 \end{pmatrix}$$

Multiply row 2 by –1:

$$\begin{pmatrix} 1 & -2 & 3 & 1 & 0 & 0 \\ 0 & 1 & -4 & 2 & -1 & 0 \\ 0 & 0 & 1 & 1 & 0 & 1 \end{pmatrix}$$

Multiply row 2 by 2 and add to row 1:

$$\begin{pmatrix} 1 & 0 & -5 & 5 & -2 & 0 \\ 0 & 1 & -4 & 2 & -1 & 0 \\ 0 & 0 & 1 & 1 & 0 & 1 \end{pmatrix}$$

Multiply row 3 by 4 and add to row 2, then multiply row 3 by 5 and add to row 1:

$$\begin{pmatrix} 1 & 0 & 0 & 10 & -2 & 5 \\ 0 & 1 & 0 & 6 & -1 & 4 \\ 0 & 0 & 1 & 1 & 0 & 1 \end{pmatrix}$$

So the inverse is $\begin{pmatrix} 10 & -2 & 5 \\ 6 & -1 & 4 \\ 1 & 0 & 1 \end{pmatrix}$.

(b) Since the system is equivalent to $A \cdot X = B$, where $A = \begin{pmatrix} 1 & -2 & 3 \\ 2 & -5 & 10 \\ -1 & 2 & -2 \end{pmatrix}$,

$X = \begin{pmatrix} x \\ y \\ z \end{pmatrix}$, and $B = \begin{pmatrix} -2 \\ -3 \\ 6 \end{pmatrix}$, then:

$$X = A^{-1} \cdot B = \begin{pmatrix} 10 & -2 & 5 \\ 6 & -1 & 4 \\ 1 & 0 & 1 \end{pmatrix}\begin{pmatrix} -2 \\ -3 \\ 6 \end{pmatrix} = \begin{pmatrix} 16 \\ 15 \\ 4 \end{pmatrix}$$

So the solution is $(16, 15, 4)$.

105. Form the matrix:

$$\begin{pmatrix} 5 & 3 & 6 & -7 & 1 & 0 & 0 & 0 \\ 3 & -4 & 0 & -9 & 0 & 1 & 0 & 0 \\ 0 & 1 & -1 & -1 & 0 & 0 & 1 & 0 \\ 2 & 2 & 3 & -2 & 0 & 0 & 0 & 1 \end{pmatrix}$$

Multiply row 4 by –2 and add to row 1 (to get a 1):

$$\begin{pmatrix} 1 & -1 & 0 & -3 & 1 & 0 & 0 & -2 \\ 3 & -4 & 0 & -9 & 0 & 1 & 0 & 0 \\ 0 & 1 & -1 & -1 & 0 & 0 & 1 & 0 \\ 2 & 2 & 3 & -2 & 0 & 0 & 0 & 1 \end{pmatrix}$$

Multiply row 1 by –3 and add to row 2, and multiply row 1 by –2 and add to row 4:

$$\begin{pmatrix} 1 & -1 & 0 & -3 & 1 & 0 & 0 & -2 \\ 0 & -1 & 0 & 0 & -3 & 1 & 0 & 6 \\ 0 & 1 & -1 & -1 & 0 & 0 & 1 & 0 \\ 0 & 4 & 3 & 4 & -2 & 0 & 0 & 5 \end{pmatrix}$$

Add row 2 to row 3, then mulitply row 2 by 4 and add to row 4:

$$\begin{pmatrix} 1 & -1 & 0 & -3 & 1 & 0 & 0 & -2 \\ 0 & -1 & 0 & 0 & -3 & 1 & 0 & 6 \\ 0 & 0 & -1 & -1 & -3 & 1 & 1 & 6 \\ 0 & 0 & 3 & 4 & -14 & 4 & 0 & 29 \end{pmatrix}$$

Multiply row 3 by 3 and add to row 4:

$$\begin{pmatrix} 1 & -1 & 0 & -3 & 1 & 0 & 0 & -2 \\ 0 & -1 & 0 & 0 & -3 & 1 & 0 & 6 \\ 0 & 0 & -1 & -1 & -3 & 1 & 1 & 6 \\ 0 & 0 & 0 & 1 & -23 & 7 & 3 & 47 \end{pmatrix}$$

Add row 4 to row 3, then multiply row 4 by 3 and add to row 1:

$$\begin{pmatrix} 1 & -1 & 0 & 0 & -68 & 21 & 9 & 139 \\ 0 & -1 & 0 & 0 & -3 & 1 & 0 & 6 \\ 0 & 0 & -1 & 0 & -26 & 8 & 4 & 53 \\ 0 & 0 & 0 & 1 & -23 & 7 & 3 & 47 \end{pmatrix}$$

Multiply row 2 by –1 and row 3 by –1:

$$\begin{pmatrix} 1 & -1 & 0 & 0 & -68 & 21 & 9 & 139 \\ 0 & 1 & 0 & 0 & 3 & -1 & 0 & -6 \\ 0 & 0 & 1 & 0 & 26 & -8 & -4 & -53 \\ 0 & 0 & 0 & 1 & -23 & 7 & 3 & 47 \end{pmatrix}$$

Add row 2 to row 1:

$$\begin{pmatrix} 1 & 0 & 0 & 0 & -65 & 20 & 9 & 133 \\ 0 & 1 & 0 & 0 & 3 & -1 & 0 & -6 \\ 0 & 0 & 1 & 0 & 26 & -8 & -4 & -53 \\ 0 & 0 & 0 & 1 & -23 & 7 & 3 & 47 \end{pmatrix}$$

So the inverse is $\begin{pmatrix} -65 & 20 & 9 & 133 \\ 3 & -1 & 0 & -6 \\ 26 & -8 & -4 & -53 \\ -23 & 7 & 3 & 47 \end{pmatrix}$.

107. Adding 2 times column 2 to column 1 and column 2 to column 3 yields:

$$D = \begin{vmatrix} 2 & -1 & 1 \\ 3 & 2 & 2 \\ 1 & -5 & -3 \end{vmatrix} = \begin{vmatrix} 0 & -1 & 0 \\ 7 & 2 & 4 \\ -9 & -5 & -8 \end{vmatrix} = -(-1)\begin{vmatrix} 7 & 4 \\ -9 & -8 \end{vmatrix} = 1(-20) = -20$$

Adding 2 times row 1 to row 3 yields:

$$D_x = \begin{vmatrix} 1 & -1 & 1 \\ 0 & 2 & 2 \\ -2 & -5 & -3 \end{vmatrix} = \begin{vmatrix} 1 & -1 & 1 \\ 0 & 2 & 2 \\ 0 & -7 & -1 \end{vmatrix} = 1\begin{vmatrix} 2 & 2 \\ -7 & -1 \end{vmatrix} = 12$$

Adding 2 times row 1 to row 3 yields:

$$D_y = \begin{vmatrix} 2 & 1 & 1 \\ 3 & 0 & 2 \\ 1 & -2 & -3 \end{vmatrix} = \begin{vmatrix} 2 & 1 & 1 \\ 3 & 0 & 2 \\ 5 & 0 & -1 \end{vmatrix} = -1\begin{vmatrix} 3 & 2 \\ 5 & -1 \end{vmatrix} = -(-13) = 13$$

Adding 2 times row 1 to row 3 yields:

$$D_z = \begin{vmatrix} 2 & -1 & 1 \\ 3 & 2 & 0 \\ 1 & -5 & -2 \end{vmatrix} = \begin{vmatrix} 2 & -1 & 1 \\ 3 & 2 & 0 \\ 5 & -7 & 0 \end{vmatrix} = 1\begin{vmatrix} 3 & 2 \\ 5 & -7 \end{vmatrix} = -31$$

So $x = \frac{D_x}{D} = -\frac{12}{20} = -\frac{3}{5}$, $y = \frac{D_y}{D} = -\frac{13}{20}$, and $z = \frac{D_z}{D} = \frac{31}{20}$.

So the solution is $\left(-\frac{3}{5}, -\frac{13}{20}, \frac{31}{20}\right)$.

109. Subtracting column 2 from column 3 and column 1 from column 2 yields:

$$D = \begin{vmatrix} 1 & 2 & 3 \\ 4 & 5 & 6 \\ 7 & 8 & 9 \end{vmatrix} = \begin{vmatrix} 1 & 1 & 1 \\ 4 & 1 & 1 \\ 7 & 1 & 1 \end{vmatrix}$$

Subtracting column 3 from column 2 yields:

$$D = \begin{vmatrix} 1 & 0 & 1 \\ 4 & 0 & 1 \\ 7 & 0 & 1 \end{vmatrix} = 0$$

So Cramer's Rule will not work. Form the augmented matrix:

$$\begin{pmatrix} 1 & 2 & 3 & -1 \\ 4 & 5 & 6 & 2 \\ 7 & 8 & 9 & -3 \end{pmatrix}$$

Add -4 times row 1 to row 2 and -7 times row 1 to row 3:

$$\begin{pmatrix} 1 & 2 & 3 & -1 \\ 0 & -3 & -6 & 6 \\ 0 & -6 & -12 & 4 \end{pmatrix}$$

Add -2 times row 2 to row 3:

$$\begin{pmatrix} 1 & 2 & 3 & -1 \\ 0 & -3 & -6 & 6 \\ 0 & 0 & 0 & -8 \end{pmatrix}$$

So $0 = -8$, which is false. The system has no solution.

111. Since $D = 0$ (See Exercise 110), we form the augmented matrix:

$$\begin{pmatrix} 3 & 2 & -2 & 1 \\ 2 & 3 & -1 & -2 \\ 8 & 7 & -5 & 0 \end{pmatrix}$$

Subtract row 2 from row 1:

$$\begin{pmatrix} 1 & -1 & -1 & 3 \\ 2 & 3 & -1 & -2 \\ 8 & 7 & -5 & 0 \end{pmatrix}$$

Add -2 times row 1 to row 2 and -8 times row 1 to row 3:

$$\begin{pmatrix} 1 & -1 & -1 & 3 \\ 0 & 5 & 1 & -8 \\ 0 & 15 & 3 & -24 \end{pmatrix}$$

Add -3 times row 2 to row 3:

$$\begin{pmatrix} 1 & -1 & -1 & 3 \\ 0 & 5 & 1 & -8 \\ 0 & 0 & 0 & 0 \end{pmatrix}$$

So we have the system:

$$x - y - z = 3$$
$$5y + z = -8$$

Solve equation 2 for z:
$$z = -8 - 5y$$
Substitute into equation 1:
$$x - y + 8 + 5y = 3$$
$$x = -5 - 4y$$
So the solution is $(-5 - 4y, y, -8 - 5y)$, for any real number y.

113. Adding 4 times column 2 to both column 1 and column 3 yields:

$$D = \begin{vmatrix} 2 & -1 & 1 & 3 \\ 1 & 2 & 0 & 2 \\ 0 & 3 & 3 & 4 \\ -4 & 1 & -4 & 0 \end{vmatrix} = \begin{vmatrix} -2 & -1 & -3 & 3 \\ 9 & 2 & 8 & 2 \\ 12 & 3 & 15 & 4 \\ 0 & 1 & 0 & 0 \end{vmatrix} = 1 \begin{vmatrix} -2 & -3 & 3 \\ 9 & 8 & 2 \\ 12 & 15 & 4 \end{vmatrix}$$

Subtracting row 2 from row 3 yields:

$$D = \begin{vmatrix} -2 & -3 & 3 \\ 9 & 8 & 2 \\ 3 & 7 & 2 \end{vmatrix}$$

Adding $\frac{2}{3}$ times column 3 to column 1, and column 3 to column 2 yields:

$$D = \begin{vmatrix} 0 & 0 & 3 \\ \frac{31}{3} & 10 & 2 \\ \frac{13}{3} & 9 & 2 \end{vmatrix} = 3 \begin{vmatrix} \frac{31}{3} & 10 \\ \frac{13}{3} & 9 \end{vmatrix} = \begin{vmatrix} 31 & 10 \\ 13 & 9 \end{vmatrix} = 149$$

Adding 11 times column 2 to column 1 and 4 times column 2 to column 3 yields:

$$D_x = \begin{vmatrix} 15 & -1 & 1 & 3 \\ 12 & 2 & 0 & 2 \\ 12 & 3 & 3 & 4 \\ -11 & 1 & -4 & 0 \end{vmatrix} = \begin{vmatrix} 4 & -1 & -3 & 3 \\ 34 & 2 & 8 & 2 \\ 45 & 3 & 15 & 4 \\ 0 & 1 & 0 & 0 \end{vmatrix} = 1 \begin{vmatrix} 4 & -3 & 3 \\ 34 & 8 & 2 \\ 45 & 15 & 4 \end{vmatrix}$$

Subtracting row 2 from row 3 yields:

$$D_x = \begin{vmatrix} 4 & -3 & 3 \\ 34 & 8 & 2 \\ 11 & 7 & 2 \end{vmatrix}$$

Subtracting row 3 from row 2 yields:

$$D_x = \begin{vmatrix} 4 & -3 & 3 \\ 23 & 1 & 0 \\ 11 & 7 & 2 \end{vmatrix}$$

Adding -23 times column 2 to column 1:

$$D_x = \begin{vmatrix} 73 & -3 & 3 \\ 0 & 1 & 0 \\ -150 & 7 & 2 \end{vmatrix} = \begin{vmatrix} 73 & 3 \\ -150 & 2 \end{vmatrix} = 596$$

Adding −2 times row 2 to row 1 and 4 times row 2 to row 4 yields:

$$D_y = \begin{vmatrix} 2 & 15 & 1 & 3 \\ 1 & 12 & 0 & 2 \\ 0 & 12 & 3 & 4 \\ -4 & -11 & -4 & 0 \end{vmatrix} = \begin{vmatrix} 0 & -9 & 1 & -1 \\ 1 & 12 & 0 & 2 \\ 0 & 12 & 3 & 4 \\ 0 & 37 & -4 & 8 \end{vmatrix} = -1 \begin{vmatrix} -9 & 1 & -1 \\ 12 & 3 & 4 \\ 37 & -4 & 8 \end{vmatrix}$$

Adding 9 times column 2 to column 1 and column 2 to column 3 yields:

$$D_y = -1 \begin{vmatrix} 0 & 1 & 0 \\ 39 & 3 & 7 \\ 1 & -4 & 4 \end{vmatrix} = 1 \begin{vmatrix} 39 & 7 \\ 1 & 4 \end{vmatrix} = 149$$

Adding −2 times row 2 and row 1 to 4 times row 2 to row 4 yields:

$$D_z = \begin{vmatrix} 2 & -1 & 15 & 3 \\ 1 & 2 & 12 & 2 \\ 0 & 3 & 12 & 4 \\ -4 & 1 & -11 & 0 \end{vmatrix} = \begin{vmatrix} 0 & -5 & -9 & -1 \\ 1 & 2 & 12 & 2 \\ 0 & 3 & 12 & 4 \\ 0 & 9 & 37 & 8 \end{vmatrix} = -1 \begin{vmatrix} -5 & -9 & -1 \\ 3 & 12 & 4 \\ 9 & 37 & 8 \end{vmatrix}$$

Adding 4 times row 1 to row 2 and 8 times row 1 to row 3:

$$D_z = -1 \begin{vmatrix} -5 & -9 & -1 \\ -17 & -24 & 0 \\ -31 & -25 & 0 \end{vmatrix} = 1 \begin{vmatrix} -17 & -24 \\ -31 & -35 \end{vmatrix} = -149$$

Adding −2 times row 2 to row 1 and 4 times row 2 to row 4:

$$D_w = \begin{vmatrix} 2 & -1 & 1 & 15 \\ 1 & 2 & 0 & 12 \\ 0 & 3 & 3 & 12 \\ -4 & 1 & -4 & -11 \end{vmatrix} = \begin{vmatrix} 0 & -5 & 1 & -9 \\ 1 & 2 & 0 & 12 \\ 0 & 3 & 3 & 12 \\ 0 & 9 & -4 & 37 \end{vmatrix} = -1 \begin{vmatrix} -5 & 1 & -9 \\ 3 & 3 & 12 \\ 9 & -4 & 37 \end{vmatrix}$$

Factoring 3 out of row 2 yields:

$$D_w = -3 \begin{vmatrix} -5 & 1 & -9 \\ 1 & 1 & 4 \\ 9 & -4 & 37 \end{vmatrix}$$

Adding −1 times row 2 to row 1 and 4 times row 2 to row 3:

$$D_w = -3 \begin{vmatrix} -6 & 0 & -13 \\ 1 & 1 & 4 \\ 13 & 0 & 53 \end{vmatrix} = -3 \begin{vmatrix} -6 & -13 \\ 13 & 53 \end{vmatrix} = -3(-149) = 447$$

So $x = \frac{D_x}{D} = \frac{596}{149} = 4$, $y = \frac{D_y}{D} = \frac{149}{149} = 1$, $z = \frac{D_z}{D} = -\frac{149}{149} = -1$, and $w = \frac{D_w}{D} = \frac{447}{149} = 3$.
So the solution is $(4, 1, -1, 3)$.

115. Substitute to obtain:
$$x^2 = 6x$$
$$x^2 - 6x = 0$$
$$x(x-6) = 0$$
$$x = 0 \text{ or } x = 6$$
When $x = 0$, $y = 0$ and when $x = 6$, $y = 36$. So the solutions are $(0,0)$ and $(6,36)$.

117. Substitute to obtain:
$$x^2 - 9 = 9 - x^2$$
$$2x^2 = 18$$
$$x^2 = 9$$
$$x = \pm 3$$
When $x = \pm 3$, $y = 0$. So the solutions are $(3,0)$ and $(-3,0)$.

119. Adding the two equations yields:
$$2x^2 = 25$$
$$x^2 = \frac{25}{2}$$
$$x = \frac{\pm 5\sqrt{2}}{2}$$
Substituting for x yields:
$$\tfrac{25}{2} + y^2 = 16$$
$$y^2 = \tfrac{7}{2}$$
$$y = \frac{\pm\sqrt{14}}{2}$$
So the solutions are $\left(\frac{5\sqrt{2}}{2}, \frac{\sqrt{14}}{2}\right)$, $\left(-\frac{5\sqrt{2}}{2}, \frac{\sqrt{14}}{2}\right)$, $\left(\frac{5\sqrt{2}}{2}, -\frac{\sqrt{14}}{2}\right)$, and $\left(-\frac{5\sqrt{2}}{2}, -\frac{\sqrt{14}}{2}\right)$.

121. Substitute to obtain:
$$x^2 + x = 1$$
$$x^2 + x - 1 = 0$$
$$x = \frac{-1 \pm \sqrt{1+4}}{2} = \frac{-1 \pm \sqrt{5}}{2}$$
Now $x = \frac{-1-\sqrt{5}}{2}$ is impossible, since $x \geq 0$. So the only solution is
$\left(\frac{-1+\sqrt{5}}{2}, \sqrt{\frac{-1+\sqrt{5}}{2}}\right)$ or $\left(\frac{-1+\sqrt{5}}{2}, \frac{\sqrt{-2+2\sqrt{5}}}{2}\right)$.

123. Substitute to obtain:
$$x^2 + \left(2x^2\right)^2 = 1$$
$$x^2 + 4x^4 = 1$$
$$4x^4 + x^2 - 1 = 0$$
$$x^2 = \frac{-1 \pm \sqrt{1+16}}{8} = \frac{-1 \pm \sqrt{17}}{8}$$

Since $x^2 \neq \frac{-1-\sqrt{17}}{8}$, we have $x^2 = \frac{-1+\sqrt{17}}{8}$, so $x = \frac{\pm\sqrt{-2+2\sqrt{17}}}{4}$. Since $y = 2x^2$, we have $y = \frac{-1+\sqrt{17}}{4}$. So the solutions are $\left(\frac{\sqrt{-2+2\sqrt{17}}}{4}, \frac{-1+\sqrt{17}}{4}\right)$ and $\left(\frac{-\sqrt{-2+2\sqrt{17}}}{4}, \frac{-1+\sqrt{17}}{4}\right)$.

125. Multiply the second equation by -3:
$$-9x^2 + 3xy - 3y^2 = -54$$
$$x^2 + 2xy + 3y^2 = 68$$
Adding, we obtain:
$$-8x^2 + 5xy = 14$$
$$-5xy = 8x^2 + 14$$
$$y = \frac{8x^2 + 14}{5x}$$
Now substitute into the second equation:
$$3x^2 - \frac{8x^2 + 14}{5} + \frac{(8x^2+14)^2}{25x^2} = 18$$
Multiply by $25x^2$:
$$75x^4 - 5x^2(8x^2+14) + (8x^2+14)^2 = 450x^2$$
$$75x^4 - 40x^4 - 70x^2 + 64x^4 + 224x^2 + 196 = 450x^2$$
$$99x^4 - 296x^2 + 196 = 0$$
$$(99x^2 - 98)(x^2 - 2) = 0$$
Therefore we have the equations:
$$x^2 = \frac{98}{99} \qquad \text{or} \qquad x^2 = 2$$
$$x = \frac{\pm 7\sqrt{22}}{33} \qquad \text{or} \qquad x = \pm\sqrt{2}$$
Now $y = \frac{8x^2 + 14}{5x}$, so:

When $x = \sqrt{2}$, $y = \frac{16+14}{5\sqrt{2}} = 3\sqrt{2}$

When $x = -\sqrt{2}$, $y = \frac{16+14}{-5\sqrt{2}} = -3\sqrt{2}$

When $x = \frac{7\sqrt{22}}{33}$, $y = \frac{31\sqrt{22}}{33}$

When $x = \frac{-7\sqrt{22}}{33}$, $y = \frac{-31\sqrt{22}}{33}$

So the solutions are $(\sqrt{2}, 3\sqrt{2})$, $(-\sqrt{2}, -3\sqrt{2})$, $\left(\frac{7\sqrt{22}}{33}, \frac{31\sqrt{22}}{33}\right)$, and $\left(\frac{-7\sqrt{22}}{33}, \frac{-31\sqrt{22}}{33}\right)$.

127. Let $u = x - 3$ and $v = y + 1$, so:
$$2u^2 - v^2 = -1$$
$$-3u^2 + 2v^2 = 6$$

Multiply the first equation by 2 and add:

$$u^2 = 4$$
$$u = \pm 2$$

Substituting, we obtain:

$$8 - v^2 = -1$$
$$-v^2 = -9$$
$$v = \pm 3$$

So the solutions (u, v) are $(2, 3)$, $(2, -3)$, $(-2, 3)$, and $(-2, -3)$. Since $u = x - 3$ and $v = y + 1$, then $x = u + 3$ and $y = v - 1$. So the solutions are $(5, 2)$, $(5, -4)$, $(1, 2)$, and $(1, -4)$.

129. Call the numbers x and y. So $x + y = s$ and $\frac{x}{y} = \frac{a}{b}$. So $y = s - x$, and substituting:

$$\frac{x}{s - x} = \frac{a}{b}$$
$$bx = as - ax$$
$$(a + b)x = as$$
$$x = \frac{as}{a + b}$$

Thus $y = s - \frac{as}{a+b} = \frac{bs}{a+b}$. So the two numbers are $\frac{as}{a+b}$ and $\frac{bs}{a+b}$.

131. We have:

$$\tfrac{1}{2}x + \tfrac{1}{3}y + \tfrac{1}{4}z = 62$$
$$\tfrac{1}{3}x + \tfrac{1}{4}y + \tfrac{1}{5}z = 47$$
$$\tfrac{1}{4}x + \tfrac{1}{5}y + \tfrac{1}{6}z = 38$$

Multiply the first equation by 12, the second equation by 60, and the third equation by 60 to clear the fractions:

$$6x + 4y + 3z = 744$$
$$20x + 15y + 12z = 2820$$
$$15x + 12y + 10z = 2280$$

We use Cramer's Rule. Subtracting row 3 from row 2 yields:

$$D = \begin{vmatrix} 6 & 4 & 3 \\ 20 & 15 & 12 \\ 15 & 12 & 10 \end{vmatrix} = \begin{vmatrix} 6 & 4 & 3 \\ 5 & 3 & 2 \\ 15 & 12 & 10 \end{vmatrix}$$

Subtracting row 2 from row 1 and adding -3 times row 2 to row 3 yields:

$$D = \begin{vmatrix} 1 & 1 & 1 \\ 5 & 3 & 2 \\ 0 & 3 & 4 \end{vmatrix}$$

Subtracting column 1 from column 2 and column 3 yields:

$$D = \begin{vmatrix} 1 & 0 & 0 \\ 5 & -2 & -3 \\ 0 & 3 & 4 \end{vmatrix} = 1 \begin{vmatrix} -2 & -3 \\ 3 & 4 \end{vmatrix} = 1(1) = 1$$

Subtracting row 3 from row 2 yields:

$$D_x = \begin{vmatrix} 744 & 4 & 3 \\ 2820 & 15 & 12 \\ 2280 & 12 & 10 \end{vmatrix} = \begin{vmatrix} 744 & 4 & 3 \\ 540 & 3 & 2 \\ 2280 & 12 & 10 \end{vmatrix}$$

Subtracting row 2 from row 1 and adding -4 times row 2 to row 3 yields:

$$D_x = \begin{vmatrix} 204 & 1 & 1 \\ 540 & 3 & 2 \\ 120 & 0 & 2 \end{vmatrix}$$

Adding -3 times row 1 to row 2 yields:

$$D_x = \begin{vmatrix} 204 & 1 & 1 \\ -72 & 0 & -1 \\ 120 & 0 & 2 \end{vmatrix} = -1 \begin{vmatrix} -72 & -1 \\ 120 & 2 \end{vmatrix} = -(-24) = 24$$

Subtracting row 3 from row 2 yields:

$$D_y = \begin{vmatrix} 6 & 744 & 3 \\ 20 & 2820 & 12 \\ 15 & 2280 & 10 \end{vmatrix} = \begin{vmatrix} 6 & 744 & 3 \\ 5 & 540 & 2 \\ 15 & 2280 & 10 \end{vmatrix}$$

Subtracting row 2 from row 1 and adding -3 times row 2 to row 3 yields:

$$D_y = \begin{vmatrix} 1 & 204 & 1 \\ 5 & 540 & 2 \\ 0 & 660 & 4 \end{vmatrix}$$

Adding -5 times row 1 to row 2 yields:

$$D_y = \begin{vmatrix} 1 & 204 & 1 \\ 0 & -480 & -3 \\ 0 & 660 & 4 \end{vmatrix} = 1 \begin{vmatrix} -480 & -3 \\ 660 & 4 \end{vmatrix} = 1(60) = 60$$

Subtracting row 3 from row 2 yields:

$$D_z = \begin{vmatrix} 6 & 4 & 744 \\ 20 & 15 & 2820 \\ 15 & 12 & 2280 \end{vmatrix} = \begin{vmatrix} 6 & 4 & 744 \\ 5 & 3 & 540 \\ 15 & 12 & 2280 \end{vmatrix}$$

Subtracting row 2 from row 1 and adding -3 times row 2 to row 3 yields:

$$D_z = \begin{vmatrix} 1 & 1 & 204 \\ 5 & 3 & 540 \\ 0 & 3 & 660 \end{vmatrix}$$

Adding -5 times row 1 to row 2 yields:

$$D_z = \begin{vmatrix} 1 & 1 & 204 \\ 0 & -2 & -480 \\ 0 & 3 & 660 \end{vmatrix} = 1 \begin{vmatrix} -2 & -480 \\ 3 & 660 \end{vmatrix} = 1(120) = 120$$

So $x = \frac{D_x}{D} = 24$, $y = \frac{D_y}{D} = 60$, and $z = \frac{D_z}{D} = 120$. So the numbers are 24, 60, and 120.

133. We have:
$$xy = m$$
$$x^2 + y^2 = n$$
Add twice the first equation to the second and get:
$$x^2 + 2xy + y^2 = n + 2m$$
$$(x + y)^2 = n + 2m$$
$$x + y = \pm\sqrt{n + 2m}$$
Subtract twice the first equation from the second to get:
$$x^2 - 2xy + y^2 = n - 2m$$
$$(x - y)^2 = n - 2m$$
$$x - y = \pm\sqrt{n - 2m}$$
We solve the systems:

$$x + y = \sqrt{n + 2m}$$
$$x - y = \sqrt{n - 2m}$$
$$\overline{2x = \sqrt{n + 2m} + \sqrt{n - 2m}}$$
$$x = \frac{\sqrt{n + 2m} + \sqrt{n - 2m}}{2}$$
$$y = \frac{\sqrt{n + 2m} - \sqrt{n - 2m}}{2}$$

$$x + y = \sqrt{n + 2m}$$
$$x - y = -\sqrt{n - 2m}$$
$$\overline{2x = \sqrt{n + 2m} - \sqrt{n - 2m}}$$
$$x = \frac{\sqrt{n + 2m} - \sqrt{n - 2m}}{2}$$
$$y = \frac{\sqrt{n + 2m} + \sqrt{n - 2m}}{2}$$

And also the systems:

$$x + y = -\sqrt{n + 2m}$$
$$x - y = \sqrt{n - 2m}$$
$$\overline{2x = \sqrt{n - 2m} - \sqrt{n + 2m}}$$
$$x = \frac{\sqrt{n - 2m} - \sqrt{n + 2m}}{2}$$
$$y = \frac{-\sqrt{n - 2m} - \sqrt{n + 2m}}{2}$$

$$x + y = -\sqrt{n + 2m}$$
$$x - y = -\sqrt{n - 2m}$$
$$\overline{2x = -\sqrt{n + 2m} - \sqrt{n - 2m}}$$
$$x = \frac{-\sqrt{n + 2m} - \sqrt{n - 2m}}{2}$$
$$y = \frac{\sqrt{n - 2m} - \sqrt{n + 2m}}{2}$$

So the possible pairs of numbers are:
$$\frac{\sqrt{n + 2m} + \sqrt{n - 2m}}{2} \text{ and } \frac{\sqrt{n + 2m} - \sqrt{n - 2m}}{2},$$
$$\frac{\sqrt{n - 2m} - \sqrt{n + 2m}}{2} \text{ and } \frac{-\sqrt{n - 2m} - \sqrt{n + 2m}}{2}$$

135. The region is neither convex nor bounded. We graph the system of inequalities:

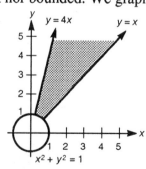

137. The region is convex and bounded. We graph the system of inequalities:

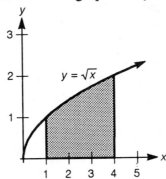

Chapter Nine Test

1. We substitute $y = x^2 + 2x + 3$ into the first equation:
$$3x + 4y = 12$$
$$3x + 4(x^2 + 2x + 3) = 12$$
$$3x + 4x^2 + 8x + 12 = 12$$
$$4x^2 + 11x = 0$$
$$x(4x + 11) = 0$$
$$x = 0, -\tfrac{11}{4}$$
When $x = 0$, $y = 3$ and when $x = -\tfrac{11}{4}$, $y = \tfrac{81}{16}$. So the solutions are $(0, 3)$ and $\left(-\tfrac{11}{4}, \tfrac{81}{16}\right)$.

2. We multiply the first equation by -3:
$$-3x + 6y = -39$$
$$3x + 5y = -16$$
Adding, we obtain:
$$11y = -55$$
$$y = -5$$
Substitute into the first equation:
$$x - 2y = 13$$
$$x - 2(-5) = 13$$
$$x + 10 = 13$$
$$x = 3$$
So the solution is $(3, -5)$.

3. (a) Adding -3 times the first equation to the second equation, and also adding -4 times the first equation to the third equation results in the system:
$$x + 4y - z = 0$$
$$-11y + 4z = -1$$
$$-20y + 9z = -7$$

To reduce coefficients, multiply the second equation by -2 and add the third equation to the second equation:

$$\begin{aligned} x + 4y - z &= 0 \\ 2y + z &= -5 \\ -20y + 9z &= -7 \end{aligned}$$

Multiply the second equation by 10 and add it to the third equation:

$$\begin{aligned} x + 4y - z &= 0 \\ 2y + z &= -5 \\ 19z &= -57 \end{aligned}$$

Solving the third equation for z yields $z = -3$. Substitute into the second equation:

$$\begin{aligned} 2y - 3 &= -5 \\ 2y &= -2 \\ y &= -1 \end{aligned}$$

Substitute into the first equation:

$$\begin{aligned} x + 4(-1) - (-3) &= 0 \\ x - 4 + 3 &= 0 \\ x &= 1 \end{aligned}$$

So the solution is $(1, -1, -3)$.

(b) Adding row 1 to row 2, and 5 times row 1 to row 3:

$$D = \begin{vmatrix} 1 & 4 & -1 \\ 3 & 1 & 1 \\ 4 & -4 & 5 \end{vmatrix} = \begin{vmatrix} 1 & 4 & -1 \\ 4 & 5 & 0 \\ 9 & 16 & 0 \end{vmatrix}$$

Now expand along column 3:

$$D = -1 \begin{vmatrix} 4 & 5 \\ 9 & 16 \end{vmatrix} = -(64 - 45) = -19$$

Adding column 1 to columns 2 and 3:

$$D_x = \begin{vmatrix} 0 & 4 & -1 \\ -1 & 1 & 1 \\ -7 & -4 & 5 \end{vmatrix} = \begin{vmatrix} 0 & 4 & -1 \\ -1 & 0 & 0 \\ -7 & -11 & -2 \end{vmatrix}$$

Now expand along row 2:

$$D_x = -(-1) \begin{vmatrix} 4 & -1 \\ -11 & -2 \end{vmatrix} = 1(-8 - 11) = -19$$

Adding column 1 to column 3:

$$D_y = \begin{vmatrix} 1 & 0 & -1 \\ 3 & -1 & 1 \\ 4 & -7 & 5 \end{vmatrix} = \begin{vmatrix} 1 & 0 & 0 \\ 3 & -1 & 4 \\ 4 & -7 & 9 \end{vmatrix}$$

Now expand along row 1:

$$D_y = 1 \begin{vmatrix} -1 & 4 \\ -7 & 9 \end{vmatrix} = -9 + 28 = 19$$

Adding -4 times column 1 to column 2:

$$D_z = \begin{vmatrix} 1 & 4 & 0 \\ 3 & 1 & -1 \\ 4 & -4 & -7 \end{vmatrix} = \begin{vmatrix} 1 & 0 & 0 \\ 3 & -11 & -1 \\ 4 & -20 & -7 \end{vmatrix}$$

Now expand along row 1:

$$D_z = 1\begin{vmatrix} -11 & -1 \\ -20 & -7 \end{vmatrix} = 77 - 20 = 57$$

So $x = \frac{D_x}{D} = \frac{-19}{-19} = 1$, $y = \frac{D_y}{D} = \frac{19}{-19} = -1$, and $z = \frac{D_z}{D} = \frac{57}{-19} = -3$. So the solution is $(1, -1, -3)$, which verifies our answer from (a).

4. (a) $2A - B = \begin{pmatrix} 2 & -6 \\ 4 & -2 \end{pmatrix} - \begin{pmatrix} 0 & 4 \\ 1 & 3 \end{pmatrix} = \begin{pmatrix} 2 & -10 \\ 3 & -5 \end{pmatrix}$

   (b) $BA = \begin{pmatrix} 0 & 4 \\ 1 & 3 \end{pmatrix}\begin{pmatrix} 1 & -3 \\ 2 & -1 \end{pmatrix} = \begin{pmatrix} 8 & -4 \\ 7 & -6 \end{pmatrix}$

5. We first graph the region:

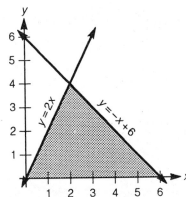

The base of the triangle is the x-intercept of $y = -x + 6$, which is 6. To find the height, we find the intersection points of the lines $y = 2x$ and $y = -x + 6$:

$$2x = -x + 6$$
$$3x = 6$$
$$x = 2$$
$$y = 4$$

So the height is 4. We find the area:

$$\text{Area} = \tfrac{1}{2}(\text{base})(\text{height}) = \tfrac{1}{2}(6)(4) = 12 \text{ sq. units}$$

6. Let $u = \frac{1}{x}$ and $v = \frac{1}{y}$, so we have the system:

$$\tfrac{1}{2}u + \tfrac{1}{3}v = 10$$
$$-5u - 4v = -4$$

Multiply the first equation by 10:
$$5u + \tfrac{10}{3}v = 100$$
$$-5u - 4v = -4$$
Adding, we obtain:
$$-\tfrac{2}{3}v = 96$$
$$v = -144$$
Substitute into the first equation to find u:
$$\tfrac{1}{2}u + \tfrac{1}{3}(-144) = 10$$
$$\tfrac{1}{2}u - 48 = 10$$
$$\tfrac{1}{2}u = 58$$
$$u = 116$$
So $\tfrac{1}{x} = 116$ thus $x = \tfrac{1}{116}$, and $\tfrac{1}{y} = -144$ thus $y = -\tfrac{1}{144}$. So the solution is $\left(\tfrac{1}{116}, -\tfrac{1}{144}\right)$.

7.   The coefficient matrix is $\begin{pmatrix} 1 & 1 & -1 \\ 2 & -1 & 2 \\ 1 & -2 & 1 \end{pmatrix}$, and the augmented matrix is $\begin{pmatrix} 1 & 1 & -1 & -1 \\ 2 & -1 & 2 & 11 \\ 1 & -2 & 1 & 10 \end{pmatrix}$.

8.   Using the augmented matrix formed in the preceeding exercise, add -2 times row 1 to row 2 and -1 times row 1 to row 3:
$$\begin{pmatrix} 1 & 1 & -1 & -1 \\ 0 & -3 & 4 & 13 \\ 0 & -3 & 2 & 11 \end{pmatrix}$$
Multiply row 2 by -1 and add to row 3:
$$\begin{pmatrix} 1 & 1 & -1 & -1 \\ 0 & -3 & 4 & 13 \\ 0 & 0 & -2 & -2 \end{pmatrix}$$
So we have the system of equations:
$$x + y - z = -1$$
$$-3y + 4z = 13$$
$$-2z = -2$$
Solving the third equation for z yields $z = 1$. Substitute into the second equation:
$$-3y + 4(1) = 13$$
$$-3y = 9$$
$$y = -3$$
Substitute into the first equation:
$$x - 3 - 1 = -1$$
$$x - 4 = -1$$
$$x = 3$$
So the solution is $(3, -3, 1)$.

9. We first find the point of intersection:
$$x + y = 11$$
$$3x + 2y = 7$$
Multiply the first equation by –2:
$$-2x - 2y = -22$$
$$3x + 2y = 7$$
Adding, we obtain:
$$x = -15$$
Substitute into the first equation:
$$-15 + y = 11$$
$$y = 26$$
So the point of intersection is $(-15, 26)$. We now find the slope of the line $2x - 4y = 7$:
$$-4y = -2x + 7$$
$$y = \tfrac{1}{2}x - \tfrac{7}{4}$$
Since this slope is $\tfrac{1}{2}$, the perpendicular slope must be -2. We use this slope and the point $(-15, 26)$ in the point-slope formula:
$$y - 26 = -2(x + 15)$$
$$y - 26 = -2x - 30$$
$$y = -2x - 4$$

10. (a)   The minor is $\begin{vmatrix} 2 & -1 \\ 0 & 4 \end{vmatrix} = 8$.

(b)   The cofactor is $-1 \begin{vmatrix} 2 & -1 \\ 0 & 4 \end{vmatrix} = -1(8) = -8$.

11. Adding 2 times row 1 to row 2, and -4 times row 1 to row 3:
$$\begin{vmatrix} 4 & -5 & 0 \\ 0 & 0 & 7 \\ 0 & 40 & 14 \end{vmatrix}$$
Expanding along column 1, we have:
$$4 \begin{vmatrix} 0 & 7 \\ 49 & 14 \end{vmatrix} = 4(0 - 280) = -1120$$

12. If we multiply the second equation by 2 we have $2xy = 10$. Adding the two equations, we obtain:
$$x^2 + 2xy + y^2 = 25$$
$$(x + y)^2 = 25$$
$$x + y = \pm 5$$
Subtracting the two equations, we obtain:
$$x^2 - 2xy + y^2 = 5$$
$$(x - y)^2 = 5$$
$$x - y = \pm\sqrt{5}$$

So we have the following four systems of equations:

$$x + y = 5 \qquad\qquad x + y = 5 \qquad\qquad x + y = -5 \qquad\qquad x + y = -5$$
$$x - y = \sqrt{5} \qquad\qquad x - y = -\sqrt{5} \qquad\qquad x - y = \sqrt{5} \qquad\qquad x - y = -\sqrt{5}$$

Adding, we have the four equations:

$$2x = 5 + \sqrt{5} \qquad\qquad 2x = 5 - \sqrt{5} \qquad\qquad 2x = -5 + \sqrt{5} \qquad\qquad 2x = -5 - \sqrt{5}$$
$$x = \frac{5 + \sqrt{5}}{2} \qquad\qquad x = \frac{5 - \sqrt{5}}{2} \qquad\qquad x = \frac{-5 + \sqrt{5}}{2} \qquad\qquad x = \frac{-5 - \sqrt{5}}{2}$$
$$y = \frac{5 - \sqrt{5}}{2} \qquad\qquad y = \frac{5 + \sqrt{5}}{2} \qquad\qquad y = \frac{-5 - \sqrt{5}}{2} \qquad\qquad y = \frac{-5 + \sqrt{5}}{2}$$

So the solutions are $\left(\frac{5+\sqrt{5}}{2}, \frac{5-\sqrt{5}}{2}\right)$, $\left(\frac{5-\sqrt{5}}{2}, \frac{5+\sqrt{5}}{2}\right)$, $\left(\frac{-5+\sqrt{5}}{2}, \frac{-5-\sqrt{5}}{2}\right)$, and $\left(\frac{-5-\sqrt{5}}{2}, \frac{-5+\sqrt{5}}{2}\right)$.

13. Multiply the first equation by -2:
$$-2A - 4B - 6C = -2$$
$$2A - B - C = 2$$

Adding, we obtain:
$$-5B - 7C = 0$$
$$-5B = 7C$$
$$B = -\tfrac{7}{5}C$$

Substitute into the (original) first equation:
$$A + 2B + 3C = 1$$
$$A - \tfrac{14}{5}C + 3C = 1$$
$$A + \tfrac{1}{5}C = 1$$
$$A = 1 - \tfrac{1}{5}C$$

So the solution is $\left(1 - \tfrac{1}{5}C, -\tfrac{7}{5}C, C\right)$, where $C = $ any real number.

14. (a) We form the matrix:
$$\begin{pmatrix} 10 & -2 & 5 & 1 & 0 & 0 \\ 6 & -1 & 4 & 0 & 1 & 0 \\ 1 & 0 & 1 & 0 & 0 & 1 \end{pmatrix}$$

Switching rows, we have:
$$\begin{pmatrix} 1 & 0 & 1 & 0 & 0 & 1 \\ 6 & -1 & 4 & 0 & 1 & 0 \\ 10 & -2 & 5 & 1 & 0 & 0 \end{pmatrix}$$

Adding -6 times row 1 to row 2, and -10 times row 1 to row 3, we have:
$$\begin{pmatrix} 1 & 0 & 1 & 0 & 0 & 1 \\ 0 & -1 & -2 & 0 & 1 & -6 \\ 0 & -2 & -5 & 1 & 0 & -10 \end{pmatrix}$$

Multiply row 2 by –1, and adding 2 times this new row 2 to row 3, we have:

$$\begin{pmatrix} 1 & 0 & 1 & 0 & 0 & 1 \\ 0 & 1 & 2 & 0 & -1 & 6 \\ 0 & 0 & -1 & 1 & -2 & 2 \end{pmatrix}$$

Adding 2 times row 3 to row 2, row 3 to row 1, then multiplying row 3 by –1, we have:

$$\begin{pmatrix} 1 & 0 & 0 & 1 & -2 & 3 \\ 0 & 1 & 0 & 2 & -5 & 10 \\ 0 & 0 & 1 & -1 & 2 & -2 \end{pmatrix}$$

So the inverse is $\begin{pmatrix} 1 & -2 & 3 \\ 2 & -5 & 10 \\ -1 & 2 & -2 \end{pmatrix}$.

(b) Calling A the coefficient matrix and $B = \begin{pmatrix} -1 \\ -2 \\ 3 \end{pmatrix}$, we know the solution will be given

by $A^{-1}B$:

$$\begin{pmatrix} u \\ v \\ w \end{pmatrix} = \begin{pmatrix} 1 & -2 & 3 \\ 2 & -5 & 10 \\ -1 & 2 & -2 \end{pmatrix}\begin{pmatrix} -1 \\ -2 \\ 3 \end{pmatrix} = \begin{pmatrix} 12 \\ 38 \\ -9 \end{pmatrix}$$

So the solution is $(12, 38, -9)$.

15. We graph $5x - 6y \geq 30$:

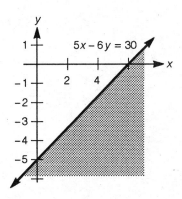

16. Substituting the two points into $y = Px^2 + Qx - 5$, we have:

$$-1 = 4P - 2Q - 5$$
$$-2 = P - Q - 5$$

So we have the system:

$$4P - 2Q = 4$$
$$P - Q = 3$$

Divide the first equation by 2, and multiply the second equation by –1:
$$2P - Q = 2$$
$$-P + Q = -3$$
Adding, we obtain $P = -1$. Substitute into the second equation:
$$-1 - Q = 3$$
$$-Q = 4$$
$$Q = -4$$
So $P = -1$ and $Q = -4$.

17. We complete the square on x:
$$x^2 - 4x + y^2 > -3$$
$$\left(x^2 - 4x + 4\right) + y^2 > -3 + 4$$
$$(x - 2)^2 + y^2 > 1$$
We now graph the inequality:

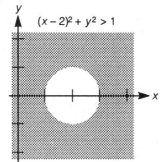

The solution set is neither bounded nor convex.

18. The vertices are $(0,0)$, $(0,7)$, $(6,10)$, $\left(\frac{261}{26}, \frac{225}{26}\right)$, and $(11,0)$. We sketch the graph:

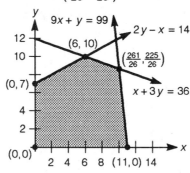

19. Multiply by $(x - 5)(x + 5)$:
$$4 = A(x + 5) + B(x - 5)$$
$$4 = (A + B)x + (5A - 5B)$$
Equating coefficients, we have the system:
$$A + B = 0$$
$$5A - 5B = 4$$

Adding 5 times the first equation to the second equation:

$$10A = 4$$
$$A = \tfrac{2}{5}$$

Substituting into the first equation yields $B = -\tfrac{2}{5}$. So $A = \tfrac{2}{5}$ and $B = -\tfrac{2}{5}$.

20. Multiply by $(x + 1)(x - 1)^2$:

$$x - 2 = A(x - 1)^2 + B(x + 1)(x - 1) + C(x + 1)$$
$$x - 2 = A(x^2 - 2x + 1) + B(x^2 - 1) + C(x + 1)$$
$$x - 2 = (A + B)x^2 + (-2A + C)x + (A - B + C)$$

Equating coefficients, we have the system:

$$\begin{aligned} A + B \quad &= 0 \\ -2A \quad + C &= 1 \\ A - B + C &= -2 \end{aligned}$$

Adding 2 times the first equation to the second equation, and -1 times the first equation to the third equation:

$$\begin{aligned} A + B \quad &= 0 \\ 2B + C &= 1 \\ -2B + C &= -2 \end{aligned}$$

Adding the second equation to the third equation:

$$\begin{aligned} A + B \quad &= 0 \\ 2B + C &= 1 \\ 2C &= -1 \end{aligned}$$

Solving the third equation for C yields $C = -\tfrac{1}{2}$. Substitute into the second equation:

$$2B - \tfrac{1}{2} = 1$$
$$B = \tfrac{3}{4}$$

Substitute into the first equation:

$$A + \tfrac{3}{4} = 0$$
$$A = -\tfrac{3}{4}$$

So $A = -\tfrac{3}{4}$, $B = \tfrac{3}{4}$, and $C = -\tfrac{1}{2}$.

Chapter Ten
Analytic Geometry

10.1 The Basic Equations

1. Using the distance formula, we have:
$$d = \sqrt{(-5-3)^2 + (-6+1)^2} = \sqrt{64+25} = \sqrt{89}$$

3. We find the slope of the line:
$$4x - 5y - 20 = 0$$
$$-5y = -4x + 20$$
$$y = \tfrac{4}{5}x - 4$$

So the perpendicular slope is $m = -\tfrac{5}{4}$. We now find the y-intercept:
$$x - y + 1 = 0$$
$$-y = -x - 1$$
$$y = x + 1$$

So $b = 1$, and the equation is $y = -\tfrac{5}{4}x + 1$. Multiplying by 4, we have:
$$4y = -5x + 4, \text{ or } 5x + 4y - 4 = 0$$

5. We find the slope of the line segment:
$$m = \tfrac{7-1}{6-2} = \tfrac{6}{4} = \tfrac{3}{2}$$

So the perpendicular slope is $m = -\tfrac{2}{3}$. We find the midpoint:
$$M = \left(\tfrac{2+6}{2}, \tfrac{1+7}{2}\right) = \left(\tfrac{8}{2}, \tfrac{8}{2}\right) = (4,4)$$

Now use the point-slope formula:

$$y - 4 = -\tfrac{2}{3}(x - 4)$$
$$y - 4 = -\tfrac{2}{3}x + \tfrac{8}{3}$$
$$y = -\tfrac{2}{3}x + \tfrac{20}{3}$$

Multiplying by 3, we have $3y = -2x + 20$, or $2x + 3y - 20 = 0$.

7.   Since the center is $(1, 0)$ and the radius is 5, the equation must be $(x - 1)^2 + y^2 = 25$. To find the x-intercepts, let $y = 0$ and solve the resulting equation for x:

$$y = 0$$
$$(x - 1)^2 = 25$$
$$x - 1 = \pm 5$$
$$x = 6, -4$$

To find the y-intercepts, let $x = 0$ and solve the resulting equation for y:

$$x = 0$$
$$(-1)^2 + y^2 = 25$$
$$y^2 = 24$$
$$y = \pm\sqrt{24} = \pm 2\sqrt{6}$$

The x-intercepts are 6 and -4, and the y-intercepts are $\pm 2\sqrt{6}$.

9.   We first find the midpoint of $\overline{AB}$:

$$M = \left(\tfrac{1+6}{2}, \tfrac{2+1}{2}\right) = \left(\tfrac{7}{2}, \tfrac{3}{2}\right)$$

Now find the slope of the line:

$$m = \frac{8 - \tfrac{3}{2}}{7 - \tfrac{7}{2}} = \frac{\tfrac{13}{2}}{\tfrac{7}{2}} = \tfrac{13}{7}$$

Using the point $(7, 8)$ in the point-slope formula:

$$y - 8 = \tfrac{13}{7}(x - 7)$$
$$y - 8 = \tfrac{13}{7}x - 13$$
$$y = \tfrac{13}{7}x - 5$$

Multiply by 7 to get $7y = 13x - 35$, or $13x - 7y - 35 = 0$.

11.   Since C is the x-intercept, we let $y = 0$ to obtain $\tfrac{x}{7} = 1$, or $x = 7$. So the coordinates of C are $(7, 0)$. Similarly, since B is the y-intercept, we let $x = 0$ to obtain $\tfrac{y}{5} = 1$, or $y = 5$. So the coordinates of B are $(0, 5)$. We find BC by the distance formula:

$$BC = \sqrt{(7 - 0)^2 + (0 - 5)^2} = \sqrt{49 + 25} = \sqrt{74}$$

The perimeter is given by:

$$P = AB + BC + AC = 5 + \sqrt{74} + 7 = 12 + \sqrt{74}$$

13.   Since $\tan \theta = \sqrt{3}$, then $\theta = \tfrac{\pi}{3}$ or $60°$.

15. (a) Since $\tan\theta = 5$, then $\theta = 1.37$ or $78.69°$.
   (b) Since $\tan\theta = -5$, then $\theta = 1.77$ or $101.31°$.

17. (a) Here $(x_0, y_0) = (1, 4)$, $m = 1$, $b = -2$, so:
$$d = \frac{|1 - 2 - 4|}{\sqrt{1 + 1}} = \frac{5}{\sqrt{2}} = \frac{5\sqrt{2}}{2}$$

   (b) Using $x - y - 2 = 0$, we have $A = 1$, $B = -1$, $C = -2$:
$$d = \frac{|1 - 4 - 2|}{\sqrt{1 + 1}} = \frac{5}{\sqrt{2}} = \frac{5\sqrt{2}}{2}$$

19. (a) Here $(x_0, y_0) = (-3, 5)$. We convert to slope-intercept form:
$$4x + 5y + 6 = 0$$
$$5y = -4x - 6$$
$$y = -\tfrac{4}{5}x - \tfrac{6}{5}$$
   So $m = -\tfrac{4}{5}$ and $b = -\tfrac{6}{5}$:
$$d = \frac{\left|\frac{12}{5} - \frac{6}{5} - 5\right|}{\sqrt{1 + \frac{16}{25}}} = \frac{\frac{19}{5}}{\frac{\sqrt{41}}{5}} = \frac{19}{\sqrt{41}} = \frac{19\sqrt{41}}{41}$$

   (b) We have $A = 4$, $B = 5$, $C = 6$:
$$d = \frac{|-12 + 25 + 6|}{\sqrt{16 + 25}} = \frac{19}{\sqrt{41}} = \frac{19\sqrt{41}}{41}$$

21. (a) The radius of the circle is the distance from the point $(-2, -3)$ to the line $2x + 3y - 6 = 0$. Using $(x_0, y_0) = (-2, -3)$, $A = 2$, $B = 3$, $C = -6$:
$$r = \frac{|-4 - 9 - 6|}{\sqrt{4 + 9}} = \frac{19}{\sqrt{13}}$$
   So the equation of the circle is $(x + 2)^2 + (y + 3)^2 = \frac{361}{13}$.

   (b) Since the radius is the distance from the point $(1, 3)$ to the line $y = \tfrac{1}{2}x + 5$, we have $(x_0, y_0) = (1, 3)$, $m = \tfrac{1}{2}$, $b = 5$:
$$r = \frac{\left|\frac{1}{2} + 5 - 3\right|}{\sqrt{1 + \frac{1}{4}}} = \frac{\frac{5}{2}}{\frac{\sqrt{5}}{2}} = \sqrt{5}$$

23. Using the suggestion, we work with $\triangle ABC$ and $\triangle CDA$. For $\triangle ABC$, we find the base using the distance formula:
$$AB = \sqrt{(8 - 0)^2 + (2 - 0)^2} = \sqrt{64 + 4} = \sqrt{68} = 2\sqrt{17}$$
Now we find the equation of the line through A and B. We find the slope:
$$m = \frac{2 - 0}{8 - 0} = \frac{1}{4}$$

So the equation is $y = \frac{1}{4}x$. We find the distance from $C(4,7)$ to this line:

$$h = \frac{|1+0-7|}{\sqrt{1+\frac{1}{16}}} = \frac{6}{\frac{\sqrt{17}}{4}} = \frac{24}{\sqrt{17}}$$

So $\triangle ABC$ will have an area of:

$$\text{Area}_{\triangle ABC} = \frac{1}{2}(\text{base})(\text{height}) = \frac{1}{2}(2\sqrt{17})\left(\frac{24}{\sqrt{17}}\right) = 24$$

For $\triangle CDA$, we find the base using the distance formula:

$$CD = \sqrt{(1-4)^2 + (6-7)^2} = \sqrt{9+1} = \sqrt{10}$$

Now we find the equation of the line through C and D. We find the slope:

$$m = \frac{6-7}{1-4} = \frac{-1}{-3} = \frac{1}{3}$$

Using $C(4,7)$, in the point-slope formula:

$$y - 7 = \frac{1}{3}(x-4)$$
$$y - 7 = \frac{1}{3}x - \frac{4}{3}$$
$$y = \frac{1}{3}x + \frac{17}{3}$$
$$3y = x + 17 \text{ or } x - 3y + 17 = 0$$

We find the distance from $A(0,0)$ to this line:

$$h = \frac{|0+0+17|}{\sqrt{1+9}} = \frac{17}{\sqrt{10}}$$

So $\triangle CDA$ will have an area of :

$$\text{Area}_{\triangle CDA} = \frac{1}{2}(\text{base})(\text{height}) = \frac{1}{2}(\sqrt{10})\left(\frac{17}{\sqrt{10}}\right) = \frac{17}{2}$$

Thus, the total combined area of quadrilateral $ABCD$ is:

$$24 + \frac{17}{2} = \frac{65}{2}$$

25. Using the point $(0,-5)$ in the point-slope formula:

$$y + 5 = m(x-0)$$
$$y = mx - 5$$

Now, since the distance from the center $(3,0)$ to this line is 2, we have:

$$2 = \frac{|3m-5-0|}{\sqrt{1+m^2}} = \frac{|3m-5|}{\sqrt{1+m^2}}$$

Squaring each side, we have:

$$4 = \frac{9m^2 - 30m + 25}{1+m^2}$$
$$4 + 4m^2 = 9m^2 - 30m + 25$$
$$0 = 5m^2 - 30m + 21$$

Using the quadratic formula:

$$m = \frac{30 \pm \sqrt{900-420}}{10} = \frac{30 \pm \sqrt{480}}{10} = \frac{30 \pm 4\sqrt{30}}{10} = \frac{15 \pm 2\sqrt{30}}{5}$$

27. We use the same approach as in the preceeding exercise. Let d_1 and d_2 represent the distances from $(0,0)$ to the lines $3x + 4y - 12 = 0$ and $3x + 4y - 24 = 0$, respectively. For d_1, we have $(x_0, y_0) = (0,0)$, $A = 3$, $B = 4$, $C = -12$:

$$d_1 = \frac{|0 + 0 - 12|}{\sqrt{9 + 16}} = \frac{12}{\sqrt{25}} = \frac{12}{5}$$

For d_2, we have $(x_0, y_0) = (0,0)$, $A = 3$, $B = 4$, $C = -24$:

$$d_2 = \frac{|0 + 0 - 24|}{\sqrt{9 + 16}} = \frac{24}{\sqrt{25}} = \frac{24}{5}$$

So the distance between the lines is:

$$d_2 - d_1 = \tfrac{24}{5} - \tfrac{12}{5} = \tfrac{12}{5}$$

29. For the described line segment to be bisected by the point $(2, 6)$, then $(2, 6)$ must be the midpoint of the x- and y-intercepts (as points) of the line. Using the point-slope formula:

$$y - 6 = m(x - 2)$$
$$y - 6 = mx - 2m$$
$$y = mx - 2m + 6$$

To find the x-intercept, let $y = 0$ and solve the resulting equation for x:

$$y = 0$$
$$mx - 2m + 6 = 0$$
$$mx = 2m - 6$$
$$x = \frac{2m - 6}{m}$$

To find the y-intercept, let $x = 0$ and solve the resulting equation for y:

$$x = 0$$
$$y = -2m + 6$$

Since $(2, 6)$ must be the midpoint of $\left(\frac{2m-6}{m}, 0\right)$ and $(0, -2m + 6)$, we have the two sets of equations:

$\dfrac{2m - 6}{2m} = 2$	$\dfrac{-2m + 6}{2} = 6$
$2m - 6 = 4m$	$-2m + 6 = 12$
$-6 = 2m$	$-2m = 6$
$-3 = m$	$m = -3$

Now use the point-slope formula with the point $(2, 6)$:

$$y - 6 = -3(x - 2)$$
$$y - 6 = -3x + 6$$
$$y = -3x + 12$$

31. Using the hint, we let d_1 and d_2 represent the distances from (x, y) to $x - y + 1 = 0$ and $x + 7y - 49 = 0$, respectively. Then:

$$d_1 = \frac{|x - y + 1|}{\sqrt{1 + 1}} = \frac{|x - y + 1|}{\sqrt{2}}$$

$$d_2 = \frac{|x + 7y - 49|}{\sqrt{1 + 49}} = \frac{|x + 7y - 49|}{\sqrt{50}}$$

Since $d_1 = d_2$, we have:

$$\frac{|x - y + 1|}{\sqrt{2}} = \frac{|x + 7y - 49|}{\sqrt{50}}$$

Multiplying each side by $\sqrt{2}$, we have:

$$|x - y + 1| = \frac{|x + 7y - 49|}{5}$$

$$5|x - y + 1| = |x + 7y - 49|$$

Rather than squaring each side, we will note that one of the two equations must hold:

$$
\begin{array}{ll}
5(x - y + 1) = x + 7y - 49 & \qquad 5(x - y + 1) = -x - 7y + 49 \\
5x - 5y + 5 = x + 7y - 49 & \qquad 5x - 5y + 5 = -x - 7y + 49 \\
4x - 12y + 54 = 0 & \qquad 6x + 2y - 44 = 0 \\
2x - 6y + 27 = 0 & \qquad 3x + y - 22 = 0 \\
y = \tfrac{1}{3}x + \tfrac{9}{2} & \qquad y = -3x + 22
\end{array}
$$

Note that the first of these lines, $y = \tfrac{1}{3}x + \tfrac{9}{2}$, is the solution as indicated in the figure. The second line we found is the second angle bisector passing through the same intersection point but bisecting the larger vertical angles.

33. (a)   The standard form for a circle is $(x - h)^2 + (y - k)^2 = r^2$. We substitute each point for (x, y):

$$(-12 - h)^2 + (1 - k)^2 = r^2$$

$$(2 - h)^2 + (1 - k)^2 = r^2$$

$$(0 - h)^2 + (7 - k)^2 = r^2$$

Setting the first two equations equal, we have:

$$(-12 - h)^2 + (1 - k)^2 = (2 - h)^2 + (1 - k)^2$$

$$144 + 24h + h^2 = 4 - 4h + h^2$$

$$28h = -140$$

$$h = -5$$

Setting the second two equations equal, we have:

$$(2 - h)^2 + (1 - k)^2 = (0 - h)^2 + (7 - k)^2$$

$$4 - 4h + h^2 + 1 - 2k + k^2 = h^2 + 49 - 14k + k^2$$

$$-4h + 12k = 44$$

$$h - 3k = -11$$

Substituting $h = -5$, we have:
$$-5 - 3k = -11$$
$$-3k = -6$$
$$k = 2$$

Now substitute into the second equation to find r:
$$(7)^2 + (-1)^2 = r^2$$
$$50 = r^2$$
$$5\sqrt{2} = r$$

So the center is $(-5, 2)$ and the radius is $5\sqrt{2}$.

(b)   We first draw the sketch:

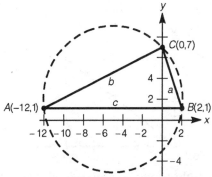

We find the lengths of three sides (using the distance formula):
$$a = \sqrt{(2-0)^2 + (1-7)^2} = \sqrt{4+36} = \sqrt{40} = 2\sqrt{10}$$
$$b = \sqrt{(-12-0)^2 + (1-7)^2} = \sqrt{144+36} = \sqrt{180} = 6\sqrt{5}$$
$$c = \sqrt{(-12-2)^2 + (1-1)^2} = \sqrt{196+0} = \sqrt{196} = 14$$

Using $R = 5\sqrt{2}$ from part (a), we have:
$$\frac{abc}{4R} = \frac{(2\sqrt{10})(6\sqrt{5})(14)}{4(5\sqrt{2})} = \frac{42\sqrt{50}}{5\sqrt{2}} = 42$$

For the area of the triangle, we have base = 14 and height = 6, so:
$$\text{Area} = \tfrac{1}{2}(\text{base})(\text{height}) = \tfrac{1}{2}(14)(6) = 42$$

This verifies the result.

35.   We will first find points P and Q. Solving for x, we have $x = 7y - 44$. Now substitute:
$$(7y - 44)^2 - 4(7y - 44) + y^2 - 6y = 12$$
$$49y^2 - 616y + 1936 - 28y + 176 + y^2 - 6y = 12$$
$$50y^2 - 650y + 2100 = 0$$
$$50(y^2 - 13y + 42) = 0$$
$$50(y - 7)(y - 6) = 0$$
$$y = 7, 6$$

When $y = 7$, $x = 49 - 44 = 5$. When $y = 6$, $x = 42 - 44 = -2$. So the two points are $P(5, 7)$ and $Q(-2, 6)$. We now use the distance formula:
$$PQ = \sqrt{(-2 - 5)^2 + (6 - 7)^2} = \sqrt{49 + 1} = \sqrt{50} = 5\sqrt{2}$$

37. Let d_1 be the distance from $(0, c)$ to $ax + y = 0$:

$$d_1 = \frac{|0 + c + 0|}{\sqrt{a^2 + 1}} = \frac{|c|}{\sqrt{a^2 + 1}}$$

Let d_2 be the distance from $(0, c)$ to $x + by = 0$:

$$d_1 = \frac{|0 + bc + 0|}{\sqrt{1 + b^2}} = \frac{|bc|}{\sqrt{1 + b^2}}$$

So the product of these distances is given by:

$$d_1 d_2 = \frac{|c|}{\sqrt{a^2 + 1}} \cdot \frac{|bc|}{\sqrt{1 + b^2}} = \frac{|bc^2|}{\sqrt{a^2 + a^2 b^2 + b^2 + 1}}$$

39. (a) We draw the triangle:

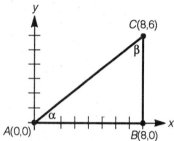

Since the triangle is a right triangle, note that $\sin\alpha = \frac{6}{10} = \frac{3}{5}$, $\cos\alpha = \frac{8}{10} = \frac{4}{5}$, $\sin\beta = \frac{8}{10} = \frac{4}{5}$ and $\cos\beta = \frac{6}{10} = \frac{3}{5}$. To find the bisector for A, note that:

$$m = \tan\frac{\alpha}{2} = \frac{\sin\alpha}{1 + \cos\alpha} = \frac{\frac{3}{5}}{1 + \frac{4}{5}} = \frac{3}{9} = \frac{1}{3}$$

Since this line passes through the point $(0, 0)$, its equation is $y = \frac{1}{3}x$. To find the bisector for B, note that its slope is -1 and that it passes through $(8, 0)$, so:

$$y - 0 = -1(x - 8)$$
$$y = -x + 8$$

To find the bisector for C, first draw the figure:

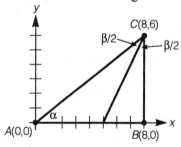

Now note that:

$$m = \tan\left(90° - \frac{\beta}{2}\right) = \cot\frac{\beta}{2} = \frac{1 + \cos\beta}{\sin\beta} = \frac{1 + \frac{3}{5}}{\frac{4}{5}} = 2$$

Using the point $(8,6)$, we use the point-slope formula:
$$y - 6 = 2(x - 8)$$
$$y - 6 = 2x - 16$$
$$y = 2x - 10$$
So the bisectors at each vertex are:

A: $y = \frac{1}{3}x$
B: $y = -x + 8$
C: $y = 2x - 10$

(b)   Setting A and B equal:
$$\frac{1}{3}x = -x + 8$$
$$\frac{4}{3}x = 8$$
$$x = 6$$
$$y = 2$$
Setting B and C equal:
$$-x + 8 = 2x - 10$$
$$18 = 3x$$
$$6 = x$$
$$2 = y$$
Setting A and C equal:
$$\frac{1}{3}x = 2x - 10$$
$$-\frac{5}{3}x = -10$$
$$x = 6$$
$$y = 2$$
So the angle bisectors are concurrent at the point $(6, 2)$.

41. The distance from the point (x, y) to $\left(0, \frac{1}{4}\right)$ is:
$$d = \sqrt{(x - 0)^2 + \left(y - \frac{1}{4}\right)^2} = \sqrt{x^2 + y^2 - \frac{1}{2}y + \frac{1}{16}}$$
The distance from the point (x, y) to $y = -\frac{1}{4}$ is:
$$d = \frac{\left|0 - \frac{1}{4} - y\right|}{\sqrt{1 + 0}} = \left|y + \frac{1}{4}\right|$$
Setting these distances equal:
$$\left|y + \frac{1}{4}\right| = \sqrt{x^2 + y^2 - \frac{1}{2}y + \frac{1}{16}}$$
Squaring, we have:
$$y^2 + \frac{1}{2}y + \frac{1}{16} = x^2 + y^2 - \frac{1}{2}y + \frac{1}{16}$$
$$y = x^2$$
Thus, the points satisfy the equation $y = x^2$.

43. (a)   We write the line in slope-intercept form:

$$Ax + By + C = 0$$
$$By = -Ax - C$$
$$y = -\frac{A}{B}x - \frac{C}{B}$$

So the slope is $-\frac{A}{B}$ and the y-intercept is $-\frac{C}{B}$.

(b)   Using the formula, we have:

$$d = \frac{\left|mx_0 + b - y_0\right|}{\sqrt{1 + m^2}} = \frac{\left|-\frac{Ax_0 + By_0 + C}{B}\right|}{\sqrt{1 + \frac{A^2}{B^2}}} = \frac{\frac{\left|Ax_0 + By_0 + C\right|}{|B|}}{\frac{\sqrt{A^2 + B^2}}{|B|}} = \frac{\left|Ax_0 + By_0 + C\right|}{\sqrt{A^2 + B^2}}$$

45.   We first find the slope of the line passing through (x_2, y_2) and (x_3, y_3):

$$m = \frac{y_2 - y_3}{x_2 - x_3}$$

Using the point-slope formula:

$$y - y_2 = \frac{y_2 - y_3}{x_2 - x_3}(x - x_2)$$
$$(x_2 - x_3)y - x_2y_2 + x_3y_2 = (y_2 - y_3)x - x_2y_2 + x_2y_3$$
$$(y_3 - y_2)x + (x_2 - x_3)y + (x_3y_2 - x_2y_3) = 0$$

We now find the distance from (x_1, y_1) to the line:

$$d = \frac{\left|x_1(y_3 - y_2) + y_1(x_2 - x_3) + x_3y_2 - x_2y_3\right|}{\sqrt{(y_3 - y_2)^2 + (x_2 - x_3)^2}}$$

$$= \frac{\left|x_1y_3 - x_1y_2 + x_2y_1 - x_3y_1 + x_3y_2 - x_2y_3\right|}{\sqrt{(x_2 - x_3)^2 + (y_2 - y_3)^2}}$$

Expanding D along the third column, we find that its terms exactly match the numerator of d.

47.   Since the center can be written as $(x, 2x - 7)$, we find the distance to each point. Let d_1 represent the distance to $(6, 3)$ and d_2 represent the distance to $(-4, -3)$, then:

$$d_1 = \sqrt{(x - 6)^2 + (2x - 7 - 3)^2}$$
$$= \sqrt{x^2 - 12x + 36 + 4x^2 - 40x + 100}$$
$$= \sqrt{5x^2 - 52x + 136}$$
$$d_2 = \sqrt{(x + 4)^2 + (2x - 7 + 3)^2}$$
$$= \sqrt{x^2 + 8x + 16 + 4x^2 - 16x + 16}$$
$$= \sqrt{5x^2 - 8x + 32}$$

Since $d_1 = d_2$ (they both represent the radius of the circle), we have:

$$\sqrt{5x^2 - 52x + 136} = \sqrt{5x^2 - 8x + 32}$$
$$5x^2 - 52x + 136 = 5x^2 - 8x + 32$$
$$-44x = -104$$
$$x = \tfrac{26}{11}$$
$$y = 2\left(\tfrac{26}{11}\right) - 7 = \tfrac{52}{11} - \tfrac{77}{11} = -\tfrac{25}{11}$$

So the center is $\left(\tfrac{26}{11}, -\tfrac{25}{11}\right)$. We find the radius using the point $(6, 3)$ in the distance formula:

$$r = \sqrt{\left(6 - \tfrac{26}{11}\right)^2 + \left(3 + \tfrac{25}{11}\right)^2} = \sqrt{\tfrac{1600}{121} + \tfrac{3364}{121}} = \frac{\sqrt{4964}}{11}$$

So the equation of the circle is $\left(x - \tfrac{26}{11}\right)^2 + \left(y + \tfrac{25}{11}\right)^2 = \tfrac{4964}{121}$.

49. Since the line $2x + 3y = 26$ has a slope of $-\tfrac{2}{3}$, the radial line will have a slope of $\tfrac{3}{2}$. Call (h, k) the center of the circle, so:

$$\frac{k - 6}{h - 4} = \tfrac{3}{2}$$
$$2k - 12 = 3h - 12$$
$$2k = 3h$$
$$k = \tfrac{3}{2}h$$

The circle has an equation of $(x - h)^2 + (y - k)^2 = 25$, so using the point $(4, 6)$ we have:

$$(4 - h)^2 + (6 - k)^2 = 25$$
$$16 - 8h + h^2 + 36 - 12k + k^2 = 25$$
$$h^2 - 8h + k^2 - 12k = -27$$

Substituting $k = \tfrac{3}{2}h$:

$$h^2 - 8h + \tfrac{9}{4}h^2 - 18h = -27$$
$$4h^2 - 32h + 9h^2 - 72h = -108$$
$$13h^2 - 104h + 108 = 0$$
$$h = \frac{104 \pm \sqrt{5200}}{26} = 4 \pm \frac{10\sqrt{13}}{13}$$
$$k = 6 \pm \frac{15\sqrt{13}}{13}$$

So the two circles are:

$$\left(x - 4 - \frac{10\sqrt{13}}{13}\right)^2 + \left(y - 6 - \frac{15\sqrt{13}}{13}\right)^2 = 25$$
$$\left(x - 4 + \frac{10\sqrt{13}}{13}\right)^2 + \left(y - 6 + \frac{15\sqrt{13}}{13}\right)^2 = 25$$

51. (a)   Replacing y with tx in the equation of the folium yields:

$$x^3 + t^3 x^3 = 6x(tx)$$
$$x^3 + t^3 x^3 = 6x^2 t \qquad \text{(since } x \neq 0 \text{ at point } Q\text{)}$$
$$x + t^3 x = 6t$$
$$x\left(1 + t^3\right) = 6t$$
$$x = \frac{6t}{1 + t^3}$$
$$y = tx = \frac{6t^2}{1 + t^3}$$

So the coordinates of point Q are $x = \dfrac{6t}{1 + t^3}$ and $y = \dfrac{6t^2}{1 + t^3}$.

(b)   The slope of $\overline{PQ}$ is given by:

$$\frac{3 - \dfrac{6t^2}{1 + t^3}}{3 - \dfrac{6t}{1 + t^3}} = \frac{1 - \dfrac{2t^2}{1 + t^3}}{1 - \dfrac{2t}{1 + t^3}} \cdot \frac{1 + t^3}{1 + t^3} = \frac{1 + t^3 - 2t^2}{1 + t^3 - 2t} = \frac{t^3 - 2t^2 + 1}{t^3 - 2t + 1}$$

We can factor both the numerator and denominator of this expression. For the numerator, add and subtract the term t^2 as follows:

$$t^3 - 2t^2 + 1 = t^3 - 2t^2 + t^2 - t^2 + 1$$
$$= \left(t^3 - t^2\right) - \left(t^2 - 1\right)$$
$$= t^2(t - 1) - (t + 1)(t - 1)$$
$$= (t - 1)\left(t^2 - t - 1\right)$$

For the denominator, add and subtract the term t^2 as follows:

$$t^3 - 2t + 1 = t^3 - t^2 + t^2 - 2t + 1$$
$$= \left(t^3 - t^2\right) + \left(t^2 - 2t + 1\right)$$
$$= t^2(t - 1) + (t - 1)^2$$
$$= (t - 1)\left(t^2 + t - 1\right)$$

So the slope of $\overline{PQ}$ is given by:

$$\frac{t^3 - 2t^2 + 1}{t^3 - 2t + 1} = \frac{(t - 1)\left(t^2 - t - 1\right)}{(t - 1)\left(t^2 + t - 1\right)} = \frac{t^2 - t - 1}{t^2 + t - 1}$$

(c)   Cross-multiplying and combining like terms, we obtain:

$$(t^2 - t - 1)(u^2 + u - 1) = (t^2 + t - 1)(u^2 - u - 1)$$

$$t^2u^2 + t^2u - t^2 - tu^2 - tu + t - u^2 - u + 1 = t^2u^2 - t^2u - t^2 + tu^2 - tu - t - u^2 + u + 1$$

$$2t^2u - 2tu^2 + 2t - 2u = 0$$

$$t^2u - tu^2 + t - u = 0$$

$$tu(t - u) + 1(t - u) = 0$$

$$(t - u)(tu + 1) = 0$$

$$t = u \ \text{ or } \ tu = -1$$

Since points R, O and Q are not collinear, $t \neq u$, and thus $tu = -1$. Therefore $\overline{OR} \perp \overline{OQ}$, as required.

10.2 The Parabola

1.   We note that $4p = 4$, so $p = 1$. So the focus is $(0, 1)$, the directrix is $y = -1$, and the focal width is 4.

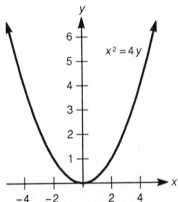

3.   We note that $4p = 8$, so $p = 2$. So the focus is $(-2, 0)$, the directrix is $x = 2$, and the focal width is 8.

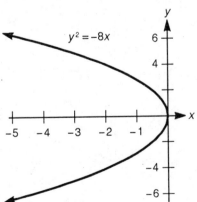

5.   We note that $4p = 20$, so $p = 5$. So the focus is $(0, -5)$, the directrix is $y = 5$, and the focal width is 20.

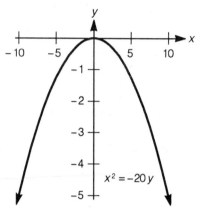

7.   Since $y^2 = -28x$, then $4p = 28$ so $p = 7$. So the focus is $(-7, 0)$, the directrix is $x = 7$, and the focal width is 28.

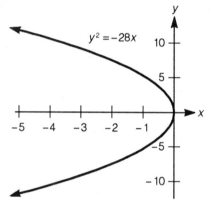

9.   We note that $4p = 6$, so $p = \frac{3}{2}$. So the focus is $\left(0, \frac{3}{2}\right)$, the directrix is $y = -\frac{3}{2}$, and the focal width is 6.

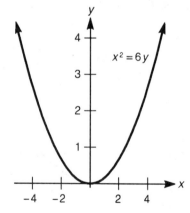

11. Since $x^2 = \frac{7}{4}y$, then $4p = \frac{7}{4}$ so $p = \frac{7}{16}$. So the focus is $\left(0, \frac{7}{16}\right)$, the directrix is $y = -\frac{7}{16}$, and the focal width is $\frac{7}{4}$.

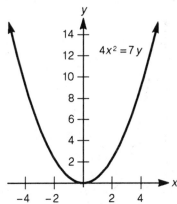

13. We first complete the square:
$$y^2 - 6y - 4x + 17 = 0$$
$$y^2 - 6y = 4x - 17$$
$$y^2 - 6y + 9 = 4x - 17 + 9$$
$$(y - 3)^2 = 4x - 8 = 4(x - 2)$$

We note that $4p = 4$, so $p = 1$. The vertex is $(2, 3)$, the focus is $(3, 3)$, the directrix is $x = 1$, and the focal width is 4.

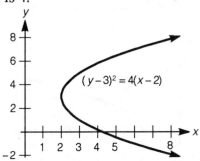

15. We first complete the square:
$$x^2 - 8x - y + 18 = 0$$
$$x^2 - 8x = y - 18$$
$$x^2 - 8x + 16 = y - 18 + 16$$
$$(x - 4)^2 = y - 2$$

We note that $4p = 1$, so $p = \frac{1}{4}$. The vertex is $(4, 2)$, the focus is $\left(4, \frac{9}{4}\right)$, the directrix is $y = \frac{7}{4}$, and the focal width is 1.

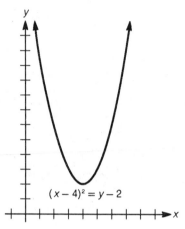

$(x - 4)^2 = y - 2$

17. We first complete the square:
$$y^2 + 2y - x + 1 = 0$$
$$y^2 + 2y = x - 1$$
$$y^2 + 2y + 1 = x - 1 + 1$$
$$(y + 1)^2 = x$$

We note that $4p = 1$, so $p = \frac{1}{4}$. The vertex is $(0, -1)$, the focus is $\left(\frac{1}{4}, -1\right)$, the directrix is $x = -\frac{1}{4}$, and the focal width is 1.

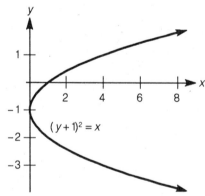

$(y + 1)^2 = x$

19. We first complete the square:
$$2x^2 - 12x - y + 18 = 0$$
$$2x^2 - 12x = y - 18$$
$$x^2 - 6x = \tfrac{1}{2}y - 9$$
$$x^2 - 6x + 9 = \tfrac{1}{2}y - 9 + 9$$
$$(x - 3)^2 = \tfrac{1}{2}y$$

We note that $4p = \frac{1}{2}$, so $p = \frac{1}{8}$. The vertex is $(3,0)$, the focus is $\left(3, \frac{1}{8}\right)$, the directrix is $y = -\frac{1}{8}$, and the focal width is $\frac{1}{2}$.

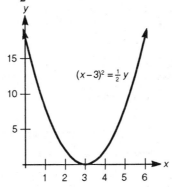

21. We first complete the square:
$$2x^2 - 16x - y + 33 = 0$$
$$2x^2 - 16x = y - 33$$
$$2(x-4)^2 = y - 1$$
$$(x-4)^2 = \tfrac{1}{2}(y-1)$$

We note that $4p = \frac{1}{2}$, so $p = \frac{1}{8}$. The vertex is $(4,1)$, the focus is $\left(4, \frac{9}{8}\right)$, the directrix is $y = \frac{7}{8}$, and the focal width is $\frac{1}{2}$.

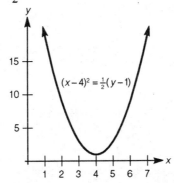

23. The line of symmetry has equation $y = 1$. The figure is graphed as:

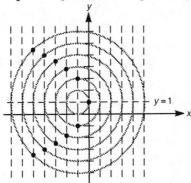

25. Given that the focus is at $(0, 3)$, we see that $p = 3$ and thus $4p = 12$. Since the parabola opens up, it has the $x^2 = 4py$ form, and its equation must be $x^2 = 12y$.

27. The directrix is $x = -32$, so $(32, 0)$ is the focus and the parabola opens to the right. So the parabola is of the form $y^2 = 4px$, where $p = 32$. Thus the equation is $y^2 = 128x$.

29. Since the parabola is symmetric about the x-axis, then $y^2 = \pm 4px$ is the form of the parabola. Since the x-coordinate of the focus is negative, then $y^2 = -4px$. Finally we are given $4p = 9$, so the equation is $y^2 = -9x$.

31. (a) Since $4p = 8$, then $p = 2$, and thus the focus is $(0, 2)$. So the slope of the line is $\frac{8-2}{8-0} = \frac{3}{4}$, and thus the equation is $y = \frac{3}{4}x + 2$.

(b) We solve the system:
$$y = \tfrac{3}{4}x + 2$$
$$x^2 = 8y$$
Substituting:
$$x^2 = 8\left(\tfrac{3}{4}x + 2\right)$$
$$x^2 = 6x + 16$$
$$x^2 - 6x - 16 = 0$$
$$(x - 8)(x + 2) = 0$$
$$x = 8, -2$$
$$y = 8, \tfrac{1}{2}$$
Since $(8, 8)$ repeats our original point, then Q must be $\left(-2, \tfrac{1}{2}\right)$.

(c) Use the distance formula:
$$PQ = \sqrt{(8+2)^2 + \left(8 - \tfrac{1}{2}\right)^2} = \sqrt{100 + \tfrac{225}{4}} = \sqrt{\tfrac{625}{4}} = \tfrac{25}{2}$$

(d) We know the radius is $\frac{25}{4}$, and the center of the circle will be the midpoint of the line segment $\overline{PQ}$:
$$\text{center} = \left(\frac{8-2}{2}, \frac{8 + \tfrac{1}{2}}{2}\right) = \left(3, \tfrac{17}{4}\right)$$
So the equation is $(x - 3)^2 + \left(y - \tfrac{17}{4}\right)^2 = \tfrac{625}{16}$.

(e) The directrix is $y = -2$. The vertical distance of the center of the circle to the directrix is:
$$\left|\tfrac{17}{4} - (-2)\right| = \left|\tfrac{17}{4} + 2\right| = \tfrac{25}{4}$$

But the radius of the circle is also $\frac{25}{4}$, which implies that the directrix must be tangent to the circle.

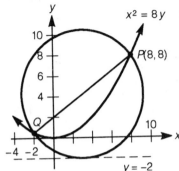

33. Call $(0,0)$ the vertex, thus the ends of the arch are at $(20,-15)$ and $(-20,-15)$. The parabola is of the form $x^2 = -4py$, so we substitute:
$$400 = -4p(-15)$$
$$-4p = -\frac{80}{3}$$
So the equation is $x^2 = -\frac{80}{3}y$. We wish to find the y-coordinate when the base is 20 ft, so we substitute $x = \pm 10$ into the equation:
$$100 = -\frac{80}{3}y$$
$$y = -\frac{15}{4}$$
So the height above the base is $15 - \frac{15}{4} = \frac{45}{4} = 11.25$ ft.

35. (a) We have $4p = 2$, so $p = \frac{1}{2}$, and thus the focus is $\left(0,\frac{1}{2}\right)$. To find A', we find the slope of the line containing A and the focus:
$$m = \frac{8-\frac{1}{2}}{4-0} = \frac{\frac{15}{2}}{4} = \frac{15}{8}$$
So $y = \frac{15}{8}x + \frac{1}{2}$ is its equation. Now substitute into the parabola:
$$x^2 = 2\left(\tfrac{15}{8}x + \tfrac{1}{2}\right)$$
$$x^2 = \tfrac{15}{4}x + 1$$
$$4x^2 = 15x + 4$$
$$4x^2 - 15x - 4 = 0$$
$$x = -\tfrac{1}{4}, 4$$
$$y = \tfrac{1}{32}, 8$$
So $A' = \left(-\frac{1}{4},\frac{1}{32}\right)$. To find B', we find the slope of the line containing B and the focus:
$$m = \frac{2-\frac{1}{2}}{-2-0} = \frac{\frac{3}{2}}{-2} = -\frac{3}{4}$$

So $y = -\frac{3}{4}x + \frac{1}{2}$ is its equation. Now substitute into the parabola:

$$x^2 = 2\left(-\frac{3}{4}x + \frac{1}{2}\right)$$
$$x^2 = -\frac{3}{2}x + 1$$
$$2x^2 + 3x - 2 = 0$$
$$x = \frac{1}{2}, -2$$
$$y = \frac{1}{8}, 2$$

So $B' = \left(\frac{1}{2}, \frac{1}{8}\right)$. We now find the slope through A and B':

$$m = \frac{8 - \frac{1}{8}}{4 - \frac{1}{2}} = \frac{\frac{63}{8}}{\frac{7}{2}} = \frac{9}{4}$$

Using $(4, 8)$ in the point-slope formula:

$$y - 8 = \frac{9}{4}(x - 4)$$
$$y - 8 = \frac{9}{4}x - 9$$
$$y = \frac{9}{4}x - 1$$

(b)   We find the slope:

$$m = \frac{2 - \frac{1}{32}}{-2 + \frac{1}{4}} = \frac{\frac{63}{32}}{-\frac{7}{4}} = -\frac{9}{8}$$

Using $(-2, 2)$ in the point-slope formula:

$$y - 2 = -\frac{9}{8}(x + 2)$$
$$y - 2 = -\frac{9}{8}x - \frac{9}{4}$$
$$y = -\frac{9}{8}x - \frac{1}{4}$$

(c)   Setting the y-coordinates equal:

$$\frac{9}{4}x - 1 = -\frac{9}{8}x - \frac{1}{4}$$
$$18x - 8 = -9x - 2$$
$$27x = 6$$
$$x = \frac{2}{9}$$
$$y = \frac{1}{2} - 1 = -\frac{1}{2}$$

So the lines intersect at the point $\left(\frac{2}{9}, -\frac{1}{2}\right)$. Since the directrix is $y = -\frac{1}{2}$, then this point lies on the directrix.

37.   (a)   Since $4p = 1$, $p = \frac{1}{4}$, thus the focus is $\left(0, \frac{1}{4}\right)$. We find the slope of the focal chord from P:

$$m = \frac{4 - \frac{1}{4}}{2 - 0} = \frac{\frac{15}{4}}{2} = \frac{15}{8}$$

So its equation is $y = \frac{15}{8}x + \frac{1}{4}$. We now find the intersection of this line with the parabola $y = x^2$:

$$x^2 = \frac{15}{8}x + \frac{1}{4}$$
$$8x^2 = 15x + 2$$
$$8x^2 - 15x - 2 = 0$$
$$(8x + 1)(x - 2) = 0$$
$$x = -\frac{1}{8}, 2$$
$$y = \frac{1}{64}, 4$$

So the coordinates of Q are $Q\left(-\frac{1}{8}, \frac{1}{64}\right)$.

(b) We find the midpoint of $\overline{PQ}$:

$$M = \left(\frac{2 - \frac{1}{8}}{2}, \frac{4 + \frac{1}{64}}{2}\right) = \left(\frac{15}{16}, \frac{257}{128}\right)$$

(c) The coordinates of S are $\left(0, \frac{257}{128}\right)$. We can find T, since the slope of this line (which is perpendicular to $\overline{PQ}$) is $-\frac{8}{15}$, and thus by the point-slope formula:

$$y - \frac{257}{128} = -\frac{8}{15}\left(x - \frac{15}{16}\right)$$
$$y - \frac{257}{128} = -\frac{8}{15}x + \frac{1}{2}$$
$$y = -\frac{8}{15}x + \frac{321}{128}$$

When $x = 0$, $y = \frac{321}{128}$, so the coordinates of T are $\left(0, \frac{321}{128}\right)$. Thus the length of $\overline{ST}$ is $\frac{321}{128} - \frac{257}{128} = \frac{64}{128} = \frac{1}{2}$. Since the focal width is $4p = 1$, this verifies that ST is one-half the focal width.

39. The sketch in the book will have the following coordinates:

$$O(0,0), B\left(x, \frac{x^2}{4p}\right), A\left(-x, \frac{x^2}{4p}\right)$$

We require that $AB = OB$. So:

$$2x = \sqrt{(x - 0)^2 + \left(\frac{x^2}{4p} - 0\right)^2} = \sqrt{x^2 + \frac{x^4}{16p^2}}$$

Squaring, we have:

$$4x^2 = x^2 + \frac{x^4}{16p^2}$$
$$0 = -3x^2 + \frac{x^4}{16p^2}$$
$$0 = -48x^2 p^2 + x^4$$
$$0 = x^2\left(-48p^2 + x^2\right)$$

So $x^2 = 0$ or $x = 0$ is one root and $x^2 = 48p^2$ or $x = \pm 4\sqrt{3}p$ is the other. The distance from A to B is therefore:

$$2(4\sqrt{3}p) = 8\sqrt{3}p \text{ units}$$

To find the area we note that this is an equilateral triangle. Half of it is a $30°$-$60°$-$90°$ triangle whose height will be $\sqrt{3}$ times its base. The height is therefore $\sqrt{3} \cdot 4\sqrt{3}p = 12p$. The area is:

$$\tfrac{1}{2}(8\sqrt{3}p)(12p) = 48\sqrt{3}p^2 \text{ sq. units}$$

41.   Since the focus is $(p, 0)$, then the slope of the focal chord is:

$$m = \frac{y_0 - 0}{\frac{y_0^2}{4p} - p} = \frac{4py_0}{y_0^2 - 4p^2}$$

Using the point-slope formula, the equation of this focal chord is:

$$y - y_0 = m\left(x - \frac{y_0^2}{4p}\right)$$

This line intersects the parabola when $x = \frac{y^2}{4p}$, so:

$$y - y_0 = m\left(\frac{y^2}{4p} - \frac{y_0^2}{4p}\right)$$

$$4p(y - y_0) = m(y + y_0)(y - y_0)$$

$$4p = m(y + y_0)$$

Substitute $m = \dfrac{4py_0}{y_0^2 - 4p^2}$:

$$4p = \frac{4py_0}{y_0^2 - 4p^2}(y + y_0)$$

$$\frac{y_0^2 - 4p^2}{y_0} = y + y_0$$

$$y = -\frac{4p^2}{y_0}$$

Therefore:

$$x = \frac{1}{4p}\left(-\frac{4p^2}{y_0}\right)^2 = \frac{1}{4p}\left(\frac{16p^4}{y_0^2}\right) = \frac{4p^3}{y_0^2}$$

But $y_0^2 = 4px_0$, so:

$$x = \frac{4p^3}{4px_0} = \frac{p^2}{x_0}$$

Thus the coordinates of Q are $\left(\frac{p^2}{x_0}, -\frac{4p^2}{y_0}\right)$.

43. Let P have coordinates (x_0, y_0). Then, from Exercise 41, Q has coordinates $\left(\frac{p^2}{x_0}, -\frac{4p^2}{y_0}\right)$. The following results summarize the calculations required for this problem. Note: At each step we have replaced the quantity y_0^2 by $4px_0$. This simplifies matters a great deal.

Coordinates of M: $\left(\dfrac{x_0^2 + p^2}{2x_0}, \dfrac{2p(x_0 - p)}{y_0}\right)$

Coordinates of S: $\left(\dfrac{p^2 + x_0^2}{2x_0}, 0\right)$

Slope of $\overline{PQ}$: $\dfrac{4px_0}{y_0(x_0 - p)}$

Slope of perpendicular: $\dfrac{y_0(p - x_0)}{4px_0}$

Since $-\frac{ST}{MS}$ is the slope, m, of the line through M and T, we have:

$$ST = -(MS)m = -\frac{2p(x_0 - p)}{y_0} \cdot \frac{4px_0}{y_0(p - x_0)} = \frac{8p^2 x_0}{y_0^2} = 2p$$

Thus ST is one-half of the focal width.

45. (a) Since the focus is $(0, p)$, then the slope of the focal chord is:

$$m = \frac{\frac{x_0^2}{4p} - p}{x_0 - 0} = \frac{x_0^2 - 4p^2}{4px_0}$$

The equation of the focal chord must be $y - p = mx$. This line intersects the parabola when $y = \frac{x^2}{4p}$, so we have:

$$\frac{x^2}{4p} - p = mx$$

$$\frac{x^2}{4p} - p = \frac{x_0^2 - 4p^2}{4px_0} \cdot x$$

$$x^2 x_0 - 4p^2 x_0 = xx_0^2 - 4p^2 x$$

$$xx_0(x - x_0) = -4p^2(x - x_0)$$

$$xx_0 = -4p^2$$

$$x = -\frac{4p^2}{x_0}$$

Therefore:

$$y = \frac{1}{4p}\left(-\frac{4p^2}{x_0}\right)^2 = \frac{1}{4p}\left(\frac{16p^4}{x_0^2}\right) = \frac{4p^3}{x_0^2}$$

But $x_0^2 = 4py_0$, so:

$$y = \frac{4p^3}{4py_0} = \frac{p^2}{y_0}$$

Thus the coordinates of Q are $\left(-\dfrac{4p^2}{x_0}, \dfrac{p^2}{y_0}\right)$.

(b) By the midpoint formula:

$$\text{Midpoint of } PQ = \left(\frac{x_0 - \frac{4p^2}{x_0}}{2}, \frac{y_0 + \frac{p^2}{y_0}}{2}\right) = \left(\frac{x_0^2 - 4p^2}{2x_0}, \frac{y_0^2 + p^2}{2y_0}\right)$$

(c) The distance from P to the directrix is $y_0 + p$. Therefore $PF = y_0 + p$. The distance from Q to the directrix is $\frac{p^2}{y_0} + p$. So we have $QF = \frac{p^2}{y_0} + p$ and consequently:

$$PQ = QF + PF = \frac{p^2}{y_0} + p + y_0 + p = \frac{p^2 + 2py_0 + y_0^2}{y_0} = \frac{(p + y_0)^2}{y_0}$$

(d) The distance from the center of the circle to the directrix is found by adding p to the y-coordinate of the midpoint of $\overline{PQ}$. This yields:

$$\frac{y_0^2 + p^2}{2y_0} + p = \frac{y_0^2 + p^2 + 2py_0}{2y_0} = \frac{(y_0 + p)^2}{2y_0}$$

Comparing this result with the expression for PQ determined in part (c), we conclude that the distance from the center of the circle to the directrix equals the radius of the circle. This implies that the circle is tangent to the directrix, as we wished to show.

Optional TI-81 Graphing Calculator Exercises for Section 10.2

1. Entering $x^2 = 8y$ as $y = \frac{1}{8}x^2$ and using the ZOOM-6 settings, we have the parabola:

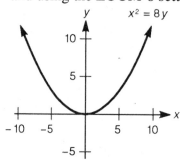

3. Entering $y^2 = 8x$ as the two functions $y = \sqrt{8x}$ and $y = -\sqrt{8x}$, and using the ZOOM-6 settings, we have the parabola:

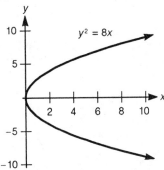

5. (a) Completing the square, we have:
$$4x + y^2 + 2y - 7 = 0$$
$$y^2 + 2y + 1 = -4x + 8$$
$$(y+1)^2 = -4(x-2)$$

(b) Entering $(y + 1)^2 = -4(x - 2)$ as the two functions $y = -1 - \sqrt{-4(x-2)}$ and $y = -1 + \sqrt{-4(x-2)}$, and using the ZOOM-6 settings, we have the parabola:

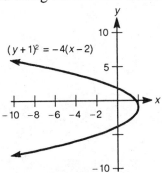

The graph is consistent with Figure 9.

7. Since the tangent line passes through the points $(4, 2)$ and $(0, -2)$, we can find the slope:
$$m = \frac{-2-2}{0-4} = \frac{-4}{-4} = 1$$
Since the y-intercept is -2, the equation of the tangent line is $y = x - 2$. Using the ZOOM-6 settings, we graph the parabola and tangent line:

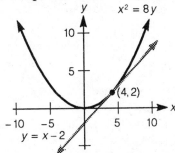

9.   Since the tangent line passes through the points $(-3, 9)$ and $(0, -9)$, we can find the slope:
$$m = \frac{-9-9}{0-(-3)} = \frac{-18}{3} = -6$$
Since the y-intercept is -9, the equation of the tangent line is $y = -6x - 9$. Using the ZOOM-6 settings, we graph the parabola and tangent line:

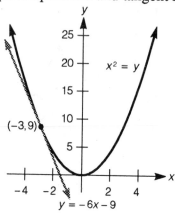

11.   We sketch the two curves:

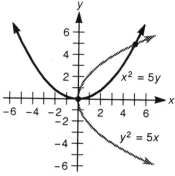

Solving $x^2 = 5y$ for y yields $y = \dfrac{x^2}{5}$, now substitute:
$$\left(\frac{x^2}{5}\right)^2 = 5x$$
$$\frac{x^4}{25} = 5x$$
$$x^4 = 125x$$
$$x^4 - 125x = 0$$
$$x\left(x^3 - 125\right) = 0$$
$$x = 0 \quad \text{or} \quad x = \sqrt[3]{125} = 5$$
$$y = 0 \quad \text{or} \quad y = 5$$
The intersection points are $(0, 0)$ and $(5, 5)$. These agree with the coordinates found using the TRACE and ARROW keys.

13. We sketch the two curves:

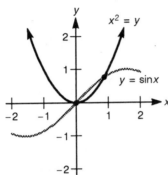

By inspection $(0,0)$ is one of the intersection points. Using the TRACE and ARROW keys, the other point is approximately $(0.88, 0.77)$.

10.3 Tangents to Parabolas (Optional)

1. Call m the slope of the tangent line, so:
$$y - 4 = m(x - 2)$$
Now substitute $y = x^2$:
$$x^2 - 4 = m(x - 2)$$
$$(x + 2)(x - 2) = m(x - 2)$$
$$(x - 2)(x + 2 - m) = 0$$
So the solutions are $x = 2$ and $x = m - 2$. But since the line is tangent to the parabola, these two x-values must be equal:
$$m - 2 = 2$$
$$m = 4$$
Thus the tangent line is given by:
$$y - 4 = 4(x - 2)$$
$$y - 4 = 4x - 8$$
$$y = 4x - 4$$
We sketch the parabola and tangent line:

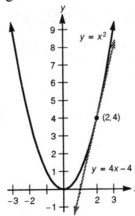

3.   Call m the slope of the tangent line, so:
$$y - 2 = m(x - 4)$$

Solving $x^2 = 8y$ for y yields $y = \dfrac{x^2}{8}$, now substitute:
$$\frac{x^2}{8} - 2 = m(x - 4)$$
$$x^2 - 16 = 8m(x - 4)$$
$$(x + 4)(x - 4) = 8m(x - 4)$$
$$(x - 4)(x + 4 - 8m) = 0$$

So the solutions are $x = 4$ and $x = 8m - 4$. But since the line is tangent to the parabola, these two x-values must be equal:
$$8m - 4 = 4$$
$$8m = 8$$
$$m = 1$$

Thus the tangent line is given by:
$$y - 2 = 1(x - 4)$$
$$y - 2 = x - 4$$
$$y = x - 2$$

We sketch the parabola and tangent line:

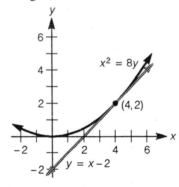

5.   Call m the slope of the tangent line, so:
$$y + 9 = m(x + 3)$$

Solving $x^2 = -y$ for y yields $y = -x^2$, now substitute:
$$-x^2 + 9 = m(x + 3)$$
$$x^2 - 9 = -m(x + 3)$$
$$(x + 3)(x - 3) = -m(x + 3)$$
$$(x + 3)(x - 3 + m) = 0$$

So the solutions are $x = -3$ and $x = 3 - m$. But since the line is tangent to the parabola, these two x-values must be equal:
$$3 - m = -3$$
$$-m = -6$$
$$m = 6$$

Thus the tangent line is given by:

$$y + 9 = 6(x + 3)$$
$$y + 9 = 6x + 18$$
$$y = 6x + 9$$

We sketch the parabola and tangent line:

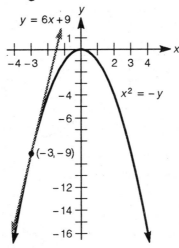

7.   Call m the slope of the tangent line, so:

$$y - 2 = m(x - 1)$$

Solving $y^2 = 4x$ for x yields $x = \dfrac{y^2}{4}$, now substitute:

$$y - 2 = m\left(\frac{y^2}{4} - 1\right)$$
$$4(y - 2) = m(y^2 - 4)$$
$$4(y - 2) = m(y + 2)(y - 2)$$
$$(y - 2)[4 - m(y + 2)] = 0$$

So the solutions are $y = 2$ and $y = \frac{4}{m} - 2$. But since the line is tangent to the parabola, these two x-values must be equal:

$$\frac{4}{m} - 2 = 2$$
$$\frac{4}{m} = 4$$
$$4 = 4m$$
$$1 = m$$

Thus the tangent line is given by:

$$y - 2 = 1(x - 1)$$
$$y - 2 = x - 1$$
$$y = x + 1$$

We sketch the parabola and tangent line:

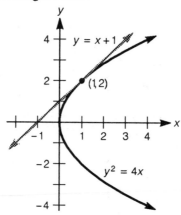

9.   Let m be the slope, so $y - 2 = m(x - 4)$. Since $y = \sqrt{x}$, then $x = y^2$:
$$y - 2 = m\left(y^2 - 4\right)$$
$$y - 2 = m(y + 2)(y - 2)$$
$$0 = (y - 2)\left[m(y + 2) - 1\right]$$
So the solutions are $y = 2$ and $y = \frac{1}{m} - 2$. But since the line is tangent to the parabola, these two x-values must be equal:
$$\frac{1}{m} - 2 = 2$$
$$\frac{1}{m} = 4$$
$$m = \tfrac{1}{4}$$

11.   Let m be the slope, so $y - 8 = m(x - 2)$. Substitute $y = x^3$:
$$x^3 - 8 = m(x - 2)$$
$$(x - 2)\left(x^2 + 2x + 4\right) = m(x - 2)$$
$$(x - 2)\left(x^2 + 2x + 4 - m\right) = 0$$
So the solutions are $x = 2$ and $m = x^2 + 2x + 4$. But since the line is tangent to the parabola, we can substitute $x = 2$ to obtain:
$$m = 2^2 + 4 + 4 = 12$$

13. Let the equation of the tangent line be $y - y_0 = m(x - x_0)$. Replacing y and y_0

by $\dfrac{x^2}{4p}$ and $\dfrac{x_0^2}{4p}$, respectively, we obtain:

$$\frac{x^2}{4p} - \frac{x_0^2}{4p} = m(x - x_0)$$
$$x^2 - x_0^2 - 4pm(x - x_0) = 0$$
$$(x - x_0)[(x + x_0) - 4pm] = 0$$
$$(x - x_0)(x + x_0 - 4pm) = 0$$

Setting the second factor equal to zero yields $x = 4pm - x_0$. But since $x = x_0$ is known

to be the unique solution, we have $x_0 = 4pm - x_0$. Thus $m = \dfrac{2x_0}{4p} = \dfrac{x_0}{2p}$. The equation

of the tangent line is therefore:

$$y - y_0 = \frac{x_0}{2p}(x - x_0)$$
$$y = \frac{x_0}{2p}x - \frac{x_0^2}{2p} + y_0$$

Now if in this last equation we replace x_0^2 by $4py_0$, we obtain $y = \dfrac{x_0}{2p}x - y_0$, as

required.

15. (a)   Let the coordinates of P be (a, b), so that $a^2 = 4pb$. Then according to Exercise 40

of Section 10.2, the coordinates of Q are $\left(-\dfrac{4p^2}{a}, \dfrac{p^2}{b}\right)$. The slope of the tangent at

P is $\dfrac{a}{2p}$. The slope of the tangent at Q is $\dfrac{-\dfrac{4p^2}{a}}{2p} = -\dfrac{2p}{a}$. Thus the two slopes are

negative reciprocals, from which it follows that the tangents are perpendicular to
one another.

(b)   The equation of the tangent at P is $y = \dfrac{a}{2p}x - b$. The equation of the tangent at Q is

found to be $y = -\dfrac{2p}{a}x - \dfrac{p^2}{b}$. By solving this system of two equations, we find that

the coordinates of the point of intersection are $x = \dfrac{a(b-p)}{2b}$ and $y = -p$. Since the

y-value here is $-p$, we conclude that the two tangents intersect at a point on the
directrix.

(c) The coordinates of the intersection point D are given in the solution to part (b). Using those coordinates, the slope of the line from D to the focus is found to be $\dfrac{4bp}{a(p-b)}$. On the other hand, the slope of $\overline{PQ}$ is found to be $\dfrac{a(b-p)}{4bp}$. (Note: In computing the slope of $\overline{PQ}$, the relationship $a^2 = 4pb$ helps in simplifying the expressions that arise.) Now by inspection we see that the two expressions for slope are negative reciprocals, thus the lines are perpendicular.

17. Summarizing our calculations, we have:

Equation of normal: $\qquad y - y_0 = -\dfrac{y_0}{2p}\left(x - x_0\right)$

Coordinates of A: $\qquad \left(2p + x_0, 0\right)$

Equation of line $\overline{AZ}$: $\qquad y = \dfrac{p - x_0}{y_0}\left(x - 2p - x_0\right)$

Equation of line $\overline{PF}$: $\qquad y = \dfrac{y_0}{x_0 - p}(x - p)$

Coordinates of Z: $\qquad x = \dfrac{2p^3 + p^2 x_0 + x_0^3}{\left(x_0 + p\right)^2} = \dfrac{2p^2 - px_0 + x_0^2}{x_0 + p}, \; y = \dfrac{y_0\left(x_0 - p\right)}{x_0 + p}$

With these coordinates for Z, the distance ZP can be determined by using the distance formula. The result is $ZP = 2p$, as indicated in the problem.

19. (a) As indicated in Exercise 14, the x-intercept of the tangent line is $-x_0$. Therefore $FA = x_0 + p$. Next, according to the definition of a parabola, FP is equal to the distance from P to the directrix. But the distance from P to the directrix is $x_0 + p$. Consequently $FP = FA$ because both equal $x_0 + p$.

(b) Since $AF = FP$, $\triangle AFP$ is isosceles, and consequently $\alpha = \gamma$. Also, since HP is parallel to the x-axis, we have $\beta = \gamma$.

(c) We have $\alpha = \gamma = \beta$ and therefore $\alpha = \beta$, as required.

21. The equation of the tangent line is $y = 2x - 4$. With $x = 3$ and $y = 2$, the equation becomes $2 = 2(3) - 4$, or $2 = 2$. Thus the point $(3, 2)$ lies on the line. The equation through $F(0, 1)$ and $P(4, 4)$ is $y = \frac{3}{4}x + 1$. The equation of the line containing AB is $y = -\frac{4}{3}x + 6$. By solving these last two equations for x and y, we find that the coordinates of B are $\left(\frac{12}{5}, \frac{14}{5}\right)$. The distance formula can now be used to find FB as well as FC. In both cases the result is 3. Thus $FB = FC$.

10.4 The Ellipse

1. Dividing by 36, the standard form is $\frac{x^2}{9} + \frac{y^2}{4} = 1$. The length of the major axis is 6 and the length of the minor axis is 4. Using $c^2 = a^2 - b^2$, we find:
$$c = \sqrt{a^2 - b^2} = \sqrt{9-4} = \sqrt{5}$$
So the foci are $(\pm\sqrt{5}, 0)$ and the eccentricity is $\frac{\sqrt{5}}{3}$.

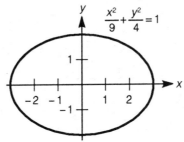

3. Dividing by 16, the standard form is $\frac{x^2}{16} + \frac{y^2}{1} = 1$. The length of the major axis is 8 and the length of the minor axis is 2. Using $c^2 = a^2 - b^2$, we find:
$$c = \sqrt{a^2 - b^2} = \sqrt{16-1} = \sqrt{15}$$
So the foci are $(\pm\sqrt{15}, 0)$ and the eccentricity is $\frac{\sqrt{15}}{4}$.

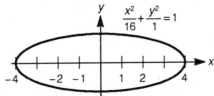

5. Dividing by 2, the standard form is $\frac{x^2}{2} + \frac{y^2}{1} = 1$. The length of the major axis is $2\sqrt{2}$ and the length of the minor axis is 2. Using $c^2 = a^2 - b^2$, we find:
$$c = \sqrt{a^2 - b^2} = \sqrt{2-1} = 1$$
So the foci are $(\pm 1, 0)$ and the eccentricity is $\frac{\sqrt{2}}{2}$.

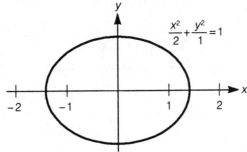

7. Dividing by 144, the standard form is $\frac{x^2}{9} + \frac{y^2}{16} = 1$. The length of the major axis is 8 and the length of the minor axis is 6. Using $c^2 = a^2 - b^2$, we find:

$$c = \sqrt{a^2 - b^2} = \sqrt{16-9} = \sqrt{7}$$

So the foci are $\left(0, \pm\sqrt{7}\right)$ and the eccentricity is $\frac{\sqrt{7}}{4}$.

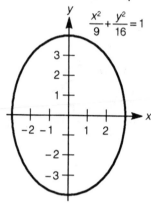

9. Dividing by 5, the standard form is $\frac{x^2}{1/3} + \frac{y^2}{5/3} = 1$. The lengths of the major and minor axes are given by:

major axis: $2\sqrt{\frac{5}{3}} = \frac{2\sqrt{5}}{\sqrt{3}} = \frac{2\sqrt{15}}{3}$

minor axis: $2\sqrt{\frac{1}{3}} = \frac{2}{\sqrt{3}} = \frac{2\sqrt{3}}{3}$

Using $c^2 = a^2 - b^2$, we find:

$$c = \sqrt{\frac{5}{3} - \frac{1}{3}} = \sqrt{\frac{4}{3}} = \frac{2}{\sqrt{3}} = \frac{2\sqrt{3}}{3}$$

So the foci are $\left(0, \pm\frac{2\sqrt{3}}{3}\right)$. The eccentricity is given by:

$$e = \frac{\frac{2\sqrt{3}}{3}}{\frac{\sqrt{15}}{3}} = \frac{2\sqrt{3}}{\sqrt{15}} = \frac{2}{\sqrt{5}} = \frac{2\sqrt{5}}{5}$$

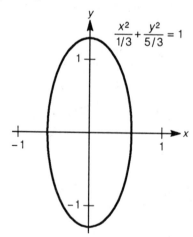

11. Dividing by 4, the standard form is $\frac{x^2}{2} + \frac{y^2}{4} = 1$. The length of the major axis is 4 and the length of the minor axis is $2\sqrt{2}$. Using $c^2 = a^2 - b^2$, we find:

$$c = \sqrt{a^2 - b^2} = \sqrt{4-2} = \sqrt{2}$$

So the foci are $\left(0, \pm\sqrt{2}\right)$ and the eccentricity is $\frac{\sqrt{2}}{2}$.

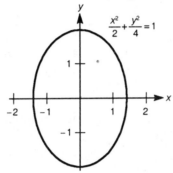

13. The equation is already in standard form with a center of $(5, -1)$. The length of the major axis is 10 and the length of the minor axis is 6. Using $c^2 = a^2 - b^2$, we find:

$$c = \sqrt{a^2 - b^2} = \sqrt{25-9} = \sqrt{16} = 4$$

So the foci are $(5+4, -1) = (9, -1)$ and $(5-4, -1) = (1, -1)$, and the eccentricity is $\frac{4}{5}$.

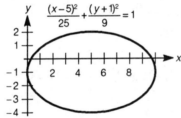

15. The equation is already in standard form with a center of $(1, 2)$. The length of the major axis is 4 and the length of the minor axis is 2. Using $c^2 = a^2 - b^2$, we find:

$$c = \sqrt{a^2 - b^2} = \sqrt{4-1} = \sqrt{3}$$

So the foci are $\left(1, 2 \pm\sqrt{3}\right)$ and the eccentricity is $\frac{\sqrt{3}}{2}$.

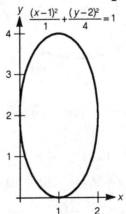

17. The equation is already in standard form with a center of $(-3,0)$. The length of the major axis is 6 and the length of the minor axis is 2. Using $c^2 = a^2 - b^2$, we find:
$$c = \sqrt{a^2 - b^2} = \sqrt{9-1} = \sqrt{8} = 2\sqrt{2}$$
So the foci are $(-3 \pm 2\sqrt{2}, 0)$ and the eccentricity is $\frac{2\sqrt{2}}{3}$.

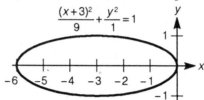

19. We complete the square to convert the equation to standard form:
$$3x^2 + 4y^2 - 6x + 16y + 7 = 0$$
$$3(x^2 - 2x) + 4(y^2 + 4y) = -7$$
$$3(x^2 - 2x + 1) + 4(y^2 + 4y + 4) = -7 + 3 + 16$$
$$3(x-1)^2 + 4(y+2)^2 = 12$$
$$\frac{(x-1)^2}{4} + \frac{(y+2)^2}{3} = 1$$

The center is $(1, -2)$, the length of the major axis is 4, and the length of the minor axis is $2\sqrt{3}$. Using $c^2 = a^2 - b^2$, we find:
$$c = \sqrt{a^2 - b^2} = \sqrt{4-3} = 1$$
So the foci are $(1 + 1, -2) = (2, -2)$ and $(1 - 1, -2) = (0, -2)$, and the eccentricity is $\frac{1}{2}$.

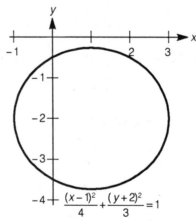

21. We complete the square to convert the equation to standard form:
$$5x^2 + 3y^2 - 40x - 36y + 188 = 0$$
$$5(x^2 - 8x) + 3(y^2 - 12y) = -188$$
$$5(x^2 - 8x + 16) + 3(y^2 - 12y + 36) = -188 + 80 + 108$$
$$5(x-4)^2 + 3(y-6)^2 = 0$$

Notice that the only solution to this is the center $(4, 6)$. This is called a degenerate ellipse, or, more commonly, a point!

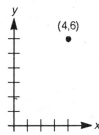

23. We complete the square to convert the equation to standard form:
$$16x^2 + 25y^2 - 64x - 100y + 564 = 0$$
$$16(x^2 - 4x) + 25(y^2 - 4y) = -564$$
$$16(x^2 - 4x + 4) + 25(y^2 - 4y + 4) = -564 + 64 + 100$$
$$16(x - 2)^2 + 25(y - 2)^2 = -400$$
Notice that there is no solution to this equation, since the left-hand side is non-negative. So there is no graph.

25. We are given $c = 3$ and $a = 5$, so we have the equation in the form:
$$\frac{x^2}{5^2} + \frac{y^2}{b^2} = 1$$
Since $c^2 = a^2 - b^2$, we find b:
$$9 = 25 - b^2$$
$$b^2 = 16$$
$$b = 4$$
So the equation is $\dfrac{x^2}{25} + \dfrac{y^2}{16} = 1$, or $16x^2 + 25y^2 = 400$.

27. We are given $a = 4$, so the equation has a form of:
$$\frac{x^2}{16} + \frac{y^2}{b^2} = 1$$
Now $\frac{c}{a} = \frac{1}{4}$, so $\frac{c}{4} = \frac{1}{4}$ and thus $c = 1$.
We find b:
$$c^2 = a^2 - b^2$$
$$1 = 16 - b^2$$
$$b^2 = 15$$
So the equation is $\dfrac{x^2}{16} + \dfrac{y^2}{15} = 1$, or $15x^2 + 16y^2 = 240$.

29. We have $c = 2$ and $a = 5$, so the equation has a form of:
$$\frac{x^2}{b^2} + \frac{y^2}{25} = 1$$
We find b:
$$c^2 = a^2 - b^2$$
$$4 = 25 - b^2$$
$$b^2 = 21$$
So the equation is $\dfrac{x^2}{21} + \dfrac{y^2}{25} = 1$, or $25x^2 + 21y^2 = 525$.

31. We know $a = 2b$ and that the equation has a form of:
$$\frac{x^2}{a^2} + \frac{y^2}{b^2} = 1$$
Using the point $\left(1, \sqrt{2}\right)$ and $a = 2b$, we have:
$$\frac{(1)^2}{(2b)^2} + \frac{\left(\sqrt{2}\right)^2}{b^2} = 1$$
$$\frac{1}{4b^2} + \frac{2}{b^2} = 1$$
Multiply by $4b^2$:
$$1 + 8 = 4b^2$$
$$b^2 = \tfrac{9}{4}$$
$$b = \tfrac{3}{2}$$
Since $a = 2b$, then $a = 3$. So the equation is $\dfrac{x^2}{9} + \dfrac{y^2}{9/4} = 1$, or $x^2 + 4y^2 = 9$.

33. (a) Using the equation $\dfrac{x_1 x}{a^2} + \dfrac{y_1 y}{b^2} = 1$, where $\left(x_1, y_1\right) = (8, 2)$, $a^2 = 76$, $b^2 = \tfrac{76}{3}$, we have:
$$\frac{8x}{76} + \frac{2y}{76/3} = 1$$
$$8x + 6y = 76$$
$$6y = -8x + 76$$
$$y = -\tfrac{4}{3}x + \tfrac{38}{3}$$

(b)   Here $(x_1, y_1) = (-7, 3)$, $a^2 = 76$, $b^2 = \frac{76}{3}$:

$$-\frac{7x}{76} + \frac{3y}{76/3} = 1$$
$$-7x + 9y = 76$$
$$9y = 7x + 76$$
$$y = \frac{7}{9}x + \frac{76}{9}$$

(c)   Here $(x_1, y_1) = (1, -5)$, $a^2 = 76$, $b^2 = \frac{76}{3}$:

$$\frac{x}{76} - \frac{5y}{76/3} = 1$$
$$x - 15y = 76$$
$$-15y = -x + 76$$
$$y = \frac{1}{15}x - \frac{76}{15}$$

35.   (a)   Using the tangent formula where $(x_1, y_1) = (4, 2)$, $a^2 = \frac{52}{3}$, $b^2 = 52$:

$$\frac{4x}{52/3} + \frac{2y}{52} = 1$$
$$12x + 2y = 52$$
$$2y = -12x + 52$$
$$y = -6x + 26$$

(b)   The y-intercept is 26 and the x-intercept is $\frac{13}{3}$, so the area is:

$$\tfrac{1}{2}(\text{base})(\text{height}) = \tfrac{1}{2} \cdot \tfrac{13}{3} \cdot 26 = \tfrac{169}{3}$$

37.   (a)   Substituting the point, we have:

$$9x^2 + 25y^2 = 9(1)^2 + 25\left(\frac{6\sqrt{6}}{5}\right)^2 = 9 + 216 = 225$$

(b)   Here $(x_1, y) = \left(1, \frac{6\sqrt{6}}{5}\right)$, $a^2 = 25$, $b^2 = 9$:

$$\frac{x}{25} + \frac{\dfrac{6\sqrt{6}}{5}y}{9} = 1$$
$$9x + 30\sqrt{6}y = 225$$
$$3x + 10\sqrt{6}y - 75 = 0$$

(c)   We have $c^2 = a^2 - b^2 = 25 - 9 = 16$, so $c = 4$. Thus the coordinates of F_1 and F_2
are $F_1(-4,0)$ and $F_2(4,0)$. We now find each distance:

$$d_1 = \frac{|-12+0-75|}{\sqrt{9+600}} = \frac{87}{\sqrt{609}}$$

$$d_2 = \frac{|12+0-75|}{\sqrt{9+600}} = \frac{63}{\sqrt{609}}$$

(d)   We compute $d_1 d_2 = \frac{87}{\sqrt{609}} \cdot \frac{63}{\sqrt{609}} = \frac{5481}{609} = 9$, which is b^2.

39. We begin by solving for y:

$$\frac{x^2}{1} + \frac{y^2}{16} = 1$$

$$\frac{y^2}{16} = 1 - x^2$$

$$y^2 = 16(1 - x^2)$$

$$y = \pm\sqrt{16(1 - x^2)} = \pm 4\sqrt{1 - x^2}$$

We use a calculator to complete the table:

x	0	0.1	0.2	0.3	0.4	0.5	0.6	0.7	0.8	0.9	1.0
y	±4	±3.98	±3.92	±3.82	±3.67	±3.46	±3.2	±2.86	±2.4	±1.74	0

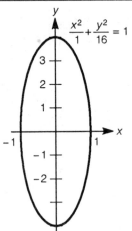

41. From Figure 4, we have $F_1 P + F_2 P > F_1 F_2$. But by definition, $F_1 P + F_2 P = 2a$, while
$F_1 F_2 = 2c$. Thus $2a > 2c$, or $a > c$, so $a^2 > c^2$ (since a and c are positive), and
consequently $a^2 - c^2 > 0$.

43. The tangent at (x_1, y_1) has equation $\dfrac{x_1 x}{a^2} + \dfrac{y_1 y}{b^2} = 1$, which has slope $\dfrac{-x_1 b^2}{y_1 a^2}$. Thus the

normal has slope $\dfrac{y_1 a^2}{x_1 b^2}$ and equation:

$$y - y_1 = \frac{y_1 a^2}{x_1 b^2}(x - x_1)$$
$$b^2 x_1 y - b^2 x_1 y_1 = y_1 a^2 x - a^2 x_1 y_1$$
$$a^2 y_1 x - b^2 x_1 y = (a^2 - b^2) x_1 y_1$$

45. If we first multiply each equation by $a^2 b^2$, we have:
$$b^2 x^2 + a^2 y^2 = a^2 b^2$$
$$a^2 x^2 + b^2 y^2 = a^2 b^2$$
Multiplying the first equation by $-a^2$ and the second equation by b^2 yields:
$$-a^2 b^2 x^2 - a^4 y^2 = -a^4 b^2$$
$$a^2 b^2 x^2 + b^4 y^2 = a^2 b^4$$
Adding, we have:
$$(b^4 - a^4) y^2 = a^2 b^2 (b^2 - a^2)$$
$$y^2 = \frac{a^2 b^2}{b^2 + a^2}$$
$$y = \pm \frac{ab}{\sqrt{a^2 + b^2}}$$
Substituting into the first equation:
$$b^2 x^2 + \frac{a^4 b^2}{a^2 + b^2} = a^2 b^2$$
$$b^2 x^2 = \frac{a^2 b^4}{a^2 + b^2}$$
$$x^2 = \frac{a^2 b^2}{a^2 + b^2}$$
$$x = \pm \frac{ab}{\sqrt{a^2 + b^2}}$$
So there are four intersection points:
$$\left(\frac{ab}{A}, \frac{ab}{A}\right), \left(\frac{ab}{A}, -\frac{ab}{A}\right), \left(-\frac{ab}{A}, \frac{ab}{A}\right), \left(-\frac{ab}{A}, -\frac{ab}{A}\right), \text{ where } A = \sqrt{a^2 + b^2}$$

We sketch the intersection:

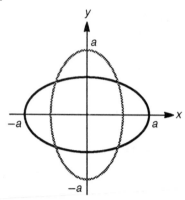

47. Using Exercise 43, we find that N is the point $\left(\dfrac{(a^2-b^2)x_1}{a^2},0\right)$. Thus:

$$FN = \frac{(a^2-b^2)x_1}{a^2}+c = \frac{c^2}{a^2}x_1+c = e^2x_1+c$$

It can be shown using $\dfrac{x_1^2}{a^2}+\dfrac{y_1^2}{b^2}=1$ that $y_1 = a^2-c^2-x_1^2+e^2x_1^2$. Now:

$$FP = \sqrt{\left(x_1+c\right)^2+y_1^2}$$
$$= \sqrt{x_1^2+2x_1c+c^2+a^2-c^2-x_1^2+e^2x_1^2}$$
$$= \sqrt{e^2x_1^2+2aex_1+a^2}$$
$$= ex_1+a$$

Thus:

$$\frac{FN}{FP} = \frac{e^2x_1+c}{ex_1+a} = \frac{e(ex_1+a)}{ex_1+a} = e$$

49. (a) We substitute the points:
$$5^2+3(1)^2 = 25+3 = 28$$
$$4^2+3(-2)^2 = 16+12 = 28$$
$$(-1)^2+3(3)^2 = 1+27 = 28$$

(b) $\overline{AB}$ has slope $\dfrac{-2-1}{4-5}=3$. The line containing C parallel to $\overline{AB}$ has equation:
$$y-3 = 3(x+1)$$
$$y = 3x+6$$

The intersections of this line with the ellipse have x-coordinates such that:
$$x^2 + 3(3x+6)^2 = 28$$
$$x^2 + 27x^2 + 108x + 108 = 28$$
$$7x^2 + 27x + 20 = 0$$
$$(7x+20)(x+1) = 0$$
$$x = -\tfrac{20}{7}, -1$$

Thus the point D we want is $\left(-\tfrac{20}{7}, -\tfrac{18}{7}\right)$.

(c) The point O is $(0,0)$. We use the suggested formula:
$$\text{Area of } \triangle OAC = \tfrac{1}{2}\left|0 - 0 + 5 \cdot 3 - (-1)(1) + 0 - 0\right| = 8$$
$$\text{Area of } \triangle OBD = \tfrac{1}{2}\left|0 - 0 + 4\left(-\tfrac{18}{7}\right) - (-2)\left(-\tfrac{20}{7}\right) + 0 - 0\right| = 8$$
Thus the areas are equal.

51. (a) We have $a\sqrt{(x+c)^2 + y^2} = a^2 + xc$. Dividing by a and noting that $e = \tfrac{c}{a}$:
$$\sqrt{(x+c)^2 + y^2} = a + xe$$
But the radical is F_1P by the distance formula, so:
$$F_1P = a + xe$$

(b) We have:
$$F_1P + F_2P = 2a$$
$$(a + xe) + F_2P = 2a$$
$$F_2P = a - xe$$

53. (a) We use the tangent formula, and multiply by a^2b^2:
$$\frac{x_1 x}{a^2} + \frac{y_1 y}{b^2} = 1$$
$$b^2 x_1 x + a^2 y_1 y = a^2 b^2$$

(b) Since (h,k) lies on this line, replacing (x,y) with (h,k) results in:
$$b^2 x_1 h + a^2 y_1 k = a^2 b^2$$

(c) Repeating part (a), we have $b^2 x_2 x + a^2 y_2 y = a^2 b^2$. Now substitute (h,k) to obtain $b^2 x_2 h + a^2 y_2 k = a^2 b^2$.

(d) Replacing (x,y) with (x_1, y_1) and (x_2, y_2), respectively, results in the equations we have proved in (b) and (c). Thus this line must pass through the points (x_1, y_1) and (x_2, y_2).

55. By Exercise 51(b), $F_2P = a - ex$ and $PR = \frac{a}{e} - x$. Thus:

$$\frac{F_2P}{PR} = \frac{a-ex}{\frac{a}{e}-x} \cdot \frac{e}{e} = \frac{e(a-ex)}{a-ex} = e$$

57. Since $(0,0)$ is the center, then $D = \frac{a}{e}$ and $d = \frac{a}{e} - c$, thus:

$$\frac{D}{d} = \frac{\frac{a}{e}}{\frac{a}{e}-c} = \frac{a}{a-ce} = \frac{a}{a-c\cdot\left(\frac{c}{a}\right)} = \frac{a^2}{a^2-c^2} = \frac{a^2}{b^2}$$

59. (a)   Substituting the point, we see that:

$$\frac{c^2}{a^2} + \frac{\left(b^2/a\right)^2}{b^2} = \frac{a^2-b^2}{a^2} + \frac{b^2}{a^2} = 1$$

So $\left(c, \dfrac{b^2}{2a}\right)$ lies on the ellipse.

(b)   The equation of the tangent is:

$$\frac{cx}{a^2} + \frac{\dfrac{b^2}{a}\cdot y}{b^2} = 1$$

$$\frac{cx}{a^2} + \frac{y}{a} = 1$$

The x-coordinate of the point of intersection is $\dfrac{a}{e}$, and the y-coordinate is the solution of:

$$\frac{c(a/e)}{a^2} + \frac{y}{a} = 1$$

$$\frac{a^2}{a^2} + \frac{y}{a} = 1$$

$$y = 0$$

Thus, the required point is $\left(\dfrac{a}{e}, 0\right)$.

61. Let P have coordinates (x_1, y_1) and Q have coordinates (x_1, y_2). The center of the circle is $(0,0)$, and the slope of the normal through P is thus $\dfrac{y_1 - 0}{x_1 - 0} = \dfrac{y_1}{x_1}$. Thus the slope of the tangent through P is $-\dfrac{x_1}{y_1}$, and its equation is:

$$y - y_1 = -\frac{x_1}{y_1}(x - x_1)$$
$$yy_1 - y_1^2 = -x_1 x + x_1^2$$
$$x_1 x = -y_1 y + x_1^2 + y_1^2$$
$$x_1 x = -y_1 y + a^2$$

The tangent of the ellipse through Q has equation $\dfrac{x_1 x}{a^2} + \dfrac{y_2 y}{b^2} = 1$. Thus the y-coordinate of the point of intersection is the solution of:

$$\frac{-y_1 y + a^2}{a^2} + \frac{y_2 y}{b^2} = 1$$
$$\left(\frac{y_2}{b^2} - \frac{y_1}{a^2}\right) y = 0$$

If $\dfrac{y_2}{b^2} - \dfrac{y_1}{a^2} = 0$ then any y-value corresponds to a point of intersection, so the lines are the same. So the points are the same, contradicting our hypothesis that the points are distinct. Otherwise, $y = 0$, as required. Note: We implicitly assumed $x_1 \neq 0$ earlier, when assuming $x_1 x \neq 0$. Note that the lines do not intersect if $x_1 = 0$.

63. Let P have coordinates (x_1, y_1). Then by Exercise 62(a), Q has coordinates $(-x_1, -y_1)$. The tangent through P has equation:

$$\frac{x_1 x}{a^2} + \frac{y_1 y}{b^2} = 1$$
$$y = \frac{b^2}{y_1}\left(\frac{-x_1 x}{a^2} + 1\right)$$

So its slope is $-\dfrac{b^2 x_1}{y_1 a^2}$. The tangent through Q has equation:

$$-\frac{x_1 x}{a^2} + \frac{-y_1 y}{b^2} = 1$$
$$y = \frac{b^2}{y_1}\left(\frac{x_1 x}{a^2} + 1\right)$$

So its slope is $\dfrac{b^2 x_1}{y_1 a^2}$, the same as that of the other tangent. Thus the two tangents are parallel, as required.

65. (a) Using the formula $d = \dfrac{\left|Ax_0 + By_0 + C\right|}{\sqrt{A^2 + B^2}}$, we have:

$$F_1A = \frac{\left|-c\left(\dfrac{x_1}{a^2}\right) + 0 - 1\right|}{\sqrt{A^2 + B^2}} = \frac{\left|\dfrac{x_1 c}{a^2} + 1\right|}{\sqrt{A^2 + B^2}},$$

where $A = \dfrac{x_1}{a^2}$ and $B = \dfrac{y_1}{b^2}$.

(b) Making the substitution $c = ae$ and $a + ex_1 = F_1P$, we have:

$$F_1A = \frac{\left|\dfrac{x_1 e}{a} + 1\right|}{\sqrt{A^2 + B^2}} = \frac{\left|x_1 e + a\right|}{a\sqrt{A^2 + B^2}} = \frac{F_1P}{a\sqrt{A^2 + B^2}}$$

(c) F_2B is the distance from the point $(c, 0)$ to $\dfrac{x_1 x}{a^2} + \dfrac{y_1 y}{b^2} = 1$, so:

$$F_2B = \frac{\left|c\left(\dfrac{x_1}{a^2}\right) + 0 - 1\right|}{\sqrt{A^2 + B^2}} = \frac{\left|\dfrac{x_1 c}{a^2} - 1\right|}{\sqrt{A^2 + B^2}}$$

Now making the substitutions $c = ae$ and $a - ex_1 = F_2P$, we have:

$$F_2B = \frac{\left|\dfrac{x_1 e}{a} - 1\right|}{\sqrt{A^2 + B^2}} = \frac{\left|x_1 e - a\right|}{a\sqrt{A^2 + B^2}} = \frac{\left|a - x_1 e\right|}{a\sqrt{A^2 + B^2}} = \frac{F_2P}{a\sqrt{A^2 + B^2}}$$

(d) We evaluate each quotient:

$$\frac{F_1A}{F_1P} = \frac{1}{a\sqrt{A^2 + B^2}} \qquad \text{and} \qquad \frac{F_2B}{F_2P} = \frac{1}{a\sqrt{A^2 + B^2}}$$

Thus they are equal.

Optional TI-81 Graphing Calculator Exercises for Section 10.4

1. (a) Solving for y, we obtain:

$$\frac{x^2}{9} + \frac{y^2}{4} = 1$$
$$4x^2 + 9y^2 = 36$$
$$9y^2 = 36 - 4x^2$$
$$y^2 = \frac{36 - 4x^2}{9}$$
$$y = \pm\tfrac{1}{3}\sqrt{36 - 4x^2}$$

Graphing the two functions with the indicated settings:

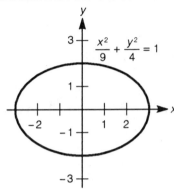

The graph appears to have symmetry with respect to the x-axis, the y-axis, and the origin.

(b)   Solving for y, we obtain:

$$\frac{x^2}{1} + \frac{y^2}{16} = 1$$
$$16x^2 + y^2 = 16$$
$$y^2 = 16 - 16x^2$$
$$y = \pm 4\sqrt{1 - x^2}$$

Graphing the two functions with the indicated settings:

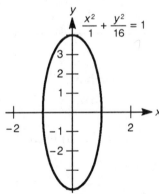

Again, the graph appears to have symmetry with respect to the x-axis, the y-axis, and the origin.

3.   (a)   Dividing by $16b^2$, we have:

$$\frac{x^2}{16} + \frac{y^2}{b^2} = 1$$

So the horizontal axis has length $2(4) = 8$.

(b)   When $b^2 = 16$, then $c^2 = 16 - 16 = 0$, so $c = 0$ and thus:

$$e = \frac{c}{a} = \frac{0}{4} = 0$$

When $b^2 = 13.44$, then $c^2 = 16 - 13.44 = 2.56$, so $c = 1.6$ and thus:
$$e = \frac{c}{a} = \frac{1.6}{4} = 0.4$$
When $b^2 = 10.24$, then $c^2 = 16 - 10.24 = 5.76$, so $c = 2.4$ and thus:
$$e = \frac{c}{a} = \frac{2.4}{4} = 0.6$$
When $b^2 = 5.76$, then $c^2 = 16 - 5.76 = 10.24$, so $c = 3.2$ and thus:
$$e = \frac{c}{a} = \frac{3.2}{4} = 0.8$$

(c)   Solving for y, we have:
$$b^2x^2 + 16y^2 = 16b^2$$
$$16y^2 = 16b^2 - b^2x^2$$
$$y^2 = \frac{b^2(16 - x^2)}{16}$$
$$y = \pm 0.25\sqrt{b^2(16 - x^2)}$$
Using the indicated settings, the graphs of all four ellipses appear as:

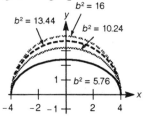

5.   Dividing the ellipse equation by 12, we have:
$$\frac{x^2}{4} + \frac{y^2}{12} = 1$$
So $a^2 = 4$ and $b^2 = 12$. Using $(x_0, y_0) = (1, -3)$ in the tangent line formula yields:
$$\frac{1x}{4} + \frac{-3y}{12} = 1$$
$$3x - 3y = 12$$
$$x - y = 4$$
$$y = x - 4$$

Now graph the ellipse $\frac{x^2}{4} + \frac{y^2}{12} = 1$ and tangent line $y = x - 4$:

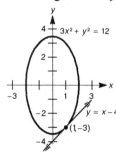

7. (a) Dividing by 12, we graph the ellipse $\frac{x^2}{12} + \frac{y^2}{4} = 1$:

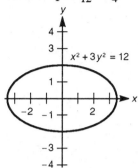

(b) Since $a^2 = 12$ and $b^2 = 4$, then $a = 2\sqrt{3}$ and $b = 2$. Therefore:
$$c^2 = a^2 - b^2 = 12 - 4 = 8, \text{ so } c = \sqrt{8} = 2\sqrt{2}$$

(c) The auxiliary circle has an equation $x^2 + y^2 = 12$. We graph the ellipse and auxiliary circle:

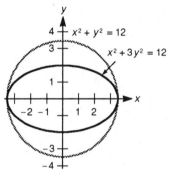

(d) We verify that the point $P(3, 1)$ lies on the ellipse:
$$(3)^2 + 3(1)^2 = 9 + 3 = 12$$
Using $(x_0, y_0) = (3, 1)$, we find the equation of the tangent line to the ellipse at P:
$$\frac{3x}{12} + \frac{1y}{4} = 1$$
$$\frac{x}{4} + \frac{y}{4} = 1$$
$$x + y = 4$$
Now we graph the upper halves of the ellipse and circle, and the line $y = -x + 4$:

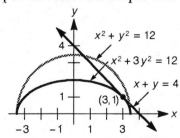

(e)   The perpendicular to the tangent at $P(3,1)$ has slope $m = 1$. Using the point $(-2\sqrt{2},0)$ and the point-slope formula, we have:
$$y - 0 = 1(x + 2\sqrt{2})$$
$$y = x + 2\sqrt{2}$$
Now we graph the upper halves of the ellipse and circle, as well as the lines $y = -x + 4$ and $y = x + 2\sqrt{2}$:

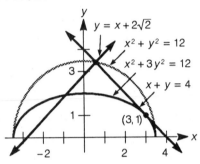

10.5 The Hyperbola

1.   Dividing by 4, the standard form is $\frac{x^2}{4} - \frac{y^2}{1} = 1$. The vertices are $(\pm 2, 0)$, the length of the transverse axis is 4, the length of the conjugate axis is 2, and the asymptotes are $y = \pm \frac{1}{2}x$. Using $c^2 = a^2 + b^2$, we find:
$$c = \sqrt{a^2 + b^2} = \sqrt{4 + 1} = \sqrt{5}$$
So the foci are $(\pm\sqrt{5}, 0)$ and the eccentricity is $\frac{\sqrt{5}}{2}$.

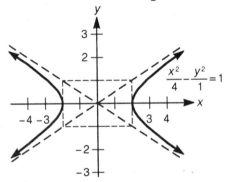

3.   Dividing by 4, the standard form is $\frac{y^2}{4} - \frac{x^2}{1} = 1$. The vertices are $(0, \pm 2)$, the length of the transverse axis is 4, the length of the conjugate axis is 2, and the asymptotes are $y = \pm 2x$. Using $c^2 = a^2 + b^2$, we find:
$$c = \sqrt{a^2 + b^2} = \sqrt{4 + 1} = \sqrt{5}$$

So the foci are $(0,\pm\sqrt{5})$ and the eccentricity is $\frac{\sqrt{5}}{2}$.

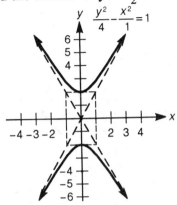

5. Dividing by 400, the standard form is $\frac{x^2}{25}-\frac{y^2}{16}=1$. The vertices are $(\pm5,0)$, the length of the transverse axis is 10, the length of the conjugate axis is 8, and the asymptotes are $y=\pm\frac{4}{5}x$. Using $c^2=a^2+b^2$, we find:
$$c=\sqrt{a^2+b^2}=\sqrt{25+16}=\sqrt{41}$$
So the foci are $(\pm\sqrt{41},0)$ and the eccentricity is $\frac{\sqrt{41}}{5}$.

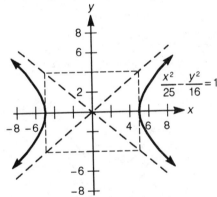

7. Rewriting the equation, the standard form is $\frac{y^2}{1/2}-\frac{x^2}{1/3}=1$. The vertices are $\left(0,\pm\sqrt{\frac{1}{2}}\right)=\left(0,\pm\frac{\sqrt{2}}{2}\right)$, the length of the transverse axis is $\sqrt{2}$, the length of the conjugate axis is $2\sqrt{\frac{1}{3}}=\frac{2\sqrt{3}}{3}$, and the asymptotes are $y=\pm\sqrt{\frac{3}{2}}x=\pm\frac{\sqrt{6}}{2}x$. Using $c^2=a^2+b^2$, we find:
$$c=\sqrt{a^2+b^2}=\sqrt{\frac{1}{2}+\frac{1}{3}}=\sqrt{\frac{5}{6}}=\frac{\sqrt{5}}{\sqrt{6}}=\frac{\sqrt{30}}{6}$$

So the foci are $\left(0,\pm\frac{\sqrt{30}}{6}\right)$ and the eccentricity is:

$$e = \frac{\frac{\sqrt{30}}{6}}{\frac{\sqrt{2}}{2}} = \frac{\sqrt{30}}{3\sqrt{2}} = \frac{\sqrt{15}}{3}$$

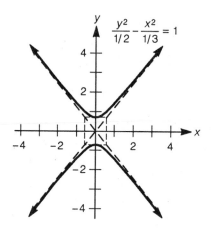

9. Dividing by 100, the standard form is $\frac{y^2}{25} - \frac{x^2}{4} = 1$. The vertices are $(0,\pm 5)$, the length of the transverse axis is 10, the length of the conjugate axis is 4, and the asymptotes are $y = \pm\frac{5}{2}x$. Using $c^2 = a^2 + b^2$, we have:

$$c = \sqrt{a^2 + b^2} = \sqrt{25 + 4} = \sqrt{29}$$

So the foci are $\left(0,\pm\sqrt{29}\right)$ and the eccentricity is $\frac{\sqrt{29}}{5}$.

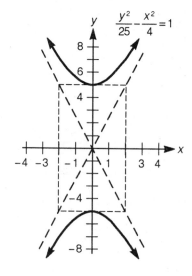

11. The equation is already in standard form with a center of $(5,-1)$. The vertices are $(5+5,-1) = (10,-1)$ and $(5-5,-1) = (0,-1)$, the length of the transverse axis is 10, and the length of the conjugate axis is 6. The asymptotes have slopes of $\pm\frac{3}{5}$, so using the point-slope formula:

$$y+1 = \tfrac{3}{5}(x-5) \qquad\qquad y+1 = -\tfrac{3}{5}(x-5)$$
$$y+1 = \tfrac{3}{5}x - 3 \qquad\qquad y+1 = -\tfrac{3}{5}x + 3$$
$$y = \tfrac{3}{5}x - 4 \qquad\qquad y = -\tfrac{3}{5}x + 2$$

Using $c^2 = a^2 + b^2$, we have:
$$c = \sqrt{a^2 + b^2} = \sqrt{25+9} = \sqrt{34}$$

So the foci are $\left(5\pm\sqrt{34}, -1\right)$ and the eccentricity is $\frac{\sqrt{34}}{5}$.

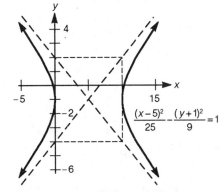

$$\frac{(x-5)^2}{25} - \frac{(y+1)^2}{9} = 1$$

13. The equation is already in standard form with a center of $(1,2)$. The vertices are $(1, 2+2) = (1,4)$ and $(1, 2-2) = (1,0)$, the length of the transverse axis is 4, and the length of the conjugate axis is 2. The asymptotes have slopes of ± 2, so using the point-slope formula:

$$y-2 = 2(x-1) \qquad\qquad y-2 = -2(x-1)$$
$$y-2 = 2x-2 \qquad\qquad y-2 = -2x+2$$
$$y = 2x \qquad\qquad y = -2x+4$$

Using $c^2 = a^2 + b^2$, we have:
$$c = \sqrt{a^2 + b^2} = \sqrt{4+1} = \sqrt{5}$$

So the foci are $\left(1, 2\pm\sqrt{5}\right)$ and the eccentricity is $\frac{\sqrt{5}}{2}$.

$$\frac{(y-2)^2}{4} - \frac{(x-1)^2}{1} = 1$$

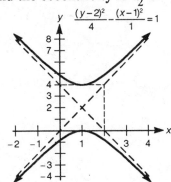

15. The equation is already in standard form with a center of $(-3, 4)$. The vertices are $(-3 + 4, 4) = (1, 4)$ and $(-3 - 4, 4) = (-7, 4)$, the length of the transverse axis is 8, and the length of the conjugate axis is 8. The asymptotes have slopes of ± 1, so using the point-slope formula:

$$y - 4 = 1(x + 3) \qquad\qquad y - 4 = -1(x + 3)$$
$$y - 4 = x + 3 \qquad\qquad\qquad y - 4 = -x - 3$$
$$y = x + 7 \qquad\qquad\qquad\quad y = -x + 1$$

Using $c^2 = a^2 + b^2$, we have:

$$c = \sqrt{a^2 + b^2} = \sqrt{16 + 16} = \sqrt{32} = 4\sqrt{2}$$

So the foci are $\left(-3 \pm 4\sqrt{2}, 4\right)$ and the eccentricity is $\frac{4\sqrt{2}}{4} = \sqrt{2}$.

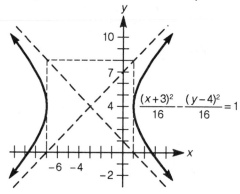

17. We complete the square to convert the equation to standard form:

$$x^2 - y^2 + 2y - 5 = 0$$
$$x^2 - \left(y^2 - 2y\right) = 5$$
$$x^2 - \left(y^2 - 2y + 1\right) = 5 - 1$$
$$x^2 - (y - 1)^2 = 4$$
$$\frac{x^2}{4} - \frac{(y - 1)^2}{4} = 1$$

The center is $(0, 1)$, the vertices are $(0 + 2, 1) = (2, 1)$ and $(0 - 2, 1) = (-2, 1)$, and the lengths of both the transverse and conjugate axes are 4. The asymptotes have slopes of ± 1, so using the point-slope formula:

$$y - 1 = 1(x - 0) \qquad\qquad y - 1 = -1(x - 0)$$
$$y - 1 = x \qquad\qquad\qquad\quad y - 1 = -x$$
$$y = x + 1 \qquad\qquad\qquad\quad y = -x + 1$$

Using $c^2 = a^2 + b^2$, we have:

$$c = \sqrt{a^2 + b^2} = \sqrt{4 + 4} = \sqrt{8} = 2\sqrt{2}$$

So the foci are $\left(\pm2\sqrt{2},1\right)$ and the eccentricity is $\frac{2\sqrt{2}}{2}=\sqrt{2}$.

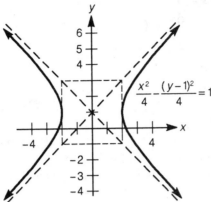

$$\frac{x^2}{4}-\frac{(y-1)^2}{4}=1$$

19. We complete the square to convert the equation to standard form:

$$x^2-y^2-4x+2y-6=0$$
$$\left(x^2-4x\right)-\left(y^2-2y\right)=6$$
$$\left(x^2-4x+4\right)-\left(y^2-2y+1\right)=6+4-1$$
$$(x-2)^2-(y-1)^2=9$$
$$\frac{(x-2)^2}{9}-\frac{(y-1)^2}{9}=1$$

The center is (2, 1), the vertices are (2 + 3, 1) = (5, 1) and (2 − 3, 1) = (−1, 1), and the lengths of both the transverse and conjugate axes are 6. The asymptotes have slopes of ±1, so using the point-slope formula:

$$y-1=1(x-2) \qquad\qquad y-1=-1(x-2)$$
$$y-1=x-2 \qquad\qquad\quad y-1=-x+2$$
$$y=x-1 \qquad\qquad\qquad y=-x+3$$

Using $c^2=a^2+b^2$, we have:

$$c=\sqrt{a^2+b^2}=\sqrt{9+9}=\sqrt{18}=3\sqrt{2}$$

So the foci are $\left(2\pm3\sqrt{2},1\right)$ and the eccentricity is $\frac{3\sqrt{2}}{3}=\sqrt{2}$.

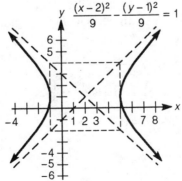

$$\frac{(x-2)^2}{9}-\frac{(y-1)^2}{9}=1$$

21. We complete the square to convert the equation to standard form:
$$y^2 - 25x^2 + 8y - 9 = 0$$
$$\left(y^2 + 8y\right) - 25x^2 = 9$$
$$\left(y^2 + 8y + 16\right) - 25x^2 = 9 + 16$$
$$(y+4)^2 - 25x^2 = 25$$
$$\frac{(y+4)^2}{25} - \frac{x^2}{1} = 1$$

The center is $(0, -4)$, the vertices are $(0, -4 + 5) = (0, 1)$ and $(0, -4 - 5) = (0, -9)$, the length of the transverse axis is 10, and the length of the conjugate axis is 2. The asymptotes have slopes of ± 5, so using the point-slope formula:

$$y + 4 = 5(x - 0) \qquad\qquad y + 4 = -5(x - 0)$$
$$y + 4 = 5x \qquad\qquad\qquad y + 4 = -5x$$
$$y = 5x - 4 \qquad\qquad\qquad y = -5x - 4$$

Using $c^2 = a^2 + b^2$, we have:
$$c = \sqrt{a^2 + b^2} = \sqrt{25 + 1} = \sqrt{26}$$

So the foci are $\left(0, -4 \pm \sqrt{26}\right)$ and the eccentricity is $\frac{\sqrt{26}}{5}$.

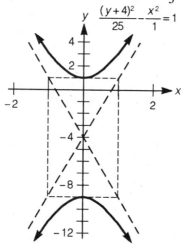

23. We complete the square to convert the equation to standard form:
$$x^2 + 7x - y^2 - y + 12 = 0$$
$$\left(x^2 + 7x\right) - \left(y^2 + y\right) = -12$$
$$\left(x^2 + 7x + \tfrac{49}{4}\right) - \left(y^2 + y + \tfrac{1}{4}\right) = -12 + \tfrac{49}{4} - \tfrac{1}{4}$$
$$\left(x + \tfrac{7}{2}\right)^2 - \left(y + \tfrac{1}{2}\right)^2 = 0$$

Notice that this is a degenerate hyperbola, and the graph consists of the "would-be" asymptotes with slopes ± 1:

$$y + \tfrac{1}{2} = 1\left(x + \tfrac{7}{2}\right) \qquad\qquad y + \tfrac{1}{2} = -1\left(x + \tfrac{7}{2}\right)$$
$$y + \tfrac{1}{2} = x + \tfrac{7}{2} \qquad\qquad\quad y + \tfrac{1}{2} = -x - \tfrac{7}{2}$$
$$y = x + 3 \qquad\qquad\qquad\quad y = -x - 4$$

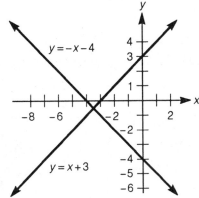

25. Since $P(x,y)$ lies on $\frac{x^2}{4} - \frac{y^2}{1} = 1$, we can find y in terms of x:

$$y^2 = \frac{x^2}{4} - 1 = \frac{x^2 - 4}{4}$$

Taking roots, we have $y = \frac{\sqrt{x^2 - 4}}{2}$, since $P(x,y)$ lies in the first quadrant. So the coordinates of P are $\left(x, \frac{\sqrt{x^2-4}}{2}\right)$. Since $Q(x,y)$ lies in the first quadrant on the asymptote, we find the equation of the asymptote:

$$y - 0 = \tfrac{1}{2}(x - 0)$$
$$y = \tfrac{1}{2}x$$

So the coordinates of Q are $\left(x, \frac{1}{2}x\right)$. Since P and Q have the same x-coordinate PQ is the difference between their y-coordinates:

$$PQ = \frac{x}{2} - \frac{\sqrt{x^2 - 4}}{2} = \frac{x - \sqrt{x^2 - 4}}{2}$$

The order of subtraction is because the asymptote lies above the hyperbola in the first quadrant, and thus $\frac{x}{2}$ is larger than $\frac{\sqrt{x^2-4}}{2}$. This proves the desired result.

27. Since the foci are $(\pm 4, 0)$ and the vertices are $(\pm 1, 0)$, then $c = 4$, $a = 1$, and the hyperbola has the form:

$$\frac{x^2}{1} - \frac{y^2}{b^2} = 1$$

We find b:
$$c^2 = a^2 + b^2$$
$$16 = 1 + b^2$$
$$b^2 = 15$$
So the equation is $\frac{x^2}{1} - \frac{y^2}{15} = 1$, or $15x^2 - y^2 = 15$.

29. The slope of the asymptotes is $\pm\frac{1}{2}$, which tells us that the ratio $\frac{b}{a} = \frac{1}{2}$ in this hyperbola. Also, since the vertices are $(\pm 2, 0)$ then $a = 2$. The required ratio is therefore:
$$\frac{b}{2} = \frac{1}{2} \text{ and } b = 1$$
The equation is $\frac{x^2}{4} - \frac{y^2}{1} = 1$, or $x^2 - 4y^2 = 4$.

31. Since the asymptotes are $y = \pm\frac{\sqrt{10}}{5}x$, then $\frac{b}{a} = \frac{\sqrt{10}}{5}$, so $b = \frac{\sqrt{10}}{5}a$. Now, since the foci are $(\pm\sqrt{7}, 0)$, then $c = \sqrt{7}$ and the hyperbola has the form:
$$\frac{x^2}{a^2} - \frac{y^2}{b^2} = 1$$
Since $b = \frac{\sqrt{10}}{5}a$ and $c = \sqrt{7}$, we have:
$$c^2 = a^2 + b^2$$
$$7 = a^2 + \left(\frac{\sqrt{10}}{5}a\right)^2$$
$$7 = a^2 + \frac{2}{5}a^2$$
$$7a^2 = 35$$
$$a^2 = 5$$
$$b^2 = \frac{2}{5}a^2 = \frac{2}{5}(5) = 2$$
So the equation is $\frac{x^2}{5} - \frac{y^2}{2} = 1$, or $2x^2 - 5y^2 = 10$.

33. The vertices are at $(0, \pm 7)$ so we know it is a "vertical" hyperbola. Its equation will be $\frac{y^2}{49} - \frac{x^2}{b^2} = 1$, but we also know that $(1, 9)$ is a point satisfying the equation. We use it to find b:
$$\frac{81}{49} - \frac{1}{b^2} = 1$$
$$81b^2 - 49 = 49b^2$$
$$32b^2 = 49$$
$$b^2 = \frac{49}{32}$$
So the equation is $\frac{y^2}{49} - \frac{x^2}{49/32} = 1$, or $y^2 - 32x^2 = 49$.

35. We have $2a = 6$, so $a = 3$. Also $2b = 2$, so $b = 1$. Since the foci are on the y-axis, the hyperbola will have the form:

$$\frac{y^2}{a^2} - \frac{x^2}{b^2} = 1$$

So the equation is $\frac{y^2}{9} - \frac{x^2}{1} = 1$, or $y^2 - 9x^2 = 9$.

37. Writing the equation as $\frac{x^2}{16} - \frac{y^2}{16} = 1$, we have $a = b = 4$. So the slopes of the asymptotes are $\pm\frac{b}{a} = \pm\frac{4}{4} = \pm 1$. But these are negative reciprocals of each other, so the asymptotes are perpendicular to each other.

39. (a) Substituting $P(5, 6)$ into $5y^2 - 4x^2 = 80$, we have:

$$5(6)^2 - 4(5)^2 = 5(36) - 4(25) = 180 - 100 = 80$$

So $P(5, 6)$ lies on the hyperbola.

(b) Dividing by 80 yields $\frac{y^2}{16} - \frac{x^2}{20} = 1$, so $a = 4$ and $b = 2\sqrt{5}$. Using $c^2 = a^2 + b^2$, we have:

$$c = \sqrt{a^2 + b^2} = \sqrt{16 + 20} = \sqrt{36} = 6$$

So $c = 6$ and the foci are $(0, \pm 6)$.

(c) We compute the distances:

$$F_1 P = \sqrt{(5-0)^2 + (6-6)^2} = 5$$
$$F_2 P = \sqrt{(5-0)^2 + (6-(-6))^2} = \sqrt{25 + 144} = 13$$

(d) We verify the result:

$$\left| F_1 P - F_2 P \right| = |5 - 13| = |-8| = 8 = 2(4) = 2a$$

41. (a) Since the asymptotes must have slopes of $\pm\frac{b}{a}$, then:

$$-\frac{b}{a} = \frac{-1}{\frac{b}{a}}$$
$$\frac{b^2}{a^2} = 1$$
$$b^2 = a^2$$
$$b = a$$

Now, since $c^2 = a^2 + b^2 = 2a^2$, then the eccentricity is:

$$\frac{c}{a} = \frac{\sqrt{2a^2}}{a} = \frac{\sqrt{2}a}{a} = \sqrt{2}$$

(b)   The slopes of the asymptotes are $\pm\frac{a}{a} = \pm 1$, so the hyperbola will have perpendicular asymptotes. The eccentricity is $\frac{\sqrt{2}a}{a} = \sqrt{2}$.

43. (a)   This equation is just $F_1P - F_2P = 2a$, the defining relation of a hyperbola. Since P is on the right-hand branch, $F_1P > F_2P$.

(b)   Squaring each side of the equation, we obtain:
$$\sqrt{(x+c)^2 + y^2} = 2a + \sqrt{(x-c)^2 + y^2}$$
$$(x+c)^2 + y^2 = 4a^2 + 4a\sqrt{(x-c)^2 + y^2} + (x-c)^2 + y^2$$
$$4xc = 4a^2 + 4a\sqrt{(x-c)^2 + y^2}$$
$$xc - a^2 = a\sqrt{(x-c)^2 + y^2}$$
$$xc - a^2 = a(F_2P)$$

(c)   Dividing by a, we have:
$$\frac{xc}{a} - a = F_2P$$
$$xe - a = F_2P$$

45.   For this hyperbola we find $a^2 = b^2 = k^2$, and $e = \sqrt{2}$.
Also $d^2 = x^2 + y^2 = x^2 + (x^2 - k^2) = 2x^2 - k^2$. Thus we want to show that $F_1P \cdot F_2P = 2x^2 - k^2$. Using the formulas for F_1P and F_2P developed in Exercises 43 and 44, we have:
$$F_1P \cdot F_2P = (xe + a)(xe - a) = x^2e^2 - a^2 = x^2(2) - k^2 = 2x^2 - k^2 = d^2$$

47.   The coordinates of D are $\left(\frac{a}{e}, \frac{b}{e}\right)$ and those of F are $(c, 0)$. Let O denote the center $(0,0)$.
We show that $\angle ODF$ is a right angle by showing $OD^2 + DF^2 = OF^2$:
$$OF^2 = c^2$$
$$OD^2 = \left(\frac{a}{e}\right)^2 + \left(\frac{b}{e}\right)^2 = \frac{a^2 + b^2}{e^2} = \frac{c^2}{e^2} = a^2$$
$$DF^2 = \left(\frac{a}{e} - c\right)^2 + \left(\frac{b}{e}\right)^2$$
$$= \frac{(a - ec)^2 + b^2}{e^2}$$
$$= \frac{a^2 + b^2 - 2aec + e^2c^2}{e^2}$$
$$= \frac{c^2 - 2c^2}{e^2} + c^2$$
$$= -\frac{c^2}{e^2} + c^2$$
$$= -a^2 + c^2$$
Thus $OD^2 + DF^2 = OF^2$, as required.

Note: An alternate approach here is to show that the slope of $\overline{FD}$ is $-\frac{a}{b}$:

$$\text{slope } = \frac{\frac{b}{e}}{\frac{a}{e}-c} = \frac{b}{a-ce} = \frac{b}{a-\frac{c^2}{a}} = \frac{ab}{a^2-c^2} = \frac{ab}{-b^2} = -\frac{a}{b}$$

Since the asymptote has a slope of $\frac{b}{a}$, the lines are perpendicular.

49. We draw the sketch:

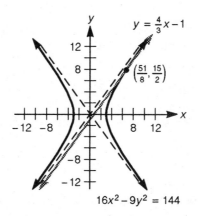

$y = \frac{4}{3}x - 1$

$\left(\frac{51}{8}, \frac{15}{2}\right)$

$16x^2 - 9y^2 = 144$

By substitution, we have:

$$16x^2 - 9\left(\tfrac{4}{3}x - 1\right)^2 = 144$$
$$16x^2 - 9\left(\tfrac{16}{9}x^2 - \tfrac{8}{3}x + 1\right) = 144$$
$$16x^2 - 16x^2 + 24x - 9 = 144$$
$$24x = 153$$
$$x = \tfrac{51}{8}$$
$$y = \tfrac{4}{3} \cdot \tfrac{51}{8} - 1 = \tfrac{17}{2} - 1 = \tfrac{15}{2}$$

Therefore $\left(\frac{51}{8}, \frac{15}{2}\right)$ is the intersection point.

51. We have $\left(x_1, y_1\right) = (4,6)$, $a^2 = 4$, $b^2 = 12$:

$$\frac{4x}{4} - \frac{6y}{12} = 1$$
$$x - \frac{y}{2} = 1$$
$$2x - y = 2$$
$$-y = -2x + 2$$
$$y = 2x - 2$$

53. The equation $\dfrac{x^2}{a^2} - \dfrac{y^2}{b^2} = 1$ is equivalent to $b^2x^2 - a^2y^2 = a^2b^2$ (1). So (x_1, y_1) is on the hyperbola if and only if:

$$b^2x_1^2 - a^2y_1^2 = a^2b^2$$

Thus if (x_1, y_1) is on the hyperbola then the equation of the hyperbola can be written as:

$$b^2(x^2 - x_1^2) - a^2(y^2 - y_1^2) = 0$$

Let $y - y_1 = m(x - x_1)$ be the equation of the tangent to the hyperbola through (x_1, y_1). Then as in Exercise 64 of Section 10.4 the system

$$b^2(x^2 - x_1^2) - a^2(y^2 - y_1^2) = 0$$
$$y - y_1 = m(x - x_1)$$

has exactly one solution, (x_1, y_1). Substituting $y = m(x - x_1) + y_1$ into the first equation, we have:

$$b^2(x^2 - x_1^2) - a^2\left[(m(x - x_1) + y_1)^2 - y_1^2\right] = 0$$
$$b^2(x^2 - x_1^2) - a^2m^2(x - x_1)^2 - 2a^2my_1(x - x_1) = 0$$
$$(x - x_1)\left[b^2(x + x_1) - a^2m^2(x - x_1) - 2a^2my_1\right] = 0$$

Since x_1 is the only solution, it is a double root, so:

$$b^2(x_1 + x_1) - a^2m^2(x_1 - x_1) - 2a^2my_1 = 0$$
$$2a^2my_1 = 2b^2x_1$$
$$m = \dfrac{b^2x_1}{a^2y_1}$$

Thus the tangent line has equation:

$$y - y_1 = \dfrac{b^2x_1}{a^2y_1}(x - x_1)$$
$$a^2y_1y - a^2y_1^2 = b^2x_1x - b^2x_1^2$$
$$b^2x_1x - a^2y_1y = b^2x_1^2 - a^2y_1^2$$
$$b^2x_1x - a^2y_1y = a^2b^2$$
$$\dfrac{x_1x}{a^2} - \dfrac{y_1y}{b^2} = 1$$

Note that $m = \dfrac{b^2 x_1}{a^2 y_1}$ cannot be the slope $\pm\dfrac{b}{a}$ of the asymtote if (x_1, y_1) lies on the hyperbola, since if so then:

$$\pm\frac{b}{a} = \frac{b^2 x_1}{a^2 y_1}$$

$$\pm\frac{a}{b} y_1 = x_1$$

Therefore:

$$\frac{\dfrac{a^2}{b^2} \cdot y_1^2}{a^2} - \frac{y_1^2}{b^2} = 0 \neq 1$$

55. By the law of sines:

$$\frac{F_2 P}{\sin \theta} = \frac{F_2 A}{\sin \alpha}$$

$$\frac{F_2 P}{F_2 A} = \frac{\sin \theta}{\sin \alpha}$$

Also by the law of sines:

$$\frac{F_1 P}{\sin(180° - \theta)} = \frac{F_1 A}{\sin \beta}$$

$$\frac{F_1 P}{F_1 A} = \frac{\sin \theta}{\sin \beta}$$

Thus if $\dfrac{F_2 P}{F_2 A} = \dfrac{F_1 P}{F_1 A}$, then $\dfrac{\sin \theta}{\sin \alpha} = \dfrac{\sin \theta}{\sin \beta}$, so $\sin \alpha = \sin \beta$. By construction α and β are acute, so $\alpha = \beta$.

57. We know the slope of the tangent line at P is $\dfrac{b^2 x_1}{a^2 y_1}$, so the slope of the normal is $\dfrac{-a^2 y_1}{b^2 x_1}$ and its equation is:

$$y - y_1 = \frac{-a^2 y_1}{b^2 x_1}(x - x_1)$$

$$b^2 x_1 y - b^2 x_1 y_1 = -a^2 y_1 x + a^2 y_1 x_1$$

$$a^2 y_1 x + b^2 x_1 y = x_1 y_1 (a^2 + b^2)$$

59. The equation of the normal line is $a^2 y_1 x + b^2 x_1 y = x_1 y_1 (a^2 + b^2)$. Letting $y = 0$ we see

that the x-intercept of the line is $x = \dfrac{x_1(a^2 + b^2)}{a^2}$. Similarily, the y-intercept is found to

be $y = \dfrac{y_1(a^2 + b^2)}{b^2}$. Thus the coordinates of Q and R are $\left(\dfrac{x_1(a^2 + b^2)}{a^2}, 0\right)$ and

$\left(0, \dfrac{y_1(a^2 + b^2)}{b^2}\right)$. The area of $\triangle OQR$ is :

$$\tfrac{1}{2} OQ \cdot OR = \frac{x_1 y_1 (a^2 + b^2)^2}{2a^2 b^2}$$

61. The equation of the tangent line at the point P with coordinates (x_1, y_1) is

$\dfrac{x_1 x}{a^2} - \dfrac{y_1 y}{b^2} = 1$. The asymtotes are $y = \pm \dfrac{b}{a} x$. Suppose the tangent intersects $y = \dfrac{b}{a} x$ at A

and $y = -\dfrac{b}{a} x$ at B. The x-coordinate of A is found by solving:

$$\frac{x_1 x}{a^2} - \frac{y_1 \cdot \frac{b}{a} x}{b^2} = 1$$

$$bx_1 x - ay_1 x = ba^2$$

$$x = \frac{ba^2}{bx_1 - ay_1}$$

So $y = \dfrac{b}{a} x = \dfrac{b^2 a}{bx_1 - ay_1}$. So A is the point $\left(\dfrac{ba^2}{bx_1 - ay_1}, \dfrac{b^2 a}{bx_1 - ay_1}\right)$. The point B is

similarly found to be $\left(\dfrac{ba^2}{bx_1 + ay_1}, \dfrac{b^2 a}{bx_1 + ay_1}\right)$. The midpoint of $\overline{AB}$ has x-coordinate:

$$\tfrac{1}{2}\left(\frac{ba^2}{bx_1 - ay_1} + \frac{ba^2}{bx_1 - ay_1}\right) = \tfrac{1}{2}\left(\frac{ba^2(2bx_1)}{b^2 x_1^2 - a^2 y_1^2}\right) = \frac{b^2 a^2 x_1}{b^2 x_1^2 - a^2 y_1^2}$$

Since $b^2 x_1^2 - a^2 y_1^2 = a^2 b^2$, this is equal to:

$$\frac{b^2 a^2 x_1}{a^2 b^2} = x_1$$

Similarly the y-coordinate of the midpoint is found to be y_1, so P is the midpoint of $\overline{AB}$ as required.

Optional TI-81 Graphing Calculator Exercises for Section 10.5

1. (a) Solving for y, we have:
$$16x^2 - 9y^2 = 144$$
$$16x^2 - 144 = 9y^2$$
$$y^2 = \frac{16x^2 - 144}{9}$$
$$y = \pm\tfrac{1}{3}\sqrt{16x^2 - 144}$$

Graphing the two functions we obtain the hyperbola:

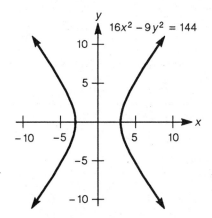

(b) Dividing by 144, the standard form of the hyperbola is $\frac{x^2}{9} - \frac{y^2}{16} = 1$, so $a = 3$ and $b = 4$. Thus the asymptotes are $y = \pm\tfrac{4}{3}x$. Now we graph the hyperbola and two asymptotes:

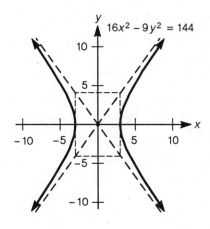

(c)   Using ZOOM-3 settings, we obtain the graphs:

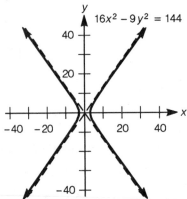

3.   (a)   Using the quadratic formula to solve $-2y^2 + 4xy + (x^2 - 6) = 0$, we have:

$$y = \frac{-4x \pm \sqrt{(4x)^2 - 4(-2)(x^2 - 6)}}{2(-2)}$$

$$= \frac{-4x \pm \sqrt{16x^2 + 8x^2 - 48}}{-4}$$

$$= \frac{-4x \pm \sqrt{24x^2 - 48}}{-4}$$

$$= \frac{-4x \pm 2\sqrt{6x^2 - 12}}{-4}$$

$$= x \pm \tfrac{1}{2}\sqrt{6x^2 - 12}$$

(b)   Entering these two functions and making the suggested range settings, we have:

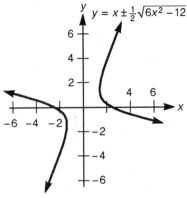

10.6 The Focus-Directrix Property of Conics

1. In order to use the formulas for the focal radii, we must find a and e. Dividing by 76, the standard form is $\frac{x^2}{76} + \frac{y^2}{76/3} = 1$. So $a = \sqrt{76} = 2\sqrt{19}$. To find the eccentricity, we first find c:

$$c = \sqrt{a^2 - b^2} = \sqrt{76 - \tfrac{76}{3}} = \sqrt{\tfrac{152}{3}} = \frac{2\sqrt{38}}{\sqrt{3}} = \frac{2\sqrt{114}}{3}$$

So the eccentricity is:

$$e = \frac{c}{a} = \frac{\frac{2\sqrt{114}}{3}}{2\sqrt{19}} = \frac{\sqrt{6}}{3}$$

So the focal radii are given by:

$$F_1P = a + ex = 2\sqrt{19} + \frac{\sqrt{6}}{3}(-8) = 2\sqrt{19} - \frac{8\sqrt{6}}{3} = \frac{6\sqrt{19} - 8\sqrt{6}}{3}$$

$$F_2P = a - ex = 2\sqrt{19} - \frac{\sqrt{6}}{3}(-8) = 2\sqrt{19} + \frac{8\sqrt{6}}{3} = \frac{6\sqrt{19} + 8\sqrt{6}}{3}$$

3. The equation is already in standard form with $a = 15$ and $b = 5$. We find c:

$$c = \sqrt{a^2 - b^2} = \sqrt{225 - 25} = \sqrt{200} = 10\sqrt{2}$$

So the eccentricity is:

$$e = \frac{c}{a} = \frac{10\sqrt{2}}{15} = \frac{2\sqrt{2}}{3}$$

So the focal radii are given by:

$$F_1P = a + ex = 15 + \frac{2\sqrt{2}}{3}(9) = 15 + 6\sqrt{2}$$

$$F_2P = a - ex = 15 - \frac{2\sqrt{2}}{3}(9) = 15 - 6\sqrt{2}$$

5. (a) The ellipse is already in standard form with $a = 4$ and $b = 3$. We find c:

$$c = \sqrt{a^2 - b^2} = \sqrt{16 - 9} = \sqrt{7}$$

So the foci are $(\pm\sqrt{7}, 0)$ and the eccentricity is $\frac{\sqrt{7}}{4}$. The directrices are given by:

$$x = \pm\frac{a}{e} = \pm\frac{4}{\frac{\sqrt{7}}{4}} = \pm\frac{16}{\sqrt{7}} = \pm\frac{16\sqrt{7}}{7}$$

(b) The hyperbola is already in standard form with $a = 4$ and $b = 3$. We find c:

$$c = \sqrt{a^2 + b^2} = \sqrt{16 + 9} = \sqrt{25} = 5$$

So the foci are $(\pm 5, 0)$ and the eccentricity is $\frac{5}{4}$. The directrices are given by:

$$x = \pm\frac{a}{e} = \pm\frac{4}{\frac{5}{4}} = \pm\frac{16}{5}$$

7. (a) Dividing by 156, the standard form for the ellipse is $\frac{x^2}{13} + \frac{y^2}{12} = 1$, and so $a = \sqrt{13}$

and $b = \sqrt{12} = 2\sqrt{3}$. We find c:
$$c = \sqrt{a^2 - b^2} = \sqrt{13 - 12} = \sqrt{1} = 1$$

So the foci are $(\pm 1, 0)$ and the eccentricity is $\frac{1}{\sqrt{13}} = \frac{\sqrt{13}}{13}$. The directrices are given by:
$$x = \pm\frac{a}{e} = \pm\frac{\sqrt{13}}{\frac{\sqrt{13}}{13}} = \pm 13$$

(b) Dividing by 156, the standard form for the hyperbola is $\frac{x^2}{13} - \frac{y^2}{12} = 1$, and so $a = \sqrt{13}$

and $b = \sqrt{12} = 2\sqrt{3}$. We find c:
$$c = \sqrt{a^2 + b^2} = \sqrt{13 + 12} = \sqrt{25} = 5$$

So the foci are $(\pm 5, 0)$ and the eccentricity is $\frac{5}{\sqrt{13}} = \frac{5\sqrt{13}}{13}$. The directrices are given by:
$$x = \pm\frac{a}{e} = \pm\frac{\sqrt{13}}{\frac{5\sqrt{13}}{13}} = \pm\frac{13}{5}$$

9. (a) Dividing by 900, the standard form for the ellipse is $\frac{x^2}{36} + \frac{y^2}{25} = 1$, and so $a = 6$ and $b = 5$. We find c:
$$c = \sqrt{a^2 - b^2} = \sqrt{36 - 25} = \sqrt{11}$$

So the foci are $(\pm\sqrt{11}, 0)$ and the eccentricity is $\frac{\sqrt{11}}{6}$. The directrices are given by:
$$x = \pm\frac{a}{e} = \pm\frac{6}{\frac{\sqrt{11}}{6}} = \pm\frac{36}{\sqrt{11}} = \pm\frac{36\sqrt{11}}{11}$$

(b) Dividing by 900, the standard form for the hyperbola is $\frac{x^2}{36} - \frac{y^2}{25} = 1$, and so $a = 6$ and $b = 5$. We find c:
$$c = \sqrt{a^2 + b^2} = \sqrt{36 + 25} = \sqrt{61}$$

So the foci are $(\pm\sqrt{61}, 0)$ and the eccentricity is $\frac{\sqrt{61}}{6}$. The directrices are given by:
$$x = \pm\frac{a}{e} = \pm\frac{6}{\frac{\sqrt{61}}{6}} = \pm\frac{36}{\sqrt{61}} = \pm\frac{36\sqrt{61}}{61}$$

11. Since the foci are $(\pm 1, 0)$, $c = 1$. Since the directrices are $x = \pm 4$, then $\frac{a}{e} = 4$, so

$a = 4e$. But since $e = \frac{c}{a} = \frac{1}{a}$, we have:
$$a = 4 \cdot \frac{1}{a}$$
$$a^2 = 4$$

Since $a^2 - b^2 = c^2$, we can find b^2:
$$2^2 - b^2 = 1^2$$
$$-b^2 = -3$$
$$b^2 = 3$$

So the equation of the ellipse is $\frac{x^2}{4} + \frac{y^2}{3} = 1$, or $3x^2 + 4y^2 = 12$.

13. Since the foci are $(\pm 2, 0)$, $c = 2$. Since the directrices are $x = \pm 1$, then $\frac{a}{e} = 1$, so

$a = e$. But since $e = \frac{c}{a} = \frac{2}{a}$, we have:
$$a = \frac{2}{a}$$
$$a^2 = 2$$
Since $a^2 + b^2 = c^2$, we can find b^2:
$$2 + b^2 = 2^2$$
$$b^2 = 2$$

So the equation of the hyperbola is $\frac{x^2}{2} - \frac{y^2}{2} = 1$, or $x^2 - y^2 = 2$.

15. (a) By the distance formula, we have:
$$d_1 = \sqrt{(x + c)^2 + (y - 0)^2} = \sqrt{(x + c)^2 + y^2}$$
$$d_2 = \sqrt{(x - c)^2 + (y - 0)^2} = \sqrt{(x - c)^2 + y^2}$$
Squaring, we have:
$$d_1^2 = (x + c)^2 + y^2$$
$$d_2^2 = (x - c)^2 + y^2$$

(b) Working from the left-hand side, we have:
$$d_1^2 - d_2^2 = (x + c)^2 - (x - c)^2 = x^2 + 2cx + c^2 - x^2 + 2cx - c^2 = 4cx$$

(c) Since d_1 and d_2 represent the distances from the foci to a point on the ellipse, then $d_1 + d_2 = 2a$ by the definition of an ellipse.

(d) Factoring, we have:
$$d_1^2 - d_2^2 = 4cx$$
$$\left(d_1 + d_2\right)\left(d_1 - d_2\right) = 4cx$$
$$2a\left(d_1 - d_2\right) = 4cx$$
$$d_1 - d_2 = \frac{2cx}{a}$$

(e)   Adding the two equations, we have:

$$2d_1 = 2a + \frac{2cx}{a}$$

$$d_1 = a + \frac{c}{a}x = a + ex$$

(f)   Substituting the result from (e), we have:

$$a + ex + d_2 = 2a$$

$$d_2 = a - ex$$

10.7 The Conics in Polar Coordinates

1.   (a)   Comparing the given equation with the four basic types, it appears this is the type associated with Figure 2. We divide both numerator and denominator by 3 to obtain:

$$r = \frac{2}{1 + \frac{2}{3}\cos\theta} = \frac{\frac{2}{3} \cdot 3}{1 + \frac{2}{3}\cos\theta}$$

Therefore $e = \frac{2}{3}$ and $d = 3$. Since $e < 1$, this confirms the given conic is an ellipse.

The eccentricity is $\frac{2}{3}$ and the directrix is $x = 3$. Computing the values of r when

$\theta = 0, \frac{\pi}{2}, \pi$ and $\frac{3\pi}{2}$, we have:

θ	0	$\frac{\pi}{2}$	π	$\frac{3\pi}{2}$
r	$\frac{6}{5}$	2	6	2

Since the major axis of this ellipse lies along the x-axis, the length of the major axis is:

$$2a = \tfrac{6}{5} + 6 = \tfrac{36}{5}, \text{ so } a = \tfrac{18}{5}$$

The endpoints of the major axis are at $\left(\frac{6}{5}, 0\right)$ and $(-6, 0)$, so the x-coordinate of the center is:

$$\tfrac{1}{2}\left(-6 + \tfrac{6}{5}\right) = \tfrac{1}{2}\left(-\tfrac{24}{5}\right) = -\tfrac{12}{5}$$

So the center is $\left(-\frac{12}{5}, 0\right)$. Finally, we calculate b:

$$b = a\sqrt{1 - e^2} = \tfrac{18}{5}\sqrt{1 - \tfrac{4}{9}} = \tfrac{18}{5}\sqrt{\tfrac{5}{9}} = \tfrac{18\sqrt{5}}{15} = \tfrac{6\sqrt{5}}{5}$$

So the endpoints of the minor axis are $\left(-\frac{12}{5}, \pm\frac{6\sqrt{5}}{5}\right)$. We graph the ellipse:

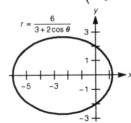

(b)   Comparing the given equation with the four basic types, it appears this is the type associated with Figure 3. We divide both numerator and denominator by 3 to obtain:

$$r = \frac{2}{1 - \frac{2}{3}\cos\theta} = \frac{\frac{2}{3} \cdot 3}{1 - \frac{2}{3}\cos\theta}$$

Therefore $e = \frac{2}{3}$ and $d = 3$. Since $e < 1$, this confirms the given conic is an ellipse.

The eccentricity is $\frac{2}{3}$ and the directrix is $x = -3$. Computing the values of r when $\theta = 0, \frac{\pi}{2}, \pi$ and $\frac{3\pi}{2}$, we have:

θ	0	$\frac{\pi}{2}$	π	$\frac{3\pi}{2}$
r	6	2	$\frac{6}{5}$	2

Since the major axis of this ellipse lies along the x-axis, the length of the major axis is:

$$2a = \frac{6}{5} + 6 = \frac{36}{5}, \text{ so } a = \frac{18}{5}$$

The endpoints of the major axis are at $(6, 0)$ and $\left(-\frac{6}{5}, 0\right)$, so the x-coordinate of the center is:

$$\frac{1}{2}\left(6 - \frac{6}{5}\right) = \frac{1}{2}\left(\frac{24}{5}\right) = \frac{12}{5}$$

So the center is $\left(\frac{12}{5}, 0\right)$. Finally, we calculate b:

$$b = a\sqrt{1 - e^2} = \frac{18}{5}\sqrt{1 - \frac{4}{9}} = \frac{18}{5}\sqrt{\frac{5}{9}} = \frac{18\sqrt{5}}{15} = \frac{6\sqrt{5}}{5}$$

So the endpoints of the minor axis are $\left(\frac{12}{5}, \pm\frac{6\sqrt{5}}{5}\right)$. We graph the ellipse:

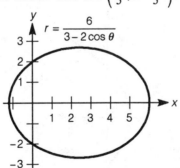

3.   (a)   Comparing the given equation with the four basic types, it appears this is the type associated with Figure 2. We divide both numerator and denominator by 2 to obtain:

$$r = \frac{\frac{5}{2}}{1 + \cos\theta} = \frac{1 \cdot \frac{5}{2}}{1 + \cos\theta}$$

Therefore $e = 1$ and $d = \frac{5}{2}$. Since $e = 1$, this confirms the given conic is a parabola.

The directrix is $x = \frac{5}{2}$. Computing the value of r when $\theta = 0$ yields $r = \frac{5}{4}$, so the vertex is $\left(\frac{5}{4}, 0\right)$. We graph the parabola:

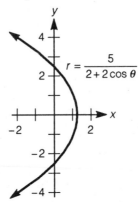

(b) Comparing the given equation with the four basic types, it appears this is the type associated with Figure 3. We divide both numerator and denominator by 2 to obtain:

$$r = \frac{\frac{5}{2}}{1 - \cos\theta} = \frac{1 \cdot \frac{5}{2}}{1 - \cos\theta}$$

Therefore $e = 1$ and $d = \frac{5}{2}$. Since $e = 1$, this confirms the given conic is a parabola. The directrix is $x = -\frac{5}{2}$. Computing the value of r when $\theta = \pi$ yields $r = \frac{5}{4}$, so the vertex is $\left(-\frac{5}{4}, 0\right)$. We graph the parabola:

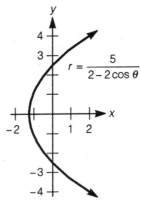

5. (a) Comparing the given equation with the four basic types, it appears this is the type associated with Figure 2. We divide both numerator and denominator by 2 to obtain:

$$r = \frac{3}{2 + 4\cos\theta} = \frac{\frac{3}{2}}{1 + 2\cos\theta} = \frac{2 \cdot \frac{3}{4}}{1 + 2\cos\theta}$$

Therefore $e = 2$ and $d = \frac{3}{4}$. Since $e > 1$, this confirms the given conic is a hyperbola.

The eccentricity is 2 and the directrix is $x = \frac{3}{4}$. Computing the values of r when $\theta = 0, \frac{\pi}{2}, \pi$ and $\frac{3\pi}{2}$, we have:

θ	0	$\frac{\pi}{2}$	π	$\frac{3\pi}{2}$
r	$\frac{1}{2}$	$\frac{3}{2}$	$-\frac{3}{2}$	$\frac{3}{2}$

Since the two vertices $\left(\frac{1}{2},0\right)$ and $\left(\frac{3}{2},0\right)$ lie on the transverse axis, then:
$$2a = \frac{3}{2} - \frac{1}{2} = 1, \text{ so } a = \frac{1}{2}$$
The center of the hyperbola is the midpoint of these two vertices, which is $(1,0)$. Since a focus is $(0,0)$, then $c = 1$. Finally, we find b:
$$b = \sqrt{c^2 - a^2} = \sqrt{1 - \frac{1}{4}} = \sqrt{\frac{3}{4}} = \frac{1}{2}\sqrt{3}$$
Notice that we could also find b from the eccentricity:
$$b = a\sqrt{e^2 - 1} = \frac{1}{2}\sqrt{4-1} = \frac{1}{2}\sqrt{3}$$
We graph the hyperbola:

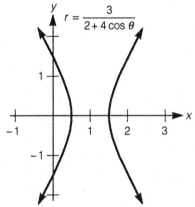

(b)   Comparing the given equation with the four basic types, it appears this is the type associated with Figure 3. We divide both numerator and denominator by 2 to obtain:
$$r = \frac{3}{2 - 4\cos\theta} = \frac{\frac{3}{2}}{1 - 2\cos\theta} = \frac{2 \cdot \frac{3}{4}}{1 - 2\cos\theta}$$
Therefore $e = 2$ and $d = \frac{3}{4}$. Since $e > 1$, this confirms the given conic is a hyperbola.

The eccentricity is 2 and the directrix is $x = -\frac{3}{4}$. Computing the values of r when $\theta = 0, \frac{\pi}{2}, \pi$ and $\frac{3\pi}{2}$, we have:

θ	0	$\frac{\pi}{2}$	π	$\frac{3\pi}{2}$
r	$-\frac{3}{2}$	$\frac{3}{2}$	$\frac{1}{2}$	$\frac{3}{2}$

Since the two vertices $\left(-\frac{3}{2},0\right)$ and $\left(-\frac{1}{2},0\right)$ lie on the transverse axis, then:
$$2a = -\frac{1}{2} + \frac{3}{2} = 1, \text{ so } a = \frac{1}{2}$$

The center of the hyperbola is the midpoint of these two vertices, which is $(-1,0)$. Since a focus is $(0,0)$, then $c = 1$. Finally, we find b:

$$b = \sqrt{c^2 - a^2} = \sqrt{1 - \tfrac{1}{4}} = \sqrt{\tfrac{3}{4}} = \tfrac{1}{2}\sqrt{3}$$

Notice that we could also find b from the eccentricity:

$$b = a\sqrt{e^2 - 1} = \tfrac{1}{2}\sqrt{4 - 1} = \tfrac{1}{2}\sqrt{3}$$

We graph the hyperbola:

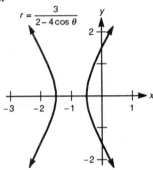

7. Comparing the given equation with the four basic types, it appears this is the type associated with Figure 3. We divide both numerator and denominator by 2 to obtain:

$$r = \frac{24}{2 - 3\cos\theta} = \frac{12}{1 - \tfrac{3}{2}\cos\theta} = \frac{\tfrac{3}{2} \cdot 8}{1 - \tfrac{3}{2}\cos\theta}$$

Since the eccentricity is $e = \tfrac{3}{2} > 1$, this conic is a hyperbola. Computing the values of r when $\theta = 0, \tfrac{\pi}{2}, \pi$ and $\tfrac{3\pi}{2}$, we have:

θ	0	$\tfrac{\pi}{2}$	π	$\tfrac{3\pi}{2}$
r	-24	12	$\tfrac{24}{5}$	12

Since the two vertices $(-24, 0)$ and $\left(-\tfrac{24}{5}, 0\right)$ lie on the transverse axis, its length must be:

$$2a = -\tfrac{24}{5} + 24 = \tfrac{96}{5}, \text{ so } a = \tfrac{48}{5}$$

The center of the hyperbola is the midpoint of these two vertices, which is $\left(-\tfrac{72}{5}, 0\right)$.

Since a focus is $(0,0)$, $c = \tfrac{72}{5}$ and thus:

$$b = \sqrt{c^2 - a^2} = \sqrt{\tfrac{5184}{25} - \tfrac{2304}{25}} = \sqrt{\tfrac{2880}{25}} = \tfrac{24\sqrt{5}}{5}$$

So the length of the conjugate axis is $2b = \tfrac{48\sqrt{5}}{5}$. We graph the hyperbola:

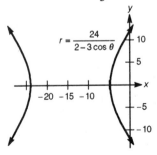

9. Comparing the given equation with the four basic types, it appears this is the type associated with Figure 4. We divide both numerator and denominator by 5 to obtain:

$$r = \frac{8}{5+3\sin\theta} = \frac{\frac{8}{5}}{1+\frac{3}{5}\sin\theta} = \frac{\frac{3}{5}\cdot\frac{8}{3}}{1+\frac{3}{5}\sin\theta}$$

Since the eccentricity is $e = \frac{3}{5} < 1$, this conic is an ellipse. Computing the values of r when $\theta = 0, \frac{\pi}{2}, \pi$ and $\frac{3\pi}{2}$, we have:

θ	0	$\frac{\pi}{2}$	π	$\frac{3\pi}{2}$
r	$\frac{8}{5}$	1	$\frac{8}{5}$	4

Since the two vertices $(0,1)$ and $(0,-4)$ lie on the major axis, its length must be:

$2a = 1+4 = 5$, so $a = \frac{5}{2}$

The center of the ellipse is the midpoint of these two vertices, which is $\left(0,-\frac{3}{2}\right)$. Since a focus is $(0,0)$, then $c = \frac{3}{2}$ and thus:

$$b = \sqrt{a^2 - c^2} = \sqrt{\frac{25}{4} - \frac{9}{4}} = \sqrt{\frac{16}{4}} = 2$$

So the length of the minor axis is $2b = 4$. We graph the ellipse:

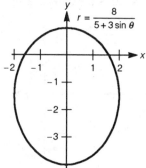

11. Comparing the given equation with the four basic types, it appears this is the type associated with Figure 5. We divide both numerator and denominator by 5 to obtain:

$$r = \frac{12}{5-5\sin\theta} = \frac{\frac{12}{5}}{1-\sin\theta} = \frac{1\cdot\frac{12}{5}}{1-\sin\theta}$$

Since the eccentricity is $e = 1$, this conic is a parabola with directrix $y = -\frac{12}{5}$. Since the focus is $(0,0)$, the vertex must be the midpoint of $(0,0)$ and $\left(0,-\frac{12}{5}\right)$, which is $\left(0,-\frac{6}{5}\right)$. We graph the parabola:

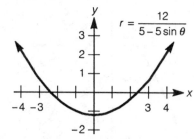

13. Comparing the given equation with the four basic types, it appears this is the type associated with Figure 2. We divide both numerator and denominator by 7 to obtain:

$$r = \frac{12}{7 + 5\cos\theta} = \frac{\frac{12}{7}}{1 + \frac{5}{7}\cos\theta} = \frac{\frac{5}{7} \cdot \frac{12}{5}}{1 + \frac{5}{7}\cos\theta}$$

Since the eccentricity is $e = \frac{5}{7} < 1$, this conic is an ellipse. Computing the values of r when $\theta = 0, \frac{\pi}{2}, \pi$ and $\frac{3\pi}{2}$, we have:

θ	0	$\frac{\pi}{2}$	π	$\frac{3\pi}{2}$
r	1	$\frac{12}{7}$	6	$\frac{12}{7}$

Since the two vertices $(1,0)$ and $(-6,0)$ lie on the major axis, its length must be:

$$2a = 1 + 6 = 7, \text{ so } a = \frac{7}{2}$$

The center of the ellipse is the midpoint of these two vertices, which is $\left(-\frac{5}{2}, 0\right)$. Since a focus is $(0,0)$, then $c = \frac{5}{2}$ and thus:

$$b = \sqrt{a^2 - c^2} = \sqrt{\frac{49}{4} - \frac{25}{4}} = \sqrt{6}$$

So the length of the minor axis is $2b = 2\sqrt{6}$. We graph the ellipse:

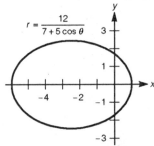

15. Comparing the given equation with the four basic types, it appears this is the type associated with Figure 4. We divide both numerator and denominator by 5 to obtain:

$$r = \frac{4}{5 + 5\sin\theta} = \frac{\frac{4}{5}}{1 + \sin\theta} = \frac{1 \cdot \frac{4}{5}}{1 + \sin\theta}$$

Since the eccentricity is $e = 1$, this conic is a parabola with directrix $y = \frac{4}{5}$. Since the focus is $(0,0)$, the vertex must be the midpoint of $(0,0)$ and $\left(0, \frac{4}{5}\right)$, which is $\left(0, \frac{2}{5}\right)$. We graph the parabola:

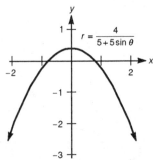

17. Comparing the given equation with the four basic types, it appears this is the type associated with Figure 3. The equation is already in standard form with eccentricity $e = 2 > 1$, so this conic is a hyperbola. Computing the values of r when $\theta = 0, \frac{\pi}{2}, \pi$ and $\frac{3\pi}{2}$, we have:

θ	0	$\frac{\pi}{2}$	π	$\frac{3\pi}{2}$
r	-9	9	3	9

Since the two vertices $(-9, 0)$ and $(-3, 0)$ lie on the transverse axis, its length must be:
$$2a = -3 + 9 = 6, \text{ so } a = 3$$
The center of the hyperbola is the midpoint of these two vertices, which is $(-6, 0)$. Since a focus is $(0, 0)$, then $c = 6$ and thus:
$$b = \sqrt{c^2 - a^2} = \sqrt{36 - 9} = \sqrt{27} = 3\sqrt{3}$$
So the length of the conjugate axis is $2b = 6\sqrt{3}$. We graph the hyperbola:

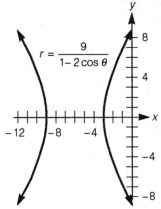

19. Since the coordinates of P are (r, θ), then the coordinates of Q are $(r, \theta + \pi)$. We now find FP and FQ, noting that $\cos(\theta + \pi) = -\cos\theta$

$$FP = r = \frac{ed}{1 - e\cos\theta}$$

$$FQ = r = \frac{ed}{1 - e\cos(\theta + \pi)} = \frac{ed}{1 + e\cos\theta}$$

Therefore:
$$\frac{1}{FP} + \frac{1}{FQ} = \frac{1 - e\cos\theta}{ed} + \frac{1 + e\cos\theta}{ed} = \frac{2}{ed}$$

This is remarkable in that $\frac{2}{ed}$ is a constant, even though P is a variable point.

21. Draw a focal chord $\overline{AB}$, with A representing the endpoint on the "left". Since $\overline{AB}$ is a 90° rotation from $\overline{PQ}$, then A corresponds to the polar coordinates $A\left(r, \theta + \frac{\pi}{2}\right)$ and B corresponds to the coordinates $B\left(r, \theta + \frac{3\pi}{2}\right)$. Using the identities $\cos\left(\theta + \frac{\pi}{2}\right) = -\sin\theta$ and $\cos\left(\theta + \frac{3\pi}{2}\right) = \sin\theta$, we have:

$$AF = r = \frac{ed}{1 - e\cos\left(\theta + \frac{\pi}{2}\right)} = \frac{ed}{1 + e\sin\theta}$$

$$FB = r = \frac{ed}{1 - e\cos\left(\theta + \frac{3\pi}{2}\right)} = \frac{ed}{1 - e\sin\theta}$$

Since $AB = AF + FB$, we have:

$$\begin{aligned}
AB &= \frac{ed}{1 + e\sin\theta} + \frac{ed}{1 - e\sin\theta} \\
&= \frac{ed(1 - e\sin\theta) + ed(1 + e\sin\theta)}{(1 + e\sin\theta)(1 - e\sin\theta)} \\
&= \frac{ed - e^2 d\sin\theta + ed + e^2 d\sin\theta}{1 - e^2\sin^2\theta} \\
&= \frac{2ed}{1 - e^2\sin^2\theta}
\end{aligned}$$

Using the result from Exercise 10, we show the required sum is constant:

$$\frac{1}{PQ} + \frac{1}{AB} = \frac{1 - e^2\cos^2\theta}{2ed} + \frac{1 - e^2\sin^2\theta}{2ed} = \frac{2 - e^2(\sin^2\theta + \cos^2\theta)}{2ed} = \frac{2 - e^2}{2ed}$$

But since e and d are constants, we have proven the desired result.

10.8 Rotation of Axes

1. For x, we have:
$$\begin{aligned}
x &= x'\cos\theta - y'\sin\theta \\
&= \sqrt{3}\cos 30° - 2\sin 30° \\
&= \sqrt{3}\cdot\frac{\sqrt{3}}{2} - 2\cdot\frac{1}{2} \\
&= \frac{3}{2} - 1 \\
&= \frac{1}{2}
\end{aligned}$$

For y, we have:
$$\begin{aligned}
y &= x'\sin\theta + y'\cos\theta \\
&= \sqrt{3}\sin 30° + 2\cos 30° \\
&= \sqrt{3}\cdot\frac{1}{2} + 2\cdot\frac{\sqrt{3}}{2} \\
&= \frac{\sqrt{3}}{2} + \sqrt{3} \\
&= \frac{3\sqrt{3}}{2}
\end{aligned}$$

So the coordinates in the x-y system are $\left(\frac{1}{2}, \frac{3\sqrt{3}}{2}\right)$.

3. For x, we have:

$$x = x'\cos\theta - y'\sin\theta$$
$$= \sqrt{2}\cos 45° + \sqrt{2}\sin 45°$$
$$= \sqrt{2}\cdot\frac{1}{\sqrt{2}} + \sqrt{2}\cdot\frac{1}{\sqrt{2}}$$
$$= 1 + 1$$
$$= 2$$

For y, we have:

$$y = x'\sin\theta + y'\cos\theta$$
$$= \sqrt{2}\sin 45° - \sqrt{2}\cos 45°$$
$$= \sqrt{2}\cdot\frac{1}{\sqrt{2}} - \sqrt{2}\cdot\frac{1}{\sqrt{2}}$$
$$= 1 - 1$$
$$= 0$$

So the coordinates in the x-y system are $(2, 0)$.

5. For x', we have:

$$x' = x\cos\theta + y\sin\theta$$
$$= -3\cos\left[\sin^{-1}\left(\tfrac{5}{13}\right)\right] + 1\sin\left[\sin^{-1}\left(\tfrac{5}{13}\right)\right]$$
$$= -3\cdot\tfrac{12}{13} + 1\cdot\tfrac{5}{13}$$
$$= -\tfrac{31}{13}$$

For y', we have:

$$y' = -x\sin\theta + y\cos\theta$$
$$= 3\sin\left[\sin^{-1}\left(\tfrac{5}{13}\right)\right] + 1\cos\left[\sin^{-1}\left(\tfrac{5}{13}\right)\right]$$
$$= 3\cdot\tfrac{5}{13} + 1\cdot\tfrac{12}{13}$$
$$= \tfrac{27}{13}$$

So the coordinates in the x'-y' system are $\left(-\tfrac{31}{13}, \tfrac{27}{13}\right)$.

7. We have:

$$\cot 2\theta = \frac{A-C}{B} = \frac{25-18}{-24} = -\tfrac{7}{24}, \text{ so } \tan 2\theta = -\tfrac{24}{7}$$

$$\sec^2 2\theta = 1 + \tan^2 2\theta = 1 + \left(-\tfrac{24}{7}\right)^2 = \tfrac{625}{49}$$

So $\sec 2\theta = -\tfrac{25}{7}$ (second quadrant, since $\cot 2\theta < 0$), and thus $\cos 2\theta = -\tfrac{7}{25}$. Since θ is in the first quadrant, we have:

$$\sin\theta = \sqrt{\frac{1-\cos 2\theta}{2}} = \sqrt{\frac{1+\frac{7}{25}}{2}} = \sqrt{\tfrac{16}{25}} = \tfrac{4}{5}$$

$$\cos\theta = \sqrt{\frac{1+\cos 2\theta}{2}} = \sqrt{\frac{1-\frac{7}{25}}{2}} = \sqrt{\tfrac{9}{25}} = \tfrac{3}{5}$$

9. We have:

$$\cot 2\theta = \frac{1-8}{-24} = \frac{-7}{-24} = \frac{7}{24}, \text{ so } \tan 2\theta = \frac{24}{7}$$

$$\sec^2 2\theta = 1 + \tan^2 2\theta = 1 + \left(\frac{24}{7}\right)^2 = \frac{625}{49}$$

So $\sec 2\theta = \frac{25}{7}$ (first quadrant, since $\cot 2\theta > 0$), and thus $\cos 2\theta = \frac{7}{25}$. Since θ is in the first quadrant, we have:

$$\sin\theta = \sqrt{\frac{1-\cos 2\theta}{2}} = \sqrt{\frac{1-\frac{7}{25}}{2}} = \sqrt{\frac{9}{25}} = \frac{3}{5}$$

$$\cos\theta = \sqrt{\frac{1+\cos 2\theta}{2}} = \sqrt{\frac{1+\frac{7}{25}}{2}} = \sqrt{\frac{16}{25}} = \frac{4}{5}$$

11. We have:

$$\cot 2\theta = \frac{A-C}{B} = \frac{1-(-1)}{-2\sqrt{3}} = -\frac{1}{\sqrt{3}}, \text{ so } \tan 2\theta = -\sqrt{3}$$

Therefore $2\theta = 120°$, and thus $\theta = 60°$. So:

$$\sin\theta = \sin 60° = \frac{\sqrt{3}}{2}$$

$$\cos\theta = \cos 60° = \frac{1}{2}$$

13. We have:

$$\cot 2\theta = \frac{A-C}{B} = \frac{0-(-240)}{161} = \frac{240}{161}, \text{ so } \tan 2\theta = \frac{161}{240}$$

$$\sec^2 2\theta = 1 + \tan^2 2\theta = 1 + \left(\frac{161}{240}\right)^2 = \frac{83521}{57600}$$

So $\sec 2\theta = \frac{289}{240}$ (first quadrant, since $\cot 2\theta > 0$), and thus $\cos 2\theta = \frac{240}{289}$. Since θ is in the first quadrant, we have:

$$\sin\theta = \sqrt{\frac{1-\cos 2\theta}{2}} = \sqrt{\frac{1-\frac{240}{289}}{2}} = \sqrt{\frac{49}{578}} = \frac{7\sqrt{2}}{34}$$

$$\cos\theta = \sqrt{\frac{1+\cos 2\theta}{2}} = \sqrt{\frac{1+\frac{240}{289}}{2}} = \sqrt{\frac{529}{578}} = \frac{23\sqrt{2}}{34}$$

15. Using the rotation equations:

$$x = x'\cos\theta - y'\sin\theta = x'\cos 45° - y'\sin 45° = \frac{\sqrt{2}}{2}x' - \frac{\sqrt{2}}{2}y'$$

$$y = x'\sin\theta + y'\cos\theta = x'\sin 45° + y'\cos 45° = \frac{\sqrt{2}}{2}x' + \frac{\sqrt{2}}{2}y'$$

So the equation $2xy = 9$ becomes:

$$2\left(\frac{\sqrt{2}}{2}x' - \frac{\sqrt{2}}{2}y'\right)\left(\frac{\sqrt{2}}{2}x' + \frac{\sqrt{2}}{2}y'\right) = 9$$

$$2\left(\frac{1}{2}x'^2 - \frac{1}{2}y'^2\right) = 9$$

$$x'^2 - y'^2 = 9$$

We graph the equation:

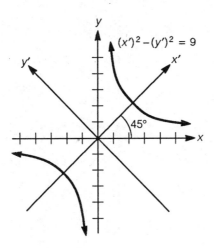

$(x')^2 - (y')^2 = 9$

17. We find:

$$\cot 2\theta = \frac{7-1}{8} = \tfrac{3}{4}, \text{ so } \tan 2\theta = \tfrac{4}{3}$$

$$\sec^2 2\theta = 1 + \tan^2 2\theta = 1 + \left(\tfrac{4}{3}\right)^2 = \tfrac{25}{9}, \text{ so } \sec 2\theta = \tfrac{5}{3} \quad (\text{since } 2\theta < 90°)$$

Thus $\cos 2\theta = \tfrac{3}{5}$, and therefore:

$$\sin\theta = \sqrt{\frac{1 - \cos 2\theta}{2}} = \sqrt{\frac{1 - \tfrac{3}{5}}{2}} = \sqrt{\tfrac{1}{5}} = \frac{\sqrt{5}}{5}$$

$$\cos\theta = \sqrt{\frac{1 + \cos 2\theta}{2}} = \sqrt{\frac{1 + \tfrac{3}{5}}{2}} = \sqrt{\tfrac{4}{5}} = \frac{2\sqrt{5}}{5}$$

Thus $\theta = \sin^{-1}\left(\frac{\sqrt{5}}{5}\right) \approx 26.6°$. Now:

$$x = x'\cos\theta - y'\sin\theta = \tfrac{2\sqrt{5}}{5}x' - \tfrac{\sqrt{5}}{5}y'$$

$$y = x'\sin\theta + y'\cos\theta = \tfrac{\sqrt{5}}{5}x' + \tfrac{2\sqrt{5}}{5}y'$$

Making the substitutions into $7x^2 + 8xy + y^2 - 1 = 0$, we have:

$$7\left(\tfrac{4}{5}x'^2 - \tfrac{4}{5}x'y' + \tfrac{1}{5}y'^2\right) + 8\left(\tfrac{2}{5}x'^2 + \tfrac{3}{5}x'y' - \tfrac{2}{5}y'^2\right) + \left(\tfrac{1}{5}x'^2 + \tfrac{4}{5}x'y' + \tfrac{4}{5}y'^2\right) - 1 = 0$$

$$\tfrac{28}{5}x'^2 - \tfrac{28}{5}x'y' + \tfrac{7}{5}y'^2 + \tfrac{16}{5}x'^2 + \tfrac{24}{5}x'y' - \tfrac{16}{5}y'^2 + \tfrac{1}{5}x'^2 + \tfrac{4}{5}x'y' + \tfrac{4}{5}y'^2 - 1 = 0$$

$$9x'^2 - y'^2 = 1$$

$$\frac{x'^2}{\tfrac{1}{9}} - \frac{y'^2}{1} = 1$$

Rotating $\theta = 26.6°$, we sketch the hyperbola:

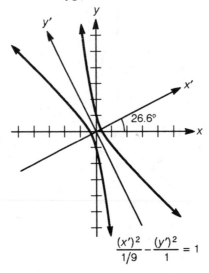

$$\frac{(x')^2}{1/9} - \frac{(y')^2}{1} = 1$$

19. We can use an alternate approach here:
$$x^2 + 4xy + 4y^2 = 1$$
$$(x + 2y)^2 = 1$$

$$x + 2y = 1 \qquad \text{or} \qquad x + 2y = -1$$
$$y = -\tfrac{1}{2}x + \tfrac{1}{2} \qquad\qquad\qquad y = -\tfrac{1}{2}x - \tfrac{1}{2}$$

The graph consists of two lines:

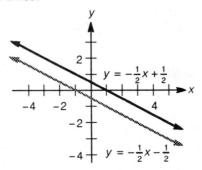

21. We find:
$$\cot 2\theta = \frac{9 - 16}{-24} = \tfrac{7}{24}, \text{ so } \tan 2\theta = \tfrac{24}{7}$$

$$\sec^2 2\theta = 1 + \tan^2 2\theta = 1 + \left(\tfrac{24}{7}\right)^2 = \tfrac{625}{49}, \text{ so } \sec 2\theta = \tfrac{25}{7} \quad (\text{since } 2\theta < 90°)$$

Thus $\cos 2\theta = \tfrac{7}{25}$, and therefore:

$$\sin\theta = \sqrt{\frac{1 - \cos 2\theta}{2}} = \sqrt{\frac{1 - \tfrac{7}{25}}{2}} = \sqrt{\tfrac{9}{25}} = \tfrac{3}{5}$$

$$\cos\theta = \sqrt{\frac{1 + \cos 2\theta}{2}} = \sqrt{\frac{1 + \tfrac{7}{25}}{2}} = \sqrt{\tfrac{16}{25}} = \tfrac{4}{5}$$

Thus $\theta = \sin^{-1}\left(\frac{3}{5}\right) \approx 36.9°$. Now:

$$x = x'\cos\theta - y'\sin\theta = \tfrac{4}{5}x' - \tfrac{3}{5}y'$$
$$y = x'\sin\theta + y'\cos\theta = \tfrac{3}{5}x' + \tfrac{4}{5}y'$$

Making the substitutions into $9x^2 - 24xy + 16y^2 - 400x - 300y = 0$ and collecting like terms, we have:

$$25y'^2 - 500x' = 0$$
$$y'^2 = 20x'$$

Rotating 36.9°, we sketch the parabola:

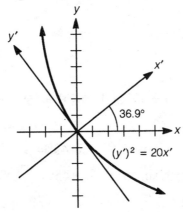

$(y')^2 = 20x'$

23. We find:

$$\cot 2\theta = \frac{0-3}{4} = -\tfrac{3}{4}, \text{ so } \tan 2\theta = -\tfrac{4}{3}$$

$$\sec^2 2\theta = 1 + \tan^2 2\theta = 1 + \left(-\tfrac{4}{3}\right)^2 = \tfrac{25}{9}$$

$$\sec 2\theta = -\tfrac{5}{3} \quad (\text{since } 2\theta > 90°)$$

Thus $\cos 2\theta = -\tfrac{3}{5}$, and therefore:

$$\sin\theta = \sqrt{\frac{1-\cos 2\theta}{2}} = \sqrt{\frac{1+\tfrac{3}{5}}{2}} = \sqrt{\tfrac{4}{5}} = \frac{2\sqrt{5}}{5}$$

$$\cos\theta = \sqrt{\frac{1+\cos 2\theta}{2}} = \sqrt{\frac{1-\tfrac{3}{5}}{2}} = \sqrt{\tfrac{1}{5}} = \frac{\sqrt{5}}{5}$$

Thus $\theta = \cos^{-1}\left(\frac{\sqrt{5}}{5}\right) \approx 63.4°$. Now:

$$x = x'\cos\theta - y'\sin\theta = \tfrac{\sqrt{5}}{5}x' - \tfrac{2\sqrt{5}}{5}y'$$
$$y = x'\sin\theta + y'\cos\theta = \tfrac{2\sqrt{5}}{5}x' + \tfrac{\sqrt{5}}{5}y'$$

Making the substitutions into $4xy + 3y^2 + 4x + 6y = 1$ and completing the square on x' and y' terms, we have:

$$\frac{\left(x' + \tfrac{2\sqrt{5}}{5}\right)^2}{1} - \frac{\left(y' + \tfrac{\sqrt{5}}{5}\right)^2}{4} = 1$$

Rotating 63.4°, we sketch the hyperbola:

$$\frac{\left(x'+\frac{2\sqrt{5}}{5}\right)^2}{1}-\frac{\left(y'+\frac{\sqrt{5}}{5}\right)^2}{4}=1$$

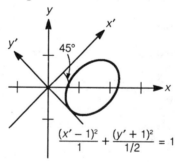

25. We find:

$$\cot 2\theta = \frac{3-3}{-2} = 0, \text{ so } 2\theta = 90° \text{ and thus } \theta = 45°$$

Therefore:

$$x = x'\cos\theta - y'\sin\theta = x'\cos 45° - y'\sin 45° = \frac{\sqrt{2}}{2}x' - \frac{\sqrt{2}}{2}y'$$
$$y = x'\sin\theta + y'\cos\theta = x'\sin 45° + y'\cos 45° = \frac{\sqrt{2}}{2}x' + \frac{\sqrt{2}}{2}y'$$

Making the substitutions into $3x^2 - 2xy + 3y^2 - 6\sqrt{2}x + 2\sqrt{2}y + 4 = 0$ and completing the square on x' and y' terms, we have:

$$\frac{(x'-1)^2}{1} + \frac{(y'+1)^2}{\frac{1}{2}} = 1$$

Rotating 45°, we sketch the ellipse:

$$\frac{(x'-1)^2}{1} + \frac{(y'+1)^2}{1/2} = 1$$

27. We first multiply out to get:

$$x^2 - 2xy + y^2 = 8y - 48$$
$$x^2 - 2xy + y^2 - 8y + 48 = 0$$

Now:

$$\cot 2\theta = \frac{1-1}{-2} = 0, \text{ so } 2\theta = 90° \text{ and thus } \theta = 45°$$

Therefore:

$$x = x'\cos\theta - y'\sin\theta = \frac{\sqrt{2}}{2}x' - \frac{\sqrt{2}}{2}y'$$

$$y = x'\sin\theta + y'\cos\theta = \frac{\sqrt{2}}{2}x' + \frac{\sqrt{2}}{2}y'$$

Making the substitutions and completing the square on x' and y' terms yields:

$$\left(y' - \sqrt{2}\right)^2 = 2\sqrt{2}\left(x' - \frac{11\sqrt{2}}{2}\right)$$

Rotating 45°, we sketch the parabola:

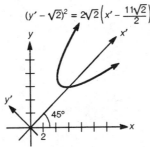

29. We find:

$$\cot 2\theta = \frac{3-6}{4} = -\frac{3}{4}, \text{ so } \tan 2\theta = -\frac{4}{3}$$

$$\sec^2 2\theta = 1 + \tan^2 2\theta = 1 + \left(-\frac{4}{3}\right)^2 = \frac{25}{9}$$

So $\sec 2\theta = -\frac{5}{3}$ (since $2\theta > 90°$), and thus $\cos 2\theta = -\frac{3}{5}$. Therefore:

$$\sin\theta = \sqrt{\frac{1 - \cos 2\theta}{2}} = \sqrt{\frac{1 + \frac{3}{5}}{2}} = \frac{2\sqrt{5}}{5}$$

$$\cos\theta = \sqrt{\frac{1 + \cos 2\theta}{2}} = \sqrt{\frac{1 - \frac{3}{5}}{2}} = \frac{\sqrt{5}}{5}$$

Thus $\theta = \cos^{-1}\left(\frac{\sqrt{5}}{5}\right) \approx 63.4°$. Now:

$$x = x'\cos\theta - y'\sin\theta = \frac{\sqrt{5}}{5}x' - \frac{2\sqrt{5}}{5}y'$$

$$y = x'\sin\theta + y'\cos\theta = \frac{2\sqrt{5}}{5}x' + \frac{\sqrt{5}}{5}y'$$

Making the substitutions into $3x^2 + 4xy + 6y^2 = 7$, we have:

$$\frac{x'^2}{1} + \frac{y'^2}{\frac{7}{2}} = 1$$

Rotating 63.4°, we sketch the ellipse:

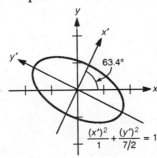

31. We find:
$$\cot 2\theta = \frac{17 - 8}{-12} = -\tfrac{3}{4}, \text{ so } \tan 2\theta = -\tfrac{4}{3}$$

As with Exercise 29 we find $\sin\theta = \frac{2\sqrt{5}}{5}$ and $\cos\theta = \frac{\sqrt{5}}{5}$, so:
$$x = \tfrac{\sqrt{5}}{5}x' - \tfrac{2\sqrt{5}}{5}y'$$
$$y = \tfrac{2\sqrt{5}}{5}x' + \tfrac{\sqrt{5}}{5}y'$$

Substituting into $17x^2 - 12xy + 8y^2 - 80 = 0$ yields:
$$\frac{x'^2}{16} + \frac{y'^2}{4} = 1$$

Rotating 63.4°, we sketch the ellipse:

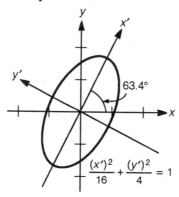

33. We find:
$$\cot 2\theta = \frac{0 + 4}{3} = \tfrac{4}{3}, \text{ so } \tan 2\theta = \tfrac{3}{4}$$
$$\sec^2 2\theta = 1 + \tan^2 2\theta = 1 + \tfrac{9}{16} = \tfrac{25}{16}, \text{ so } \sec 2\theta = \tfrac{5}{4}$$

Then $\cos 2\theta = \tfrac{4}{5}$, and thus:
$$\sin\theta = \sqrt{\frac{1 - \cos 2\theta}{2}} = \sqrt{\frac{1 - \tfrac{4}{5}}{2}} = \frac{\sqrt{10}}{10}$$
$$\cos\theta = \sqrt{\frac{1 + \cos 2\theta}{2}} = \sqrt{\frac{1 + \tfrac{4}{5}}{2}} = \frac{3\sqrt{10}}{10}$$

Then $\theta = \sin^{-1}\left(\frac{\sqrt{10}}{10}\right) \approx 18.4°$, and:
$$x = \tfrac{3\sqrt{10}}{10}x' - \tfrac{\sqrt{10}}{10}y'$$
$$y = \tfrac{\sqrt{10}}{10}x' + \tfrac{3\sqrt{10}}{10}y'$$

Substituting into $3xy - 4y^2 + 18 = 0$ results in:
$$\frac{y'^2}{4} - \frac{x'^2}{36} = 1$$

Rotating 18.4°, we sketch the hyperbola:

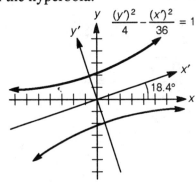

$$\frac{(y')^2}{4} - \frac{(x')^2}{36} = 1$$

35. We first multiply out terms to obtain:

$$x^2 + 2xy + y^2 + 4\sqrt{2}x - 4\sqrt{2}y = 0$$

We find:

$$\cot 2\theta = \frac{1-1}{2} = 0, \text{ so } 2\theta = 90° \text{ and thus } \theta = 45°$$

Now:

$$x = \frac{\sqrt{2}}{2}x' - \frac{\sqrt{2}}{2}y'$$
$$y = \frac{\sqrt{2}}{2}x' + \frac{\sqrt{2}}{2}y'$$

Substituting into $x^2 + 2xy + y^2 + 4\sqrt{2}x - 4\sqrt{2}y = 0$ results in:

$$x'^2 = 4y'$$

Rotating 45°, we sketch the parabola:

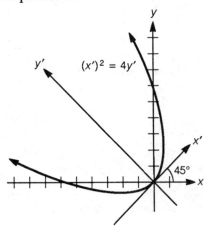

$(x')^2 = 4y'$

37. We find:

$$\cot 2\theta = \frac{3-2}{-\sqrt{15}} = -\frac{1}{\sqrt{15}}, \text{ so } \tan 2\theta = -\sqrt{15}$$
$$\sec^2 2\theta = 1 + \tan^2 2\theta = 1 + 15 = 16, \text{ so } \sec 2\theta = -4 \quad (\text{since } 2\theta > 90°)$$

Thus $\cos 2\theta = -\frac{1}{4}$ and we find:

$$\sin\theta = \sqrt{\frac{1-\cos 2\theta}{2}} = \sqrt{\frac{1+\frac{1}{4}}{2}} = \sqrt{\frac{5}{8}} = \frac{\sqrt{10}}{4}$$

$$\cos\theta = \sqrt{\frac{1+\cos 2\theta}{2}} = \sqrt{\frac{1-\frac{1}{4}}{2}} = \sqrt{\frac{3}{8}} = \frac{\sqrt{6}}{4}$$

Then $\theta = \cos^{-1}\left(\frac{\sqrt{6}}{4}\right) \approx 52.2°$, and:

$$x = \frac{\sqrt{6}}{4}x' - \frac{\sqrt{10}}{4}y'$$
$$y = \frac{\sqrt{10}}{4}x' + \frac{\sqrt{6}}{4}y'$$

Substituting into $3x^2 - \sqrt{15}xy + 2y^2 = 3$ results in:

$$\frac{x'^2}{6} + \frac{y'^2}{\frac{2}{3}} = 1$$

Rotating $52.2°$, we sketch the ellipse:

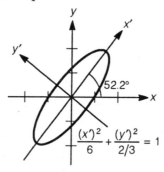

39. We find:

$$\cot 2\theta = \frac{3-3}{-2} = 0, \text{ so } 2\theta = 90° \text{ and thus } \theta = 45°$$

Now:

$$x = \frac{\sqrt{2}}{2}x' - \frac{\sqrt{2}}{2}y'$$
$$y = \frac{\sqrt{2}}{2}x' + \frac{\sqrt{2}}{2}y'$$

Substituting into $3x^2 - 2xy + 3y^2 + 2 = 0$ results in:

$$x'^2 + 2y'^2 = -1$$

But clearly this is impossible, so there is no graph.

41. Multiplying the first equation by $\sin\theta$ and the second equation by $\cos\theta$ yields:

$$(\sin\theta\cos\theta)x + (\sin^2\theta)y = x'\sin\theta$$
$$(-\sin\theta\cos\theta)x + (\cos^2\theta)y = y'\cos\theta$$

Adding, we get:

$$(\sin^2\theta + \cos^2\theta)y = x'\sin\theta + y'\cos\theta$$
$$y = x'\sin\theta + y'\cos\theta$$

Multiplying the first equation by $\cos\theta$ and the second equation by $-\sin\theta$ yields:

$$(\cos^2\theta)x + (\sin\theta\cos\theta)y = x'\cos\theta$$
$$(\sin^2\theta)x - (\sin\theta\cos\theta)y = -y'\sin\theta$$

Adding, we get:

$$(\sin^2\theta + \cos^2\theta)x = x'\cos\theta - y'\sin\theta$$
$$x = x'\cos\theta - y'\sin\theta$$

43. (a) Since $\cot 2\theta = -\frac{7}{24}$, $\tan 2\theta = -\frac{24}{7}$ so:

$$\sec^2 2\theta = 1 + \left(\frac{24}{7}\right)^2 = \frac{49}{49} + \frac{576}{49} = \frac{625}{49}$$

Thus $\sec 2\theta = \pm\frac{25}{7}$. Since 2θ is in either the first or second quadrant, and $\cot 2\theta < 0$, then 2θ is in the second quadrant, so $\sec 2\theta = -\frac{25}{7}$ and thus $\cos 2\theta = -\frac{7}{25}$.

(b) We compute $\sin\theta$ and $\cos\theta$:

$$\sin\theta = \sqrt{\frac{1-\cos 2\theta}{2}} = \sqrt{\frac{1+\frac{7}{25}}{2}} = \sqrt{\frac{32}{50}} = \sqrt{\frac{16}{25}} = \frac{4}{5}$$

$$\cos\theta = \sqrt{\frac{1+\cos 2\theta}{2}} = \sqrt{\frac{1-\frac{7}{25}}{2}} = \sqrt{\frac{18}{50}} = \sqrt{\frac{9}{25}} = \frac{3}{5}$$

(c) Making the substitutions and simplifying, we have:

$$16\left(\tfrac{1}{5}(3x'-4y')\right)^2 - 24\left(\tfrac{1}{5}(3x'-4y')\right)\left(\tfrac{1}{5}(4x'+3y')\right)$$
$$+9\left(\tfrac{1}{5}(4x'+3y')\right)^2 + 110\left(\tfrac{1}{5}(3x'-4y')\right) - 20\bullet\tfrac{1}{5}(4x'+3y') + 100 = 0$$
$$\tfrac{16}{25}\left(9x'^2 - 24x'y' + 16y'^2\right) - \tfrac{24}{25}\left(12x'^2 - 7x'y' - 12y'^2\right)$$
$$+\tfrac{9}{25}\left(16x'^2 + 24x'y' + 9y'^2\right) + 22(3x'-4y') - 4(4x'+3y') + 100 = 0$$
$$\frac{144-288+144}{25}x'^2 + \frac{-384+168+216}{25}x'y' + \frac{256+288+81}{25}y'^2$$
$$+ (66-16)x' + (-88-12)y' + 100 = 0$$
$$25y'^2 + 50x' - 100y' + 100 = 0$$
$$y'^2 + 2x' - 4y' + 4 = 0$$

45. Working from the right-hand side, we have:
$$A' + C' = A\cos^2\theta + B\sin\theta\cos\theta + C\sin^2\theta + A\sin^2\theta - B\sin\theta\cos\theta + C\cos^2\theta$$
$$= (A+C)\left(\cos^2\theta + \sin^2\theta\right)$$
$$= A+C$$

Chapter Ten Review Exercises

1.   The x-axis contains $\overline{AB}$ and has equation $y = 0$. The slope of the line containing $\overline{BC}$ is $-\frac{1}{b}$ and the equation is:

$$y = -\frac{1}{b}(x - 6b)$$
$$by = 6b - x$$
$$x + by = 6b$$

The slope of the line containing $\overline{AC}$ is $-\frac{1}{a}$ and the equation is:

$$y = -\frac{1}{a}(x - 6a)$$
$$x + ay = 6a$$

3.   Using the equations from Exercise 2, we show $G(2a + 2b, 2)$ lies on all three medians:

$$2(2a + 2b) + (a + b)(2) = 4a + 4b + 2a + 2b = 6(a + b)$$
$$2a + 2b - (b - 2a)(2) = 2a + 2b - 2b + 4a = 6a$$
$$2a + 2b - (a - 2b)(2) = 2a + 2b - 2a + 4b = 6b$$

5.   Using the equations from Exercise 4, we show $H(0, -6ab)$ lies on each altitude:

$$x = 0$$
$$-6ab = b \cdot 0 - 6ab = -6ab$$
$$-6ab = a \cdot 0 - 6ab = -6ab$$

7.   Using the equations from Exercise 6, we show $O(3a + 3b, 3ab + 3)$ lies on each perpendicular bisector:

$$x = 3a + 3b$$
$$b(3a + 3b) - (3ab + 3) = 3ab + 3b^2 - 3ab - 3 = 3b^2 - 3$$
$$a(3a + 3b) - (3ab + 3) = 3a^2 + 3ab - 3ab - 3 = 3a^2 - 3$$

9.   We find each distance and use the result from Exercise 8:

$$p = \sqrt{(6b)^2 + 6^2} = 6\sqrt{b^2 + 1}$$
$$q = \sqrt{(6a)^2 + 6^2} = 6\sqrt{a^2 + 1}$$
$$r = 6(a - b)$$
$$R = 3\sqrt{(a^2 + 1)(b^2 + 1)}$$

Thus:

$$\frac{pqr}{4R} = \frac{\left(6\sqrt{b^2 + 1}\right)\left(6\sqrt{a^2 + 1}\right)(6(a - b))}{4 \cdot 3\sqrt{(a^2 + 1)(b^2 + 1)}} = 18(a - b)$$

Area of $\triangle ABC = \frac{1}{2}(6(a - b))(6) = 18(a - b)$

So the area is $\frac{pqr}{4R}$, as required.

11. We compute each side of the identity:

$$OH^2 = (3a + 3b - 0)^2 + (3ab + 3 + 6ab)^2$$
$$= 9(a^2 + 2ab + b^2) + 9(3ab + 1)^2$$
$$= 9(a^2 + 2ab + b^2 + 9a^2b^2 + 6ab + 1)$$
$$= 81a^2b^2 + 9a^2 + 9b^2 + 72ab + 9$$

$$9R^2 - (p^2 + q^2 + r^2)$$
$$= 9[9(a^2 + 1)(b^2 + 1)] - 36(b^2 + 1) - 36(a^2 + 1) - 36(a^2 - 2ab + b^2)$$
$$= 81a^2b^2 + 81a^2 + 81b^2 + 81 - 36b^2 - 36 - 36a^2 - 36 - 36a^2 + 72ab - 36b^2$$
$$= 81a^2b^2 + 9a^2 + 9b^2 + 72ab + 9$$

Thus $OH^2 = 9R^2 - (p^2 + q^2 + r^2)$.

13. We first compute the squares:

$$HA^2 = (6a)^2 + (-6ab)^2 = 36a^2 + 36a^2b^2$$
$$HB^2 = (6b)^2 + (-6ab)^2 = 36b^2 + 36a^2b^2$$
$$HC^2 = 0^2 + (6 + 6ab)^2 = 36 + 72ab + 36a^2b^2$$

So $HA^2 + HB^2 + HC^2 = 108a^2b^2 + 36a^2 + 36b^2 + 72ab + 36$.

Now compute the right-hand side:

$$12R^2 - (p^2 + q^2 + r^2)$$
$$= 12[9(a^2 + 1)(b^2 + 1)] - [36(b^2 + 1) + 36(a^2 + 1) + 36(a^2 - 2ab + b^2)]$$
$$= 108a^2b^2 + 108a^2 + 108b^2 + 108 - 36b^2 - 36 - 36a^2 - 36 - 36a^2 + 72ab - 36b^2$$
$$= 108a^2b^2 + 36a^2 + 36b^2 + 72ab + 36$$

Thus $HA^2 + HB^2 + HC^2 = 12R^2 - (p^2 + q^2 + r^2)$.

15. In Exercise 12 we saw that:

$$GH^2 = 4(9a^2b^2 + a^2 + b^2 + 8ab + 1)$$

Therefore:

$$GH = 2\sqrt{9a^2b^2 + a^2 + b^2 + 8ab + 1}$$

Now:

$$2GO = 2\sqrt{(2a + 2b - 3a - 3b)^2 + (2 - 3ab - 3)^2}$$
$$= 2\sqrt{a^2 + 2ab + b^2 + 1 + 6ab + 9a^2b^2}$$
$$= 2\sqrt{9a^2b^2 + a^2 + b^2 + 8ab + 1}$$

Thus $GH = 2GO$.

17. Since $\tan\theta = -\frac{2}{3}$, then $\theta \approx 146.3°$.

19. We have $(x_0, y_0) = (-1, -3)$, $A = 5$, $B = 6$ and $C = -30$, so using the distance formula from a point to a line yields:

$$d = \frac{|5(-1) + 6(-3) - 30|}{\sqrt{5^2 + 6^2}} = \frac{53}{\sqrt{61}} = \frac{53\sqrt{61}}{61}$$

21. Label $A(-6,0)$, $B(6,0)$ and $C(0,6\sqrt{3})$. The height of the triangle is $6\sqrt{3}$. Since the line $\overline{AB}$ is the x-axis, the distance from $(1,2)$ to $\overline{AB}$ is 2. The line containing $\overline{AC}$ has slope $\sqrt{3}$ and equation:

$$y - 0 = \sqrt{3}(x+6)$$
$$y = \sqrt{3}x + 6\sqrt{3}$$

The distance from $(1,2)$ to $\overline{AC}$ is thus:

$$\frac{\left|\sqrt{3}(1) + 6\sqrt{3} - 2\right|}{\sqrt{1^2 + \left(-\sqrt{3}\right)^2}} = \frac{7\sqrt{3} - 2}{2}$$

The line containing $\overline{BC}$ has slope $-\sqrt{3}$ and equation:

$$y - 0 = -\sqrt{3}(x-6)$$
$$y = -\sqrt{3}x + 6\sqrt{3}$$

The distance from $(1,2)$ to $\overline{BC}$ is thus:

$$\frac{\left|-\sqrt{3}(1) + 6\sqrt{3} - 2\right|}{\sqrt{1^2 + \left(\sqrt{3}\right)^2}} = \frac{5\sqrt{3} - 2}{2}$$

The sum of these distances is $6\sqrt{3}$, which is also the height.

23. (a) We have the form $y^2 = 4px$, where $p = 4$. Thus the equation is $y^2 = 16x$.
(b) We have the form $x^2 = 4py$, where $p = 4$. Thus the equation is $x^2 = 16y$.

25. Since the parabola is symmetric about the positive y-axis, its equation must be of the form $x^2 = 4py$, where $p > 0$. Now the focal width is 12, so $4p = 12$. Thus the equation is $x^2 = 12y$.

27. We have $c = 2$ and $a = 8$, and the ellipse must have the form:

$$\frac{x^2}{8^2} + \frac{y^2}{b^2} = 1$$

Since $c^2 = a^2 - b^2$, we can find b:

$$4 = 64 - b^2$$
$$b^2 = 60$$

So the equation is $\frac{x^2}{64} + \frac{y^2}{60} = 1$, or $15x^2 + 16y^2 = 960$.

29. Since one end of the minor axis is $(-6,0)$, then $b = 6$ and the ellipse has a form of:

$$\frac{x^2}{36} + \frac{y^2}{a^2} = 1$$

Now $\frac{c}{a} = \frac{4}{5}$, so $c = \frac{4}{5}a$. We find a:

$$c^2 = a^2 - b^2$$
$$\left(\tfrac{4}{5}a\right)^2 = a^2 - 36$$
$$\tfrac{16}{25}a^2 = a^2 - 36$$
$$36 = \tfrac{9}{25}a^2$$
$$100 = a^2$$

So the equation is $\frac{x^2}{36} + \frac{y^2}{100} = 1$, or $25x^2 + 9y^2 = 900$.

31. Since the foci are $(\pm 6, 0)$ and the vertices are $(\pm 2, 0)$, then $c = 6$, $a = 2$, and the equation has the form:

$$\frac{x^2}{4} - \frac{y^2}{b^2} = 1$$

We find b:

$$c^2 = a^2 + b^2$$
$$36 = 4 + b^2$$
$$32 = b^2$$

So the equation is $\frac{x^2}{4} - \frac{y^2}{32} = 1$, or $8x^2 - y^2 = 32$.

33. Since the foci are $(\pm 3, 0)$, then $c = 3$ and the equation has the form:

$$\frac{x^2}{a^2} - \frac{y^2}{b^2} = 1$$

Now $\frac{c}{a} = 4$, so $\frac{3}{a} = 4$, thus $a = \frac{3}{4}$. We substitute to find b:

$$c^2 = a^2 + b^2$$
$$9 = \tfrac{9}{16} + b^2$$
$$144 = 9 + 16b^2$$
$$135 = 16b^2$$
$$\tfrac{135}{16} = b^2$$

So the equation is $\dfrac{x^2}{\frac{9}{16}} - \dfrac{y^2}{\frac{135}{16}} = 1$, or $240x^2 - 16y^2 = 135$.

35. We note that $4p = 10$, so $p = \frac{5}{2}$. So the vertex is $(0,0)$, the focus is $\left(0, \frac{5}{2}\right)$, the directrix is $y = -\frac{5}{2}$, and the focal width is 10.

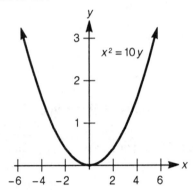

37. We note that $4p = 12$, so $p = 3$. So the vertex is $(0,3)$, the focus is $(0,0)$, the directrix is $y = 6$, and the focal width is 12.

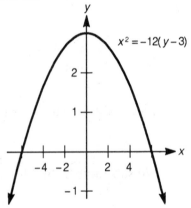

39. We note that $4p = 4$, so $p = 1$. So the vertex is $(1,1)$, the focus is $(0,1)$, the directrix is $x = 2$, and the focal width is 4.

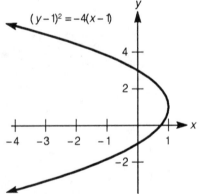

41. Dividing by 4, the standard form is $\frac{x^2}{4} + \frac{y^2}{2} = 1$. The center is $(0,0)$, the length of the major axis is 4, and the length of the minor axis is $2\sqrt{2}$. Using $c^2 = a^2 - b^2$, we find:
$$c = \sqrt{a^2 - b^2} = \sqrt{4 - 2} = \sqrt{2}$$
So the foci are $\left(\pm\sqrt{2}, 0\right)$ and the eccentricity is $\frac{\sqrt{2}}{2}$.

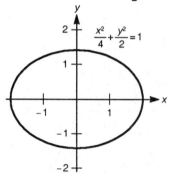

43. Dividing by 441, the standard form is $\frac{x^2}{9} + \frac{y^2}{49} = 1$. The center is $(0,0)$, the length of the major axis is 14, and the length of the minor axis is 6. Using $c^2 = a^2 - b^2$, we find:
$$c = \sqrt{a^2 - b^2} = \sqrt{49 - 9} = \sqrt{40} = 2\sqrt{10}$$
So the foci are $\left(0, \pm 2\sqrt{10}\right)$ and the eccentricity is $\frac{2\sqrt{10}}{7}$.

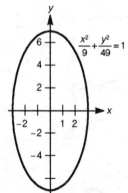

45. The equation is already in standard form where the center is $(1, -2)$, the length of the major axis is 10, and the length of the minor axis is 6. Using $c^2 = a^2 - b^2$, we find:
$$c = \sqrt{a^2 - b^2} = \sqrt{25 - 9} = \sqrt{16} = 4$$
So the foci are $(1 + 4, -2) = (5, -2)$ and $(1 - 4, -2) = (-3, -2)$, and the eccentricity is $\frac{4}{5}$.

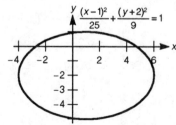

47. Dividing by 4, the standard form is $\frac{x^2}{4} - \frac{y^2}{2} = 1$. The center is $(0,0)$, the vertices are $(\pm2,0)$, and the asymptotes are $y = \pm\frac{\sqrt{2}}{2}x$. Using $c^2 = a^2 + b^2$, we have:

$$c = \sqrt{a^2 + b^2} = \sqrt{4+2} = \sqrt{6}$$

So the foci are $(\pm6,0)$ and the eccentricity is $\frac{\sqrt{6}}{2}$.

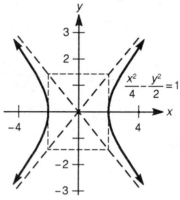

49. Dividing by 441, the standard form is $\frac{y^2}{9} - \frac{x^2}{49} = 1$. The center is $(0,0)$, the vertices are $(0,\pm3)$, and the asymptotes are $y = \pm\frac{3}{7}x$. Using $c^2 = a^2 + b^2$, we have:

$$c = \sqrt{a^2 + b^2} = \sqrt{9+49} = \sqrt{58}$$

So the foci are $\left(0,\pm\sqrt{58}\right)$ and the eccentricity is $\frac{\sqrt{58}}{3}$.

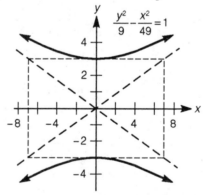

51. The equation is already in standard form where the center is $(1,-2)$ and the vertices are $(1 + 5,-2) = (6,-2)$ and $(1 - 5,-2) = (-4,-2)$. The asymptotes have slopes of $\pm\frac{3}{5}$, so using the point-slope formula:

$$y-(-2) = \tfrac{3}{5}(x-1) \qquad\qquad y-(-2) = -\tfrac{3}{5}(x-1)$$
$$y+2 = \tfrac{3}{5}x - \tfrac{3}{5} \qquad\qquad\quad y+2 = -\tfrac{3}{5}x + \tfrac{3}{5}$$
$$y = \tfrac{3}{5}x - \tfrac{13}{5} \qquad\qquad\quad\;\; y = -\tfrac{3}{5}x - \tfrac{7}{5}$$

Using $c^2 = a^2 + b^2$, we have:
$$c = \sqrt{a^2 + b^2} = \sqrt{25 + 9} = \sqrt{34}$$
So the foci are $\left(1 \pm \sqrt{34}, -2\right)$ and the eccentricity is $\frac{\sqrt{34}}{5}$.

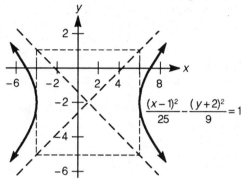

53. We complete the square to convert the equation to standard form:
$$3x^2 - 6x + 4y^2 + 16 = -7$$
$$3\left(x^2 - 2x + 1\right) + 4\left(y^2 + 4y + 4\right) = -7 + 3 + 16$$
$$3(x - 1)^2 + 4(y + 2)^2 = 12$$
$$\frac{(x - 1)^2}{4} + \frac{(y + 2)^2}{3} = 1$$
This is the equation of an ellipse. Its center is $(1, -2)$, the length of the major axis is 4, and the length of the minor axis is $2\sqrt{3}$. Using $c^2 = a^2 - b^2$, we find:
$$c = \sqrt{a^2 - b^2} = \sqrt{4 - 3} = 1$$
So the foci are $(1 + 1, -2) = (2, -2)$ and $(1 - 1, -2) = (0, -2)$.

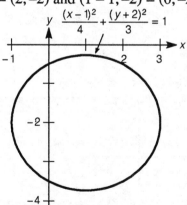

55. We complete the square to convert the equation to standard form:
$$y^2 + 2y = -4x + 15$$
$$y^2 + 2y + 1 = -4x + 15 + 1$$
$$(y + 1)^2 = -4(x - 4)$$

This is the equation of a parabola. Since $4p = 4$, then $p = 1$. So the vertex is $(4, -1)$, the axis of symmetry is $y = -1$, the focus is $(3, -1)$, and the directrix is $x = 5$.

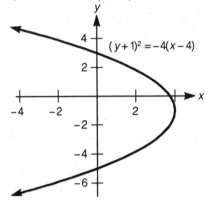

57. We complete the square to convert the equation to standard form:
$$16(x^2 - 2x) - 9(y^2 - 10y) = 353$$
$$16(x^2 - 2x + 1) - 9(y^2 - 10y + 25) = 353 + 16 - 225$$
$$16(x - 1)^2 - 9(y - 5)^2 = 144$$
$$\frac{(x-1)^2}{9} - \frac{(y-5)^2}{16} = 1$$

This is the equation of a hyperbola. Its center is $(1, 5)$ and its vertices are $(1 + 3, 5) = (4, 5)$ and $(1 - 3, 5) = (-2, 5)$. The asymptotes have slopes of $\pm\frac{4}{3}$, so using the point-slope formula:

$$y - 5 = \frac{4}{3}(x - 1) \qquad\qquad y - 5 = -\frac{4}{3}(x - 1)$$
$$y - 5 = \frac{4}{3}x - \frac{4}{3} \qquad\qquad y - 5 = -\frac{4}{3}x + \frac{4}{3}$$
$$y = \frac{4}{3}x + \frac{11}{3} \qquad\qquad y = -\frac{4}{3}x + \frac{19}{3}$$

Using $c^2 = a^2 + b^2$, we have:
$$c = \sqrt{a^2 + b^2} = \sqrt{9 + 16} = \sqrt{25} = 5$$

So the foci are $(1 + 5, 5) = (6, 5)$ and $(1 - 5, 5) = (-4, 5)$.

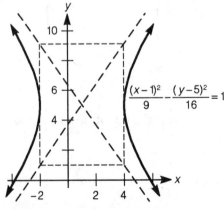

59. We complete the square to convert the equation to standard form:
$$5x^2 + 3y^2 - 40x - 36y + 188 = 0$$
$$5(x^2 - 8x) + 3(y^2 - 12y) = -188$$
$$5(x - 4)^2 + 3(y - 6)^2 = -188 + 80 + 108$$
$$5(x - 4)^2 + 3(y - 6)^2 = 0$$
This equation has only one solution, namely the point $(4, 6)$.

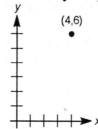

61. We complete the square to convert the equation to standard form:
$$9x^2 - 90x - 16y^2 + 32y = -209$$
$$9(x^2 - 10x) - 16(y^2 - 2y) = -209$$
$$9(x - 5)^2 - 16(y - 1)^2 = -209 + 225 - 16$$
$$9(x - 5)^2 - 16(y - 1)^2 = 0$$
$$9(x - 5)^2 = 16(y - 1)^2$$
$$\pm 3(x - 5) = 4(y - 1)$$
The graph of this equation consists of just two lines:

$3(x - 5) = 4(y - 1)$	$-3(x - 5) = 4(y - 1)$
$3x - 15 = 4y - 4$	$-3x + 15 = 4y - 4$
$3x - 4y = 11$	$3x + 4y = 19$

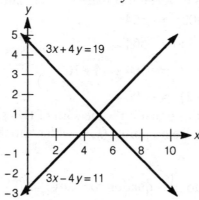

63. We complete the square to convert the equation to standard form:

$$y^2 + 8y - 25x^2 = 9$$

$$(y+4)^2 - 25x^2 = 25$$

$$\frac{(y+4)^2}{25} - \frac{x^2}{1} = 1$$

This is the equation of a hyperbola. Its center is $(0, -4)$ and its vertices are $(0, -4+5) = (0, 1)$ and $(0, -4-5) = (0, -9)$. The asymptotes have slopes of ± 5, so using the point-slope formula:

$$y - (-4) = 5(x - 0) \qquad\qquad y - (-4) = -5(x - 0)$$

$$y + 4 = 5x \qquad\qquad\qquad y + 4 = -5x$$

$$y = 5x - 4 \qquad\qquad\qquad y = -5x - 4$$

Using $c^2 = a^2 + b^2$, we have:

$$c = \sqrt{a^2 + b^2} = \sqrt{25 + 1} = \sqrt{26}$$

So the foci are $\left(0, -4 \pm \sqrt{26}\right)$.

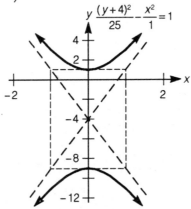

65. We complete the square to convert the equation to standard form:

$$16x^2 - 64x + 25y^2 - 100y = -564$$

$$16\left(x^2 - 4x\right) + 25\left(y^2 - 4y\right) = -564$$

$$16(x-2)^2 + 25(y-2)^2 = -564 + 64 + 100$$

$$16(x-2)^2 + 25(y-2)^2 = -400$$

Since the left-hand side of this equation is the sum of two squares, it must be non-negative and thus there are no points (x, y) which satisfy the equation. Thus there is no graph for this equation.

67. (a) Substituting $P(6, 5)$ into the equation $5x^2 - 4y^2 = 80$:

$$5(6)^2 - 4(5)^2 = 5(36) - 4(25) = 180 - 100 = 80$$

So $P(6, 5)$ lies on the hyperbola.

(b)   Since P lies on the hyperbola with foci F_1, F_2, then the quantity $F_1P - F_2P = \pm 2a$ (by the definition of the hyperbola). Thus $(F_1P - F_2P)^2 = 4a^2$. It remains to find a:

$$5x^2 - 4y^2 = 80$$

$$\frac{x^2}{16} - \frac{y^2}{20} = 1$$

So $a^2 = 16$, thus $(F_1P - F_2P)^2 = 4a^2 = 4(16) = 64$.

69.   Completing the square, we obtain:

$$A\left(x^2 + \frac{D}{A}x + \frac{D^2}{4A^2}\right) + C\left(y^2 + \frac{E}{C}y + \frac{E^2}{4C^2}\right) = \frac{D^2}{4A} + \frac{E^2}{4C} - F$$

$$A\left(x + \frac{D}{2A}\right)^2 + C\left(y + \frac{E}{2C}\right)^2 = \frac{CD^2 + AE^2 - 4ACF}{4AC}$$

Now if the curve represents an ellipse or a hyperbola, this last equation shows that the center is $\left(-\frac{D}{2A}, -\frac{E}{2C}\right)$.

71.   For the parabola, the vertex is $(5,0)$, so we know $y^2 = -4p(x - 5)$. Since the focus is $(3,0)$, then $p = 2$, so the equation is $y^2 = -8(x - 5)$. For the ellipse, $a = 5$ and $c = 3$, so $b^2 = a^2 - c^2 = 25 - 9 = 16$, so $b = 4$. Thus the equation of the ellipse is $\frac{x^2}{25} + \frac{y^2}{16} = 1$.

73.   We have:

$$\cot 2\theta = \frac{4-1}{4} = \tfrac{3}{4}, \text{ so } \tan 2\theta = \tfrac{4}{3}$$

$$\sec^2 2\theta = 1 + \tan^2 2\theta = 1 + \tfrac{16}{9} = \tfrac{25}{9}$$

So $\sec 2\theta = \tfrac{5}{3}$ and $\cos 2\theta = \tfrac{3}{5}$. Thus:

$$\sin\theta = \sqrt{\frac{1 - \cos 2\theta}{2}} = \sqrt{\frac{1 - \frac{3}{5}}{2}} = \sqrt{\tfrac{1}{5}} = \frac{\sqrt{5}}{5}$$

$$\cos\theta = \sqrt{\frac{1 + \cos 2\theta}{2}} = \sqrt{\frac{1 + \frac{3}{5}}{2}} = \sqrt{\tfrac{4}{5}} = \frac{2\sqrt{5}}{5}$$

So $\theta = \sin^{-1}\left(\frac{\sqrt{5}}{5}\right) \approx 26.6°$, and:

$$x = \frac{2\sqrt{5}}{5}x' - \frac{\sqrt{5}}{5}y'$$

$$y = \frac{\sqrt{5}}{5}x' + \frac{2\sqrt{5}}{5}y'$$

Substituting, we obtain:

$$\left(x' + \frac{3\sqrt{5}}{5}\right)^2 = \frac{8\sqrt{5}}{5}\left(y' + \frac{9\sqrt{5}}{40}\right)$$

Rotating 26.6°, we graph the parabola:

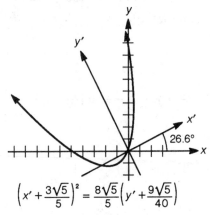

$$\left(x' + \frac{3\sqrt{5}}{5}\right)^2 = \frac{8\sqrt{5}}{5}\left(y' + \frac{9\sqrt{5}}{40}\right)$$

75. We have:
$$\cot 2\theta = \frac{13-13}{10} = 0, \text{ so } 2\theta = 90° \text{ and } \theta = 45°$$
Therefore:
$$\sin \theta = \frac{\sqrt{2}}{2} \text{ and } \cos \theta = \frac{\sqrt{2}}{2}$$
$$x = \frac{\sqrt{2}}{2}x' - \frac{\sqrt{2}}{2}y'$$
$$y = \frac{\sqrt{2}}{2}x' + \frac{\sqrt{2}}{2}y'$$
Substituting, we obtain:
$$\frac{x'^2}{16} + \frac{\left(y' - \sqrt{2}\right)^2}{36} = 1$$
Rotating 45°, we graph the ellipse:

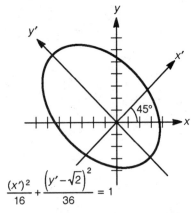

$$\frac{(x')^2}{16} + \frac{\left(y' - \sqrt{2}\right)^2}{36} = 1$$

77. Comparing the given equation with the four basic types, it appears this is the type associated with Figure 3 (Section 10.7). We divide both numerator and denominator by 2 to obtain:
$$r = \frac{3}{2 - 4\cos\theta} = \frac{\frac{3}{2}}{1 - 2\cos\theta}$$

Since the eccentricity is $e = 2 > 1$, this conic is a hyperbola. Computing values of r when $\theta = 0, \frac{\pi}{2}, \pi$ and $\frac{3\pi}{2}$, we have:

θ	0	$\frac{\pi}{2}$	π	$\frac{3\pi}{2}$
r	$-\frac{3}{2}$	$\frac{3}{2}$	$\frac{1}{2}$	$\frac{3}{2}$

So the vertices are $\left(-\frac{3}{2},0\right)$ and $\left(-\frac{1}{2},0\right)$, and thus the center is $(-1,0)$. Note that $a = \frac{1}{2}$ and since a focus is $(0,0)$, then $c = 1$ and thus:
$$b = \sqrt{c^2 - a^2} = \sqrt{1-\frac{1}{4}} = \sqrt{\frac{3}{4}} = \frac{\sqrt{3}}{2}$$
We graph the hyperbola:

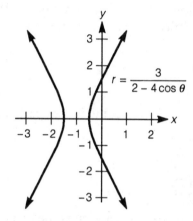

$$r = \frac{3}{2 - 4\cos\theta}$$

79. Comparing the given equation with the four basic types, it appears this is the type associated with Figure 3 (Section 10.7). We divide both numerator and denominator by 4 to obtain:
$$r = \frac{4}{4 - \cos\theta} = \frac{1}{1 - \frac{1}{4}\cos\theta}$$
Since the eccentricity is $e = \frac{1}{4} < 1$, this conic is an ellipse. Computing the values of r when $\theta = 0, \frac{\pi}{2}, \pi$ and $\frac{3\pi}{2}$, we have:

θ	0	$\frac{\pi}{2}$	π	$\frac{3\pi}{2}$
r	$\frac{4}{3}$	1	$\frac{4}{5}$	1

So the vertices on the major axis are $\left(\frac{4}{3},0\right)$ and $\left(-\frac{4}{5},0\right)$, and thus the center is $\left(\frac{4}{15},0\right)$.
The major axis has length:
$$2a = \frac{4}{3} - \left(-\frac{4}{5}\right) = \frac{32}{15}, \text{ so } a = \frac{16}{15}$$
Since a focus is $(0,0)$, then $c = \frac{4}{15}$ and thus:
$$b = \sqrt{a^2 - c^2} = \sqrt{\left(\frac{16}{15}\right)^2 - \left(\frac{4}{15}\right)^2} = \sqrt{\frac{240}{225}} = \frac{4\sqrt{15}}{15}$$

We graph the ellipse:

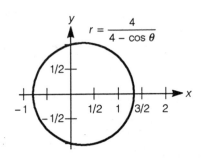

81. Comparing the given equation with the four basic types, it appears this is the type associated with Figure 5 (Section 10.7). We divide both numerator and denominator by 3 to obtain:

$$r = \frac{20}{3 - 5\sin\theta} = \frac{\frac{20}{3}}{1 - \frac{5}{3}\sin\theta}$$

Since the eccentricity is $e = \frac{5}{3} > 1$, this conic is a hyperbola. Computing the values of r when $\theta = 0, \frac{\pi}{2}, \pi$ and $\frac{3\pi}{2}$, we have:

θ	0	$\frac{\pi}{2}$	π	$\frac{3\pi}{2}$
r	$\frac{20}{3}$	-10	$\frac{20}{3}$	$\frac{5}{2}$

So the vertices are $(0,-10)$ and $\left(0,-\frac{5}{2}\right)$, and thus the center is $\left(0,-\frac{25}{4}\right)$. The transverse axis has length:

$$2a = -\frac{5}{2} + 10 = \frac{15}{2}, \text{ so } a = \frac{15}{4}$$

Since a focus is $(0,0)$, then $c = \frac{25}{4}$ and thus:

$$b = \sqrt{c^2 - a^2} = \sqrt{\left(\frac{25}{4}\right)^2 - \left(\frac{15}{4}\right)^2} = \sqrt{\frac{400}{16}} = 5$$

We graph the hyperbola:

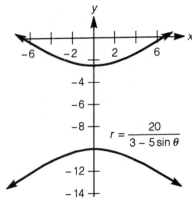

83. We have $\tan\theta = \dfrac{y}{a+x}$ and $\tan\beta = \dfrac{y}{a-x}$, thus:

$$\tan\theta\tan\beta = 2$$
$$\frac{y}{a+x}\cdot\frac{y}{a-x} = 2$$
$$\frac{y^2}{a^2-x^2} = 2$$
$$y^2 = 2a^2 - 2x^2$$
$$2x^2 + y^2 = 2a^2$$

85. We use $a = 5$ and $b = 3$ in each formula:

	Approximation Obtained	Percentage Error
C_1	25.531776	0.019
C_2	25.526986	0.000049
C_3	25.519489	0.029

Since C_2 has the smallest percentage error, it would be the best approximation of the circumference of the three.

87. (a) The first two lines intersect where:

$$(2p-2q)x = r(q-p)$$
$$2x = -r$$
$$x = -\frac{r}{2}$$

Thus one vertex is $A\left(-\frac{r}{2}, -\frac{r}{2}(p+q)\right)$. Similarly, we find that $B\left(-\frac{q}{2}, -\frac{q}{2}(p+r)\right)$ and $C\left(-\frac{p}{2}, -\frac{p}{2}(q+r)\right)$ are the other vertices.

(b) We use the formula $A = \frac{1}{2}\left| x_1 y_2 - x_2 y_1 + x_2 y_3 - x_3 y_2 + x_3 y_1 - x_1 y_3 \right|$ for the area of a triangle with vertices (x_1,y_1), (x_2,y_2), and (x_3,y_3):

$$A = \frac{1}{2}\left| \frac{rq(p+r)}{4} - \frac{rq(p+q)}{4} + \frac{pq(q+r)}{4} - \frac{pq(p+r)}{4} + \frac{pr(p+q)}{4} - \frac{pr(q+r)}{4}\right|$$
$$= \frac{1}{8}\left| r^2q - rq^2 + pq^2 - p^2q + p^2r - pr^2 \right|$$

Therefore:

$$\frac{\left|(p-q)(q-r)(r-p)\right|}{8}$$
$$= \frac{\left| pqr - p^2q - pr^2 + p^2r - q^2r + q^2p + qr^2 - prq \right|}{8}$$
$$= A$$

Chapter Ten Test

1.   Since $4p = 12$, then $p = 3$. So the focus is $(-3, 0)$ and the directrix is $x = 3$.

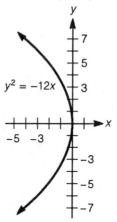

2.   Dividing by 4, the standard form is $\frac{x^2}{4} - \frac{y^2}{1} = 1$. So the asymptotes are $y = \pm\frac{1}{2}x$. Using $c^2 = a^2 + b^2$, we have:
$$c = \sqrt{a^2 + b^2} = \sqrt{4+1} = \sqrt{5}$$
So the foci are $\left(\pm\sqrt{5}, 0\right)$.

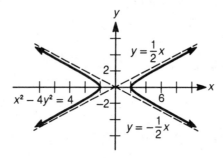

3.   (a)   We have:
$$\cot 2\theta = \frac{A-C}{B} = \frac{1-3}{2\sqrt{3}} = -\frac{1}{\sqrt{3}}, \text{ so } \tan 2\theta = -\sqrt{3}$$
Thus $2\theta = 120°$, or $\theta = 60°$.

     (b)   Applying the rotation formulas, we have:
$$x = x'\cos 60° - y'\sin 60° = \frac{1}{2}x' - \frac{\sqrt{3}}{2}y'$$
$$y = x'\sin 60° + y'\cos 60° = \frac{\sqrt{3}}{2}x' + \frac{1}{2}y'$$
     Substituting into $x^2 + 2\sqrt{3}xy + 3y^2 - 12\sqrt{3}x + 12y = 0$ yields:
$$x'^2 = -6y'$$

Rotating 60°, we graph the parabola:

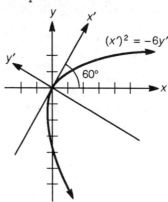

4. Since $\tan\theta = \frac{1}{\sqrt{3}}$, then $\theta = 30°$.

5. Since $e = \frac{c}{a}$, then $\frac{c}{a} = \frac{1}{2}$, so $a = 2c$. Since the foci are $(0, \pm 2)$, then $c = 2$ and thus $a = 4$. Now we find b^2:

$$c^2 = a^2 - b^2$$
$$4 = 16 - b^2$$
$$b^2 = 12$$

Thus the equation of the ellipse is $\frac{x^2}{12} + \frac{y^2}{16} = 1$.

6. Call m the slope of the tangents, so:

$$y - 0 = m(x + 4)$$
$$y = mx + 4m$$

We now find the distance from the center of the circle $(0, 0)$ to this line:

$$r = \frac{|0 + 0 - 4m|}{\sqrt{1 + m^2}} = \frac{|4m|}{\sqrt{1 + m^2}}$$

Since this radius is 1, we have:

$$1 = \frac{|4m|}{\sqrt{1 + m^2}}$$

Squaring, we have:

$$1 = \frac{16m^2}{1 + m^2}$$
$$1 + m^2 = 16m^2$$
$$1 = 15m^2$$
$$m^2 = \tfrac{1}{15}$$
$$m = \pm\frac{\sqrt{15}}{15}$$

7.   The slope is given by $m = \tan 60° = \sqrt{3}$. Using the point $(2,0)$ in the point-slope formula:
$$y - 0 = \sqrt{3}(x - 2)$$
$$y = \sqrt{3}x - 2\sqrt{3}$$
$$\sqrt{3}x - y - 2\sqrt{3} = 0$$

8.   Since the foci are $(\pm 2, 0)$, $c = 2$. Also $\frac{b}{a} = \frac{1}{\sqrt{3}}$, so $a = b\sqrt{3}$. So:
$$c^2 = a^2 + b^2$$
$$4 = \left(b\sqrt{3}\right)^2 + b^2$$
$$4 = 4b^2$$
$$1 = b^2$$
$$1 = b$$
So $a = 1\sqrt{3} = \sqrt{3}$, thus the equation is $\frac{x^2}{3} - \frac{y^2}{1} = 1$.

9.   (a)   Substituting $P(6, 5)$ into the equation $5x^2 - 4y^2 = 80$, we have:
$$5(6)^2 - 4(5)^2 = 5(36) - 4(25) = 180 - 100 = 80$$
So $P(6, 5)$ lies on the hyperbola.

      (b)   The quantity $(F_1P - F_2P)^2$ can be computed without determining the coordinates of F_1 and F_2. By definition we have $|F_1P - F_2P| = 2a$ for any point P on the hyperbola. Squaring both sides here yields $(F_1P - F_2P)^2 = 4a^2$. Now to compute a, we convert the equation $5x^2 - 4y^2 = 80$ to standard form. The result is $\frac{x^2}{16} - \frac{y^2}{20} = 1$. Therefore $a = 4$ and we obtain $(F_1P - F_2P)^2 = 4a^2 = 4(16) = 64$.

10.   Dividing by 100 yields the standard form $\frac{x^2}{25} + \frac{y^2}{4} = 1$. The length of the major axis is 10 and the length of the minor axis is 4. Using $c^2 = a^2 - b^2$, we find:
$$c = \sqrt{a^2 - b^2} = \sqrt{25 - 4} = \sqrt{21}$$
So the foci are $\left(\pm\sqrt{21}, 0\right)$.

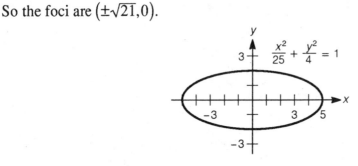

11. Here $(x_0, y_0) = (-1, 0)$, $A = 2$, $B = -1$ and $C = -1$, so using the distance formula from a point to a line yields:

$$d = \frac{|Ax_0 + By_0 + C|}{\sqrt{A^2 + B^2}} = \frac{|-2 + 0 - 1|}{\sqrt{4 + 1}} = \frac{3}{\sqrt{5}} = \frac{3\sqrt{5}}{5}$$

12. We complete the square to convert the equation to standard form:

$$16x^2 + y^2 - 64x + 2y + 65 = 0$$
$$16(x^2 - 4x) + (y^2 + 2y) = -65$$
$$16(x^2 - 4x + 4) + (y^2 + 2y + 1) = -65 + 64 + 1$$
$$16(x - 2)^2 + (y + 1)^2 = 0$$

Since the only solution to this equation is the point $(2, -1)$, the graph consists of a single point.

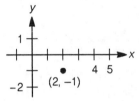

13. The equation represents a hyperbola with center $(-4, 4)$. We graph the hyperbola:

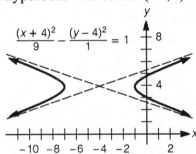

14. Comparing the given equation with the four basic types, it appears this is the type associated with Figure 3 (Section 10.7). We divide both numerator and denominator by 5 to obtain:

$$r = \frac{9}{5 - 4\cos\theta} = \frac{\frac{9}{5}}{1 - \frac{4}{5}\cos\theta}$$

Since the eccentricity is $e = \frac{4}{5} < 1$, this conic is an ellipse. Computing the values of r when $\theta = 0, \frac{\pi}{2}, \pi$ and $\frac{3\pi}{2}$, we have:

θ	0	$\frac{\pi}{2}$	π	$\frac{3\pi}{2}$
r	9	$\frac{9}{5}$	1	$\frac{9}{5}$

So the vertices on the major axis are $(9,0)$ and $(-1,0)$, and thus the center is $(4,0)$ and the length of the major axis is $2a = 10$, so $a = 5$. Since a focus is $(0,0)$, then $c = 4$ and thus:

$$b = \sqrt{a^2 - c^2} = \sqrt{25 - 16} = \sqrt{9} = 3$$

We graph the ellipse:

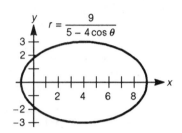

15. Since $4p = 8$, then $p = 2$. The focal width is 8 and the vertex is $(1,2)$.

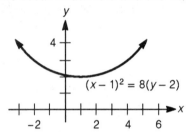

16. (a) Since $a = 6$ and $b = 5$, we find c:

$$c = \sqrt{a^2 - b^2} = \sqrt{36 - 25} = \sqrt{11}$$

So the eccentricity is $e = \frac{c}{a} = \frac{\sqrt{11}}{6}$. Thus the directrices are given by:

$$x = \pm\frac{a}{e} = \pm\frac{6}{\frac{\sqrt{11}}{6}} = \pm\frac{36}{\sqrt{11}} = \pm\frac{36\sqrt{11}}{11}$$

(b) The focal radii are given by:

$$F_1P = a + ex = 6 + \frac{\sqrt{11}}{6}(3) = 6 + \frac{\sqrt{11}}{2} = \frac{12 + \sqrt{11}}{2}$$

$$F_2P = a - ex = 6 - \frac{\sqrt{11}}{6}(3) = 6 - \frac{\sqrt{11}}{2} = \frac{12 - \sqrt{11}}{2}$$

17. The equation of the tangent line is $\dfrac{x_1x}{a^2} + \dfrac{y_1y}{b^2} = 1$. With $x_1 = -2$ and $y_1 = 4$, this

becomes $\dfrac{-2x}{a^2} + \dfrac{4y}{b^2} = 1$. Now to determine a^2 and b^2, we divide both sides of the

equation $x^2 + 3y^2 = 52$ by 52. This yields $\dfrac{x^2}{52} + \dfrac{y^2}{52/3} = 1$. Thus $a^2 = 52$ and $b^2 = \frac{52}{3}$,

and the equation of the tangent becomes $\dfrac{-2x}{52} + \dfrac{4y}{52/3} = 1$. When we simplify and solve

for y, the result is $y = \frac{1}{6}x + \frac{13}{3}$. This is the equation of the tangent line, as required.

18. Let m denote the slope for the required tangent line. Then the equation of the line is $y - 8 = m(x - 4)$. Now because this is the tangent line, the system

$$y - 8 = m(x - 4)$$
$$x^2 = 2y$$

must have exactly one solution, namely $(4, 8)$. Solving the second equation yields $y = \dfrac{x^2}{2}$. Now substitute into the first equation:

$$\frac{x^2}{2} - 8 = m(x - 4)$$
$$x^2 - 16 = 2m(x - 4)$$
$$(x - 4)(x + 4) = 2m(x - 4)$$
$$(x - 4)(x + 4 - 2m) = 0$$

So the solutions are $x = 4$ and $x = 2m - 4$. But since the line is tangent to the parabola, these two x-values must be equal:

$$2m - 4 = 4$$
$$2m = 8$$
$$m = 4$$

With this value for m, the equation $y - 8 = m(x - 4)$ becomes:

$$y - 8 = 4(x - 4)$$
$$y - 8 = 4x - 16$$
$$y = 4x - 8$$

This is the required tangent line.

Chapter Eleven
Roots of Polynomial Equations

11.1 The Complex Number System

1. We complete the table:

i^2	i^3	i^4	i^5	i^6	i^7	i^8
-1	$-i$	1	i	-1	$-i$	1

3. (a) The real part is 4 and the imaginary part is 5.
    (b) The real part is 4 and the imaginary part is -5.
    (c) The real part is $\frac{1}{2}$ and the imaginary part is -1.
    (d) The real part is 0 and the imaginary part is 16.

5. Equating the real parts gives $2c = 8$, and therefore $c = 4$. Similarly, equating the imaginary parts yields $d = -3$.

7. (a) $(5 - 6i) + (9 + 2i) = (5 + 9) + (-6 + 2)i = 14 - 4i$
    (b) $(5 - 6i) - (9 + 2i) = (5 - 9) + (-6 - 2)i = -4 - 8i$

9. (a) $(3 - 4i)(5 + i) = 15 - 17i - 4i^2 = 19 - 17i$

    (b) $(5 + i)(3 - 4i) = 19 - 17i$, from part (a)

(c) We compute the quotient:
$$\frac{3-4i}{5+i} \cdot \frac{5-i}{5-i} = \frac{15-23i+4i^2}{25-i^2} = \frac{11-23i}{26} = \tfrac{11}{26} - \tfrac{23}{26}i$$

(d) We compute the quotient:
$$\frac{5+i}{3-4i} \cdot \frac{3+4i}{3+4i} = \frac{15+23i+4i^2}{9-16i^2} = \frac{11+23i}{25} = \tfrac{11}{25} + \tfrac{23}{25}i$$

11. (a) $z+w = (2+3i)+(9-4i) = 11-i$

(b) $\bar{z}+w = (2-3i)+(9-4i) = 11-7i$

(c) $z+\bar{z} = (2+3i)+(2-3i) = 4$

13. We evaluate the expression:
$$(z+w)+w_1 = \left[(2+3i)+(9-4i)\right]+(-7-i) = (11-i)+(-7-i) = 4-2i$$

15. We compute the product:
$$zw = (2+3i)(9-4i) = 18+19i-12i^2 = 30+19i$$

17. $z\bar{z} = (2+3i)(2-3i) = 4-9i^2 = 13$

19. We first compute the product:
$$ww_1 = (9-4i)(-7-i) = -63+19i+4i^2 = -67+19i$$
Therefore:
$$z(ww_1) = (2+3i)(-67+19i) = -134-163i+57i^2 = -191-163i$$

21. We first compute the sum:
$$w+w_1 = (9-4i)+(-7-i) = 2-5i$$
Therefore:
$$z(w+w_1) = (2+3i)(2-5i) = 4-4i-15i^2 = 19-4i$$

23. We first compute the powers:
$$z^2 = (2+3i)(2+3i) = 4+12i+9i^2 = -5+12i$$
$$w^2 = (9-4i)(9-4i) = 81-72i+16i^2 = 65-72i$$
Therefore:
$$z^2-w^2 = (-5+12i)-(65-72i) = -70+84i$$

25. Since $zw = 30 + 19i$ (from Exercise 15), then:
$$(zw)^2 = (30+19i)(30+19i) = 900+1140i+361i^2 = 539+1140i$$

27. Since $z^2 = -5 + 12i$ (from Exercise 23), then:
$$z^3 = z \cdot z^2 = (2+3i)(-5+12i) = -10+9i+36i^2 = -46+9i$$

29. We compute the quotient:
$$\frac{z}{w} = \frac{2+3i}{9-4i} \cdot \frac{9+4i}{9+4i} = \frac{18+35i+12i^2}{81-16i^2} = \frac{6+35i}{97} = \frac{6}{97} + \frac{35}{97}i$$

31. Using the values of $\bar{z}$ and $\bar{w}$, we have:
$$\frac{\bar{z}}{\bar{w}} = \frac{2-3i}{9+4i} \cdot \frac{9-4i}{9-4i} = \frac{18-35i+12i^2}{81-16i^2} = \frac{6-35i}{97} = \frac{6}{97} - \frac{35}{97}i$$

33. Using the value of $\bar{z}$, we have:
$$\frac{z}{\bar{z}} = \frac{2+3i}{2-3i} \cdot \frac{2+3i}{2+3i} = \frac{4+12i+9i^2}{4-9i^2} = \frac{-5+12i}{13} = -\frac{5}{13} + \frac{12}{13}i$$

35. Since $w - \bar{w} = (9-4i) - (9+4i) = -8i$, then:
$$\frac{w-\bar{w}}{2i} = \frac{-8i}{2i} = -4$$

37. We compute the quotient:
$$\frac{i}{5+i} \cdot \frac{5-i}{5-i} = \frac{5i-i^2}{25-i^2} = \frac{1+5i}{26} = \frac{1}{26} + \frac{5}{26}i$$

39. We compute the quotient:
$$\frac{1}{i} \cdot \frac{i}{i} = \frac{i}{i^2} = \frac{i}{-1} = -i$$

41. By writing the numbers in complex form, we have:
$$\sqrt{-49} + \sqrt{-9} + \sqrt{-4} = 7i + 3i + 2i = 12i$$

43. By writing the numbers in complex form, we have:
$$\sqrt{-20} - 3\sqrt{-45} + \sqrt{-80} = \sqrt{4}\sqrt{-5} - 3\sqrt{9}\sqrt{-5} + \sqrt{16}\sqrt{-5}$$
$$= 2\sqrt{5}i - 9\sqrt{5}i + 4\sqrt{5}i$$
$$= -3\sqrt{5}i$$

45. By writing the numbers in complex form, we have:
$$1 + \sqrt{-36}\sqrt{-36} = 1 + (6i)(6i) = 1 + 36i^2 = -35$$

47. By writing the numbers in complex form, we have:
$$3\sqrt{-128} - 4\sqrt{-18} = 3\sqrt{-64}\sqrt{2} - 4\sqrt{-9}\sqrt{2}$$
$$= 3(8i)\sqrt{2} - 4(3i)\sqrt{2}$$
$$= 24\sqrt{2}i - 12\sqrt{2}i$$
$$= 12\sqrt{2}i$$

49.   (a) Notice that the result does agree with the definition:
$$z + w = (a + bi) + (c + di) = (a + c) + (b + d)i$$

(b) Notice that the result does agree with the definition:
$$z - w = (a + bi) - (c + di) = (a - c) + (b - d)i$$

(c) Notice that the result does agree with the definition:
$$zw = (a + bi)(c + di) = ac + bci + adi + bdi^2 = (ac - bd) + (bc + ad)i$$

(d) Notice that the result does agree with the definition:
$$\frac{z}{w} = \frac{a + bi}{c + di} \cdot \frac{c - di}{c - di} = \frac{ac + bci - adi - bdi^2}{c^2 - d^2i^2} = \frac{ac + bd}{c^2 + d^2} + \frac{bc - ad}{c^2 + d^2}i$$

51.   (a) We compute each power:

$$z^3 = \left(\frac{-1 + \sqrt{3}i}{2}\right)^3$$
$$= \frac{(-1 + \sqrt{3}i)^3}{8}$$
$$= \frac{(-1 + \sqrt{3}i)(-1 + \sqrt{3}i)^2}{8}$$
$$= \frac{(-1 + \sqrt{3}i)(1 - 2\sqrt{3}i + 3i^2)}{8}$$
$$= \frac{(-1 + \sqrt{3}i)(-2 - 2\sqrt{3}i)}{8}$$
$$= \frac{2 - 2\sqrt{3}i + 2\sqrt{3}i - 6i^2}{8}$$
$$= \frac{8}{8}$$
$$= 1$$

$$w^3 = \left(\frac{-1 - \sqrt{3}i}{2}\right)^3$$
$$= \frac{(-1 - \sqrt{3}i)^3}{8}$$
$$= \frac{(-1 - \sqrt{3}i)(-1 - \sqrt{3}i)^2}{8}$$
$$= \frac{(-1 - \sqrt{3}i)(1 + 2\sqrt{3}i + 3i^2)}{8}$$
$$= \frac{(-1 - \sqrt{3}i)(-2 + 2\sqrt{3}i)}{8}$$
$$= \frac{2 + 2\sqrt{3}i - 2\sqrt{3}i - 6i^2}{8}$$
$$= \frac{8}{8}$$
$$= 1$$

(b) We compute the product:
$$zw = \left(\frac{-1 + \sqrt{3}i}{2}\right)\left(\frac{-1 - \sqrt{3}i}{2}\right) = \frac{1 - \sqrt{3}i + \sqrt{3}i - 3i^2}{4} = \frac{4}{4} = 1$$

(c) We compute each power:
$$w^2 = \left(\frac{-1 - \sqrt{3}i}{2}\right)^2 = \frac{1 + 2\sqrt{3}i + 3i^2}{4} = \frac{-2 + 2\sqrt{3}i}{4} = \frac{-1 + \sqrt{3}i}{2} = z$$
$$z^2 = \left(\frac{-1 + \sqrt{3}i}{2}\right)^2 = \frac{1 - 2\sqrt{3}i + 3i^2}{4} = \frac{-2 - 2\sqrt{3}i}{4} = \frac{-1 - \sqrt{3}i}{2} = w$$

Note: Another approach (using parts (a) and (b)) is to recognize:

$$z^3 = zw, \text{ so } z^2 = \frac{z^3}{z} = \frac{zw}{z} = w$$

$$w^3 = zw, \text{ so } w^2 = \frac{w^3}{w} = \frac{zw}{w} = z$$

(d) We compute the product:
$$\left(1 - z + z^2\right)\left(1 + z - z^2\right) = (1 - z + w)(1 + z - w)$$
$$= \left(\frac{2}{2} - \frac{-1 + \sqrt{3}i}{2} + \frac{-1 - \sqrt{3}i}{2}\right)\left(\frac{2}{2} + \frac{-1 + \sqrt{3}i}{2} - \frac{-1 - \sqrt{3}i}{2}\right)$$
$$= \left(\frac{2 - 2\sqrt{3}i}{2}\right)\left(\frac{2 + 2\sqrt{3}i}{2}\right)$$
$$= \left(1 - \sqrt{3}i\right)\left(1 + \sqrt{3}i\right)$$
$$= 1 - 3i^2$$
$$= 4$$

53.   (a) Let $z = a + bi$, then:
$$0 + z = (0 + 0i) + (a + bi) = a + bi = z$$
$$z + 0 = (a + bi) + (0 + 0i) = a + bi = z$$

   (b) Let $z = a + bi$, then:
$$0 \cdot z = (0 + 0i)(a + bi) = 0 + 0i = 0$$
$$z \cdot 0 = (a + bi)(0 + 0i) = 0$$

55.   (a) We compute each sum:
$$z + w = (a + bi) + (c + di) = (a + c) + (b + d)i$$
$$w + z = (c + di) + (a + bi) = (c + a) + (d + b)i = z + w$$

   (b) We compute each product:
$$zw = (a + bi)(c + di) = ac + bci + adi + bdi^2 = (ac - bd) + (bc + ad)i$$
$$wz = (c + di)(a + bi) = ac + adi + bci + bdi^2 = (ac - bd) + (bc + ad)i = zw$$

57.   We compute each quotient then add the complex numbers:
$$\frac{a + bi}{a - bi} + \frac{a - bi}{a + bi} = \frac{a + bi}{a - bi} \cdot \frac{a + bi}{a + bi} + \frac{a - bi}{a + bi} \cdot \frac{a - bi}{a - bi}$$
$$= \frac{(a + bi)^2}{a^2 + b^2} + \frac{(a - bi)^2}{a^2 + b^2}$$
$$= \frac{a^2 + 2abi - b^2 + a^2 - 2abi - b^2}{a^2 + b^2}$$
$$= \frac{2a^2 - 2b^2}{a^2 + b^2}$$

Thus the real part is $\frac{2a^2 - 2b^2}{a^2 + b^2}$, and the imaginary part is 0.

59.   We compute each quotient then add the complex numbers:

$$\frac{(a+bi)^2}{a-bi} - \frac{(a-bi)^2}{a+bi} = \frac{(a+bi)^3 - (a-bi)^3}{(a-bi)(a+bi)}$$

$$= \frac{\left(a^3 + 3a^2bi + 3ab^2i^2 + b^3i^3\right) - \left(a^3 - 3a^2bi + 3ab^2i^2 - b^3i^3\right)}{a^2 - b^2i^2}$$

$$= \frac{\left(a^3 - 3ab^2\right) + \left(3a^2b - b^3\right)i - \left(a^3 - 3ab^2\right) + \left(3a^2b - b^3\right)}{a^2 + b^2}$$

$$= \frac{6a^2b - 2b^3}{a^2 + b^2}i$$

Thus the real part is 0.

61.   We will first compute zw:

$$zw = (a+bi)(c+di) = ac + bci + adi + bdi^2 = (ac - bd) + (bc + ad)i$$

Since $zw = 0$, then $ac - bd = 0$ and $bc + ad = 0$. Now assuming $a \neq 0$, we can solve the first equation for c to obtain $c = \frac{bd}{a}$. Substituting into the second equation yields:

$$b\left(\frac{bd}{a}\right) + ad = 0$$
$$b^2d + a^2d = 0$$
$$\left(b^2 + a^2\right)d = 0$$

But if $a \neq 0$, then $b^2 + a^2 \neq 0$ and thus $d = 0$. Since $c = \frac{bd}{a}$, then $c = 0$. Thus $w = 0$.

11.2 Division of Polynomials

1.   We use long division:

```
              x − 5
      ┌──────────────────
x − 3 │  x² − 8x + 4
         x² − 3x
         ─────────
              − 5x + 4
              − 5x + 15
              ─────────
                  −11
```

The quotient is $x - 5$ and the remainder is -11. We write the equation:
$$x^2 - 8x + 4 = (x - 3)(x - 5) - 11$$

3.   We use long division:

```
              x − 11
      ┌──────────────────
x + 5 │  x² − 6x − 2
         x² + 5x
         ─────────
              − 11x − 2
              − 11x − 55
              ─────────
                   53
```

The quotient is $x - 11$ and the remainder is 53. We write the equation:
$$x^2 - 6x - 2 = (x + 5)(x - 11) + 53$$

5.   We use long division:

$$
\begin{array}{r}
3x^2 - \frac{3}{2}x - \frac{1}{4} \\
2x+1\overline{\smash{\big)}\,6x^3 + 0x^2 - 2x + 3} \\
\underline{6x^3 + 3x^2} \\
-3x^2 - 2x \\
\underline{-3x^2 - \frac{3}{2}x} \\
-\frac{1}{2}x + 3 \\
\underline{-\frac{1}{2}x - \frac{1}{4}} \\
\frac{13}{4}
\end{array}
$$

The quotient is $3x^2 - \frac{3}{2}x - \frac{1}{4}$ and the remainder is $\frac{13}{4}$. We write the equation:

$$6x^3 - 2x + 3 = (2x+1)\left(3x^2 - \frac{3}{2}x - \frac{1}{4}\right) + \frac{13}{4}$$

7.   We use long division:

$$
\begin{array}{r}
x^4 - 3x^3 + 9x^2 - 27x + 81 \\
x+3\overline{\smash{\big)}\,x^5 + 0x^4 + 0x^3 + 0x^2 + 0x + 2} \\
\underline{x^5 + 3x^4} \\
-3x^4 + 0x^3 \\
\underline{-3x^4 - 9x^3} \\
9x^3 + 0x^2 \\
\underline{9x^3 + 27x^2} \\
-27x^2 + 0x \\
\underline{-27x^2 - 81x} \\
81x + 2 \\
\underline{81x + 243} \\
-241
\end{array}
$$

The quotient is $x^4 - 3x^3 + 9x^2 - 27x + 81$ and the remainder is -241. We write the equation:

$$x^5 + 2 = (x+3)\left(x^4 - 3x^3 + 9x^2 - 27x + 81\right) - 241$$

9.   We use long division:

$$
\begin{array}{r}
x^5 + 2x^4 + 4x^3 + 8x^2 + 16x + 32 \\[2pt]
x-2 \,\overline{\big)\; x^6 + 0x^5 + 0x^4 + 0x^3 + 0x^2 + 0x - 64} \\
\underline{x^6 - 2x^5} \\
2x^5 + 0x^4 \\
\underline{2x^5 - 4x^4} \\
4x^4 + 0x^3 \\
\underline{4x^4 - 8x^3} \\
8x^3 + \; 0x^2 \\
\underline{8x^3 - 16x^2} \\
16x^2 + \; 0x \\
\underline{16x^2 - 32x} \\
32x - 64 \\
\underline{32x - 64} \\
0
\end{array}
$$

The quotient is $x^5 + 2x^4 + 4x^3 + 8x^2 + 16x + 32$ and the remainder is 0. We write the equation:

$$x^6 - 64 = (x-2)\left(x^5 + 2x^4 + 4x^3 + 8x^2 + 16x + 32\right) + 0$$

11.   We use long division:

$$
\begin{array}{r}
5x^2 + 15x + 17 \\[2pt]
x^2 - 3x + 5 \,\overline{\big)\; 5x^4 + \; 0x^3 - \; 3x^2 + \; 0x + 2} \\
\underline{5x^4 - 15x^3 + 25x^2} \\
15x^3 - 28x^2 + \; 0x \\
\underline{15x^3 - 45x^2 + 75x} \\
17x^2 - 75x + 2 \\
\underline{17x^2 - 51x + 85} \\
-24x - 83
\end{array}
$$

The quotient is $5x^2 + 15x + 17$ and the remainder is $-24x - 83$. We write the equation:

$$5x^4 - 3x^2 + 2 = \left(x^2 - 3x + 5\right)\left(5x^2 + 15x + 17\right) + (-24x - 83)$$

13.   We use long division:

$$
\begin{array}{r}
3y - 19 \\[2pt]
y^2 + 5y + 2 \,\overline{\big)\; 3y^3 - \; 4y^2 + \; 0y - 3} \\
\underline{3y^3 + 15y^2 + \; 6y} \\
-19y^2 - \; 6y - 3 \\
\underline{-19y^2 - 95y - 38} \\
89y + 35
\end{array}
$$

The quotient is $3y - 19$ and the remainder is $89y + 35$. We write the equation:

$$3y^3 - 4y^2 - 3 = (y^2 + 5y + 2)(3y - 19) + (89y + 35)$$

15. We use long division:

$$
\begin{array}{r}
t^2 - 2t - 4 \\
t^2 - 2t + 4 \overline{\smash{\big)}\ t^4 - 4t^3 + 4t^2 + 0t - 16} \\
\underline{t^4 - 2t^3 + 4t^2} \\
-2t^3 + 0t^2 + 0t \\
\underline{-2t^3 + 4t^2 - 8t} \\
-4t^2 + 8t - 16 \\
\underline{-4t^2 + 8t - 16} \\
0
\end{array}
$$

The quotient is $t^2 - 2t - 4$ and the remainder is 0. We write the equation:

$$t^4 - 4t^3 + 4t^2 - 16 = (t^2 - 2t + 4)(t^2 - 2t - 4) + 0$$

17. We use long division:

$$
\begin{array}{r}
z^4 + z^3 + z^2 + z + 1 \\
z - 1 \overline{\smash{\big)}\ z^5 + 0z^4 + 0z^3 + 0z^2 + 0z - 1} \\
\underline{z^5 - 1z^4} \\
z^4 + 0z^3 \\
\underline{z^4 - 1z^3} \\
z^3 + 0z^2 \\
\underline{z^3 - 1z^2} \\
z^2 + 0z \\
\underline{z^2 - 1z} \\
z - 1 \\
\underline{z - 1} \\
0
\end{array}
$$

The quotient is $z^4 + z^3 + z^2 + z + 1$ and the remainder is 0. We write the equation:

$$z^5 - 1 = (z - 1)(z^4 + z^3 + z^2 + z + 1) + 0$$

19. We use long division:

$$
\begin{array}{r}
ax + (b + ar) \\
x - r \overline{\smash{\big)}\ ax^2 + bx + c} \\
\underline{ax^2 - arx} \\
(b + ar)x + c \\
\underline{(b + ar)x - r(b + ar)} \\
c + r(b + ar)
\end{array}
$$

The quotient is $ax + (b + ar)$ and the remainder is $c + r(b + ar) = ar^2 + br + c$. We write the equation:

$$ax^2 + bx + c = (x - r)(ax + (b + ar)) + (ar^2 + br + c)$$

21. Using synthetic division:

$$
\begin{array}{r|rrr}
5 & 1 & -6 & -2 \\
  &   & 5 & -5 \\
\hline
  & 1 & -1 & -7
\end{array}
$$

The quotient is $x - 1$ and the remainder is -7. So $x^2 - 6x - 2 = (x - 5)(x - 1) - 7$.

23. Using synthetic division:

$$
\begin{array}{r|rrr}
-1 & 4 & -1 & -5 \\
   &   & -4 & 5 \\
\hline
   & 4 & -5 & 0
\end{array}
$$

The quotient is $4x - 5$ and the remainder is 0. So $4x^2 - x - 5 = (x + 1)(4x - 5) + 0$.

25. Using synthetic division:

$$
\begin{array}{r|rrrr}
4 & 6 & -5 & 2 & 1 \\
  &   & 24 & 76 & 312 \\
\hline
  & 6 & 19 & 78 & 313
\end{array}
$$

The quotient is $6x^2 + 19x + 78$ and the remainder is 313.
So $6x^3 - 5x^2 + 2x + 1 = (x - 4)(6x^2 + 19x + 78) + 313$.

27. Using synthetic division:

$$
\begin{array}{r|rrrr}
2 & 1 & 0 & 0 & -1 \\
  &   & 2 & 4 & 8 \\
\hline
  & 1 & 2 & 4 & 7
\end{array}
$$

The quotient is $x^2 + 2x + 4$ and the remainder is 7. So $x^3 - 1 = (x - 2)(x^2 + 2x + 4) + 7$.

29. Using synthetic division:

$$
\begin{array}{r|rrrrrr}
-2 & 1 & 0 & 0 & 0 & 0 & -1 \\
   &   & -2 & 4 & -8 & 16 & -32 \\
\hline
   & 1 & -2 & 4 & -8 & 16 & -33
\end{array}
$$

The quotient is $x^4 - 2x^3 + 4x^2 - 8x + 16$ and the remainder is -33.
So $x^5 - 1 = (x + 2)(x^4 - 2x^3 + 4x^2 - 8x + 16) - 33$.

31. Using synthetic division:

$$
\begin{array}{r|rrrrr}
-4 & 1 & -6 & 0 & 0 & 2 \\
   &   & -4 & 40 & -160 & 640 \\
\hline
   & 1 & -10 & 40 & -160 & 642
\end{array}
$$

The quotient is $x^3 - 10x^2 + 40x - 160$ and the remainder is 642.
So $x^4 - 6x^3 + 2 = (x + 4)(x^3 - 10x^2 + 40x - 160) + 642$.

33. Using synthetic division:

$$
\begin{array}{r|rrrr}
10 & 1 & -4 & -3 & 6 \\
   &   & 10 & 60 & 570 \\
\hline
   & 1 & 6 & 57 & 576
\end{array}
$$

The quotient is $x^2 + 6x + 57$ and the remainder is 576.
So $x^3 - 4x^2 - 3x + 6 = (x - 10)(x^2 + 6x + 57) + 576$.

35. Using synthetic division:

$$\begin{array}{r|rrrr} -5 & 1 & -1 & 0 & 0 \\ & & -5 & 30 & -150 \\ \hline & 1 & -6 & 30 & -150 \end{array}$$

The quotient is $x^2 - 6x + 30$ and the remainder is -150.
So $x^3 - x^2 = (x + 5)(x^2 - 6x + 30) - 150$.

37. Using synthetic division:

$$\begin{array}{r|rrrr} 2/3 & 54 & -27 & -27 & 14 \\ & & 36 & 6 & -14 \\ \hline & 54 & 9 & -21 & 0 \end{array}$$

The quotient is $54x^2 + 9x - 21$ and the remainder is 0.
So $54x^3 - 27x^2 - 27x + 14 = \left(x - \frac{2}{3}\right)\left(54x^2 + 9x - 21\right) + 0$.

39. Using synthetic division:

$$\begin{array}{r|rrrrr} 3 & 1 & 0 & 3 & 0 & 12 \\ & & 3 & 9 & 36 & 108 \\ \hline & 1 & 3 & 12 & 36 & 120 \end{array}$$

The quotient is $x^3 + 3x^2 + 12x + 36$ and the remainder is 120.
So $x^4 + 3x^2 + 12 = (x - 3)(x^3 + 3x^2 + 12x + 36) + 120$.

41. Since $\sqrt[5]{32} = 2$ is a root, we use synthetic division:

$$\begin{array}{r|rrrrrr} 2 & 1 & 0 & 0 & 0 & 0 & -32 \\ & & 2 & 4 & 8 & 16 & 32 \\ \hline & 1 & 2 & 4 & 8 & 16 & 0 \end{array}$$

So $x^5 - 32 = (x - 2)\left(x^4 + 2x^3 + 4x^2 + 8x + 16\right)$.

43. Since $\sqrt[4]{81} = 3$ is a root, we use synthetic division:

$$\begin{array}{r|rrrrr} 3 & 1 & 0 & 0 & 0 & -81 \\ & & 3 & 9 & 27 & 81 \\ \hline & 1 & 3 & 9 & 27 & 0 \end{array}$$

So $z^4 - 81 = (z - 3)\left(z^3 + 3z^2 + 9z + 27\right)$.

45. Since the two quotients are the same, we can now perform synthetic division:

$$\begin{array}{r|rrr} 4/3 & 2 & -8/3 & 1/3 \\ & & 8/3 & 0 \\ \hline & 2 & 0 & 1/3 \end{array}$$

The quotient is $2x$ and the remainder is $\frac{1}{3}$. So the original problem has a quotient of $2x$ and a remainder of $3\left(\frac{1}{3}\right) = 1$.

47. Adapting the hint from Exercise 45, we first divide the numerator and denominator by 2 to form the quotient:

$$\frac{3x^3 + \frac{1}{2}}{x + \frac{1}{2}}$$

Using synthetic division:

$$
\begin{array}{r|rrrr}
-1/2 & 3 & 0 & 0 & 1/2 \\
     &   & -3/2 & 3/4 & -3/8 \\
\hline
     & 3 & -3/2 & 3/4 & 1/8
\end{array}
$$

The quotient is $3x^2 - \frac{3}{2}x + \frac{3}{4}$ and the remainder is $\frac{1}{8}$. So the original problem has a quotient of $3x^2 - \frac{3}{2}x + \frac{3}{4}$ and a remainder of $2\left(\frac{1}{8}\right) = \frac{1}{4}$.

49. Using synthetic division:

$$
\begin{array}{r|rrrr}
-1 & 1 & 0 & k & 1 \\
   &   & -1 & 1 & -k-1 \\
\hline
   & 1 & -1 & k+1 & -k
\end{array}
$$

So $-k = -4$, and thus $k = 4$.

51. Using synthetic division:

$$
\begin{array}{r|rrr}
p & 1 & 2p & -3q^2 \\
  &   & p & 3p^2 \\
\hline
  & 1 & 3p & 3p^2 - 3q^2
\end{array}
$$

Since the remainder is 0, we have:
$$3p^2 - 3q^2 = 0$$
$$3p^2 = 3q^2$$
$$p^2 = q^2$$

53. Using synthetic division:

$$
\begin{array}{r|rrr}
i & 1 & -4 & 1 \\
  &   & i & -1-4i \\
\hline
  & 1 & -4+i & -4i
\end{array}
$$

The quotient is $x + (-4 + i)$ and the remainder is $-4i$.

55. Using synthetic division:

$$
\begin{array}{r|rrr}
1+i & 1 & -2 & 2 \\
    &   & 1+i & -2 \\
\hline
    & 1 & -1+i & 0
\end{array}
$$

The quotient is $x + (-1 + i)$ and the remainder is 0.

57. We know that:
$$f(x) = d(x) \cdot q(x) + R(x)$$
Since $d(\sqrt{3}) = 3 - 3 = 0$, then:
$$f(\sqrt{3}) = R(\sqrt{3}) = -6\sqrt{3} + 57$$

59. Using synthetic division:

$$\begin{array}{r|rrrrrr} a & 1 & 0 & 0 & 0 & -5a^4 & 4a^5 \\ & & a & a^2 & a^3 & a^4 & -4a^5 \\ \hline & 1 & a & a^2 & a^3 & -4a^4 & 0 \end{array}$$

The remainder is 0.

11.3 The Remainder Theorem and the Factor Theorem

1. We substitute $x = 10$:
$$12(10) - 8 = 112$$
$$120 - 8 = 112$$
Yes, it is a root.

3. We substitute $x = 1 - \sqrt{5}$:
$$\left(1 - \sqrt{5}\right)^2 - 2\left(1 - \sqrt{5}\right) - 4 = 0$$
$$1 - 2\sqrt{5} + 5 - 2 + 2\sqrt{5} - 4 = 0$$
$$4 - 4 = 0$$
Yes, it is a root.

5. We substitute $x = \frac{1}{2}$:
$$2\left(\tfrac{1}{2}\right)^2 - 3\left(\tfrac{1}{2}\right) + 1 = 0$$
$$\tfrac{1}{2} - \tfrac{3}{2} + 1 = 0$$
$$-1 + 1 = 0$$
Yes, it is a root.

7. We compute $f\left(\tfrac{2}{3}\right)$:
$$f\left(\tfrac{2}{3}\right) = 3\left(\tfrac{2}{3}\right) - 2 = 2 - 2 = 0$$
So $x = \tfrac{2}{3}$ is a zero of $f(x)$.

9. We compute $h(-1)$:
$$h(-1) = 5(-1)^3 - (-1)^2 + 2(-1) + 8 = -5 - 1 - 2 + 8 = 0$$
So $x = -1$ is a zero of $h(x)$.

11. We compute $f(2)$:
$$f(2) = 1 + 2(2) + (2)^3 - (2)^5 = 1 + 4 + 8 - 32 = -19$$
So $t = 2$ is not a zero of $f(t)$.

13. (a) We compute $f\left(\frac{\sqrt{3}-1}{2}\right)$:

$$f\left(\frac{\sqrt{3}-1}{2}\right) = 2\left(\frac{\sqrt{3}-1}{2}\right)^3 - 3\left(\frac{\sqrt{3}-1}{2}\right) + 1 = \frac{3\sqrt{3}-5}{2} - \frac{3\sqrt{3}-3}{2} + 1 = -1+1 = 0$$

So $x = \frac{\sqrt{3}-1}{2}$ is a zero of $f(x)$.

(b) We compute $f\left(\frac{\sqrt{3}+1}{2}\right)$:

$$f\left(\frac{\sqrt{3}+1}{2}\right) = 2\left(\frac{\sqrt{3}+1}{2}\right)^3 - 3\left(\frac{\sqrt{3}+1}{2}\right) + 1 = \frac{5+3\sqrt{3}}{2} - \frac{3\sqrt{3}-3}{2} + 1 = 4+1 = 5$$

So $x = \frac{\sqrt{3}+1}{2}$ is not a zero of $f(x)$.

15. (a) 1, 2 (multiplicity 3), 3
    (b) 1 (multiplicity 3)
    (c) 5 (multiplicity 6), −1 (multiplicity 4)
    (d) 0 (multiplicity 5), 1

17. Using synthetic division:

$$\begin{array}{r|rrrr}
-3 & 4 & -6 & 1 & -5 \\
   &   & -12 & 54 & -165 \\
\hline
   & 4 & -18 & 55 & -170
\end{array}$$

So $f(-3) = -170$.

19. Using synthetic division:

$$\begin{array}{r|rrrrr}
1/2 & 6 & 5 & -8 & -10 & -3 \\
    &   & 3 & 4 & -2 & -6 \\
\hline
    & 6 & 8 & -4 & -12 & -9
\end{array}$$

So $f\left(\tfrac{1}{2}\right) = -9$.

21. Using synthetic division:

$$\begin{array}{r|rrr}
-\sqrt{2} & 1 & 3 & -4 \\
          &   & -\sqrt{2} & -3\sqrt{2}+2 \\
\hline
          & 1 & 3-\sqrt{2} & -3\sqrt{2}-2
\end{array}$$

So $f\left(-\sqrt{2}\right) = -3\sqrt{2} - 2$.

23. Using synthetic division:

$$\begin{array}{r|rrrr}
12 & 1/2 & -5 & -13 & -10 \\
   &   & 6 & 12 & -12 \\
\hline
   & 1/2 & 1 & -1 & -22
\end{array}$$

So $f(12) = -22$.

25. Using synthetic division:

$$
\begin{array}{r|rrrr}
-3 & 1 & -4 & -9 & 36 \\
   &   & -3 & 21 & -36 \\
\hline
   & 1 & -7 & 12 & 0
\end{array}
$$

So $x^3 - 4x^2 - 9x + 36 = (x+3)(x^2 - 7x + 12) = (x+3)(x-4)(x-3)$.

So the roots are $\pm 3, 4$.

27. Using synthetic division:

$$
\begin{array}{r|rrrr}
1 & 1 & 1 & -7 & 5 \\
  &   & 1 & 2 & -5 \\
\hline
  & 1 & 2 & -5 & 0
\end{array}
$$

So $x^3 + x^2 - 7x + 5 = (x-1)(x^2 + 2x - 5)$. So $x = 1$ is a root. We use the quadratic formula:

$$
x = \frac{-2 \pm \sqrt{4 + 20}}{2} = \frac{-2 \pm 2\sqrt{6}}{2} = -1 \pm \sqrt{6}
$$

So the roots are $1, -1 \pm \sqrt{6}$.

29. Using synthetic division:

$$
\begin{array}{r|rrrr}
-2 & 3 & -5 & -16 & 12 \\
   &   & -6 & 22 & -12 \\
\hline
   & 3 & -11 & 6 & 0
\end{array}
$$

So $3x^3 - 5x^2 - 16x + 12 = (x+2)(3x^2 - 11x + 6) = (x+2)(3x-2)(x-3)$.

So the roots are $-2, \frac{2}{3}$, and 3.

31. Using synthetic division:

$$
\begin{array}{r|rrrr}
-3/2 & 2 & 1 & -5 & -3 \\
     &   & -3 & 3 & 3 \\
\hline
     & 2 & -2 & -2 & 0
\end{array}
$$

So $2x^3 + x^2 - 5x - 3 = \left(x + \frac{3}{2}\right)(2x^2 - 2x - 2) = 2\left(x + \frac{3}{2}\right)(x^2 - x - 1)$.

So $x = -\frac{3}{2}$ is a root. We use the quadratic formula:

$$
x = \frac{1 \pm \sqrt{1+4}}{2} = \frac{1 \pm \sqrt{5}}{2}
$$

So the roots are $-\frac{3}{2}, \frac{1 \pm \sqrt{5}}{2}$.

33. Using synthetic division:

$$
\begin{array}{r|rrrrr}
5 & 1 & -15 & 75 & -125 & 0 \\
  &   & 5   & -50 & 125 & 0 \\
\hline
  & 1 & -10 & 25 & 0 & 0
\end{array}
$$

Therefore:

$$x^4 - 15x^3 + 75x^2 - 125x = (x-5)(x^3 - 10x^2 + 25x)$$
$$= x(x-5)(x^2 - 10x + 25)$$
$$= x(x-5)^3$$

So the roots are 0 and 5.

35. Using synthetic division:

```
-4 |  1     2    -23    -24    144
   |       -4      8     60   -144
   ----------------------------------
      1    -2    -15     36      0
```

Use synthetic division again:

```
3 |  1    -2    -15     36
  |        3      3    -36
  --------------------------
     1     1    -12      0
```

So $x^4 + 2x^3 - 23x^2 - 24x + 144 = (x+4)(x-3)(x^2 + x - 12) = (x+4)^2(x-3)^2$.
So the roots are -4 and 3.

37. Using synthetic division:

```
-9 |  1     7    -19    -9
   |       -9     18     9
   -------------------------
      1    -2     -1     0
```

So $x^3 + 7x^2 - 19x - 9 = (x+9)(x^2 - 2x - 1)$. So $x = -9$ is a root. Use the quadratic formula:

$$x = \frac{2 \pm \sqrt{4+4}}{2} = \frac{2 \pm 2\sqrt{2}}{2} = 1 \pm \sqrt{2}$$

So the roots are -9 and $1 \pm \sqrt{2}$.

39. Using long division:

$$
\begin{array}{r}
2x^2 - 3x - 1 \\
2x^2 - 3x - 1 \overline{\big)\, 4x^4 - 12x^3 + 5x^2 + 6x + 1} \\
\underline{4x^4 - 6x^3 - 2x^2} \\
-6x^3 + 7x^2 + 6x \\
\underline{-6x^3 + 9x^2 + 3x} \\
-2x^2 + 3x + 1 \\
\underline{-2x^2 + 3x + 1} \\
0
\end{array}
$$

We use the quadratic formula to solve $2x^2 - 3x - 1 = 0$:

$$x = \frac{3 \pm \sqrt{9+8}}{4} = \frac{3 \pm \sqrt{17}}{4}$$

The solutions are $\frac{3 \pm \sqrt{17}}{4}$ (each of multiplicity 2).

41. (a) $R\left(\frac{1}{2}\right) = f\left(\frac{1}{2}\right) = 1.125$

    (b) $R(1.25) = f(1.25) = -0.046875$

    (c) Since $f(1) = 0$ then $t - 1$ is a linear factor of $f(t)$.

    (d) Since 1 is a solution, we use synthetic division:

$$\begin{array}{r|rrrr} 1 & 1 & 0 & -4 & 3 \\ & & 1 & 1 & -3 \\ \hline & 1 & 1 & -3 & 0 \end{array}$$

Using the quadratic formula to solve $t^2 + t - 3 = 0$:
$$t = \frac{-1 \pm \sqrt{1+12}}{2} = \frac{-1 \pm \sqrt{13}}{2}$$
So the solutions are $1, \frac{-1\pm\sqrt{13}}{2}$.

43. We must have $f(x) = (x - 3)(x - 5)(x + 4) = 0$. Multiplying out, we have
$x^3 - 4x^2 - 17x + 60 = 0$.

45. We must have $f(x) = (x + 1)^2(x + 6) = 0$. Multiplying out, we have $x^3 + 8x^2 + 13x + 6 = 0$.

47. Since $\frac{1}{2}$ is not a root of $x^2 - 3x - 4$, we know that such an equation must have
$f(x) = \left(x^2 - 3x - 4\right)\left(x - \frac{1}{2}\right)^3$, which has degree 5. Thus no such polynomial of degree 4 exists.

49. We must have $x^2 - 3x + 1$ and $x + 6$ as factors, and thus we can write
$f(x) = (x^2 - 3x + 1)(x + 6)(ax + b) = 0$, for some real numbers a and b, where $a \neq 0$.
Multiplying out, we have $ax^4 + (3a + b)x^3 + (-17a + 3b)x^2 + (6a - 17b)x + 6b = 0$.

51. Using synthetic division:

$$\begin{array}{r|rrrr} 1.16 & 1 & -3 & 12 & 9 \\ & & 1.16 & -2.1344 & 11.44 \\ \hline & 1 & -1.84 & 9.8656 & 20.44 \end{array}$$

So $f(1.16)$ is approximately equal to 20.44.

53. (a) Using synthetic division:

$$\begin{array}{r|rrrr} 2.41 & 1 & 0 & -5 & -2 \\ & & 2.41 & 5.8081 & 1.9475 \\ \hline & 1 & 2.41 & 0.8081 & -0.05 \end{array}$$

So $f(2.41)$ is approximately equal to -0.05.

(b) Using synthetic division:

$$
\begin{array}{r|cccc}
2.42 & 1 & 0 & -5 & -2 \\
& & 2.42 & 5.8564 & 2.0725 \\
\hline
& 1 & 2.42 & 0.8564 & 0.07
\end{array}
$$

So $f(2.42)$ is approximately equal to 0.07.

55. (a) Since -5 is a zero, we use synthetic division:

$$
\begin{array}{r|cccc}
-5 & 1 & 1 & -18 & 10 \\
& & -5 & 20 & -10 \\
\hline
& 1 & -4 & 2 & 0
\end{array}
$$

So the remaining factor is $x^2 - 4x + 2 = 0$. Using the quadratic formula:

$$x = \frac{4 \pm \sqrt{16 - 4(2)}}{2(1)} = \frac{4 \pm \sqrt{8}}{2} = \frac{4 \pm 2\sqrt{2}}{2} = 2 \pm \sqrt{2}$$

So $x_2 = 2 - \sqrt{2}$ and $x_3 = 2 + \sqrt{2}$.

(b) The values are the same.

57. We evaluate the function:

$$
\begin{aligned}
f\left(-b - \sqrt{b^2 - 2c}\right) &= \tfrac{1}{2}\left(-b - \sqrt{b^2 - 2c}\right)^2 + b\left(-b - \sqrt{b^2 - 2c}\right) + c \\
&= \tfrac{1}{2}\left(b^2 + 2b\sqrt{b^2 - 2c} + b^2 - 2c\right) - b^2 - b\sqrt{b^2 - 2c} + c \\
&= b^2 + b\sqrt{b^2 - 2c} - c - b^2 - b\sqrt{b^2 - 2c} + c \\
&= 0
\end{aligned}
$$

So $x = -b - \sqrt{b^2 - 2c}$ is a zero of $f(x)$.

59. We evaluate the function:

$$
\begin{aligned}
F\left(\frac{-\sqrt{2} + \sqrt{2\sqrt{2} - 2}}{2}\right) &= 2\left(\frac{-\sqrt{2} + \sqrt{2\sqrt{2} - 2}}{2}\right)^4 + 4\left(\frac{-\sqrt{2} + \sqrt{2\sqrt{2} - 2}}{2}\right) + 1 \\
&= -1 + 2\sqrt{2} - 2\sqrt{2\sqrt{2} - 2} - 2\sqrt{2} + \sqrt{2\sqrt{2} - 2} + 1 \\
&= 0
\end{aligned}
$$

Yes, it is a zero.

61. Using synthetic division:

$$
\begin{array}{r|cccc}
1 & 1 & 1 & a & b \\
& & 1 & 2 & a+2 \\
\hline
& 1 & 2 & a+2 & a+b+2
\end{array}
$$

Using synthetic division again:

$$
\begin{array}{r|cccc}
1 & 1 & -1 & -a & b \\
& & 1 & 0 & -a \\
\hline
& 1 & 0 & -a & -a+b
\end{array}
$$

If 1 is a root, then:

$$a + b + 2 = 0$$
$$-a + b = 0$$

Adding these, we get $2b + 2 = 0$, so $b = -1$. Substituting we get $a = -1$.
So $a = -1$ and $b = -1$.

63. Let r_1 and r_2 be the roots, so $r_2 = 2r_1$. Then:
$$x^2 + bx + 1 = (x - r_1)(x - 2r_1) = x^2 - 3r_1 x + 2r_1^2$$
Since r_1 is a constant, then $-3r_1 = b$ and $2r_1^2 = 1$, so $r_1^2 = \frac{1}{2}$ and thus $r_1 = \frac{\pm\sqrt{2}}{2}$.
So $b = \frac{\pm 3\sqrt{2}}{2}$.

65. Let r denote the root with multiplicity 2. Using synthetic division:

$$
\begin{array}{r|cccc}
r & 1 & 0 & -12 & 16 \\
  &   & r & r^2 & r^3 - 12r \\
\hline
  & 1 & r & r^2 - 12 & r^3 - 12r + 16
\end{array}
$$

Since r is a double root, then the equation $x^2 + rx + (r^2 - 12) = 0$ must also have r as a root. Substituting $x = r$, we have:
$$r^2 + r^2 + r^2 - 12 = 0$$
$$3r^2 = 12$$
$$r^2 = 4$$
$$r = \pm 2$$
But $r = -2$ does not check in the original equation, thus $r = 2$. Thus the quadratic equation becomes $x^2 + 2x - 8 = 0$, which factors to $(x + 4)(x - 2) = 0$, and thus $x = -4$ is also a solution. So the solutions are 2 (multiplicity 2) and -4.

11.4 The Fundamental Theorem of Algebra

1. (a) yes
    (b) yes
    (c) yes
    (d) no--not a polynomial equation

3. $x^2 - 2x - 3 = (x + 1)(x - 3) = [x - (-1)](x - 3)$

5. $4x^2 + 23x - 6 = (4x - 1)(x + 6) = 4\left(x - \frac{1}{4}\right)[x - (-6)]$

7. $x^2 - 5 = (x + \sqrt{5})(x - \sqrt{5}) = [x - (-\sqrt{5})](x - \sqrt{5})$

9. Using the quadratic formula for $x^2 - 10x + 26 = 0$, we have:
$$x = \frac{10 \pm \sqrt{100 - 104}}{2} = \frac{10 \pm 2i}{2} = 5 \pm i$$
So $x^2 - 10x + 26 = [x - (5 + i)][x - (5 - i)]$

11. $f(x) = (x - 1)^2(x + 3) = x^3 + x^2 - 5x + 3$

13. $f(x) = (x - 2)(x + 2)(x - 2i)(x + 2i) = x^4 - 16$

15. $f(x) = (x - \sqrt{3})^2 [x - (-\sqrt{3})]^2 (x - 4i)[x - (-4i)] = x^6 + 10x^4 - 87x^2 + 144$

17. We can write $f(x)$ as $f(x) = a(x + 4)(x - 9)$, for some constant a. Since $f(3) = 5$, then:
$$5 = a(7)(-6)$$
$$a = -\tfrac{5}{42}$$
So $f(x) = -\tfrac{5}{42}(x^2 - 5x - 36) = -\tfrac{5}{42}x^2 + \tfrac{25}{42}x + \tfrac{30}{7}$.

19. We can write $f(x)$ as $f(x) = a(x + 5)(x - 2)(x - 3)$, for some constant a. Since $f(0) = 1$, then:
$$1 = a(5)(-2)(-3)$$
$$a = \tfrac{1}{30}$$
So $f(x) = \tfrac{1}{30}(x^3 - 19x + 30) = \tfrac{1}{30}x^3 - \tfrac{19}{30}x + 1$.

21. We have $-b = -i - \sqrt{3}$, so $b = i + \sqrt{3}$. Also $c = -i(-\sqrt{3}) = i\sqrt{3}$.
So $x^2 + (i + \sqrt{3})x + i\sqrt{3} = 0$.

23. We have $-b = 3$, so $b = -3$. Also $c = (9)(-6) = -54$. So $x^2 - 3x - 54 = 0$.

25. We have $-b = 2$, so $b = -2$. Also $c = (1 + \sqrt{5})(1 - \sqrt{5}) = -4$. So $x^2 - 2x - 4 = 0$.

27. We have $-B = 2a$, so $B = -2a$. Also $C = (a + \sqrt{b})(a - \sqrt{b}) = a^2 - b$.
So $x^2 - 2ax + a^2 - b = 0$.

29. Using the hint, we have:
$$x^4 + 64 = x^4 + 16x^2 + 64 - 16x^2$$
$$= (x^2 + 8)^2 - (4x)^2$$
$$= (x^2 + 8 + 4x)(x^2 + 8 - 4x)$$
$$= (x^2 + 4x + 8)(x^2 - 4x + 8)$$
Now $x^2 + 4x + 8 = 0$ when $x = \dfrac{-4 \pm \sqrt{16 - 32}}{2} = \dfrac{-4 \pm 4i}{2} = -2 \pm 2i$.

Also $x^2 - 4x + 8 = 0$ when $x = \dfrac{4 \pm \sqrt{16 - 32}}{2} = \dfrac{4 \pm 4i}{2} = 2 \pm 2i$.

Therefore:
$$x^4 + 64 = (x^2 + 4x + 8)(x^2 - 4x + 8)$$
$$= [x - (-2 + 2i)][x - (-2 - 2i)][x - (2 + 2i)][x - (2 - 2i)]$$
$$= (x + 2 - 2i)(x + 2 + 2i)(x - 2 - 2i)(x - 2 + 2i)$$

31. We know:
$$x^3 + bx^2 + cx + d = (x - r_1)(x - r_2)(x - r_3)$$
$$= x^3 - (r_1 + r_2 + r_3)x^2 + (r_1r_2 + r_1r_3 + r_2r_3)x - r_1r_2r_3$$

Therefore:
$$r_1 + r_2 + r_3 = -b$$
$$r_1r_2 + r_1r_3 + r_2r_3 = c$$
$$r_1r_2r_3 = -d$$

33. Let r_1, r_2 and r_3 be the roots and assume $r_2 = -r_1$ (their sum is therefore 0). The identities from Exercise 31 give us:

$$r_1 + r_2 + r_3 = 4 \qquad\qquad r_1r_2r_3 = -36$$
$$r_1 - r_1 + r_3 = 4 \qquad\qquad -r_1^2 \cdot 4 = -36$$
$$r_3 = 4 \qquad\qquad r_1^2 = 9$$
$$r_1 = \pm 3$$

So $r_1 = 3$ and $r_2 = -3$ (or switch them, it doesn't matter). So the roots are 4, 3, and –3.

35. (a) Using the hint:
$$\alpha^2 + \beta^2 = (\alpha + \beta)^2 - 2\alpha\beta = (-b)^2 - 2c = b^2 - 2c$$

(b) Using our result from (a):
$$\left(\frac{1}{\alpha}\right)^2 + \left(\frac{1}{\beta}\right)^2 = \frac{\alpha^2 + \beta^2}{(\alpha\beta)^2} = \frac{b^2 - 2c}{c^2}$$

(c) Using our result from (a):
$$\alpha^3 + \beta^3 = (\alpha + \beta)(\alpha^2 - \alpha\beta + \beta^2)$$
$$= (-b)(\alpha^2 + \beta^2 - \alpha\beta)$$
$$= (-b)(b^2 - 2c - c)$$
$$= -b(b^2 - 3c)$$

37. (a) Suppose $A \neq 0$ or $B \neq 0$. Then $f(x) = Ax^2 + Bx + C = 0$ can have at most two distinct roots by the Linear Factor Theorem. So $A = B = 0$. But this implies $C = 0$, and thus $f(x) = 0$ for all values of x.

(b) Using the hint, let:
$$f(x) = \frac{a^2 - x^2}{(a-b)(a-c)} + \frac{b^2 - x^2}{(b-c)(b-a)} + \frac{c^2 - x^2}{(c-a)(c-b)} - 1$$

$$f(a) = 0 + \frac{b^2 - a^2}{(b-c)(b-a)} + \frac{c^2 - a^2}{(c-a)(c-b)} - 1$$

$$= \frac{b+a}{b-c} + \frac{c+a}{c-b} - 1$$

$$= \frac{-b-a+c+a}{c-b} - 1$$

$$= \frac{c-b}{c-b} - 1$$

$$= 0$$

$$f(b) = \frac{a^2 - b^2}{(a-b)(a-c)} + 0 + \frac{c^2 - b^2}{(c-a)(c-b)} - 1$$

$$= \frac{a+b}{a-c} + \frac{c+b}{c-a} - 1$$

$$= \frac{-a-b+c+b}{c-a} - 1$$

$$= \frac{c-a}{c-a} - 1$$

$$= 0$$

$$f(c) = \frac{a^2 - c^2}{(a-b)(a-c)} + \frac{b^2 - c^2}{(b-c)(b-a)} + 0 - 1$$

$$= \frac{a+c}{a-b} + \frac{b+c}{b-a} - 1$$

$$= \frac{-a-c+b+c}{b-a} - 1$$

$$= \frac{b-a}{b-a} - 1$$

$$= 0$$

Since a, b and c are distinct (for denominators to be nonzero), and all three are roots, then by part (a) $f(x) = 0$ for all values of x. Thus the equation is an identity.

39. (a) Using $r_1 = \cos 72°$ and $r_2 = \cos 144°$, we are given that $r_1 + r_2 = -\frac{1}{2}$ and $r_1 r_2 = -\frac{1}{4}$, so $b = \frac{1}{2}$ and $c = -\frac{1}{4}$. Thus a quadratic equation is $x^2 + \frac{1}{2}x - \frac{1}{4} = 0$. A better form (with integer coefficients) is $4x^2 + 2x - 1 = 0$.

(b) Solving the equation $4x^2 + 2x - 1 = 0$ using the quadratic formula, we have:
$$x = \frac{-2 \pm \sqrt{4 - 4(4)(-1)}}{2(4)} = \frac{-2 \pm \sqrt{20}}{8} = \frac{-2 \pm 2\sqrt{5}}{8} = \frac{1}{4}\left(-1 \pm \sqrt{5}\right)$$
The smaller of these two roots is $\cos 144°$ (since it is negative), thus $\cos 144° = \frac{1}{4}\left(-1 - \sqrt{5}\right)$ and $\cos 72° = \frac{1}{4}\left(-1 + \sqrt{5}\right)$.

(c) Using a calculator verifies that $\cos 72° \approx 0.3090$ and $\cos 144° \approx -0.8090$.

11.5 Rational and Irrational Roots

1. p-factors: $\pm 1, \pm 3$
    q-factors: $\pm 1, \pm 2, \pm 4$
    possible rational roots $\frac{p}{q}$: $\pm 1, \pm \frac{1}{2}, \pm \frac{1}{4}, \pm 3, \pm \frac{3}{2}, \pm \frac{3}{4}$

3. p-factors: $\pm 1, \pm 3, \pm 9$
    q-factors: $\pm 1, \pm 2, \pm 4, \pm 8$
    possible rational roots $\frac{p}{q}$: $\pm 1, \pm \frac{1}{2}, \pm \frac{1}{4}, \pm \frac{1}{8}, \pm 3, \pm \frac{3}{2}, \pm \frac{3}{4}, \pm \frac{3}{8}, \pm 9, \pm \frac{9}{2}, \pm \frac{9}{4}, \pm \frac{9}{8}$

5. First multiply by 3 (we need integer coefficients) to get:
    $$2x^3 - 3x^2 - 15x + 6 = 0$$
    p-factors: $\pm 1, \pm 2, \pm 3, \pm 6$
    q-factors: $\pm 1, \pm 2$

    possible rational roots $\frac{p}{q}$: $\pm 1, \pm \frac{1}{2}, \pm 2, \pm 3, \pm \frac{3}{2}, \pm 6$

7. The possible rational roots are ± 1. Using synthetic division:

    $$\begin{array}{r|rrrr} 1 & 1 & 0 & -3 & 1 \\ & & 1 & 1 & -2 \\ \hline & 1 & 1 & -2 & -1 \end{array}$$

    So $x = 1$ is not a root.

    $$\begin{array}{r|rrrr} -1 & 1 & 0 & -3 & 1 \\ & & -1 & 1 & 2 \\ \hline & 1 & -1 & -2 & 3 \end{array}$$

    So $x = -1$ is not a root. There are no rational roots.

9. The possible rational roots are ± 1. Using synthetic division:

    $$\begin{array}{r|rrrr} 1 & 1 & 1 & -1 & 1 \\ & & 1 & 2 & 1 \\ \hline & 1 & 2 & 1 & 2 \end{array}$$

    So $x = 1$ is not a root.

    $$\begin{array}{r|rrrr} -1 & 1 & 1 & -1 & 1 \\ & & -1 & 0 & 1 \\ \hline & 1 & 0 & -1 & 2 \end{array}$$

    So $x = -1$ is not a root. There are no rational roots.

11. The possible rational roots are $\pm 1, \pm 2, \pm 4, \pm 8, \pm 16$. Using synthetic division:

    $$\begin{array}{r|rrrrr} 1 & 1 & 4 & 4 & 0 & -16 \\ & & 1 & 5 & 9 & 9 \\ \hline & 1 & 5 & 9 & 9 & -7 \end{array}$$

    So $x = 1$ is not a root.

$$
\begin{array}{r|rrrrr}
2 & 1 & 4 & 4 & 0 & -16 \\
  &   & 2 & 12 & 32 & 64 \\
\hline
  & 1 & 6 & 16 & 32 & 48
\end{array}
$$

So $x = 2$ is not a root. Since this row is all positive, we can also exclude $x = 4$, $x = 8$ and $x = 16$.

$$
\begin{array}{r|rrrrr}
-1 & 1 & 4 & 4 & 0 & -16 \\
   &   & -1 & -3 & -1 & 1 \\
\hline
   & 1 & 3 & 1 & -1 & -15
\end{array}
$$

So $x = -1$ is not a root.

$$
\begin{array}{r|rrrrr}
-2 & 1 & 4 & 4 & 0 & -16 \\
   &   & -2 & -4 & 0 & 0 \\
\hline
   & 1 & 2 & 0 & 0 & -16
\end{array}
$$

So $x = -2$ is not a root.

$$
\begin{array}{r|rrrrr}
-4 & 1 & 4 & 4 & 0 & -16 \\
   &   & -4 & 0 & -16 & 64 \\
\hline
   & 1 & 0 & 4 & -16 & 48
\end{array}
$$

So $x = -4$ is not a root. Since these signs alternate, we can also exclude $x = -8$ and $x = -16$. So there are no rational roots.

13. The possible rational roots are $\pm 1, \pm 3$. Using synthetic division:

$$
\begin{array}{r|rrrr}
1 & 1 & 3 & -1 & -3 \\
  &   & 1 & 4 & 3 \\
\hline
  & 1 & 4 & 3 & 0
\end{array}
$$

So $x^3 + 3x^2 - x - 3 = (x - 1)(x^2 + 4x + 3) = (x - 1)(x + 1)(x + 3)$.
So the roots are 1, -1, and -3.

15. The possible rational roots are $\pm 1, \pm \frac{1}{2}, \pm \frac{1}{4}, \pm 5, \pm \frac{5}{2}, \pm \frac{5}{4}$. Using synthetic division:

$$
\begin{array}{r|rrrr}
-1/4 & 4 & 1 & -20 & -5 \\
     &   & -1 & 0 & 5 \\
\hline
     & 4 & 0 & -20 & 0
\end{array}
$$

Therefore:
$$
\begin{aligned}
4x^3 + x^2 - 20x - 5 &= \left(x + \tfrac{1}{4}\right)\left(4x^2 - 20\right) \\
&= 4\left(x + \tfrac{1}{4}\right)\left(x^2 - 5\right) \\
&= 4\left(x + \tfrac{1}{4}\right)\left(x + \sqrt{5}\right)\left(x - \sqrt{5}\right)
\end{aligned}
$$

So the roots are $-\frac{1}{4}, -\sqrt{5}$ and $\sqrt{5}$.

17. The possible rational roots are $\pm 1, \pm \frac{1}{3}, \pm \frac{1}{9}, \pm 2, \pm \frac{2}{3}, \pm \frac{2}{9}$. Using synthetic division:

$$
\begin{array}{r|rrrr}
1 & 9 & 18 & 11 & 2 \\
  &   & 9 & 27 & 38 \\
\hline
  & 9 & 27 & 38 & 40
\end{array}
$$

So $x = 2$ is excluded also (the row is positive).

$$\begin{array}{r|rrrr} -1 & 9 & 18 & 11 & 2 \\ & & -9 & -9 & -2 \\ \hline & 9 & 9 & 2 & 0 \end{array}$$

So $9x^3 + 18x^2 + 11x + 2 = (x+1)(9x^2 + 9x + 2) = (x+1)(3x+2)(3x+1)$.

So the roots are $-1, -\frac{2}{3}$ and $-\frac{1}{3}$.

19. The possible rational roots are $\pm 1, \pm 2, \pm 3, \pm 4, \pm 6, \pm 8, \pm 12, \pm 24$. Using synthetic division:

$$\begin{array}{r|rrrrr} 1 & 1 & 1 & -25 & -1 & 24 \\ & & 1 & 2 & -23 & -24 \\ \hline & 1 & 2 & -23 & -24 & 0 \end{array}$$

So $x = 1$ is a root. Using synthetic division again:

$$\begin{array}{r|rrrr} -1 & 1 & 2 & -23 & -24 \\ & & -1 & -1 & 24 \\ \hline & 1 & 1 & -24 & 0 \end{array}$$

So $x = -1$ is also a root. So $x^4 + x^3 - 25x^2 - x + 24 = (x-1)(x+1)(x^2 + x - 24)$. Use the quadratic formula:

$$x = \frac{-1 \pm \sqrt{1+96}}{2} = \frac{-1 \pm \sqrt{97}}{2}$$

So the roots are $1, -1, \frac{-1+\sqrt{97}}{2}$ and $\frac{-1-\sqrt{97}}{2}$.

21. The possible rational roots are ± 1. Using synthetic division:

$$\begin{array}{r|rrrrr} 1 & 1 & -4 & 6 & -4 & 1 \\ & & 1 & -3 & 3 & -1 \\ \hline & 1 & -3 & 3 & -1 & 0 \end{array}$$

Using synthetic division again:

$$\begin{array}{r|rrrr} 1 & 1 & -3 & 3 & -1 \\ & & 1 & -2 & 1 \\ \hline & 1 & -2 & 1 & 0 \end{array}$$

So $x^4 - 4x^3 + 6x^2 - 4x + 1 = (x-1)^2(x^2 - 2x + 1) = (x-1)^4$.

So the only root is 1 (with multiplicity 4).

23. First multiply by 2 to get integer coefficients:
$$2x^3 - 5x^2 - 46x + 24 = 0$$

The possible rational roots are $\pm 1, \pm \frac{1}{2}, \pm 2, \pm 3, \pm \frac{3}{2}, \pm 4, \pm 6, \pm 8, \pm 12, \pm 24$. Using synthetic division:

$$\begin{array}{r|rrrr} 1/2 & 2 & -5 & -46 & 24 \\ & & 1 & -2 & -24 \\ \hline & 2 & -4 & -48 & 0 \end{array}$$

Therefore:
$$2x^3 - 5x^2 - 46x + 24 = \left(x - \tfrac{1}{2}\right)\left(2x^2 - 4x - 48\right)$$
$$= 2\left(x - \tfrac{1}{2}\right)\left(x^2 - 2x - 24\right)$$
$$= 2\left(x - \tfrac{1}{2}\right)(x - 6)(x + 4)$$

So the roots are $\tfrac{1}{2}, 6$ and -4.

25. First multiply by 20 to get integer coefficients:
$$40x^4 - 18x^3 - 58x^2 + 27x - 3 = 0$$

The possible rational roots are $\pm 1, \pm \tfrac{1}{2}, \pm \tfrac{1}{4}, \pm \tfrac{1}{5}, \pm \tfrac{1}{8}, \pm \tfrac{1}{10}, \pm \tfrac{1}{20}, \pm \tfrac{1}{40}, \pm 3, \pm \tfrac{3}{2}, \pm \tfrac{3}{4}, \pm \tfrac{3}{5},$
$\pm \tfrac{3}{8}, \pm \tfrac{3}{10}, \pm \tfrac{3}{20}, \pm \tfrac{3}{40}$. Using synthetic division:

```
1/4│  40    -18    -58     27    -3
              10    - 2    -15     3
      40    - 8    -60     12     0
```

Using synthetic division again:

```
1/5│  40    -8    -60     12
             8      0    -12
      40     0    -60      0
```

Therefore:
$$40x^4 - 18x^3 - 58x^2 + 27x - 3 = \left(x - \tfrac{1}{4}\right)\left(x - \tfrac{1}{5}\right)\left(40x^2 - 60\right)$$
$$= 10\left(x - \tfrac{1}{4}\right)\left(x - \tfrac{1}{5}\right)\left(4x^2 - 6\right)$$
$$= 10\left(x - \tfrac{1}{4}\right)\left(x - \tfrac{1}{5}\right)\left(2x + \sqrt{6}\right)\left(2x - \sqrt{6}\right)$$

So the roots are $\tfrac{1}{4}, \tfrac{1}{5}, -\tfrac{\sqrt{6}}{2}$ and $\tfrac{\sqrt{6}}{2}$.

27. (a) The possible rational roots are $\pm 1, \pm \tfrac{1}{5}, \pm 2, \pm \tfrac{2}{5}, \pm 3, \pm \tfrac{3}{5}, \pm 4, \pm \tfrac{4}{5}, \pm 6, \pm \tfrac{6}{5}, \pm 12,$
$\pm \tfrac{12}{5}$. Using synthetic division:

```
2│  5     0      0    -10    -12
          10    20     40     60
    5     10    20     30     48
```

Since this row is all positive, $x = 2$ is an upper bound.

```
-1│  5     0     0    -10    -12
          -5     5    - 5     15
     5    -5     5    -15      3
```

Since this row alternates signs, $x = -1$ is a lower bound.

(b) The possible rational roots are $\pm 1, \pm \tfrac{1}{3}, \pm 2, \pm \tfrac{2}{3}, \pm 4, \pm \tfrac{4}{3}$. Using synthetic division:

```
4/3│  3    -4     5     - 2      - 4
            4     0    20/3     56/9
      3     0     5    14/3     20/9
```

So $x = \tfrac{4}{3}$ is the upper bound.

$$\begin{array}{r|rrrrr} -2/3 & 3 & -4 & 5 & -2 & -4 \\ & & -2 & 4 & -6 & 16/3 \\ \hline & 3 & -6 & 9 & -8 & 4/3 \end{array}$$

So $x = -\frac{2}{3}$ is the lower bound. Since the question asked for the integral upper and lower bounds, the upper bound is 2 and the lower bound is –1.

(c) The possible rational roots are $\pm 1, \pm \frac{1}{2}, \pm 2, \pm 3, \pm \frac{3}{2}, \pm 4, \pm 6, \pm 12$. Using synthetic division:

$$\begin{array}{r|rrrrr} 6 & 2 & -7 & -5 & 28 & -12 \\ & & 12 & 30 & 150 & 1068 \\ \hline & 2 & 5 & 25 & 178 & 1056 \end{array}$$

So $x = 6$ is an upper bound.

$$\begin{array}{r|rrrrr} -2 & 2 & -7 & -5 & 28 & -12 \\ & & -4 & 22 & -34 & 12 \\ \hline & 2 & -11 & 17 & -6 & 0 \end{array}$$

Actually –2 is a root, but also a lower bound.

29. Let $f(x) = x^3 + x - 1$.
   $f(0) = -1$
   $f(1) = 1$
So the root lies between 0 and 1.
   $f(0.5) = -0.375$
   $f(0.7) = 0.043$
   $f(0.6) = -0.373$
So the root lies between 0.6 and 0.7.
   $f(0.65) = -0.075$
   $f(0.66) = -0.052$
   $f(0.67) = -0.029$
   $f(0.68) = -0.005$
   $f(0.69) = 0.018$
So the root lies between 0.68 and 0.69.

31. Let $f(x) = x^5 - 200$.
   $f(2) = -168$
   $f(3) = 43$
So the root lies between 2 and 3.
   $f(2.5) = -102$
   $f(2.8) = -27.9$
   $f(2.9) = 5.11$
So the root lies between 2.8 and 2.9.
   $f(2.87) = -5.28$
   $f(2.88) = -1.86$
   $f(2.89) = 1.60$
So the root lies between 2.88 and 2.89.

33. Let $f(x) = x^3 - 8x^2 + 21x - 22$.

$$f(4) = -2$$
$$f(5) = 8$$

So the root lies between 4 and 5.

$$f(4.5) = 1.63$$
$$f(4.4) = 0.704$$
$$f(4.3) = -0.113$$

So the root lies between 4.3 and 4.4.

$$f(4.33) = 0.12$$
$$f(4.32) = 0.04$$
$$f(4.31) = -0.03$$

So the root lies between 4.31 and 4.32.

35. Let $f(x) = x^3 + x^2 - 2x + 1$.

$$f(-2) = 1$$
$$f(-3) = -11$$

So the root lies between -3 and -2.

$$f(-2.1) = 0.35$$
$$f(-2.2) = -0.41$$

So the root lies between -2.2 and -2.1.

$$f(-2.15) = -0.02$$
$$f(-2.14) = 0.05$$

So the root lies between -2.15 and -2.14.

37. Let $f(x) = x^3 + 2x^2 + 2x + 101$.

$$f(-5) = 16$$
$$f(-6) = -55$$

So the root lies between -6 and -5.

$$f(-5.4) = -8.94$$
$$f(-5.3) = -2.30$$
$$f(-5.2) = 4.07$$

So the root lies between -5.3 and -5.2.

$$f(-5.26) = 0.28$$
$$f(-5.27) = -0.36$$

So the root lies between -5.27 and -5.26.

39. (a) Since 2 is a factor of $8 \cdot 5 = 40$, and 2 is not a factor of 5, then 2 must be a factor of 8, which is true.

(b) The result guarantees only that A is a factor of B in the case where A and C have no factor in common. Here A and C have a common factor of 5, so the result does not apply.

(c) Since $x = \frac{p}{q}$ is a root of the equation, this statement must be true.

(d)   Subtract a_0 and multiply by q^n to get:

$$a_n p^n + a_{n-1} q p^{n-1} + a_{n-2} q^2 p^{n-2} + \ldots + a_1 q^{n-1} p = -a_0 q^n$$

Therefore:

$$p\left(a_n p^{n-1} + a_{n-1} q p^{n-2} + a_{n-2} q^2 p^{n-3} + \ldots + a_1 q^{n-1}\right) = -a_0 q^n$$

41.   We sketch the graph:

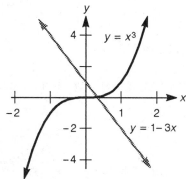

We find where $x^3 = 1 - 3x$, so $x^3 + 3x - 1 = 0$.

Let $f(x) = x^3 + 3x - 1$.

$$f(0) = -1$$
$$f(1) = 3$$

So the root lies between 0 and 1.

$$f(0.3) = -0.07$$
$$f(0.4) = 0.264$$

So the root lies between 0.3 and 0.4.

$$f(0.32) = -0.007$$
$$f(0.33) = 0.02$$

So the root lies between 0.32 and 0.33.

$$f(0.325) = 0.009$$

So the root lies between 0.32 and 0.325, thus the x-coordinate is $x = 0.32$, accurate to two decimal places.

43.   We find where:

$$\sqrt{x} = x^2 - 1$$
$$x = x^4 - 2x^2 + 1$$
$$x^4 - 2x^2 - x + 1 = 0$$

Let $f(x) = x^4 - 2x^2 - x + 1$.

$$f(1) = -1$$
$$f(2) = 7$$

So the root lies between 1 and 2.

$$f(1.4) = -0.48$$
$$f(1.5) = 0.06$$

So the root lies between 1.4 and 1.5

$$f(1.48) = -0.06$$
$$f(1.49) = -0.001$$

So the root lies between 1.49 and 1.50.

$$f(1.495) = 0.03$$

So the root lies between 1.49 and 1.495, thus the x-coordinate is $x = 1.49$, accurate to two decimal places.

45. (a) Here $a_n = 9$, $a_0 = 27$ and $f(1) = 29$. Since these three numbers are odd, then the equation has no rational roots.

   (b) Here $a_n = 5$, $a_0 = -25$ and $f(1) = -29$. Since these three numbers are odd, then the equation has no rational roots.

47. Since p is a prime number, the only possible rational roots are $\pm 1, \pm p$.
   Let $f(x) = x^3 + x^2 + x - p$
   $$f(1) = 1 + 1 + 1 - p = 3 - p$$
   So 1 is a root only if $p = 3$.
   $$f(-1) = -1 + 1 - 1 - p = -1 - p$$
   So -1 is a root only if $p = -1$, but -1 is not prime.
   $$f(p) = p^3 + p^2 + p - p = p^3 + p^2 = p^2(p + 1)$$
   So p is a root only if $p = 0$ or $p = -1$, neither of which is prime.
   $$f(-p) = -p^3 + p^2 - p - p = -p^3 + p^2 - 2p = -p(p^2 - p + 2)$$
   So $-p$ is a root only if $p = 0$ or $p = \frac{1 \pm \sqrt{1-8}}{2}$, neither of which is prime.
   Thus $p = 3$ is the only prime number such that $f(x) = 0$ will have rational roots.
   When $p = 3$, we have $x^3 + x^2 + x - 3 = 0$. We know $x = 1$ is a root. Now using synthetic division:

   $$\begin{array}{r|rrrr} 1 & 1 & 1 & 1 & -3 \\ & & 1 & 2 & 3 \\ \hline & 1 & 2 & 3 & 0 \end{array}$$

   Since $x^2 + 2x + 3 = 0$ has no real roots, $x = 1$ is the only root.

49. Following the suggestion, we substitute:
   $x = 1$: $1 + 1 - pq = 0$, so $pq = 2$, which is impossible for prime numbers p and q.
   $x = -1$: $1 - 1 - pq = 0$, so $pq = 0$, which is impossible.
   $x = p$: $p^2 + p - pq = 0$
              $p(p + 1 - q) = 0$
              Clearly $p \neq 0$, so $p = q - 1$, which occurs when $p = 2$ and $q = 3$. Thus the equation is $x^2 + x - 6 = 0$, which factors to $(x + 3)(x - 2) = 0$, thus the solutions are -3 and 2.
   $x = -p$: $p^2 - p - pq = 0$
              $p(p - 1 - q) = 0$
              Clearly $p \neq 0$, so $p = q + 1$, which occurs when $p = 3$ and $q = 2$. This results in identical solutions as when $x = p$.
   $x = q$: $q^2 + q - pq = 0$
              $q(q + 1 - p) = 0$
              Clearly $q \neq 0$, so $p = q + 1$, as above.

$x = -q$: $q^2 - q - pq = 0$
$\qquad q(q - 1 - p) = 0$
$\qquad$ Clearly $q \neq 0$, so $q = p + 1$ again, as above.

$x = pq$: $p^2q^2 + pq - pq = 0$
$\qquad\qquad p^2q^2 = 0$
$\qquad$ Clearly this is impossible.

$x = -pq$: $p^2q^2 - pq - pq = 0$
$\qquad\qquad pq(pq - 2) = 0$
$\qquad$ Since $pq \neq 0$, then $pq = 2$. But this is impossible for prime numbers p and q.
Thus if the equation has rational roots they must be -3 and 2.

51. The possible rational roots are $\pm 1, \pm 2, \pm 4$. Using synthetic division:

$\underline{1\rfloor}$	1	$-b^2$	$3b$	-4
		1	$-b^2+1$	$-b^2+3b+1$
	1	$-b^2+1$	$-b^2+3b+1$	$-b^2+3b-3$

Setting the remainder equal to 0:
$$-3 + 3b - b^2 = 0$$
$$b^2 - 3b + 3 = 0$$
$$b = \frac{3 \pm \sqrt{9 - 12}}{2}, \text{ which is not an integer}$$

We use synthetic division again:

$\underline{-1\rfloor}$	1	$-b^2$	$3b$	-4
		-1	b^2+1	$-b^2-3b-1$
	1	$-b^2-1$	b^2+3b+1	$-b^2-3b-5$

Setting the remainder equal to 0:
$$-b^2 - 3b - 5 = 0$$
$$b^2 + 3b + 5 = 0$$
$$b = \frac{-3 \pm \sqrt{9 - 20}}{2}, \text{ which is not an integer}$$

Using synthetic division again:

$\underline{2\rfloor}$	1	$-b^2$	$3b$	-4
		2	$-2b^2+4$	$-4b^2+6b+8$
	1	$-b^2+2$	$-2b^2+3b+4$	$-4b^2+6b+4$

Setting the remainder equal to 0:
$$4 + 6b - 4b^2 = 0$$
$$2b^2 - 3b - 2 = 0$$
$$(2b + 1)(b - 2) = 0$$
$$b = -\tfrac{1}{2} \text{ or } b = 2$$

Hence, if $b = 2$, the given equation has $x = 2$ as a rational root. Using synthetic division again:

$\underline{-2\rfloor}$	1	$-b^2$	$3b$	-4
		-2	$2b^2+4$	$-4b^2-6b-8$
	1	$-b^2-2$	$2b^2+3b+4$	$-4b^2-6b-12$

Setting the remainder equal to 0:

$-4b^2 - 6b - 12 = 0$

$2b^2 + 3b + 6 = 0$

$$b = \frac{-3 \pm \sqrt{9-48}}{4}, \text{ which is not an integer}$$

Using synthetic division again:

$$\begin{array}{r|cccc}
4 & 1 & -b^2 & 3b & -4 \\
  &   & 4   & -4b^2+16 & -16b^2+12b+64 \\
\hline
  & 1 & -b^2+4 & -4b^2+3b+16 & -16b^2+12b+60
\end{array}$$

Setting the remainder equal to 0:

$60 + 12b - 16b^2 = 0$

$4b^2 - 3b - 15 = 0$

$$b = \frac{3 \pm \sqrt{9+240}}{8}, \text{ which is not an integer}$$

Using synthetic division again:

$$\begin{array}{r|cccc}
-4 & 1 & -b^2 & 3b & -4 \\
   &   & -4   & 4b^2+16 & -16b^2-12b-64 \\
\hline
   & 1 & -b^2-4 & 4b^2+3b+16 & -16b^2-12b-68
\end{array}$$

Setting the remainder equal to 0:

$-16b^2 - 12b - 68 = 0$

$4b^2 + 3b + 17 = 0$

$$b = \frac{-3 \pm \sqrt{9-272}}{8}, \text{ which is not an integer}$$

So $b = 2$ is the only integer value of b for which the given equation has a rational root.

11.6 Conjugate Roots and Descartes's Rule of Signs

1. The other root must be $7 + 2i$.

3. One other root must be $5 - 2i$. So $[x - (5 - 2i)][x - (5 + 2i)]$ are factors, or $x^2 - 10x + 29$. Using long division:

$$\begin{array}{r}
x - 3 \\
x^2 - 10x + 29 \overline{\smash{)}\, x^3 - 13x^2 + 59x - 87} \\
\underline{x^3 - 10x^2 + 29x} \\
-3x^2 + 30x - 87 \\
\underline{-3x^2 + 30x - 87} \\
0
\end{array}$$

Since $x - 3$ is the other factor, then 3 is the other root. So the other two roots are $5 - 2i$ and 3.

5. One other root must be $-2 - i$, so $[x - (-2 + i)][x - (-2 - i)]$ are factors, or $x^2 + 4x + 5$. Using long division:

$$
\begin{array}{r}
x^2 + 6x + 9 \\
x^2 + 4x + 5 \overline{\smash{\big)}\ x^4 + 10x^3 + 38x^2 + 66x + 45} \\
\underline{x^4 + 4x^3 + 5x^2} \\
6x^3 + 33x^2 + 66x \\
\underline{6x^3 + 24x^2 + 30x} \\
9x^2 + 36x + 45 \\
\underline{9x^2 + 36x + 45} \\
0
\end{array}
$$

Since $x^2 + 6x + 9 = (x + 3)^2$, the other root is -3. So the remaining roots are $-2 - i$ and -3 (multiplicity 2).

7. One other root must be $6 + 5i$, so $[x - (6 - 5i)][x - (6 + 5i)]$ are factors, or $x^2 - 12x + 61$. Using long division:

$$
\begin{array}{r}
4x + 1 \\
x^2 - 12x + 61 \overline{\smash{\big)}\ 4x^3 - 47x^2 + 232x + 61} \\
\underline{4x^3 - 48x^2 + 244x} \\
x^2 - 12x + 61 \\
\underline{x^2 - 12x + 61} \\
0
\end{array}
$$

Since $4x + 1$ is the other factor, then $-\frac{1}{4}$ is the other root. So the remaining roots are $6 + 5i$ and $-\frac{1}{4}$.

9. One other root must be $4 - \sqrt{2}i$, so $\left[x - (4 + \sqrt{2}i)\right]\left[x - (4 - \sqrt{2}i)\right]$ are factors, or $x^2 - 8x + 18$. Using long division:

$$
\begin{array}{r}
4x^2 + 9 \\
x^2 - 8x + 18 \overline{\smash{\big)}\ 4x^4 - 32x^3 + 81x^2 - 72x + 162} \\
\underline{4x^4 - 32x^3 + 72x^2} \\
9x^2 - 72x + 162 \\
\underline{9x^2 - 72x + 162} \\
0
\end{array}
$$

Since $4x^2 + 9 = (2x + 3i)(2x - 3i)$, the other roots are $-\frac{3i}{2}$ and $\frac{3i}{2}$. So the remaining roots are $4 - \sqrt{2}i$, $-\frac{3i}{2}$ and $\frac{3i}{2}$.

11. One other root must be $10 - 2i$, so $[x - (10 + 2i)][x - (10 - 2i)]$ are factors, or $x^2 - 20x + 104$. Using long division:

$$
\begin{array}{r}
x^2 - 2x - 4 \\
x^2 - 20x + 104 \overline{\smash{\big)}\ x^4 - 22x^3 + 140x^2 - 128x - 416} \\
\underline{x^4 - 20x^3 + 104x^2} \\
-2x^3 + 36x^2 - 128x \\
\underline{-2x^3 + 40x^2 - 208x} \\
-4x^2 + 80x - 416 \\
\underline{-4x^2 + 80x - 416} \\
0
\end{array}
$$

So the other factor is $x^2 - 2x - 4$. We use the quadratic formula:

$$x = \frac{2 \pm \sqrt{4 + 16}}{2} = \frac{2 \pm 2\sqrt{5}}{2} = 1 \pm \sqrt{5}$$

So the remaining roots are $10 - 2i$, $1 + \sqrt{5}$ and $1 - \sqrt{5}$.

13. One other root must be $\frac{1 - \sqrt{2}i}{3}$ so $\left(x - \frac{1 + \sqrt{2}i}{3}\right)\left(x - \frac{1 - \sqrt{2}i}{3}\right)$ are factors, or $x^2 - \frac{2}{3}x + \frac{1}{3}$. Using long division:

$$
\begin{array}{r}
15x - 6 \\
x^2 - \frac{2}{3}x + \frac{1}{3} \overline{\smash{\big)}\ 15x^3 - 16x^2 + 9x - 2} \\
\underline{15x^3 - 10x^2 + 5x} \\
-6x^2 + 4x - 2 \\
\underline{-6x^2 + 4x - 2} \\
0
\end{array}
$$

So the other factor is $15x - 6$, so $x = \frac{2}{5}$ is a root. The remaining roots are $\frac{1 - i\sqrt{2}}{3}$ and $\frac{2}{5}$.

15. We know $3 + 2i$ is a root, so $[x - (3 - 2i)][x - (3 + 2i)]$ are factors, which is $x^2 - 6x + 13$. Using long division:

$$
\begin{array}{r}
x^5 + 3x^4 + x^3 - 3x^2 - 4x + 2 \\
x^2 - 6x + 13 \overline{\smash{\big)}\ x^7 - 3x^6 - 4x^5 + 30x^4 + 27x^3 - 13x^2 - 64x + 26} \\
\underline{x^7 - 6x^6 + 13x^5} \\
3x^6 - 17x^5 + 30x^4 \\
\underline{3x^6 - 18x^5 + 39x^4} \\
x^5 - 9x^4 + 27x^3 \\
\underline{x^5 - 6x^4 + 13x^3} \\
-3x^4 + 14x^3 - 13x^2 \\
\underline{-3x^4 + 18x^3 - 39x^2} \\
-4x^3 + 26x^2 - 64x \\
\underline{-4x^3 + 24x^2 - 52x} \\
2x^2 - 12x + 26 \\
\underline{2x^2 - 12x + 26} \\
0
\end{array}
$$

We know $x = -1 - i$ will be a root, so $[x - (-1 + i)][x - (-1 - i)]$ are factors, which is $x^2 + 2x + 2$. Using long division:

$$
\begin{array}{r}
x^3 + x^2 - 3x + 1 \\
x^2 + 2x + 2 \overline{\smash{\big)}\ x^5 + 3x^4 + x^3 - 3x^2 - 4x + 2} \\
\underline{x^5 + 2x^4 + 2x^3} \\
x^4 - x^3 - 3x^2 \\
\underline{x^4 + 2x^3 + 2x^2} \\
-3x^3 - 5x^2 - 4x \\
\underline{-3x^3 - 6x^2 - 6x} \\
x^2 + 2x + 2 \\
\underline{x^2 + 2x + 2} \\
0
\end{array}
$$

Finally, we know $x = 1$ is a root, so we synthetically divide:

$$
\begin{array}{r|rrrr}
1 & 1 & 1 & -3 & 1 \\
  &   & 1 & 2 & -1 \\
\hline
  & 1 & 2 & -1 & 0
\end{array}
$$

So we are left with $x^2 + 2x - 1 = 0$. Using the quadratic formula:

$$
x = \frac{-2 \pm \sqrt{4+4}}{2} = \frac{-2 \pm 2\sqrt{2}}{2} = -1 \pm \sqrt{2}
$$

So the remaining roots are $3 + 2i$, $-1 - i$, $-1 + \sqrt{2}$, and $-1 - \sqrt{2}$.

17. Here $r_1 = 1 + \sqrt{6}$ and $r_2 = 1 - \sqrt{6}$, so $[x - (1 + \sqrt{6})][x - (1 - \sqrt{6})] = 0$. Multiplied out, we obtain $x^2 - 2x - 5 = 0$.

19. Here $r_1 = \frac{2 + \sqrt{10}}{3}$ and $r_2 = \frac{2 - \sqrt{10}}{3}$, so $\left(x - \frac{2 + \sqrt{10}}{3}\right)\left(x - \frac{2 - \sqrt{10}}{3}\right) = 0$. Multiplied out, we obtain $x^2 - \frac{4}{3}x - \frac{2}{3} = 0$.

21. Let $f(x) = x^3 + 5$. Since there are no sign changes, there will be no positive roots. Now $f(-x) = -x^3 + 5$. Since there is one sign change, there is 1 negative root. So the equation has 2 complex roots and 1 negative real root.

23. Let $f(x) = 2x^5 + 3x + 4$. Since there are no sign changes, there will be no positive roots. Now $f(-x) = -2x^5 - 3x + 4$. Since there is one sign change, there is 1 negative root. So the equation has 4 complex roots and 1 negative real root.

25. Let $f(x) = 5x^4 + 2x - 7$. Since there is one sign change, there will be 1 positive root. Now $f(-x) = 5x^4 - 2x - 7$. Since there is one sign change, there will be 1 negative root. So the equation has 2 complex roots, 1 positive real root and 1 negative real root.

27. Let $f(x) = x^3 - 4x^2 - x - 1$. Since there is one sign change, there will be 1 positive root. Now $f(-x) = -x^3 - 4x^2 + x - 1$. Since there are two sign changes, there will be 0 or 2 negative roots. So the equation has either 1 positive real root and 2 negative real roots, or 1 positive real root and 2 complex roots.

29. Let $f(x) = 3x^8 + x^6 - 2x^2 - 4$. Since there is one sign change, there is 1 positive root. Now $f(-x) = 3x^8 + x^6 - 2x^2 - 4$. Since there is one sign change, there is 1 negative root. So the equation has 1 positive real root, 1 negative real root and 6 complex roots.

31. Let $f(x) = x^9 - 2$. Since there is one sign change, there is 1 positive root. Now $f(-x) = -x^9 - 2$. Since there are no sign changes, there are no negative roots. So the equation has 1 positive real root and 8 complex roots.

33. Let $f(x) = x^8 - 2$. Since there is one sign change, there is 1 positive root. Now $f(-x) = x^8 - 2$. Since there is one sign change, there is 1 negative root. So the equation has 1 positive real root, 1 negative real root and 6 complex roots.

35. Let $f(x) = x^6 + x^2 - x - 1$. Since there is one sign change, there is 1 positive root. Now $f(-x) = x^6 + x^2 + x - 1$. Since there is one sign change, there is 1 negative root. So the equation has 1 positive real root, 1 negative real root and 4 complex roots.

37. Let $f(x) = x^4 + cx^2 + dx - e$. Since there is one sign change, there is 1 positive root. Now $f(-x) = x^4 + cx^2 - dx - e$. Since there is one sign change, there is 1 negative root. So the equation has 1 positive real root, 1 negative real root and 2 complex roots.

39. The two roots $\sqrt{3} + 2i$ and $\sqrt{3} - 2i$ must be included, so:
$$f(x) = \left[x - (\sqrt{3} + 2i)\right]\left[x - (\sqrt{3} - 2i)\right] = x^2 - 2\sqrt{3} + 7 = 0$$
Unfortunately, not all coefficients are rational:
$$x^2 + 7 = 2\sqrt{3}x$$
Squaring, we obtain:
$$x^4 + 14x^2 + 49 = 12x^2$$
$$x^4 + 2x^2 + 29 = 0$$
So $f(x) = x^4 + 2x^2 + 49$ is the desired polynomial.

41. (a) If $b = 0$, then $a + b\sqrt{c} = a = a - b\sqrt{c}$, so $a - b\sqrt{c}$ is also a root.

   (b) We compute $d(a + b\sqrt{c})$:
   $$d(a + b\sqrt{c}) = \left[a + b\sqrt{c} - (a + b\sqrt{c})\right]\left[a + b\sqrt{c} - (a - b\sqrt{c})\right] = 0(2b\sqrt{c}) = 0$$

   (c) We factor $d(x) = x^2 - 2ax + a^2 - b^2c = (x - a)^2 - b^2c$.

   (d) If $x = a + b\sqrt{c}$, we have:
   $$f(a + b\sqrt{c}) = d(a + b\sqrt{c})Q(a + b\sqrt{c}) + C(a + b\sqrt{c}) + D$$
   $$0 = 0 \cdot Q(a + b\sqrt{c}) + Ca + D + bC\sqrt{c}$$
   $$0 = Ca + D + bC\sqrt{c}$$
   But if C and D are rational, then $C = 0$ (otherwise $\sqrt{c}$ will not be "cancelled" out). If $C = 0$, then $D = 0$. So $C = D = 0$.

(e)   We compute $f(a - b\sqrt{c})$:

$$f(a - b\sqrt{c}) = \left[a - b\sqrt{c} - (a + b\sqrt{c})\right]\left[a - b\sqrt{c} - (a - b\sqrt{c})\right]Q(x)$$
$$= (-2b\sqrt{c})(0)Q(x)$$
$$= 0$$

So $x = a - b\sqrt{c}$ is also a root of $f(x) = 0$.

Optional TI-81 Graphing Calculator Exercises for Sections 11.3-11.6

1.   (a)   We graph the function $y = f(x)$:

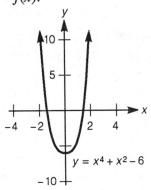

$$y = x^4 + x^2 - 6$$

   (b)   Using the ZOOM-TRACE technique, the roots are ± 1.414. This agrees with the roots $\pm\sqrt{2}$ found in the text.

3.   (a)   We graph the function $y = f(x)$:

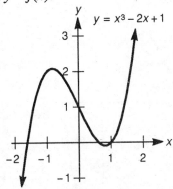

$$y = x^3 - 2x + 1$$

   (b)   Using the ZOOM-TRACE technique, the roots are -1.618, 0.618, and 1. This agrees with the roots 1, $\frac{1}{2}(-1 \pm \sqrt{5})$ found in the text.

5. (a) We graph the function $y = f(x)$:

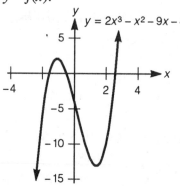

$y = 2x^3 - x^2 - 9x - 4$

(b) Using the ZOOM-TRACE technique, the roots are -1.562, -0.5, and 2.562. This agrees with the roots $-\frac{1}{2}, \frac{1}{2}(1 \pm \sqrt{17})$ found in the text.

7. (a) We graph the function $y = f(x)$:

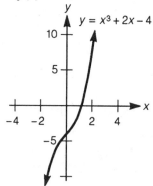

$y = x^3 + 2x - 4$

(b) Using the ZOOM-TRACE technique, the root is 1.180. This agrees with the information obtained in the text.

9. (a) We graph the function $y = f(x)$:

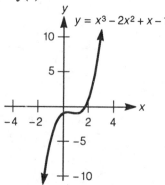

$y = x^3 - 2x^2 + x - 1$

(b) Using the ZOOM-TRACE technique, the root is 1.755. This agrees with the information obtained in the text.

11. We graph the function $f(x)$:

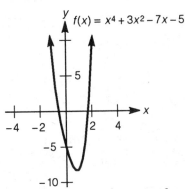

$f(x) = x^4 + 3x^2 - 7x - 5$

Notice that this function has one negative real root and one positive real root.

11.7 DeMoivre's Theorem

1. The complex number $4 + 2i$ is identified with the point $(4, 2)$:

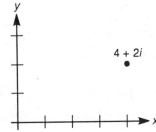

$4 + 2i$

3. The complex number $-5 + i$ is identified with the point $(-5, 1)$:

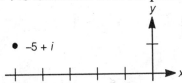

$-5 + i$

5. The complex number $1 - 4i$ is identified with the point $(1, -4)$:

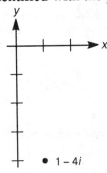

$1 - 4i$

7.   The complex number $-i$, or $0 - 1i$, is identified with the point $(0, -1)$:

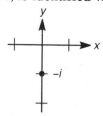

9.   We convert the complex number to rectangular form:
$$2\left[\cos\tfrac{\pi}{4} + i\sin\tfrac{\pi}{4}\right] = 2\left[\tfrac{\sqrt{2}}{2} + \tfrac{\sqrt{2}}{2}i\right] = \sqrt{2} + \sqrt{2}i$$

11.   We convert the complex number to rectangular form:
$$4\left[\cos\tfrac{5\pi}{6} + i\sin\tfrac{5\pi}{6}\right] = 4\left[-\tfrac{\sqrt{3}}{2} + \tfrac{1}{2}i\right] = -2\sqrt{3} + 2i$$

13.   We convert the complex number to rectangular form:
$$\sqrt{2}\left[\cos 225° + i\sin 225°\right] = \sqrt{2}\left[-\tfrac{\sqrt{2}}{2} - \tfrac{\sqrt{2}}{2}i\right] = -1 - i$$

15.   We convert the complex number to rectangular form:
$$\sqrt{3}\left[\cos\tfrac{\pi}{2} + i\sin\tfrac{\pi}{2}\right] = \sqrt{3}(0 + 1i) = \sqrt{3}i$$

17.   We use the hint to find $\cos 75°$ and $\sin 75°$:
$$\cos 75° = \cos(30° + 45°) = \cos 30°\cos 45° - \sin 30°\sin 45° = \tfrac{\sqrt{3}}{2} \cdot \tfrac{\sqrt{2}}{2} - \tfrac{1}{2} \cdot \tfrac{\sqrt{2}}{2} = \tfrac{\sqrt{6} - \sqrt{2}}{4}$$
$$\sin 75° = \sin(30° + 45°) = \sin 30°\cos 45° + \cos 30°\sin 45° = \tfrac{1}{2} \cdot \tfrac{\sqrt{2}}{2} + \tfrac{\sqrt{3}}{2} \cdot \tfrac{\sqrt{2}}{2} = \tfrac{\sqrt{6} + \sqrt{2}}{4}$$
Therefore:
$$4\left[\cos 75° + i\sin 75°\right] = 4\left[\tfrac{\sqrt{6} - \sqrt{2}}{4} + \tfrac{\sqrt{6} + \sqrt{2}}{4}i\right] = \left(\sqrt{6} - \sqrt{2}\right) + \left(\sqrt{6} + \sqrt{2}\right)i$$

19.   Here $a = \tfrac{\sqrt{3}}{2}$ and $b = \tfrac{1}{2}$, so:
$$r = \sqrt{a^2 + b^2} = \sqrt{\tfrac{3}{4} + \tfrac{1}{4}} = 1$$
We now find θ such that $\cos\theta = \tfrac{a}{r} = \tfrac{\sqrt{3}}{2}$ and $\sin\theta = \tfrac{b}{r} = \tfrac{1}{2}$. Such a θ is $\theta = \tfrac{\pi}{6}$. Thus:
$$\tfrac{\sqrt{3}}{2} + \tfrac{1}{2}i = \cos\tfrac{\pi}{6} + i\sin\tfrac{\pi}{6}$$

21.   Here $a = -1$ and $b = \sqrt{3}$, so:
$$r = \sqrt{a^2 + b^2} = \sqrt{1 + 3} = \sqrt{4} = 2$$
We now find θ such that $\cos\theta = \tfrac{a}{r} = -\tfrac{1}{2}$ and $\sin\theta = \tfrac{b}{r} = \tfrac{\sqrt{3}}{2}$. Such a θ is $\theta = \tfrac{2\pi}{3}$. Thus:
$$-1 + \sqrt{3}i = 2\left[\cos\tfrac{2\pi}{3} + i\sin\tfrac{2\pi}{3}\right]$$

23. Here $a = -2\sqrt{3}$ and $b = -2$, so:

$$r = \sqrt{a^2 + b^2} = \sqrt{12 + 4} = \sqrt{16} = 4$$

We now find θ such that $\cos\theta = \frac{a}{r} = -\frac{2\sqrt{3}}{4} = -\frac{\sqrt{3}}{2}$ and $\sin\theta = \frac{b}{r} = -\frac{2}{4} = -\frac{1}{2}$. Such a θ is $\theta = \frac{7\pi}{6}$. Thus:

$$-2\sqrt{3} - 2i = 4\left[\cos\frac{7\pi}{6} + i\sin\frac{7\pi}{6}\right]$$

25. Here $a = 0$ and $b = -6$, so:

$$r = \sqrt{a^2 + b^2} = \sqrt{0 + 36} = 6$$

We now find θ such that $\cos\theta = \frac{a}{r} = \frac{0}{6} = 0$ and $\sin\theta = \frac{b}{r} = -\frac{6}{6} = -1$. Such a θ is $\theta = \frac{3\pi}{2}$. Thus:

$$-6i = 6\left[\cos\frac{3\pi}{2} + i\sin\frac{3\pi}{2}\right]$$

27. Here $a = \frac{\sqrt{3}}{4}$ and $b = -\frac{1}{4}$, so:

$$r = \sqrt{a^2 + b^2} = \sqrt{\frac{3}{16} + \frac{1}{16}} = \sqrt{\frac{1}{4}} = \frac{1}{2}$$

We now find θ such that $\cos\theta = \frac{a}{r} = \frac{\sqrt{3}/4}{1/2} = \frac{\sqrt{3}}{2}$ and $\sin\theta = \frac{b}{r} = \frac{-1/4}{1/2} = -\frac{1}{2}$. Such a θ is $\theta = \frac{11\pi}{6}$. Thus:

$$\frac{\sqrt{3}}{4} - \frac{1}{4}i = \frac{1}{2}\left[\cos\frac{11\pi}{6} + i\sin\frac{11\pi}{6}\right]$$

29. Performing the multiplication and adding angles, we have:

$$2[\cos 22° + i\sin 22°] \cdot 3[\cos 38° + i\sin 38°] = 6[\cos 60° + i\sin 60°]$$
$$= 6\left(\frac{1}{2} + \frac{\sqrt{3}}{2}i\right)$$
$$= 3 + 3\sqrt{3}i$$

31. Performing the multiplication and adding angles, we have:

$$\sqrt{2}\left[\cos\frac{\pi}{3} + i\sin\frac{\pi}{3}\right] \cdot \sqrt{2}\left[\cos\frac{4\pi}{3} + i\sin\frac{4\pi}{3}\right] = 2\left[\cos\frac{5\pi}{3} + i\sin\frac{5\pi}{3}\right] = 2\left(\frac{1}{2} - \frac{\sqrt{3}}{2}i\right) = 1 - \sqrt{3}i$$

33. Performing the multiplication and adding angles, we have:

$$3\left[\cos\frac{\pi}{7} + i\sin\frac{\pi}{7}\right] \cdot \sqrt{2}\left[\cos\frac{\pi}{7} + i\sin\frac{\pi}{7}\right] = 3\sqrt{2}\left[\cos\frac{2\pi}{7} + i\sin\frac{2\pi}{7}\right]$$
$$= 3\sqrt{2}\cos\frac{2\pi}{7} + \left(3\sqrt{2}\sin\frac{2\pi}{7}\right)i$$

35. Performing the division and subtracting angles, we have:

$$6[\cos 50° + i\sin 50°] \div 2[\cos 5° + i\sin 5°] = 3[\cos 45° + i\sin 45°]$$
$$= 3\left(\frac{\sqrt{2}}{2} + \frac{\sqrt{2}}{2}i\right)$$
$$= \frac{3\sqrt{2}}{2} + \frac{3\sqrt{2}}{2}i$$

37. Performing the division and subtracting angles, we have:
$$2^{4/3}\left[\cos\tfrac{5\pi}{12}+i\sin\tfrac{5\pi}{12}\right]\div 2^{1/3}\left[\cos\tfrac{\pi}{4}+i\sin\tfrac{\pi}{4}\right]=2\left[\cos\tfrac{\pi}{6}+i\sin\tfrac{\pi}{6}\right]=2\left(\tfrac{\sqrt{3}}{2}+\tfrac{1}{2}i\right)=\sqrt{3}+i$$

39. Performing the division and subtracting angles, we have:
$$\left[\cos\tfrac{2\pi}{5}+i\sin\tfrac{2\pi}{5}\right]\div\left[\cos\tfrac{2\pi}{5}+i\sin\tfrac{2\pi}{5}\right]=\cos 0+i\sin 0=1+0i=1$$

41. Using DeMoivre's theorem, we have:
$$\left[3\left(\cos\tfrac{\pi}{3}+i\sin\tfrac{\pi}{3}\right)\right]^5=3^5\left(\cos\tfrac{5\pi}{3}+i\sin\tfrac{5\pi}{3}\right)=243\left(\tfrac{1}{2}-\tfrac{\sqrt{3}}{2}i\right)=\tfrac{243}{2}-\tfrac{243\sqrt{3}}{2}i$$

43. Using DeMoivre's theorem, we have:
$$\left[\tfrac{1}{2}\left(\cos\tfrac{\pi}{24}+i\sin\tfrac{\pi}{24}\right)\right]^6=\left(\tfrac{1}{2}\right)^6\left(\cos\tfrac{\pi}{4}+i\sin\tfrac{\pi}{4}\right)=\tfrac{1}{64}\left(\tfrac{\sqrt{2}}{2}+\tfrac{\sqrt{2}}{2}i\right)=\tfrac{\sqrt{2}}{128}+\tfrac{\sqrt{2}}{128}i$$

45. Using DeMoivre's theorem, we have:
$$\left[2^{1/5}\left(\cos 63°+i\sin 63°\right)\right]^{10}=\left(2^{1/5}\right)^{10}\left(\cos 630°+i\sin 630°\right)=4(0-1i)=-4i$$

47. Performing the multiplication and adding angles, we have:
$$2(\cos 100°+i\sin 200°)\times\sqrt{2}(\cos 20°+i\sin 20°)\times\tfrac{1}{2}(\cos 5°+i\sin 5°)$$
$$=\sqrt{2}(\cos 225°+i\sin 225°)$$
$$=\sqrt{2}\left(-\tfrac{\sqrt{2}}{2}-\tfrac{\sqrt{2}}{2}i\right)$$
$$=-1-i$$

49. We first write $\tfrac{1}{2}-\tfrac{\sqrt{3}}{2}i$ in trigonometric form:
$$\tfrac{1}{2}-\tfrac{\sqrt{3}}{2}i=\cos\tfrac{5\pi}{3}+i\sin\tfrac{5\pi}{3}$$
Now using DeMoivre's theorem, we have:
$$\left(\tfrac{1}{2}-\tfrac{\sqrt{3}}{2}i\right)^5=\cos\tfrac{25\pi}{3}+i\sin\tfrac{25\pi}{3}=\tfrac{1}{2}+\tfrac{\sqrt{3}}{2}i$$

51. We first write $-2-2i$ in trigonometric form:
$$-2-2i=2\sqrt{2}\left(-\tfrac{\sqrt{2}}{2}-\tfrac{\sqrt{2}}{2}i\right)=2\sqrt{2}\left(\cos\tfrac{5\pi}{4}+i\sin\tfrac{5\pi}{4}\right)$$
Now using DeMoivre's theorem, we have:
$$(-2-2i)^5=\left(2\sqrt{2}\right)^5\left(\cos\tfrac{25\pi}{4}+i\sin\tfrac{25\pi}{4}\right)=128\sqrt{2}\left(\tfrac{\sqrt{2}}{2}+\tfrac{\sqrt{2}}{2}i\right)=128+128i$$

53. We first write $-2\sqrt{3}-2i$ in trigonometric form:
$$-2\sqrt{3}-2i=4\left(-\tfrac{\sqrt{3}}{2}-\tfrac{1}{2}i\right)=4\left(\cos\tfrac{7\pi}{6}+i\sin\tfrac{7\pi}{6}\right)$$
Now using DeMoivre's theorem, we have:
$$(-2\sqrt{3}-2i)^4=4^4\left[\cos\tfrac{14\pi}{3}+i\sin\tfrac{14\pi}{3}\right]=256\left(-\tfrac{1}{2}+\tfrac{\sqrt{3}}{2}i\right)=-128+128\sqrt{3}i$$

55. We begin by writing $-27i$ in trigonometric form:
$$-27i = 27(0 - 1i) = 27\left[\cos\tfrac{3\pi}{2} + i\sin\tfrac{3\pi}{2}\right]$$
Now let $z = r(\cos\theta + i\sin\theta)$ denote a cube root of $-27i$. Then:
$$z^3 = r^3(\cos 3\theta + i\sin 3\theta) = 27\left[\cos\tfrac{3\pi}{2} + i\sin\tfrac{3\pi}{2}\right]$$
Then $r^3 = 27$ so $r = 3$, and $3\theta = \tfrac{3\pi}{2} + 2\pi k$, so $\theta = \tfrac{\pi}{2} + \tfrac{2\pi}{3}k$.
When $k = 0$, we have:
$$z_1 = 3\left[\cos\tfrac{\pi}{2} + i\sin\tfrac{\pi}{2}\right] = 3(0 + i) = 3i$$
When $k = 1$, we have:
$$z_2 = 3\left[\cos\tfrac{7\pi}{6} + i\sin\tfrac{7\pi}{6}\right] = 3\left(-\tfrac{\sqrt{3}}{2} - \tfrac{1}{2}i\right) = -\tfrac{3\sqrt{3}}{2} - \tfrac{3}{2}i$$
When $k = 2$, we have:
$$z_3 = 3\left[\cos\tfrac{11\pi}{6} + i\sin\tfrac{11\pi}{6}\right] = 3\left(\tfrac{\sqrt{3}}{2} - \tfrac{1}{2}i\right) = \tfrac{3\sqrt{3}}{2} - \tfrac{3}{2}i$$
So the cube roots of $-27i$ are $3i$, $-\tfrac{3\sqrt{3}}{2} - \tfrac{3}{2}i$ and $\tfrac{3\sqrt{3}}{2} - \tfrac{3}{2}i$.

57. We begin by writing 1 in trigonometric form:
$$1 = 1(1 + 0i) = 1[\cos 0 + i\sin 0]$$
Now let $z = r(\cos\theta + i\sin\theta)$ denote an eighth root of 1. Then:
$$z^8 = r^8(\cos 8\theta + i\sin 8\theta) = 1[\cos 0 + i\sin 0]$$
Then $r^8 = 1$ so $r = 1$, and $8\theta = 0 + 2\pi k$, so $\theta = \tfrac{\pi}{4}k$.
When $k = 0$, we have:
$$z_1 = 1[\cos 0 + i\sin 0] = 1 + 0i = 1$$
When $k = 1$, we have:
$$z_2 = 1\left[\cos\tfrac{\pi}{4} + i\sin\tfrac{\pi}{4}\right] = \tfrac{\sqrt{2}}{2} + \tfrac{\sqrt{2}}{2}i$$
When $k = 2$, we have:
$$z_3 = 1\left[\cos\tfrac{\pi}{2} + i\sin\tfrac{\pi}{2}\right] = 0 + 1i = i$$
When $k = 3$, we have:
$$z_4 = 1\left[\cos\tfrac{3\pi}{4} + i\sin\tfrac{3\pi}{4}\right] = -\tfrac{\sqrt{2}}{2} + \tfrac{\sqrt{2}}{2}i$$
When $k = 4$, we have:
$$z_5 = 1[\cos\pi + i\sin\pi] = -1 + 0i = -1$$
When $k = 5$, we have:
$$z_6 = 1\left[\cos\tfrac{5\pi}{4} + i\sin\tfrac{5\pi}{4}\right] = -\tfrac{\sqrt{2}}{2} - \tfrac{\sqrt{2}}{2}i$$
When $k = 6$, we have:
$$z_7 = 1\left[\cos\tfrac{3\pi}{2} + i\sin\tfrac{3\pi}{2}\right] = 0 - 1i = -i$$
When $k = 7$, we have:
$$z_8 = 1\left[\cos\tfrac{7\pi}{4} + i\sin\tfrac{7\pi}{4}\right] = \tfrac{\sqrt{2}}{2} - \tfrac{\sqrt{2}}{2}i$$
So the eighth roots of 1 are 1, $\tfrac{\sqrt{2}}{2} + \tfrac{\sqrt{2}}{2}i$, i, $-\tfrac{\sqrt{2}}{2} + \tfrac{\sqrt{2}}{2}i$, -1, $-\tfrac{\sqrt{2}}{2} - \tfrac{\sqrt{2}}{2}i$, $-i$, and $\tfrac{\sqrt{2}}{2} - \tfrac{\sqrt{2}}{2}i$.
These final results were obtained using the half-angle formulas for sine and cosine.

59. We begin by writing 64 in trigonometric form:
$$64 = 64(1 + 0i) = 64[\cos 0 + i \sin 0]$$
Now let $z = r(\cos \theta + i \sin \theta)$ denote a cube root of 64. Then:
$$z^3 = r^3(\cos 3\theta + i \sin 3\theta) = 64[\cos 0 + i \sin 0]$$
Then $r^3 = 64$ so $r = 4$, and $3\theta = 0 + 2\pi k$, so $\theta = 0 + \frac{2\pi}{3}k$.
When $k = 0$, we have:
$$z_1 = 4[\cos 0 + i \sin 0] = 4(1 + 0i) = 4$$
When $k = 1$, we have:
$$z_2 = 4\left[\cos \frac{2\pi}{3} + i \sin \frac{2\pi}{3}\right] = 4\left(-\frac{1}{2} + \frac{\sqrt{3}}{2}i\right) = -2 + 2\sqrt{3}i$$
When $k = 2$, we have:
$$z_3 = 4\left[\cos \frac{4\pi}{3} + i \sin \frac{4\pi}{3}\right] = 4\left(-\frac{1}{2} - \frac{\sqrt{3}}{2}i\right) = -2 - 2\sqrt{3}i$$
So the cube roots of 64 are 4, $-2 + 2\sqrt{3}i$, and $-2 - 2\sqrt{3}i$.

61. We first write 729 in trigonometric form:
$$729 = 729(1 + 0i) = 729[\cos 0 + i \sin 0]$$
Now let $z = r(\cos \theta + i \sin \theta)$ denote a sixth root of 729. Then:
$$z^6 = r^6[\cos 6\theta + i \sin 6\theta] = 729[\cos 0 + i \sin 0]$$
Then $r^6 = 729$ so $r = 3$, and $6\theta = 0 + 2\pi k$, so $\theta = 0 + \frac{\pi}{3}k$.
When $k = 0$, we have:
$$z_1 = 3[\cos 0 + i \sin 0] = 3(1 + 0i) = 3$$
When $k = 1$, we have:
$$z_2 = 3\left[\cos \frac{\pi}{3} + i \sin \frac{\pi}{3}\right] = 3\left(\frac{1}{2} + \frac{\sqrt{3}}{2}i\right) = \frac{3}{2} + \frac{3\sqrt{3}}{2}i$$
When $k = 2$, we have:
$$z_3 = 3\left[\cos \frac{2\pi}{3} + i \sin \frac{2\pi}{3}\right] = 3\left(-\frac{1}{2} + \frac{\sqrt{3}}{2}i\right) = -\frac{3}{2} + \frac{3\sqrt{3}}{2}i$$
When $k = 3$, we have:
$$z_4 = 3[\cos \pi + i \sin \pi] = 3(-1 + 0i) = -3$$
When $k = 4$, we have:
$$z_5 = 3\left[\cos \frac{4\pi}{3} + i \sin \frac{4\pi}{3}\right] = 3\left(-\frac{1}{2} - \frac{\sqrt{3}}{2}i\right) = -\frac{3}{2} - \frac{3\sqrt{3}}{2}i$$
When $k = 5$, we have:
$$z_6 = 3\left[\cos \frac{5\pi}{3} + i \sin \frac{5\pi}{3}\right] = 3\left(\frac{1}{2} - \frac{\sqrt{3}}{2}i\right) = \frac{3}{2} - \frac{3\sqrt{3}}{2}i$$
So the sixth roots of 729 are 3, $\frac{3}{2} + \frac{3\sqrt{3}}{2}i$, $-\frac{3}{2} + \frac{3\sqrt{3}}{2}i$, -3, $-\frac{3}{2} - \frac{3\sqrt{3}}{2}i$ and $\frac{3}{2} - \frac{3\sqrt{3}}{2}i$.

63. We first write $7 - 7i$ in trigonometric form:
$$7 - 7i = 7\sqrt{2}\left(\frac{1}{\sqrt{2}} - \frac{1}{\sqrt{2}}i\right) = 7\sqrt{2}\left[\cos \frac{7\pi}{4} + i \sin \frac{7\pi}{4}\right]$$
Using DeMoivre's theorem we have:
$$(7 - 7i)^8 = (7\sqrt{2})^8[\cos 14\pi + i \sin 14\pi] = (7^8)(2^4)(1 + 0i) = 92,236,816$$

65. We begin by writing i in trigonometric form:
$$i = 1(0+1i) = 1\left[\cos\tfrac{\pi}{2} + i\sin\tfrac{\pi}{2}\right]$$
Now let z be a fifth root of i, where $z = r(\cos\theta + i\sin\theta)$. Then:
$$z^5 = r^5[\cos 5\theta + i\sin 5\theta] = 1\left[\cos\tfrac{\pi}{2} + i\sin\tfrac{\pi}{2}\right]$$
So $r^5 = 1$ thus $r = 1$, and $5\theta = \tfrac{\pi}{2} + 2\pi k = 90°+360°k$, so $\theta = 18° + 72°k$.

When $k = 0$: $z_1 = 1[\cos 18°+i\sin 18°] = 0.95 + 0.31i$

When $k = 1$: $z_2 = 1[\cos 90°+i\sin 90°] = i$

When $k = 2$: $z_3 = 1[\cos 162°+i\sin 162°] = -0.95 + 0.31i$

When $k = 3$: $z_4 = 1[\cos 234°+i\sin 234°] = -0.59 - 0.81i$

When $k = 4$: $z_5 = 1[\cos 306°+i\sin 306°] = 0.59 - 0.81i$

So the fifth roots of i are $0.95 + 0.31i$, i, $-0.95 + 0.31i$, $-0.59 - 0.81i$ and $0.59 - 0.81i$. Each of the real and imaginary parts here has been rounded off to two decimal places.

67. We begin by writing $8 - 8\sqrt{3}i$ in trigonometric form:
$$8 - 8\sqrt{3}i = 16\left(\tfrac{1}{2} - \tfrac{\sqrt{3}}{2}i\right) = 16\left[\cos\tfrac{5\pi}{3} + i\sin\tfrac{5\pi}{3}\right]$$
Now let $z = r(\cos\theta + i\sin\theta)$ denote a fourth root of $8 - 8\sqrt{3}i$. Then:
$$z^4 = r^4[\cos 4\theta + i\sin 4\theta] = 16\left[\cos\tfrac{5\pi}{3} + i\sin\tfrac{5\pi}{3}\right]$$
Then $r^4 = 16$ so $r = 2$, and $4\theta = \tfrac{5\pi}{3} + 2\pi k$, so $\theta = \tfrac{5\pi}{12} + \tfrac{\pi}{2}k$.

When $k = 0$, we have:
$$z_1 = 2\left[\cos\tfrac{5\pi}{12} + i\sin\tfrac{5\pi}{12}\right] = 2\left[\tfrac{\sqrt{6}-\sqrt{2}}{4} + \tfrac{\sqrt{6}+\sqrt{2}}{4}i\right] = \tfrac{\sqrt{6}-\sqrt{2}}{2} + \tfrac{\sqrt{6}+\sqrt{2}}{2}i$$

When $k = 1$, we have:
$$z_2 = 2\left[\cos\tfrac{11\pi}{12} + i\sin\tfrac{11\pi}{12}\right] = 2\left[\tfrac{-\sqrt{2}-\sqrt{6}}{4} + \tfrac{\sqrt{6}-\sqrt{2}}{4}i\right] = \tfrac{-\sqrt{2}-\sqrt{6}}{2} + \tfrac{\sqrt{6}-\sqrt{2}}{2}i$$

When $k = 2$, we have:
$$z_3 = 2\left[\cos\tfrac{17\pi}{12} + i\sin\tfrac{17\pi}{12}\right] = 2\left[\tfrac{\sqrt{2}-\sqrt{6}}{4} + \tfrac{-\sqrt{2}-\sqrt{6}}{4}i\right] = \tfrac{\sqrt{2}-\sqrt{6}}{2} - \tfrac{\sqrt{2}+\sqrt{6}}{2}i$$

When $k = 3$, we have:
$$z_4 = 2\left[\cos\tfrac{23\pi}{12} + i\sin\tfrac{23\pi}{12}\right] = 2\left[\tfrac{\sqrt{2}+\sqrt{6}}{4} + \tfrac{\sqrt{2}-\sqrt{6}}{4}i\right] = \tfrac{\sqrt{2}+\sqrt{6}}{2} + \tfrac{\sqrt{2}-\sqrt{6}}{2}i$$

So the fourth roots of $8 - 8\sqrt{3}i$ are $\tfrac{\sqrt{6}-\sqrt{2}}{2} + \tfrac{\sqrt{6}+\sqrt{2}}{2}i$, $\tfrac{-\sqrt{2}-\sqrt{6}}{2} + \tfrac{\sqrt{6}-\sqrt{2}}{2}i$,
$\tfrac{\sqrt{2}-\sqrt{6}}{2} + \tfrac{-\sqrt{2}-\sqrt{6}}{2}i$ and $\tfrac{\sqrt{2}+\sqrt{6}}{2} + \tfrac{\sqrt{2}-\sqrt{6}}{2}i$. These final results were obtained using the addition formulas for sine and cosine.
Note: If you used half-angle formulas, your answers, though identical in value, may "look" vastly different. Those answers (which are correct) are $\sqrt{2-\sqrt{3}} + \sqrt{2+\sqrt{3}}i$, $-\sqrt{2+\sqrt{3}} + \sqrt{2-\sqrt{3}}i$, $-\sqrt{2-\sqrt{3}} - \sqrt{2+\sqrt{3}}i$ and $\sqrt{2+\sqrt{3}} - \sqrt{2-\sqrt{3}}i$.

69. (a) Let $z = r(\cos\theta + i\sin\theta)$, so $z^3 = r^3(\cos 3\theta + i\sin 3\theta)$. Since $1 = 1(\cos 0 + i\sin 0)$, we have $r^3 = 1$ so $r = 1$, and $3\theta = 0 + 2\pi k$, so $\theta = 0 + \tfrac{2\pi}{3}k$.

When $k = 0$, we have:
$$z_1 = 1[\cos 0 + i\sin 0] = 1(1 + 0i) = 1$$
When $k = 1$, we have:
$$z_2 = 1\left[\cos\frac{2\pi}{3} + i\sin\frac{2\pi}{3}\right] = 1\left(-\frac{1}{2} + \frac{\sqrt{3}}{2}i\right) = -\frac{1}{2} + \frac{\sqrt{3}}{2}i$$
When $k = 2$, we have:
$$z_3 = 1\left[\cos\frac{4\pi}{3} + i\sin\frac{4\pi}{3}\right] = 1\left(-\frac{1}{2} - \frac{\sqrt{3}}{2}i\right) = -\frac{1}{2} - \frac{\sqrt{3}}{2}i$$
So the cube roots of 1 are 1, $-\frac{1}{2} + \frac{\sqrt{3}}{2}i$ and $-\frac{1}{2} - \frac{\sqrt{3}}{2}i$.

(b) We compute the sum:
$$z_1 + z_2 + z_3 = 1 + \left(-\frac{1}{2} + \frac{\sqrt{3}}{2}i\right) + \left(-\frac{1}{2} - \frac{\sqrt{3}}{2}i\right) = 0 + 0i = 0$$
Now compute the products:
$$z_1 z_2 = 1\left[\cos\frac{2\pi}{3} + i\sin\frac{2\pi}{3}\right] = -\frac{1}{2} + \frac{\sqrt{3}}{2}i \qquad \text{(Note this is } z_2\text{)}$$
$$z_2 z_3 = 1[\cos 2\pi + i\sin 2\pi] = 1 \qquad \text{(Note this is } z_1\text{)}$$
$$z_3 z_1 = 1\left[\cos\frac{4\pi}{3} + i\sin\frac{4\pi}{3}\right] = -\frac{1}{2} - \frac{\sqrt{3}}{2}i \qquad \text{(Note this is } z_3\text{)}$$
So $z_1 z_2 + z_2 z_3 + z_3 z_1 = z_1 + z_2 + z_3 = 0$.

71. Using trigonometric forms and DeMoivre's theorem, we compute the powers:
$$\left[\frac{-1 + i\sqrt{3}}{2}\right]^5 = \left[-\frac{1}{2} + \frac{\sqrt{3}}{2}i\right]^5 = \left[\cos\frac{2\pi}{3} + i\sin\frac{2\pi}{3}\right]^5 = \cos\frac{10\pi}{3} + i\sin\frac{10\pi}{3} = -\frac{1}{2} - \frac{\sqrt{3}}{2}i$$
$$\left[\frac{-1 - i\sqrt{3}}{2}\right]^5 = \left[-\frac{1}{2} - \frac{\sqrt{3}}{2}i\right]^5 = \left[\cos\frac{4\pi}{3} + i\sin\frac{4\pi}{3}\right]^5 = \cos\frac{20\pi}{3} + i\sin\frac{20\pi}{3} = -\frac{1}{2} + \frac{\sqrt{3}}{2}i$$
Now compute the sum:
$$\left[\frac{-1 + i\sqrt{3}}{2}\right]^5 + \left[\frac{-1 - i\sqrt{3}}{2}\right]^5 = \left(-\frac{1}{2} - \frac{\sqrt{3}}{2}i\right) + \left(-\frac{1}{2} + \frac{\sqrt{3}}{2}i\right) = -1$$

73. Since the multiplication results in a difference of squares, we have:
$$(\cos\theta + i\sin\theta)(\cos\theta - i\sin\theta) = \cos^2\theta - i^2\sin^2\theta$$
$$= \cos^2\theta + \sin^2\theta \quad (\text{since } i^2 = -1)$$
$$= 1$$

75. Using the hint, we have:
$$\frac{r(\cos\alpha + i\sin\alpha)}{R(\cos\beta + i\sin\beta)} \cdot \frac{(\cos\beta - i\sin\beta)}{(\cos\beta - i\sin\beta)}$$
$$= \frac{r(\cos\alpha\cos\beta + i\sin\alpha\cos\beta - i\cos\alpha\sin\beta - i^2\sin\alpha\sin\beta)}{R(\cos^2\beta - i^2\sin^2\beta)}$$
$$= \frac{r[(\cos\alpha\cos\beta + \sin\alpha\sin\beta) + i(\sin\alpha\cos\beta - \cos\alpha\sin\beta)]}{R(\cos^2\beta + \sin^2\beta)}$$

Using the difference identities for sine and cosine:
$$\frac{r\left[\cos(\alpha-\beta)+i\sin(\alpha-\beta)\right]}{R}=\frac{r}{R}\left[\cos(\alpha-\beta)+i\sin(\alpha-\beta)\right]$$

77. Using the hint and the identities $\cos(-\theta)=\cos\theta$ and $\sin(-\theta)=-\sin\theta$, we have:
$$\frac{1}{z}=\frac{1(\cos 0+i\sin 0)}{r(\cos\theta+i\sin\theta)}$$
$$=\frac{1}{r}\left[\cos(0-\theta)+i\sin(0-\theta)\right]\qquad\text{(by Exercise 75)}$$
$$=\frac{1}{r}\left[\cos(-\theta)+i\sin(-\theta)\right]$$
$$=\frac{1}{r}(\cos\theta-i\sin\theta)$$

Chapter Eleven Review Exercises

1. Using synthetic division:

$$\begin{array}{r|rrrrr}
-2 & 1 & 3 & -1 & -5 & 1 \\
   &   & -2 & -2 & 6 & -2 \\
\hline
   & 1 & 1 & -3 & 1 & -1
\end{array}$$

So $q(x)=x^3+x^2-3x+1$ and $R(x)=-1$.

3. Using synthetic division:

$$\begin{array}{r|rrrrr}
3 & 1 & 0 & -2 & 0 & 8 \\
  &   & 3 & 9 & 21 & 63 \\
\hline
  & 1 & 3 & 7 & 21 & 71
\end{array}$$

So the quotient is $x^3+3x^2+7x+21$ and the remainder is 71.

5. Using synthetic division:

$$\begin{array}{r|rrrr}
-4 & 2 & -5 & -6 & -3 \\
   &   & -8 & 52 & -184 \\
\hline
   & 2 & -13 & 46 & -187
\end{array}$$

So the quotient is $2x^2-13x+46$ and the remainder is -187.

7. Using synthetic division:

$$\begin{array}{r|rrr}
-0.2 & 5 & -19 & -4 \\
     &   & -1 & 4 \\
\hline
     & 5 & -20 & 0
\end{array}$$

So the quotient is $5x-20$ and the remainder is zero.

9. Using synthetic division:

$$
\begin{array}{r|rrrrrr}
10 & 1 & 0 & 0 & 0 & -10 & 4 \\
   &   & 10 & 100 & 1000 & 10000 & 99900 \\
\hline
   & 1 & 10 & 100 & 1000 & 9990 & 99904
\end{array}
$$

So $f(10) = 99{,}904$.

11. Using synthetic division:

$$
\begin{array}{r|rrrr}
1/10 & 1 & -10 & 1 & -1 \\
     &   & 1/10 & -99/100 & 1/1000 \\
\hline
     & 1 & -99/10 & 1/100 & -999/1000
\end{array}
$$

So $f\left(\frac{1}{10}\right) = -\frac{999}{1000}$.

13. Using synthetic division:

$$
\begin{array}{r|rrrr}
a-1 & 1 & 3 & 3 & 1 \\
    &   & a-1 & a^2+a-2 & a^3-1 \\
\hline
    & 1 & a+2 & a^2+a+1 & a^3
\end{array}
$$

So $f(a - 1) = a^3$.

15. (a) Using synthetic division:

$$
\begin{array}{r|rrrrr}
-0.3 & 1 & 4 & -6 & -8 & -2 \\
     &   & -0.3 & -1.11 & 2.133 & 1.7601 \\
\hline
     & 1 & 3.7 & -7.11 & -5.867 & -0.24
\end{array}
$$

So $f(-0.3) \approx -0.24$.

(b) Using synthetic division:

$$
\begin{array}{r|rrrrr}
-0.39 & 1 & 4 & -6 & -8 & -2 \\
      &   & -0.39 & -1.4079 & 2.8891 & 1.9933 \\
\hline
      & 1 & 3.61 & -7.4079 & -5.1109 & -0.007
\end{array}
$$

So $f(-0.39) \approx -0.007$.

(c) Using synthetic division:

$$
\begin{array}{r|rrrrr}
-0.394 & 1 & 4 & -6 & -8 & -2 \\
       &   & -0.394 & -1.420764 & 2.92378 & 2.00003 \\
\hline
       & 1 & 3.606 & -7.420764 & -5.07622 & 0.00003
\end{array}
$$

So $f(-0.394) \approx 0.00003$

17. Using synthetic division:

$$
\begin{array}{r|rrrr}
3 & 1 & -4 & -a & -6 \\
  &   & 3 & -3 & -3a-9 \\
\hline
  & 1 & -1 & -a-3 & -3a-15
\end{array}
$$

So if 3 is a root, then:

$$-3a - 15 = 0$$
$$-3a = 15$$
$$a = -5$$

19. Using synthetic division:

$$
\begin{array}{r|cccc}
1 & a^2 & 3a & 0 & 2 \\
  &   & a^2 & a^2+3a & a^2+3a \\
\hline
  & a^2 & a^2+3a & a^2+3a & a^2+3a+2
\end{array}
$$

So if $x - 1$ is a factor, then:

$$a^2 + 3a + 2 = 0$$
$$(a+1)(a+2) = 0$$
$$a = -1, -2$$

So $a = -1$ or $a = -2$.

21. (a) Let $f(x + h) = 0$, and let $x = r - h$. Then $f(r - h + h) = f(r) = 0$, since r is a root of $f(x) = 0$.

(b) Let $f(-x) = 0$, and let $x = -r$. Then $f(-(-r)) = f(r) = 0$, since r is a root of $f(x) = 0$.

(c) Let $f\left(\frac{x}{k}\right) = 0$, and let $x = kr$. Then $f\left(\frac{kr}{k}\right) = f(r) = 0$, since r is a root of $f(x) = 0$.

23. p-factors: $\pm1, \pm2, \pm3, \pm6, \pm9, \pm18$
q-factors: ±1
possible rational roots: $\pm1, \pm2, \pm3, \pm6, \pm9, \pm18$

25. p-factors: $\pm1, \pm2, \pm4, \pm8$
q-factors: $\pm1, \pm2$
possible rational roots: $\pm1, \pm\frac{1}{2}, \pm2, \pm4, \pm8$

27. p-factors: $\pm p, \pm1$
q-factors: ±1
possible rational roots: $\pm p, \pm1$

29. The possible rational roots are $\pm1, \pm\frac{1}{2}, \pm2, \pm3, \pm\frac{3}{2}, \pm6$. Using synthetic division:

$$
\begin{array}{r|cccc}
2 & 2 & 1 & -7 & -6 \\
  &   & 4 & 10 & 6 \\
\hline
  & 2 & 5 & 3 & 0
\end{array}
$$

So $2x^3 + x^2 - 7x - 6 = (x - 2)(2x^2 + 5x + 3) = (x - 2)(2x + 3)(x + 1)$.
So the roots are 2, $-\frac{3}{2}$ and -1.

31. The possible rational roots are $\pm1, \pm\frac{1}{2}, \pm2, \pm5, \pm\frac{5}{2}, \pm10$. Using synthetic division:

$$
\begin{array}{r|cccc}
5/2 & 2 & -1 & -14 & 10 \\
    &   & 5 & 10 & -10 \\
\hline
    & 2 & 4 & -4 & 0
\end{array}
$$

So $2x^3 - x^2 - 14x + 10 = \left(x - \frac{5}{2}\right)(2x^2 + 4x - 4) = 2\left(x - \frac{5}{2}\right)(x^2 + 2x - 2)$.

Using the quadratic formula:
$$x = \frac{-2 \pm \sqrt{4+8}}{2} = \frac{-2 \pm 2\sqrt{3}}{2} = -1 \pm \sqrt{3}$$
So the roots are $\frac{5}{2}$, $-1 + \sqrt{3}$ and $-1 - \sqrt{3}$.

33. We first multiply by 2 to obtain $3x^3 + x^2 + x - 2 = 0$. The possible rational roots are $\pm 1, \pm \frac{1}{3}, \pm 2, \pm \frac{2}{3}$. Using synthetic division:

$$
\begin{array}{r|rrrr}
2/3 & 3 & 1 & 1 & -2 \\
    &   & 2 & 2 & 2 \\
\hline
    & 3 & 3 & 3 & 0
\end{array}
$$

So $3x^3 + x^2 + x - 2 = \left(x - \frac{2}{3}\right)\left(3x^2 + 3x + 3\right) = 3\left(x - \frac{2}{3}\right)\left(x^2 + x + 1\right)$.
Using the quadratic formula:
$$x = \frac{-1 \pm \sqrt{1-4}}{2} = \frac{-1 \pm i\sqrt{3}}{2}$$
So the roots are $\frac{2}{3}$, $\frac{-1+i\sqrt{3}}{2}$, and $\frac{-1-i\sqrt{3}}{2}$.

35. The possible rational roots are ± 1, ± 7, ± 49. Using synthetic division:

$$
\begin{array}{r|rrrrrr}
-1 & 1 & 1 & -14 & -14 & 49 & 49 \\
   &   & -1 & 0 & 14 & 0 & -49 \\
\hline
   & 1 & 0 & -14 & 0 & 49 & 0
\end{array}
$$

Therefore:
$$x^5 + x^4 - 14x^3 - 14x^2 + 49x + 49 = (x+1)\left(x^4 - 14x^2 + 49\right)$$
$$= (x+1)\left(x^2 - 7\right)^2$$
$$= (x+1)\left(x + \sqrt{7}\right)^2\left(x - \sqrt{7}\right)^2$$
So the roots are -1, $-\sqrt{7}$ (multiplicity 2), and $\sqrt{7}$ (multiplicity 2).

37. Since $x^3 - 9x^2 + 24x - 20 = 0$ has a root r with multiplicity 2, we can use synthetic division by r twice, each time the remainder must be 0:

$$
\begin{array}{r|rrrr}
r & 1 & -9 & 24 & -20 \\
  &   & r & r^2 - 9r & r^3 - 9r^2 + 24r \\
\hline
  & 1 & r-9 & r^2 - 9r + 24 & r^3 - 9r^2 + 24r - 20
\end{array}
$$

We continue the division:

$$
\begin{array}{r|rrr}
r & 1 & r-9 & r^2 - 9r + 24 \\
  &   & r & 2r^2 - 9r \\
\hline
  & 1 & 2r-9 & 3r^2 - 18r + 24
\end{array}
$$

So $3r^2 - 18r + 24 = 3(r^2 - 6r + 8) = 3(r-4)(r-2) = 0$, thus $r = 4$ or $r = 2$. We check these in $r^3 - 9r^2 + 24r - 20 = 0$:

$r = 4$: $(4)^3 - 9(4)^2 + 24(4) - 20 = 64 - 144 + 96 - 20 = -4 \neq 0$

So $r = 4$ cannot be a root.

$r = 2$: $(2)^3 - 9(2)^2 + 24(2) - 20 = 8 - 36 + 48 - 20 = 0$

So $r = 2$ is the root with multiplicity 2. Since the synthetic division resulted in $x + (2r - 9) = 0$, then $x + (-5) = 0$ and thus $x = 5$. So the roots are 2 (multiplicity 2) and 5. Note: Actually, an easier (and more direct) approach is to find the roots directly. The possible rational roots are $\pm 1, \pm 2, \pm 4, \pm 5, \pm 10, \pm 20$. Note that there are 3 sign changes, so there can be 1 or 3 positive real roots. Since $f(-x)$ has no sign changes, all 3 roots must be positive real numbers for one of them to have multiplicity of 2 (note that it cannot have a radical or be complex--why?). We use synthetic division:

$$
\begin{array}{r|rrrr}
2 & 1 & -9 & 24 & -20 \\
  &   & 2 & -14 & 20 \\
\hline
  & 1 & -7 & 10 & 0 \\
\end{array}
$$

So $x^3 - 9x^2 + 24x - 20 = (x - 2)(x^2 - 7x + 10) = (x - 2)^2(x - 5)$.

So the roots are 2 (with multiplicity 2) and 5.

39. (a) Let $p(x)$ and $d(x)$ be the polynomials where $d(x) \neq 0$. Then there are unique polynomials $q(x)$ and $R(x)$ such that
$$p(x) = d(x) \cdot q(x) + R(x)$$
where either $R(x) = 0$ or the degree of $R(x)$ is less than the degree of $d(x)$.

(b) When a polynomial $f(x)$ is divided by $x - r$, the remainder is $f(r)$.

(c) Let $f(x)$ be a polynomial. If $f(r) = 0$, then $x - r$ is a factor of $f(x)$. Conversely, if $x - r$ is a factor of $f(x)$, then $f(r) = 0$.

(d) Every polynomial equation of the form
$$a_n x^n + a_{n-1} x^{n-1} + \ldots + a_1 x + a_0 = 0 \quad (n \geq 1, a_n \neq 0)$$
has at least one root among the complex numbers. (This root may be a real number.)

41. We factor:
$$6x^2 + 7x - 20 = (3x - 4)(2x + 5) = 6\left(x - \tfrac{4}{3}\right)\left[x - \left(-\tfrac{5}{2}\right)\right]$$

43. We factor:
$$
\begin{aligned}
x^4 - 4x^3 + 5x - 20 &= x^3(x - 4) + 5(x - 4) \\
&= (x - 4)(x^3 + 5) \\
&= (x - 4)\left(x + \sqrt[3]{5}\right)\left(x^2 - \sqrt[3]{5}x + \sqrt[3]{25}\right)
\end{aligned}
$$

We solve $x^2 - \sqrt[3]{5}x + \sqrt[3]{25} = 0$ using the quadratic formula, which yields:
$$x = \frac{\sqrt[3]{5} \pm i\sqrt{3\sqrt[3]{25}}}{2}$$

So $x^4 - 4x^3 + 3x - 20 = (x - 4)\left(x - \left(-\sqrt[3]{5}\right)\right)\left(x - \dfrac{\sqrt[3]{5} + i\sqrt{3\sqrt[3]{25}}}{2}\right)\left(x - \dfrac{\sqrt[3]{5} - i\sqrt{3\sqrt[3]{25}}}{2}\right)$.

45. One other root is $2 + 3i$, so $\left[x - (2 - 3i)\right]\left[x - (2 + 3i)\right]$ are factors, which is $x^2 - 4x + 13$. Using long division:

$$
\begin{array}{r}
x - 3 \\
x^2 - 4x + 13 \overline{\smash{\big)}\ x^3 - 7x^2 + 25x - 39} \\
\underline{x^3 - 4x^2 + 13x} \\
-3x^2 + 12x - 39 \\
\underline{-3x^2 + 12x - 39} \\
0
\end{array}
$$

So $x - 3$ is the other factor. So the roots are $2 - 3i$, $2 + 3i$, and 3.

47. One other root is $1 - i\sqrt{2}$, so $\left[x - (1 + i\sqrt{2})\right]\left[x - (1 - i\sqrt{2})\right]$ are factors, which is $x^2 - 2x + 3$. Using long division:

$$
\begin{array}{r}
x^2 - 4 \\
x^2 - 2x + 3 \overline{\smash{\big)}\ x^4 - 2x^3 - 4x^2 + 14x - 21} \\
\underline{x^4 - 2x^3 + 3x^2} \\
-7x^2 + 14x - 21 \\
\underline{-7x^2 + 14x - 21} \\
0
\end{array}
$$

So $x^2 - 7 = (x + \sqrt{7})(x - \sqrt{7})$ is the other factor. So the roots are $1 + i\sqrt{2}$, $1 - i\sqrt{2}$, $\sqrt{7}$, and $-\sqrt{7}$.

49. Let $f(x) = x^3 + 8x - 7$. There is one sign change, so there is 1 positive root. Now $f(-x) = -x^3 - 8x - 7$. Since there are no sign changes, there are no negative roots. So the equation has 1 positive real root and 2 complex roots.

51. Let $f(x) = x^3 + 3x + 1$. There are no sign changes, so there are no positive roots. Now $f(-x) = -x^3 - 3x + 1$. Since there is one sign change, there is 1 negative root. So the equation has 1 negative real root and 2 complex roots.

53. Let $f(x) = x^4 - 10$. There is one sign change, so there is 1 positive root. Since $f(-x) = f(x)$, there is also 1 negative root. So the equation has 1 positive real root, 1 negative real root and 2 complex roots.

55. (a) Let $f(x) = x^3 + x^2 + x + 1$, so $f(-x) = -x^3 + x^2 - x + 1$. There are three sign changes, so there are either 1 or 3 negative roots.

   (b) Using the hint, we have $(x - 1)(x^3 + x^2 + x + 1) = x^4 - 1$. Let $g(x) = x^4 - 1$. Then $g(-x) = x^4 - 1$, so this can only have 1 negative root. Since $x = 1$ is a positive root, then $x^3 + x^2 + x + 1 = 0$ can only have 1 negative root also.

(c) Factoring, we have $x^3 + x^2 + x + 1 = x^2(x+1) + (x+1) = (x+1)(x^2+1)$.
So $x = -1$ is a root. Also:
$$x^2 + 1 = 0$$
$$x^2 = -1$$
$$x = \pm i$$
So the roots are -1, i, and $-i$.

57. Substituting $y = x^3$, we have $x^2 + \left(x^3\right)^2 = 1$, so $x^6 + x^2 - 1 = 0$.
Let $f(x) = x^6 + x^2 - 1$.
$$f(0) = -1$$
$$f(1) = 1$$
So there is a root between 0 and 1.
$$f(0.8) = -0.09$$
$$f(0.9) = 0.34$$
So there is a root between 0.8 and 0.9.
$$f(0.82) = -0.02$$
$$f(0.83) = 0.01$$
So there is a root between 0.82 and 0.83.
So the x-coordinate lies between 0.82 and 0.83.

59. (a) Let $f(x) = x^3 - 36x - 84$. Since there is only one sign change, there is 1 positive real root.

(b) Using synthetic division:

$$
\begin{array}{r|rrrr}
7 & 1 & 0 & -36 & -84 \\
  &   & 7 & 49 & 91 \\
\hline
  & 1 & 7 & 13 & 7 \\
\end{array}
$$

Since the row is all positive, there are no real roots to the equation greater than 7.

(c) Here $f(6) = -84$ and $f(7) = 7$, so there is a root between 6 and 7.
$f(6.8) = -14.4$ and $f(6.9) = -3.89$, so there is a root between 6.9 and 7.0.
$f(6.93) = -0.67$ and $f(6.94) = 0.42$, so there is a root between 6.93 and 6.94.

61. Another root will be $4 + \sqrt{5}$, so:
$$\left[x - (4 - \sqrt{5})\right]\left[x - (4 + \sqrt{5})\right] = 0$$
$$x^2 - 8x + 11 = 0$$

63. Other roots will be $6 + 2i$ and $-\sqrt{5}$:
$$\left(x - \sqrt{5}\right)\left[x - (-\sqrt{5})\right]\left[x - (6 - 2i)\right]\left[x - (6 + 2i)\right] = 0$$
$$\left(x^2 - 5\right)\left(x^2 - 12x + 40\right) = 0$$
$$x^4 - 12x^3 + 35x^2 + 60x - 200 = 0$$

65. We write $x - 1 = \sqrt{2} + \sqrt{3}$.
Squaring both sides, we obtain:
$$x^2 - 2x + 1 = 5 + 2\sqrt{6}$$
$$x^2 - 2x - 4 = 2\sqrt{6}$$
Squaring both sides again, we obtain:
$$x^4 - 4x^3 - 4x^2 + 16x + 16 = 24$$
$$x^4 - 4x^3 - 4x^2 + 16x - 8 = 0$$

67. Factoring, we have $y = x^3 - 2x^2 - 3x = x(x^2 - 2x - 3) = x(x - 3)(x + 1)$. The zeros are 0, 3, and –1.

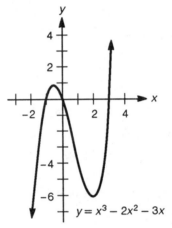
$$y = x^3 - 2x^2 - 3x$$

69. Factoring, we have $y = x^4 - 4x^2 = x^2(x^2 - 4) = x^2(x + 2)(x - 2)$. The zeros are 0, –2, and 2.

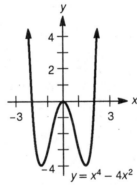
$$y = x^4 - 4x^2$$

71. Performing the indicated operations yields:
$$(3 - 2i)(3 + 2i) + (1 + 3i)^2 = 9 - 4i^2 + 1 + 6i + 9i^2 = 9 + 4 + 1 + 6i - 9 = 5 + 6i$$

73. Performing the indicated operations yields:
$$(1 + i\sqrt{2})(1 - i\sqrt{2}) + (\sqrt{2} + i)(\sqrt{2} - i) = 1 - 2i^2 + 2 - i^2 = 1 + 2 + 2 + 1 = 6$$

75. Performing the indicated operations yields:
$$\frac{3-i\sqrt{3}}{3+i\sqrt{3}}\cdot\frac{3-i\sqrt{3}}{3-i\sqrt{3}}=\frac{9-6i\sqrt{3}-3}{9+3}=\frac{6-6i\sqrt{3}}{12}=\frac{1-i\sqrt{3}}{2}=\frac{1}{2}-\frac{\sqrt{3}}{2}i$$

77. Writing in complex form, we have:
$$-\sqrt{-2}\sqrt{-9}+\sqrt{-8}-\sqrt{-72}=-(i\sqrt{2})(3i)+2i\sqrt{2}-6i\sqrt{2}=3\sqrt{2}-4i\sqrt{2}=3\sqrt{2}-4\sqrt{2}i$$

79. Let $z=a+bi$, so $\bar{z}=a-bi$. Therefore:
$$\frac{z+\bar{z}}{2}=\frac{a+bi+a-bi}{2}=\frac{2a}{2}=a=Re(z)$$

81. (a) We compute the absolute values:
$$|6+2i|=\sqrt{6^2+2^2}=\sqrt{36+4}=\sqrt{40}=2\sqrt{10}$$
$$|6-2i|=\sqrt{6^2+(-2)^2}=\sqrt{36+4}=\sqrt{40}=2\sqrt{10}$$

(b) Using the new definition:
$$|-3|=|-3+0i|=\sqrt{(-3)^2+0^2}=\sqrt{9}=3$$

(c) Let $z=a+bi$, so $\bar{z}=a-bi$. We now compute:
$$z\bar{z}=(a+bi)(a-bi)=a^2-b^2i^2=a^2+b^2$$
$$|z|^2=\left(\sqrt{a^2+b^2}\right)^2=a^2+b^2$$
Thus $z\bar{z}=|z|^2$.

83. Combining the fractions, we have:
$$\frac{1}{a-bi}-\frac{1}{a+bi}=\frac{(a+bi)-(a-bi)}{(a-bi)(a+bi)}=\frac{2bi}{a^2-b^2i^2}=\frac{2bi}{a^2+b^2}$$

85. Combining the fractions, we have:
$$\frac{a+bi}{a-bi}-\frac{a-bi}{a+bi}=\frac{(a+bi)^2-(a-bi)^2}{(a-bi)(a+bi)}$$
$$=\frac{a^2+2abi-b^2-a^2+2abi+b^2}{a^2+b^2}$$
$$=\frac{4abi}{a^2+b^2}$$

87. We convert to rectangular form:
$$3\left[\cos\frac{\pi}{3}+i\sin\frac{\pi}{3}\right]=3\left(\frac{1}{2}+\frac{\sqrt{3}}{2}i\right)=\frac{3}{2}+\frac{3\sqrt{3}}{2}i$$

89. We convert to rectangular form:
$$2^{1/4}\left[\cos\frac{7\pi}{4}+i\sin\frac{7\pi}{4}\right]=2^{1/4}\left(\frac{\sqrt{2}}{2}-\frac{\sqrt{2}}{2}i\right)=2^{-1/4}-2^{-1/4}i$$

91. We convert to trigonometric form:
$$\tfrac{1}{2}+\tfrac{\sqrt{3}}{2}i = 1\left[\cos\tfrac{\pi}{3}+i\sin\tfrac{\pi}{3}\right]$$

93. We convert to trigonometric form:
$$-3\sqrt{2}-3\sqrt{2}i = 6\left(-\tfrac{\sqrt{2}}{2}-\tfrac{\sqrt{2}}{2}i\right) = 6\left[\cos\tfrac{5\pi}{4}+i\sin\tfrac{5\pi}{4}\right]$$

95. We compute the product:
$$5\left[\cos\tfrac{\pi}{7}+i\sin\tfrac{\pi}{7}\right]\cdot 2\left[\cos\tfrac{3\pi}{28}+i\sin\tfrac{3\pi}{28}\right] = 10\left[\cos\tfrac{7\pi}{28}+i\sin\tfrac{7\pi}{28}\right]$$
$$= 10\left[\cos\tfrac{\pi}{4}+i\sin\tfrac{\pi}{4}\right]$$
$$= 10\left(\tfrac{\sqrt{2}}{2}+\tfrac{\sqrt{2}}{2}i\right)$$
$$= 5\sqrt{2}+5\sqrt{2}i$$

97. We compute the quotient:
$$8\left[\cos\tfrac{\pi}{12}+i\sin\tfrac{\pi}{12}\right]\div 4\left[\cos\tfrac{\pi}{3}+i\sin\tfrac{\pi}{3}\right] = 2\left[\cos\left(-\tfrac{3\pi}{12}\right)+i\sin\left(-\tfrac{3\pi}{12}\right)\right]$$
$$= 2\left[\cos\left(-\tfrac{\pi}{4}\right)+i\sin\left(-\tfrac{\pi}{4}\right)\right]$$
$$= 2\left(\tfrac{\sqrt{2}}{2}-\tfrac{\sqrt{2}}{2}i\right)$$
$$= \sqrt{2}-\sqrt{2}i$$

99. We compute the power using DeMoivre's theorem:
$$\left[3^{1/4}\left(\cos\tfrac{\pi}{36}+i\sin\tfrac{\pi}{36}\right)\right]^{12} = 3^{3}\left[\cos\tfrac{12\pi}{36}+i\sin\tfrac{12\pi}{36}\right]$$
$$= 27\left[\cos\tfrac{\pi}{3}+i\sin\tfrac{\pi}{3}\right]$$
$$= 27\left(\tfrac{1}{2}+\tfrac{\sqrt{3}}{2}i\right)$$
$$= \tfrac{27}{2}+\tfrac{27\sqrt{3}}{2}i$$

101. We first write $\sqrt{3}+i$ in trigonometric form:
$$\sqrt{3}+i = 2\left(\tfrac{\sqrt{3}}{2}+\tfrac{1}{2}i\right) = 2\left[\cos\tfrac{\pi}{6}+i\sin\tfrac{\pi}{6}\right]$$
Now compute the power using DeMoivre's theorem:
$$(\sqrt{3}+i)^{10} = \left[2\left(\cos\tfrac{\pi}{6}+i\sin\tfrac{\pi}{6}\right)\right]^{10}$$
$$= 2^{10}\left[\cos\tfrac{5\pi}{3}+i\sin\tfrac{5\pi}{3}\right]$$
$$= 2^{10}\left(\tfrac{1}{2}-\tfrac{\sqrt{3}}{2}i\right)$$
$$= 2^{9}-2^{9}\sqrt{3}i$$
$$= 512-512\sqrt{3}i$$

103. Let $z = r(\cos\theta + i\sin\theta)$ and $1 = 1[\cos 0 + i\sin 0]$, so:
$$z^6 = r^6(\cos 6\theta + i\sin 6\theta) = 1(\cos 0 + i\sin 0)$$
Thus $r = 1$ and $6\theta = 0 + 2\pi k$, so $\theta = 0 + \frac{\pi}{3}k$.
When $k = 0$, we have:
$$z_1 = 1[\cos 0 + i\sin 0] = 1(1 + 0i) = 1$$
When $k = 1$, we have:
$$z_2 = 1\left[\cos\frac{\pi}{3} + i\sin\frac{\pi}{3}\right] = \frac{1}{2} + \frac{\sqrt{3}}{2}i$$
When $k = 2$, we have:
$$z_3 = 1\left[\cos\frac{2\pi}{3} + i\sin\frac{2\pi}{3}\right] = -\frac{1}{2} + \frac{\sqrt{3}}{2}i$$
When $k = 3$, we have:
$$z_4 = 1[\cos\pi + i\sin\pi] = -1 + 0i = -1$$
When $k = 4$, we have:
$$z_5 = 1\left[\cos\frac{4\pi}{3} + i\sin\frac{4\pi}{3}\right] = -\frac{1}{2} - \frac{\sqrt{3}}{2}i$$
When $k = 5$, we have:
$$z_6 = 1\left[\cos\frac{5\pi}{3} + i\sin\frac{5\pi}{3}\right] = \frac{1}{2} - \frac{\sqrt{3}}{2}i$$
So the sixth roots of 1 are $1, \frac{1}{2} + \frac{\sqrt{3}}{2}i, -\frac{1}{2} + \frac{\sqrt{3}}{2}i, -1, -\frac{1}{2} - \frac{\sqrt{3}}{2}i$ and $\frac{1}{2} - \frac{\sqrt{3}}{2}i$.

105. Let $z = r(\cos\theta + i\sin\theta)$ and write $\sqrt{2} - \sqrt{2}i$ in trigonometric form:
$$\sqrt{2} - \sqrt{2}i = 2\left(\frac{\sqrt{2}}{2} - \frac{\sqrt{2}}{2}i\right) = 2\left[\cos\frac{7\pi}{4} + i\sin\frac{7\pi}{4}\right]$$
Then $z^2 = r^2(\cos 2\theta + i\sin 2\theta) = 2\left[\cos\frac{7\pi}{4} + i\sin\frac{7\pi}{4}\right]$.
So $r^2 = 2$ and $r = \sqrt{2}$, and also $2\theta = \frac{7\pi}{4} + 2\pi k$, so $\theta = \frac{7\pi}{8} + \pi k$.
When $k = 0$, we have:
$$z_1 = \sqrt{2}\left[\cos\frac{7\pi}{8} + i\sin\frac{7\pi}{8}\right]$$
$$= \sqrt{2}\left[-\frac{\sqrt{2+\sqrt{2}}}{2} + i\frac{\sqrt{2-\sqrt{2}}}{2}\right]$$
(by the half-angle identities)
$$= -\frac{\sqrt{4+2\sqrt{2}}}{2} + \frac{\sqrt{4-2\sqrt{2}}}{2}i$$
When $k = 1$, we have:
$$z_2 = \sqrt{2}\left[\cos\frac{15\pi}{8} + i\sin\frac{15\pi}{8}\right] = \sqrt{2}\left[\frac{\sqrt{2+\sqrt{2}}}{2} - i\frac{\sqrt{2-\sqrt{2}}}{2}\right] = \frac{\sqrt{4+2\sqrt{2}}}{2} - \frac{\sqrt{4-2\sqrt{2}}}{2}i$$
So the square roots of $\sqrt{2} - \sqrt{2}i$ are $-\frac{\sqrt{4+2\sqrt{2}}}{2} + \frac{\sqrt{4-2\sqrt{2}}}{2}i$ and $\frac{\sqrt{4+2\sqrt{2}}}{2} - \frac{\sqrt{4-2\sqrt{2}}}{2}i$.

107. Let $z = r(\cos\theta + i\sin\theta)$, and write $1 + i$ in trigonometric form:
$$1 + i = \sqrt{2}\left(\frac{\sqrt{2}}{2} + \frac{\sqrt{2}}{2}i\right) = \sqrt{2}\left[\cos\frac{\pi}{4} + i\sin\frac{\pi}{4}\right]$$
Then $z^5 = r^5[\cos 5\theta + i\sin 5\theta] = \sqrt{2}[\cos 45° + i\sin 45°]$.
So $r^5 = \sqrt{2}$ and $r = 2^{1/10} \approx 1.0718$, and $5\theta = 45° + 360°k$, so $\theta = 9° + 72°k$.

When $k = 0$: $z_1 = 1.0718[\cos 9° + i \sin 9°] \approx 1.06 + 0.17i$

When $k = 1$: $z_2 = 1.0718[\cos 81° + i \sin 81°] \approx 0.17 + 1.06i$

When $k = 2$: $z_3 = 1.0718[\cos 153° + i \sin 153°] \approx -0.95 + 0.49i$

When $k = 3$: $z_4 = 1.0718[\cos 225° + i \sin 225°] \approx -0.76 - 0.76i$

When $k = 4$: $z_5 = 1.0718[\cos 297° + i \sin 297°] \approx 0.49 - 0.95i$

So the fifth roots of $1 + i$ are $1.06 + 0.17i$, $0.17 + 1.06i$, $-0.95 + 0.49i$, $-0.76 - 0.76i$ and $0.49 - 0.95i$.

Chapter Eleven Test

1. Since $f\left(\frac{1}{2}\right)$ will be the remainder after synthetic division, we synthetically divide by $\frac{1}{2}$:

$$
\begin{array}{r|rrrrr}
1/2 & 6 & -5 & 7 & -2 & -2 \\
    &   & 3 & -1 & 3 & 1/2 \\
\hline
    & 6 & -2 & 6 & 1 & -3/2
\end{array}
$$

So $f\left(\frac{1}{2}\right) = -\frac{3}{2}$.

2. Using synthetic division:

$$
\begin{array}{r|rrrr}
-3 & 1 & 1 & -11 & -15 \\
   &   & -3 & 6 & 15 \\
\hline
   & 1 & -2 & -5 & 0
\end{array}
$$

So $x^3 + x^2 - 11x - 15 = (x + 3)(x^2 - 2x - 5)$. Using the quadratic formula, we have:

$$x = \frac{2 \pm \sqrt{4 + 20}}{2} = \frac{2 \pm 2\sqrt{6}}{2} = 1 \pm \sqrt{6}$$

So the roots of the equation are -3, $1 + \sqrt{6}$, and $1 - \sqrt{6}$.

3. $\pm 1, \pm \frac{1}{2}, \pm 2, \pm 3, \pm \frac{3}{2}, \pm 6$

4. Such a function would be $y = a(x - 1)(x + 8)$. Now $y = -24$ when $x = 0$, so $-24 = a(-1)(8)$, so $a = 3$. So the function is $y = 3(x - 1)(x + 8)$, or $f(x) = 3x^2 + 21x - 24$.

5. Using synthetic division:

$$
\begin{array}{r|rrrr}
-1 & 4 & 1 & -8 & 3 \\
   &   & -4 & 3 & 5 \\
\hline
   & 4 & -3 & -5 & 8
\end{array}
$$

The quotient is $4x^2 - 3x - 5$ and the remainder is 8.

6. (a) Let $f(x)$ be a polynomial. If $f(r) = 0$, then $x - r$ is a factor of $f(x)$. Conversely, if $x - r$ is a factor of $f(x)$, then $f(r) = 0$.

(b) Every polynomial of the form
$$a_n x^n + a_{n-1} x^{n-1} + \ldots + a_1 x + a_0 = 0 \ (n \geq 1, a_n \neq 0)$$
has at least one root among the complex numbers. (This root may be a real number.)

7. (a) Using synthetic division with 1:

$$\begin{array}{r|rrrr} 1 & 1 & -2 & 0 & -1 \\ & & 1 & -1 & -1 \\ \hline & 1 & -1 & -1 & -2 \end{array}$$

Now we use 2:

$$\begin{array}{r|rrrr} 2 & 1 & -2 & 0 & -1 \\ & & 2 & 0 & 0 \\ \hline & 1 & 0 & 0 & -1 \end{array}$$

Now use 3:

$$\begin{array}{r|rrrr} 3 & 1 & -2 & 0 & -1 \\ & & 3 & 3 & 9 \\ \hline & 1 & 1 & 3 & 8 \end{array}$$

So 3 is an upper bound for the roots.

(b) Since $f(2.2) = -0.032$ and $f(2.3) = 0.587$, the root lies between 2.2 and 2.3.

8. Two other roots are $1 - i$ and $3 + 2i$, so $x^2 - 2x + 2$ and $x^2 - 6x + 13$ are factors. Using long division:

$$x^2 - 2x + 2 \overline{\smash{\big)}\ x^5 - 6x^4 + 11x^3 + 16x^2 - 50x + 52}$$

quotient $x^3 - 4x^2 + x + 26$

$$\begin{aligned}
& x^5 - 2x^4 + 2x^3 \\
& -4x^4 + 9x^3 + 16x^2 \\
& -4x^4 + 8x^3 - 8x^2 \\
& \ x^3 + 24x^2 - 50x \\
& \ x^3 - 2x^2 + 2x \\
& 26x^2 - 52x + 52 \\
& 26x^2 - 52x + 52 \\
& 0
\end{aligned}$$

Using long division again:

$$x^2 - 6x + 13 \overline{\smash{\big)}\ x^3 - 4x^2 + x + 26}$$

quotient $x + 2$

$$\begin{aligned}
& x^3 - 6x^2 + 13x \\
& \ 2x^2 - 12x + 26 \\
& \ 2x^2 - 12x + 26 \\
& 0
\end{aligned}$$

So the roots are $1 \pm i$, $3 \pm 2i$, and -2.

9.   Using long division:

$$x^2 + 1 \enclose{longdiv}{x^4 + 2x^3 + 0x^2 - x + 6}$$

with quotient $x^2 + 2x - 1$:

$$
\begin{array}{r}
x^4 + x^2 \\ \hline
2x^3 - x^2 - x \\
2x^3 + 2x \\ \hline
-x^2 - 3x + 6 \\
-x^2 - 1 \\ \hline
-3x + 7
\end{array}
$$

So $q(x) = x^2 + 2x - 1$ and $R(x) = -3x + 7$.

10.   We first find the roots by the quadratic formula:

$$x = \frac{6 \pm \sqrt{36 - 40}}{4} = \frac{6 \pm 2i}{4} = \frac{3}{2} \pm \frac{1}{2}i$$

So $2x^2 - 6x + 5 = 2\left(x^2 - 3x + \frac{5}{2}\right) = 2\left[x - \left(\frac{3}{2} + \frac{1}{2}i\right)\right]\left[x - \left(\frac{3}{2} - \frac{1}{2}i\right)\right]$.

11.   (a)   The possible rational roots are $\pm 1, \pm 2, \pm 3, \pm 4, \pm 6, \pm 8, \pm 12$, and ± 24.

(b)   Using synthetic division:

$$
\begin{array}{r|rrrrr}
2 & 1 & -1 & 0 & 0 & 24 \\
  &   & 2 & 2 & 4 & 8 \\ \hline
  & 1 & 1 & 2 & 4 & 32
\end{array}
$$

Since this last row consists of all positive numbers, we know that 2 is an upper bound for the roots of this equation.

(c)   Only $x = 1$, since $x = 2$ is an upper bound and not a root.

(d)   Using synthetic division:

$$
\begin{array}{r|rrrrr}
1 & 1 & -1 & 0 & 0 & 24 \\
  &   & 1 & 0 & 0 & 1 \\ \hline
  & 1 & 0 & 0 & 0 & 25
\end{array}
$$

Since $x = 1$ was the only possibility, there are no positive rational roots.

12.   (a)   The possible rational roots are $\pm 1, \pm \frac{1}{2}, \pm 3, \pm \frac{3}{2}$.
Using synthetic division:

$$
\begin{array}{r|rrrr}
3/2 & 2 & -1 & -1 & -3 \\
    &   & 3 & 3 & 3 \\ \hline
    & 2 & 2 & 2 & 0
\end{array}
$$

So $2x^3 - x^2 - x - 3 = \left(x - \frac{3}{2}\right)\left(2x^2 + 2x + 2\right) = 2\left(x - \frac{3}{2}\right)\left(x^2 + x + 1\right)$. Using the quadratic equation, we have:

$$x = \frac{-1 \pm \sqrt{1 - 4}}{2} = \frac{-1 \pm i\sqrt{3}}{2}$$

So the only rational root is $\frac{3}{2}$.

(b)   All solutions are $\frac{3}{2}, \frac{-1+i\sqrt{3}}{2}$, and $\frac{-1-i\sqrt{3}}{2}$.

13.   Let $f(x) = 3x^4 + x^2 - 5x - 1$. Since there is one sign change, there is 1 positive root. $f(-x) = 3x^4 + x^2 + 5x - 1$. Since there is one sign change, there is 1 negative root. So the equation has 1 positive real root, 1 negative real root, and 2 complex roots.

14.   The factors are $x + 2$, $x - (1 - 3i)$, and $x - (1 + 3i)$, so:
$$(x+2)[x-(1-3i)][x-(1+3i)] = 0$$
$$(x+2)(x^2 - 2x + 10) = 0$$
$$x^3 + 6x + 20 = 0$$

15.   Here $f(x) = (x-2)(x-3i)^3[x-(1+\sqrt{2})]^2$. Note that had a restriction of rational coefficients been added, we would also have $-3i$ (multiplicity 3) and $1 - \sqrt{2}$ (multiplicity 2) as roots.

16.   We simplify the expression:
$$(4+2i)(4-2i) + \frac{3+i}{1+2i} = 16 - 4i^2 + \frac{3+i}{1+2i} \cdot \frac{1-2i}{1-2i}$$
$$= 16 + 4 + \frac{3 - 5i - 2i^2}{1 - 4i^2}$$
$$= 20 + \frac{5 - 5i}{5}$$
$$= 20 + 1 - i$$
$$= 21 - i$$

17.   We convert to rectangular form:
$$z = 2\left(\cos\tfrac{2\pi}{3} + i\sin\tfrac{2\pi}{3}\right) = 2\left(-\tfrac{1}{2} + \tfrac{\sqrt{3}}{2}i\right) = -1 + \sqrt{3}i$$

18.   We convert to trigonometric form:
$$\sqrt{2} - \sqrt{2}i = 2\left(\tfrac{\sqrt{2}}{2} - \tfrac{\sqrt{2}}{2}i\right) = 2\left(\cos\tfrac{7\pi}{4} + i\sin\tfrac{7\pi}{4}\right)$$

19.   We compute the product:
$$zw = 3\left(\cos\tfrac{2\pi}{9} + i\sin\tfrac{2\pi}{9}\right) \bullet 5\left(\cos\tfrac{\pi}{9} + i\sin\tfrac{\pi}{9}\right)$$
$$= 15\left(\cos\tfrac{\pi}{3} + i\sin\tfrac{\pi}{3}\right)$$
$$= 15\left(\tfrac{1}{2} + \tfrac{\sqrt{3}}{2}i\right)$$
$$= \tfrac{15}{2} + \tfrac{15\sqrt{3}}{2}i$$

20.   Let $z = r(\cos\theta + i\sin\theta)$ be a cube root of $64i$. We write $64i$ in trigonometric form:
$$64i = 64(0 + 1i) = 64\left(\cos\tfrac{\pi}{2} + i\sin\tfrac{\pi}{2}\right)$$

Thus:
$$z^3 = r^3(\cos 3\theta + i \sin 3\theta) = 64\left(\cos \tfrac{\pi}{2} + i \sin \tfrac{\pi}{2}\right)$$

So $r^3 = 64$ and thus $r = 4$, and $3\theta = \tfrac{\pi}{2} + 2\pi k$ thus $\theta = \tfrac{\pi}{6} + \tfrac{2\pi}{3}k$.

When $k = 0$, we have:
$$z_1 = 4\left(\cos \tfrac{\pi}{6} + i \sin \tfrac{\pi}{6}\right) = 4\left(\tfrac{\sqrt{3}}{2} + \tfrac{1}{2}i\right) = 2\sqrt{3} + 2i$$

When $k = 1$, we have:
$$z_2 = 4\left(\cos \tfrac{5\pi}{6} + i \sin \tfrac{5\pi}{6}\right) = 4\left(-\tfrac{\sqrt{3}}{2} + \tfrac{1}{2}i\right) = -2\sqrt{3} + 2i$$

When $k = 2$, we have:
$$z_3 = 4\left(\cos \tfrac{3\pi}{2} + i \sin \tfrac{3\pi}{2}\right) = 4(0 - i) = -4i$$

So the cube roots of $64i$ are $2\sqrt{3} + 2i$, $-2\sqrt{3} + 2i$, and $-4i$.

Chapter Twelve
Additional Topics in Algebra

12.1 Mathematical Induction

1. Here P_n denotes the statement:
$$1 + 2 + 3 + \ldots + n = \frac{n(n+1)}{2}$$

Since $1 = \frac{1(1+1)}{2} = 1$, then P_1 is true. It remains to show that P_k implies P_{k+1}. Assume P_k is true:
$$1 + 2 + \ldots + k = \frac{k(k+1)}{2}$$

Thus:
$$1 + 2 + \ldots + k + k + 1 = \frac{k(k+1)}{2} + k + 1$$
$$= (k+1)\left(\frac{k}{2} + 1\right)$$
$$= \frac{(k+1)(k+2)}{2}$$
$$= \frac{(k+1)[(k+1)+1]}{2}$$

So P_{k+1} is true and the induction is complete.

3. Here P_n denotes the statement:

$$1 + 4 + 7 + \ldots + (3n - 2) = \frac{n(3n-1)}{2}$$

Since $1 = \frac{1(3(1)-1)}{2} = 1$, then P_1 is true. Assume P_k is true:

$$1 + 4 + \ldots + (3k - 2) = \frac{k(3k-1)}{2}$$

Thus:

$$
\begin{aligned}
1 + 4 + \ldots + (3k - 2) + [3(k+1) - 2] &= \frac{k(3k-1)}{2} + 3(k+1) - 2 \\
&= \tfrac{1}{2}\big[k(3k-1) + 6(k+1) - 4\big] \\
&= \tfrac{1}{2}\big(3k^2 - k + 6k + 2\big) \\
&= \tfrac{1}{2}\big(3k^2 + 5k + 2\big) \\
&= \frac{(k+1)(3k+2)}{2} \\
&= \frac{(k+1)\big[3(k+1) - 1\big]}{2}
\end{aligned}
$$

So P_{k+1} is true and the induction is complete.

5. Here P_n denotes:

$$1^2 + 2^2 + 3^2 + \ldots + n^2 = \frac{n(n+1)(2n+1)}{6}$$

Since $1^2 = 1 = \frac{1(1+1)[2(1)+1]}{6} = \frac{1(2)(3)}{6} = 1$, then P_1 is true. Assume P_k is true:

$$1^2 + 2^2 + \ldots + k^2 = \frac{k(k+1)(2k+1)}{6}$$

Thus:

$$
\begin{aligned}
1^2 + 2^2 + \ldots + k^2 + (k+1)^2 &= \frac{k(k+1)(2k+1)}{6} + (k+1)^2 \\
&= \frac{(k+1)[k(2k+1)] + 6(k+1)^2}{6} \\
&= \frac{(k+1)\big(2k^2 + k + 6k + 6\big)}{6} \\
&= \frac{(k+1)\big(2k^2 + 7k + 6\big)}{6} \\
&= \frac{(k+1)(k+2)(2k+3)}{6} \\
&= \frac{(k+1)[(k+1)+1][2(k+1)+1]}{6}
\end{aligned}
$$

So P_{k+1} is true and the induction is complete.

7. Here P_n denotes:

$$1^2 + 3^2 + 5^2 + ... + (2n-1)^2 = \frac{n(2n-1)(2n+1)}{3}$$

As $1^2 = 1 = \dfrac{1[2(1)-1][2(1)+1]}{3} = 1$, then P_1 is true. Assume P_k is true:

$$1^2 + 3^2 + ... + (2k-1)^2 = \frac{k(2k-1)(2k+1)}{3}$$

Thus:

$$
\begin{aligned}
1^2 + 3^2 + ... + (2k-1)^2 + [2(k+1)-1]^2 &= \frac{k(2k-1)(2k+1)}{3} + [2(k+1)-1]^2 \\
&= \frac{k(2k-1)(2k+1)}{3} + \frac{3(2k+1)^2}{3} \\
&= \frac{2k+1}{3}[k(2k-1) + 3(2k+1)] \\
&= \frac{[2(k+1)-1]}{3}(2k^2 + 5k + 3) \\
&= \frac{(k+1)[2(k+1)-1](2k+3)}{3} \\
&= \frac{(k+1)[2(k+1)-1][2(k+1)]+1}{3}
\end{aligned}
$$

So P_{k+1} is true and the induction is complete.

9. Here P_n denotes:

$$3 + 3^2 + 3^3 + ... + 3^n = \tfrac{1}{2}(3^{n+1} - 3)$$

Since $3 = \tfrac{1}{2}(3^2 - 3) = \tfrac{1}{2}(6) = 3$, then P_1 is true. Assume P_k is true:

$$3 + 3^2 + ... + 3^k = \tfrac{1}{2}(3^{k+1} - 3)$$

Thus:

$$
\begin{aligned}
3 + 3^2 + ... + 3^k + 3^{k+1} &= \tfrac{1}{2}(3^{k+1} - 3) + 3^{k+1} \\
&= \tfrac{3}{2}(3^{k+1}) - \tfrac{3}{2} \\
&= \tfrac{1}{2}[3(3^{k+1}) - 3] \\
&= \tfrac{1}{2}(3^{k+2} - 3) \\
&= \tfrac{1}{2}[3^{(k+1)+1} - 3]
\end{aligned}
$$

So P_{k+1} is true and the induction is complete.

11. Here P_n denotes:

$$1^3 + 2^3 + 3^3 + ... + n^3 = \left[\frac{n(n+1)}{2}\right]^2$$

Since $1^3 = 1 = \left[\frac{1(2)}{2}\right]^2 = 1$, then P_1 is true. Assume P_k is true:

$$1^3 + 2^3 + ... + k^3 = \left[\frac{k(k+1)}{2}\right]^2$$

Thus:

$$1^3 + 2^3 + ... + k^3 + (k+1)^3 = \left[\frac{k(k+1)}{2}\right]^2 + (k+1)^3$$

$$= (k+1)^2\left[\frac{k^2}{4} + (k+1)\right]$$

$$= \frac{(k+1)^2}{4}\left[k^2 + 4k + 4\right]$$

$$= \frac{(k+1)^2(k+2)^2}{4}$$

$$= \frac{(k+1)^2[(k+1)+1]^2}{2^2}$$

$$= \left(\frac{(k+1)[(k+1)+1]}{2}\right)^2$$

So P_{k+1} is true and the induction is complete.

13. Here P_n denotes:

$$1^3 + 3^3 + 5^3 + ... + (2n-1)^3 = n^2(2n^2 - 1)$$

Since $1^3 = 1(1)^2[2(1)^2 - 1] = (1)(1) = 1$, then P_1 is true. Assume P_k is true:

$$1^3 + 3^3 + ... + (2k-1)^3 = k^2(2k^2 - 1)$$

Thus:

$$1^3 + 3^3 + ... + (2k-1)^3 + [2(k+1) - 1]^3 = k^2(2k^2 - 1) + (2k+1)^3$$

$$= 2k^4 - k^2 + 8k^3 + 12k^2 + 6k + 1$$

$$= 2k^4 + 8k^3 + 11k^2 + 6k + 1$$

$$= (k+1)^2(2k^2 + 4k + 1)$$

$$= (k+1)^2\left[2(k^2 + 2k + 1) - 1\right]$$

$$= (k+1)^2\left[2(k+1)^2 - 1\right]$$

So P_{k+1} is true and the induction is complete.

15. Here P_n denotes:

$$1 \times 3 + 3 \times 5 + 5 \times 7 + ... + (2n-1)(2n+1) = \frac{n(4n^2 + 6n - 1)}{3}$$

Since $1 \times 3 = 3 = \dfrac{(1)\left[4(1)^2 + 6(1) - 1\right]}{3} = \frac{9}{3} = 3$, then P_1 is true. Assume P_k is true:

$$1 \times 3 + 3 \times 5 + \ldots + (2k - 1)(2k + 1) = \dfrac{k\left(4k^2 + 6k - 1\right)}{3}$$

Thus:

$$1 \times 3 + 3 \times 5 + \ldots + (2k - 1)(2k + 1) + [2(k + 1) - 1][2(k + 1) + 1]$$

$$= \dfrac{k\left(4k^2 + 6k - 1\right)}{3} + (2k + 1)(2k + 3)$$

$$= \dfrac{4k^3 + 6k^2 - k + 12k^2 + 24k + 9}{3}$$

$$= \dfrac{4k^3 + 18k^2 + 23k + 9}{3}$$

Noting that -1 is a root:

$$= \dfrac{(k + 1)\left(4k^2 + 14k + 9\right)}{3}$$

$$= \dfrac{(k + 1)\left[4(k + 1)^2 + 6(k + 1) - 1\right]}{3}$$

So P_{k+1} is true and the induction is complete.

17. Here P_n denotes:

$$1 + \frac{3}{2} + \frac{5}{2^2} + \ldots + \frac{2n - 1}{2^{n-1}} = 6 - \frac{2n + 3}{2^{n-1}}$$

Since $1 = 6 - \dfrac{[2(1) + 3]}{2^0} = 6 - 5 = 1$, then P_1 is true. Assume P_k is true:

$$1 + \frac{3}{2} + \ldots + \frac{2k - 1}{2^{k-1}} = 6 - \frac{2k + 3}{2^{k-1}}$$

Thus:

$$1 + \frac{3}{2} + \ldots + \frac{2k - 1}{2^{k-1}} + \frac{2(k + 1) - 1}{2^k} = 6 - \frac{2k + 3}{2^{k-1}} + \frac{2k + 1}{2^k}$$

$$= 6 - \frac{4k + 6 - 2k - 1}{2^k}$$

$$= 6 - \frac{2k + 5}{2^k}$$

$$= 6 - \frac{2(k + 1) + 3}{2^{(k+1)-1}}$$

So P_{k+1} is true and the induction is complete.

19. Let P_n denote the statement:
$$n \le 2^{n-1}$$
Since $2^{1-1} = 2^0 = 1 \ge 1$, then P_1 is true. Assume P_k is true:
$$k \le 2^{k-1}$$
Hence:
$$k + 1 \le 2^{k-1} + 1$$
Since $1 \le 2^{k-1}$ (use induction to prove this), then:
$$k + 1 \le 2^{k-1} + 1 \le 2^{k-1} + 2^{k-1} = 2\left(2^{k-1}\right) = 2^k = 2^{(k+1)-1}$$
So P_{k+1} is true and the induction is complete.

21. For $n = 2$:
$$2^2 + 4 = 8 < (2+1)^2$$
Assuming:
$$k^2 + 4 < (k+1)^2$$
Then:
$$k^2 + 2k + 4 < (k+1)^2 + 2k$$
$$(k+1)^2 + 3 < k^2 + 4k + 1$$
$$(k+1)^2 + 4 < k^2 + 4k + 2$$
$$(k+1)^2 + 4 < k^2 + 4k + 4$$
$$(k+1)^2 + 4 < (k+2)^2$$
This completes the induction for $n \ge 2$.

23. For $n = 7$, we have:
$$(1.5)^7 \approx 17.09 > 2(7), \text{ so } P_1 \text{ is true}$$
Assume P_k is true, so:
$$(1.5)^k > 2k$$
Then for $k \ge 7$:
$$\begin{aligned}
(1.5)^{k+1} &= 1.5(1.5)^k \\
&> 1.5(2k) \\
&= 3k \\
&= 2k + k \\
&> 2k + 2 \quad \text{(since } k > 2\text{)} \\
&= 2(k+1)
\end{aligned}$$
So P_{k+1} is true and the induction is complete.

25. For $n = 39$, we have:
$$(1.1)^{39} \approx 41.1 > 39, \text{ so } P_{39} \text{ is true}$$
Assume P_k is true for some $k \ge 39$, so:
$$(1.1)^k > k$$
Then:
$$(1.1)^{k+1} = 1.1(1.1)^k > 1.1k = k + 0.1k > k + 1 \quad \text{(since } k > 10\text{)}$$
So P_{k+1} is true and the induction is complete.

27. (a) We complete the table:

n	1	2	3	4	5
$f(n)$	$\frac{1}{2}$	$\frac{2}{3}$	$\frac{3}{4}$	$\frac{4}{5}$	$\frac{5}{6}$

(b) $\frac{6}{7}$, $f(6) = f(5) + \frac{1}{42} = \frac{5}{6} + \frac{1}{42} = \frac{36}{42} = \frac{6}{7}$

(c) $f(n) = \dfrac{1}{1 \times 2} + \dfrac{1}{2 \times 3} + \ldots + \dfrac{1}{n(n+1)} = \dfrac{n}{n+1}$

$f(1) = \frac{1}{2} = \dfrac{1}{1+1} = \frac{1}{2}$

Assuming $f(k) = \dfrac{k}{k+1}$, then:

$$f(k+1) = f(k) + \frac{1}{(k+1)(k+2)}$$
$$= \frac{k}{k+1} + \frac{1}{(k+1)(k+2)}$$
$$= \frac{k(k+2)+1}{(k+1)(k+2)}$$
$$= \frac{k^2 + 2k + 1}{(k+1)(k+2)}$$
$$= \frac{(k+1)^2}{(k+1)(k+2)}$$
$$= \frac{k+1}{k+2}$$

This completes the induction.

29. (a) We complete the table:

n	1	2	3	4	5
$f(n)$	1	4	9	16	25

(b) 36, $f(6) = f(5) + 2\sqrt{f(5)} + 1 = 25 + 2\sqrt{25} + 1 = 36$

(c) $f(n) = n^2$

$f(1) = 1 = 1^2$

Assume that for some $k \geq 1$ that $f(k) = k^2$. Then we have:

$$f(k+1) = f(k) + 2\sqrt{f(k)} + 1 = k^2 + 2\sqrt{k^2} + 1 = k^2 + 2k + 1 = (k+1)^2$$

The induction is complete.

31. For $n = 1$: 13 is prime and P_1 is true
For $n = 2$: 17 is prime and P_2 is true
For $n = 3$: 23 is prime and P_3 is true
For $n = 4$: 31 is prime and P_4 is true
For $n = 5$: 41 is prime and P_5 is true
For $n = 6$: 53 is prime and P_6 is true
For $n = 7$: 67 is prime and P_7 is true
For $n = 8$: 83 is prime and P_8 is true
For $n = 9$: 101 is prime and P_9 is true
However, for $n = 10$, we have $(10)^2 + 10 + 11 = 121 = (11)(11)$ and so P_{10} is false.

33. Let P_n denote:

$$1 + r + r^2 + \ldots + r^{n-1} = \frac{r^n - 1}{r - 1} \text{ when } r \neq 1$$

Then since $1 = \dfrac{r - 1}{r - 1} = 1$, then P_1 is true. Assuming P_k is true, then:

$$1 + r + \ldots + r^{k-1} = \frac{r^k - 1}{r - 1}$$

Thus:

$$1 + r + \ldots + r^{k-1} + r^k = \frac{r^k - 1}{r - 1} + r^k = \frac{r^k - 1 + r^k(r - 1)}{r - 1} = \frac{r^{k+1} - 1}{r - 1}$$

So P_{k+1} is true and the induction is complete.

35. Let P_n denote:
$n^5 - n = 5R$ for some natural number R
Since $2^5 - 2 = 32 - 2 = 30 = 5(6)$, then P_2 is true. Since:

$$(k + 1)^5 - (k + 1) = k^5 + 5k^4 + 10k^3 + 10k^2 + 4k = k^5 - k + 5(k^4 + 2k^3 + 2k^2 + k)$$

Assuming P_k is true results in:

$$(k + 1)^5 - (k + 1) = 5(R + k^4 + 2k^3 + 2k^2 + k)$$

So P_{k+1} is true and the induction is complete.

37. For $n = 0$, we have:

$$2^1 + 3^1 = 5 = 5(1), \text{ so } P_0 \text{ is true}$$

Assume P_k is true, so:

$$2^{2k+1} + 3^{2k+1} = 5A, \text{ for some natural number } A$$

Then:
$$2^{2k+3} + 3^{2k+3} = 4 \cdot 2^{2k+1} + 9 \cdot 3^{2k+1}$$
$$= 4 \cdot 2^{2k+1} + 9\left(5A - 2^{2k+1}\right)$$
$$= 4 \cdot 2^{2k+1} + 45A - 9 \cdot 2^{2k+1}$$
$$= 45A - 5 \cdot 2^{2k+1}$$
$$= 5\left(9A - 2^{2k+1}\right)$$

Since 5 is a factor of this expression, P_{k+1} is true and the induction is complete.

39. For $n = 0$, we have:
$$2^1 + (-1)^0 = 2 + 1 = 3 = 3(1), \text{ so } P_0 \text{ is true}$$
Assume P_k is true, so:
$$2^{k+1} + (-1)^k = 3A, \text{ for some natural number } A$$
Then:
$$2^{k+2} + (-1)^{k+1} = 2 \cdot 2^{k+1} - (-1)^k$$
$$= 2\left[3A - (-1)^k\right] - (-1)^k$$
$$= 6A - 2(-1)^k - (-1)^k$$
$$= 6A - 3(-1)^k$$
$$= 3\left[2A - (-1)^k\right]$$

Since 3 is a factor of this expression, P_{k+1} is true and the induction is complete.

41. Let P_n denote:
$$x^n - y^n = (x - y)(Q(x,y)) \text{ where } Q(x,y) \text{ is a polynomial in } x \text{ and } y.$$
Then:
$$x - y = (x - y)(1), \text{ so } P_1 \text{ is true}$$
While using the suggested identity:
$$x^{k+1} - y^{k+1} = x^k(x - y) + (x^k - y^k)y$$
So assuming P_k, we have $x^k - y^k = (x - y)Q(x,y)$:
$$x^{k+1} - y^{k+1} = \left(x^k + yQ(x,y)\right)(x - y)$$
So P_{k+1} is true and the induction is complete.

43. For $n = 1$:
$$(1 + p) \geq (1 + p), \text{ so } P_1 \text{ is true}$$
When $p > -1$ then $p + 1 > 0$. Assuming $(1 + p)^k \geq 1 + kp$ then:
$$(1 + p)^{k+1} \geq (1 + kp)(1 + p) \geq 1 + p + kp + kp^2 = 1 + (k+1)p + kp^2 \geq 1 + (k+1)p$$
The induction is complete.

45. For $n = 1$, we have:
$$1^5 = \frac{1(2)^2(3)}{12} = \frac{12}{12} = 1$$

Assume P_k is true, so:

$$1^5 + 2^5 + \ldots + k^5 = \frac{k^2(k+1)^2(2k^2+2k-1)}{12}$$

Therefore:

$$1^5 + 2^5 + \ldots + k^5 + (k+1)^5 = \frac{k^2(k+1)^2(2k^2+2k-1)}{12} + (k+1)^5$$

$$= \frac{(k+1)^2}{12}\left(2k^4 + 14k^3 + 35k^2 + 36k + 12\right)$$

$$= \frac{(k+1)^2}{12}(k+2)^2\left(2k^2 + 6k + 3\right)$$

$$= \frac{(k+1)^2(k+2)^2\left[2(k+1)^2 + 2(k+1) - 1\right]}{12}$$

So P_{k+1} is true and the induction is complete.

47. For $n = 1$, we have:

$$\frac{0+1}{1+1} = \tfrac{1}{4} + \tfrac{1}{4}, \text{ so } \tfrac{1}{2} = \tfrac{1}{2}$$

Assume P_k is true, so:

$$\frac{0^3 + 1^3 + \ldots + k^3}{(k+1)k^3} = \frac{1}{4} + \frac{1}{4k}$$

Therefore:

$$\frac{0^3 + 1^3 + \ldots + k^3 + (k+1)^3}{(k+2)(k+1)^3} = \frac{(k+1)k^3\left(\frac{1}{4} + \frac{1}{4k}\right) + (k+1)^3}{(k+2)(k+1)^3}$$

$$= \frac{\frac{k^3}{4}\left(1 + \frac{1}{k}\right) + (k+1)^2}{(k+2)(k+1)^2}$$

$$= \frac{\frac{k^2}{4}(k+1) + (k+1)^2}{(k+2)(k+1)^2}$$

$$= \frac{\frac{k^2}{4} + (k+1)}{(k+2)(k+1)}$$

$$= \frac{k^2 + 4k + 4}{4(k+2)(k+1)}$$

$$= \frac{(k+2)^2}{4(k+2)(k+1)}$$

$$= \frac{k+2}{4(k+1)}$$

$$= \frac{k+1}{4(k+1)} + \frac{1}{4(k+1)}$$

$$= \frac{1}{4} + \frac{1}{4(k+1)}$$

So P_{k+1} is true and the induction is complete.

12.2 The Binomial Theorem

1. (a)   We compute:
$$(a+b)^2 = (a+b)(a+b) = a^2 + ab + ba + b^2 = a^2 + 2ab + b^2$$

(b)   We compute:
$$\begin{aligned}(a+b)^3 &= (a+b)(a+b)^2 \\ &= (a+b)\left(a^2 + 2ab + b^2\right) \\ &= a^3 + 2a^2b + ab^2 + ba^2 + 2ab^2 + b^3 \\ &= a^3 + 3a^2b + 3ab^2 + b^3\end{aligned}$$

3. Using the binomial theorem, we expand:
$$\begin{aligned}(a+b)^9 &= \binom{9}{0}a^9 + \binom{9}{1}a^8b + \binom{9}{2}a^7b^2 + \binom{9}{3}a^6b^3 + \binom{9}{4}a^5b^4 + \binom{9}{5}a^4b^5 \\ &\quad + \binom{9}{6}a^3b^6 + \binom{9}{7}a^2b^7 + \binom{9}{8}ab^8 + \binom{9}{9}b^9 \\ &= a^9 + 9a^8b + 36a^7b^2 + 84a^6b^3 + 126a^5b^4 + 126a^4b^5 + 84a^3b^6 \\ &\quad + 36a^2b^7 + 9ab^8 + b^9\end{aligned}$$

5. Using the binomial theorem, we expand:
$$\begin{aligned}(2A+B)^3 &= \binom{3}{0}(2A)^3 + \binom{3}{1}(2A)^2B + \binom{3}{2}(2A)B^2 + \binom{3}{3}B^3 \\ &= 8A^3 + 12A^2B + 6AB^2 + B^3\end{aligned}$$

7. Using the binomial theorem, we expand:
$$\begin{aligned}(1-2x)^6 &= \binom{6}{0}1^6 - \binom{6}{1}1^5(2x) + \binom{6}{2}1^4(2x)^2 - \binom{6}{3}1^3(2x)^3 \\ &\quad + \binom{6}{4}1^2(2x)^4 - \binom{6}{5}1(2x)^5 + \binom{6}{6}(2x)^6 \\ &= 1 - 12x + 60x^2 - 160x^3 + 240x^4 - 192x^5 + 64x^6\end{aligned}$$

9. Using the binomial theorem, we expand:
$$\begin{aligned}\left(\sqrt{x} + \sqrt{y}\right)^4 \\ = \binom{4}{0}\left(\sqrt{x}\right)^4 + \binom{4}{1}\left(\sqrt{x}\right)^3\left(\sqrt{y}\right) + \binom{4}{2}\left(\sqrt{x}\right)^2\left(\sqrt{y}\right)^2 + \binom{4}{3}\left(\sqrt{x}\right)\left(\sqrt{y}\right)^3 + \binom{4}{4}\left(\sqrt{y}\right)^4 \\ = x^2 + 4x\sqrt{xy} + 6xy + 4y\sqrt{xy} + y^2\end{aligned}$$

11. Using the binomial theorem, we expand:

$$\left(x^2 + y^2\right)^5 = \binom{5}{0}\left(x^2\right)^5 + \binom{5}{1}\left(x^2\right)^4 y^2 + \binom{5}{2}\left(x^2\right)^3\left(y^2\right)^2 + \binom{5}{3}\left(x^2\right)^2\left(y^2\right)^3$$

$$+ \binom{5}{4}\left(x^2\right)\left(y^2\right)^4 + \binom{5}{5}\left(y^2\right)^5$$

$$= x^{10} + 5x^8 y^2 + 10x^6 y^4 + 10x^4 y^6 + 5x^2 y^8 + y^{10}$$

13. Using the binomial theorem, we expand:

$$\left(1 - \frac{1}{x}\right)^6 = \binom{6}{0} - \binom{6}{1}\frac{1}{x} + \binom{6}{2}\left(\frac{1}{x}\right)^2 - \binom{6}{3}\left(\frac{1}{x}\right)^3 + \binom{6}{4}\left(\frac{1}{x}\right)^4 - \binom{6}{5}\left(\frac{1}{x}\right)^5 + \binom{6}{6}\left(\frac{1}{x}\right)^6$$

$$= 1 - \frac{6}{x} + \frac{15}{x^2} - \frac{20}{x^3} + \frac{15}{x^4} - \frac{6}{x^5} + \frac{1}{x^6}$$

15. Using the binomial theorem, we expand:

$$\left(\frac{x}{2} - \frac{y}{3}\right)^3 = \binom{3}{0}\left(\frac{x}{2}\right)^3 - \binom{3}{1}\left(\frac{x}{2}\right)^2\left(\frac{y}{3}\right) + \binom{3}{2}\left(\frac{x}{2}\right)\left(\frac{y}{3}\right)^2 - \binom{3}{3}\left(\frac{y}{3}\right)^3$$

$$= \frac{x^3}{8} - \frac{x^2 y}{4} + \frac{xy^2}{6} - \frac{y^3}{27}$$

17. Using the binomial theorem, we expand:

$$\left(ab^2 + c\right)^7 = \binom{7}{0}\left(ab^2\right)^7 + \binom{7}{1}\left(ab^2\right)^6 c + \binom{7}{2}\left(ab^2\right)^5 c^2 + \binom{7}{3}\left(ab^2\right)^4 c^3 + \binom{7}{4}\left(ab^2\right)^3 c^4$$

$$+ \binom{7}{5}\left(ab^2\right)^2 c^5 + \binom{7}{6}\left(ab^2\right)c^6 + \binom{7}{7}c^7$$

$$= a^7 b^{14} + 7a^6 b^{12} c + 21a^5 b^{10} c^2 + 35a^4 b^8 c^3 + 35a^3 b^6 c^4 + 21a^2 b^4 c^5$$
$$+ 7ab^2 c^6 + c^7$$

19. Using the binomial theorem, we expand:

$$\left(x + \sqrt{2}\right)^8 = \binom{8}{0}x^8 + \binom{8}{1}x^7\sqrt{2} + \binom{8}{2}x^6\left(\sqrt{2}\right)^2 + \binom{8}{3}x^5\left(\sqrt{2}\right)^3 + \binom{8}{4}x^4\left(\sqrt{2}\right)^4$$

$$+ \binom{8}{5}x^3\left(\sqrt{2}\right)^5 + \binom{8}{6}x^2\left(\sqrt{2}\right)^6 + \binom{8}{7}x\left(\sqrt{2}\right)^7 + \binom{8}{8}\left(\sqrt{2}\right)^8$$

$$= x^8 + 8\sqrt{2}x^7 + 56x^6 + 112\sqrt{2}x^5 + 280x^4 + 224\sqrt{2}x^3 + 224x^2$$
$$+ 64\sqrt{2}x + 16$$

21. Using the binomial theorem, we expand:

$$\left(\sqrt{2} - 1\right)^3 = \binom{3}{0}\left(\sqrt{2}\right)^3 - \binom{3}{1}\left(\sqrt{2}\right)^2 + \binom{3}{2}\left(\sqrt{2}\right) - \binom{3}{3} = 2\sqrt{2} - 6 + 3\sqrt{2} - 1 = 5\sqrt{2} - 7$$

23. Using the binomial theorem, we expand:

$$(\sqrt{2}+\sqrt{3})^5 = \binom{5}{0}(\sqrt{2})^5 + \binom{5}{1}(\sqrt{2})^4\sqrt{3} + \binom{5}{2}(\sqrt{2})^3(\sqrt{3})^2 + \binom{5}{3}(\sqrt{2})^2(\sqrt{3})^3$$

$$+ \binom{5}{4}(\sqrt{2})(\sqrt{3})^4 + \binom{5}{5}(\sqrt{3})^5$$

$$= 4\sqrt{2} + 20\sqrt{3} + 60\sqrt{2} + 60\sqrt{3} + 45\sqrt{2} + 9\sqrt{3}$$

$$= 89\sqrt{3} + 109\sqrt{2}$$

25. Using the binomial theorem, we expand:

$$\left(2\sqrt[3]{2}-\sqrt[3]{4}\right)^3 = \binom{3}{0}\left(2\sqrt[3]{2}\right)^3 - \binom{3}{1}\left(2\sqrt[3]{2}\right)^2\left(\sqrt[3]{4}\right) + \binom{3}{2}\left(2\sqrt[3]{2}\right)\left(\sqrt[3]{4}\right)^2 - \binom{3}{3}\left(\sqrt[3]{4}\right)^3$$

$$= 16 - 12\sqrt[3]{16} + 6\sqrt[3]{32} - 4$$

$$= 12 - 24\sqrt[3]{2} + 12\sqrt[3]{4}$$

27. Using the binomial theorem, we expand:

$$\left[x^2 - (2x+1)\right]^5$$

$$= \binom{5}{0}(x^2)^5 - \binom{5}{1}(x^2)^4(2x+1) + \binom{5}{2}(x^2)^3(2x+1)^2 - \binom{5}{3}(x^2)^2(2x+1)^3$$

$$+ \binom{5}{4}(x^2)(2x+1)^4 - \binom{5}{5}(2x+1)^5$$

$$= x^{10} - \left[5x^8(2x+1)\right] + \left[10x^6(4x^2+4x+1)\right] - \left[10x^4(8x^3+12x^2+6x+1)\right]$$

$$+ \left[5x^2(16x^4+32x^3+24x^2+8x+1)\right]$$

$$- \left(32x^5+80x^4+80x^3+40x^2+10x+1\right)$$

$$= x^{10} - \left(10x^9+5x^8\right) + \left(40x^8+40x^7+10x^6\right) - \left(80x^7+120x^6+60x^5+10x^4\right)$$

$$+ \left(80x^6+160x^5+120x^4+40x^3+5x^2\right)$$

$$- \left(32x^5+80x^4+80x^3+40x^2+10x+1\right)$$

$$= x^{10} - 10x^9 + 35x^8 - 40x^7 - 30x^6 + 68x^5 + 30x^4 - 40x^3 - 35x^2 - 10x - 1$$

29. $5! = (5)(4)(3)(2)(1) = 120$

31. We compute:

$$\binom{7}{3}\binom{3}{2} = \frac{7!}{3!(4!)} \cdot \frac{3!}{2!(1!)} = \frac{7 \cdot 6 \cdot 5}{3 \cdot 2} \cdot 3 = 105$$

33. (a) We compute:

$$\binom{5}{3} = \frac{5!}{3!(5-3)!} = \frac{5(4)}{2} = 10$$

(b) We compute:
$$\binom{5}{4} = \frac{5!}{4!(5-4)!} = 5$$

35. We simplify:
$$\frac{(n+2)!}{n!} = \frac{(n+2)(n+1)n!}{n!} = n^2 + 3n + 2$$

37. We compute:
$$\binom{6}{4} + \binom{6}{3} - \binom{7}{4} = \frac{6!}{4!2!} + \frac{6!}{3!3!} - \frac{7!}{4!3!}$$
$$= \frac{(6)(5)}{2} + \frac{(6)(5)(4)}{6} - \frac{(7)(6)(5)}{6}$$
$$= 15 + 20 - 35$$
$$= 0$$

39. The fifteenth term will be given by:
$$\binom{16}{14}a^2b^{14} = 120a^2b^{14}$$

41. The one-hundredth term will be given by:
$$\binom{100}{99}x^{99} = 100x^{99}$$

43. Here $n - r + 1 = 8$, so $10 - r + 1 = 8$, so $r = 3$. Hence the coefficient is:
$$\binom{10}{2} = 45$$

45. Here $r - 1 = 8$, thus $r = 9$. Hence the coefficient is:
$$\frac{1}{2}(-4)^8\binom{9}{8} = 9 \cdot \frac{1}{2}(65536) = 294912$$

47. Here $r - 1 = 6$. Hence $r = 7$, and the coefficient is:
$$\binom{8}{6} = \frac{8(7)}{2} = 28$$

49. Here $(12 - r + 1)(-1) + 2(r - 1) = 0$, so $-12 + r - 1 + 2r - 2 = 0$, then $3r = 15$, so $r = 5$ and the coefficient is:
$$(3)^4\binom{12}{4} = \frac{(12)(11)(10)(9)}{(4)(3)(2)}(3)^4 = 495(81) = 40095$$

51. Here $r - 1 = n$, so $r = n + 1$, so the coefficient is:

$$\binom{2n}{n} = \frac{(2n)!}{n!n!} = \frac{(2n)!}{(n!)^2}$$

53. (a) We complete the table:

k	0	1	2	3	4	5	6	7	8
$\binom{8}{k}$	1	8	28	56	70	56	28	8	1

(b) We have $1 + 8 + 28 + 56 + 70 + 56 + 28 + 8 + 1 = 256$, while $2^8 = 256$.

(c) Since a^L and b^L equal 1 for any L since $a = b = 1$, then from the binomial theorem:

$$2^n = (1+1)^n = \binom{n}{0} + \binom{n}{1} + \ldots + \binom{n}{n}$$

55. (a) We simplify:

$$(1+x)^n \left(1 + \frac{1}{x}\right)^n = (1+x)^n \left(\frac{x+1}{x}\right)^n = (1+x)^n \cdot \frac{(1+x)^n}{x^n} = \frac{(1+x)^{2n}}{x^n}$$

(b) The nth term of expansion on the right is:

$$\frac{\binom{2n}{n}(1)^n(x)^n}{x^n} = \binom{2n}{n}$$

This verifies the result.

(c) Using the binomial theorem:

$$(1+x)^n = \binom{n}{0} + \binom{n}{1}x + \binom{n}{2}x^2 + \binom{n}{3}x^3 + \ldots + \binom{n}{n}x^n$$

$$\left(1 + \frac{1}{x}\right)^n = \binom{n}{0} + \binom{n}{1}x^{-1} + \binom{n}{2}x^{-2} + \binom{n}{3}x^{-3} + \ldots + \binom{n}{n}x^{-n}$$

When multiplied, the terms that do not contain x come from terms with corresponding positive and negative exponents:

$$\binom{n}{1}x \cdot \binom{n}{1}x^{-1} + \binom{n}{2}x^2 \cdot \binom{n}{2}x^{-2} + \ldots + \binom{n}{n}x^n \cdot \binom{n}{n}x^{-n}$$

$$= \binom{n}{1}^2 + \binom{n}{2}^2 + \ldots + \binom{n}{n}^2$$

This verifies the identity.

12.3 Introduction to Sequences and Series

1. Since $a_n = \frac{n}{n+1}$, we can write out the first five terms by setting n equal to the natural numbers 1 through 5:

$$a_1 = \frac{1}{1+1} = \frac{1}{2}, \quad a_2 = \frac{2}{2+1} = \frac{2}{3}, \quad a_3 = \frac{3}{3+1} = \frac{3}{4}, \quad a_4 = \frac{4}{4+1} = \frac{4}{5}, \quad a_5 = \frac{5}{5+1} = \frac{5}{6}$$

So the first five terms are $\frac{1}{2}, \frac{2}{3}, \frac{3}{4}, \frac{4}{5}, \frac{5}{6}$.

3. Again, letting n be successively 1 through 5, we obtain:

$$a_1 = (1-1)^2, \quad a_2 = (2-1)^2, \quad a_3 = (3-1)^2, \quad a_4 = (4-1)^2, \quad a_5 = (5-1)^2$$

So the first five terms are 0, 1, 4, 9, 16.

5. Here $b_n = \left(1 + \frac{1}{n}\right)^n$, so $b_1 = 2, b_2 = \frac{9}{4}, b_3 = \frac{64}{27}, b_4 = \frac{625}{256}$, and $b_5 = \frac{7776}{3125}$.

7. Here $u_n = (-1)^n$, so $u_1 = -1, u_2 = 1, u_3 = -1, u_4 = 1$, and $u_5 = -1$.

9. Here $a_n = \frac{(-1)^{n+1}}{n!}$, so $a_1 = 1, a_2 = -\frac{1}{2}, a_3 = \frac{1}{6}, a_4 = -\frac{1}{24}$, and $a_5 = \frac{1}{120}$.

11. Here $a_n = 3n$, so $a_1 = 3, a_2 = 6, a_3 = 9, a_4 = 12$, and $a_5 = 15$.

13. Here $a_1 = 1$, so $a_n = \left(1 + a_{n-1}\right)^2$ for $n \geq 2$, so:

$$a_1 = 1$$
$$a_2 = \left(1 + a_1\right)^2 = 2^2 = 4$$
$$a_3 = \left(1 + a_2\right)^2 = 5^2 = 25$$
$$a_4 = \left(1 + a_3\right)^2 = 26^2 = 676$$
$$a_5 = \left(1 + a_4\right)^2 = 677^2 = 458329$$

The first five terms are 1, 4, 25, 676, 458329.

15. Here $a_1 = 2, a_2 = 2$, and $a_n = a_{n-1}a_{n-2}$ for $n \geq 3$, so $a_1 = 2, a_2 = 2, a_3 = 4, a_4 = 8$, and $a_5 = 32$.

17. Here $a_1 = 1, a_{n+1} = na_n$ for $n \geq 1$, so $a_1 = 1, a_2 = 1, a_3 = 2, a_4 = 6$, and $a_5 = 24$.

19. Here $a_1 = 0, a_n = 2^{a_{n-1}}$ for $n \geq 2$, so $a_1 = 0, a_2 = 1, a_3 = 2, a_4 = 4$, and $a_5 = 16$.

21. First we compute the required values of a_n as shown in the table:

n	1	2	3	4
a_n	-1	1	-1	1

Now graph these values:

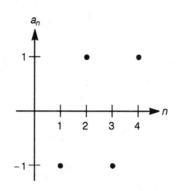

23. First we compute the required values of a_n as shown in the table:

n	1	2	3	4
a_n	1	$\frac{1}{2}$	$\frac{1}{3}$	$\frac{1}{4}$

Now graph these values:

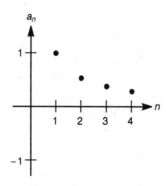

25. First we compute the required values of a_n as shown in the table:

n	1	2	3
a_n	0	$\frac{1}{3}$	$\frac{1}{2}$

Now graph these values:

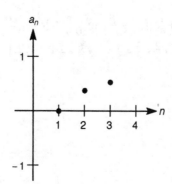

27. Since $a_n = 2^n$, the sum is $2 + 4 + 8 + 16 + 32 = 62$.

29. Since $a_n = n^2 - n$, the sum is $0 + 2 + 6 + 12 + 20 = 40$.

31. Since $a_n = \dfrac{(-1)^n}{n!}$, the sum is $-1 + \frac{1}{2} - \frac{1}{6} + \frac{1}{24} - \frac{1}{120} = -\frac{19}{30}$.

33. Here $a_1 = 1$, $a_2 = 2$, and $a_n = a_{n-1}^2 + a_{n-2}^2$ for $n \geq 3$, so the sum is
$1 + 2 + 5 + 29 + 866 = 903$.

35. Here $a_1 = 2$ and $a_n = \left(a_{n-1}\right)^2$ for $n \geq 2$, so the sum is $2 + 4 + 16 + 256 = 278$.

37. We expand the sum:
$$\sum_{k=1}^{3} (k-1) = 0 + 1 + 2 = 3$$

39. We expand the sum:
$$\sum_{k=4}^{5} k^2 = 16 + 25 = 41$$

41. We expand the sum:
$$\sum_{n=1}^{3} x^n = x + x^2 + x^3$$

43. We expand the sum:
$$\sum_{n=1}^{4} \frac{1}{n} = 1 + \frac{1}{2} + \frac{1}{3} + \frac{1}{4} = \frac{25}{12}$$

45. We expand the sum:
$$\sum_{j=1}^{9} \log_{10} \frac{j}{j+1} = \log_{10} \frac{1}{2} + \log_{10} \frac{2}{3} + \log_{10} \frac{3}{4} + \log_{10} \frac{4}{5} + \log_{10} \frac{5}{6} + \log_{10} \frac{6}{7}$$
$$+ \log_{10} \frac{7}{8} + \log_{10} \frac{8}{9} + \log_{10} \frac{9}{10}$$
$$= \log_{10} \left(\frac{1}{2} \cdot \frac{2}{3} \cdot \frac{3}{4} \cdot \frac{4}{5} \cdot \frac{5}{6} \cdot \frac{6}{7} \cdot \frac{7}{8} \cdot \frac{8}{9} \cdot \frac{9}{10} \right)$$
$$= \log_{10} \frac{1}{10}$$
$$= -1$$

47. We expand the sum:

$$\sum_{j=1}^{6}\left(\frac{1}{j}-\frac{1}{j+1}\right)=\left(1-\tfrac{1}{2}\right)+\left(\tfrac{1}{2}-\tfrac{1}{3}\right)+\left(\tfrac{1}{3}-\tfrac{1}{4}\right)+\left(\tfrac{1}{4}-\tfrac{1}{5}\right)+\left(\tfrac{1}{5}-\tfrac{1}{6}\right)+\left(\tfrac{1}{6}-\tfrac{1}{7}\right)=1-\tfrac{1}{7}=\tfrac{6}{7}$$

49. We write the sum in sigma notation:

$$5+5^2+5^3+5^4=\sum_{j=1}^{4}5^j$$

51. We write the sum in sigma notation:

$$x+x^2+x^3+x^4+x^5+x^6=\sum_{j=1}^{6}x^j$$

53. We write the sum in sigma notation:

$$1+\tfrac{1}{2}+\tfrac{1}{3}+\ldots+\tfrac{1}{12}=\sum_{k=1}^{12}\frac{1}{k}$$

55. We write the sum in sigma notation:

$$2-2^2+2^3-2^4+2^5=\sum_{j=1}^{5}(-1)^{j+1}2^j$$

57. We write the sum in sigma notation:

$$1-2+3-4+5=\sum_{j=1}^{5}(-1)^{j+1}j$$

59. (a) Here $F_1=1$, $F_2=1$, and $F_{n+2}=F_n+F_{n+1}$ for $n\geq 1$.
Evaluating the first 10 terms, we obtain the table:

F_1	F_2	F_3	F_4	F_5	F_6	F_7	F_8	F_9	F_{10}
1	1	2	3	5	8	13	21	34	55

(b) We complete the chart:

n	$F_1+F_2+\ldots+F_n$	$F_{n+2}-1$
1	1	1
2	2	2
3	4	4
4	7	7
5	12	12

(c)   The chart completed in (b) shows the hypothesis to be true for $n = 1$. Assuming:
$$F_1 + F_2 + \ldots + F_k = F_{k+2} - 1$$
Therefore:
$$F_1 + F_2 + \ldots + F_k + F_{k+1} = F_{k+2} - 1 + F_{k+1} = F_{k+3} - 1 = F_{(k+1)+2} - 1$$
This completes the induction.

(d)   From (a), $F_5 = 5 \geq 5$, so the hypothesis is true for $n = 5$. For some $k \geq 5$, assume:
$$F_k \geq k$$
Then:
$$F_{k+1} = F_k + F_{k-1} \geq F_k + 1 \geq k + 1$$
This completes the induction.

(e)   Since $F_1^2 = 1^2 = 1 = F_1 F_2 = 1(1) = 1$, then the hypothesis is true for $n = 1$.
Assuming:
$$F_1^2 + F_2^2 + \ldots + F_k^2 = F_k F_{k+1}$$
Then:
$$F_1^2 + F_2^2 + \ldots + F_k^2 + F_{k+1}^2 = F_k F_{k+1} + F_{k+1}^2 = F_{k+1}(F_k + F_{k+1}) = F_{k+1}(F_{k+2})$$
This completes the induction.

(f)   Since $F_2^2 = 1^2 = 1 = F_1 F_3 + (-1)^1 = 2 - 1 = 1$, the hypothesis is true for $n = 1$.
Assume:
$$F_{k+1}^2 = F_k F_{k+2} + (-1)^k$$
Then:
$$F_{k+1}F_{k+2} + F_{k+1}^2 = F_k F_{k+2} + F_{k+1}F_{k+2} + (-1)^k$$
$$F_{k+1}(F_{k+2} + F_{k+1}) = F_{k+2}(F_k + F_{k+1}) + (-1)^k$$
$$F_{k+1}F_{k+3} = F_{k+2}^2 + (-1)^k$$
$$F_{k+2}^2 = F_{k+1}F_{k+3} + (-1)^{k+1}$$
This completes the induction.

(g)   Since $\begin{pmatrix} 1 & 1 \\ 1 & 0 \end{pmatrix}^2 = \begin{pmatrix} 2 & 1 \\ 1 & 1 \end{pmatrix} = \begin{pmatrix} F_3 & F_2 \\ F_2 & F_1 \end{pmatrix}$ then the hypothesis is true for $n = 2$. Assuming:
$$\begin{pmatrix} 1 & 1 \\ 1 & 0 \end{pmatrix}^k = \begin{pmatrix} F_{k+1} & F_k \\ F_k & F_{k-1} \end{pmatrix}$$
Then:
$$\begin{pmatrix} 1 & 1 \\ 1 & 0 \end{pmatrix}^{k+1} = \begin{pmatrix} 1 & 1 \\ 1 & 0 \end{pmatrix}\begin{pmatrix} F_{k+1} & F_k \\ F_k & F_{k-1} \end{pmatrix}$$
$$= \begin{pmatrix} F_k + F_{k+1} & F_k + F_{k-1} \\ F_{k+1} & F_k \end{pmatrix}$$
$$= \begin{pmatrix} F_{k+2} & F_{k+1} \\ F_{k+1} & F_k \end{pmatrix}$$
This completes the induction.

12.4 Arithmetic Sequences and Series

1. (a) To find the common difference d, we subtract any term from the succeeding term. Here that can be:

$$3 - 1 = 2$$
$$5 - 3 = 2$$
$$7 - 5 = 2$$

So 2 is the required common difference.

(b) Again we subtract terms:

$$6 - 10 = -4$$
$$2 - 6 = -4$$
$$-2 - 2 = -4$$

The common difference is -4. It is not necessary to try all three pairs, but this helps to verify that we have an arithmetic sequence.

(c) We subtract:

$$1 - \tfrac{2}{3} = \tfrac{4}{3} - 1 = \tfrac{5}{3} - \tfrac{4}{3} = \tfrac{1}{3}$$

So $\tfrac{1}{3}$ is the common difference.

(d) We subtract:

$$1 + \sqrt{2} - 1 = \sqrt{2} \text{ or } \left(1 + 2\sqrt{2}\right) - \left(1 + \sqrt{2}\right) = \sqrt{2}$$

Here $\sqrt{2}$ is the common difference.

3. Since $a = 10$ and $d = 11$, then using $a_n = a + (n-1)d$ we find:

$$a_{12} = 10 + (12 - 1)11 = 131$$

5. $a_{100} = 6 + (100 - 1)(5) = 6 + 495 = 501$

7. $a_{1000} = -1 + (1000 - 1)(1) = 998$

9. Here $a_4 = -6$ and $a_{10} = 5$, thus:

$$-6 = a + 3d$$
$$5 = a + 9d$$

Subtracting the second equation from the first yields:

$$-6d = -11$$
$$d = \tfrac{11}{6}$$

Therefore:

$$a = 5 - 9\left(\tfrac{11}{6}\right) = 5 - \tfrac{33}{2} = -\tfrac{23}{2}$$

11. Here $a_{60} = 105$ and $d = 5$, so:
$$105 = a + 59(5)$$
$$105 = a + 295$$
$$a = -190$$

13. Since $a_{15} = a + 14d$ and $a_7 = a + 6d$, then:
$$a_{15} - a_7 = 8d$$
$$-1 = 8d$$
$$d = -\tfrac{1}{8}$$

15. Since $a = 1$ and $d = 1$, then:
$$S_{1000} = \tfrac{1000}{2}(2 + 999) = 500500$$

17. For $\tfrac{\pi}{3} + \tfrac{2\pi}{3} + \pi + \tfrac{4\pi}{3} + \ldots + \tfrac{13\pi}{3}$, $a = \tfrac{\pi}{3}$, $d = \tfrac{\pi}{3}$, so the sum is:
$$S_{13} = \frac{13}{2}\left(\frac{2\pi}{3} + \frac{12\pi}{3}\right) = \frac{14(13)(\pi)}{6} = \frac{91\pi}{3}$$

19. Since $d = 5$ and $S_{38} = 3534$, then:
$$3534 = \tfrac{38}{2}[2a + 37(5)]$$
$$3534 = 38a + 3515$$
$$19 = 38a$$
$$a = \tfrac{1}{2}$$

21. Here $a = 4$, $a_{16} = -100$, therefore:
$$S_{16} = 16\left(\tfrac{4-100}{2}\right) = -768$$
So we have:
$$-768 = \tfrac{16}{2}(2(4) + 15d)$$
$$-768 = 64 + 120d$$
$$-832 = 120d$$
$$d = -\tfrac{104}{15}$$

23. Since $a_8 = 5$ and $S_{10} = 20$, then:
$$5 = a + 7d$$
$$20 = 5(2a + 9d)$$
$$4 = 2a + 9d$$
We solve the system:
$$a + 7d = 5$$
$$2a + 9d = 4$$

Multiplying the first equation by -2 and adding to the second equation:

$$-2a - 14d = -10$$
$$2a + 9d = 4$$
$$-5d = -6$$
$$d = \tfrac{6}{5}$$

Therefore:

$$a = 5 - 7\left(\tfrac{6}{5}\right) = -\tfrac{17}{5}$$

25. $S = S_{20} = \displaystyle\sum_{k=1}^{20}(4k + 3)$ so $a = 7$ and $a_{20} = 83$, thus:

$$S = 20\left(\tfrac{7+83}{2}\right) = 900$$

27. Let x denote the middle term, so the three terms are $x - d, x, x + d$:

$$x - d + x + x + d = 30$$
$$3x = 30$$
$$x = 10$$

Therefore:

$$x(x - d)(x + d) = 360$$
$$10(100 - d^2) = 360$$
$$100 - d^2 = 36$$
$$-d^2 = -64$$
$$d = \pm 8$$

So the terms are 2, 10, 18 or 18, 10, 2.

29. Using equations as in Exercise 27:

$$3x = 6, \text{ so } x = 2$$
$$(x - d)^3 + x^3 + (x + d)^3 = 132$$
$$(2 - d)^3 + 8 + (2 + d)^3 = 132$$
$$(2 - d)^3 + (2 + d)^3 = 124$$
$$16 + 12d^2 = 124$$
$$d^2 = 9$$
$$d = \pm 3$$

So the terms are $-1, 2, 5$ or $5, 2, -1$.

31. (a) Subtracting, we have:

$$a_2 - a_1 = -1 - \frac{1}{1 + \sqrt{2}} = \frac{-2 - \sqrt{2}}{1 + \sqrt{2}} = \frac{(1 - \sqrt{2})(-2 - \sqrt{2})}{-1} = -\sqrt{2}$$

$$a_3 - a_2 = \frac{1}{1 - \sqrt{2}} + 1 = \frac{2 - \sqrt{2}}{1 - \sqrt{2}} = \frac{(2 - \sqrt{2})(1 + \sqrt{2})}{-1} = -\sqrt{2}$$

So $a_2 - a_1 = a_3 - a_2 = -\sqrt{2}$.

(b)   Using $a = \frac{1}{1+\sqrt{2}}$ and $d = -\sqrt{2}$, we have:

$$S_6 = 3\left[2\left(\frac{1}{1+\sqrt{2}}\right)+5(-\sqrt{2})\right]$$

$$= 3\left(\frac{2}{1+\sqrt{2}}-\frac{10+5\sqrt{2}}{1+\sqrt{2}}\right)$$

$$= 3\left(\frac{-8-5\sqrt{2}}{1+\sqrt{2}}\right)$$

$$= -\frac{24+15\sqrt{2}}{1+\sqrt{2}}$$

$$= (24+15\sqrt{2})(1-\sqrt{2})$$

$$= 24-30-9\sqrt{2}$$

$$= -6-9\sqrt{2}$$

33. Using $a = \frac{1}{1+\sqrt{b}}$, we have:

$$d = \frac{1}{2}\left(\frac{1}{1-\sqrt{b}}-\frac{1}{1+\sqrt{b}}\right) = \frac{1}{2}\left(\frac{2\sqrt{b}}{1-b}\right) = \frac{\sqrt{b}}{1-b}$$

Therefore:

$$S_n = \frac{n}{2}\left[\frac{2}{1+\sqrt{b}}+(n-1)\frac{\sqrt{b}}{1-b}\right]$$

$$= \frac{n}{2}\left[\frac{(2-2\sqrt{b})+(n-1)\sqrt{b}}{1-b}\right]$$

$$= \frac{n}{2(1-b)}\left[2+(n-3)\sqrt{b}\right]$$

35. Using the given ratio, we have:

$$\frac{n^2}{m^2} = \frac{\frac{n}{2}[2a+(n-1)d]}{\frac{m}{2}[2a+(m-1)d]}$$

$$\frac{2a+(n-1)d}{2a+(m-1)d} = \frac{n}{m}$$

$$m[2a+(n-1)d] = n[2a+(m-1)d]$$

$$2am+m(n-1)d = 2an+n(m-1)d$$

$$2a(m-n) = [n(m-1)-m(n-1)]d$$

$$d = \frac{2a(m-n)}{m-n}$$

$$d = 2a \quad (\text{assuming } m \neq n)$$

Consequently:

$$\frac{a_n}{a_m} = \frac{a+(n-1)(2a)}{a+(m-1)(2a)} = \frac{1+2n-2}{1+2m-2} = \frac{2n-1}{2m-1}$$

37. Let the three consecutive terms be $x - d$, x, and $x + d$, then using the Pythagorean theorem:

$$(x-d)^2 + x^2 = (x+d)^2$$
$$x^2 - 2xd + d^2 + x^2 = x^2 + 2xd + d^2$$
$$x^2 - 4xd = 0$$
$$x = 0 \quad \text{or} \quad x = 4d$$

Since the side of the triangle cannot have zero length, then $x = 4d$ and the three terms are $3d$, $4d$, and $5d$, which is clearly similar to a 3-4-5 right triangle.

39. We have $\frac{1}{b} - \frac{1}{a} = \frac{1}{c} - \frac{1}{b}$, from which it follows that $b = \frac{2ac}{a+c}$. Therefore:

$$\ln(a+c) + \ln(a - 2b + c) = \ln(a+c) + \ln\left(a - \frac{4ac}{a+c} + c\right)$$
$$= \ln(a+c) + \ln\left(\frac{a^2 + ac - 4ac + ac + c^2}{a+c}\right)$$
$$= \ln(a+c) + \ln\frac{(a-c)^2}{a+c}$$
$$= \ln(a-c)^2$$
$$= 2\ln(a-c)$$

12.5 Geometric Sequences and Series

1. The first three terms have a common ratio, hence:

$$\frac{x}{9} = \frac{4}{x}, \text{ therefore } x^2 = 36 \text{ and } x = \pm 6$$

Since the ratio is positive the second term is 6.

3. Letting the common ratio be r, we have:

$$4, 4r, \text{ and } 4r^2 \text{ so } 64r^3 = 8000$$

Therefore, $r^3 = 125$, so $r = 5$, and the second and third terms are 20 and 100, respectively.

5. Here $a_1 = -1$ and $r = -1$, so $a_{100} = -1(-1)^{99} = 1$.

7. Here $a_1 = \frac{2}{3}$ and $r = \frac{2}{3}$, so $a_8 = \frac{2}{3}\left(\frac{2}{3}\right)^7 = \frac{256}{6561}$.

9. Here $4096 = a_7 = r^6$, so $r = \pm\sqrt[6]{4096} = \pm 4$.

11. We compute the sum:

$$S_{10} = \frac{7(1 - 2^{10})}{1 - 2} = -7(1 - 1024) = 7161$$

13. We compute the sum:

$$1+\sqrt{2}+2+\ldots+32 = S_{11} = \frac{1\left[1-\left(\sqrt{2}\right)^{11}\right]}{1-\sqrt{2}} = \frac{1-32\sqrt{2}}{1-\sqrt{2}} = 63+31\sqrt{2}$$

15. We compute the sum:

$$\sum_{k=1}^{6}\left(\tfrac{3}{2}\right)^{k} = S_{6} = \frac{\tfrac{3}{2}\left[1-\left(\tfrac{3}{2}\right)^{6}\right]}{1-\tfrac{3}{2}} = -3\left(1-\tfrac{729}{64}\right) = \tfrac{1995}{64}$$

17. We compute the sum:

$$\sum_{k=2}^{6}\left(\tfrac{1}{10}\right)^{k} = S_{6} - a_{1}$$

$$= \frac{\tfrac{1}{10}\left[1-\left(\tfrac{1}{10}\right)^{6}\right]}{1-\tfrac{1}{10}} - \tfrac{1}{10}$$

$$= \tfrac{1}{9}\left(1-\tfrac{1}{1000000}\right) - \tfrac{1}{10}$$

$$= \tfrac{1}{9}\cdot\tfrac{999999}{1000000} - \tfrac{1}{10}$$

$$= \tfrac{111111}{1000000} - \tfrac{1}{10}$$

$$= \tfrac{11111}{1000000}$$

This is equivalent is 0.011111.

19. We compute the sum:

$$\tfrac{2}{3}-\tfrac{4}{9}+\tfrac{8}{27}-\ldots = \frac{\tfrac{2}{3}}{1+\tfrac{2}{3}} = \tfrac{2}{5}$$

21. We compute the sum:

$$1+\frac{1}{1.01}+\frac{1}{(1.01)^{2}}+\ldots = \frac{1}{1-\tfrac{1}{1.01}} = 101$$

23. Writing as a geometric sum, we have:

$$0.555\ldots = \tfrac{5}{10}+\tfrac{5}{100}+\ldots = \frac{\tfrac{5}{10}}{1-\tfrac{1}{10}} = \tfrac{5}{9}$$

25. Writing as a geometric sum, we have:

$$0.12323\ldots = \tfrac{1}{10}+\tfrac{23}{1000}+\tfrac{23}{100000}+\ldots = \tfrac{1}{10}+\frac{\tfrac{23}{1000}}{1-\tfrac{1}{100}} = \tfrac{1}{10}+\tfrac{23}{990} = \tfrac{122}{990} = \tfrac{61}{495}$$

27. Writing as a geometric sum, we have:

$$0.432\ldots = \frac{432}{1000} + \frac{432}{1000000} + \ldots = \frac{\frac{432}{1000}}{1 - \frac{1}{1000}} = \frac{432}{999} = \frac{16}{37}$$

29. For $\frac{a}{r}$, a, ar we have:

$$\frac{a}{r}(a)(ar) = -1000$$
$$a^3 = -1000$$
$$a = -10$$

So we have:

$$\frac{a}{r} + a + ar = 15$$
$$\frac{-10}{r} - 10 - 10r = 15$$
$$2r^2 + 5r + 2 = 0$$
$$(2r + 1)(r + 2) = 0$$
$$r = -\tfrac{1}{2}, -2$$

31. We compute the sum:

$$\frac{\sqrt{3}}{\sqrt{3}+1} + \frac{\sqrt{3}}{\sqrt{3}+3} + \ldots = \frac{\frac{\sqrt{3}}{\sqrt{3}+1}}{1 - \frac{\sqrt{3}+1}{\sqrt{3}+3}} = \frac{\frac{\sqrt{3}}{\sqrt{3}+1}}{\frac{2}{\sqrt{3}+3}} = \frac{\sqrt{3}(\sqrt{3}+3)}{2(\sqrt{3}+1)} = \frac{3+3\sqrt{3}}{2(1+\sqrt{3})} = \frac{3}{2}$$

33. For $a_1, a_2, \ldots$ a geometric sequence, with ratio r:

$$S = a_1 + a_2 + \ldots + a_n = \frac{a_1(1-r^n)}{1-r}$$

Also:

$$T = \frac{1}{a_1} + \frac{1}{a_2} + \ldots + \frac{1}{a_n} = \frac{\frac{1}{a_1}\left(1 - \frac{1}{r^n}\right)}{1 - \frac{1}{r}}$$

Therefore:

$$\frac{S}{T} = \frac{\frac{a_1(1-r^n)}{1-r}}{\frac{\frac{1}{a_1}\left(1 - \frac{1}{r^n}\right)}{\frac{r-1}{r}}} = \frac{a_1(1-r^n)a_1(r-1)r^n}{(1-r)r(r^n-1)} = a_1^2 r^{n-1} = a_1\left(a_1 r^{n-1}\right) = a_1 a_n$$

35. The total distance traveled by the ball consists of two infinite geometric series:

down: $6 + \tfrac{1}{3}(6) + \tfrac{1}{9}(6) + \ldots$

up: $2 + \tfrac{1}{3}(2) + \tfrac{1}{9}(2) + \ldots$

We now use the formula $S = \frac{a}{1-r}$ to find each sum:

down: $a = 6, r = \frac{1}{3}$

$$S = \frac{6}{1-\frac{1}{3}} = \frac{6}{\frac{2}{3}} = 9 \text{ ft.}$$

up: $a = 1, r = \frac{1}{3}$

$$S = \frac{2}{1-\frac{1}{3}} = \frac{2}{\frac{2}{3}} = 3 \text{ ft.}$$

So the total distance traveled is 12 ft.

Chapter Twelve Review Exercises

1. When $n = 1$:
$$5 = \tfrac{5}{2}(1)(1+1) = 5$$
Assuming:
$$5 + 10 + \ldots + 5k = \tfrac{5}{2}k(k+1)$$
Then:
$$5 + 10 + \ldots + 5k + 5(k+1) = \tfrac{5}{2}k(k+1) + 5(k+1) = (k+1)\left(\tfrac{5}{2}k+5\right) = \tfrac{5}{2}(k+1)(k+2)$$
This completes the induction.

3. When $n = 1$:
$$1 \cdot 2 = 2 = \tfrac{1}{3}(1)(1+1)(1+2) = 2$$
Assuming:
$$1 \cdot 2 + 2 \cdot 3 + \ldots + k(k+1) = \tfrac{1}{3}k(k+1)(k+2)$$
Then:
$$1 \cdot 2 + 2 \cdot 3 + \ldots + k(k+1) + (k+1)(k+2)$$
$$= \tfrac{1}{3}k(k+1)(k+2) + (k+1)(k+2)$$
$$= \left(\tfrac{1}{3}k+1\right)(k+1)(k+2)$$
$$= \tfrac{1}{3}(k+1)(k+2)(k+3)$$
This completes the induction.

5. For $n = 1$:
$$1 = 3 + (2-3)2 = 3 - 2 = 1$$
Assuming:
$$1 + 3 \cdot 2 + 5 \cdot 2^2 + \ldots + (2k-1) \cdot 2^{k-1} = 3 + (2k-3) \cdot 2^k$$

Then:
$$1+3\cdot2+...+(2k-1)2^{k-1}+(2k+1)2^k$$
$$=3+(2k-3)2^k+(2k+1)2^k$$
$$=3+(4k-2)2^k$$
$$=3+(2k-1)2^{k+1}$$
$$=3+(2(k+1)-3)\cdot2^{k+1}$$
The induction is complete.

7. For $n=1$:
$$1=(1^2-2+3)2-3=1$$
Assuming:
$$1+2^2\cdot2+3^2\cdot2^2+4^2\cdot2^3+...+k^2\cdot2^{k-1}=(k^2-2k+3)2^k-3$$
Then:
$$1+2^2\cdot2+...+(k+1)^22^k=(k^2-2k+3)2^k-3+(k+1)^22^k$$
$$=2^k\left[k^2-2k+3+(k+1)^2\right]-3$$
$$=2^k(2k^2+4)-3$$
$$=2^{k+1}(k^2+2)-3$$
$$=2^{k+1}\left[(k+1)^2-2k+1\right]-3$$
$$=2^{k+1}\left[(k+1)^2-2(k+1)+3\right]-3$$
The induction is complete.

9. For $n=1$:
$$7^1-1=6=3(2)$$
Assuming:
$$7^k-1=3L$$
Then:
$$7^{k+1}-1=3L+7^{k+1}-7^k=3L+7^k(7-1)=3L+6(7^k)=3(L+2\cdot7^k)$$
The induction is complete.

11. We expand:
$$(3a+b^2)^4=\binom{4}{0}(3a)^4+\binom{4}{1}(3a)^3(b^2)+\binom{4}{2}(3a)^2(b^2)^2+\binom{4}{3}(3a)(b^2)^3+\binom{4}{4}(b^2)^4$$
$$=81a^4+108a^3b^2+54a^2b^4+12ab^6+b^8$$

13. We expand:
$$(x+\sqrt{x})^4=\binom{4}{0}x^4+\binom{4}{1}x^3(\sqrt{x})+\binom{4}{2}x^2(\sqrt{x})^2+\binom{4}{3}x(\sqrt{x})^3+\binom{4}{4}(\sqrt{x})^4$$
$$=x^4+4x^3\sqrt{x}+6x^3+4x^2\sqrt{x}+x^2$$

15. We expand:
$$\left(x^2 - 2y^2\right)^5 = \binom{5}{0}\left(x^2\right)^5 - \binom{5}{1}\left(x^2\right)^4\left(2y^2\right) + \binom{5}{2}\left(x^2\right)^3\left(2y^2\right)^2 - \binom{5}{3}\left(x^2\right)^2\left(2y^2\right)^3$$
$$+ \binom{5}{4}\left(x^2\right)\left(2y^2\right)^4 - \binom{5}{5}\left(2y^2\right)^5$$
$$= x^{10} - 10x^8y^2 + 40x^6y^4 - 80x^4y^6 + 80x^2y^8 - 32y^{10}$$

17. We expand:
$$\left(1 + \frac{1}{x}\right)^5 = \binom{5}{0} + \binom{5}{1}\frac{1}{x} + \binom{5}{2}\left(\frac{1}{x}\right)^2 + \binom{5}{3}\left(\frac{1}{x}\right)^3 + \binom{5}{4}\left(\frac{1}{x}\right)^4 + \binom{5}{5}\left(\frac{1}{x}\right)^5$$
$$= 1 + \frac{5}{x} + \frac{10}{x^2} + \frac{10}{x^3} + \frac{5}{x^4} + \frac{1}{x^5}$$

19. We expand:
$$\left(a\sqrt{b} - b\sqrt{a}\right)^4 = a^2b^2\left(\sqrt{a} - \sqrt{b}\right)^4$$
$$= a^2b^2\left[\left(\sqrt{a}\right)^4 - 4\left(\sqrt{a}\right)^3\sqrt{b} + 6\left(\sqrt{a}\right)^2\left(\sqrt{b}\right)^2 - 4\left(\sqrt{a}\right)\left(\sqrt{b}\right)^3 + \left(\sqrt{b}\right)^4\right]$$
$$= a^2b^2\left(a^2 - 4a\sqrt{ab} + 6ab - 4b\sqrt{ab} + b^2\right)$$
$$= a^4b^2 - 4a^3b^2\sqrt{ab} + 6a^3b^3 - 4a^2b^3\sqrt{ab} + a^2b^4$$

21. The fifth term is given by:
$$\binom{5}{5-1}(3x)\left(y^2\right)^4 = 5(3x)y^8 = 15xy^8$$

23. Here $7 - r + 1 = 5$, so $r = 3$, and the coefficient of the third term is:
$$\binom{7}{2}(2)^2 = (4)\frac{7!}{5!2!} = 84$$

25. Here $r - 1 = 6$, so $r = 7$, and the coefficient of the seventh term is:
$$\binom{8}{6} = \frac{8!}{6!2!} = 28$$

27. Computing each side:
$$\binom{2}{0}^2 + \binom{2}{1}^2 + \binom{2}{2}^2 = (1)^2 + (2)^2 + (1)^2 = 6$$
$$\binom{4}{2} = \frac{4!}{2!2!} = 6$$

29. Computing each side:

$$\binom{4}{0}^2 + \binom{4}{1}^2 + \binom{4}{2}^2 + \binom{4}{3}^2 + \binom{4}{4}^2 = 1^2 + 4^2 + 6^2 + 4^2 + 1^2 = 70$$

$$\binom{8}{4} = \frac{8!}{4!4!} = 70$$

31. Computing values:

$$\binom{3}{0} + \binom{3}{1} + \binom{3}{2} + \binom{3}{3} = 1 + 3 + 3 + 1 = 8 = 2^3$$

33. Computing values, we have:

$$a_1 = \frac{2}{1+1} = 1, \quad a_2 = \frac{4}{2+1} = \frac{4}{3}, \quad a_3 = \frac{6}{3+1} = \frac{3}{2}, \quad a_4 = \frac{8}{4+1} = \frac{8}{5}$$

We graph these points:

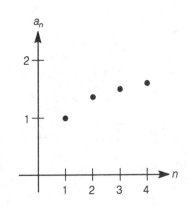

35. Computing values, we have:

$$a_1 = (-1)\left(1 - \tfrac{1}{2}\right) = -\tfrac{1}{2}$$
$$a_2 = (1)\left(1 - \tfrac{1}{3}\right) = \tfrac{2}{3}$$
$$a_3 = (-1)\left(1 - \tfrac{1}{4}\right) = -\tfrac{3}{4}$$
$$a_4 = (1)\left(1 - \tfrac{1}{5}\right) = \tfrac{4}{5}$$

We graph these points:

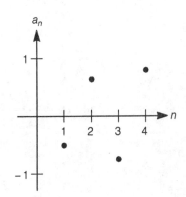

37. Computing values, we have:
$$a_0 = -3, \quad a_1 = -12, \quad a_2 = -48, \quad a_3 = -192$$
The first four terms are $-3, -12, -48,$ and -192.

39. (a) Expanding the sum, we have:
$$\sum_{k=1}^{3}(-1)^k(2k+1) = -3+5-7 = -5$$

(b) Expanding the sum, we have:
$$\sum_{k=0}^{8}\left(\frac{1}{k+1} - \frac{1}{k+2}\right) = \left(1-\tfrac{1}{2}\right)+\left(\tfrac{1}{2}-\tfrac{1}{3}\right)+\left(\tfrac{1}{3}-\tfrac{1}{4}\right)+\left(\tfrac{1}{4}-\tfrac{1}{5}\right)+\left(\tfrac{1}{5}-\tfrac{1}{6}\right)$$
$$+\left(\tfrac{1}{6}-\tfrac{1}{7}\right)+\left(\tfrac{1}{7}-\tfrac{1}{8}\right)+\left(\tfrac{1}{8}-\tfrac{1}{9}\right)+\left(\tfrac{1}{9}-\tfrac{1}{10}\right)$$
$$= 1 - \frac{1}{10}$$
$$= \frac{9}{10}$$

41. The sum can be written as:
$$\frac{5}{3} + \frac{5}{3^2} + \frac{5}{3^3} + \frac{5}{3^4} + \frac{5}{3^5} = \sum_{k=1}^{5}\frac{5}{3k}$$

43. Here $a_n = 1 + 4n$, so $a_{18} = 1 + 4(18) = 73$.

45. Here $a_n = \dfrac{10}{2^{n-1}}$, so $a_{12} = \dfrac{10}{2^{11}} = \dfrac{5}{1024}$.

47. The sum is given by:
$$S_{12} = (12)\left(\frac{8 + \frac{43}{2}}{2}\right) = \tfrac{59}{2}(6) = 177$$

49. By inspection $S_1 = 7$, $S_2 = 77$, ... so $S_{10} = 7,777,777,777$.

51. We compute:
$$a_3 = 4$$
$$a_5 = 10$$
$$\frac{a_5}{a_3} = \tfrac{10}{4} = \tfrac{5}{2} = r^2 \text{ and } r = \pm\sqrt{\tfrac{5}{2}}$$
Since r is known to be negative, $r = -\sqrt{\tfrac{5}{2}}$ and $a_6 = -\sqrt{\tfrac{5}{2}}(10) = -5\sqrt{10}$.

53. The sum is given by:

$$S = \frac{\frac{3}{5}}{1-\frac{1}{5}} = \frac{3}{4}$$

55. The sum is given by:

$$S = \frac{\frac{1}{9}}{1+\frac{1}{9}} = \frac{1}{10}$$

57. We have $0.\overline{45} = 0.45 + 0.0045 + 0.000045 + \dots$. This is a geometric sequence with $a = 0.45$, $r = \frac{1}{100}$, so:

$$0.\overline{45} = \frac{0.45}{1-\frac{1}{100}} = \frac{45}{99} = \frac{15}{33} = \frac{5}{11}$$

59. The sum is given by:

$$S_n = \frac{n}{2}\left[2(1)+(n-1)^1\right]$$
$$= \frac{n}{2}(2+n-1)$$
$$= \frac{n}{2}(1+n)$$
$$= \frac{n^2}{2}+\frac{n}{2}$$
$$= n+\frac{n^2}{2}-\frac{n}{2}$$
$$= n+\frac{n(n-1)}{2}$$

61. The sum is given by:

$$S_n = \frac{n}{2}[2(1)+(n-1)3]$$
$$= \frac{n}{2}(2+3n-3)$$
$$= \frac{n}{2}(3n-1)$$
$$= \frac{3n^2}{2}-\frac{n}{2}$$
$$= n+\frac{3n^2}{2}-\frac{3n}{2}$$
$$= n+\frac{3n(n-1)}{2}$$

63. (a)   Using formula (1), we have:

$$1^2 + 2^2 + \ldots + 50^2 \approx \frac{\left(50 + \frac{1}{2}\right)^{2+1}}{2+1} = \frac{\left(\frac{101}{2}\right)^3}{3} = \frac{1030301}{24} \approx 42929$$

(b)   The exact sum is given by:

$$\frac{50(51)(101)}{6} = 42925$$

The percent error is:

$$100 \cdot \frac{4}{42925} \approx 0.00932\%$$

(c)   Using formula (1), we have:

$$1^4 + 2^4 + \ldots + 200^4 \approx \frac{(200.5)^5}{5} \approx 6.48040 \times 10^{10}$$

(d)   We compute the sum:

$$1^4 + 2^4 + \ldots + 200^4 \approx \frac{200(201)(401)(120599)}{30} \approx 6.48027 \times 10^{10}$$

The percent error is:

$$100 \cdot \frac{0.00013 \times 10^{10}}{6.48027 \times 10^{10}} = 2 \times 10^{-3}\%$$

65.   Call $b = ra$ and $c = r^2a$, so $a + ar + ar^2 = 70$. Since $4a$, $5b$, $4c$ are consecutive terms in an arithmetic sequence, then their common difference is the same. So:

$$5b - 4a = 4c - 5b$$
$$10b = 4a + 4c$$
$$5b = 2a + 2c$$
$$5(ra) = 2a + 2\left(r^2a\right)$$
$$5ra = 2a + 2r^2a$$

Since a is nonzero, we divide by a:

$$5r = 2 + 2r^2$$
$$2r^2 - 5r + 2 = 0$$
$$(2r - 1)(r - 2) = 0$$
$$r = \tfrac{1}{2}, 2$$

If $r = \frac{1}{2}$, we find a:

$$a + a \cdot \tfrac{1}{2} + a \cdot \tfrac{1}{4} = 70$$
$$7a = 280$$
$$a = 40$$

So the terms are 40, 20, 10.
If $r = 2$, we find a:

$$a + 2a + 4a = 70$$
$$7a = 70$$
$$a = 10$$

So the terms are 10, 20, 40.

67. We assume:
$$\frac{1}{c+a} - \frac{1}{b+c} = \frac{1}{a+b} - \frac{1}{c+a}$$
Our goal is to prove that:
$$b^2 - a^2 = c^2 - b^2$$
Multiply the first equation by $(c+a)(b+c)(a+b)$:
$$(b+c)(a+b) - (c+a)(a+b) = (c+a)(b+c) - (b+c)(a+b)$$
$$\left(ab+ac+bc+b^2\right) - \left(ac+bc+ab+a^2\right) = \left(bc+ab+ac+c^2\right) - \left(ab+ac+bc+b^2\right)$$
$$b^2 - a^2 = c^2 - b^2$$
This proves the desired result.

69. (a) We must have $4+x = r(3+x)$ and $5+x = r^2(3+x)$, so $r = \frac{4+x}{3+x}$. Substituting into the second equality:
$$5+x = \left(\frac{4+x}{3+x}\right)^2 (3+x)$$
$$5+x = \frac{(4+x)^2}{3+x}$$
Multiplying each side by $3+x$:
$$15 + 8x + x^2 = 16 + 8x + x^2$$
$$15 = 16$$
So no such value for x exists.

(b) Here $b+x = r(a+x)$ and $c+x = r^2(a+x)$, so $r = \frac{b+x}{a+x}$. Substituting:
$$c+x = \left(\frac{b+x}{a+x}\right)^2 (a+x)$$
$$(c+x)(a+x) = (b+x)^2$$
$$ac + (a+c)x + x^2 = b^2 + 2bx + x^2$$
$$(a+c-2b)x = b^2 - ac$$
$$x = \frac{b^2 - ac}{a+c-2b}$$
(Note that $a+c-2b = 0$ for (a), explaining why no x could be found.)

71. (a) Let $S = a_1 + ra_2 + r^2 a_3 + \ldots + r^{n-1} a_n$, so:
$$rS = ra_1 + r^2 a_2 + r^3 a_3 + \ldots + r^n a_n$$
Then:
$$S - rS = a_1 + r(a_2 - a_1) + r^2(a_3 - a_2) + \ldots + r^{n-1}(a_n - a_{n-1}) - r^n a_n$$
$$= a_1 + rd + r^2 d + \ldots + r^{n-1} d - r^n a_n$$
$$= a_1 + \frac{d(r - r^n)}{1 - r} - r^n a_n$$

(b) Continuing from (a), we have:

$$S(1-r) = a_1 - r^n a_n + \frac{d(r-r^n)}{1-r}$$

$$S = \frac{a_1 - r^n a_n}{1-r} + \frac{d(r-r^n)}{(1-r)^2}$$

73. (a) We start by rationalizing denominators on the left-hand side of the equation:

$$\frac{1}{\sqrt{a_1} + \sqrt{a_2}} \cdot \frac{\sqrt{a_1} - \sqrt{a_2}}{\sqrt{a_1} - \sqrt{a_2}} = \frac{\sqrt{a_1} - \sqrt{a_2}}{a_1 - a_2} = \frac{\sqrt{a_1} - \sqrt{a_2}}{-d}$$

$$\frac{1}{\sqrt{a_2} + \sqrt{a_3}} \cdot \frac{\sqrt{a_2} - \sqrt{a_3}}{\sqrt{a_2} - \sqrt{a_3}} = \frac{\sqrt{a_2} - \sqrt{a_3}}{a_2 - a_3} = \frac{\sqrt{a_2} - \sqrt{a_3}}{-d}$$

Adding, we have:

$$\frac{1}{\sqrt{a_1} + \sqrt{a_2}} + \frac{1}{\sqrt{a_2} + \sqrt{a_3}} = \frac{\sqrt{a_1} - \sqrt{a_3}}{-d} \cdot \frac{\sqrt{a_1} + \sqrt{a_3}}{\sqrt{a_1} + \sqrt{a_3}}$$

$$= \frac{a_1 - a_3}{-d\left(\sqrt{a_1} + \sqrt{a_3}\right)}$$

$$= \frac{-2d}{-d\sqrt{a_1} + \sqrt{a_3}}$$

$$= \frac{2}{\sqrt{a_1} + \sqrt{a_3}}$$

(b) Proceeding as in part (a), we rationalize denominators on the left-hand side of the equation:

$$\frac{1}{\sqrt{a_3} + \sqrt{a_4}} \cdot \frac{\sqrt{a_3} - \sqrt{a_4}}{\sqrt{a_3} - \sqrt{a_4}} = \frac{\sqrt{a_3} - \sqrt{a_4}}{a_3 - a_4} = \frac{\sqrt{a_3} - \sqrt{a_4}}{-d}$$

Adding the three fractions, we have:

$$\frac{1}{\sqrt{a_1} + \sqrt{a_2}} + \frac{1}{\sqrt{a_2} + \sqrt{a_3}} + \frac{1}{\sqrt{a_3} + \sqrt{a_4}} = \frac{\sqrt{a_1} - \sqrt{a_4}}{-d} \cdot \frac{\sqrt{a_1} + \sqrt{a_4}}{\sqrt{a_1} + \sqrt{a_4}}$$

$$= \frac{a_1 - a_4}{-d\left(\sqrt{a_1} + \sqrt{a_4}\right)}$$

$$= \frac{-3d}{-d\sqrt{a_1} + \sqrt{a_4}}$$

$$= \frac{3}{\sqrt{a_1} + \sqrt{a_4}}$$

(c) We will prove this by mathematical induction. Part (a) shows the statement to be true when $n = 3$. Now assume P_k is true, so:

$$\frac{1}{\sqrt{a_1} + \sqrt{a_2}} + \dots + \frac{1}{\sqrt{a_{k-1}} + \sqrt{a_k}} = \frac{k-1}{\sqrt{a_1} + \sqrt{a_k}}$$

We must show P_{k+1} is true. We have:

$$\frac{1}{\sqrt{a_1} + \sqrt{a_2}} + \dots + \frac{1}{\sqrt{a_{k-1}} + \sqrt{a_k}} + \frac{1}{\sqrt{a_k} + \sqrt{a_{k+1}}} = \frac{k-1}{\sqrt{a_1} + \sqrt{a_k}} + \frac{1}{\sqrt{a_k} + \sqrt{a_{k+1}}}$$

Rationalize each denominator:

$$\frac{k-1}{\sqrt{a_1} + \sqrt{a_k}} \cdot \frac{\sqrt{a_1} - \sqrt{a_k}}{\sqrt{a_1} - \sqrt{a_k}} = \frac{(k-1)\left(\sqrt{a_1} - \sqrt{a_k}\right)}{a_1 - a_k}$$

$$= \frac{(k-1)\left(\sqrt{a_1} - \sqrt{a_k}\right)}{-(k-1)d}$$

$$= \frac{\sqrt{a_1} - \sqrt{a_k}}{-d}$$

$$\frac{1}{\sqrt{a_k} + \sqrt{a_{k+1}}} \cdot \frac{\sqrt{a_k} - \sqrt{a_{k+1}}}{\sqrt{a_k} - \sqrt{a_{k+1}}} = \frac{\sqrt{a_k} - \sqrt{a_{k+1}}}{a_k - a_{k+1}} = \frac{\sqrt{a_k} - \sqrt{a_{k+1}}}{-d}$$

Adding these two fractions yields:

$$\frac{\sqrt{a_1} - \sqrt{a_{k+1}}}{-d} \cdot \frac{\sqrt{a_1} + \sqrt{a_{k+1}}}{\sqrt{a_1} + \sqrt{a_{k+1}}} = \frac{a_1 - a_{k+1}}{-d\left(\sqrt{a_1} - \sqrt{a_{k+1}}\right)}$$

$$= \frac{-kd}{-d\left(\sqrt{a_1} + \sqrt{a_{k+1}}\right)}$$

$$= \frac{k}{\sqrt{a_1} + \sqrt{a_{k+1}}}$$

$$= \frac{(k+1) - 1}{\sqrt{a_1} + \sqrt{a_{k+1}}}$$

So P_{k+1} is true. The identity is true for all natural numbers n.

Chapter Twelve Test

1. For $n = 1$, we have:

$$1^2 = \frac{1(1+1)(2+1)}{6} = \frac{6}{6} = 1$$

Assume P_k is true, so:

$$1^2 + 2^2 + 3^2 + \dots + k^2 = \frac{k(k+1)(2k+1)}{6}$$

Then:

$$1^2 + 2^2 + \ldots + k^2 + (k+1)^2 = \frac{k(k+1)(2k+1)}{6} + (k+1)^2$$

$$= (k+1)\left[\frac{k(2k+1)}{6} + k + 1\right]$$

$$= (k+1)\left(\frac{2k^2 + k + 6k + 6}{6}\right)$$

$$= \frac{(k+1)(2k^2 + 7k + 6)}{6}$$

$$= \frac{(k+1)(k+2)(2k+3)}{6}$$

$$= \frac{(k+1)(k+2)[2(k+1)+1]}{6}$$

So P_{k+1} is true and the induction is complete.

2. (a) Expanding the sum:

$$\sum_{k=0}^{2} (10k - 1) = -1 + 9 + 19 = 27$$

(b) Expanding the sum:

$$\sum_{k=1}^{3} (-1)^k k^2 = -1 + 4 - 9 = -6$$

3. (a) $S_n = \dfrac{a(1 - r^n)}{1 - r}$

(b) Here $a = \frac{3}{2}, r = \frac{3}{2}$, and $n = 10$, so:

$$S_{10} = \frac{\frac{3}{2}\left[1 - \left(\frac{3}{2}\right)^{10}\right]}{1 - \frac{3}{2}} = \frac{\frac{3}{2}\left(1 - \frac{59049}{1024}\right)}{-\frac{1}{2}} = -3\left(-\frac{58025}{1024}\right) = \frac{174075}{1024}$$

4. (a) Here $11 - r + 1 = 3$, so $r = 9$, and the coefficient is:

$$\binom{11}{8}(-2)^8 = \frac{(11)(10)(9)}{(3)(2)}(256) = 42240$$

(b) The fifth term is:

$$\binom{11}{4}(a)^7(-2b^3)^4 = \frac{(11)(10)(9)(8)}{(4)(3)(2)}a^7(16b^{12}) = 5280a^7b^{12}$$

5. Using the binomial theorem, we have:

$$(3x^2 + y^3)^5 = \binom{5}{0}(3x^2)^5 + \binom{5}{1}(3x^2)^4(y^3) + \binom{5}{2}(3x^2)^3(y^3)^2 + \binom{5}{3}(3x^2)^2(y^3)^3$$
$$+ \binom{5}{4}(3x^2)(y^3)^4 + \binom{5}{5}(y^3)^5$$
$$= 243x^{10} + 405x^8 y^3 + 270x^6 y^6 + 90x^4 y^9 + 15x^2 y^{12} + y^{15}$$

6. The sum is:

$$S_{12} = 12\left(\frac{8 + \frac{43}{2}}{2}\right) = \tfrac{59}{2}(6) = 177$$

7. The sum is:

$$S = \frac{a}{1-r} = \frac{\frac{7}{10}}{1 - \frac{1}{10}} = \tfrac{7}{9}$$

8. Here $a_1 = 1$ and $a_2 = 1$, so:

$$a_3 = a_2^2 + a_1 = 1 + 1 = 2$$
$$a_4 = a_3^2 + a_2 = 2^2 + 1 = 5$$
$$a_5 = a_4^2 + a_3 = 5^2 + 2 = 27$$

9. Since $a_3 = 4$ and $a_5 = 10$, then $a_5 = a_3 r^2$, so:

$$10 = 4r^2$$
$$\tfrac{5}{2} = r^2$$
$$r = -\sqrt{\tfrac{5}{2}} = -\tfrac{\sqrt{10}}{2}$$

Then $a_6 = ra_5 = -\tfrac{\sqrt{10}}{2}(10) = -5\sqrt{10}$.

10. Here $a_n = -61 + 15(n-1)$, so:

$$a_{20} = -61 + 15(19) = 224$$

Appendix

A.5 Integer Exponents and nth Roots

1. (a) Using the properties of exponents, we have:
$$x^3 x^{12} = x^{3+12} = x^{15}$$

(b) Using the properties of exponents, we have:
$$\left(x^3\right)^{12} = x^{3 \cdot 12} = x^{36}$$

(c) Using the properties of exponents, we have:
$$(x+1)^3 (x+1)^{12} = (x+1)^{3+12} = (x+1)^{15}$$

(d) Using the properties of exponents, we have:
$$\left[(x+1)^3\right]^{12} = (x+1)^{3 \cdot 12} = (x+1)^{36}$$

3. (a) Using the properties of exponents, we have:
$$\frac{a^{15}}{a^9} = a^{15-9} = a^6$$

(b) Using the properties of exponents, we have:
$$\frac{(a+1)^{15}}{(a+1)^9} = (a+1)^{15-9} = (a+1)^6$$

5.   (a)   Since $x^0 = 1$ for any value of $x \neq 0$, then $64^0 = 1$.

   (b)   Since $x^0 = 1$ for any value of $x \neq 0$, then $\left(64^3\right)^0 = 1$.

7.   (a)   Evaluating negative exponents, we have:
$$10^{-1} + 10^{-2} = \tfrac{1}{10} + \tfrac{1}{100} = \tfrac{10}{100} + \tfrac{1}{100} = \tfrac{11}{100}$$

   (b)   Using the result from part (a), we have:
$$\left(10^{-1} + 10^{-2}\right)^{-1} = \left(\tfrac{11}{100}\right)^{-1} = \tfrac{100}{11}$$

   (c)   Using the properties of exponents, we have:
$$\left(10^{-1}\right)^{-2} = 10^{(-1)(-2)} = 10^2 = 100$$

9.   (a)   Using properties of exponents, we have:
$$\left(a^2 b c^0\right)^{-3} = \frac{1}{\left(a^2 b\right)^3} = \frac{1}{a^6 b^3}$$

   (b)   Using properties of exponents, we have:
$$\left(a^3 b\right)^3 \left(a^2 b^4\right)^{-1} = a^9 b^3 a^{-2} b^{-4} = a^7 b^{-1} = \frac{a^7}{b}$$

   (c)   Using properties of exponents, we have:
$$\left(a^{-3} b^{-1} c^3\right)^{-2} = a^6 b^2 c^{-6} = \frac{a^6 b^2}{c^6}$$

11.   Evaluating negative exponents, we have:
$$\left(2^{-2} + 2^{-1} + 2^0\right)^{-2} = \left(\tfrac{1}{4} + \tfrac{1}{2} + 1\right)^{-2} = \left(\tfrac{7}{4}\right)^{-2} = \left(\tfrac{4}{7}\right)^2 = \tfrac{16}{49}$$

13.   Using properties of exponents, we have:
$$\left(\frac{x^3 y^{-2} z}{x y^2 z^{-3}}\right)^{-3} = \frac{x^{-9} y^6 z^{-3}}{x^{-3} y^{-6} z^9} = \frac{y^{6-(-6)}}{x^{-3+9} z^{9+3}} = \frac{y^{12}}{x^6 z^{12}}$$

15.   Using properties of exponents, we have:
$$\left(\frac{a^{-2} b^{-3} c^{-4}}{a^2 b^3 c^4}\right)^2 = \frac{a^{-4} b^{-6} c^{-8}}{a^4 b^6 c^8} = \frac{1}{a^{4+4} b^{6+6} c^{8+8}} = \frac{1}{a^8 b^{12} c^{16}}$$

17.   Using properties of exponents, we have:
$$\frac{x^2}{y^{-3}} \div \frac{x^2}{y^3} = \frac{x^2}{y^{-3}} \cdot \frac{y^3}{x^2} = \frac{x^2 y^3}{x^2 y^{-3}} = y^6$$

19. (a) The statement is false, since $\sqrt{81} = 9$.

    (b) The statement is true.

    (c) The statement is true.

21. (a) The statement is false, since:
$$\sqrt{9+16} = \sqrt{25} = 5$$
$$\sqrt{9} + \sqrt{16} = 3 + 4 = 7$$

    (b) The statement is true, since:
$$\sqrt{(9)(16)} = \sqrt{144} = 12$$
$$\sqrt{9}\sqrt{16} = 3 \cdot 4 = 12$$

23. (a) The statement is false, since $\sqrt{(-5)^2} = \sqrt{25} = 5$.

    (b) The statement is false, since $\sqrt{x^2} = |x| = -x$ if $x < 0$.

25. (a) $\sqrt[3]{-64} = -4$

    (b) $\sqrt[4]{-64}$ is undefined

27. (a) $\sqrt[3]{\frac{8}{125}} = \frac{2}{5}$

    (b) $\sqrt[3]{-\frac{8}{125}} = -\frac{2}{5}$

29. (a) $\sqrt{-16}$ is undefined

    (b) $\sqrt[4]{-16}$ is undefined

31. (a) $\sqrt[4]{\frac{256}{81}} = \frac{4}{3}$

    (b) $\sqrt[3]{-\frac{27}{125}} = -\frac{3}{5}$

33. (a) Factoring $18 = 9 \cdot 2$, we have:
$$\sqrt{18} = \sqrt{9}\sqrt{2} = 3\sqrt{2}$$

    (b) Factoring $54 = 27 \cdot 2$, we have:
$$\sqrt[3]{54} = \sqrt[3]{27}\sqrt[3]{2} = 3\sqrt[3]{2}$$

35. (a) Factoring $98 = 49 \cdot 2$, we have:
$$\sqrt{98} = \sqrt{49}\sqrt{2} = 7\sqrt{2}$$

    (b) Factoring $-64 = -32 \cdot 2$, we have:
$$\sqrt[5]{-64} = \sqrt[5]{-32}\sqrt[5]{2} = -2\sqrt[5]{2}$$

37. (a) $\sqrt{\frac{25}{4}} = \frac{5}{2}$

  (b) $\sqrt[4]{\frac{16}{625}} = \frac{2}{5}$

39. (a) Factoring $8 = 4 \cdot 2$, we have:
$$\sqrt{2} + \sqrt{8} = \sqrt{2} + \sqrt{4}\sqrt{2} = \sqrt{2} + 2\sqrt{2} = 3\sqrt{2}$$

  (b) Factoring $16 = 8 \cdot 2$, we have:
$$\sqrt[3]{2} + \sqrt[3]{16} = \sqrt[3]{2} + \sqrt[3]{8}\sqrt[3]{2} = \sqrt[3]{2} + 2\sqrt[3]{2} = 3\sqrt[3]{2}$$

41. (a) Factoring $50 = 25 \cdot 2$ and $128 = 64 \cdot 2$, we have:
$$4\sqrt{50} - 3\sqrt{128} = 4\sqrt{25}\sqrt{2} - 3\sqrt{64}\sqrt{2} = 20\sqrt{2} - 24\sqrt{2} = -4\sqrt{2}$$

  (b) Factoring $32 = 16 \cdot 2$ and $162 = 81 \cdot 2$, we have:
$$\sqrt[4]{32} + \sqrt[4]{162} = \sqrt[4]{16}\sqrt[4]{2} + \sqrt[4]{81}\sqrt[4]{2} = 2\sqrt[4]{2} + 3\sqrt[4]{2} = 5\sqrt[4]{2}$$

43. $\sqrt{\sqrt{64}} = \sqrt{8} = \sqrt{4}\sqrt{2} = 2\sqrt{2}$

45. (a) $\sqrt{36x^2} = 6x$, since $x > 0$
  (b) $\sqrt{36y^2} = -6y$, since $y < 0$

47. (a) Multiplying radicals yields:
$$\sqrt{ab^2}\sqrt{a^2b} = \sqrt{a^3b^3} = ab\sqrt{ab}$$

  (b) Multiplying radicals yields:
$$\sqrt{ab^3}\sqrt{a^3b} = \sqrt{a^4b^4} = a^2b^2$$

49. Simplifying the radical (remember that $a < 0$) yields:
$$\sqrt[4]{16a^4b^5} = \sqrt[4]{16a^4b^4} \cdot \sqrt[4]{b} = -2ab\sqrt[4]{b}$$

51. Simplifying the radical yields:
$$\sqrt[3]{\frac{16a^{12}b^2}{c^9}} = \frac{\sqrt[3]{8a^{12}}\sqrt[3]{2b^2}}{\sqrt[3]{c^9}} = \frac{2a^4\sqrt[3]{2b^2}}{c^3}$$

53. Simplifying the radical (remember that $b < 0$) yields:
$$\sqrt[6]{\frac{5a^7}{a^{-5}b^6}} = \sqrt[6]{\frac{5a^{12}}{b^6}} = \frac{\sqrt[6]{5}\sqrt[6]{a^{12}}}{\sqrt[6]{b^6}} = -\frac{a^2\sqrt[6]{5}}{b}$$

55. (a) Rationalizing denominators, we have:
$$\frac{4}{\sqrt{7}} \cdot \frac{\sqrt{7}}{\sqrt{7}} = \frac{4\sqrt{7}}{7}$$

(b) Rationalizing denominators, we have:

$$\frac{3}{\sqrt{3}} \cdot \frac{\sqrt{3}}{\sqrt{3}} = \frac{3\sqrt{3}}{3} = \sqrt{3}$$

(c) Rationalizing denominators, we have:

$$\frac{\sqrt{2}}{\sqrt{5}} \cdot \frac{\sqrt{5}}{\sqrt{5}} = \frac{\sqrt{10}}{5}$$

57. (a) Rationalizing denominators, we have:

$$\frac{1}{1+\sqrt{5}} \cdot \frac{1-\sqrt{5}}{1-\sqrt{5}} = \frac{1-\sqrt{5}}{1-5} = \frac{1-\sqrt{5}}{-4} = \frac{\sqrt{5}-1}{4}$$

(b) Rationalizing denominators, we have:

$$\frac{1}{1-\sqrt{5}} \cdot \frac{1+\sqrt{5}}{1+\sqrt{5}} = \frac{1+\sqrt{5}}{1-5} = \frac{1+\sqrt{5}}{-4} = -\frac{1+\sqrt{5}}{4}$$

(c) Rationalizing denominators, we have:

$$\frac{1+\sqrt{5}}{1-\sqrt{5}} \cdot \frac{1+\sqrt{5}}{1+\sqrt{5}} = \frac{1+\sqrt{5}+\sqrt{5}+5}{1-5} = \frac{6+2\sqrt{5}}{-4} = -\frac{3+\sqrt{5}}{2}$$

59. Rationalizing denominators and simplifying, we have:

$$\frac{1}{\sqrt{5}} + 4\sqrt{45} = \frac{1}{\sqrt{5}} \cdot \frac{\sqrt{5}}{\sqrt{5}} + 4\sqrt{9}\sqrt{5} = \frac{\sqrt{5}}{5} + 12\sqrt{5} = \frac{\sqrt{5}}{5} + \frac{60\sqrt{5}}{5} = \frac{61\sqrt{5}}{5}$$

61. Rationalizing denominators, we have:

$$\frac{1}{\sqrt[3]{25}} \cdot \frac{\sqrt[3]{5}}{\sqrt[3]{5}} = \frac{\sqrt[3]{5}}{\sqrt[3]{125}} = \frac{\sqrt[3]{5}}{5}$$

63. Rationalizing denominators, we have:

$$\frac{1}{\sqrt[4]{2ab^5}} \cdot \frac{\sqrt[4]{8a^3b^3}}{\sqrt[4]{8a^3b^3}} = \frac{\sqrt[4]{8a^3b^3}}{\sqrt[4]{16a^4b^8}} = \frac{\sqrt[4]{8a^3b^3}}{2ab^2}$$

65. Rationalizing denominators, we have:

$$\frac{3}{\sqrt[5]{16a^4b^9}} \cdot \frac{\sqrt[5]{2ab}}{\sqrt[5]{2ab}} = \frac{3\sqrt[5]{2ab}}{\sqrt[5]{32a^5b^{10}}} = \frac{3\sqrt[5]{2ab}}{2ab^2}$$

67. Rationalizing denominators, we have:

$$\frac{x}{\sqrt{x}-2} \cdot \frac{\sqrt{x}+2}{\sqrt{x}+2} = \frac{x\left(\sqrt{x}+2\right)}{x-4}$$

69. Rationalizing denominators, we have:

$$\frac{\sqrt{x}-\sqrt{a}}{\sqrt{x}+\sqrt{a}}\cdot\frac{\sqrt{x}-\sqrt{a}}{\sqrt{x}-\sqrt{a}}=\frac{x-2\sqrt{ax}+a}{x-a}$$

71. Rationalizing numerators, we have:

$$\frac{\sqrt{x}-\sqrt{5}}{x-5}\cdot\frac{\sqrt{x}+\sqrt{5}}{\sqrt{x}+\sqrt{5}}=\frac{x-5}{(x-5)\left(\sqrt{x}+\sqrt{5}\right)}=\frac{1}{\sqrt{x}+\sqrt{5}}$$

73. Rationalizing numerators, we have:

$$\frac{\sqrt{2+h}-\sqrt{2}}{h}\cdot\frac{\sqrt{2+h}+\sqrt{2}}{\sqrt{2+h}+\sqrt{2}}=\frac{2+h-2}{h\left(\sqrt{2+h}+\sqrt{2}\right)}=\frac{h}{h\left(\sqrt{2+h}+\sqrt{2}\right)}=\frac{1}{\sqrt{2+h}+\sqrt{2}}$$

75. (a) To six decimal places, both values are 1.645751.

   (b) Following the hint, we square each side:
$$\left(\sqrt{7}-1\right)^2=7-2\sqrt{7}+1=8-2\sqrt{7}$$

77. Using properties of exponents, we have:

$$\frac{x^{3a+2b-c}}{\left(x^{2a}\right)\left(x^{b}\right)}\cdot x^{3c-a-b}=\frac{x^{2a+b+2c}}{x^{2a+b}}=x^{2c}$$

A.6 Rational Exponents

1. $a^{3/5}=\sqrt[5]{a^3}=\left(\sqrt[5]{a}\right)^3$

3. $5^{2/3}=\sqrt[3]{5^2}=\left(\sqrt[3]{5}\right)^2$

5. $\left(x^2+1\right)^{3/4}=\sqrt[4]{\left(x^2+1\right)^3}=\left(\sqrt[4]{x^2+1}\right)^3$

7. $2^{xy/3}=\sqrt[3]{2^{xy}}=\left(\sqrt[3]{2}\right)^{xy}$

9. $\sqrt[3]{p^2}=p^{2/3}$

11. $\sqrt[7]{(1+u)^4}=(1+u)^{4/7}$

13. $\sqrt[p]{\left(a^2+b^2\right)^3}=\left(a^2+b^2\right)^{3/p}$

15. $16^{1/2} = \sqrt{16} = 4$

17. $\left(\frac{1}{36}\right)^{1/2} = \sqrt{\frac{1}{36}} = \frac{1}{6}$

19. $(-16)^{1/2} = \sqrt{-16}$, which is undefined

21. $625^{1/4} = \sqrt[4]{625} = 5$

23. $8^{1/3} = \sqrt[3]{8} = 2$

25. $8^{2/3} = \left(\sqrt[3]{8}\right)^2 = (2)^2 = 4$

27. $(-32)^{1/5} = \sqrt[5]{-32} = -2$

29. $(-1000)^{1/3} = \sqrt[3]{-1000} = -10$

31. $49^{-1/2} = \left(\sqrt{49}\right)^{-1} = 7^{-1} = \frac{1}{7}$

33. $(-49)^{-1/2} = \left(\sqrt{-49}\right)^{-1}$, which is undefined

35. $(36)^{-3/2} = \left(\sqrt{36}\right)^{-3} = 6^{-3} = \frac{1}{6^3} = \frac{1}{216}$

37. $125^{2/3} = \left(\sqrt[3]{125}\right)^2 = (5)^2 = 25$

39. $(-1)^{3/5} = \left(\sqrt[5]{-1}\right)^3 = (-1)^3 = -1$

41. Evaluating the rational exponents:
$$32^{4/5} - 32^{-4/5} = \left(\sqrt[5]{32}\right)^4 - \left(\sqrt[5]{32}\right)^{-4} = 2^4 - 2^{-4} = 16 - \frac{1}{16} = \frac{255}{16}$$

43. Evaluating the rational exponents:
$$\begin{aligned}
\left(\frac{9}{16}\right)^{-5/2} - \left(\frac{1000}{27}\right)^{4/3} &= \left(\frac{16}{9}\right)^{5/2} - \left(\frac{1000}{27}\right)^{4/3} \\
&= \left(\sqrt{\frac{16}{9}}\right)^5 - \left(\sqrt[3]{\frac{1000}{27}}\right)^4 \\
&= \left(\frac{4}{3}\right)^5 - \left(\frac{10}{3}\right)^4 \\
&= \frac{1024}{243} - \frac{10000}{81} \\
&= -\frac{28976}{243}
\end{aligned}$$

45. $\left(2a^{1/3}\right)\left(3a^{1/4}\right) = 6a^{7/12}$, since $\frac{1}{3} + \frac{1}{4} = \frac{4}{12} + \frac{3}{12} = \frac{7}{12}$

47. Writing with rational exponents, then simplifying:

$$\sqrt[4]{\frac{64a^{2/3}}{a^{1/3}}} = \sqrt[4]{64a^{1/3}} = \left(2^6\,a^{1/3}\right)^{1/4} = 2^{3/2}\,a^{1/12}$$

49. Subtracting exponents yields:

$$\frac{\left(x^2+1\right)^{3/4}}{\left(x^2+1\right)^{-1/4}} = \left(x^2+1\right)^{4/4} = x^2+1$$

51. (a) $\sqrt{3}\,\sqrt[3]{6} = 3^{1/2}\,6^{1/3} = 3^{1/2}2^{1/3}3^{1/3} = 2^{1/3}\,3^{5/6}$

  (b) $3^{1/2}\,6^{1/3} = 3^{3/6}\,6^{2/6} = \sqrt[6]{3^3\,6^2} = \sqrt[6]{972}$

53. (a) $\sqrt[3]{6}\,\sqrt[4]{2} = 6^{1/3}\,2^{1/4} = 2^{1/3}3^{1/3}2^{1/4} = 2^{7/12}\,3^{1/3}$

  (b) $6^{1/3}\,2^{1/4} = 6^{4/12}\,2^{3/12} = \sqrt[12]{6^4\,2^3} = \sqrt[12]{10368}$

55. (a) $\sqrt[3]{x^2}\,\sqrt[5]{y^4} = x^{2/3}\,y^{4/5}$

  (b) $x^{2/3}\,y^{4/5} = x^{10/15}\,y^{12/15} = \sqrt[15]{x^{10}\,y^{12}}$

57. (a) Using rational exponents, we have:

$$\sqrt[4]{x^a}\,\sqrt[3]{x^b}\,\sqrt{x^{a/6}} = x^{a/4}\,x^{b/3}\,x^{a/12} = x^{3a/12}\,x^{4b/12}\,x^{a/12} = x^{(4a+4b)/12} = x^{(a+b)/3}$$

  (b) $x^{(a+b)/3} = \sqrt[3]{x^{a+b}}$

59. $\sqrt[3]{(x+1)^2} = (x+1)^{2/3}$

61. $\left(\sqrt[5]{x+y}\right)^2 = (x+y)^{2/5}$

63. Using rational exponents, we have:

$$\sqrt[3]{\sqrt{x}} + \sqrt{\sqrt[3]{x}} = \left(x^{1/2}\right)^{1/3} + \left(x^{1/3}\right)^{1/2} = x^{1/6} + x^{1/6} = 2x^{1/6}$$

65. Using rational exponents, we have:

$$\sqrt{\sqrt[3]{x}\,\sqrt[4]{y}} = \left(x^{1/3}\,y^{1/4}\right)^{1/2} = x^{1/6}\,y^{1/8}$$

67. Using $a = 9$ and $b = 16$, we have:

$$(a+b)^{1/2} = (9+16)^{1/2} = 25^{1/2} = 5$$
$$a^{1/2} + b^{1/2} = 9^{1/2} + 16^{1/2} = 3+4 = 7$$

So the formula is not valid.

69. Using $u = 1$ and $v = 8$, we have:

$$(u+v)^{1/3} = (1+8)^{1/3} = 9^{1/3} \approx 2.08$$
$$u^{1/3} + v^{1/3} = 1^{1/3} + 8^{1/3} = 1 + 2 = 3$$

So the formula is not valid.

71. Using $x = 4$ and $m = 2$, we have:

$$x^{1/m} = 4^{1/2} = 2$$
$$\frac{1}{x^m} = \frac{1}{4^2} = \tfrac{1}{16}$$

So the formula is not valid.

73. (a) Note that $2^{2/3}$ is less than 2 and $2^{3/2}$ is greater than 2, thus $2^{3/2}$ is the larger number.

(b) Note that $5^{1/2} = \sqrt{5}$ and $5^{-2} = \tfrac{1}{25}$, thus $5^{1/2}$ is the larger number.

(c) $2^{1/2}$ is larger. (One way to see this is to raise both numbers to the sixth power.)

(d) $\left(\tfrac{1}{2}\right)^{1/3}$ is larger. (One way to see this is to raise both numbers to the sixth power.)

(e) Note that $10^{1/10}$ is larger than 1, but $\left(\tfrac{1}{10}\right)^{10}$ is much less than 1, thus $10^{1/10}$ is the larger number.

75. Using rational exponents, we have:

$$\frac{a-b}{a+b}\sqrt{\frac{a+b}{a-b}} = \frac{a-b}{a+b} \cdot \frac{(a+b)^{1/2}}{(a-b)^{1/2}} = \frac{a-b}{(a-b)^{1/2}} \cdot \frac{(a+b)^{1/2}}{a+b} = \frac{(a-b)^{1/2}}{(a+b)^{1/2}} = \left(\frac{a-b}{a+b}\right)^{1/2}$$